Classics in Mathematics

Arthur L. Besse

T0254763

Arthur L. Besse

Besides his personal work in Riemannian Geometry, Marcel Berger is well known for his persistent and untiring propaganda for the problems he considers to be (and which actually are) natural and fundamental.

In 1975, he convinced his students to organise a workshop about one of his favorite problems, namely to understand manifolds, all of whose geodesics are closed. The workshop took place in Besse-en-Chandesse, a very pleasant village in the centre of France, and turned out to be so successful that a consensus emerged to write a book about this topic. Arthur Besse was born.

At that time, such a first name seemed old-fashioned and funny in France. But why not ? Besides, the association with King Arthur could not be overlooked, since this type of meeting was denoted, by the CNRS, as a "Table Ronde"!

The experience was so enjoyable that Arthur did not stop there, and settled down to write another book.

A preliminary workshop took place in another village, even lovelier than the first: Espalion, in the South-West of France. This second book, Einstein Manifolds, was eventually published in 1987.

Years have passed. Arthur's friends (the list of which can be found in the beginning of his books) haved scattered to various places.

For Arthur himself, who never laid any claim to immortality, it may be time for retirement.

One FAQ. What do Bourbaki and Besse have in common? Hardly anything. Simply that both are mathematicians, of course, and share a taste for working in pleasant and quiet places.

Arthur L. Besse

Einstein Manifolds

Reprint of the 1987 Edition

With 22 Figures

Arthur L. Besse

Originally published as Vol. 10 of the
Ergebnisse der Mathematik und ihrer Grenzgebiete, 3rd series

ISBN 978-3-540-74120-6 e-ISBN 978-3-540-74311-8

DOI 10.1007/978-3-540-74311-8

Classics in Mathematics ISSN 1431-0821

Library of Congress Control Number: 2007938035

Mathematics Subject Classification (2000): 54C25, 53-02, 53C21, 53C30, 34C55, 58D17, 58E11

© 2008, 2002, 1987 Springer-Verlag Berlin Heidelberg

This work is subject to copyright. All rights are reserved, whether the whole or part of the material is concerned, specifically the rights of translation, reprinting, reuse of illustrations, recitation, broadcasting, reproduction on microfilm or in any other way, and storage in data banks. Duplication of this publication or parts thereof is permitted only under the provisions of the German Copyright Law of September 9, 1965, in its current version, and permission for use must always be obtained from Springer. Violations are liable to prosecution under the German Copyright Law.

The use of general descriptive names, registered names, trademarks, etc. in this publication does not imply, even in the absence of a specific statement, that such names are exempt from the relevant protective laws and regulations and therefore free for general use.

Cover design: WMX Design GmbH, Heidelberg

Printed on acid-free paper

9 8 7 6 5 4 3 2 1

springer.com

Arthur L. Besse

Einstein Manifolds

With 22 Figures

Springer-Verlag
Berlin Heidelberg New York
London Paris Tokyo

Mathematics Subject Classification (1980): 53 C 25, 53 C 55, 53 C 30, 83 C . . . 83 E 50

First Reprint 2002

ISBN 3-540-15279-2 Springer-Verlag Berlin Heidelberg New York
ISBN 0-387-15279-2 Springer-Verlag New York Berlin Heidelberg

Library of Congress Cataloging-in-Publication Data
Besse, A. L.
Einstein manifolds.
(Ergebnisse der Mathematik und ihrer Grenzgebiete ; 3. Folge, Bd. 10)
Bibliography: p.
Includes indexes.
1. Einstein manifolds. 2. Relativity (Physics) I. Title. II. Series. QA649.B49 1987
530.1'1 86-15411
ISBN 0-387-15279-2 (U.S.)

This work is subject to copyright. All rights are reserved, whether the whole or part of the material is concerned, specifically those of translation, reprinting, re-use of illustrations, broadcasting, reproduction by photocopying machine or similar means, and storage in data banks. Under § 54 of the German Copyright Law where copies are made for other than private use a fee is payable to "Verwertungsgesellschaft Wort", Munich.

Springer-Verlag Berlin Heidelberg New York
Springer-Verlag is a part of Springer Science+Business Media
springeronline.com

© Springer-Verlag Berlin Heidelberg 1987
Printed in Germany

Typesetting: Asco Trade Typesetting, Ltd., Hong Kong

SPIN 10972151 44/3111 – 543

Acknowledgements

Pour rassembler les éléments un peu disparates qui constituent ce livre, j'ai dû faire appel à de nombreux amis, heureusement bien plus savants que moi. Ce sont, entre autres, Geneviève Averous, Lionel Bérard-Bergery, Marcel Berger, Jean-Pierre Bourguignon, Andrei Derdzinski, Dennis M. DeTurck, Paul Gauduchon, Nigel J. Hitchin, Josette Houillot, Hermann Karcher, Jerry L. Kazdan, Norihito Koiso, Jacques Lafontaine, Pierre Pansu, Albert Polombo, John A. Thorpe, Liane Valère.

Les institutions suivantes m'ont prêté leur concours matériel, et je les en remercie: l'UER de mathématiques de Paris 7, le Centre de Mathématiques de l'Ecole Polytechnique, Unités Associées du CNRS, l'UER de mathématiques de Chambéry et le Conseil Général de Savoie.

Enfin, qu'il me soit permis de saluer ici mon prédecesseur et homonyme Jean Besse, de Zürich, qui s'est illustré dans la théorie des fonctions d'une variable complexe (voir par exemple [Bse]).

Vôtre,

Arthur Besse

Le Faux, le 15 septembre 1986

Table of Contents

Chapter 0. Introduction 1

A. Brief Definitions and Motivation 1
B. Why Write a Book on Einstein Manifolds? 5
C. Existence 6
D. Examples
   1. Algebraic Examples 6
   2. Examples from Analysis 7
   3. Sporadic Examples 8
E. Uniqueness and Moduli 9
F. A Brief Survey of Chapter Contents 10
G. Leitfaden 14
H. Getting the Feel of Ricci Curvature 15
I. The Main Problems Today 18

Chapter 1. Basic Material 20

A. Introduction 20
B. Linear Connections 22
C. Riemannian and Pseudo-Riemannian Manifolds 29
D. Riemannian Manifolds as Metric Spaces 35
E. Riemannian Immersions, Isometries and Killing Vector Fields 37
F. Einstein Manifolds 41
G. Irreducible Decompositions of Algebraic Curvature Tensors 45
H. Applications to Riemannian Geometry 48
I. Laplacians and Weitzenböck Formulas 52
J. Conformal Changes of Riemannian Metrics 58
K. First Variations of Curvature Tensor Fields 62

Chapter 2. Basic Material (Continued): Kähler Manifolds 66

0. Introduction 66
A. Almost Complex and Complex Manifolds 66
B. Hermitian and Kähler Metrics 69
C. Ricci Tensor and Ricci Form 73
D. Holomorphic Sectional Curvature 75
E. Chern Classes 78

F. The Ricci Form as the Curvature Form of a Line Bundle 81
G. Hodge Theory 83
H. Holomorphic Vector Fields and Infinitesimal Isometries 86
I. The Calabi-Futaki Theorem 92

Chapter 3. Relativity 94

A. Introduction 94
B. Physical Interpretations 94
C. The Einstein Field Equation 96
D. Tidal Stresses 97
E. Normal Forms for Curvature 98
F. The Schwarzschild Metric 101
G. Planetary Orbits 105
H. Perihelion Precession 107
I. Geodesics in the Schwarzschild Universe 108
J. Bending of Light 110
K. The Kruskal Extension 111
L. How Completeness May Fail 113
M. Singularity Theorems 115

Chapter 4. Riemannian Functionals 116

A. Introduction 116
B. Basic Properties of Riemannian Functionals 117
C. The Total Scalar Curvature: First Order Properties 119
D. Existence of Metrics with Constant Scalar Curvature 122
E. The Image of the Scalar Curvature Map 124
F. The Manifold of Metrics with Constant Scalar Curvature 126
G. Back to the Total Scalar Curvature: Second Order Properties 129
H. Quadratic Functionals 133

Chapter 5. Ricci Curvature as a Partial Differential Equation 137

A. Pointwise (Infinitesimal) Solvability 137
B. From Pointwise to Local Solvability: Obstructions 138
C. Local Solvability of $\mathrm{Ric}(g) = r$ for Nonsingular r 140
D. Local Construction of Einstein Metrics 142
E. Regularity of Metrics with Smooth Ricci Tensors 143
F. Analyticity of Einstein Metrics and Applications 145
G. Einstein Metrics on Three-Manifolds 146
H. A Uniqueness Theorem for Ricci Curvature 152
I. Global Non-Existence 153

Chapter 6. Einstein Manifolds and Topology 154

A. Introduction 154
B. Existence of Einstein Metrics in Dimension 2 155
C. The 3-Dimensional Case 157

D. The 4-Dimensional Case ... 161
E. Ricci Curvature and the Fundamental Group ... 165
F. Scalar Curvature and the Spinorial Obstruction ... 169
G. A Proof of the Cheeger-Gromoll Theorem on Complete Manifolds with Non-Negative Ricci Curvature ... 171

Chapter 7. Homogeneous Riemannian Manifolds ... 177

A. Introduction ... 177
B. Homogeneous Riemannian Manifolds ... 178
C. Curvature ... 181
D. Some Examples of Homogeneous Einstein Manifolds ... 186
E. General Results on Homogeneous Einstein Manifolds ... 189
F. Symmetric Spaces ... 191
G. Standard Homogeneous Riemannian Manifolds ... 196
H. Tables ... 200
I. Remarks on Homogeneous Lorentz Manifolds ... 205

Chapter 8. Compact Homogeneous Kähler Manifolds ... 208

0. Introduction ... 208
A. The Orbits of a Compact Lie Group for the Adjoint Representation ... 209
B. The Canonical Complex Structure ... 212
C. The *G*-Invariant Ricci Form ... 215
D. The Symplectic Structure of Kirillov-Kostant-Souriau ... 220
E. The Invariant Kähler Metrics on the Orbits ... 221
F. Compact Homogeneous Kähler Manifolds ... 224
G. The Space of Orbits ... 227
H. Examples ... 229

Chapter 9. Riemannian Submersions ... 235

A. Introduction ... 235
B. Riemannian Submersions ... 236
C. The Invariants *A* and *T* ... 238
D. O'Neill's Formulas for Curvature ... 241
E. Completeness and Connections ... 244
F. Riemannian Submersions with Totally Geodesic Fibres ... 249
G. The Canonical Variation ... 252
H. Applications to Homogeneous Einstein Manifolds ... 256
I. Further Examples of Homogeneous Einstein Manifolds ... 263
J. Warped Products ... 265
K. Examples of Non-Homogeneous Compact Einstein Manifolds with Positive Scalar Curvature ... 272

Chapter 10. Holonomy Groups ... 278

A. Introduction ... 278
B. Definitions ... 280

C. Covariant Derivative Vanishing Versus Holonomy Invariance. Examples ... 282
D. Riemannian Products Versus Holonomy ... 285
E. Structure I ... 288
F. Holonomy and Curvature ... 290
G. Symmetric Spaces; Their Holonomy ... 294
H. Structure II ... 300
I. The Non-Simply Connected Case ... 307
J. Lorentzian Manifolds ... 309
K. Tables ... 311

Chapter 11. Kähler-Einstein Metrics and the Calabi Conjecture ... 318

A. Kähler-Einstein Metrics ... 318
B. The Resolution of the Calabi Conjecture and its Consequences ... 322
C. A Brief Outline of the Proofs of the Aubin-Calabi-Yau Theorems ... 326
D. Compact Complex Manifolds with Positive First Chern Class ... 329
E. Extremal Metrics ... 333

Chapter 12. The Moduli Space of Einstein Structures ... 340

A. Introduction ... 340
B. Typical Examples: Surfaces and Flat Manifolds ... 342
C. Basic Tools ... 345
D. Infinitesimal Einstein Deformations ... 346
E. Formal Integrability ... 348
F. Structure of the Premoduli Spaces ... 351
G. The Set of Einstein Constants ... 352
H. Rigidity of Einstein Structures ... 355
I. Dimension of the Moduli Space ... 358
J. Deformations of Kähler-Einstein Metrics ... 361
K. The Moduli Space of the Underlying Manifold of K3 Surfaces ... 365

Chapter 13. Self-Duality ... 369

A. Introduction ... 369
B. Self-Duality ... 370
C. Half-Conformally Flat Manifolds ... 372
D. The Penrose Construction ... 379
E. The Reverse Penrose Construction ... 385
F. Application to the Construction of Half-Conformally Flat Einstein Manifolds ... 390

Chapter 14. Quaternion-Kähler Manifolds ... 396

A. Introduction ... 396
B. Hyperkählerian Manifolds ... 398
C. Examples of Hyperkählerian Manifolds ... 400

D. Quaternion-Kähler Manifolds 402
E. Symmetric Quaternion-Kähler Manifolds 408
F. Quaternionic Manifolds 410
G. The Twistor Space of a Quaternionic Manifold 412
H. Applications of the Twistor Space Theory 415
I. Examples of Non-Symmetric Quaternion-Kähler Manifolds 419

Chapter 15. A Report on the Non-Compact Case 422

A. Introduction 422
B. A Construction of Nonhomogeneous Einstein Metrics 423
C. Bundle Constructions 424
D. Bounded Domains of Holomorphy 428

Chapter 16. Generalizations of the Einstein Condition 432

A. Introduction 432
B. Natural Linear Conditions on Dr 433
C. Codazzi Tensors 436
D. The Case $Dr \in C^\infty(Q \oplus S)$: Riemannian Manifolds with Harmonic Weyl Tensor 440
E. Condition $Dr \in C^\infty(S)$: Riemannian Manifolds with Harmonic Curvature 443
F. The Case $Dr \in C^\infty(Q)$ 447
G. Condition $Dr \in C^\infty(A)$: Riemannian Manifolds such that $(D_X r)(X, X) = 0$ for all Tangent Vectors X 450
H. Oriented Riemannian 4-Manifolds with $\delta W^+ = 0$ 451

Appendix. Sobolev Spaces and Elliptic Operators 456

A. Hölder Spaces 456
B. Sobolev Spaces 457
C. Embedding Theorems 457
D. Differential Operators 459
E. Adjoint 460
F. Principal Symbol 460
G. Elliptic Operators 461
H. Schauder and L^p Estimates for Linear Elliptic Operators 463
I. Existence for Linear Elliptic Equations 464
J. Regularity of Solutions for Elliptic Equations 466
K. Existence for Nonlinear Elliptic Equations 467

Addendum 471

A. Infinitely Many Einstein Constants on $S^2 \times S^{2m+1}$ 471
B. Explicit Metrics with Holonomy G_2 and Spin(7) 472
C. Inhomogeneous Kähler-Einstein Metrics with Positive Scalar Curvature 474

D. Uniqueness of Kähler-Einstein Metrics with Positive Scalar Curvature 475
E. Hyperkählerian Quotients........ 477

Bibliography 479

Notation Index 500

Subject Index........ 505

Errata 511

Chapter 0. Introduction

A. Brief Definitions and Motivation

0.1. "Un espace de Riemann est au fond formé d'une infinité de petits morceaux d'espaces euclidiens" (E. Cartan) (A Riemannian manifold is really made up of an infinity of small pieces of Euclidean spaces). In modern language, a Riemannian manifold (M, g) consists of the following data: a compact C^∞ manifold M and a metric tensor field g which is a positive definite bilinear symmetric differential form on M. In other words, we associate with every point p of M a Euclidean structure g_p on the tangent space T_pM of M at p and require the association $p \mapsto g_p$ to be C^∞. We say that g is a Riemannian *metric* on M.

0.2. In contrast, a Riemannian *structure* (see Chapters 4 and 12 for more on this point) is a class of isometric Riemannian manifolds (two Riemannian manifolds (M, g), (M', g') are isometric if there exists some diffeomorphism $f: M \to M'$ which transfers g' into g, i.e., $f^*g' = g$). In other words, if $\mathscr{M}$ (or $\mathscr{M}(M)$ if appropriate) denotes the set of Riemannian metrics on M, the set of Riemannian structures on M is the quotient $\mathscr{M}/\mathfrak{D}$ of $\mathscr{M}$ by the group $\mathfrak{D} = \mathfrak{D}(M)$ of diffeomorphisms of M.

0.3. As we have said Riemannian manifolds are generalizations of Euclidean spaces. They also occur naturally in mechanics, see [Ab-Ma] and [CB-DW-DB]. Further generalizations are Finsler manifolds. These occur when we associate (always in a C^∞ manner of course) to each point p of M, a normed (Banach) structure in the tangent space T_pM. Although Finsler manifolds are found naturally in mechanics and physics, they turn out to be less important than Riemannian manifolds and good published material on them is rare. (See however E. Cartan [Car 10], H. Busemann [Bus 1], H. Rund [Run], M. Gromov [Gro 3]).

0.4. We now start with a given compact C^∞-differentiable manifold M of dimension n, and ask ourselves the following question:

> ARE THERE ANY BEST (OR NICEST, OR DISTINGUISHED) RIEMANNIAN STRUCTURES ON M?

This question was put to us by a good friend, René Thom, in the Strasbourg mathematics library in 1958. For the sake of simplicity we shall restrict ourselves to the *compact* case. If one looks on Einstein manifolds as critical points of the U functional (see 0.12 or Chapter 4) then it is reasonable to hope for an existence result in the compact case (even if this approach is disappointing in the end, see 4.1). The non-compact case remains interesting nevertheless (see Chapter 15).

0.5. For surfaces ($n = 2$) the answer to the above question is known. The best Riemannian structures on a given compact surface M are those of *constant curvature*. In dimension 2, there is only one notion of curvature, namely the Gauss curvature, which is a function $K: M \to \mathbb{R}$. If M is embedded in $\mathbb{R}^3$ with the induced natural Riemannian metric, the Gauss curvature is the product of the principal curvatures, or the inverse of the product of the principal radii of curvature. There exists at least one Riemannian metric of constant curvature on any compact surface and the constant can be normalized by scaling to $+1, 0, -1$. Moreover, on a given M, the Riemannian structures of constant given curvature 1, 0 or -1 form a nice finite dimensional submanifold—with singularities—of $\mathcal{M}/\mathfrak{D}$, the so-called *Moduli Space*. For example, on S^2 (the 2-dimensional sphere) and on $\mathbb{R}P^2$ (the real projective plane) there exists exactly one Riemannian structure of constant curvature equal to 1. On a compact surface M, orientable or not, with Euler characteristic $\chi(M) < 0$, the Riemannian structures with constant curvature -1 depend on $-3\chi(M)$ real parameters. For example, $6\gamma - 6$ parameters when M is orientable and of genus γ. For more detail, see 12.B.

0.6. What would a natural generalization of the concept of constant curvature be, for a Riemannian manifold, when the dimension is greater than 2? We claim that a good generalization would be the notion of *constant Ricci curvature*.

0.7. For Ricci curvature the reader can refer to Chapter 1. Here we shall simply say this. The Ricci curvature of a Riemannian manifold (M, g) is a quadratic differential form (alias a bilinear symmetric form), *denoted by* r (or r_g if the metric needs to be specified). By diagonal restriction r defines a function (still denoted by r) on the unit tangent bundle UM of M. This function r is the trace of the curvature endomorphism defined as follows by the curvature 4-tensor R: $r(x) = \text{trace}(z \mapsto R(x, z)x)$. Equivalently, for x in UM, the value $r(x)$ is the sum $\sum_{i=2}^{n} K(x, x_i)$ of the sectional curvatures $K(x, x_i)$ for any completion of x to an orthonormal basis $(x, x_2, \ldots, x_n)$.

0.8. A Riemannian manifold has three main notions of curvature. i) The Riemann *curvature tensor* R (equivalent to the *sectional curvature* function, K, defined on tangent planes) which is a biquadratic form giving complete information, at the curvature level, on g. ii) The Ricci curvature above, r, the trace of R with respect to g. iii) Finally, the scalar curvature, s, a scalar function on M, which is the trace $s = \text{trace}_g r$ with respect to g of the quadractic form r. Note first that when $n = 2$ the three curvatures are equivalent. When $n = 3$ the Ricci curvature contains as much information as the Riemann curvature tensor. This is the main reason why our book is devoted to manifolds of dimension greater than or equal to 4.

0.9. An easy argument shows that the Ricci quadratic form $r\colon UM \to \mathbb{R}$ is a constant λ if and only if $r = \lambda g$. Such a Riemannian manifold is called an *Einstein manifold*. By normalization one can always assume to be in one of the following three cases:

$$r = g \qquad \text{(when } \lambda \text{ is positive)}$$

$$r = 0 \qquad \text{(when } \lambda \text{ is zero)}$$

$$r = -g \qquad \text{(when } \lambda \text{ is negative).}$$

The corresponding number $+1$, 0, or -1 will be called the *sign* of the Einstein manifold.

0.10. We give a *first justification* of our claim that an Einstein metric is a good candidate for privileged metric on a given manifold. It might be natural to consider as "best" metrices those of constant curvature (homogeneity at the curvature level). For the sectional curvature (viewed as a function on the Grassmann bundle of tangent 2-planes to M), its constancy makes the Riemannian manifold, after normalization, locally isometric to a unique model space, with the same constant curvature, namely the standard sphere (S^n, can) with its canonical structure "can" (inherited for example from the standard imbedding $S^n \subset \mathbb{R}^{n+1}$) when the constant is positive, or the Euclidean space ($\mathbb{R}^n$, can) (viewed as a Riemannian manifold) when the constant is zero or finally the hyperbolic space (H^n, can) when the constant is negative. In particular, there is one and only one Riemannian simply connected normalized structure of constant sectional curvature $+1$, 0, or -1. The corresponding manifolds are diffeomorphic to $\mathbb{R}^n$ or S^n. Consequently, many manifolds of dimension $n \geqslant 4$ cannot admit such a metric. For $n = 3$, the situation is still not clear (see 6.C).

On the other hand, on any compact manifold of any dimension there exist Riemannian metrics of constant scalar curvature and they form an infinite dimensional family for $n \geqslant 3$ (see 4.F). So there are too many of them to be legitimately called privileged metrics. In short, constancy of sectional curvature is too strong, constancy of scalar curvature is too weak and we are left with constancy of Ricci curvature.

From a naive analytic viewpoint, the study of such metrics looks reasonable since both the metric and the Ricci curvature depend apparently on the same number of parameters, namely $n(n + 1)/2$. In fact, we shall see in the book that they form a finite dimensional family. Existence is another problem. But at least we will meet many examples.

0.11. For the amateur of linear representations (say for the algebraically minded reader) let us mention here that the curvature tensor R of a Riemannian manifold is acted upon by the orthogonal group and then gives rise to a decomposition

$$R = U + Z + W$$

into pieces corresponding to the irreducible components of this action. The part W is the famous H. Weyl conformal curvature tensor, U is equivalent to the scalar curvature s and Z is the traceless part of the Ricci curvature. So the condition $Z = 0$

is equivalent to the Einstein condition. (See more on this point and on conditions $U = 0$ and $W = 0$ in 1.G and H).

0.12. We now give a second justification. It is a double one under one cover. Roughly speaking the Einstein condition for a Riemannian manifold (M, g_0) is equivalent to the following: on the space $\mathscr{M}_1(M)$ of metrics of volume one, the functional

$$\mathbf{S}: g \mapsto \int_M s_g \mu_g$$

(the total scalar curvature) admits g_0 as a critical point. The latter assertion means that, for any variation $g_0 + th$ where h belongs to $\mathscr{S}^2 M$, the space of bilinear symmetric differential forms on M, the derivative

$$\frac{d}{dt}\mathbf{S}(g_0 + th)|_{t=0}$$

vanishes. The simplicity of the functional S, and the naturality of the critical point condition, favour the Einstein metrics as "best".

0.13. Further justification comes from history and from physics. Working on relativity and considering Lorentzian manifolds instead of Riemannian manifolds (Lorentzian manifolds are manifolds equipped with a tensor field g of signature $(-, -, \ldots, -, +)$ in place of positive definite $(+, \ldots, +)$, see Chapter 3), A. Einstein proposed in 1913 that the field equations for the interaction of gravitation and other fields take the form $r - \frac{1}{2}sg = T$ where T is the energy-momentum tensor. In particular, "no mass" leads to Ricci flat manifolds: $r = 0$. A. Einstein derived this condition as the Euler-Lagrange equation of a variational problem. In fact computation shows (see 4.17) that $r - \frac{1}{2}sg$ is the gradient of the functional S. If one considers only metrics of volume equal to one, one gets the "Ricci constant" condition, which we have already considered and which defines Einstein metrics. We have to do this because Ricci flatness is too strong a condition on most manifolds. For example, due to Bochner's theorem 6.56, a compact manifold M with first Betti number $b_1(M) > \dim M$ does not admit any Ricci flat metric.

0.14. Let us be clear about our aim. We use the term "Einstein manifolds" for Riemannian manifolds of constant Ricci curvature, because it has been admitted among mathematicians for a long time. We do not claim to work for mathematical physics. However, in their views on this, theoretical physicists fall into two groups. The first believes that what we do is just rubbish. The second thinks that Riemannian manifolds (for example Einstein Riemannian manifolds) may be of some help to them, if only by way of inspiration. Or perhaps even truly helpful, by putting everything into a bigger setting by complexification, so that differences in sign disappear. The same considerations apply to Yang-Mills theory.

0.15. A third justification is the fact that constant Ricci curvature is equivalent to constant curvature in dimension 2, and here things work pretty well (see 0.5 and

12.B). For any compact surface, there exist metrics with constant curvature and they form a nice finite dimensional set. For dimension 3, see 6.C.

0.16. This book is devoted to Einstein manifolds of dimension $n \geqslant 4$. In Sections D, E, F we will describe the contents as a whole, as opposed to the detailed description by chapters in G. For simplicity's sake, we have put things under three headings: existence, examples and uniqueness. We hope this will be useful for a first approach to the subject, (see the reviews [Ber 6] and [Bou 5, 10]).

B. Why Write a Book on Einstein Manifolds?

0.17. In September 1979, a symposium on Einstein manifolds was held at Espalion, France. It was there we realized that a book on the subject could be worthwhile. The subject seemed ripe enough: certainly most basic questions were still open, but good progress had been made, due in particular to the solution of Calabi's conjecture (by T. Aubin for the negative sign and S.T. Yau for both negative and zero signs) and to N. Koiso's results on the moduli space. The subject is in full growth at present. Moreover a fair number of examples of Einstein manifolds of various types are now available.

0.18. Einstein manifolds are not only interesting in themselves but are also related to many important topics of Riemannian geometry. For example: Riemannian submersions, homogeneous Riemannian spaces, Riemannian functionals and their critical points, Yang-Mills theory, self-dual manifolds of dimension four, holonomy groups, quaternionic manifolds, algebraic geometry via K3 surfaces. The study of these topics is flourishing today. On the other hand, there seem to be no links established between Einstein manifolds and geodesics or the spectrum of the Laplacian.

0.19. The book we present here is intended to be a complete reference book, even including related material mentioned above. We confess to having used Einstein manifolds also as a partial pretext for treating some questions of geometry which we hold dear. Some very difficult proofs (e.g. 11.C) are only sketched in order to keep the book down to size for the Ergebnisse but only when excellent proofs are easily found elsewhere.

0.20. The various chapters are self-contained. Consequently, many things appear several times, but repetition is after all a pedagogic quality. Due to our lack of competence on some of the material, we had to ask many of our friends to contribute to this book. Numerous meetings were held in order to get as much coherence as possible. Agreeing on notation was particularly difficult (see 0.39 below). Credit is given in detail in the various chapters, so we shall not dwell on it here. Should there be any complaints, in this respect "les tribunaux seuls compétents sont ceux du domicile de l'auteur, i.e. Seine, France".

C. Existence

0.21. Today things look completely different when $n = 4$ and when $n \geqslant 5$. Ridiculous as it may seem, when $n \geqslant 5$ we do not know the answer to the simple question: "Does every compact manifold carry at least one Einstein metric?". In some sense we do not know if to be an Einstein manifold in dimension $n \geqslant 5$ is a *strong* condition or not. Of course we do not prescribe a given sign in the Einstein condition; this sign can be arbitrary a priori. Otherwise the existence of an Einstein metric on M with a positive sign is known to imply finiteness of the fundamental group of M (see 6.52). Let us also mention here that if we look for *Kähler* Einstein manifolds then we do have topological consequences in all complex dimensions (see 11.A).

0.22. The situation of manifolds of dimension 4 is, on the contrary, somewhat better understood. The Einstein condition means in this dimension that the curvature operator commutes with the Hodge $*$ operator acting on exterior 2-forms. This commutation relation implies strong restrictions on the two integrands involved in Chern's formulas for the Euler class and the Pontryagin class (in dimension 4 there is only one Pontryagin class, hence number, to consider). If $\chi(M)$ denotes the Euler characteristic of M and $p_1(M)$ its Pontryagin number, then the commutation relation above implies the inequality

$$\chi(M) \geqslant \tfrac{1}{2}|p_1(M)|.$$

Not every compact manifold of dimension 4 can therefore carry an Einstein metric: for instance the Euler characteristic has to be non-negative. A more sophisticated example where the above condition fails is the connected sum of five or more complex projective planes, or two or more tori. But whether the above condition is sufficient remains an open question. The reader should be aware that when $n \geqslant 5$ the Einstein condition on the curvature tensor does not imply any condition on the various integrands involved in Chern's formulas for the Euler and Pontryagin numbers (see 6.41).

D. Examples

0.23. Despite the simplicity of the condition $r = \lambda g$ the reader should not imagine that examples are easy to find. If you are not convinced, try to find one yourself which is not in our book. And if you succeed, please write to us immediately. Ricci flat compact manifolds are even harder to come by. The author will be happy to stand you a meal in a starred restaurant in exchange of one of these!

1. Algebraic Examples

0.24. What is the easiest way to find Riemannian manifolds of constant Ricci curvature? Let M be some homogeneous Riemannian manifold, i.e. we assume that the isometry group $G = \mathrm{Isom}(M)$ acts transitively on M. In particular M can be written as a coset space $M = G/H$ where G is a Lie group and H is a closed, in fact compact, subgroup of G. In particular, G will preserve every Riemannian invariant of M. To begin with, the scalar curvature s will be constant. If we now want the Ricci curvature to be a constant function on the unit tangent bundle UM, it will suffice for G to be transitive on UM (one then says that $M = G/H$ is isotropic). Of course it is enough to require transitivity at one point, i.e., that the linear isotropy representation $\operatorname{Ad} H$ act transitively on the unit sphere at the base point $p(e) = p_0$. But this condition is extremely strong. It says that such a metric space is two-point homogeneous. The class of these spaces is known to consist only of the symmetric spaces of rank one. Namely, Euclidean spaces, spheres, the various projective spaces and the non-compact duals of these spaces.

0.25. There is however the following basic remark due to Elie Cartan ([Car 12] or [Wol 3] p. 137). The homogeneous Riemannian space $M = G/H$ will be automatically Einstein as soon as this isotropy representation $\operatorname{Ad} H$ in $T_{p_0}M$ is irreducible. For $\operatorname{Ad} H$ will leave invariant two quadratic forms on $T_{p_0}M$, the metric g and the Ricci curvature r. Since one is positive definite, the reduction theory and the fact that $\operatorname{Ad} H$ is irreducible imply that these two quadratic forms are proportional. Note that by the same token, when $\operatorname{Ad} H$ is linear irreducible, then $M = G/H$, as an a priori only homogeneous manifold, possesses (up to a scalar) exactly one homogeneous Riemannian metric, which is automatically Einstein.

0.26. The naive approach of 0.24 yields only the symmetric spaces of rank one. Elie Cartan's remark above yields many more. First all irreducible Riemannian symmetric spaces are isotropy irreducible. This includes spaces of constant curvature, the various projective spaces and also the Grassmann manifolds. In 7.51 it will be seen that there are roughly twice as many G/H's with $\operatorname{Ad} H$ acting irreducibly as there are irreducible symmetric spaces. Some of them are extremely interesting in the sense that the dimension of H is very small compared with that of G (see tables in 7.106, 7.107).

0.27. To find all homogeneous Riemannian compact Einstein manifolds is an algebraic problem but not a simple one. In theory it can be solved by analysing the pair (g, h) of the Lie algebras of G and H respectively. The more reducible the representation $\operatorname{Ad} H$, the harder the problem becomes, because there are more and more different homogeneous Riemannian metrics on G/H. At this very moment, great progress is being made in this classification (see Chapter 7 and [DA-Zi] together with [Wa-Zi]). To give some idea of how "non-canonical" a homogeneous Einstein space can be, here are some examples. On S^{4n+3} and $\mathbb{C}P^{2n+1}$ there are non-canonical Einstein metrics and on S^{15} at least two non-canonical ones. On the group $SO(n)$ there are at least n distinct Einstein metrics (if $n \geqslant 12$). There are Einstein metrics on $S^2 \times S^3$, $S^3 \times S^3$, $S^6 \times S^7$ and $S^7 \times S^7$ which are not product metrics.

2. Examples from Analysis

0.28. Up to 1980 there was no general series of compact Einstein manifolds other than the homogeneous ones. Back in 1954 E. Calabi made a conjecture—now known to be true—which yields a large class of compact manifolds with vanishing Ricci curvature, i.e., *Ricci flat* Riemannian manifolds, that is to say Einstein with 0 sign.

We consider here compact Kähler manifolds. In this case the Ricci curvature, thanks to Chern, can be interpreted in a very nice way, as follows. Via the complex structure the Ricci curvature is first transformed into a closed exterior 2-form ρ (see chapter 2 for all Kählerian considerations). Now, by the de Rham theorem, this 2-form yields a 2-dimensional real cohomology class, which (up to 2π) is simply the first Chern class $c_1(M)$ of our manifold M. E. Calabi conjectured that, if the first Chern class of a compact Kähler manifold vanishes, then this manifold admits some Kähler metric which is Ricci flat.

0.29. In 1955 E. Calabi proved the uniqueness of such a metric (within any Kähler class). For a proof of existence we had to wait until 1976. S.T. Yau's existence theorem rests on a non-linear partial differential equation of Monge-Ampère type, which has also excited analysts for a long time. In oral communications Calabi extended his conjecture to the case where $c_1(M)$ is negative definite and for Einstein metrics with sign -1. This conjecture was then proved by T. Aubin and independently by S.T. Yau in 1976.

0.30. Now there are many examples of compact Kähler manifolds with vanishing first Chern class or with negative first Chern class, for which we refer to 11.A. Here let us just mention that algebraic hypersurfaces of degree d of the complex projective space $\mathbb{C}P^{m+1}$ have first Chern class which is zero if $d = m + 2$ and negative if $d > m + 2$. But in order to appreciate the difficulty of Calabi's conjecture, and at the same time that of finding Einstein manifolds, the lay reader should be aware that such a hypersurface will never be Einstein for the induced Kähler metric from $\mathbb{C}P^{m+1}$. Analysis provides us with an Einstein metric in a non constructive way. Note also that homogeneous Riemannian manifolds, if non flat, can never be Ricci flat. The above examples are the only compact Ricci flat manifolds known today.

3. Sporadic Examples

0.31. Besides the preceding examples only two other constructions of Einstein manifolds are known and they provide only a few new compact examples. The first series comes from a result of S. Kobayashi. Consider a compact Kähler Einstein manifold with sign $+1$ and over it an S^1-principal bundle whose cohomology class is a multiple of the Kähler class of the base manifold. Then, on this bundle (viewed as a manifold) there exists an Einstein metric. Unfortunately, all examples obtained by this method are homogeneous.

0.31a. In 1979 Don Page found a completely new example of an Einstein metric on the connected sum of two complex projective planes (this manifold can also be obtained by blowing up one point of the complex projective plane, or else viewed as the non-trivial S^2-bundle over S^2). D. Page's construction was turned into a systematic technique by L. Bérard-Bergery. However at present this technique yields few new compact Einstein manifolds. The non-compact case is much easier.

E. Uniqueness and Moduli

0.32. By uniqueness and moduli we mean the problem of classifying all Einstein structures on a given compact manifold M. In dimensions 2 and 3 things are clear enough (see 12.B and G). But starting at $n = 4$ and then for $n \geqslant 5$, little is known. Note first that no manifold is known to admit exactly one Einstein structure, not even S^4 or $\mathbb{C}P^2$ for example.

0.33. The following are the basic known results in the uniqueness direction.

First: on S^n the canonical metric is isolated among Einstein metrics, and unique within an interval of pinching for sectional curvature of $3n/(7n - 4)$. This is probably not the best possible bound. For $n = 4$ we know the sharp bound, namely 1/4.

Second: a compact complex homogeneous space admits exactly one Kähler-Einstein metrio (Y. Matsushima, see Chapter 8). However there may be other Einstein metrics, non-Kähler, or Kähler with respect to other complex structures. This is the case for the manifold obtained by blowing up one point on $\mathbb{C}P^2$ (see 0.31).

Third: there is definitely no uniqueness on the spheres S^{4n+3} on which G. Jensen found Einstein non-canonical structures for every n. They are homogeneous. There are examples of this type for other classical manifolds (see 8.2 and 9.82).

0.34. Are there manifolds on which all Einstein structures are known? This is now true for some 4-manifolds, namely the torus T^4 (and its quotients) on which every Einstein metric has to be flat (vanishing curvature tensor) and the K3 surface (and its quotients by $\mathbb{Z}/2$) on which every Einstein metric has to be Kähler with respect to some K3 complex structure. The set of Einstein structures forms in $\mathcal{M}/\mathcal{D}$ a finite dimensional submanifold of dimension 6 for T^4 and 57 for the K3 surface (see 12.B and 12.K).

0.35. Now let M be a compact manifold and let us *denote* by $\mathscr{E}(M)$ the subset of $\mathcal{M}/\mathfrak{D}$ made up by the Einstein structures. We call $\mathscr{E}(M)$ the *Moduli Space* of M. What is known about $\mathscr{E}(M)$? The first thing to do is to look in a naive way at the tangent space of $\mathscr{E}(M)$. We consider a curve $g(t)$ in $\mathscr{E}(M)$, i.e., $r_{g(t)} = \lambda(t)g(t)$ for every t and $g(0) = g$. Taking the derivative with respect to t at $t = 0$ and setting $h = \left.\frac{dg(t)}{dt}\right|_{t=0}$, we first prove that $\lambda(t)$ cannot change signs. So let us set $\lambda = 1, 0$ or -1. It can now be seen that h should obey an elliptic partial differential equation. It follows that $\mathscr{E}(M)$ is everywhere infinitesimally finite-dimensional. So one might expect that, roughly speaking, $\mathscr{E}(M)$ would be made up of connected components each of finite dimension.

0.36. This is a vast programme. The difficulty is not of an analytic nature. It lies rather in the algebraic complexity of the global system of PDE's $r_g = \lambda g$.

Recently, N. Koiso carried out part of this programme. The first thing he discovered was that the naive tangent space above is not in general the tangent space of $\mathscr{E}(M)$ at the point g. He gave examples where some solution h of this equation is not the derivative at $t = 0$ of any curve $g(t)$ in $\mathscr{E}(M)$. However he proved that $\mathscr{E}(M)$ is always an analytic subset of a smooth submanifold of $\mathscr{M}/\mathfrak{D}$. Thus in some sense $\mathscr{E}(M)$ is finite dimensional. Moreover, it is Hausdorff.

0.37. A fascinating question is that of the possible set of Einstein constants of a given compact manifold. By this, we mean the set of real numbers λ with $r_g = \lambda g$ for some metric g on M with (normalized) volume$(g) = 1$. One knows only that this set is countable. One does not know whether it is a discrete subset of the real line, nor whether it is bounded below. It can easily be shown to have an upper bound (see 12.G).

F. A Brief Survey of Chapter Contents

Chapters 1 and 2. Basic Material

0.38. Over the years we have often been requested to write a treatise on Riemannian geometry. Having resisted the temptation, we present in these chapters mainly things which are essential for the present book. Examples of not quite standard notions are to be found in 0.H, 1.G, 1.I, 1.K, 2.H, 2.I.

0.39. We pay specific attention to the question of *notation*, subject of much animated discussion. Curvature, and especially the scalar curvature, posed the biggest problem. We finally agreed—not unanimously—on the following:

K for the sectional curvature,
R for the curvature tensor,
r for the Ricci curvature,
s for the scalar curvature.

We did not use notations like ric, or ricci, or scal, because in certain cases equations would have become impossible to read, e.g. $\Delta_L(\text{ricci}) = \text{scal} \cdot h - \text{ricci}(\delta h)$.

Chapter 3. Relativity

0.40. We emphasize again that our book deals with truly Riemannian manifolds, i.e., our metrics g are positive definite. Nevertheless the subject has some relations with relativity (not just because of the name) and Lorentzian manifolds. So we thought it fair to present briefly a few notions and facts on relativity theory: the energy-momentum tensor etc... Since Lorentzian geometry is considerably expanding as a mathematical discipline in itself, we had to make a choice, which we hope will please the non-expert reader.

0.41. It would seem that Riemannian and Lorentzian geometry have much in common: canonical connections, geodesics, curvature tensor, etc... But in fact this common part is only a common disposition at the onset: one soon enters different realms. For example, looking at the Einstein condition, the Riemannian geometer will proceed as in the present book: existence, uniqueness, moduli... On the other hand the physicist starts from a spacelike hypersurface and propagates the induced Riemannian metric to get a Lorentzian metric on space-time using Einstein's equation. The Riemannian will meet elliptic PDE's and the Lorentzian hyperbolic ones. Moreover, the latter will work essentially on non compact objects and meet singularities. The interaction between the two fields is best illustrated by the following picture from C.N. Yang in [Yan]:

Fig. 0.43

There are wonderful books on Lorentzian geometry: [Be-Eh], [Ha-El], [Mi-Th-Wh], [ONe 3], [Sa-Wu].

Chapter 4. Riemannian Functionals

0.42. This chapter stems directly from 0.12: the Riemannian manifold (M, g) is Einstein if and only if g is a critical point of the functional $\mathbf{S}(g) = \int_M s_g \mu_g$ on the space $\mathcal{M}_1(M)$. This fact leads the mathematician to many natural questions. Is there a good Morse theory to compute the second variation (index form)? Are there some other functionals and what are the corresponding critical metrics? Chapter 4 will tell you everything known about these aspects, but there remain more open questions than results. Note that in this chapter we will meet metric variations, the set $\mathcal{M}/\mathfrak{D}$ of Riemannian structures, and also the recent theory on the possible signs for scalar curvature.

Chapter 5. Ricci Curvature as a Partial Differential Equation

0.43. If we forget the metric g then its Ricci curvature r_g appears just like some poor orphan r in the space $\mathcal{S}^2M$ of bilinear symmetric differential forms on a given compact manifold M. Many questions come to mind. In particular the existence and uniqueness of metrics g such that $r_g = r$. Chapter 5 deals with this kind of question, to which many good answers have recently been put forward. The difficulty is the following: the PED $r_g = r$, with g unknown, is highly nonlinear, nonelliptic, and of a quite original type. Its main feature is that it is overdetermined because of Bianchi's identity.

0.44. In this chapter the reader will find many nice results. There will be real orphans, i.e., r forms with no metric they could come from, but there will also be real children, and even only children. In the same vein, it will turn out that any Einstein metric g is real analytic (in suitable coordinates of course) because of the character of the equation $r_g = \lambda g$. This has nice corollaries. Finally Hamilton's result on the existence of Einstein metrics on certain 3-manifolds is given with a sketch of the proof.

Chapter 6. Einstein Manifolds and Topology

0.45. This chapter contains the material announced in C above. The problem of positive scalar curvature treated here is at the heart of present day research.

Chapter 7. Homogeneous Riemannian Manifolds

0.46. As explained in D above, homogeneous spaces provide many examples of Einstein manifolds. There is a huge amount of journal literature on Riemannian homogeneous manifolds, a few monographs but not too many systematic accounts in books (except Kobayashi-Nomizu [Ko-No 2]). We therefore thought it worthwhile to present here a systematic introduction. We present *tables* because we like them and there are not many in the literature.

Chapter 9. Riemannian Submersions

0.47. The idea of Riemannian submersions comes from the study of Riemannian metrics on a fibre bundle $E \to B$. In a way, they are the closest possible metrics to product metrics and roughly speaking, for Riemannian submersions (especially for those with totally geodesic fibres), one can compute the curvature of the total space E in terms of the curvatures of F and B. Their most striking property is that the sectional curvature of a plane tangent to B is always larger than or equal to that of its horizontal lift. This method has produced a wealth of examples (see 9.G to 9.K) and so it seemed to us that a systematic exposition of the theory would be useful.

Chapter 10. Holonomy Groups

0.48. The holonomy group of a Riemannian manifold (M, g) (strictly speaking at the point $p \in M$ but different points have isomorphic groups and representations) is the subgroup $H \subset O(T_pM)$ of the orthogonal group at p which is built up of all parallel transports along various loops based at p. This object seems to have no relation with Einstein metrics, but, because of Bianchi's identities and the fact that the holonomy group is generated by the curvature tensor of (M, g), there are few possibilities for the orthogonal representation $H \subset O(n)$. One can always assume that (M, g) is irreducible, since, if not, one proceeds component by component. Then for irreducible Riemannian manifolds either $H = SO(n)$, or $H = U(n/2)$ and we are necessarily in the Kähler case, or H can be taken in a *very short list*. But, again because of Bianchi's identities, every Riemannian manifold with holonomy in this list turns out to be automatically Einstein.

0.49. We shall add the following remarks. *First*, research on the holonomy group practically stopped from 1955 to 1975. It was then revived and most open problems were solved. In particular the solution of Calabi's conjecture (the Ricci flat case) produced examples of compact Riemannian manifolds with holonomy group $H = SU(n/2)$ and finally with $H = Sp(n/4)$. *Second*, there has not been a single textbook exposition on holonomy since Lichnerowicz [Lic 3]. *Third*, in the short list quoted above one finds $H = Sp(1)Sp(n/4)$. Manifolds with this holonomy group are called *quaternion-Kähler* and are the subject of Chapter 14.

Chapter 8 and 11. The Kähler Story

0.50. These are the central chapters of the book. They are quite detailed except for the proof of the existence theorem. The solution of the Calabi conjecture yields a large class of Einstein manifolds and in particular Ricci flat ones. These can at present be obtained in no other way, since Riemannian homogeneous manifolds are never Ricci flat except if already flat.

0.51. In passing we detail the theory of compact complex homogeneous spaces. They are at the intersection of algebra and analysis. There are not too many accounts of these spaces and in particular of their property of being exactly the orbits of the adjoint representation of compact Lie groups (Kirillov-Souriau theory). We have tried to present a modern exposition.

Chapter 12. The Moduli Space of Einstein Structure

0.52. A key chapter. The problem is the basic one on Einstein manifolds. We give a detailed account of Koiso's recent results, outlined in *E* above.

Chapter 13. Self-Dual Manifolds

0.53. For an oriented Riemannian manifold of dimension 4 the Weyl curvature tensor W splits into two parts W^+ and W^- under the action of $SO(4)$. If one requires that W^- be identically zero, one has by definition the notion of a half-conformally flat manifold. We discuss these manifolds since, thanks to R. Penrose, there is some hope of classifying Einstein half-conformally-flat manifolds or of obtaining new examples through 3-dimensional complex manifolds. By way of indication, we will prove that the only half-conformally-flat Einstein manifolds with $s \geqslant 0$ are the canonical S^4, CP^2 and the Ricci flat K3 surfaces. Unfortunately complex manifolds have given only non-compact examples. This chapter is quite detailed since these recent results have not appeared in book form.

Chapter 14. Quaternion-Kähler Manifolds

0.54. These are the Riemannian manifolds of dimension $4m$ whose holonomy group is contained in $Sp(1)Sp(m) \subset O(4m)$. Apparently they are very special Riemannian manifolds and we do not have many examples. Firstly, the group $Sp(1)Sp(m)$ appears in the short list quoted in 0.51 and therefore the underlying manifold is automatically Einstein. Secondly, the quaternion—Kähler manifolds are those for which a twistor construction, but based on the quaternions instead of the complex numbers, works. This chapter also treats Riemannian manifolds with holonomy contained in $Sp(m) \subset O(4m)$, called *hyperkählerian.* The solution of the Calabi conjecture yields a large class of these.

Chapter 15. A Report on the Non-Compact Case

0.55. We have chosen to put emphasis on compact manifolds, but there are some nice results in the non-compact case, including a deep existence theorem for complete Kähler-Einstein manifolds. We present a brief account and a complete bibliography of this rapidly growing field.

Chapter 16. Generalizations of the Einstein Condition

0.56. The product of two Einstein manifolds is not usually Einstein but it is still a "good" metric. In fact, its Ricci curvature is parallel: $Dr = 0$. Conversely, if $Dr = 0$, then the metric has to be locally a product of Einstein metrics.

0.57. $Dr = 0$ is a natural generalization of the Einstein condition. One can see that it implies that the curvature tensor is harmonic. This and analogous conditions are the subject of a considerable amount of literature. We have tried to present a complete survey of this material.

Appendix. Sobolev Spaces and Elliptic Operators

0.58. Since many tools from analysis are needed for the partial differential relations involved in this book, we have given some of the relevant material in this appendix.

G. Leitfaden

You are looking for *or* *you are interested in*	*Then look here*
Lie algebras and Lie groups	1, 2, 7, 8, 10
algebraic geometry	11, 12.K, 13, 14
analysis	4, 5, 9, 11, 15, App.
counter-examples	6, 11
existence	5, 7, 9, 11, 15
geometry	0.H, 3, 8, 9, 10, 11
Kähler geometry	2, 8, 11
Lorentz manifolds	1, 3, 6.D, 7.I, 10.J
motivation	0

open problems	0.I
a survey	0
topology	6
uniqueness	9, 11, 12
other subjects	index
hiking	follow the path: 0 + Section A of each chapter
trekking	try to follow the path: 0, 3, 5, 6, 15, 10, 7, 9

H. Getting the Feel of Ricci Curvature

Throughout our long life we have found the Ricci curvature quite hard to FEEL. We think, therefore, that the reader will appreciate some of our experiences. Historically G.C. Ricci introduced his curvature for the following reason. When $M \subset \mathbb{R}^{n+1}$ is a hypersurface embedded in Euclidean space it carries the second fundamental form, which is a quadratic differential form on M. In particular its eigenvalues (with respect to the metric g) are the principal curvature of M and its eigendirections yield the lines of curvature of M. Now if (M, g) is an abstract Riemannian manifold, there is no such quadratic form, neither privileged directions nor real numbers. In order to remedy this state of affairs, Ricci introduced the trace of the curvature tensor as in 0.7 above. Yet he does not seem to have done anything with it. This is not surprising. The Ricci eigendirections in abstract Riemannian manifolds do not appear to have geometric properties, with the exception of conformal Riemannian geometry (see [Fer] and [Eis], p. 181) and of special manifolds which are discussed in Chapter 16.

0.60. To try to discover the geometric meaning of the Ricci curvature, one can think of *volume*. The general idea is that the Ricci curvature is a trace and that a trace appears as the derivative of a determinant. And determinants are related to volume. Consider a geodesic ball $B(p, t)$, the set of points in (M, g) at a distance from the point p smaller than t. We are going to see that Ricci curvature appears when one considers the asymptotic expansion of the volume $\operatorname{vol}(B(p, t))$ with respect to t, and we are then going to discover that it is even possible to get an estimate valid for every t. The asymptotic expansion is:

$$\operatorname{vol}(B(p,t)) = \omega(n)t^n\left(1 - \frac{1}{(6(n+2)}s(p)t^2 + o(t^2)\right).$$

Here $\omega(n)$ denotes the volume of the unit ball of $\mathbb{R}^n$ and $s(p)$ is the scalar curvature at p. Ricci curvature plays a role in the proof of this formula.

0.61. To work with a measure on (M, g) we transfer things to the tangent space T_pM by means of the exponential map. More precisely we introduce the function $f(x, t)$ defined on the product $U_pM \times \mathbb{R}_+$ by the equation

$$\exp_p^*(\mu_g) = d_p x \otimes dt$$

where $d_p x$ denotes the canonical measure on the unit sphere $U_p M \subset T_p M$ and where dt is the Lebesgue measure on $\mathbb{R}$. Then we prove (see [He-Ka] or convince yourself by playing with Jacobi fields and determinants) that:

$$\theta(x,t) = t^{n-1}(1 - \tfrac{1}{3}r(x)t^2 + o(t^2))$$

Here $r(x)$ is the Ricci curvature (viewed as a quadratic form) of the vector x. Roughly speaking one can say that Ricci curvature measures the defect of the solid angle in (M, g) from being equal to that of Euclidean space. From this the above formula follows directly since $\operatorname{vol}(B(p,t)) = \int_{x \in U_p M} \theta(x,t)\, d_p x$.

0.62. An extremely important discovery was made by R. Bishop [Bis], namely that the Ricci curvature yields a global inequality. The story goes as follows. Assume that $r \geqslant (n-1)kg$ for some real number k. The best such k is simply the lowest eigenvalue of r with respect to g. Bishop's theorem states that

$$\text{the function } t \mapsto \frac{\theta(x,t)}{\mathbf{s}_k^{n-1}(t)} \text{ is non increasing}$$

where $\mathbf{s}_k$ denotes the value that $(\theta(x,t))^{1/n-1}$ takes on the model space of constant sectional curvature equal to k (namely t when $k = 0$, $\sin\sqrt{k}t/\sqrt{k}$ when $k > 0$ and $\sinh\sqrt{-k}t/\sqrt{-k}$ when $k < 0$) and under the restriction that $\theta(x,s)$ is non zero on the interval $[0,t]$ (a condition equivalent to saying that t is smaller than the first conjugate-value of x). In particular one has $\theta(x,t) \leqslant \mathbf{s}_k^{n-1}(t)$ for every x and t (restricted by the above condition).

0.63. Bishop's theorem turns out to be basic in Riemannian geometry. It means that a fundamental numerical invariant of a compact Riemannian manifolds (M, g) is $k(g)$, the lowest eigenvalue of the Ricci curvature. Assume for example that $k(g) = 0$, namely that the Ricci curvature of (M, g) is non-negative: $r \geqslant 0$. Then for every point p and every t one has $\operatorname{vol}(B(p,t)) \leqslant \omega(n)t^n$. From this Milnor deduced that the fundamental group of M has polynomial growth ([Mil 1], see also 6.61).

0.64. Bishop's inequality can be integrated as was discovered by Gromov. If we *denote* by S_k^n the model space of constant sectional curvature equal to k then

$$\text{the function } t \mapsto \frac{\operatorname{vol}(B(p,t))}{\operatorname{vol}(B(S_k^n,t))} \text{ is nonincreasing}$$

for every p in M and t going up to the diameter $\operatorname{diam}(g)$ of (M, g). This implies that, together with the diameter, the invariant k gives considerable information on Riemannian manifolds. Gromov's basic result in this direction is the following. One can endow the set of all Riemannian manifolds with a metric structure of Hausdorff type. Given an integer n, a positive number D and a real number K, the subset of Riemannian manifolds of dimension n, whose diameter is bounded above by D and whose Ricci curvature is bounded from below by K, is a precompact subset in the huge metric space above.

0.65. This precompactness statement leads to a big programme which consists in giving bounds for every Riemannian invariant in terms of only dimension, diameter and $k(g)$. Recently various authors have succeeded in doing this for many invariants. In particular: lower and upper bounds for every eigenvalue of the Laplacian on (M, g); upper bounds for the best constants in Sobolev inequalities; lower bounds for Cheeger's constant $h(g)$ and for the isoperimetric constant $C_1(g)$; bounds for the heat kernel and for the eigenfunctions of the Laplacian. Gromov conjectures that one can bound the Betti numbers of M using only these three invariants. On the other hand, we know it is impossible to bound the number of homotopy types in these terms only (see [Wal]).

0.66. To obtain the above bounds one needs more than just the volume of balls and Bishop's estimate. One has to use a trick of Gromov's, based on a deep existence and regularity theorem of Almgren for extremal hypersurfaces in (M, g). Gallot's ideas also help in reducing everything to the following. Let H be some hypersurface in (M, g), equipped with some unit normal vector field $h \mapsto x(h)$. This gives us an exponential map from the product $H \times \mathbb{R}$ into M defined as $F(h, t) = \exp_h(t \cdot x(h))$, and again one can define a function θ which enables us to compute with measures $F^*(\mu_g) = \theta(h, t)\mu_H \otimes dt$, where μ_H is the canonical Riemannian measure on H. E. Heintze and H. Karcher, and independently M. Maeda, proved an inequality dual to Bishop's:

$$\theta(h, t) \leqslant (\mathbf{s}_k'(t) + v(h)\mathbf{s}_k(t))^{n-1},$$

where $v(h)$ is the mean curvature of H at h and $\mathbf{s}'$ is the derivative with respect to t of the comparison function $\mathbf{s}$ introduced in 0.62 above. This is valid for any h in H and any t before the first focal point at h. We denoted by $k = k(g)$ the smallest eigenvalue of the Ricci curvature of (M, g).

0.67. Ricci curvature cannot do everything you would like it to do. To begin with, in the last section, H cannot be replaced by a submanifold of any dimension: to get an upper bound for θ in general, one needs a lower bound not only for Ricci curvature but also for the sectional curvature (see Heintze-Karcher [He-Ka]). Ricci curvature is appropriate only when the submanifold is either a point or a hypersurface. Then again, Bishop's inequality can work in one direction only (an upper bound for volume given a lower bound for Ricci). It is easy to prove this impossibility: with both inequalities true, one would get the following equality for the volume of balls in manifolds of constant Ricci curvature, alias Einstein manifolds: $\mathrm{vol}(B(p, t)) = \mathrm{vol}(B(S_k^n, t))$ for every p and t. However this implies constant sectional curvature (see [Gr-Va], p. 183), and we know there are Einstein manifolds with non constant sectional curvautre. There is one exception: the case of 3-dimensional manifolds, where Eschenburg and O'Sullivan proved the desired lower-upper inequality ([Es-OS]). And we know that in this dimension the Einstein condition is equivalent to the sectional curvature being constant.

0.68. There is another connection between Ricci curvature and volume. In the case of Kähler manifolds the Ricci curvature, transformed by the complex structure into a 2-form of type $(1,1)$, is nothing but the $d'd''$ second derivative of the volume function. The relation with the above inequalities is still somewhat mysterious (see 2.100).

0.69. Bishop's inequality implies Myers' theorem on the diameter and fundamental group of Riemannian manifolds with positive Ricci curvature. There seems to be no implication in the case of negative Ricci curvature. For manifolds which are non compact and of non-negative Ricci curvature there is a nice geometric result of J. Cheeger and D. Gromoll. If such a manifold possesses a *line* (a geodesic which minimizes distance from $-\infty$ to $+\infty$) then it is a product manifold $M = N \times \mathbb{R}$ endowed with a product metric (see 6.65).

I. The Main Problems Today

We focus attention on problems related directly to the material presented in this book. (For a broader point of view, see for example the Problem Section in [Yau 7]).

For the moment, a "general" example of *a globally defined Einstein metric is sadly lacking*. We have of course the global Kählerian and homogeneous examples, and we also know that the local problem is well posed.

Some very symmetric solutions have been found, sometimes even by an elementary algebraic manipulation as in the case of homogeneous spaces with irreducible linear isotropy. The situation is indeed less satisfactory with cohomogeneity 1 examples which we seem to get hold of almost by chance and not many of which are compact. This leads us to set as a first good problem *the search for a cohomogeneity* 2 *Einstein metric on a compact manifold*. Solving the 2-dimensional system that one gets in this case from the local point of view makes good sense, but, in spite of several serious attempts, the (overdetermined) boundary conditions have so far not been matched. It could very well be that cases of higher cohomogeneity will turn out to be easier, since the codimensions of the singular components of the orbit space are related to the ways in which the generic orbits degenerate to the special ones.

The problem can also be approached from the other end, by trying to show that many manifolds do not admit any Einstein metric. In other words, *are there obstructions of a topological* (or other) *nature to the existence of Einstein metrics*? So far it is only in dimension 4 that an obstruction has been found. This leaves wide open the basic question:

Are manifolds admitting an Einstein metric rather scarce or numerous?

It might be that a distinction between positive Einstein metrics and negative Einstein metrics has to be made in a way similar to that used in the Kähler set-up. In any dimension only finitely many examples of positive Kähler-Einstein manifolds are known, whereas infinitely many occur with a negative Einstein constant. Also, the positive cases seem to be tied up with the existence of non trivial isometries. So far, *there is no known example of a positive Einstein metric without any isometry*. So that it would be interesting to *find some*!

The functional approach to the Euclidean Einstein equation, in particular the interpretation put forward by physicists in terms of classical approximation of quantum phenomena (cf. [Jackiw] or [Perry]) suggests that the total scalar curvature functional **S** (see Chapter 4) should have some critical points in many cases. It would thus capture some topology of $B\mathfrak{D}(M)$ the classifying space of the diffeomorphism group $\mathfrak{D}(M)$ of the manifold which can be identified with the space $\mathscr{M}(M)/\mathfrak{D}(M)$ of Riemannian structures on M, a space to which the functional **S** descends.

Among the preliminary questions to be studied along these lines is a version of the Palais-Smale condition (C) which we can formulate as follows. *Does a sequence of metrics for which the deviation to being Einstein goes to zero* (measured for example as the square norm of the traceless Ricci tensor in a fixed background metric) *contain a subsequence converging to an Einstein metric*? Another more difficult question is the following. *Does there exist a bound depending only on the dimension so that any manifold admitting a metric whose deviation has a norm smaller than this bound has an Einstein metric*? The known 4-dimensional obstruction leaves room for such a statement (see [Pol 1]).

More specific questions can also be asked:

on Ricci-flat manifolds. The only known compact examples have special holonomy. Is this a general fact? If so, the classification of Ricci flat manifolds could be pursued.

on the different Einstein metrics existing on an Einstein manifold. What are the possible values of the Einstein constant (after normalization of the total volume to 1)? More specifically, *can Einstein metrics with Einstein constants of opposite signs exist on the same manifold*? If this is impossible, it would add weight to the remark made earlier to the effect that positive and negative Einstein metrics belong to essentially distinct families.

Chapter 1. Basic Material

A. Introduction

1.1. In this first chapter, we collect the basic material which will be used throughout the book. In particular, we recall the definitions of the main notions of Riemannian (and pseudo-Riemannian) geometry. This is mainly intended to fix the definitions and notations that we will use in the book. As a consequence, many fundamental theorems will be quoted without proofs because these are available in classical textbooks on Riemannian geometry such as [Ch-Eb], [Hel 1], [Ko-No 1 and 2], [Spi].

We also give results which, though standard, are not available in the above-mentioned textbooks. For them we give complete proofs. This is the case in particular for the canonical irreducible decompositions of curvature tensors (§ G, H), the Weitzenböck formulas (§ I) and the first variations of curvature tensor fields under changes of metrics (§ K).

We emphasize that this chapter is for reference, and is not a complete course in Riemannian geometry. The whole chapter should be skipped by experts and used only when needed (non-experts may find it useful to read one of the above-mentioned textbooks).

1.2. We assume that the reader is familiar with all basic notions of differential topology, differential geometry, algebraic topology or Lie group theory such as manifolds and differentiable maps, fibre bundles (especially principal and vector bundles), tangent bundles, tensors and tensor fields, differential forms, spinors and spinor fields, Lie groups (and differentiable actions), Lie algebras and exponential maps (of Lie groups), differential operators and their symbols, singular homology and cohomology, Cartan-de Rham cohomology, homotopy groups, ...

1.3. Unless otherwise stated,

all manifolds are assumed to be smooth (i.e., C^∞), finite dimensional, Hausdorff, paracompact and (usually) connected; all maps are smooth (in particular, all actions of Lie groups are smooth).

In the particular case of differential operators, the smooth theory is often not the most convenient one, and we will have to work with Sobolev spaces. A theory of these spaces and some additional tools from analysis have been collected in an Appendix.

1.4 Notation. Given an n-dimensional manifold M, we *denote* by $T^{r,s}M$ the vector bundle of r-times covariant and s-times contravariant tensors on M, with two exceptions: the *tangent bundle* $T^{(0,1)}M$ will be denoted by TM and the *cotangent bundle* $T^{(1,0)}M$ by T^*M. Notice that $T^{(r,s)}M = \bigotimes^r T^*M \otimes \bigotimes^s TM$.

We *denote* by $\bigwedge^s M$ the vector bundle of s-forms on M. Hence $\bigwedge^s M = \bigwedge^s(T^*M)$. More generally, $\bigwedge$ will denote any alternating procedure and similarly S or $\odot$ will denote symmetrization.

It will sometimes be convenient to denote with the corresponding script letter the space of *sections* of a fibre bundle. For example, $\mathscr{T}^{(r,s)}M$ will be the vector space of (smooth) (r, s)-tensor fields on M (sections of $T^{(r,s)}M$). In particular $\mathscr{T}M$ is the space of *vector fields* on M. Also $\mathscr{C}M$ will be the space of C^∞*-functions* on M, and we denote by $\Omega^p M$ the space of *differential p-forms* on M.

In differential geometry, we often use quantities which are not tensor fields (such as connections or differential operators), from which we may deduce some tensor fields (like the Riemann curvature field for example). So, in order to characterize a tensor field, we will frequently use the following result.

1.5 Theorem (Fundamental lemma of differential geometry). *Let $p\colon E \to M$ be a vector bundle with finite rank over a manifold M. We denote by $\mathscr{E}$ the vector space of its sections. Then, given a linear map $A\colon \mathscr{E} \to \mathscr{C}M$, there exists a (necessarily unique) section α of the dual vector bundle $E^* \to M$ such that, for any point x in M and any element X in $\mathscr{E}$,*

$$A(X)(x) = \alpha(X(x))$$

if and only if A is $\mathscr{C}M$-linear, i.e., if and only if, for any function f in $\mathscr{C}M$, $A(fX) = fA(X)$.

This theorem will be mainly applied to $\mathscr{C}M$-multilinear maps $A\colon (\mathscr{T}M)^r \to (\mathscr{T}M)^s$. We then say that such a map A "defines" an (r, s)-tensor field. In order to illustrate this procedure, we recall two well-known examples:

(a) The *exterior differential* $d\alpha$ of a differential p-form α may be defined by its values on $p+1$ vector fields $X_0, X_1, \ldots, X_p$ through the formula

$$\textbf{(1.5a)}\quad d\alpha(X_0, X_1, \ldots, X_p) = \sum_{i=0}^{p} (-1)^i \alpha(X_0, \ldots, \hat{X}_i, \ldots, X_p) + \sum_{0 \leqslant i < j \leqslant p} (-1)^{i+j} \alpha([X_i, X_j], X_0, \ldots, \hat{X}_i, \ldots, \hat{X}_j, \ldots, X_p)$$

(where $\hat{X}_i$ means that X_i has to be deleted).

One easily sees that the right-hand side is $\mathscr{C}M$-multilinear.

(b) The *Lie derivative* $L_X A$ of a (r, s)-tensor field A along a vector field X may be defined by its values on r vector fields $X_1, \ldots, X_r$ through the following formulas

(1.5b) if A is a $(r,0)$-tensor

$$L_X A(X_1,\ldots,X_r) = X \cdot A(X_1,\ldots,X_r) - \sum_{i=1}^{r} A(X_1,\ldots,[X,X_i],\ldots,X_r);$$

(1.5c) if A is a $(r,1)$-tensor

$$L_X A(X_1,\ldots,X_r) = [X, A(X_1,\ldots,X_r)] - \sum_{i=1}^{r} A(X_1,\ldots,[X,X_i],\ldots,X_r);$$

and the rule

(1.5d) L_X is linear and a derivation with respect to the tensor product $\otimes$.

Notice that $L_X A$ is $\mathscr{C}M$-linear with respect to $X_1, \ldots, X_r$, but that it is *not* $\mathscr{C}M$-linear with respect to X.

(c) Finally, we recall that for a differential p-form α, these derivatives are related through the formula

(1.5e) $$L_X \alpha = i_X d\alpha + d(i_X \alpha),$$

where $i_X\colon \Omega^p M \to \Omega^{p-1} M$ denotes the *interior product* (with a vector field X), given by the formula

(1.5f) $$(i_X \beta)(X_2,\ldots,X_p) = \beta(X,X_2,\ldots,X_p) \quad \text{if } p > 0,$$
$$\text{and} \quad i_X = 0 \quad \text{if } p = 0.$$

B. Linear Connections

1.6. The notion of a connection, fundamental in differential geometry, has various aspects and many equivalent definitions (see the comment by M. Spivak [Spi] volume 5 p. 604). In Chapter 9, we need the general (geometric) notion of an Ehresmann-connection which we recall there, together with its links to the usual notions of principal and linear connections. In many textbooks (see for example [Ko-No 1]) principal connections are introduced first. Here we only recall the notion of a linear connection from the *covariant derivative point of view* (this approach is due to J.L. Koszul).

1.7 Definition. Let $p\colon E \to M$ be a vector bundle over a manifold M. A *linear connection* or a *covariant derivative* on E is a map

$$\nabla\colon \mathscr{T}M \times \mathscr{E} \to \mathscr{E}$$
$$(X,s) \mapsto \nabla_X s$$

which, for any vector fields X and Y in $\mathscr{T}M$, any sections s and t in $\mathscr{E}$, and any functions f and h in $\mathscr{C}M$, satisfies

(1.7)
$$\nabla_{fX+hY}s = f\nabla_X s + h\nabla_Y s,$$
$$\nabla_X(s+t) = \nabla_X s + \nabla_X t,$$
$$\nabla_X(fs) = X(f)s + f\nabla_X s.$$

1.8 Remarks. a) Note that, for any section s of E, the map $\nabla s\colon \mathcal{T}M \to \mathcal{E}$ defined by $(\nabla s)(X) = \nabla_X s$ is $\mathcal{C}M$-linear, so by Theorem 1.5, ∇s is a differential 1-form on M with values in E. In particular, we may define $\nabla_X s$ for any tangent vector X in TM. The map $s \mapsto \nabla_X s$ is on the other hand, not $\mathcal{C}M$-linear (in fact, ∇ is a first order differential operator, see Section I).

b) From Theorem 1.5, it follows easily that, given two connections ∇ and ∇' on the same vector bundle E, the difference

$$A(X,s) = \nabla'_X s - \nabla_X s$$

is $\mathcal{C}M$-linear, hence defines a section of $\bigwedge^1 M \otimes E^* \otimes E$.

Conversely, given any connection ∇ on E, and any section A of $\bigwedge^1 M \otimes E^* \otimes E$, the map $\nabla'_X s = \nabla_X s + A(X,s)$ is a connection on E.

c) One example of a connection is the following. Let $B \times F$ be a *trivialized* vector bundle over B. Then there exists a unique connection, called the *trivial connection*, such that the *constant* sections (i.e., sections s such that $s(b) = (b,\xi)$ with a constant ξ) satisfy $\nabla_X s = 0$ for any X.

As a corollary, let ∇ be any connection on a vector bundle E on B and let (U,φ) be a local trivialization of E, i.e., U is an open subset of B and φ a (C^∞)-fibered isomorphism

$$p^{-1}(U) \to U \times F,$$

then $U \times F$ admits the trivial connection ∇^F as defined above, and we may consider the connection ∇^φ on $p^{-1}(U)$ such that φ interchanges ∇^φ and ∇^F. The difference Γ between ∇ and ∇^φ is a section of $\bigwedge^1 U \otimes E^* \otimes E$. It is called the *Christoffel tensor of ∇ with respect to the trivialization φ*.

Moreover, if U is the domain of a chart on M with coordinates (x^i) and if we identify the fibre F with $\mathbb{R}^p$, then the components $\Gamma^\alpha_{i\beta}$ of Γ in the canonical basis $\left(\dfrac{\partial}{\partial x^i}\right)$ of TU and (ε_α) of $\mathbb{R}^p$ are the classical *Christoffel symbols* of ∇.

1.9. Let

$$\begin{array}{ccc} F & \xrightarrow{h} & E \\ \downarrow q & & \downarrow p \\ N & \xrightarrow{f} & M \end{array}$$

be a vector bundle homomorphism (i.e., p and q are vector bundles, $p \circ h = f \circ q$ and h is linear when restricted to any fibre of q).

Given any linear connection ∇ on E, there exists a unique linear connection ∇' on F such that, for any tangent vector X of N, and any sections s of p and t of q satisfying $s \circ f = h \circ t$, we have

$$h(\nabla'_X t) = \nabla_{f_* X} s.$$

In particular, for any map $f: N \to M$, any linear connection ∇ on E (vector bundle over M) induces a linear connection ∇^f, called the *induced connection* on the induced vector bundle f^*E on N.

1.10. Let E_1, E_2 be two vector bundles over M. Then any linear connection ∇ on the direct sum vector bundle $E = E_1 \oplus E_2$ induces naturally two linear connections ∇^1 on E_1 and ∇^2 on E_2 in the following way. For any vector field X in $\mathscr{T}M$ and any section s_1 in $\mathscr{E}_1$, we define ∇^1 as

$$\nabla^1_X s_1 = pr_1(\nabla_X s_1)$$

where, on the right hand side, s_1 is viewed as a section of E and pr_1 is the projection onto the first factor in the direct sum $E = E_1 \oplus E_2$.

Note that an immediate consequence of Theorem 1.5 is the following. Keeping the same notations, the map $(X, s_1) \mapsto A_X s_1 = pr_2(\nabla_X s_1)$ defines a section of $\bigwedge^1 M \otimes E_1^* \otimes E_2$ (where of course pr_2 is the projection onto the second factor). Later we will encounter geometric situations where this construction is relevant (see for example Chapter 9).

1.11 Definition. The *curvature* R^∇ of a linear connection ∇ on a vector bundle E is the 2-form on M with values in $E^* \otimes E$ defined by

$$R^\nabla_{X,Y}s = \nabla_{[X,Y]}s - [\nabla_X, \nabla_Y]s \tag{1.11}$$

for any vector fields X, Y in $\mathscr{T}M$ and any section s in $\mathscr{E}$.

The fact that R^∇ is a tensor field follows easily from Theorem 1.5.

Since the curvature is the most important invariant attached to a connection, it is worth mentioning that there are many other interpretations of R (see for example (9.53b) for Ehresmann's point of view). One of them is the following.

1.12. We consider the differential p-forms on M with values in E, i.e., the sections of $\bigwedge^p M \otimes E$. We define the exterior differential d^∇ associated with ∇ by the following formula. For any section α of $\bigwedge^p M \otimes E$, $d^\nabla \alpha$ is the section of $\bigwedge^{p+1} M \otimes E$ such that for $X_0, \ldots, X_p$ in $T_x M$, extended to vector fields $\tilde{X}_0, \ldots, \tilde{X}_p$ in a neighborhood,

$$(d^\nabla \alpha)(X_0, \ldots, X_p) = \sum_i (-1)^i \nabla_{X_i}(\alpha(\tilde{X}_0, \ldots, \hat{\tilde{X}}_i, \ldots, \tilde{X}_p))$$
$$+ \sum_{i \neq j} (-1)^{i+j} \alpha([\tilde{X}_i, \tilde{X}_j], X_0, \ldots, \hat{X}_i, \ldots, \hat{X}_j, \ldots, X_p).$$

Then

$$R^\nabla_{X,Y}s = -d^\nabla(d^\nabla s)(X, Y). \tag{1.12}$$

Note that $d^\nabla \circ d^\nabla$ is not zero in general unlike the case of the ordinary exterior differential.

1.13 Examples. a) The curvature of the trivial connection of a trivialized bundle vanishes.

b) If $\tilde{\nabla} = \nabla + A$, with A a section of $\bigwedge^1 M \otimes E^* \otimes E$, then we obtain easily

$$R^{\tilde{\nabla}}_{X,Y}s - R^{\nabla}_{X,Y}s = -(d^{\nabla}A)_{X,Y}s - A_X(A_Y s) + A_Y(A_X s)$$

where A is considered as a 1-form with values in $E^* \otimes E$.

This shows that R^{∇} is not zero in general since any A gives a connection. Moreover this formula permits the computation of R^{∇} in a local trivialization by comparing ∇ with the trivial connection in terms of its Christoffel tensor.

1.14 Theorem (Differential Bianchi identity). *Let ∇ be a connection on a vector bundle E over M. Then*

$$d^{\nabla}R^{\nabla} = 0. \tag{1.14}$$

Proof. This follows from the second definition and the fact that $(d^{\nabla} \circ d^{\nabla}) \circ d^{\nabla} = d^{\nabla} \circ (d^{\nabla} \circ d^{\nabla})$. □

1.15 Some more definitions. A section s of a vector bundle E equipped with a connection ∇ is called *parallel* if $\nabla s = 0$. Note that, given a point x in M, there always exists a neighbourhood U and a finite family $(s_i)_{i \in I}$ of sections of E under U such that $(s_i(y))_{i \in I}$ is a basis of the fibre E_y for each y in U and $\nabla s_i(x) = 0$. (Take a local trivialization of E.)

If the vector bundle E is equipped with some additional structure (such as a Euclidean fibre metric h, an imaginary map J, a symplectic form ω or a Hermitian triple (g, J, ω)), then a linear connection ∇ on E is called respectively *Euclidean* (or metric), *complex*, *symplectic* or *Hermitian*, if respectively g, J, ω or (g, J, ω) are parallel. In such a case, the curvature R^{∇} of ∇ satisfies additional properties, namely, for each X, Y in T_xM the linear map $R^{\nabla}_{X,Y}$ on E_x is respectively skewsymmetric, complex, skewsymplectic or skew-Hermitian.

In the particular case where E is the tangent bundle of the base manifold M, further considerations can be developed.

1.16 Definition. A *linear connection D on a manifold M* is a linear connection on the tangent bundle TM of M.

1.17. Let ∇ be a linear connection on a vector bundle E over M. For any section s in $\mathscr{E}$, ∇s is a section of $T^*M \otimes E$ (see Remark (1.8a)). Now let D be a linear connection on M. Then D and ∇ induce a linear connection (that we still denote by ∇) on $T^*M \otimes E$, so we may define $\nabla(\nabla s)$, and we denote it by $\nabla^2 s$. It is the section of $T^{(2,0)}M \otimes E$ defined by

$$(\nabla^2 s)_{X,Y} = \nabla_X(\nabla_Y s) - \nabla_{(D_X Y)} s.$$

Instead of $(\nabla^2 s)_{X,Y}$ we may write $(\nabla\nabla s)(X, Y)$, or $\nabla^2_{X,Y}s$, or even $(\nabla^2 s)(X, Y)$.

Now, using an obvious induction, we may define the *iterated covariant derivative* $\nabla^p s$ as the section of $T^{(p,0)}M \otimes E$ such that

$$(\nabla^p s)_{X_1,\ldots,X_p} = (\nabla_{X_1}(\nabla^{p-1}s))_{X_2,\ldots,X_p}.$$

The main point to notice here is that the order in which $X_1, \ldots, X_p$ are written is important, because the covariant derivatives (unlike usual derivatives) do not commute in general.

1.18 Definition. The *torsion tensor* T of a linear connection D on a manifold M is the (2,1)-tensor field defined by

$$T_{X,Y} = D_X Y - D_Y X - [X, Y]. \tag{1.18}$$

The fact that T is a tensor field follows easily from Theorem 1.5.

1.19 Remarks. a) Note that the torsion is, by definition, skew-symmetric in its covariant variables.

b) For any (1,2)-tensor field A on M, then the connection $\tilde{D}$ defined by $\tilde{D}_X Y = D_X Y + A_{X,Y}$ is also a connection. Its torsion tensor $\tilde{T}$ is given by

$$\tilde{T}_{X,Y} = T_{X,Y} + A_{X,Y} - A_{Y,X}.$$

In particular, for any connection D on M with torsion tensor T, the connection $(X, Y) \mapsto D_X Y - \frac{1}{2} T_{X,Y}$ is *torsion-free* (i.e., its torsion tensor vanishes).

c) The assumption that the connection D be *torsion-free* enables one to express brackets of vector fields in terms of D:

$$[X, Y] = D_X Y - D_Y X$$

Dually, the exterior differential of differential forms may also be expressed in terms of D. For $\alpha \in \Omega^p M$, and vectors $X_0, \ldots, X_p$,

$$d\alpha(X_0, \ldots, X_p) = \sum_{i=0}^{p} (-1)^i (D_{X_i}\alpha)(X_0, \ldots, \hat{X}_i, \ldots, X_p).$$

1.20. We now come back to the relations between the iterated covariant derivatives. The simplest one is the following so-called *Ricci formula.*

$$\nabla^2_{X,Y} s - \nabla^2_{Y,X} s = -R^\nabla_{X,Y} s - \nabla_{T_{X,Y}} s \tag{1.21}$$

where T is the torsion field of D and R^∇ the curvature of ∇.

Note that if D is torsion-free (i.e. *symmetric*), then the right hand side does not involve ∇s.

Note that, if $\tilde{\nabla}_X s = \nabla_X s + B_X s$, then

$$\tilde{\nabla}^2_{X,Y} s = \nabla^2_{X,Y} s + (\nabla_X B)_Y s + B_Y(\nabla_X s) + B_X(\nabla_Y s) + B_X(B_Y s)$$

and in this way we may obtain the formula for $R^{\tilde{\nabla}}$.

At order 3 we obtain many formulas of which we give only two.

1.22 Corollary. *Further Ricci formulas read*

$$\nabla^3_{X,Y,Z} s - \nabla^3_{Y,X,Z} s = -R^\nabla_{X,Y}(\nabla_Z s) + \nabla_{R_{X,Y}Z} s - \nabla^2_{T_{X,Y},Z} s \tag{1.22a}$$

$$\nabla^3_{X,Y,Z} s - \nabla^3_{X,Z,Y} s = -(\nabla_X R^\nabla)_{Y,Z} s - R^\nabla_{Y,Z}(\nabla_X s) - \nabla^2_{X,T_{Y,Z}} s - \nabla_{(D_X T)_{Y,Z}} s. \tag{1.22b}$$

Proof. a) Apply 1.21 to ∇s.
b) Apply ∇_X to 1.21. □

1.23. We recall that a linear connection D on a manifold M induces a connection on any tensor bundle and we may apply the preceding formulas in this case. In particular, since the torsion T and the curvature R of D are tensor fields on M, we may define their covariant derivatives of any order D^pT or D^pR. These are still tensor fields on M. There are many relations between these tensor fields. The simplest ones involve T, R and their first covariant derivatives. They are called Bianchi identities. We have already met one of them under the name of the differential Bianchi identity. This one is valid for a general bundle over a manifold equipped with any connection.

1.24 Theorem. *Let D be a linear connection on M. Then its torsion field T and curvature field R satisfy*

$$\mathfrak{S}_{X,Y,Z}(R_{X,Y}Z + T_{T_{X,Y}}Z + (D_XT)_{Y,Z}) = 0$$

Proof. We compute in a local chart φ. Then the trivial connection D^φ has neither torsion nor curvature. Hence the Christoffel tensor $\Gamma = \nabla - D^\varphi$ satisfies

$$T_{X,Y} = \Gamma_{X,Y} - \Gamma_{Y,X}.$$

And the formula computing R from R^φ gives the result after one has taken the cyclic sum. □

1.25 Proposition. *The differential Bianchi identity for a connection ∇ on a general vector bundle E can be given the following expression (if M is equipped with a linear connection D with torsion T)*

(1.25) $$\mathfrak{S}_{X,Y,Z}(\nabla_X R^\nabla)_{Y,Z} + R^\nabla_{T_{X,Y},Z} = 0.$$

In particular, if D is torsion-free, we have

(1.25′) $$\mathfrak{S}_{X,Y,Z}(\nabla_X R^\nabla)_{Y,Z} = 0.$$

Proof. It follows directly from expressing 1.14 in terms of iterated covariant derivatives. □

1.26 Definition. Given a linear connection D on a manifold, a *geodesic* (for D) is a smooth curve $c: I \to M$ such that

$$D_{\dot c}\dot c = 0$$

$\left(\text{i.e., the vector field } \frac{d}{dt} \text{ on the interval } I \text{ is parallel for the induced connection } c^*D \text{ on the induced bundle } c^*TM \text{ on } I\right)$.

We recall that the general existence theorem for solutions of a differential equation implies that, for any tangent vector X in T_xM, there exists a unique geodesic $c_X: I \to M$ such that $c_X(0) = x$ and $\dot{c}_X(0) = T_0 c\left(\frac{d}{dt}\right) = X$ and I is a maximal open interval.

For any x in M, we denote by $\mathscr{D}_x$ the set of tangent vectors X in $T_x M$ such that 1 belongs to the interval I of definition for c_X. And we denote by $\mathscr{D}$ the union of all $\mathscr{D}_x$ for all x in M. Then $\mathscr{D}_x$ is open in $T_x M$ and $\mathscr{D}$ is open in TM.

1.27 Definition. The *exponential map* of D is the map $\exp: \mathscr{D} \to M$ defined by $\exp(X) = c_X(1)$.

We *denote by* $\exp_x$ the restriction of exp to $\mathscr{D}_x = \mathscr{D} \cap T_x M$.

1.28 Theorem. *The tangent map to* $\exp_x$ *at the origin* 0_x *of* $T_x M$ *is the identity of* $T_x M$ *(if we identify* $T_{0_x} T_x M$ *with* $T_x M$*).*

In particular the implicit function theorem implies that there exists a neighbourhood U_x of 0_x in $T_x M$ and a neighbourhood V_x of x in M such that the restriction $\exp_x \restriction U_x$ is a diffeomorphism from U_x onto V_x. Such a diffeomorphism gives in particular a set of local coordinates around x in M.

1.29 Definition. A linear connection D on a manifold M is called *complete* if the domain $\mathscr{D}$ of its exponential map is all of TM.

The tangent map to $\exp_x$ at other points of its domain $\mathscr{D}_x$ is described by deforming infinitesimally a geodesic by geodesics. This gives rise to special "vector fields" along the geodesic.

1.30 Definition. Given a connection D on a manifold M and c a geodesic of D, a *Jacobi field along* c is a vector field J along c (i.e., the image by Tc of a section of c^*TM) satisfying

$$D_{\dot{c}} D_{\dot{c}} J + R_{\dot{c},J}\dot{c} + D_{T_{\dot{c},\dot{c}}} J = 0. \tag{1.30}$$

Note that (1.30) is a second order differential equation along c, hence J is well defined as soon as, at one point $c(t)$, we know both $J_{c(t)}$ and $(D_{\dot{c}}J)_{c(t)}$.

1.31 Proposition. *For any* Y *in* $T_x M$ *and* X *in* $\mathscr{D}_x$, $(T_X \exp_x)(Y)$ *is the value at* $c_X(1)$ *of the unique Jacobi field along* c_X *with initial data* $J(0) = 0$ *and* $(D_{\dot{c}}J)(0) = Y$.

It follows from Proposition 1.31 that the differential of $\exp_x$ is singular precisely when a Jacobi field vanishing at the origin vanishes again for some time t.

1.32 Definition. We say that $c(0)$ and $c(t)$ are *conjugate points* along the geodesic c if and only if there exists a non-zero Jacobi field along c such that $J(0) = 0$ and $J(t) = 0$.

The study of conjugate points plays an important role in differential geometry. See for example [Ch-Eb] for the Riemannian case.

C. Riemannian and Pseudo-Riemannian Manifolds

1.33 Definition. a) A *pseudo-Riemannian metric* of signature (p, q) on a smooth manifold M of dimension $n = p + q$ is a smooth symmetric differential 2-form g on M such that, at each point x of M, g_x is non-degenerate on T_xM, with signature (p, q). We call (M, g) a *pseudo-Riemannian manifold.*

b) In the particular case where $q = 0$ (i.e., where g_x is positive definite), we call g a *Riemannian metric* and (M, g) a *Riemannian manifold;*

c) in the particular case where $p = 1$ (and $q > 0$), we call g a *Lorentz metric* and (M, g) a *Lorentz manifold.*

1.34. *The First Examples are the Flat Model Spaces.* Let g_0 be a non-degenerate symmetric linear 2-form on $\mathbb{R}^n$ with signature (p, q) with $p + q = n$. The vector space structure of $\mathbb{R}^n$ induces a canonical trivialization of $T\mathbb{R}^n = \mathbb{R}^n \times \mathbb{R}^n$ (using the translations). We define the canonical pseudo-Riemannian metric g on $\mathbb{R}^n$ (associated to g_0) to be such that, for each x in $\mathbb{R}^n$, g_x is identified with g_0 when we identify $T_x\mathbb{R}^n$ with $\mathbb{R}^n$.

1.35. We obtain many more examples through the following general construction.

Let $i: N \to M$ be an immersion, and g a pseudo-Riemannian metric on M. We *assume* that, for any x in N, $(T_xi)(T_xN)$ is a non-isotropic subspace of $T_{i(x)}M$ (i.e., the induced form i^*g is non-degenerate). Then i^*g is a pseudo-Riemannian metric on N.

Notice that if g is a Riemannian metric, then i^*g is always non-degenerate and so is a Riemannian metric on N.

1.36. *Other Model Spaces.* By way of example, let g_p be the canonical 2-form with signature $(p, n + 1 - p)$ on $\mathbb{R}^{n+1}$ (i.e., $g_p = dx_1^2 + \cdots + dx_p^2 - dx_{p+1}^2 - \cdots - dx_{n+1}^2$). Then g_p induces a pseudo-Riemannian metric on $\mathbb{R}^{n+1}$ as in 1.34. We consider the imbedded submanifolds

$$S_p^n = \{x \in \mathbb{R}^{n+1}; g_{p+1}(x, x) = +1\}$$

and

$$H_p^n = \{x \in \mathbb{R}^{n+1}; g_p(x, x) = -1\}.$$

Then these imbeddings i satisfy the assumption made in 1.35, so i^*g is a pseudo-Riemannian metric and (S_p^n, i^*g_p) and (H_p^n, i^*g_{p+1}) are two pseudo-Riemannian manifolds with signatures $(p, n - p)$.

In the particular case when $n = p$, the Riemannian manifolds S_n^n and the connected component of $(0, \ldots, 0, 1)$ in H_n^n (which has two connected components corresponding to $x_{n+1} > 0$ and $x_{n+1} < 0$) are called respectively the *canonical sphere* S^n and the *canonical hyperbolic space* H^n.

For more details, see for example [Wol 4] p. 67.

1.37. Let (M, g) and (M', g') be two pseudo-Riemannian manifolds with signatures (p, q) and (p', q'). The *product manifold* $M \times M'$ admits a canonical splitting

$T(M \times M') = TM \oplus TM'$ of its tangent space. For each (x, x') in $M \times M'$, we define the symmetric 2-form $g \oplus g'$ on $T_{(x,x')}(M \times M') = T_xM \oplus T_{x'}M'$ as the direct sum of g_x on T_xM and $g_{x'}$ on $T_{x'}M'$. Then $g \oplus g'$ is obviously a pseudo-Riemannian metric on $M \times M'$ with signature $(p + p', q + q')$. It is called the *product metric.*

1.38. Let (M, g) be a pseudo-Riemannian manifold. Then at each point x of M, the non-degenerate quadratic form g_x induces a canonical isomorphism $T_xM \to T_x^*M$ and more generally, a canonical isomorphism between any $T_x^{(p,q)}M$ and $T_x^{(p+1,q-1)}M$ (hence onto any $T^{(r,s)}M$ with $r + s = p + q$). This isomorphism is often denoted by $\flat$ ("flat") and its inverse by $\sharp$ ("sharp") since in classical tensor notation, they correspond to lowering (resp. raising) indices, see below 1.42.

By composition of the isomorphism $T^{(p,q)}M \to T^{(q,p)}M$ with the "evaluation map" (pairing any vector space with its dual), we get a non-degenerate quadratic form (still denoted by g_x) on any $T^{(p,q)}M$ and consequently on any subspace of $T^{(p,q)}M$ such as $\bigwedge^p M$ or $S^p(T^*M)$. Note that if g is positive definite on TM, it is positive definite on any $T^{(p,q)}M$.

1.39 Theorem (Fundamental Theorem of (Pseudo-) Riemannian Geometry). *Given a pseudo-Riemannian manifold* (M, g), *there exists a* unique *linear connection* D *on* M, *called the* Levi-Civita connection (*of* g), *such that*

a) D *is metric* (*i.e.*, $Dg = 0$);
b) D *is torsion-free* (*i.e.*, $T = 0$).

1.40 Definition. On a pseudo-Riemannian manifold (M, g), the curvature tensor field R of the Levi-Civita connection is called the *Riemann curvature tensor* of (M, g).

Note that, since D is torsion free, the Bianchi identities 1.24, (1.25′) and the Ricci formula (1.21) take their simplified forms.

1.41. For future use, we compute the arguments of all these tensors in local coordinates.

Let $\varphi\colon U \to V$ be a chart on M, i.e., let φ be a diffeomorphism from some open subset U of $\mathbb{R}^n$ onto some open subset V of M. Using the coordinate functions x^i on $\mathbb{R}^n$, we get coordinate functions $x^i \circ \varphi^{-1}$ on V, which we still denote by x^i. Then the differential 1-forms (dx^i) give a basis of T^*V, and we denote by (∂_i) the dual basis of TV. Also the tensor fields $dx^i \otimes dx^j$ are a basis of $T^{(2,0)}V$ at each point of the chart, so we may write the restriction of g to V as

$$g = \sum_{i,j=1}^{n} g_{ij}\, dx^i \otimes dx^j,$$

where the g_{ij}'s are functions on V satisfying $g_{ij} = g_{ji}$.

Now we may characterize the Levi-Civita connection D through its values on the basis (∂_i). We get

$$D_{\partial_i}\partial_j = \sum_{k=1}^{n} \Gamma_{ij}^k \partial_k,$$

where the "*Christoffel symbols*" (Γ_{ij}^k) are given by

$$\Gamma_{ij}^k = \frac{1}{2}\sum_{l=1}^n g^{kl}(\partial_i g_{jl} + \partial_j g_{il} - \partial_l g_{ij}),$$

(at each point x of V, (g^{kl}) is the inverse matrix of (g_{ij})). These Γ_{ij}^k's are the components of the difference tensor between D and the trivial connection on V (compare Remark 1.8c).

Finally, the curvature R has components R_{ijk}^l given by

$$R(\partial_i, \partial_j)\partial_k = \sum_{l=1}^n R_{ijk}^l \partial_l,$$

where

$$R_{ijk}^l = \partial_j(\Gamma_{ik}^l) - \partial_i(\Gamma_{jk}^l) + \sum_{m=1}^n (\Gamma_{ik}^m \Gamma_{jm}^l - \Gamma_{jk}^m \Gamma_{im}^l).$$

1.42. In classical tensor calculus, a convention is used to avoid too many summation signs ($\sum$). Any index which is repeated has to be summed (usually from 1 to $n = \dim M$). For example, we write

$$\Gamma_{ij}^k = \tfrac{1}{2} g^{kl}(\partial_i g_{jl} + \partial_j g_{il} - \partial_l g_{ij}),$$

and

$$R_{ijk}^l = \partial_j \Gamma_{ik}^l - \partial_i \Gamma_{jk}^l + \Gamma_{ik}^m \Gamma_{jm}^l - \Gamma_{jk}^m \Gamma_{im}^l.$$

Another convention avoids some g_{ij} and g^{kl}; this is the convention of "raising and lowering indices". Given any tensor A in $T^{(r,s)}M$ whose components in a local basis are $A_{i_1 \ldots i_r}^{j_1 \ldots j_s}$, we

denote by
$$A_{i_1 \ldots i_r l}^{j_2 \ldots j_s} = g_{lk} A_{i_1 \ldots i_r}^{k j_2 \ldots j_s},$$

and
$$A_{i_1 \ldots i_{r-1}}^{l j_1 \ldots j_s} = g^{lk} A_{i_1 \ldots i_{r-1} k}^{j_1 \ldots j_s}$$

(with the "summation of repeated indices" convention), and so on.

There are numerous other conventions, more or less widely used. For example, δ_i^j is the Kronecker symbol, defined as

$$\delta_i^i = 1 \text{ for any } i, \quad \text{and} \quad \delta_i^j = 0 \text{ for any } i \neq j.$$

Also a bracket [ ] around two indices means alternating them, or { } summing cyclicly, but if we use these conventions, we will remind the reader of what is meant.

1.43. Given a pseudo-Riemannian manifold (M, g), the geodesics, the exponential map and the Jacobi fields of its Levi-Civita connection D are called the *geodesics*, the *exponential map* and the *Jacobi fields* of (M, g). Furthermore, (M, g) is called *complete* if and only if D is complete.

Note that in local coordinates (x^i) a geodesic $c(t) = (x^i(t))_{i=1,\ldots,n}$ satisfies the following system of n second order differential equations (for $i = 1, \ldots, n$)

$$\ddot{x}^i + \sum_{j,k=1}^n \Gamma_{jk}^i \dot{x}^j \dot{x}^k = 0, \tag{1.43}$$

where the dot denotes the usual derivative in the variable t.

1.44. Using the exponential map, we may construct some special types of local coordinates around each point x of M in the following way. We choose some orthonormal frame $(X_1,\dots,X_n)$ of T_xM, which induces a linear isomorphism α: $\mathbb{R}^n \to T_xM$. Let U be a neighborhood of 0 in T_xM and V a neighborhood of x in M such that $\exp_x$ is a diffeomorphism from U onto V (see 1.28). Now

$$\exp_x \circ \alpha \colon \alpha^{-1}(U) \to \mathrm{V}$$

is a local chart for M around x. The corresponding coordinates are called *normal coordinates (centered at x)*.

Note that (∂_i) coincides with (X_i) *at* x, but that the basis (∂_i) is not necessarily orthonormal at other points of V.

Since $\exp_x$ maps a radial curve $(t \to tX)$ of T_xM onto a geodesic c_X, the geodesics issued from x become the radial curves in normal coordinates (centered at x).

We may characterize normal coordinates as follows.

1.45 Theorem (Folklore; see D.B.A. Epstein [Eps]). *Local coordinates (x^i) on a pseudo-Riemannian manifold (M, g), defined in an open disk centered at the origin, are normal coordinates (centered at x) if and only if the expression (g_{ij}) of g in these coordinates satisfies*

$$\sum_{j=1}^{n} g_{ij}(x^1,\dots,x^n)x^j = x^i.$$

In fact, this theorem is no more than the classical "*Gauss Lemma*". It is usually stated more intrinsically as follows.

1.46 Theorem ("Gauss Lemma", see for example [Ch-Eb]). *Let (M, g) be a pseudo-Riemannian manifold, x a point in M and $X \in \mathcal{D}_x \subset T_xM$. Then*

a) $g_{c_X(1)}((T_X \exp_x)(X), (T_X \exp_x)(X)) = g_x(X, X)$,

b) *For any Y in T_xM such that $g_x(X, Y) = 0$, we have*

$$g_{c_X(1)}((T_X \exp_x)(X), (T_X \exp_x)(Y)) = 0.$$

1.47 Definition. The *volume element* μ_g of a pseudo-Riemannian manifold (M, g) is the unique density (i.e., locally the absolute value of an n-form) such that, for any orthonormal basis (X_i) of T_xM,

$$\mu_g(X_1,\dots,X_n) = 1.$$

Obviously, in local coordinates (x^i), we have

$$\mu_g = \sqrt{|\det(g_{ij})|}\,|dx^1 \wedge \cdots \wedge dx^n|,$$

so that μ_g is locally "equivalent" to the Lebesgue measure in any set of coordinates.

In normal coordinates, there is a formula for μ_g involving the values of Jacobi fields.

1.48 Definition. Given normal coordinates (x^i) on M, we *define* the function $\theta = \sqrt{|\det(g_{ij})|}$.

Note that θ does *not* depend on the particular basis (X_i) that we choose to define the normal coordinates (because they are all orthonormal at the center).

For any *non-isotropic* vector X in $\mathscr{D}_x \subset T_xM$, we consider an orthonormal basis (X_i) such that X is proportional to X_n, and then the $n-1$ Jacobi fields $J_1, \ldots, J_{n-1}$ along c_X with initial data $J_i(0) = 0$, $(D_{\dot{c}_X} J_i)(0) = X_i$.

1.49 Proposition. *If c_X lies in the domain of normal coordinates,*

$$\theta(\exp_x X) = |\det(J_1, \ldots, J_{n-1})|(\exp_x X).$$

This follows directly from 1.31 and 1.46. The determinant is taken with respect to an orthonormal basis.

1.50. When the manifold (M, g) is *oriented*, we denote by ω_g the canonical n-form, called the *volume form* of (M, g), such that $\mu_g = |\omega_g|$ and ω_g is in the class of the given orientation. Note that $g(\omega_g, \omega_g) = (-1)^s$ where s is the number of -1 in the signature of g (i.e., g has signature $(n-s, s)$). The following definition gives a generalization.

1.51 Definition. For any p with $0 \leqslant p \leqslant n$, we define the *Hodge* operator $*$ to be the unique vector-bundle *isomorphism*

$$*: \Lambda^p M \to \Lambda^{n-p} M$$

such that

$$\alpha \wedge (*\beta) = g(\alpha, \beta)\omega_g$$

for any α and β in $\Lambda^p_x M$, and any x in M.

This operator $*$ satisfies the following properties.

1.52 Proposition. a) $*1 = \omega_g$ *and* $*\omega_g = (-1)^s$;

b) *for any α in $\Lambda^p M$ and β in $\Lambda^{n-p} M$, we have*

$$g(\alpha, *\beta) = (-1)^{p(n-p)} g(*\alpha, \beta);$$

c) *on $\Lambda^p M$, we have*

$$*^2 = (-1)^{p(n-p)+s} Id_{\Lambda^p M}.$$

1.53 Remark. In even dimensions $n = 2m$, $*$ will induce an automorphism of $\Lambda^m M$. In the Riemannian case ($s = 0$), this automorphism is

—an involution if m is even,

—a complex structure if m is odd.

These facts have strong geometric consequences. For example, for a 4-dimensional Riemannian manifold, the splitting of $\Lambda^2 M$ into two eigenspaces relative to $*$ gives rise to the notion of self-duality, which is developed in Chapter 13.

This contrasts with the fact that, for a 4-dimensional Lorentz manifold (with $s = 1$), $*$ induces a complex structure on $\Lambda^2 M$. For an application to the classification of curvature tensors of space-times, see 3.14.

1.54. *Some more notation.* We recall that a pseudo-Riemannian metric induces canonical isomorphisms ($\flat$ and $\sharp$) between tensor spaces. But for some very useful objects, we prefer not to use these isomorphisms and we introduce a special notation.

Given a smooth function f on M,

a) the *gradient* of f is the vector field $Df = \sharp df$ (or $df^\sharp$), i.e., Df satisfies $g(Df, X) = X(f) = df(X)$ for any X in TM;

b) the *Hessian* of f is the covariant derivative of df, i.e., Ddf (we also *denote* it by D^2f); it satisfies $Ddf(X, Y) = X\dot{}Y\dot{}f - (D_X Y)f$ (notice that Ddf is symmetric);

c) the *Laplacian* of f is the opposite of the trace of its Hessian with respect to g, i.e., $\Delta f = -\operatorname{tr}_g(Ddf) = -g(g, Ddf)$.

Note that Δ is an elliptic operator if and only if g is *Riemannian.*

1.55. Since g *induces a pseudo-Euclidean structure on each tensor bundle, any differential operator* A from tensor fields to tensor fields admits a canonical formal adjoint A^*. For example, the covariant derivative

$$D\colon \mathscr{T}^{(r,s)}M \to \Omega^1 M \otimes \mathscr{T}^{(r,s)}M$$

admits a formal adjoint

$$D^*\colon \Omega^1 M \otimes \mathscr{T}^{(r,s)}M \to \mathscr{T}^{(r,s)}M.$$

For vector fields $X_1, \ldots, X_r$ and α in $\Omega^1 M \otimes \mathscr{T}^{(r,s)}M$, $(D^*\alpha)(X_1, \ldots, X_r)$ is the opposite of the trace (with respect to g), of the $\bigotimes^s TM$-valued 2-form

$$(X, Y) \to (D_X\alpha)(Y, X_1, \ldots, X_r).$$

This also holds for natural subbundles of $T^{(r,s)}M$. In the Riemannian case, with an orthonormal basis $(Y_i)_{i=1,\ldots,n}$, we have

$$(D^*\alpha)(X_1, \ldots, X_r) = -\sum_{i=1}^{n} (D_{Y_i}\alpha)(Y_i, X_1, \ldots, X_r).$$

For the most useful cases, we use some special notation.

1.56 Definition. Let $d\colon \Omega^p M \to \Omega^{p+1} M$ denote the exterior differential on p-forms on M. We *denote* by δ its formal adjoint, and we call it the *codifferential.*

We may compute δ in a number of ways.

a) Take a local orientation of M and the corresponding Hodge operator $*_g$; then

$$\delta = -*_g \circ d \circ *_g;$$

b) we may consider $\bigwedge^{p+1} M$ as a subspace of $\bigwedge^1 M \otimes \bigwedge^p M$; then δ is simply the restriction of D^* to $\bigwedge^{p+1} M$;

c) in the Riemannian case, if $(Y_i)_{i=1,\ldots,n}$ is a local orthonormal basis of vector fields,

$$(\delta\alpha)(X_1, \ldots, X_p) = -\sum_{i=1}^{n} (D_{Y_i}\alpha)(Y_i, X_1, \ldots, X_p).$$

1.57. The operator $\Delta = d\delta + \delta d\colon \Omega^p M \to \Omega^p M$ is the *Hodge-de Rham Laplacian* on p-forms.

1.58. For any vector field X on M, its *divergence* $\operatorname{div} X$ (or δX) is the codifferential of the dual 1-form, i.e., $\operatorname{div} X = \delta(X^\flat)$. In the Riemannian case, we get

$$\operatorname{div} X = -\sum_{i=1}^{n} g(D_{Y_i}X, Y_i).$$

1.59. Instead of forms, we may also consider symmetric tensors. If we consider the covariant derivative

$$D\colon \mathscr{S}^pM \to \Omega^1M \otimes \mathscr{S}^pM = \mathscr{S}^1M \otimes \mathscr{S}^pM,$$

and compose with the symmetrization

$$\mathscr{S}^1M \otimes \mathscr{S}^pM \to \mathscr{S}^{p+1}M,$$

we obtain a differential operator, *denoted* by δ^*,

$$\delta^*\colon \mathscr{S}^pM \to \mathscr{S}^{p+1}M,$$

whose formal adjoint is called the *divergence*, and denoted by δ,

$$\delta\colon \mathscr{S}^{p+1} \to \mathscr{S}^pM.$$

Notice that δ is nothing but the $\bigotimes^{p+1} TM$ restriction of D^* to $\mathscr{S}^{p+1}M$ included in $\mathscr{S}^1M \otimes \mathscr{S}^pM$.

1.60 Lemma. *On* 1*-forms, the operator* $\delta^*\colon \Omega^1M \to \mathscr{S}^2M$ *satisfies*

$$\delta^*\alpha = -\tfrac{1}{2}L_{\alpha^\sharp}g,$$

where $L_{\alpha^\sharp}$ *denotes the Lie derivative of the vector field* $\alpha^\sharp$ *(dual of the* 1*-form* α*).*

In particular, $\delta^*\alpha = 0$ if and only if $\alpha^\sharp$ is a Killing vector field.

Proof.
$$\begin{aligned}\delta^*\alpha(X,Y) &= \tfrac{1}{2}((D_X\alpha)(Y) + (D_Y\alpha)(X))\\ &= \tfrac{1}{2}\{X\alpha(Y) - \alpha(D_XY) + Y\alpha(X) - \alpha(D_YX)\}\\ &= \tfrac{1}{2}\{X\cdot g(\alpha^\sharp, Y) - g(\alpha^\sharp, D_XY) + Y\cdot g(\alpha^\sharp, X) - g(\alpha^\sharp, D_YX)\}\\ &= \tfrac{1}{2}\{g(D_X\alpha^\sharp, Y) + g(D_Y\alpha^\sharp, X)\}\\ &= -\tfrac{1}{2}(L_{\alpha^\sharp}g)(X,Y)\end{aligned}$$

(compare the proof of Theorem 1.81). □

D. Riemannian Manifolds as Metric Spaces

In the particular case of a Riemannian manifold, there is another very important invariant, the *distance*, which is defined in the following way.

> Throughout section D, (M, g) is assumed to be Riemannian.

1.61 Definitions. Let (M, g) be a Riemannian manifold.

(a) Given a piecewise smooth curve $c: [a, b] \to M$, the *length* of c is

$$L(c) = \int_a^b \sqrt{g(\dot{c}, \dot{c})}\, dt.$$

(b) For each pair of points x and y in M, we *denote* by $d(x, y)$ the infimum of the lengths of all piecewise smooth curves starting from x and ending at y.

Note that in (a), if c is a geodesic, then $g(\dot{c}, \dot{c})$ is constant and $L(c) = (b - a)\sqrt{g(\dot{c}, \dot{c})}$; in (b), the infimum $d(x, y)$ may or may not be realized by a curve.

1.62 Theorem. *Given a Riemannian manifold (M, g), the function d is a distance on M, and the topology of the metric space (M, d) is the same as the manifold topology of M.*

A corollary of the Gauss Lemma 1.46 is that the distance is realized (at least locally) by geodesics. More precisely

1.63 Theorem. *For each x in M, there is a neighborhood U_x in M such that, for any y in U_x, the distance $d(x, y)$ is the length of the unique geodesic from x to y in U_x.*

1.64 Corollary. *Any geodesic minimizes the length between any pair of sufficiently near points on it; conversely, any curve having this property is (up to reparameterization) a geodesic.*

Also there is a notion of completeness of a metric space. Fortunately, these notions are as compatible as they can be. This is the content of the following theorem.

1.65 Theorem (H. Hopf-W. Rinow). *For a Riemannian manifold (M, g), the following conditions are equivalent.*

a) *(M, g) is complete for the Levi-Civita connection;*

b) *(M, d) is a complete metric space;*

c) *the bounded subsets of M are relatively compact.*

And these properties imply that

d) *for any two points x, y in M, there exists at least a geodesic starting at x and ending at y.*

Note however that there may be many more than one geodesic connecting x and y and that property d) does not imply the completeness of (M, g).

1.66 Corollary. *A compact Riemannian manifold is complete.*

(This need not be true for pseudo-Riemannian manifolds, even in the homogeneous case, see Chapter 7).

Note that a geodesic connecting two points does not necessarily realize the distance between them. The study of "minimizing" geodesics is a very important tool in Riemannian geometry. A key point is the fact that a limit of minimizing geodesics is a minimizing geodesic.

1.67 Lemma. *Let (c_k) be a sequence of geodesics and (t_k) a sequence of real numbers such that, for each k,*

$$d(c_k(0), c_k(t_k)) = t_k.$$

Assume that the vectors $(\dot{c}_k(0))$ converge in TM towards some vector X and (t_k) converge to t when k goes to infinity. Then the geodesic c such that $\dot{c}(0) = X$ satisfies

$$d(c(0), c(t)) = t.$$

Here are a few elementary applications. The *diameter* of a Riemannian manifold (M, g) is the supremum of the distances of any two points in M. A ray (respectively, a *line*) is an infinite geodesic c: $[0, +\infty[\to M$ (respectively c: $]-\infty, +\infty[\to M$) such that, for any two points x, y on c, the distance $d(x, y)$ is exactly the length of c between x and y (i.e., c minimizes the length between any two of its points).

1.68 Theorem. *If (M, g) is compact, the diameter of (M, g) is finite, and there exists x and y in M such that $d(x, y)$ is the diameter.*

If (M, g) is complete, non-compact, the diameter is infinite. For any x in M, there exists a ray c with $c(0) = x$.

Note that there is not always a line on a non-compact Riemannian manifold. But as soon as M has two "ends", there exists a line connecting them.

Finally, we want to mention that the notion of a conjugate point (see Definition 1.32) enters into the problem of distance through the following.

1.69 Theorem. *Let c be a geodesic on a Riemannian manifold (M, g); and let t_0 be such that $c(0)$ and $c(t_0)$ are* conjugate points *along c. Then, for any $t > t_0$, the geodesic c does not minimize the distance between $c(0)$ and $c(t)$.*

Note that the "*first cut point along c*" (i.e., the point $c(t_1)$ such that c fails to minimize the distance between $c(0)$ and $c(t)$ for any $t > t_1$) may appear before the first conjugate point to $c(0)$ along c.

E. Riemannian Immersions, Isometries and Killing Vector Fields

1.70 Definition. Let (M, g) and (N, h) be two pseudo-Riemannian manifolds. A smooth map $f: M \to N$ is a *pseudo-Riemannian immersion* if it satisfies $f^*h = g$ or, equivalently, if, for any x in M, the tangent map $T_x f$ satisfies

$$h((T_x f)X, (T_x f)Y) = g(X, Y)$$

for any X, Y in $T_x M$.

Note that such an f is obviously an immersion, and that the restriction of h to $(T_x f)(T_x M)$ is non-degenerate.

Conversely, given a smooth immersion $f\colon M \to N$ and a pseudo-Riemannian metric h on M, which is *non-degenerate* on $(T_x f)(T_x M)$ for *each* x in M, the map f is a pseudo-Riemannian immersion from (M, f^*h) into (N, h). Note that $f(M)$ does not need to be a submanifold of N; this happens only if f is an imbedding.

1.71. Let $f\colon (M,g) \to (N,h)$ be a pseudo-Riemannian immersion. Then we may consider the tangent bundle TM to M as a subbundle of the induced vector bundle $f^*(TN)$, which we endow with the pseudo-Euclidean structure induced from h, and the linear connection $\bar{D}$ induced from the Levi-Civita connection of h (see 1.9).

Let NM be the orthogonal complement of TM in $f^*(TN)$. We call it the *normal bundle* (of the immersion). Using 1.10, $\bar{D}$ induces a connection on TM and NM, together with a tensor. Obviously, the induced connection on TM is nothing but the Levi-Civita connection of g. We *denote* by ∇ the connection induced on NM, and we define the *second fundamental form* of f to be the unique "tensor"

$$\mathrm{II}\colon TM \otimes TM \to NM$$

such that, for two vector fields U and V on M,

$$\mathrm{II}(U, V) = \mathcal{N}(\bar{D}_U V)$$

where $\mathcal{N}$ is the orthogonal projection onto NM. We *define* also the tensor B: $TM \otimes NM \to TM$ such that for U, V in $T_x M$ and X in $N_x M$, $g(B_U X, V) = -g(\mathrm{II}(U, V), X)$. Then one easily proves

1.72 Theorem. *Let $f\colon (M,g) \to (N,h)$ be a pseudo-Riemannian immersion. Let U, V, W be vector fields on M, and X, Y be sections of NM. Then*

a) $\bar{D}_U V = D_U V + \mathrm{II}(U, V)$ *(Gauss Formula)*,

b) $\bar{D}_U X = B_U X + \nabla_U X$ *(Weingarten Equation)*,

c) $(\bar{R}(U, V)U, V) = (R(U, V)U, V) + |\mathrm{II}(U, V)|^2 - (\mathrm{II}(U, U), \mathrm{II}(V, V))$ *(Gauss Equation)*,

d) $(\bar{R}(U, V)W, X) = -((\nabla_U \mathrm{II})(V, W), X) + ((\nabla_V \mathrm{II})(U, W), X)$ *(Codazzi-Mainardi Equation)*,

e) $(\bar{R}(U, V)X, Y) = (R^\nabla(U, V)X, Y) - (B_U X, B_V Y) + (B_V X, B_U Y)$ *(Ricci Equation)*,

where R, R, R^∇ are the curvatures of $\bar{D}$, D and ∇ respectively, $\nabla\mathrm{II}$ *is the (covariant) derivative of* II *with respect to ∇, and we have omitted g.*

1.73 Definitions. Let $f\colon (M,g) \to (N,h)$ be a pseudo-Riemannian immersion.

a) The *mean curvature vector* of f at $x \in M$ is the normal vector

$$H_x = \operatorname{tr} \mathrm{II} = \sum_{i=1}^{n} \mathrm{II}(X_i, X_i),$$

where $X_1, \ldots, X_n$ is an orthonormal basis of $T_x M$.

b) A point x M is said to be *umbilic* if there exists a normal vector $v \in N_x M$ such that $\mathrm{II}(U, V) = g_x(U, V)v$ for any U, V in $T_x M$.

c) f has *constant mean curvature* if the normal vector field H is parallel, i.e., $\nabla H \equiv 0$;

d) f is *totally umbilic* if every point of M is umbilic;

e) f is *minimal* if $H \equiv 0$;

f) f is *totally geodesic* if $\mathrm{II} \equiv 0$.

Note that d) does not imply c) in general.

In the special case where the dimensions of M and N are equal, a pseudo-Riemannian immersion is *locally* a diffeomorphism (but not necessarily globally) and the signatures of (M, g) and (N, h) are the same.

1.74 Definition. An *isometry* is a pseudo-Riemannian immersion which is also a diffeomorphism.

In the special case of *Riemannian* manifolds, there is a characterization of isometries which involves distances.

1.75 Theorem. *A* surjective *smooth map $f: (M, g) \to (N, h)$ between two* Riemannian *manifolds is an isometry if and only if it preserves the distance, i.e., $d_h(f(x), f(y)) = d_g(x, y)$ for any x, y in M.*

Obviously, the composition of two isometries is an isometry and, for any isometry f, the inverse diffeomorphism f^{-1} is an isometry. As a consequence, the set of all isometries from one pseudo-Riemannian manifold (M, g) into itself is a group. We call it the *isometry group* of (M, g) and *denote* it by $I(M, g)$. As a subgroup of diffeomorphisms of M, it has a natural topology ("compact-open" topology).

1.76 Examples. a) Given any pseudo-Riemannian manifold (M, g) and any diffeomorphism $\alpha: N \to M$, then α is an isometry of $(N, \alpha^* g)$ onto (M, g).

b) On the flat model space $(\mathbb{R}^n, g_0)$ of 1.34, any translation is an isometry. More generally, the isometry group is exactly the semidirect product $\mathbb{R}^n \rtimes O(g_0)$ of the group of translations $\mathbb{R}^n$ by the orthogonal group $O(g_0)$ of g_0.

c) The groups $O(g_p)$ (respectively $O(g_{p+1})$) acting on $\mathbb{R}^{n+1}$ as in 1.36 preserve the submanifolds S^n_p (respectively H^n_p) and induce on them isometries of the induced metric. One may show that they induce in fact the whole isometry group.

d) More generally, if (M, g) is a pseudo-Riemannian submanifold of (N, h) (as in 1.35), any isometry of (N, h) which preserves M (i.e., such that $f(M) = M$) induces an isometry of (M, g).

The basic result on $I(M, g)$ is the following theorem.

1.77 Theorem (S.B. Myers-N. Steenrod [My-St]). *Let (M, g) be a pseudo-Riemannian manifold.*

a) *The group $I(M, g)$ of all isometries of (M, g) is a Lie group and acts differentiably on (M, g);*

b) *for any x in M, the* isotropy subgroup

$$I_x(M, g) = \{f \in I(M, g); f(x) = x\}$$

is a closed subgroup of $I(M,g)$. *Moreover, if we denote by*

$$\rho\colon I_x(M,g) \to Gl(T_xM),\ f \to \rho(f) = T_xf$$

the isotropy representation, *then* ρ *defines an isomorphism of* $I_x(M,g)$ *onto a closed subgroup of* $O(T_xM, g_x) \subset Gl(T_xM)$.

1.78 Corollary. *If* (M,g) *is a* Riemannian *manifold,* $I_x(M,g)$ *is a compact subgroup of* $I(M,g)$. *Moreover, if* (M,g) *is compact,* $I(M,g)$ *is compact.*

1.79 Remarks. a) More generally, $I(M,g)$ acts *properly* on any *Riemannian* manifold (M,g) (see S.T. Yau [Yau 6]).

b) Note that $I(M,g)$ may be compact (e.g., trivial), even if (M,g) is non-compact or non-Riemannian.

c) One may show that $\dim(I(M,g)) \leqslant \dfrac{n(n+1)}{2}$ with equality only if (M,g) has constant sectional curvature.

Of course, since an isometry preserves g, it preserves the Levi-Civita connection, the geodesics, the volume element and the different types of curvature (defined in § F). We now examine the corresponding infinitesimal notion.

1.80 Definition. Let (M,g) be a pseudo-Riemannian manifold. A vector field X on M is called a *Killing vector field* if the (local) 1-parameter group of diffeomorphisms associated to X consists in (local) isometries.

1.81 Theorem. *For a vector field* X, *the following properties are equivalent.*

a) X *is a Killing vector field;*

b) *the Lie derivative of* g *by* X *vanishes, i.e.,* $L_Xg = 0$;

c) *the covariant derivative* DX *is skewsymmetric with respect to* g, *i.e.,* $g(D_YX,Z) + g(D_ZX,Y) = 0$;

Moreover, any Killing vector field satisfies also

d) *the Lie derivative of* D *by* X *vanishes, i.e.,* $L_XD = 0$;

e) *the restriction of* X *along any geodesic is a Jacobi field;*

f) *the second covariant derivative* D^2X *satisfies* $D^2_{U,V}X = R(X,U)V$.

Proof. We recall that, for any tensor (or connection) A,

$$L_XA = \frac{d}{dt}(\varphi_t^*A)|_{t=0},$$

where φ_t is the (local) 1-parameter group of diffeomorphisms generated by X.

The equivalence of a) and b) follows easily, together with d). Now

$$\begin{aligned}(L_Xg)(Y,Z) &= X\cdot g(Y,Z) - g([X,Y],Z) - g(Y,[X,Z])\\ &= g(D_XY,Z) + g(Y,D_XZ) - g(D_XY,Z) + g(D_YX,Z)\\ &\quad - g(Y,D_XZ) + g(Y,D_ZX)\\ &= g(D_YX,Z) + g(Y,D_ZX),\end{aligned}$$

hence c) is equivalent to b). Then e) follows easily from the definition of Jacobi fields, since isometries preserve geodesics. Finally f) follows from e) through polarization and the algebraic Bianchi identity. □

1.82 Remarks. a) Conditions d), e), f), are *not* characteristic of Killing vector fields.

b) The bracket of two Killing vector fields is a Killing vector field, so the space of all Killing vector fields of (M, g) is a Lie subalgebra of the Lie algebra of all vector fields.

1.83 Theorem. *If (M, g) is complete, then any Killing vector field of (M, g) is complete, i.e., generates a 1-parameter group of isometries. Consequently, the Lie algebra of Killing vector fields is the Lie algebra of the Lie group $I(M, g)$.*

We finish with a vanishing theorem, due to S. Bochner [Boc 1], which involves the Ricci curvature defined in 1.90 below.

1.84 Theorem. *Let (M, g) be a compact Riemannian manifold, with Ricci curvature r.*

a) *If r is negative, i.e., if $r(U, U) < 0$ for any non-zero tangent vector U, then there are no non-zero Killing vector fields and the isometry group $I(M, g)$ is finite.*

b) *If r is nonpositive, i.e., if $r(U, U) \leqslant 0$ for any tangent vector U, then any Killing vector field on M is parallel, and the connected component of the identity in $I(M, g)$ is a torus.*

c) *If r vanishes identically, then the space of Killing vector fields has dimension exactly the first Betti number $b_1(M, \mathbb{R})$.*

We only sketch the starting point of the proof, which follows the same lines as 1.155. The relevant Weitzenböck formula is the following consequence of (1.81f): $D^*DX = \operatorname{Ric}(X)$ for any Killing vector field X. By evaluating against X and integrating over M, we get

$$\int_M |DX|^2 \mu_g = \int_M r(X, X) \mu_g$$

and the theorem follows. □

F. Einstein Manifolds

We first collect various properties of the Riemann curvature tensor R that we have met before.

1.85 Proposition. *The curvature tensor field R of a pseudo-Riemannian manifold (M, g) satisfies the following properties:*

(1.85a) *R is a $(3, 1)$-tensor;*

(1.85b) *R is skewsymmetric with respect to its first two arguments, i.e.,*

$$R(X, Y) = -R(Y, X);$$

(1.85c) *$R(X, Y)$ is skewsymmetric with respect to g, i.e.,*

$$g(R(X,Y)Z,U) = -g(R(X,Y)U,Z);$$

(1.85d) *(algebraic Bianchi identity)*

$$R(X,Y)Z + R(Y,Z)X + R(Z,X)Y = 0;$$

(1.85e) $g(R(X,Y)Z,U) = g(R(Z,U)X,Y)$;

(1.85f) *(differential Bianchi identity)*

$$(DR)(X,Y,Z) + (DR)(Y,Z,X) + (DR)(Z,X,Y) = 0.$$

Property (1.85e) follows from repeated applications of b), c) and d) (exercise!).

1.86. Using the metric g, we may also consider the curvature as a (4,0)-tensor, namely

$$(X,Y,Z,U) \to g(R(X,Y)Z,U).$$

We will also use the (2,2)-tensor deduced from R, that we *denote* by $\mathscr{R}$. Due to the symmetries, $\mathscr{R}$ may be considered as a linear map from $\bigwedge^2 M$ to $\bigwedge^2 M$, satisfying

$$g(\mathscr{R}(X \wedge Y), Z \wedge U) = g(R(X,Y)Z,U)$$

for any vectors X, Y, Z, U.

Notice that e) now states that $\mathscr{R}$ is a symmetric map with respect to the pseudo-Euclidean structure induced by g on $\bigwedge^2 M$ (see also 1.106).

1.87 Definition ($n \geqslant 2$). Given a non-isotropic 2-plane σ in T_xM, the *sectional curvature* of σ is given by

$$K(\sigma) = g(R(X,Y)X,Y)/(g(X,X)g(Y,Y) - g(X,Y)^2)$$

for any basis $\{X, Y\}$ of σ.

1.88 Proposition. *The sectional curvature of (M,g) is a constant k for any tangent two-plane at x if and only if the curvature tensor at x satisfies*

$$R(X,Y)Z = k(g(Y,Z)X - g(X,Z)Y). \tag{1.88}$$

Note that for the canonical example $\mathbb{R}^n$ then $R \equiv 0$, and that the examples S_p^n and H_p^n (see 1.36) have constant sectional curvature, respectively $+1$ and -1.

The differential Bianchi identity does not allow the curvature to vary too simply.

1.89 Theorem (F. Schur). *Let $n \geqslant 3$. If for each x in M, the sectional curvatures of the 2-planes through x are all equal (to a number that might a priori depend on x), then (M,g) has constant curvature.*

Because of the various algebraic symmetries of R only one contraction gives an interesting result:

1.90 Definition. The *Ricci curvature* tensor (or *Ricci tensor*) r of a pseudo-Riemannian manifold (M,g) is the 2-tensor

$$r(X,Y) = \mathrm{tr}(Z \to R(X,Z)Y),$$

where tr denotes the trace of the linear map $Z \to R(X,Z)Y$.

Note that the Ricci tensor is symmetric (this follows from the fact that the Levi-Civita connection has no torsion).

1.91 Remark. We will sometimes view r as a $(1,1)$-tensor and then *denote* it by Ric: $TM \to TM$; it satisfies

$$r(X,Y) = g(\mathrm{Ric}(X), Y).$$

Notice that Ric is a g-symmetric map. In particular, in the *Riemannian* case, its eigenvalues are real (this is not always true in the pseudo-Riemannian case).

Also, still in the Riemannian case, the Ricci tensor is related to the "local" function θ (defined in 1.48) through the following inequality.

1.92 Proposition. *Let (M,g) be a Riemannian manifold, c a geodesic with $c(0) = x$ and θ the function defined in* 1.48 *for normal coordinates centered at x. If we define φ by $\varphi(c(t)) = \frac{1}{t}(\theta(c(t)))^{1/(n-1)}$, then*

$$(n-1)\dot{c}\dot{c}\varphi + r(\dot{c},\dot{c})\varphi \leqslant 0. \tag{1.92}$$

One may deduce from proposition 1.92 that if M is complete and there exists $a > 0$ such that $r(X,X) \geqslant (n-1)a^2 g(X,X)$ for any X, then any geodesic c has a conjugate point $c(t_0)$ to $c(0)$ for some $t_0 \leqslant \frac{\pi}{a}$; together with 1.69, this gives Myers' theorem 6.51. More generally, a lower bound for r gives an upper bound for the growth of the volume of "distance balls" $B(x,t) = \{y \in M \text{ such that } d(x,y) \leqslant t\}$, see 0.62. In the case of *Lorentz* manifolds, similar formulas give the so-called "incompleteness theorems", see [Ha-El].

Of course, the trace of r with respect to g is also an interesting invariant.

1.93 Definition. The *scalar curvature* of a pseudo-Riemannian manifold (M,g) is the function $s = \mathrm{tr}_g r$ on M.

The scalar curvature will be studied in more detail in Chapter 4. Note that, on any manifold M, there are metrics with constant scalar curvature, see Chapter 4 below.

The derivatives of r and s are related by the following formula

1.94 Proposition. $\delta r = -\frac{1}{2}ds$, *where the divergence δ of a symmetric 2-tensor field has been defined in* 1.59.

Proof. Take suitable traces of the differential Bianchi identity. □

1.95 Definition. A pseudo-Riemannian manifold (M, g) is *Einstein* if there exists a real constant λ such that

$$r(X, Y) = \lambda g(X, Y) \tag{1.95}$$

for each X, Y in T_xM and each x in M.

1.96 Remarks. a) This notion is relevant only if $n \geqslant 4$. Indeed, if $n = 1$, $r = 0$. If $n = 2$, then at each x in M, we have $r(X, Y) = \frac{1}{2}sg(X, Y)$, so a 2-dimensional pseudo-Riemannian manifold is Einstein if and only if it has constant (sectional or scalar) curvature. If $n = 3$, then (M, g) is Einstein if and only if it has constant (sectional) curvature (exercise!).

b) If we replace g by t^2g for some positive constant t, then r does not change, hence for an Einstein manifold, the number λ in 1.95 changes to λt^{-2}.

1.97 Theorem. *Assume $n \geqslant 3$. Then an n-dimensional pseudo-Riemannian manifold is Einstein if and only if, for each x in M, there exists a constant λ_x such that*

$$r_x = \lambda_x g_x. \tag{1.97}$$

Proof. The "only if" part is trivial. In the other direction, applying the divergence δ to both sides of (1.97), we get

$$\delta r = -\tfrac{1}{2}ds = -d\lambda.$$

So $\lambda - \frac{1}{2}s$ is a constant. Taking the trace of (1.97) with respect to g, we get $n\lambda = s$. So finally λ (and s) are constant. □

1.98. Einstein manifolds (especially in the Riemannian case) are the subject of this book, so many examples will be described in the coming chapters (see in particular Chapter 7). Here we give only the obvious examples. Any manifold with constant sectional curvature is Einstein. In particular, $\mathbb{R}^n$, S^n_p, H^n_p with the metrics described in 1.34 and 1.36 are Einstein.

1.99 Proposition. *The product of two pseudo-Riemannian manifolds which are Einstein with the same constant λ is an Einstein manifold with the same constant λ.*

Note that the product of two Einstein manifolds with different constants is *not* Einstein. In fact, we may describe these products in the following way.

1.100 Theorem. *If the Ricci tensor r of a Riemannian manifold (M, g) is parallel* (*i.e.*, $Dr = 0$), *then at least locally, (M, g) is the product* (as in 1.37) *of a finite number of Einstein manifolds.*

This is a consequence of the de Rham decomposition theorem, see 10.43.

In 7.117, there is an example of a *Lorentz* manifold with $Dr = 0$, but which is not, even locally, a product, and is *not* Einstein.

G. Irreducible Decompositions of Algebraic Curvature Tensors

1.101. In this paragraph, we show that the curvature tensor of a Riemannian manifold splits naturally into three components, involving respectively its scalar curvature, the trace-free part of its Ricci tensor and its Weyl curvature tensor (if $n \geqslant 4$). This basic fact does not appear in most textbooks on Riemannian geometry, but it is classical and was well known in past times.

The key point here is that the bundle in which the curvature tensor lives naturally (according to the symmetry properties 1.85) is not irreducible under the action of the orthogonal group, and consequently has a natural decomposition into irreducible components. These give a canonical decomposition of the curvature tensor.

1.102. We begin with a review of pseudo-Euclidean geometry from the point of view of the representations of the orthogonal group on higher tensor spaces. We also consider the special orthogonal group in dimension 4, which gives an extra decomposition of the Weyl tensor for oriented four dimensional Riemannian manifolds.

These results give better insight into the role played by scalar and Ricci curvature, and provide very simple proofs of some classical results.

1.103. Let E be an n-dimensional real vector space ($n > 1$). Then each tensor space $T^{(k,l)}E = \bigotimes^k E^* \otimes \bigotimes^l E$ is a representation space for the linear group $Gl(E)$. For any γ in $Gl(E)$, $\xi_1, \ldots, \xi_k$ in E^* and $x_1, \ldots, x_l$ in E, the natural action of $Gl(E)$ satisfies

$$\gamma(\xi_1 \otimes \cdots \otimes \xi_k \otimes x_1 \otimes \cdots \otimes x_l) = ({}^t\gamma^{-1}\xi_1) \otimes \cdots \otimes ({}^t\gamma^{-1})\xi_k \otimes (\gamma x_1) \otimes \cdots \otimes (\gamma x_l).$$

Let q be a *non-degenerate* quadratic form on E. Then q induces a canonical identification between E and E^*. Moreover, if γ belongs to the orthogonal group $O(q)$ of q, we have ${}^t\gamma^{-1} = \gamma$, so E and E^* are isomorphic as $O(q)$-modules, and we may consider tensor products of E only.

We recall that we write S^kE for the k-th symmetric power of E, and $\bigwedge^l E$ for the l-th exterior power of E. Also, we denote by $\circ$ the symmetric product of two tensors, with the convention that $x \otimes x = x \circ x$.

1.104. Of course, the $O(q)$-module E is irreducible. It is well-known that $\bigotimes^2 E$ is not irreducible (even as a $Gl(E)$-module). We *denote* by $S_0^2 E$ the space of traceless symmetric 2-tensors. We recall that q induces a *trace*, which may be considered as a linear map $\mathrm{tr}_q\colon S^2E \to \mathbb{R}$.

1.105 Proposition. *The irreducible orthogonal decomposition of the $O(q)$-module $\bigotimes^2 E$ is the following*

$$\bigotimes^2 E = \bigwedge^2 E \oplus S_0^2 E \oplus \mathbb{R}q.$$

Proof. It is obvious that for any k in $\bigotimes^2 E$ $(= \bigotimes^2 E^*)$, we have

$$k = \Lambda^2 k + S_0^2 k + (\mathrm{tr}_q k/n)q,$$

where
$$\Lambda^2 k(x,y) = \tfrac{1}{2}(k(x,y) - k(y,x)),$$
$$S^2 k(x,y) = \tfrac{1}{2}(k(x,y) + k(y,x))$$
and
$$S_0^2 k = S^2 k - (\mathrm{tr}_q k/n)q.$$

The non-trivial point is that $\bigwedge^2 E$ and $S_0^2 E$ are irreducible. One proof of that follows from invariant theory; it suffices to check that the vector space of $O(q)$-invariant quadratic forms on $\bigotimes^2 E$ is 3-dimensional (cf. [Wey] II.9 and II.17, [Bes 2] Exposé IX or [Be-Ga-Ma] p. 82–83 for more details). □

1.106. We now study the tensors satisfying the same algebraic identities as the curvature tensor of a pseudo-Riemannian manifold at one point. The properties 1.85a), b), c) and e) mean that, if $E = T_x^*M$ and $g = g_x$, through the identification $\bigotimes^3 E \otimes E^* = \bigotimes^4 E$, the curvature tensor lies inside the subspace $S^2\bigwedge^2 E$.

Moreover, we know that $\bigotimes^k E$ is not irreducible, already as a $Gl(E)$-module. Indeed, the symmetric group $\mathfrak{S}_k$ (and its algebra $\mathbb{R}(\mathfrak{S}_k)$) provides natural $Gl(E)$-morphisms. The $Gl(E)$-irreducible components of $\bigotimes^k E$ appear as kernels of certain idempotents of $\mathbb{R}(\mathfrak{S}_k)$, the so-called *Young symmetrizers* (cf [Wey] Chap. IV or [Nai] II.3).

In particular, the algebraic Bianchi identity 1.85d) corresponds to the following Young symmetrizer.

1.107 Definition. We define the *Bianchi map* b to be the following idempotent of $\bigotimes^4 E$
$$b(R)(x,y,z,t) = \tfrac{1}{3}(R(x,y,z,t) + R(y,z,x,t) + R(z,x,y,t))$$
for any R in $\bigotimes^4 E$ and x, y, z, t in E^*.

Obviously, b is $Gl(E)$-equivariant, $b^2 = b$, and b maps $S^2\bigwedge^2 E$ into itself. So we have the $Gl(E)$-equivariant decomposition
$$S^2\textstyle\bigwedge^2 E = \operatorname{Ker} b \oplus \operatorname{Im} b.$$

Moreover, an easy calculation shows that $\operatorname{Im} b$ is precisely $\bigwedge^4 E$ (note that this fact implies that $b = 0$ on $S^2\bigwedge^2 E$ if $n = 2$ or 3. In other words, the Bianchi identity follows from the other ones in these dimensions).

1.108 Definition. We let $\mathscr{C}E = \operatorname{Ker} b$ (in $S^2\bigwedge^2 E$) and we call it the vector space of "*algebraic curvature tensors*".

Of course, the curvature tensor of a pseudo-Riemannian manifold lies inside $\mathscr{C}T_x^*M$ at each point x. But the key point of the whole paragraph is that this space $\mathscr{C}E$ may be decomposed as a $O(q)$-module. For algebraic curvature tensors, we introduce the notions equivalent to the Ricci and scalar curvatures for curvature tensors.

1.109 Definition. The *Ricci contraction* is the $O(q)$-equivariant map
$$c\colon S^2\textstyle\bigwedge^2 E \to S^2 E$$

defined, for any R in $S^2\Lambda^2E$ and any x, y in E^*, by

$$c(R)(x,y) = \operatorname{tr} R(x,.,y,.).$$

Conversely, there is a canonical way to build an element of $S^2\Lambda^2E$ from two elements of S^2E.

1.110 Definition. The *Kulkarni-Nomizu product* of two symmetric 2-tensors h and k is the 4-tensor $h \owedge k$ given by

$$(h \owedge k)(x,y,z,t) = h(x,z)k(y,t) + h(y,t)k(x,z) - h(x,t)k(y,z) - h(y,z)k(x,t),$$

for any x, y, z, t in E^*.

An easy computation gives

1.111 Proposition.
a) *$h \owedge k = k \owedge h$;*
b) *$h \owedge k$ belongs to $\mathscr{C}E$;*
c) *$q \owedge q$ is twice the identity of Λ^2E (through the identification* $\operatorname{End}(\Lambda^2E) = \bigotimes^2\Lambda^2E$).

Note that in the Riemannian case, the curvature tensor of a manifold with constant sectional curvature k is exactly $\frac{k}{2} g_x \owedge g_x$.

1.112 Remark. The Kulkarni-Nomizu product is a special case of the natural product of the graded algebra $\sum_{p=0}^{n} S^2(\Lambda^pE)$ where

$$(\alpha \circ \beta)\cdot(\mu \circ \nu) = (\alpha \wedge \mu)\circ(\beta \wedge \nu).$$

We recall that q also induces quadratic forms (still denoted by q) on $\bigotimes^k E$, and on subspaces as S^2E and $S^2\Lambda^2E$. If we identify S^2E with $\operatorname{End}E$ and $S^2(\Lambda^2E)$ with $\operatorname{End}(\Lambda^2E)$, then, in both cases, $q(h,k) = \operatorname{tr}(h \circ k)$ where $\circ$ is here the composition of linear maps. Then a straightforward computation gives

1.113 Lemma. *If $n > 2$, the map $q \owedge \cdot$ from S^2E into $\mathscr{C}E$ defined by $k \mapsto q \owedge k$ is* injective *and its adjoint is precisely the restriction to $\mathscr{C}E$ of the Ricci contraction.*

Now we come to the fundamental result.

1.114 Theorem. *If $n \geqslant 4$, the $O(q)$-module $\mathscr{C}E$ has the following orthogonal decomposition into (unique) irreducible subspaces*

$$\mathscr{C}E = \mathscr{U}E \oplus \mathscr{Z}E \oplus \mathscr{W}E \tag{1.114}$$

where

$$\mathscr{U}E = \mathbb{R}q \owedge q,$$
$$\mathscr{Z}E = q \owedge (S_0^2E)$$
$$\mathscr{W}E = \operatorname{Ker}(c \restriction \mathscr{C}E) = \operatorname{Ker} c \cap \operatorname{Ker} b.$$

Proof. The existence of the decomposition is clear from Lemma 1.113 (with $S^2E = \mathbb{R}q \oplus S_0^2E$). Here again the key point is the irreducibility of the factor $\mathscr{W}E$. This follows from invariant theory, since the vector space of $O(q)$-invariant quadratic forms on $\mathscr{C}E$ is 3-dimensional. More precisely, this vector space is generated by $q(R,R)$, $q(c(R), c(R))$ and $(\operatorname{tr} c(R))^2$, cf. [Bes 2] Exposé IX and [Be-Ga-Ma] pp. 82–83. □

1.115 Remark. In terms of representation theory, one may show that S_0^2E and $\mathscr{W}E$ (or more precisely their complexifications) are the irreducible $O(q)$-modules whose highest weights are twice the highest weights of E and $\bigwedge^2 E$ respectively.

1.116 Definitions. a) $\mathscr{W}E$ is called the space of *Weyl tensors*;

b) for any algebraic curvature tensor R, we *denote* by W (or $W(R)$) its component in $\mathscr{W}E$ and we call W the *Weyl part* of R.

One can compute W explicitly, using the Ricci contraction and the trace. Indeed, for any h in S^2E, an easy computation gives

$$c(q \owedge h) = (n-2)h + (\operatorname{tr} h)q.$$

So, if we write $r = c(R)$ and $s = \operatorname{tr} r$, we get the formula

$$R = \frac{s}{2n(n-1)} q \owedge q + \frac{1}{n-2} z \owedge q + W, \tag{1.116}$$

where we have *denoted* $r - \frac{s}{n} q$ by z.

In some sense, W appears as a remainder after successive "divisions" by q.

H. Applications to Riemannian Geometry

Let (M, g) be a pseudo-Riemannian manifold. For any x in M the curvature tensor R_x (through the identification $T_xM = T_x^*M$) belongs to $\mathscr{C}T_x^*M$ (with $q = g_x^*$). Of course $r = c(R)$ is the Ricci curvature and $s = \operatorname{tr} r$ the scalar curvature. Formula (1.116) gives the following

1.117 Definition. The *Weyl tensor* W of an n-dimensional pseudo-Riemannian manifold (M, g) (with $n \geqslant 4$) is the Weyl part (considered as a $(3,1)$-tensor) of its curvature tensor.

Of course an explicit computation of W is given by Formula (1.116) (as a $(4,0)$-tensor).

In section J below, we will show that W depends only on the conformal structure defined by g; in particular, $W = 0$ if and only if (M, g) is conformally flat (if $n \geqslant 4$), see 1.164 below.

1.118. From Definition 1.95, we see that (M, g) is Einstein if and only if the component of R in $\mathscr{Z}(T_xM)$ is zero for each x in M. Note that a manifold has constant sectional curvature if and only if the components of R in $\mathscr{Z}(T_xM)$ and $\mathscr{W}(T_xM)$ both vanish.

These various components of R are not completely independent as fields, since there are relations among their derivatives. The first basic relation is $\delta r = -\frac{1}{2}ds$ (see 1.94). We shall encounter three other similar relations, see 4.72 and 16.3.

1.119. When n is 2 or 3, the situation is simpler.

a) *If* $n = 2$, we have $S^2\bigwedge^2 E = \mathbb{R}q \owedge q$ and $R = \frac{s}{4} q \owedge q$, $r = \frac{s}{2} q$. Moreover, $s = 2\kappa$ where κ is the *Gauss curvature* (i.e., the sectional curvature of the 2-plane T_xM).

b) *If* $n = 3$, we have $S^2\bigwedge^2 E = \mathbb{R}q \owedge q \oplus S_0^2 E \owedge q$ and $R = \frac{s}{12} q \owedge q + \left(r - \frac{s}{3} q\right) \owedge q$.

In particular, the Ricci curvature determines the full curvature tensor and we have

1.120 Proposition. *A* 2 *or* 3-*dimensional pseudo-Riemannian manifold is Einstein if and only if it has constant sectional curvature.*

1.121. If we consider subgroups of $O(q)$, then more refined decompositions may appear. In the geometric case, this happens when M has additional structures (Hermitian or quaternionic, for example) and the theory is particularly interesting when the holonomy group of (M, g) is smaller than $O(n)$ (see Chapter 10).

Here we consider only the case of the special orthogonal group $SO(q)$, which corresponds geometrically to the case where M is oriented. It is known that the decomposition (1.114) is also $SO(q)$-irreducible if $n \neq 4$ (see [Kir] or [Bes 2] $n°$ 9). But new phenomena occur when $n = 4$. This is related to the non-simplicity of $SO(q)$ (or its complexification if the signature of q is $(1, 3)$.).

1.122. Throughout this section, the space E is an oriented 4-dimensional vector space and q a *positive definite* quadratic form on E. We restrict ourselves to this case because the phenomena are different for other signatures (see below 1.130).

We consider the Hodge operator $*$ defined in 1.51. First note that $*$ induces an isomorphism from $\bigwedge^4 E$ to $\bigwedge^0 E = \mathbb{R}$ as $SO(4)$-modules (we recall that they are *not* isomorphic as $O(4)$-modules). Also $*$ induces an automorphism of $\bigwedge^2 E$, which is selfadjoint, so we may consider it as well as an element of $S^2\bigwedge^2 E$. The main point here is that, as an automorphism of $\bigwedge^2 E$, $*$ is an involution. Let

$$\textstyle\bigwedge^+ E = \{\alpha \in \bigwedge^2 E;\ *\alpha = \alpha\}$$

$$\textstyle\bigwedge^- E = \{\alpha \in \bigwedge^2 E;\ *\alpha = -\alpha\}.$$

Then we have of course $\bigwedge^2 E = \bigwedge^+ E \oplus \bigwedge^- E$ and more precisely

1.123 Proposition. *The $SO(4)$-modules $\Lambda^+ E$ and $\Lambda^- E$ are both irreducible and 3-dimensional but they are* not *$SO(4)$-isomorphic. The $SO(4)$-module $\bigotimes^2 E$ admits the decomposition into (unique) irreducible subspaces*

$$\bigotimes^2 E = \Lambda^+ E \oplus \Lambda^- E \oplus S_0^2 E \oplus \mathbb{R}q.$$

For a proof, of this statement and of the next one, see [Ast] or [Bes 2] $n°$ IX. □

Another important property of these modules is the following

1.124 Lemma. *The subspace $\Lambda^+ E \otimes \Lambda^- E$ of $\bigotimes^4 E$ is exactly $\mathscr{Z}E$, so the map $q \owedge \cdot$ induces a canonical isomorphism of $S_0^2 E$ with $\Lambda^+ E \otimes \Lambda^- E$.*

1.125 Remark. As an $SO(4)$-module, $\Lambda^2 E$ is isomorphic to the Lie algebra $\mathfrak{so}(4)$ of $SO(4)$ (with the adjoint representation). This isomorphism maps $\Lambda^+ E$ and $\Lambda^- E$ onto the two 3-dimensional commuting ideals in $\mathfrak{so}(4)$, isomorphic to $\mathfrak{so}(3)$ (cf [Bes 2] appendix 1).

We now study the irreducible decomposition of $S^2\Lambda^2 E$ as an $SO(4)$-module. Of course $*$ generates $\Lambda^4 E$ and we see easily that for any R in $S^2\Lambda^2 E$, $b(R) = q(R, *)$.

Now the decomposition of $\Lambda^2 E$ immediately yields the following

1.126 Theorem. *The $SO(4)$-module $S^2\Lambda^2 E$ admits the orthogonal decomposition into irreducible subspaces*

$$S^2\Lambda^2 E = \mathbb{R}Id_{\Lambda^+ E} \oplus \mathbb{R}Id_{\Lambda^- E} \oplus (\Lambda^+ E \otimes \Lambda^- E) \oplus S_0(\Lambda^+ E) \oplus S_0(\Lambda^- E).$$

We easily see that $Id_{\Lambda^+ E} + Id_{\Lambda^- E} = Id_{\Lambda^2 E} = \frac{1}{2} q \owedge q$,

and $$Id_{\Lambda^+ E} - Id_{\Lambda^- E} = *;$$

thus $$\mathbb{R}Id_{\Lambda^+ E} \oplus \mathbb{R}Id_{\Lambda^- E} = \mathbb{R}q \owedge q \oplus \mathbb{R}*,$$

and $$\mathbb{R}* = \Lambda^4 E.$$

1.127. Let

$$\mathscr{W}^+ E = S_0(\Lambda^+ E)$$

and $$\mathscr{W}^- E = S_0(\Lambda^- E).$$

Since $\mathscr{Z}E = \Lambda^+ E \otimes \Lambda^- E$, we have $\mathscr{W}E = \mathscr{W}^+ E \oplus \mathscr{W}^- E$, so the irreducible decomposition of $\mathscr{C}E$ as an $SO(4)$-module is the following

$$\mathscr{C}E = \mathscr{U}E \oplus \mathscr{Z}E \oplus \mathscr{W}^+ E \oplus \mathscr{W}^- E.$$

The only difference between this and 1.114 is that the Weyl tensor may be split into two parts, which we have denoted by W^+ and W^-.

1.128. We may characterize the various subspaces by the behaviour of their elements under composition with $*$ as selfadjoint linear maps of $\Lambda^2 E$. Namely, we have

$$\mathscr{Z}E = \{R \in S^2\Lambda^2 E;\ *R = -*R\}$$
$$\mathscr{W}^+E = \{R \in S_0^2\Lambda^2 E;\ *R = R* = R\}$$
$$\mathscr{W}^-E = \{R \in S_0^2\Lambda^2 E;\ *R = R* = -R\}.$$

As a corollary, if we consider R in $S^2\Lambda^2 E$ as a linear map of $\Lambda^2 E$ and if we decompose $\Lambda^2 E = \Lambda^+ E \oplus \Lambda^- E$, we get a matrix

$$R = \left(\begin{array}{c|c} W^+ + \frac{s}{12} Id & Z \\ \hline {}_tZ & W^- + \frac{s}{12} Id \end{array}\right).$$

In dimension 4, we obtain a characterization of Einstein manifolds.

1.129 Corollary. *For a 4-dimensional Riemannian manifold, the following properties are equivalent:*

a) *(M, g) is Einstein;*

b) *for any local orientation, $R* = *R$;*

c) *for any x in M and any 2-plane σ in T_xM, its sectional curvature is the same as the sectional curvature of its orthogonal 2-plane $\sigma^\perp$ in T_xM.*

(All these results from 1.126 to 1.129 are due to I.M. Singer and J. Thorpe [Si-Th]).

Note that, given a point x in M, there always exists an orientation in a neighborhood of x in M.

1.130 Remark. a) If q has signature $(2, 2)$, then the same proof gives an analogous decomposition, since in this case also $*^2 = 1$.

b) Now if q has signature $(1, 3)$, the decomposition is quite different since $*^2 = -1$ (so $\Lambda^2 E$ has a natural complex structure). The decomposition in this case (due to A. Petrov) is studied in Chapter 3.

The difference in decomposition arises from the fact that $SO(3, 1)$ is simple, unlike $SO(4)$ or $SO(2, 2)$, (although they all have the same complexification).

As shown in 1.119, the curvature tensor has special properties in dimension $n = 2$ or 3 and its decomposition differs from the "general case" $n \geqslant 4$. In fact, there are also some special properties in dimension 4 (besides the preceding ones) and we will need them in Chapters 4 and 12. We begin with some general definitions.

1.131 Definitions. Let R be an algebraic curvature tensor.

a) We *denote* by $\check{R}$ the symmetric 2-tensor such that

$$\check{R}(x, y) = \sum_{i,j,k=1}^{n} \varepsilon_i\varepsilon_j\varepsilon_k R(x, e_i, e_j, e_k) R(y, e_i, e_j, e_k)$$

where x, y are in E^* and (e_i) is an orthonormal basis of E^* with $q(e_i, e_j) = \delta_{ij}\varepsilon_i$.

b) We *denote* by $\mathring{R}$ the linear map of S^2E into itself such that

$$(\mathring{R}h)(x, y) = \sum_{i=1}^{n} \varepsilon_i h(R(x, e_i)y, e_i)$$

for any h in S^2E and x, y, e_i as above, where R is then viewed as an element of $\bigotimes^2 E \otimes E^*$.

Easy calculations give

1.132 Proposition. a) $\operatorname{tr} \check{R} = q(R, R)$;

b) $\mathring{R}$ *is symmetric*;

c) R *is Einstein* $\left(\text{i.e., } r = \frac{s}{n} q\right)$ *if and only if* $\mathring{R}$ *maps* S_0^2E *into itself.*

See [Bo-Ka] for a further study of $\mathring{R}$. This operator appears naturally in the infinitesimal variation of the Ricci curvature, hence in the study of moduli of Einstein manifolds (see 12.30).

1.133 Remark. As a corollary, if R is Einstein and $n = 4$, then $\check{R} = \frac{1}{4}q(R, R)R$, but the converse is not true. For example, the curvature tensor of the Riemannian product $S^2 \times H^2$ (where the two spaces have opposite constant curvature) satisfies $\check{R} = \frac{1}{4}|R|^2 q$ since here $s = W = 0$ but $S^2 \times H^2$ is not Einstein.

I. Laplacians and Weitzenböck Formulas

1.134. On a Riemmanian manifold (M, g), the differential operators giving information on the geometry of M must be tied to the tensor field g. Among differential operators of order 1 acting on tensor fields, we have already seen that the Levi-Civita connection D plays a very special role. It maps a tensor field of order k to a tensor field of order $k + 1$ (via the metric we ignore the variance). Its principal symbol σ_D: $T^*M \otimes (\bigotimes^k TM) \to \bigotimes^{k+1} TM$ is the identity. (Here, we use the identification of T^*M with TM given by the metric.)

1.135. We will be interested in a special class of second order differential operators from which many geometric statements can be derived. We first identify this family.

The principal symbol σ_A of a second order differential operator A acting on sections of a vector bundle E maps $S^2T^*M \otimes E$ to E. We say for convenience that A is a *Laplace operator* if, for ξ in T^*M and e in E, $\sigma_A(\xi \otimes \xi \otimes e) = -g(\xi, \xi)e$. (We still denote the metric on covectors by g.) Other definitions would be possible.

Of special importance is the *rough Laplacian* $\nabla^*\nabla$ of a connection ∇ on the vector bundle E. From now on, *we suppose that E is a hermitian vector bundle.* We denote the formal adjoint of ∇ by ∇^*. Notice that $\nabla^*\nabla$ can be written as $-\operatorname{Trace} \nabla^1 \nabla$ where ∇^1 is the covariant derivative on the bundle $T^*M \otimes E$ defined by $\nabla^1 = \nabla \otimes 1_{T^*M} + 1_E \otimes D$.

1.136. The main property of an operator of the class we want to consider is the following. A Laplace operator is said to admit *a Weitzenböck decomposition* if

$$A = \nabla^*\nabla + K(R^\nabla)$$

where $K(R^\nabla)$ depends linearly on the curvature $R^\nabla \in \Omega^2(M, \mathrm{End}(E))$ of the covariant derivative ∇ on the bundle E.

1.137. Weitzenböck decompositions are important because of the following. There is a method, due to Bochner [Boc 1], of proving vanishing theorems for the null space of a Laplace operator admitting a Weitzenböck decomposition (and further of estimating its lowest eigenvalue). This method works mostly on *compact* manifolds. If $A = \nabla^*\nabla + K(R^\nabla)$,

$$\int_M (A\varphi, \varphi)v_g = \int_M \{(\nabla^*\nabla\varphi, \varphi) + (K(R^\nabla)\varphi, \varphi)\}v_g$$

$$= \int_M |\nabla\varphi|^2 v_g + \int_M (K(R^\nabla)\varphi, \varphi)v_g.$$

When $K(R^\nabla)$ is a non-negative endomorphism at each point, then the right hand side is the sum of two non-negative terms. Hence $A\varphi = 0$ implies both $\nabla\varphi = 0$ and $K(R^\nabla)\varphi = 0$.

If $K(R^\nabla)$ is strictly positive, then clearly $\varphi = 0$. This argument is by now classical in differential geometry and will be used several times in this book for various Laplace operators A. Part of the game is often to manufacture the appropriate second order operator. The Weitzenböck formula is then just a verification. To work with an auxiliary coefficient bundle turns out to be useful. This can be the cotangent bundle as in 12.69, 16.9 or 16.54.

1.138. A few comments are in order. When a vector bundle E comes equipped with a covariant derivative, a sequence (a_k) of relative symbols can be attached to a differential operator A of order k mapping sections of E to sections of another vector bundle F, so that $A = \sum_{i=0}^k a_i \nabla^i$ where $a_i \in C^\infty(S^i T^*M \otimes E, F)$. Of course a_k coincides with the principal symbol of A. Thanks to the covariant derivative, one can give a more algebraic description of differential operators.

In this context a Laplace operator admits a Weitzenböck decomposition if its relative symbol of order 1 vanishes and if its relative symbol of order 0 depends linearly on the curvature. These conditions are automatically satisfied by natural Riemannian differential operators on subbundles of the tangent bundle (cf [Eps], [Str] or [Ter] for a complete theory).

1.139. For *any* representation ρ of $O(n)$ (or $SO(n)$ or $\mathrm{Spin}(n)$ if the manifold is orientable or spin respectively), the associated vector bundle E_ρ and the representation on the Lie algebra defines a section c_ρ with $c_\rho \in \Omega^2(M, \mathrm{End}(E))$.

Here we have used the metric to identify the exterior power representation with the adjoint representation of $O(n)$ on its Lie algebra of skewadjoint matrices. We also use the Levi-Civita connection on E.

If $\mathscr{R} \in C^\infty(\bigwedge^2 T^*M \otimes \bigwedge^2 T^*M)$ is the Riemann curvature tensor, then we may form

$$c_\rho \otimes c_\rho(R) \in C^\infty(\mathrm{End}(E) \otimes \mathrm{End}(E)),$$

and, composing the endomorphisms, a section $c_\rho^2(R) \in C^\infty(\mathrm{End}(E))$. If $(\omega_\alpha)\left(1 \leqslant \alpha \leqslant \frac{n(n-1)}{2}\right)$ is an orthonormal basis for $\bigwedge^2 T^*M$ and $\mathscr{R} = \sum_{\alpha\beta} R_{\alpha\beta}\omega_\alpha \otimes \omega_\beta$, then

(1.140)
$$c_\rho^2(R) = \sum_{\alpha,\beta} R_{\alpha\beta} c_\rho(\omega_\alpha) c_\rho(\omega_\beta).$$

As we know (cf. 1.86), $R_{\alpha\beta} = R_{\beta\alpha}$.

Since $c_\rho(\omega_\alpha)$ is skew-adjoint, $c_\rho^2(R)$ is then self-adjoint.

1.141. An operator of the form

(1.142)
$$A = D^*D + \mu c_\rho^2(R)$$

(where μ is a constant) has the important property that when the holonomy group of the Levi-Civita connection reduces to $G \subseteq O(n)$, the operator A commutes with the orthogonal projection onto subbundles of E_ρ determined by the decomposition of the representation into irreducible components under the action of G.

1.143. An illustration of this construction is given by the so-called *Lichnerowicz Laplacian* on p-tensor fields. In [Lic 2], A. Lichnerowicz introduced the operator Δ_L defined for T in $C^\infty(\bigotimes^p T^*M)$ by

$$\Delta_L T = D^*DT + \Gamma T$$

where

$$(\Gamma T)_{i_1 \ldots i_p} = \sum_k r_{i_k j} T_{i_1 \ldots}{}^{j}{}_{\ldots i_p} - \sum_{k \neq l} R_{i_k j i_l h} T_{i_1 \ldots}{}^{j}{}_{\ldots}{}^{h}{}_{\ldots i_p}.$$

It is a straightforward matter to check that

(1.144)
$$\Gamma R = -2c_\rho^2(R)$$

for the natural representation of $O(n)$ on p-tensors.

1.145. The second order differential operator which defines first order deformations of Einstein metrics is the Lichnerowicz Laplacian for symmetric 2-tensor fields (see [Be-Eb] and Chapter 12). The curvature term may of course be described in the invariant manner given above.

1.146. We now list the Weitzenböck decompositions for the following operators from the above viewpoint:

i) the Hodge Laplacian $\Delta_H = dd^* + d^*d$ acting on exterior differential p-forms;

ii) the complex Laplacian $\bar{\partial}\bar{\partial}^* + \bar{\partial}^*\bar{\partial}$ acting on exterior differential forms of type $(0, p)$ on a Kähler manifold (see Chapter 2);

iii) the Dirac Laplacian $\mathscr{D}^2$ acting on the (complex) spinor bundle ΣM (see next alinea);

iv) the Dirac Laplacian $(\mathscr{D}^\nabla)^2$ acting on sections of $\Sigma M \otimes E$ for a coefficient bundle E with connection ∇.

1.147. The Dirac operator $\mathscr{D}: C^\infty(\Sigma M) \to C^\infty(\Sigma M)$ on a spin manifold is defined by

$$\mathscr{D} = \sigma \circ D,$$

where $\sigma: C^\infty(T^*M \otimes \Sigma M) \to C^\infty(\Sigma M)$ is the Clifford multiplication map

$$\sigma(\alpha \otimes \varphi) = \alpha \cdot \varphi.$$

Thus

$$\mathscr{D}^2\varphi = \sigma^2 D^2\varphi$$

where $\mathscr{D}^2\varphi \in C^\infty(T^*M \otimes T^*M \otimes \Sigma M)$ and

$$\sigma^2(\alpha \otimes \beta \otimes \varphi) = \alpha \cdot \beta \cdot \varphi. \tag{1.148}$$

But, under Clifford multiplication,

$$\alpha \cdot \beta + \beta \cdot \alpha = -2g(\alpha, \beta)Id,$$

so that

$$\mathscr{D}^2\varphi = D^*D\varphi + c_\sigma^2(R)\varphi. \tag{1.149}$$

Now on the Lie algebra the spin representation δ is given [Hit 2] for $a = \sum a_{ij} e_i \wedge e_j$ by

$$c_\delta(a) = -\frac{1}{4}\sum a_{ij} e_i \cdot e_j$$

so

$$\begin{aligned}\sum \sigma^2(e_i \otimes e_j \otimes a\varphi) &= \sum a_{ij} e_i \cdot e_j \cdot \varphi \\ &= -4\delta(a)\varphi.\end{aligned}$$

Thus, from (1.149),

$$\mathscr{D}^2 = D^*D - 4c_\delta^2(R).$$

It is a straightforward computation involving the symmetries of the curvature tensor to show that

$$\sum R_{ijkl} e_i \cdot e_j \cdot e_k \cdot e_l = -2sId \in C^\infty(\mathrm{End}(\Sigma M))$$

and hence one obtains the Lichnerowicz formula [Lic 4].

$$\mathscr{D}^2 = D^*D + \tfrac{1}{4}s. \tag{1.150}$$

1.151. It is often convenient to work with a coefficient bundle.

The Dirac operator $\mathscr{D}^\nabla$: $C^\infty(\Sigma M \otimes E) \to C^\infty(\Sigma M \otimes E)$ for a coefficient bundle E with connection ∇ is defined by

$$\mathscr{D}^\nabla = (\sigma \otimes 1)(D \otimes 1 + 1 \otimes \nabla).$$

Thus

(1.151)
$$(\mathscr{D}^\nabla)^2 = \nabla^{1*}\nabla^1 + c_\sigma^2(R \otimes 1 + 1 \otimes R^\nabla).$$

Hence, from I.147,

(1.152)
$$(\mathscr{D}^\nabla)^2 = \nabla^{1*}\nabla^1 + \tfrac{1}{4}s + c_\sigma^2(1 \otimes R^\nabla)$$

where

$$c_\sigma^2(1 \otimes R^\nabla)(\varphi \otimes e) = \sum e_i \cdot e_j \cdot \varphi \otimes R_{ij}(e).$$

In the special case (which will be important in Chapter 13) where (E, ∇) is itself associated with the principal frame bundle with the Levi-Civita connection by a representation ρ, we have

$$c_\sigma^2(1 \otimes R^\nabla) = -4c_{\delta\otimes\rho}(R) \in C^\infty(\mathrm{End}(\Sigma M \otimes E)),$$

with

(1.153)
$$c_{\delta\otimes\rho}(R) = \sum_{\alpha,\beta} R_{\alpha\beta}\delta(\omega_\alpha) \otimes \rho(\omega_\beta).$$

Let us take the coefficient bundle $E = \Sigma M$, the spinor bundle. Then (cf [At-Bo-Sh], [Hit 2]), if n is even,

$$\Sigma M \otimes \Sigma M \simeq \bigoplus_{p=0}^{n} \Lambda^p T_{\mathbb{C}}^* M,$$

and, if n is odd,

$$\Sigma M \otimes \Sigma M \simeq \bigoplus_{p=0}^{[\frac{n}{2}]} \Lambda^{2p} T_{\mathbb{C}}^* M \simeq \bigoplus_{p=0}^{[\frac{n}{2}]} \Lambda^{2p+1} T_{\mathbb{C}}^* M.$$

Furthermore, Clifford multiplication on the left is just, for n even, $e(\alpha) - i(\alpha)$, and, for n odd, $\pm *e(\alpha) \pm *i(\alpha)$, where $e(\alpha)$ denotes exterior multiplication by a 1-form α and $i(\alpha)$ its adjoint, interior multiplication.

Consequently,

$$\mathscr{D}^D = d + d^* \qquad \text{for } n \text{ even},$$

$$\mathscr{D}^D = \pm * d \pm d* \qquad \text{for } n \text{ odd},$$

and in either case

$$(\mathscr{D}^D)^2 = dd^* + d^*d, \text{ the Hodge Laplacian } \Delta.$$

We may therefore use the result of 1.151 to write

$$\Delta = dd^* + d^*d = D^*D - 4(c_\delta^2 \otimes 1 + c_\delta \otimes c_\delta)(R).$$

Now for the tensor product representation ρ on $\Sigma M \otimes \Sigma M$, the section $c_\rho \in C^\infty(\mathrm{Hom}(\Lambda^2 T^*M, \Sigma M \otimes \Sigma M)$ is

$$c_\rho = c_\delta \otimes 1 + 1 \otimes c_\delta$$

and then

$$c_\rho^2(R) = (c_\delta^2 \otimes 1 + 2c_\delta \otimes c_\delta + 1 \otimes c_\delta^2)(R).$$

But $c_\rho^2(R) = -\frac{1}{16}s$ is a scalar, hence

$$(c_\delta^2 \otimes 1 + c_\delta \otimes c_\delta)(R) = \tfrac{1}{2}c_\rho^2(R),$$

and we obtain

(1.154) $$dd^* + d^*d = D^*D - 2c_\rho^2(R),$$

which is a Lichnerowicz Laplacian (cf. 1.143).

1.155. An even more special case consists of the representation on 1-forms, the standard n-dimensional representation of $O(n)$. Here

$$\begin{aligned} dd^* + d^*d &= D^*D - 2\sum \tfrac{1}{2}R_{ijjl}e_i \otimes e_l \\ &= D^*D + r \end{aligned}$$

which yields Bochner's vanishing theorem [Boc 1] (cf. Chapter 6): if $r > 0$, then the first Betti number of a compact manifold vanishes.

1.156. A Kähler manifold can be characterized as a Riemannian manifold whose holonomy reduces to $U(m) \subseteq SO(2m)$. From [At-Bo-Sh], [Hit 2], the spinor bundle ΣM is given by

(1.157) $$\Sigma M \simeq \bigoplus_{p=0}^{n} \Lambda^{0,p} T_{\mathbb{C}}^* M \otimes K^{1/2}$$

where $K^{1/2}$ is the line bundle associated with the (2-valued) representation

$$(\det)^{-1/2}\colon U(m) \to U(1).$$

(Note that in computing Weitzenböck formulas, the global problems of orientation and spin play no part.)

Thus, from (1.157), we obtain

$$\bigoplus \Lambda^{0,p} T_{\mathbb{C}}^* M \simeq \Sigma M \otimes K^{-1/2}$$

where, moreover, the Dirac operator

$$\mathscr{D}^D = \sqrt{2}(\bar\partial + \bar\partial^*) \qquad \text{[Hit 2].}$$

Thus

$$\bar\partial\bar\partial^* + \bar\partial^*\bar\partial = \tfrac{1}{2}(\mathscr{D}^D)^2.$$

From (1.152), we find

$$\bar\partial\bar\partial^* + \bar\partial^*\bar\partial = D^*D + \tfrac{1}{4}s - 4(c_\delta \otimes c_\rho)(R).$$

In this case the curvature of the coefficient bundle is half the Ricci form and the eigenvalues of $-4(c_\delta \otimes c_\rho)(R)$ are $\frac{1}{2}\sum_{i=1}^{m}(\pm\lambda_i)$ where the Ricci form has eigenvalues λ_i $(1 \leqslant i \leqslant m)$. Since

$$s = 2\sum_{i=1}^{m} \lambda_i,$$

we obtain from the vanishing argument quoted as Bochner's method (cf. 1.137) and the Dolbeault isomorphism the theorem of Bochner and Yano [Bo-Ya]: if on a Kähler manifold we have $r > 0$, then $H^p(M, \mathcal{O}) = 0$ for $p > 0$ where $\mathcal{O}$ is the sheaf of holomorphic functions on M.

For further vanishing theorems deduced from a spinorial approach, see [Mic], [Hsi].

J. Conformal Changes of Riemannian Metrics

We consider here the simplest non-homothetic deformations of a (pseudo-) Riemannian metric, namely the conformal ones. They are obtained by changing at each point the lengths of all vectors by a scaling factor (depending on the point) without changing the "angles". More precisely, we set

1.158 Definition. Two pseudo-Riemannian metrics g and g^1 on a manifold M are said to be

a) *(pointwise) conformal* if there exists a C^∞ function f on M such that $g^1 = e^{2f}g$;

b) *conformally equivalent* if there exists a diffeomorphism α of M such that α^*g^1 and g are pointwise conformal.

Note that, if g and g^1 are conformally equivalent, then α is an *isometry* from $e^{2f}g$ onto g^1. So we will only study below the case $g^1 = e^{2f}g$.

In the following, we compute the various invariants of g^1 in terms of those of g and the derivatives of f (with respect to the Levi-Civita connection D of g). We denote as before R, r, s, W the various curvature tensors of g.

We recall that, for a function f, Df is the gradient, Δf the Laplacian and Ddf the Hessian with respect to g, see 1.54.

1.159 Theorem. *Let (M, g) be a pseudo-Riemannian manifold and f a function on M. Then the pseudo-Riemannian metric $g^1 = e^{2f}g$ has the following invariants:*

a) *Levi-Civita connection*

$$D^1_X Y = D_X Y + df(X)Y + df(Y)X - g(X, Y)Df$$

b) (4, 0) *curvature tensor*

$$R^1 = e^{2f}(R - g \owedge (Ddf - df \circ df + \tfrac{1}{2}|df|^2 g))$$

c) (3, 1) *Weyl tensor*

$$W^1 = W$$

d) *Ricci tensor*

$$r^1 = r - (n-2)(Ddf - df \circ df) + (\Delta f - (n-2)|df|^2)g$$

e) *trace-free part of the Ricci tensor*

$$Z^1 = Z - (n-2)(Ddf - df \circ df) - \frac{n-2}{n}(\Delta f + |df|^2)g$$

f) *scalar curvature*

$$s^1 = e^{-2f}(s + 2(n-1)\Delta f - (n-2)(n-1)|df|^2)$$

g) *volume element*

$$\mu_{g^1} = e^{nf}\mu_g$$

h) *Hodge operator on p-forms (if M is oriented)*

$$*_{g^1} = e^{(n-2p)f} *_g$$

i) *codifferential on p-forms*

$$\delta^1\beta = e^{-2f}(\delta\beta - (n-2p)i_{Df}\beta)$$

j) (*pseudo-*) *Laplacian on p-forms*

$$\begin{aligned}\Delta^1\alpha = e^{-2f}(\Delta\alpha - (n-2p)d(i_{Df}\alpha) - (n-2p-2)i_{Df}d\alpha + \\ 2(n-2p)df \wedge i_{Df}\alpha - 2df \wedge \delta\alpha).\end{aligned} \quad \square$$

1.160 Remarks. a) The canonical isomorphisms ♯ and ♭ between TM and T^*M are not the same for g and g^1. In particular, viewed as a (4, 0)-tensor, the Weyl tensor would satisfy $W^1 = e^{2f}W$ and conversely, if we view R^1 as a (3, 1)-tensor, there is no factor e^{2f}.

b) If we are interested in only one of these invariants, it is often useful to choose the scaling factor in a different form in order to simplify the expression. In particular we have

1.161 Corollary. a) *If* $g^1 = \psi^{4/n-2}g$ $(n \geqslant 3, \psi > 0)$, *then*

$$\psi^{n+2/n-2}s^1 = 4\frac{n-1}{n-2}\Delta\psi + s\psi. \tag{1.161a}$$

b) *If* $g^1 = \varphi^{-2}g$ $(\varphi > 0)$, *then*

$$Z^1 = Z + \frac{n-2}{\varphi}\left(Dd\varphi + \frac{\Delta\varphi}{n}g\right). \tag{1.161b}$$

c) *If* $g^1 = \tau^{-1}g$ $(\tau > 0)$, *then*

$$\Delta^1\alpha = \tau\Delta\alpha + \frac{n-2p-2}{2}i_{D\tau}d\alpha + \frac{n-2p}{2}d(i_{D\tau}\alpha) + d\tau \wedge \delta\alpha. \tag{1.161c}$$

Formula (1.161a) yields the equation of the Yamabe conjecture (see Chapter 4 and references there). Formula (1.161b) was studied in particular by M. Obata [Oba].

We notice that some other invariants do not vary under conformal changes. For example, for an even-dimensional oriented manifold, we have $*_{g^1} = *_g$ on $\frac{n}{2}$-forms. We will use this fact in particular for 4-manifolds. As a consequence, we have the following

1.162 Corollary. *Let M be a compact even dimensional manifold. For any function f, the harmonic $\frac{n}{2}$-forms are the same for g and $g^1 = e^{2f}g$.*

1.163. Another example is the following. Let c be a *null* geodesic for a pseudo-Riemannian manifold (M,g), i.e., c satisfies $g(\dot{c}(t),\dot{c}(t)) = 0$ (this implies that g is not Riemannian). Then, up to some change of parametrization, c is still a null geodesic for the pseudo-Riemannian metrics $e^{2f}g$ on M, for any function f. Indeed,

$$D^1_{\dot{c}}\dot{c} = 2df(\dot{c})\dot{c}$$

so if we choose $c_1 = c \circ \varphi$ where the real function $\varphi(t)$ satisfies $\varphi'' + 2\varphi' df(\dot{c})(c(t)) = 0$, we get $D^1_{\dot{c}_1}\dot{c}_1 = 0$.

Finally, we see that the second fundamental form of an immersion does not vary too badly under a (point-wise) conformal change of the ambient metric. If II is the second fundamental form of the Riemannian immersion $i: (M,g) \to (N,h)$, then the second fundamental form II^1 of the Riemannian immersion

$$i: (M, e^{2(f \circ i)}g) \to (N, e^{2f}h)$$

is given by

$$\mathrm{II}^1(U,V) = \mathrm{II}(U,V) - g(U,V)\mathcal{N}(Df) \tag{1.163}$$

where $\mathcal{N}(Df)$ denotes the normal component of the gradient of f.

The main invariant under conformal changes is the Weyl tensor. In particular, since $W = 0$ for a flat metric, the Weyl tensor is seen to be an obstruction for a pseudo-Riemannian metric to be locally conformal to a flat metric.

1.164 Definition. A pseudo-Riemannian manifold (M,g) is *conformally flat* if, for any x in M, there exists a neighborhood V of x and a C^∞ function f on V such that $(V, e^{2f}g)$ is flat.

Note that we do not require f to be defined on the whole of M. It turns out that, for $n \geqslant 4$, the Weyl tensor is the only obstruction to conformal flatness.

1.165 Theorem. *If $n \geqslant 4$, an n-dimensional pseudo-Riemannian manifold (M,g) is conformally flat if and only if its Weyl tensor vanishes.*

For a detailed proof, see [Eis] p. 85 or [Grr] p. 188. Following Theorem 1.159, this amounts to solving locally the second order overdetermined system

$$Ddf - df \circ df + \left(|df|^2 - \frac{\Delta f}{n-2} \right) g = r.$$

The main point is that for $n \geqslant 4$, if $W = 0$, there are no integrability conditions on higher derivatives. □

1.166 Remark. A necessary but not sufficient condition for conformal flatness is that $\mathscr{R}$ has a basis of decomposable 2-forms as eigenvectors. This is clear since, if $W = 0$,

$$R = \left(Z - \frac{s}{2n-2} g \right) \mathbin{\bigcirc\!\!\!\!\wedge}\, g.$$

1.167 Examples. 1) Any manifold with constant sectional curvature is conformally flat.

2) The product of a manifold with constant sectional curvature with S^1 or $\mathbb{R}$ is conformally flat.

3) The product of two Riemannian manifolds, one with sectional curvature 1, and the other with curvature -1, is conformally flat. Here is a short proof of 3): we (abusively) write

$$g = g_M + g_N \quad \text{on} \quad T(M \times N) = TM \times TN.$$

Then $\quad R_M = g_M \mathbin{\bigcirc\!\!\!\!\wedge}\, g_M \quad \text{and} \quad R_N = -g_N \mathbin{\bigcirc\!\!\!\!\wedge}\, g_N.$

So

$$\begin{aligned} R = R_M + R_N &= g_M \mathbin{\bigcirc\!\!\!\!\wedge}\, g_M - g_N \mathbin{\bigcirc\!\!\!\!\wedge}\, g_N \\ &= (g_M + g_N) \mathbin{\bigcirc\!\!\!\!\wedge}\, (g_M - g_N) \\ &= g \mathbin{\bigcirc\!\!\!\!\wedge}\, (g_M - g_N) \quad \text{and so} \quad W = 0. \end{aligned}$$

□

Indeed, it can be checked that 2) and 3) provide the only cases where a Riemannian product is conformally flat.

1.168. The same formalism also gives a simple proof of the following result due to E. Cartan [Car 4]: a hypersurface of a conformally flat Riemannian manifold (with $n \geqslant 5$) is conformally flat if and only if its second fundamental form either has a unique eigenvalue, or has two eigenvalues, one of which has multiplicity one. (The "if" part is related to phenomena studied in Chapter 16, for Codazzi tensors).

Note that Theorem 1.165 does not apply to dimensions 2 and 3, since the Weyl tensor disappears from the decomposition of R. The results here are quite different.

1.169 Theorem. *Any 2-dimensional pseudo-Riemannian manifold is conformally flat.*

In fact, this theorem is usually known under an equivalent formulation, namely the existence of "isothermal coordinates" (i.e., local coordinates (x, y) such that $g = e^{2f}(dx^2 + dy^2)$), which was proved first by C.F. Gauss [Gau] in the analytic case and then by A. Korn [Kor] and L. Lichtenstein [Lit] in the C^∞ case (see also the proof by S.S. Chern [Chr 1] or the presentation [Spi] volume 4 p. 455. Using Theorem 1.159, this amounts to solving locally the equation $2\Delta f + s = 0$. Since Δ is elliptic if g is Riemannian, the local solvability in this case is an immediate consequence of Appendix Theorem 45. □

1.170. In dimension 3, there is another obstruction to conformal flatness, which lives in the decomposition of DR into irreducible components. This obstruction is called the *Weyl-Schouten tensor field* and vanishes if and only if (M, g) is conformally flat. Since R is determined by r ($n = 3$), this Weyl-Schouten tensor may be given by a component of the decomposition of Dr; in 16.4.(v)d) we will see that (M, g) is conformally flat if and only if Dr is *symmetric*.

Finally, let us mention that conformally flat manifolds are not classified in full generality, although we have

1.171 Theorem (N. Kuiper [Kui]). *A conformally flat compact simply connected Riemannian manifold is conformally equivalent to the canonical sphere* (*of the same dimension*).

K. First Variations of Curvature Tensor Fields

1.172. If M is a compact manifold, the set $\mathcal{M}$ of all Riemannian metrics is open in $\mathcal{S}^2 M$, for the compact open topology or for the L^p_k-topologies as well. Recall that the definition of the L^p_k-norm (cf. Appendix. 3) requires the choice of a metric on M. But when M is compact, the topology defined by such a norm does not depend on the metric. (When studying geometric problems, one has to consider Riemannian structures, i.e., Riemannian metrics modulo the group of diffeomorphisms. We refer to Chapters 4 and 12 for this point of view. Here it appears only as a remark.)

1.173. The formulas for various curvature tensors in local coordinates are not very pleasing (see 1.42). However, they prove that the maps $g \mapsto R_g$ (resp. $g \mapsto r_g$, $g \mapsto s_g$) from $\mathcal{M}$ into $C^\infty(\Lambda^2 M \otimes \Lambda^2 M)$ (resp. $\mathcal{S}^2 M$, $\mathcal{C}M$), are quasilinear second order differential operators. In particular, they are differentiable if $\mathcal{M}$ is equipped with some L^2_k-norm ($k \geqslant 2$) and the target space with the corresponding L^2_{k-2}-norm. Moreover, for a given g, the differentials at the point g are linear second order differential operators, which will be *denoted* by R'_g, r'_g and s'_g. They have been computed in local coordinates by many relativists (see [Bln], [Lic 2]). Here we give a synthetic treatment, close to [Ble].

1.174 Theorem. *Let (M, g) be a pseudo-Riemannian manifold and h be in $\mathcal{S}^2 M$. Then the differentials at g, in the direction of h, of the Levi-Civita connection and the various curvature tensors are given by the following formulas*:

(a) *Levi-Civita connection*

$$g(D'_g h(X, Y), Z) = \tfrac{1}{2}\{D_X h(Y, Z) + D_Y h(X, Z) - D_Z h(X, Y)\};$$

(b) (3, 1)-*curvature tensor*

$$R'_g h(X, Y)Z = (D_Y D'_g h)(X, Z) - (D_X D'_g h)(Y, Z);$$

(c) (4, 0)-*curvature tensor*

$$R'_g h(X,Y,Z,U) = \tfrac{1}{2}\{D^2_{Y,Z}h(X,U) + D^2_{X,U}h(Y,Z) - D^2_{X,Z}h(Y,U) - D^2_{Y,U}h(X,Z) + h(R(X,Y)Z,U) - h(R(X,Y)U,Z)\};$$

(d) *Ricci tensor*

$$r'_g h = \tfrac{1}{2}\Delta_L h - \delta^*_g(\delta_g h) - \tfrac{1}{2}D_g d(\mathrm{tr}_g h);$$

(e) *scalar curvature*

$$s'_g h = \Delta_g(\mathrm{tr}_g h) + \delta_g(\delta_g h) - g(r_g, h).$$

Proof. Since the difference between two connections is a tensor field, the differential of the Levi-Civita connection is a symmetric (2, 1)-tensor field. Now the Definition 1.39 gives the formula

(1.175) $$g(D_X Y, Z) = \tfrac{1}{2}\{Xg(Y,Z) + Yg(Z,X) - Zg(X,Y) - g(X,[Y,Z]) + g(Y,[Z,X]) + g(Z,[X,Y])\},$$

for any vector fields X, Y and Z. And (1.174a) follows easily. □

1.176 Notation. It will sometimes be convenient to consider the (3, 0)-tensor field

$$C_g h(X,Y,Z) = g(D'_g h(X,Y), Z),$$

instead of $D'_g h$.

1.177. In order to compute the differential of R, we differentiate the Formula (1.11), which gives

$$R'_g h(X,Y)Z = D'_g h(Y, D_X Z) + D_Y(D'_g h(X,Z)) - D'_g h(X, D_Y Z) - D_X(D'_g h(Y,Z)) + D'_g h([X,Y],Z)$$

and (1.174b) follows immediately.

1.178 Remark. We will not prove (1.174c), since it will not be used in the book. We have mentioned it mainly for the sake of completeness. Recall that differentiation does *not* commute with changing the type of a tensor field, nor with taking the trace of a (2, 0)-tensor, since these operations do involve the metric. But we want to point out the following facts.

(a) Using the Ricci identity (1.21), it can be checked that this tensor field is symmetric with respect to the pairs (X, Z) and (Y, U).

(b) The symmetries of $R'_g(h)$ can be viewed concisely at the symbol level (see Appendix 15). We have

$$\sigma_t R'_g(h) = -\tfrac{1}{2}(t \circ t) \owedge h.$$

1.179. In order to compute r'_g, we first notice that taking the trace of a linear map does commute with differentiation. Hence, if (X_i) is any orthonormal frame, we get

$$r'_g h(X,Y) = \sum_{i=1}^{n} \varepsilon_i (D_{X_i}(C_g h)(X,Y,X_i) - D_X(C_g h)(X_i,Y,X_i)).$$

The second term is easily seen to be $\frac{1}{2}D_g d(\operatorname{tr}_g h)$, and the first term may be written

$$\frac{1}{2}\sum_{i=1}^{n} \varepsilon_i (D^2_{X_i,X} h(Y,X_i) + D^2_{X_i,Y} h(X,X_i) - D^2_{X_i,X_i} h(X,Y)).$$

Now the Ricci identity (1.21) gives

$$D^2_{X_i,X} h(Y,X_i) = D^2_{X,X_i} h(Y,X_i) + h(R(X_i,X)Y,X_i) + h(R(X_i,X)X_i,Y).$$

We put all these terms together and we note that $D^*Dh = -\sum_{i=1}^n D^2_{X_i,X_i} h$ (where D^* is the formal adjoint of D and has been defined in 1.55). We get

(1.180a) $$r'_g h = \tfrac{1}{2}(D^*Dh + r_g \circ h + h \circ r_g - 2\mathring{R}_g h - 2\delta_g^*\delta_g h - D_g d(\operatorname{tr}_g h)).$$

In this formula, δ_g denotes the divergence and δ_g^* its formal adjoint, both defined in 1.59; also $h \circ k$ is the (2, 0)-tensor associated by the metric to the composition of h and k viewed as (1, 1)-tensors, i.e., as linear maps of TM into itself; and finally $\mathring{R}_g$ denotes the action of the curvature on symmetric 2-tensors, and has been defined in 1.131.

Now we recognize the operator

(1.180b) $$\Delta_L h = D^*Dh + r_g \circ h + h \circ r_g - 2\mathring{R}_g h.$$

It is nothing but the Lichnerowicz Laplacian (restricted to symmetric (2, 0)-tensors) which we have already defined in 1.143. This gives (1.174d).

1.181. In order to compute s'_g, the simplest procedure is to differentiate the formula $s_g = \operatorname{tr}_g r_g$. We get

$$s'_g h = -g(h, r_g) + \operatorname{tr}_g(r'_g h).$$

We notice that $\Delta_g(\operatorname{tr}_g h) = \operatorname{tr}_g(\Delta_L h)$, that $\operatorname{tr}_g(\delta_g^*(\delta_g h)) = -\delta_g(\delta_g h)$ and that $\Delta_g f = -\operatorname{tr}_g(D_g df)$. This gives immediately (1.174e). □

1.182 Remark. The symbol of the operator s'_g is given by

(1.182) $$\sigma_t s'_g(h) = -|t|^2 \operatorname{tr}_g h - h(t^\sharp, t^\sharp).$$

It is clearly surjective when the covector t is not zero. Therefore, the formal adjoint $(s'_g)^*$ of s'_g is an overdetermined elliptic operator from $\mathscr{C}M$ to $\mathscr{S}^2 M$. It is given by the following formula

(1.183) $$(s'_g)^* f = D_g df + (\Delta_g f)g - f r_g$$

for any function f on M.

For future use (see 4.33 below), we will need the following

1.184 Proposition. *The differential Δ'_g of the Laplacian acting on functions is given by the formula*

(1.184) $$\Delta'_g h(f) = g(D_g df, h) - g(df, \delta_g h + \tfrac{1}{2}d(\operatorname{tr}_g h)).$$

Proof. We first compute the differential δ'_g of the codifferential δ_g acting on differential 1-forms. Since $\delta_g \alpha = D_g^* \alpha = -\operatorname{tr}_g(D_g \alpha)$, we have immediately $\delta'_g h(\alpha) = g(h, D_g \alpha) -$

$\mathrm{tr}_g(D'_g h(\alpha))$, and formula (1.174a) gives

(1.185) $$\delta'_g h(\alpha) = g(h, D_g\alpha) - g(\alpha, \delta_g h + \tfrac{1}{2}d(\mathrm{tr}_g h)).$$

And we get (1.184) by applying (1.185) to $\alpha = df$. □

Finally, an easy computation gives the differential of the volume and the volume element.

1.186 Proposition. *Let (M, g) be a pseudo-Riemannian manifold.*

(a) *The differential μ'_g of the volume element is given by the formula*

$$\mu'_g h = -\tfrac{1}{2}(\mathrm{tr}_g h)\mu_g.$$

(b) *We assume furthermore that M is* compact *and we* denote $\int_M \mu_g$ *by* $\mathrm{Vol}(M)(g)$. *Then we have*

$$\mathrm{Vol}(M)'_g h = \int_M \mu'_g h = -\frac{1}{2}\int_M (\mathrm{tr}_g h)\mu_g.$$

Note that, given a volume element μ_g, any other volume element μ_{g_1} may be written $\mu_{g_1} = f\mu_g$ for some positive function f on M.

1.187. As an application of these computations, we give an easy consequence of formulas (1.174e) and (1.186b) for compact Einstein manifolds.

If (M, g) is Einstein, then $r_g = \frac{1}{n}s_g g$ (with a constant s_g), hence $g(r_g, h) = \frac{1}{n}s_g g(g, h) = \frac{1}{n}s_g\,\mathrm{tr}_g h$. Moreover, if M is compact, the integral over M of a divergence (and a Laplacian) is always zero, and formula (1.174e) gives (in the case where (M, g) is Einstein and compact)

$$\int_M (s'_g h)\mu_g = -\frac{1}{n}s_g\int_M (\mathrm{tr}_g h)\mu_g = -\frac{2}{n}s_g\int_M \mu'_g h,$$ and we have proved

1.188 Proposition (M. Ville [Vie]). *Let (M, g) be a compact Einstein manifold, and h be in $\mathcal{S}^2 M$. Then*

$$\int_M (s'_g h)\mu_g = -\frac{2}{n}s_g\,\mathrm{Vol}(M)'_g h.$$

In particular, if $s_g = 0$ or if $\mathrm{Vol}(M)'_g h = 0$, *then s'_g cannot have a constant sign unless it is identically* 0.

Be careful to note that $\int_M (s'_g h)\mu_g$ is *not* the derivative (in h) of the total scalar curvature $\int_M (s_g)\mu_g$, which will be studied in greater detail in Chapter 4 below.

Chapter 2. Basic Material (Continued): Kähler Manifolds

0. Introduction

In this introductory chapter, we develop in a self-contained way, the elements of Kähler geometry—as a part of Riemannian geometry—which are to be of use in the book, especially in Chapters 8, 11, 12, 13 and 14.

We have tried, as far as possible, to overcome the frequent reluctance of Riemannian geometers to deal with complex geometry. For this purpose, we decided to work only with real objects, almost avoiding the use of complex coordinates. This brought us to re-write some proofs, especially in Paragraph H.

Priority was of course given to points where the Ricci tensor plays a prominent part. For further aspects of the theory, that we ignore here, the reader is referred to the considerable literature available, on the subject, such as [Chr 2], [Gr-Ha], [Hir], [Ko-No 2], [Ko-Mo], [Wei 1], [Wel] ... and many others.

A. Almost Complex and Complex Manifolds

2.2 Definition. An *almost complex structure* on a (connected) manifold is a (smooth) field of automorphisms J of the tangent bundle TM satisfying

(2.2) $$J_X^2 = -I_X, \qquad x \in M$$

where I denotes the identity

Equivalently, an almost complex structure on M is an action of the field $\mathbb{C}$ of complex numbers on TM giving to it a structure of complex vector bundle over M.

2.3. The dimension n of an *almost complex manifold* M is even. We set

$$n = 2m,$$

where m is the *complex dimension* of M, and the *complex operator* J induces a preferred orientation on M, for which *adapted* orthonormal frames $\{e_1, Je_1, \ldots, e_m, Je_m\}$ are positively oriented.

2.4. The almost-complex structure J induces a *splitting* of the complexified tangent bundle $T_{\mathbb{C}}M \simeq TM \otimes_{\mathbb{R}} \mathbb{C}$ into two complementary complex sub-bundles, conjugate to each other:

(2.4) $$T_{\mathbb{C}}M \simeq T'M \oplus T''M,$$

where at each point x of M, the fiber T'_xM (resp. T''_xM) is the eigen-space of J_x relative to the eigen-value $+i$ (resp. $-i$). Conversely, each complex sub-bundle $T'M$ of $T_{\mathbb{C}}M$ with complementary conjugate determines an almost-complex structure J on M.

2.5. The elements of $T'M$ (resp. $T''M$) are the (complex) *vectors of type* (1,0) (resp. *of type* (0,1)).

Each real tangent vector X can be expressed—in a unique way—as a sum:

$$X = U + \bar{U},$$

where U (resp. $\bar{U}$) is its *part of type* (1,0) (resp. *of type* (0,1)). The complex components U and $\bar{U}$ are, in turn, determined by:

$$U = \tfrac{1}{2}(X - iJX) \qquad \bar{U} = \tfrac{1}{2}(X + iJX).$$

We have, in this way, a $\mathbb{C}$-linear isomorphism of TM onto $T'M$ and a $\mathbb{C}$-skew linear isomorphism of TM onto $T''M$.

The above splitting (2.4) of the complexified tangent bundle $T_{\mathbb{C}}M$ induces a *splitting into types* of the whole complex tensor bundle. In particular, we have

(2.6) $$\textstyle\bigwedge^r_{\mathbb{C}} M = \sum_{p+q=r} \bigwedge^p(T'M)^* \otimes \bigwedge^q(T''M)^*,$$

where $\bigwedge^r_{\mathbb{C}} M = \bigwedge^r \otimes \mathbb{C}$ is the bundle of the $\mathbb{C}$-valued r-forms and where $\bigwedge^p(T'M)^*$ (resp. $\bigwedge^q(T''M)^*$) denotes the bundle of the $\mathbb{C}$-linear alternate p-forms (resp. q-forms) on $T'M$ (resp. $T''M$).

2.7. The bundle $\bigwedge^p(T'M)^* \otimes \bigwedge^q(T''M)^*$ will be denoted by $\bigwedge^{p,q}_{\mathbb{C}} M$. Its elements are called (complex) *forms of type* (p,q).

When p is equal to q, we denote by $\bigwedge^{p,p} M$ the bundle of the *real* forms of type (p,p).

The complex operator J induces a complex operator, also denoted by J, on the cotangent bundle $T^*M \simeq \bigwedge^1 M$, defined by

(2.8) $$J\xi(X) = -\xi(JX) \quad \forall \xi \in T^*_xM, \quad \forall X \in T_xM, \quad \forall x \in M.$$

2.9. *Note*: The choice of the sign is due to the fact that the action of J has to be compatible with the identification of the tangent and cotangent bundles via a Hermitian metric (see below 2.19).

This operator J extends to the whole of $\bigwedge^*_{\mathbb{C}} M$: we have, for any form φ of type (p,q),

(2.10) $$J\varphi = i^{(q-p)}\varphi$$

2.11. *Integrability.* The almost-complex structure J is *integrable* if one of the equivalent following conditions holds:

i) The exterior differential $d\theta$ of any 1-form θ of type $(1,0)$ belongs to the ideal generated by the 1-forms of type $(1,0)$ (equivalently, $d\theta$ has no component of type $(0,2)$).

ii) The *complex torsion tensor* or *Nijenhuis tensor* N defined by

$$4N(X,Y) = [X,Y] + J[JX,Y] + J[X,JY] - [JX,JY] \quad \forall X, Y \in T_xM, \quad \forall x \in M, \tag{2.11}$$

is equal to zero. On the right, X and Y denote any local extensions of the two vectors X and Y.

iii) The bracket of any two vector fields of type $(1,0)$ is likewise a vector field of type $(1,0)$.

2.12. The equivalence of the above three conditions is easy to establish. They all imply that *the almost-complex structure J is induced by a* (unique) *complex structure.* (By the Theorem of Newlander-Nirenberg) [Ne-Ni]. See also [Koh], [Hör 1], [Mal 1].

2.13. Recall that a *complex manifold* of (complex) dimension m is a (paracompact, Hausdorff) topological space admitting a covering by open subsets, homeomorphic with open subsets of $\mathbb{C}^m$ and with holomorphic transition functions (see [Chr 2], p. 1, for instance). A complex manifold is then a (smooth) manifold whose tangent bundle is naturally endowed with a complex structure J which is easily seen to be integrable in the above sense.

On the over hand, an almost-complex structure J on a manifold cannot be induced by two non-equivalent complex structures (see [Ko-No 2] p. 123).

Observe that the tangent bundle, considered as a complex vector bundle, of a complex manifold is *holomorphic*, that is admits holomorphic transition functions.

2.14. When J is integrable, the exterior differential $d\varphi$ of any form φ of type (p,q) is the sum of a form of type $(p+1,q)$ and a form of type $(p,q+1)$ denoted respectively by $d'\varphi$ and $d''\varphi$. Thus we obtain two differential operators d' and d'' such that the exterior differential d is equal to the sum

$$d = d' + d'' \tag{2.15}$$

and satisfying

$$d'd'' + d''d' = 0 \qquad d'^2 = 0 \qquad d''^2 = 0.$$

We define a new (real) differential operator d^c, acting on forms, by

$$d^c\varphi = -J^{-1}dJ\varphi = (-1)^r JdJ\varphi \tag{2.16}$$

for any r-form φ. We have

$$d^c = i(d'' - d')$$

and, equivalently

$$d'' = \tfrac{1}{2}(d - id^c) \qquad d' = \tfrac{1}{2}(d + id^c).$$

The square of d^c is equal to zero, and the two operators d and d^c anti-commute:

$$dd^c + d^c d = 0 \qquad (d^c)^2 = 0.$$

On the other hand, we have:

(2.17) $$dd^c = 2id'd''.$$

2.18. Let E be any complex vector bundle over M. A *complex E-valued r-form* is a section of the bundle $\bigwedge^r_{\mathbb{C}} M \otimes_{\mathbb{C}} E$.

If E is equipped with a $\mathbb{C}$-linear connection ∇, we define, as in the real case (see 1.7), an exterior differential d^∇ on the space of the E-valued forms.

In the case where M is endowed with an almost-complex structure J, the bundle $\bigwedge^r_{\mathbb{C}} M \otimes_{\mathbb{C}} E$ splits into types in an obvious way, and if J is integrable d^∇ splits into the sum

$$d^\nabla = d^{\nabla'} + d^{\nabla''}$$

just as in the scalar case. We also define the operator $d^{\nabla c}$ just as in 2.16. But, of course, the squares of these operators are no longer equal to zero and involve the curvature of ∇.

From now on, M is a complex manifold or, equivalently, a manifold endowed with an integrable almost-complex structure J, of dimension $n = 2m$.

B. Hermitian and Kähler Metrics

2.19 Definition. A *Hermitian metric* on M is a Riemannian metric g such that

(2.19) $$g(JX, JY) = g(X, Y) \quad \forall X, Y \in T_x M, \quad \forall x \in M.$$

Equivalently, the metric tensor is of type (1, 1), or J-invariant.

2.20. We denote by the *same symbols* the various tools of Riemannian geometry and their $\mathbb{C}$-linear or multilinear extensions to the complexified bundles.

In particular, the scalar product $(.,.)$ of complex tensors or forms is $\mathbb{C}$-*bilinear*. It induces $\mathbb{C}$-linear isomorphisms $\sharp$ from $T^*_{\mathbb{C}}M$ onto $T_{\mathbb{C}}M$ and $\flat$ from $T_{\mathbb{C}}M$ onto $T^*_{\mathbb{C}}M$ which permute the types. Observe that the sub-bundles $T'M$ and $T''M$ are *completely isotropic* for the scalar product.

2.21. The *Hodge-star* operator $*$ (cf. 1.51) transforms a form of type (p, q) into a form of type $(m - q, m - p)$.

The two operators $*$ and J, acting on forms, commute.

2.22. With the help of the Hodge-star operator $*$ we define the *codifferential operators* corresponding to the differential operators d, d', d'', d^c by:

$$\delta = -*d* \quad \delta' = -*d''* \quad \delta'' = -*d'* \quad \delta^c = -*d^c*$$

(compare 1.56). The pairs (d,δ), (d',δ''), (d'',δ') and (d^c,δ^c) are (formally) *adjoint* relatively to the global scalar product $\langle .,. \rangle$.

2.23. The Hodge-star operator and the various co-differential operators above can also be defined for complex E-valued forms, where E is any complex vector bundle equipped with a $\mathbb{C}$-linear connexion ∇.

2.24. We obtain a *Hermitian scalar product* h on $T_{\mathbb{C}}M$ by setting

$$h(U,V) = g(U,\bar{V}) \quad \forall U,V \in (T_{\mathbb{C}}M)_x, \quad \forall x \in M. \tag{2.24}$$

The restriction of h to the sub-bundle $T'M$ induces, via the above identification 2.5 of $T'M$ with TM, a *Hermitian scalar product* H on TM. We get:

2.25. $$H(X,Y) = \tfrac{1}{2}[g(X,Y) - ig(JX,Y)] \quad \forall X,Y \in T_xM \quad \forall x \in M.$$

2.26. We obtain a *linear one-to-one correspondence between the vector bundle* $S^{1,1}M$ *of the real symmetric J-invariant bilinear forms and* $\bigwedge^{1,1}M$, by associating with any section b of $S^{1,1}M$ the 2-form β defined by

$$\beta(X,Y) = b(JX,Y) \quad \forall X,Y \in T_xM, \quad \forall x \in M. \tag{2.26}$$

The J-invariant bilinear forms b and β are said to be *associated with each other*. The 2-form β is said to be *positive* (resp. *negative*) if b is positive-definite (resp. negative-definite).

2.27 Definition. The *Kähler form* ω of a Hermitian metric is the 2-form associated with the metric tensor.

The Kähler-form ω is equal, up to a factor $-\frac{1}{2}$, to the imaginary part of the Hermitian scalar product H.

2.28 Definition. A Hermitian metric is *Kähler* if the complex operator J is parallel for the Levi-Civita connexion:

$$DJ = 0 \quad \Leftrightarrow \quad D_X JY = JD_X Y \quad \forall X,Y \in T_xM, \quad \forall x \in M. \tag{2.28}$$

2.29 Proposition. *A Hermitian metric is Kähler if and only if the Kähler form ω is closed.*

Proof. If the metric is Kähler, g and J are parallel so that ω itself is parallel and closed a fortiori. The converse is more delicate (see [Ko-No 2] p. 148). □

2.30 Proposition. *A Hermitian metric is Kähler if and only if there exists in the neighborhood of any point x of M a geodesic complex coordinate system, that is a complex coordinate system, such that the metric osculates to order 2 to the corresponding flat metric, i.e.*

(2.30) $$g(\partial/\partial z_\alpha, \overline{\partial/\partial z_\beta}) = \partial_{\alpha\beta} \quad \text{and} \quad D(\partial/\partial z_\alpha) = 0$$

at the point x, for any $\alpha, \beta = 1, \ldots, m$.

Proof. See [Wel] p. 196, [Gr-Ha] p. 108, [D-G-M-S] p. 265. □

Note that the ordinary exponential map (see 1.27) is geodesic complex only in flat Kähler manifolds.

2.31. With the help of the Kähler form, we construct two linear operators L and Λ acting on forms, where L is defined by:

(2.31) $$L\varphi = \omega \wedge \varphi$$

for any form φ, and Λ is the (formal) adjoint of L, that is the *contraction* by ω.

A form φ is said to be *primitive* if $\Lambda\varphi$ is equal to zero. When φ is a 2-form, the scalar $\Lambda\varphi$, equal to the scalar product (φ, ω) of φ with the Kähler form ω, is the *trace* of φ, denoted by $\operatorname{tr}\varphi$.

2.32. Proposition. *If a Hermitian metric is Kähler, we have*

(2.32) $$[\Lambda, d] = -\delta^c \quad [\Lambda, d^c] = \delta.$$

Proof. The two relations follow easily from each other. In view of Proposition 2.30, it is sufficient to prove them for the canonical flat metric in $\mathbb{C}^m$. The computation is somewhat messy but not difficult (see [Gr-Ha] p. 111–114). □

2.33. Applying (2.32) to a closed 2-form φ of type (1, 1) we get

(2.33) $$\delta\varphi = -d^c(\operatorname{tr}\varphi)$$

In particular, *φ is harmonic if and only if its trace is constant*. On the other hand, 2.32 holds also for E-forms where E is a vector bundle endowed with a linear connection ∇. For any 2-form Φ of type (1, 1) with values in E the above relation becomes

(2.34) $$\delta^\nabla \Phi = -J\nabla(\operatorname{tr}\Phi)$$

where $\operatorname{tr}\Phi$ is the trace of the 2-form part of Φ. Applying this relation to the curvature of a Kähler metric R—viewed as a (closed) 2-form with values in $\bigwedge^{1,1}M$—we infer at once that *the curvature of a Kähler manifold is harmonic if and only if the Ricci tensor is parallel* [Mat 3] (see a refinement of this result in Chapter 16.)

2.35 Remark. It is easily checked that the commutator $B = [\Lambda, L]$ is the operator defined by

$$B(\varphi) = (m - r)\varphi$$

for any r-form φ. On the other hand, we have

$$[B, L] = -2L \qquad [B, \Lambda] = 2\Lambda.$$

We thus obtain a linear action of the complex Lie algebra $\mathfrak{sl}(s, \mathbb{C})$ of the special linear group of order 2 on the space of complex forms at each point of M. Using the well-known theory of linear representations of $\mathfrak{sl}(2, \mathbb{C})$, we obtain without effort the following Lefschetz decomposition theorem:

$$\varphi = \sum_{p \geqslant (r-m)^+} L^p \varphi_p \qquad (r-m)^+ = \max(0, r-m)$$

where φ is any r-form and φ_p are primitive $(r-2p)$-forms. When the Kähler form is closed, that action induces an action of $\mathfrak{sl}(2, \mathbb{C})$ on the complex cohomology and we get a similar decomposition on the cohomology spaces involving primitive cohomology. (For details on related results including the hard Lefschetz theorem, we refer the reader to [Wel] Chap. V or [Gr-Ha] Chap. 0.7).

2.36. Any real bilinear form ψ may be interpreted as a homomorphism from TM into T^*M by setting

$$\psi(X) = i_X \psi$$

or, using the metric, as an endomorphism of TM, defined by

$$(\psi(X), Y) = \psi(X, Y).$$

According to the context, $\psi(X)$ will then denote a vector or its dual 1-form. If β is a real 2-form of type $(1, 1)$ and b the associated symmetric form as in 2.26, the corresponding endomorphisms of TM are related by

(2.36) $$\beta(X) = b(JX).$$

The endomorphism β (resp. b) is skew-symmetric (resp. symmetric) and commutes with J, that is, is skew-Hermitian (resp. Hermitian) in the real sense. Extending β (resp. b) $\mathbb{C}$-linearly to $T_{\mathbb{C}}M$ and, restricting it to $T'M$, we obtain a skew-Hermitian (resp. a Hermitian) endomorphism, in the complex sense, of the Hermitian holomorphic tangent bundle $(T'M, h)$.

2.37. The mixing up of the real and complex points of view may be a source of trouble. In particular, the trace of β, viewed as a (real) endomorphism of TM, is equal to zero, while its trace, when viewed as a $\mathbb{C}$-linear endomorphism of $T'M$, is a complex number—purely imaginary—, non-zero in general. The latter is called the *complex trace* of β, denoted by $\operatorname{tr}_{\mathbb{C}} \beta$, related to the (real) trace $\operatorname{tr} \beta$, as defined in 2.31, by

(2.37) $$\operatorname{tr}_{\mathbb{C}} \varphi = i \cdot \operatorname{tr} \varphi.$$

2.38. For a 2-form φ of type $(1, 1)$, the Lefschetz decomposition reduces to

(2.38)
$$\varphi = \frac{\operatorname{tr}\varphi}{m}\cdot\omega + \varphi_0$$

where the primitive part φ_0 of φ is also its trace-free component.

2.39. Observe that the endomorphism associated with the Kähler form ω is J, and $\operatorname{tr}\omega$ is equal to m.

2.40. The fiber $\bigwedge^{1,1}$ of $\bigwedge^{1,1}M$ is identified, through any of the above interpretations, with the Lie algebra $\mathfrak{u}(m)$, of the unitary group $U(m)$, and, in fact, the vector bundle $\bigwedge^{1,1}M$ is the vector bundle associated, via the adjoint representation $\mathrm{Ad}_{U(m)}$ of $U(m)$ on $\mathfrak{u}(m)$, to the $U(m)$-principal bundle attached to the Hermitian metric.

Observe that the latter may be viewed, alternatively, as the bundle of adapted orthonormal frames of TM (see 2.3) or else as the bundle of orthonormal frames of $T'M$ relative to the Hermitian scalar product h.

C. Ricci Tensor and Ricci Form

2.42. In addition to the usual symmetries of the Riemannian curvature, the curvature R of the Levi-Civita connection of a Kähler metric satisfies the following relation:

(2.42)
$$R(X, Y)(JZ) = J(R(X, Y)(Z)) \quad \forall X, Y, Z \in T_xM, \quad \forall x \in M,$$

so that the Kähler curvature R is a (real) 2-form—of type (1, 1)—with values in the vector bundle of the skew-Hermitian endomorphisms of TM (in the real sense or in the complex sense as well (see 2.36), that is in $\bigwedge^{1,1}M$.

Alternatively, the Kähler curvature may be viewed as a (symmetric) section of the vector bundle $\bigwedge^{1,1}M \otimes \bigwedge^{1,1}M$, or as a (symmetric) endomorphism of the vector bundle $\bigwedge^{1,1}M$.

All these interpretations of the Kähler curvature are useful.

2.43. From the last point of view, the Kähler curvature is called the *Kähler curvature operator*. It is the same as the Riemannian curvature operator but viewed as an endomorphism of $\bigwedge^{1,1}M$ only, instead of $\bigwedge^2 M$.

2.44 Definition. The *Ricci form* ρ is the image $R(\omega)$ of the Kähler form ω by the Kähler curvature operator.

2.45 Proposition. *The Ricci tensor r of a Kähler metric is a (real, symmetric) bilinear form of type* (1, 1) *and the associated 2-form is the Ricci form ρ.*

Proof. Let $\{e_1, Je_1, \ldots, e_m, Je_m\}$ be an orthonormal adapted basis of T_xM at any point x of M, and X, Y any two elements of T_xM. We get

$$\begin{aligned} r(X, Y) &= \sum_{\alpha=1}^{m} (R(e_\alpha, X)e_\alpha, Y) + \sum_{\alpha=1}^{m} (R(Je_\alpha, X)Je_\alpha, Y) \\ &= \sum_{\alpha=1}^{m} (R(e_\alpha, X)Je_\alpha, JY) - \sum_{\alpha=1}^{m} (R(Je_\alpha, X)e_\alpha, JY) \\ &= \sum_{\alpha=1}^{m} (R(e_\alpha, Je_\alpha)X, JY). \end{aligned}$$

The second equality is obtained by using (2.42) and the last equality is the algebraic Bianchi identity 1.24.

Now, for any adapted orthonormal basis $\{e_\alpha, Je_\alpha\}$ $\alpha = 1, \ldots, m$, the Kähler form is equal to:

(2.46)
$$\omega = \sum_{\alpha=1}^{m} e_\alpha \wedge Je_\alpha. \qquad \square$$

2.47 Proposition. *The Ricci form ρ is closed.*

Proof. Because of the symmetry of the Kähler curvature R, ρ is equal to the contracted product of R, considered as a $\bigwedge^{1,1}M$-valued 2-form by the Kähler form ω considered as a $\bigwedge^{1,1}M$-valued 0-form:

$$\rho = R \wedge \omega.$$

More generally, if φ and ψ are $\bigwedge^{1,1}M$-valued p-forms and q-forms respectively we get, by contracting via the scalar product the fiber parts, a scalar $(p+q)$-form denoted by $\varphi \wedge \psi$. We get easily:

(2.48)
$$d(\varphi \wedge \psi) = d^D\varphi \wedge \psi + (-1)^{pq}\varphi \wedge d^D\psi.$$

Applying the above relation to $R \wedge \omega$ where R and ω are both closed as $\bigwedge^{1,1}M$-valued forms, R by the differential Bianchi identity (see 1.14) and ω because it is parallel, we get at once the result. $\square$

2.49 Remark. More generally, it follows from the proof above that the image $R(\varphi)$ of any *parallel* 2-form φ of type $(1, 1)$ by the Kähler curvature operator is closed.

2.50. The Ricci tensor r, which is, in Riemannian geometry, the zero-order term in the well-known *Bochner formula* 1.155

(2.51)
$$\Delta\xi = \delta D\xi + r(X),$$

where ξ is any 1-form and X its dual vector field (see 2.36 for the notations), still appears, in Kählerian geometry, in another Weitzenböck-type formula involving two new operators D^+ and D^- operating on 1-forms (and vector fields as well by metric duality) as follows. For any 1-form ξ the bilinear form $D\xi$ splits into

(2.52)
$$D\xi = D^+\xi + D^-\xi,$$

where $D^+\xi$ (resp. $D^-\xi$) is J-invariant (resp. J-skew-invariant). Alternatively $D^+\xi(D^-\xi)$ is the part of type $(1, 1)$ (resp. of type $(2, 0) + (0, 2)$) of $D\xi$.

2.53 Proposition. *For any 1-form ξ we have*

$$\delta D^+\xi - \delta D^-\xi = r(X), \tag{2.53}$$

where X is the dual vector field of ξ.

Proof. For any vector field X, $R(.,.)X$ is a TM-valued 2-form equal to

$$R(.,.)X = -d^D DX \qquad \text{(see 1.12)}.$$

By 2.36, 2.33 and Remark 2.34, we get

$$r(X) = -\rho(JX) = \Lambda d^D DJX = -(\delta^D)^C DJX = \delta^D(J \cdot DJX),$$

where $J \cdot DJX$ denotes the action (2.8) of J on DJX considered as a (TM-valued) 1-form. Then $J \cdot DJX$ is clearly equal to the difference $D^+\xi - D^-\xi$. □

2.54 Remark. The restriction of δ to the space of real bilinear forms of type (1, 1) (resp. of type (2, 0) + (0, 2)) is the (formal) adjoint of D^+ (resp. D^-).

2.55 Remark. Combining (2.53) with the Bochner formula we have

$$\tfrac{1}{2}\Delta\xi = \delta D^+\xi \tag{2.55}$$

for any 1-form ξ.

In particular, *the 1-form ξ is harmonic on a compact Kähler manifold if and only if it belongs to the kernel of D^+.*

Observe that $D^+\xi$ does not depend on the Kähler metric. In fact ξ is harmonic if and only if its part of type (1, 0) is holomorphic (see 2.102).

2.56 Remark. Combining (2.55) with the decomposition

$$D\xi = \frac{1}{2}d\xi + \frac{1}{2}(L_X g)_0 - \frac{1}{2m}\delta\xi \cdot g \tag{2.56}$$

of the bilinear form $D\xi$ under the action of the orthogonal group $O(2m)$—where X denotes as usual the dual vector field of ξ—we immediately get the relation

$$\delta D^-\xi = \frac{1}{2}\delta(L_X g)_0 - \frac{1}{2}\frac{(m-1)}{m} d\delta\xi \tag{2.56a}$$

or, equivalently (see 1.60), for any 1-form ξ,

$$\delta D^-\xi = \delta\delta^*\xi - \tfrac{1}{2}d\delta\xi. \tag{2.56b}$$

D. Holomorphic Sectional Curvature

2.57 Definition. The *holomorphic sectional curvature* is the function H defined on the unitary tangent bundle UM by

(2.57) $$H(X) = (R(X, JX)X, JX) \quad \forall X \in U_x M, \quad \forall x \in M.$$

Alternatively, $H(X)$ is the sectional curvature of the complex line determined by X, that is the 2-plane generated by X and JX.

2.57. The holomorphic sectional curvature *determines the whole Kähler curvature* just as the sectional curvature determines the whole Riemannian curvature. (The former fact is easily deduced from the latter).

As in the Riemannian case, if H is constant on any fiber of UM, it is constant everywhere (unless $m = 1$).

The Kähler curvature operator corresponding to constant holomorphic sectional curvature equal to c, is given by:

(2.58) $$R = \frac{c}{2}((m+1)Id_{|\mathbb{R}\cdot\omega} + Id_{|\Lambda_0^{1,1}M}),$$

where m is the (complex) dimension of M, $\mathbb{R}\cdot\omega$ the trivial (real) line sub-bundle of $\Lambda^{1,1}M$ generated by the Kähler form ω, $\Lambda_0^{1,1}M$ the vector bundle of real primitive 2-forms of type $(1, 1)$, Id the identity.

Note that the identity of the whole vector bundle $\Lambda^{1,1}M$ is *not* the Kähler curvature operator of any Kähler structure as it does not satisfy the first Bianchi identity.

2.59 Example. The *complex projective space* $\mathbb{C}P^m$ is the manifold of complex lines through the origin of $\mathbb{C}^{m+1}$; equivalently, $\mathbb{C}P^m$ is the quotient of $\mathbb{C}^{m+1} - \{0\}$ by the natural action of $\mathbb{C}^*$. Let U_i be the image on $\mathbb{C}P^m$ in the open subset of $\mathbb{C}^{m+1} - \{0\}$ determined by the condition: $z^i \neq 0$, where $(z^1, \ldots, z^{m+1})$ denotes a generic point of $\mathbb{C}^{m+1}$. On each U_i we get complex coordinates $\{{}_ix^\alpha\}$, $\alpha = 1, \ldots, m$ by

$${}_ix^\alpha = \frac{z^\alpha}{z^i} \quad \alpha < i, \qquad {}_ix^\alpha = \frac{z^{\alpha+1}}{z^i} \quad \alpha \geqslant i.$$

We obtain in this manner a complex structure on $\mathbb{C}P^m$. Consider now the 2-form ω_c of type $(1, 1)$ defined, on each open subset U_i, by:

(2.59) $$\omega_c = \frac{1}{c} dd^c \log\left(1 + \sum_{\alpha=1}^{m} |{}_ix^\alpha|^2\right),$$

where c is a positive number. This 2-form ω_c is the Kähler form of a Kähler metric, called the *Fubini-Study metric* (with parameter c).

The holomorphic sectional curvature of this Kähler metric is constant, equal to c.

The *usual Fubini-Study metric* corresponds to c being equal to 4.

2.60 Example. A *complex torus* of (complex) dimension m is the quotient space of $\mathbb{C}^m$ by a lattice Γ. The flat metric of $\mathbb{C}^m$ induces a flat Kähler metric on $\mathbb{C}^m/\Gamma$ for any lattice Γ. Conversely, *any compact flat Kähler manifold M is holomorphically covered by a complex torus* (see [Bie]).

2.61 Example. Let D^m be the unit ball in $\mathbb{C}^m$. The 2-forms ω_c defined by

$$\omega_c = \frac{1}{c} dd^c \log\left(1 - \sum_{\alpha=1}^{m} |z^\alpha|^2\right) \tag{2.61}$$

are the Kähler forms of (complete) Kähler metric on D^m parametrized by the positive number c, called the *Bergmann metric*.

Again, the holomorphic sectional curvature of the Kähler metric parametrized by c is constant, equal to $-c$.

The *usual Bergmann metric* corresponds to c being equal to 4.

2.62. *Conversely, any compact Kähler manifold with constant holomorphic sectional curvature c is isomorphic to $\mathbb{C}P^m$ with a Fubini-Study metric if c is positive, or to a (compact) quotient of D^m with a Bergmann metric by a discrete group of isometries, if c is negative.* (see [Ko-No 2] p. 170).

2.63. *Decomposition of the Kähler curvature tensor.*

The Lefschetz decomposition 2.35 of $\Lambda^{1,1}M$ induces a decomposition of the Kähler curvature tensor, viewed as a section of $\Lambda^{1,1}M \otimes \Lambda^{1,1}M$. We get

$$R = \frac{s}{2m^2}\omega \otimes \omega + \frac{1}{m}\omega \otimes \rho_0 + \frac{1}{m}\rho_0 \otimes \omega + B, \tag{2.63}$$

where ρ_0 is the primitive part of the Ricci form and B is a section of the sub-bundle $\Lambda_0^{1,1}M \otimes \Lambda_0^{1,1}M$.

Viewed as an endomorphism of $\Lambda_0^{1,1}M$, the tensor B decomposes as follows

$$B = \frac{\operatorname{tr} B}{(m^2 - 1)} Id_{|\Lambda^{1,1}M} + B_0, \tag{2.64}$$

where the trace of B_0 is null, and $\operatorname{tr} B$ is the trace of B.

Since the trace of the Kähler curvature operator is equal to $s/2$, we infer at once

$$\operatorname{tr} B = \frac{s}{2}(1 - 1/m). \tag{2.65}$$

2.66. In view of 2.57, we see that *the holomorphic sectional curvature is constant if and only if ρ_0 and B_0 are both equal to zero*. Moreover, the scalar curvature is then constant, equal to

$$s = m(m-1)c \tag{2.66}$$

if c is the value of the constant holomorphic sectional curvature.

2.67. Comparing the two decompositions 1.114 and (2.63) of the Riemannian curvature, we obtain the following relations between norms:

$$\begin{aligned} |B_0|^2 &= -\frac{3(m-1)}{(m+1)}|U|^2 - \frac{(m-2)}{m}|Z|^2 + |W|^2 \\ |\rho_0|^2 &= (m-1)|Z|^2 \\ s^2 &= 4m(2m-1)|U|^2 \end{aligned} \tag{2.67}$$

which implies

2.68 Proposition. *The norms of the components U and W of the curvature of a Kähler manifold of (complex) dimension $m > 1$ are related by the inequality*

$$|W|^2 \geqslant 3\frac{m-1}{m+1}|U|^2. \tag{2.68}$$

If $m > 2$, equality holds if and only if the holomorphic sectional curvature is constant. In particular, in dimension $m > 2$, any conformally flat Kähler metric is flat.

2.69 Remark. If $m = 2$, inequality (2.68) reduces to the obvious inequality

$$|W^+|^2 \leqslant |W|^2,$$

and equality holds if and only if the Kähler metric is self-dual or half conformally flat ($W^- = 0$).

If, in addition, the metric is conformally flat, the curvature reduces to its component Z. In particular, the curvature is harmonic (cf. 16.E), and the Ricci tensor is parallel, by 2.34. It follows easily that any compact conformally flat Kähler manifold of complex dimension 2 is locally covered by a product of Riemann surfaces endowed with opposite, constant curvature metrics.

E. Chern Classes

2.70. The characteristic classes of a complex manifold M are obtained from the Kähler curvature R in the following way. Let P be any real $\mathrm{Ad}_{U(m)}$-symmetric p-multilinear form on the Lie algebra $\mathfrak{u}(m)$. On account of the $\mathrm{Ad}_{U(m)}$-invariance, P induces a (symmetric) p-multilinear form, also denoted by P, on the vector bundle $\bigwedge^{1,1}M$. Then, P induces a (real) $2p$-form of type (p, p) on M defined by:

$$c_p(R) = \frac{1}{(2\pi)^p} P(R, \ldots, R), \tag{2.70}$$

where, of course, R is considered as a $\bigwedge^{1,1}M$-valued 2-form.

It can be proved that *$c_p(R)$ is closed and that its cohomology class in $H^{2p}(M, R)$ does not depend on the choice of the Kähler metric.* (see [Chr 3] pp. 113–115). In this way we get a homomorphism—the so-called *Chern-Weil homomorphism*—from the space $I(m)$ of $\mathrm{Ad}_{U(m)}$-invariant symmetric r-multilinear forms on $\mathfrak{u}(m)$ into $H^*(M, \mathbb{R})$, whose image is the space of *characteristic classes* of M.

2.71 Remark. We obtain the same homomorphism by using the curvature of any complex connection, where a *complex* connection is a linear connection relatively to which J is parallel (see 1.15).

Observe that each element P of $I(m)$ is perfectly well determined by the associated polynomial function $\tilde{P}$ on $\mathfrak{u}(m)$. P is then the *complete polarized form* of $\tilde{P}$. Then $I(m)$ has the natural structure of a ring and the Chern-Weil homomorphism is itself a ring homomorphism. As a ring, $I(m)$ is generated by 1 and the m elementary symmetric polynomials $\tilde{\sigma}_p$ defined by

$$\text{(2.72)}\quad \tilde{\sigma}_p(A) = \tilde{\sigma}_p\left[\begin{pmatrix} i\lambda_1 & & 0 \\ & \ddots & \\ 0 & & i\lambda_m \end{pmatrix}\right] = \sigma_p(\lambda_1, \ldots, \lambda_m) \qquad p = 1, \ldots, m,$$

where $\begin{pmatrix} i\lambda_1 & & 0 \\ & \ddots & \\ 0 & & i\lambda_m \end{pmatrix}$ is the diagonal form of any skew-Hermitian matrix A and σ_p is the pth *elementary symmetric function:*

$$\sigma_p(\lambda_1, \ldots, \lambda_m) = \sum_{i_1 < \cdots < i_p} \lambda_{i_1} \cdots\cdots \lambda_{i_p}.$$

2.73 Definition. The pth *real Chern class* $c_p(M)$ is the image under the Chern-Weil homomorphism of the polynomial function $\tilde{\sigma}_p$. The corresponding p-form $\tilde{\sigma}_p(R)$ is the *canonical representative* of $c_p(M)$ associated with the Kähler metric.

2.74 Lemma. *Let Σ_p be the complete polarized form of $\tilde{\sigma}_p$ acting on the vector bundle $\bigwedge^{1,1}M$. We have:*

$$\text{(2.74)}\qquad \Sigma_p(\varphi_1, \ldots, \varphi_p) = \frac{1}{(p!)^2} \Lambda^p(\varphi_1 \wedge \cdots \wedge \varphi_p).$$

For $p = 1$, we get the trace $\Lambda\varphi = \Sigma_1(\varphi)$. From the symmetry of the curvature operator we deduce the following basic fact

2.75 Proposition. *The first real Chern class $c_1(M)$ is represented by the 2-form $\dfrac{1}{2\pi}\rho$.*

2.76 Corollary. *The real cohomology class of the Ricci form ρ depends only on the complex structure of M, as we shall see, directly, later on* (see 2.101).

For $p = 2$, we get, using the square Λ^2 of the operator Λ of 2.32, the

2.77 Lemma.

$$\text{(2.77)}\qquad \Sigma_2(\varphi, \psi) = \frac{1}{4}\Lambda^2(\varphi \wedge \psi) = \frac{1}{2}(\operatorname{tr}\varphi \cdot \operatorname{tr}\psi - (\varphi, \psi))$$

$$= \frac{1}{2}\left[\frac{(m-1)}{m} \cdot \operatorname{tr}\varphi \cdot \operatorname{tr}\psi - (\varphi_0, \psi_0)\right]$$

where φ_0, ψ_0 are the primitive parts of φ, ψ respectively (see 2.32).

Proof. (exercise). □

Consider the symmetric bilinear form q defined on $\bigwedge^{1,1}M$ by

(2.78) $$q(\varphi,\psi)=\frac{\Lambda^2}{2}(\varphi\wedge\psi)=\left[\frac{(m-1)}{m}\operatorname{tr}\varphi\cdot\operatorname{tr}\psi-(\varphi_0,\psi_0)\right]\quad\forall\varphi,\psi\in{\textstyle\bigwedge}_x^{1,1}M,\quad\forall x\in M.$$

It is a non-degenerate form of signature $(1,m^2-1)$.

It induces a non-degenerate symmetric bilinear form on the tensor product $\bigwedge^{1,1}M\otimes\bigwedge^{1,1}M$, also denoted by q, of signature $(m^4-2m^2+2,2m^2-2)$. Let $q(R,R)$ denote the square of the "norm" of the Kähler curvature R relative to q.

Let $c_2(R)=\frac{1}{4\pi^2}\Sigma_2(R,R)$ be the canonical representative of the second Chern class attached to R. We easily obtain from (2.77)

(2.79) $$\frac{\Lambda^2}{2}(c_2(R))=\frac{1}{8\pi^2}q(R,R).$$

Using the decomposition 2.63 of R we infer the *Apte formula* [Apt]

(2.80) $$\begin{aligned}\frac{1}{2}\Lambda^2(c_2(R))&=\frac{1}{8\pi^2}\left(\frac{(m-1)}{4(m+1)}s^2-2\frac{(m-1)}{m}|\rho_0|^2+|B_0|^2\right)\\&=\frac{1}{8\pi^2}((2m-3)(m-1)|U|^2-(2m-3)|Z|^2+|W|^2),\end{aligned}$$

while

(2.81) $$\frac{1}{2}\Lambda^2((c_1^2(R))=\frac{1}{4\pi^2}\left(\frac{(m-1)}{4m}s^2-|\rho_0|^2\right).$$

Hence

(2.82) $$\frac{1}{2}\Lambda^2\left(\frac{2(m+1)}{m}c_2(R)-c_1^2(R)\right)=\frac{1}{4\pi^2}\left(-\left(1-\frac{2}{m^2}\right)|\rho_0|^2+\frac{(m+1)}{m}|B_0|^2\right).$$

Integrating over M, we get

(2.80a) $$\begin{aligned}&\frac{1}{(m-2)!}(c_2\cup[\omega]^{m-2})M\\&=\frac{1}{8\pi^2}\int_M\left(\frac{(m-1)}{4(m+1)}s^2-\frac{2(m-1)}{m}|\rho_0|^2+|B_0|^2\right)\mu_g\\&=\frac{1}{8\pi^2}\int_M((2m-3)(m-1)|U|^2-(2m-3)|Z|^2+|W|^2)\mu_g,\end{aligned}$$

(2.81a) $$\frac{1}{(m-2)!}(c_1^2\cup[\omega]^{m-2})M=\frac{1}{4\pi^2}\int_M\left(\frac{(m-1)}{4m}s^2-|\rho_0|^2\right)\mu_g,$$

(2.82a) $$\begin{aligned}&\frac{1}{(m-2)!}\left(\left(\frac{2(m+1)}{m}c_2-c_1^2\right)\cup[\omega]^{m-2}\right)M\\&=\frac{1}{4\pi^2}\int_M\left(-\left(1-\frac{2}{m^2}\right)|\rho_0|^2+\frac{(m+1)}{m}|B_0|^2\right)\mu_g.\end{aligned}$$

2.83 Remark. The left-hand members of (2.80a), (2.81a), (2.82a) depend only on the Chern classes c_1 and c_2 and on the Kähler class $[\omega]$ (see below 2.114).

Moreover, when m is equal to 2, the two Chern numbers c_1^2 and c_2 depend only on the underlying oriented manifold. They are related to the signature σ and the Euler-Poincaré characteristic χ by

(2.84) $$c_1^2 = 3\tau + 2\chi \qquad c_2 = \chi.$$

In particular, when $m = 2$, (2.82a) becomes

(2.85) $$3c_2 - c_1^2 = \chi - 3\tau = \frac{1}{4\pi^2}\int_M \left(-\frac{1}{2}|\rho_0|^2 + |B_0|^2\right)\mu_g.$$

F. The Ricci Form as the Curvature Form of a Line Bundle

2.86. Let L be any *holomorphic line bundle* over M, that is a complex vector bundle of (complex) rank one admitting holomorphic transition functions. We assume L endowed with a Hermitian fibered scalar product h and a connection ∇ compatible with h. Let l be a local holomorphic "frame" of L, that is a (local) holomorphic section without zero. The connection ∇ is then determined by a local 1-form η defined by

(2.87) $$\nabla_X l = \eta(X) l \qquad \forall X \in T_x M, \quad \forall x \in M.$$

The compatibility of ∇ with h is expressed by

$$X \cdot h(l, l) = h(\nabla_X l, l) + h(l, \nabla_X l) = (\eta + \bar{\eta})(X) \cdot h(l, l)$$

or, more simply, by

(2.88) $$\eta + \bar{\eta} = d \log |l|^2,$$

where $|l|^2$ denotes the square $h(l, l)$ of the norm of s.

On the other hand, the curvature R^∇ is defined, in the usual way, by:

(2.89) $$R^\nabla(X, Y)l = \nabla_Y(\nabla_X l) - \nabla_X(\nabla_Y l) - \nabla_{[Y,X]} l = -d\eta(X, Y)l,$$

so that the curvature R^∇ appears as a (globally defined) scalar 2-form, locally equal to $-d\eta$. In particular, R^∇ is closed and, in view of 2.88, is a pure imaginary scalar form.

2.90 Proposition. *The curvature R^∇ is zero if and only if there exists a parallel section in the neighborhood of any point of M.*

Proof. Proposition 2.90 is just a special case of a general fact in the theory of linear connections but the direct verification here is easy. Indeed, if R^∇ is zero, the local connection form η relative to any (local) holomorphic section l is equal—possibly on a smaller neighborhood of a fixed point x of M—to $d(\log f)$ by the Poincaré lemma so that $f \cdot l$ is parallel. □

Among the Hermitian connections, we distinguished a preferred one satisfying the additional condition

2.91. *$\nabla_{\bar{U}} l$ is equal to zero for any holomorphic section l and any vector $\bar{U}$ of type* (1, 0).

For such a connection, we infer easily that

$$\eta = d' \log |l|^2 \tag{2.92}$$

hence

$$R^{\nabla} = -d''d' \log |l|^2 \tag{2.93}$$

relatively to any *holomorphic* local section l without a zero.

In particular, the curvature R^{∇} is of type (1, 1).

Observe that (2.92) ensures both the existence and the uniqueness of the preferred connection, which is called *the* Hermitian connection, or *Chern connection*, of (L, h). (see [Chr 2] § 6).

2.94. For this connection, *any (local) parallel section is holomorphic*, for, if $f \cdot l$ is parallel and l holomorphic, the 1-form df has to be of type (1, 0). This implies that f is holomorphic.

2.95. On a complex manifold, we have two distinguished holomorphic line bundles (apart from the trivial one): the *canonical line bundle* K and its *dual* K^*.

The canonical line bundle K is the bundle $\bigwedge^m (T'M)^*$ of forms of type $(m, 0)$, while the *anti-canonical line bundle* K^* is the bundle $\bigwedge^m T'M$ of alternate contravariant tensors of type $(m, 0)$.

Both K and K^* inherit a fibered Hermitian metric from the fibered Hermitian metric h of $T'M$ and the connections induced on K and K^* by the Levi-Civita connection are clearly the associated Hermitian connections.

Now, the associated curvature tensor of K^* (resp. K) is just equal to the trace (resp. minus the trace)—in the complex sense—of the fiber part of the Kähler curvature R.

We infer immediately, by the symmetry of the Kähler curvature operator, the following

2.96 Proposition. *The Ricci form ρ is equal to i times (resp. $-i$ times) the curvature tensor of the canonical line bundle K (resp. the anti-canonical line bundle K^*).*

2.97 Corollary. *The Ricci form ρ is equal to zero if and only if there exists a parallel—hence holomorphic—form of type $(m, 0)$ in a neighborhood of any point of M.*

Proof. This follows at once from Propositions 2.90 and 2.96. □

2.98 Corollary. *Let $\{\varepsilon_\alpha\}$, $\alpha = 1, \ldots, m$, by any local holomorphic frame of $T'M$, $(g_{\alpha\bar{\beta}})$ the Hermitian matrix whose generic element is equal to the scalar product $(\varepsilon_\alpha, \bar{\varepsilon}_\beta)$, $\det((g_{\alpha\bar{\beta}}))$ the determinant of this matrix. The Ricci form ρ is expressed locally as*

$$\boxed{\rho = -id'd'' \log \det((g_{\alpha\bar{\beta}}))}\,. \tag{2.98}$$

Proof. By proposition 2.93 and 2.96, ρ is equal to

(2.99) $$\rho = -id'd'' \log |l|^2$$

where l is any local holomorphic section without zero of K^*. This relation holds in particular for $l = \varepsilon_1 \wedge \cdots \wedge \varepsilon_m$ whose squared norm is clearly equal to $\det((g_{\alpha\bar{\beta}}))$. □

2.100 Corollary. *The Ricci form ρ depends only on the complex structure and the volume form μ_g of the Kähler metric*

Proof. Obvious from Corollary 2.98. □

2.101 Remark. Let g_0 be any fixed Kähler metric on a complex manifold M, v_{g_0} and ρ_0 its volume form and Ricci form respectively. For any change, from g_0 to g, of the Kähler metric, the volume form is changed into $\mu_g = f \cdot \mu_{g_0}$, where f is a positive scalar function, while the Ricci form is changed into $\rho_g = \rho_{g_0} - id'd'' \log f$ by, 2.98.

We recover the fact that *the Ricci form of a Kähler metric on a complex manifold M varies in a given cohomology class.*

Conversely, we may ask the question: *is any 2-form $\rho_{g_0} - id'd'' \log f$ the Ricci form of a Kähler metric for any positive scalar function?*

Equivalently, *is any volume form μ—that is any 2m-form without zero inducing the orientation of M—the volume form of a Kähler metric?*

That is the famous *Calabi problem.* The answer is *yes* when M is compact (see Chapter 11).

G. Hodge Theory

Consider the following *Laplace operators*

$$\Delta = d\delta + \delta d \quad \Delta' = d'\delta' + \delta' d' \quad \Delta'' = d''\delta'' + \delta'' d'' \quad \Delta^c = d^c\delta^c + \delta^c d^c.$$

One of the main features of a Kähler metric is the following

2.102 Proposition. *The Laplace operators of a Kähler metric satisfy the relations*

(2.102) $$\Delta = \Delta^c = 2\Delta' = 2\Delta''.$$

In particular, the Riemannian Laplacian Δ commutes with J and preserves the form types.

Proof. Easy consequence of proposition 2.33. □

Assume now that the complex manifold M is compact.

2.103. It is well known that the pth cohomology space with complex coefficients $H^p(M, \mathbb{C})$ is isomorphic to the space of harmonic (complex) p-forms. More precisely

$H^*(M, \mathbb{C})$ is isomorphic to the cohomology space associated to the exterior differential d acting on the (complex) exterior forms (*De Rham theorem*) and in each class of d-cohomology there is one and only one harmonic form (*Hodge theory*).

In the same way, consider the d''-cohomology associated with d'' acting on the complex exterior forms. Again, it can be proved (*Dolbeault theory* see [Hir] § 15) that each d''-cohomology class contains one and only one Δ''-harmonic form. For each type (p, q) we thus obtain a cohomology space $H^{p,q}(M, \mathbb{C})$ isomorphic to the space of forms of type (p, q) annihilated by Δ''.

In the Kähler case, where Δ'' is equal, up to a factor 1/2, to the Riemannian Laplacian, *the spaces $H^{p,q}(M, \mathbb{C})$ appear as sub-spaces of $H^{p+q}(M, \mathbb{C})$* and we get the following decomposition into types of $H^r(M, \mathbb{C})$:

(2.104) $$H^r(M, \mathbb{C}) = \sum_{p+q=r} H^{p,q}(M, \mathbb{C}).$$

We denote by $h^{p,q}$ the (complex) dimension of $H^{p,q}(M, \mathbb{C})$, while b_r, the rth Betti number, denotes as usual the (complex) dimension of $H^r(M, \mathbb{C})$.

Since Δ'' is real, conjugation induces a (skew-linear) isomorphism between $H^{p,q}(M, \mathbb{C})$ and $H^{q,p}(M, \mathbb{C})$ (*Serre duality*). In particular

(2.105) $$h^{p,q} = h^{q,p} \quad \forall p, q = 0, 1, \ldots, m.$$

We infer at once that *the Betti numbers of odd order are even* (possibly zero).

2.106. On the other hand, the Kähler form and its powers are harmonic so that $h^{p,q}$ is non-null for every $p = 0, 1, \ldots, m$. This implies that *the Betti numbers of even order are non-null.*

We thus obtain obstructions of a topological nature to the existence of a Kähler metric on a given compact complex manifold.

2.107. For example, the so-called *Calabi-Eckmann manifolds*, whose underlying manifold is the product $S^{2p+1} \times S^{2q+1}$ of two spheres of odd dimensions ($S^1 \times S^1$ excluded), do not admit any Kähler metric (see [Chr 2] pp. 4–6).

2.108 Remark. The isomorphism between $H^{p,q}(M, \mathbb{C})$ and the kernel of Δ'' acting on forms of type (p, q) still holds for any Hermitian metric. In general the spaces $H^{p,q}(M, \mathbb{C})$ are related to the ordinary cohomology by a spectral sequence (*Fröhlicher spectral sequence*)—which degenerates in the Kähler case—, and from which we deduce the inequality

(2.108) $$b_r \leqslant \sum_{p+q=r} h^{p,q}.$$

But we get, *in general*, the following *equality*

(2.108 bis) $$\chi = \sum_{p,q=0}^{m} (-1)^{p+q} h^{p,q},$$

where χ denotes the Euler-Poincaré number. (see [Gr-Ha] p. 444).

Another important consequence of 2.102 is the following

2.109 Proposition (dd^c-lemma). *The Hodge decomposition of any real dd^c-closed p-form φ may be written*

$$\varphi = d\delta\alpha + d^c\delta\beta + \varphi_H \tag{2.109}$$

where α and β are real p-forms and φ_H is the harmonic part of φ.

Proof. see [D-G-M-S] 5.11. □

2.110 Corollary. *Any real d-exact 2-form $\varphi = d\xi$ of type* (1, 1) *is dd^c-exact.*

Proof. Apply 2.109 to the 1-form ξ which is dd^c-closed since $\varphi = J\varphi$. □

2.111 Definition. The *Levi form* of a (real) function f is the (real) 2-form of type (1, 1) $\frac{1}{2}dd^cf$. A (local) function whose Levi form is zero is called *pluriharmonic*. The *complex Laplacian*, that is the opposite of the trace of the Levi form, is related, *on a Kahler manifold*, to the (Riemannian) Laplacian Δ by

$$\Delta f = -(dd^cf, \omega). \tag{2.112}$$

2.113 Corollary. 2.115 is a global theorem which holds on a compact *Kähler* manifold. Now it is easily proved (see, for example [Wei 1]) that, *on any complex manifold, any closed 2-form of type* (1, 1) *is locally the Levi form of a* (*real*) *function.*

In particular, the Kähler form of a Kähler manifold is locally the Levi form of a real function called *Kähler potential*, determined up to a (local) pluriharmonic function.

2.114 Definition. The *Kähler class* of a Kähler form ω is the set of Kähler forms belonging to the cohomology class $[\omega]$ of ω.

By Corollary 2.110 each element ω' of the Kähler class of ω is deduced from ω by

$$\omega' = \omega + dd^cf, \tag{2.114}$$

where f is a real function determined up to a constant.

2.115. Let (E, h) be a *holomorphic Hermitian bundle*, that is a holomorphic vector bundle E endowed with a fibered Hermitian metric h. With the help of the Chern connection ∇ associated with h, we construct the usual operators d, d'', δ'', Δ'' etc... (here we drop the subscript ∇). In fact, it is easily checked that d'' does not depend on h and is a cohomology operator acting on E-valued forms. We thus obtain as above a cohomology space $H^{p,q}(M, E)$ which is still *isomorphic to the space of E-valued forms of type* (p, q) *annihilated by* Δ'' (which depends on h).

For $p = 0$, the space $H^{0,q}(M, E)$ is also isomorphic to the qth cohomology space of M with values in the sheaf $\mathscr{E}$ of germs of holomorphic sections of E, obtained, for instance, by the Čech method (see [Hir] § 15).

Let $h^{p,q}(E)$ be the (complex) dimension of $H^{p,q}(M,E)$.

We define the *Euler-number* $\chi(M,\mathscr{E})$ of M relative to $\mathscr{E}$ by

$$\chi(M,\mathscr{E}) = \sum_{q=0}^{m} (-1)^q h^{0,q}(E). \tag{2.116}$$

The celebrated *Riemann-Roch-Hirzebruch Theorem* asserts that $\chi(M,E)$ can be expressed by the integral on M of a universal polynomial, depending only on the dimension of M and the rank of E, in the Chern classes of M and E. (see [Hir] § 21).

2.117. In particular, the *Euler number* $\chi(M,\mathcal{O})$ *relative to the structure sheaf* $\mathcal{O}$ (associated to the trivial line bundle), which is equal to

$$\chi(M,\mathcal{O}) = \sum_{q=0}^{m} (-1)^q h^{0,q} \tag{2.117}$$

can be expressed as the integral on M of a universal polynomial (depending only on m) in the Chern classes of M, and consequently, by 2.70, *in the Kähler curvature.*

2.118. Examples of Hermitian holomorphic vector bundles over M are given by the bundles $\bigwedge^p(T'M)$ and their duals $\bigwedge^{p,0}M \simeq \bigwedge^p(T'M)^*$, $p = 1,\ldots,m$.

Holomorphic sections of $\bigwedge^{p,0}M$ are called *holomorphic p-forms.*

The space of holomorphic p-forms is identified with $H^{p,0}(M;\mathbb{C})$. In particular, *holomorphic p-forms are closed on a compact Kähler manifold.*

In view of Serre duality, the dimension of the space of holomorphic p-forms is also equal to $h^{0,p}$, $p = 1,\ldots,m$.

In particular, *if the (compact) Kähler manifold M admits no non-trivial holomorphic p-forms the Euler number* $\chi(M;\mathcal{O})$ *relative to the structure sheaf* $\mathcal{O}$ *is equal to* 1.

H. Holomorphic Vector Fields and Infinitesimal Isometries

Let M be a compact almost-complex manifold.

2.119 Definition. An *infinitesimal automorphism of the almost-complex structure* is a (real) vector field X which preserves the complex operator J:

$$L_X J = 0 \iff [X,JY] = J[X,Y] \quad \forall Y \in \mathscr{T}M. \tag{2.119}$$

The space of these vector fields forms a finite-dimensional Lie algebra for the usual bracket of vector field. More precisely, we recall the following well-known result (see [Kob 4] 1–4):

2.120. *The group of automorphisms of a compact almost-complex manifold is a Lie transformation group* $\mathfrak{A}(M)$. *The Lie algebra of right-invariant vector fields on* $\mathfrak{A}(M)$ *is naturally identified with the Lie algebra of infinitesimal automorphisms of the almost complex structure.*

We assume now that M is a compact complex manifold equipped with a Kähler metric g.

We then have the following

2.121 Proposition. *A (real) vector field X is an infinitesimal automorphism of the complex structure if and only if its part of type $(1,0)$ is holomorphic.*

Proof. It is sufficient to prove that, for any (local) holomorphic 1-form θ of type $(1,0)$, $\theta(X)$ is holomorphic. For any vector field Y we have

$$d\theta(X, Y) = X \cdot \theta(Y) - Y \cdot \theta(X) - \theta([X, Y]),$$

while

$$\begin{aligned} d\theta(X, JY) &= X \cdot \theta(JY) - JY \cdot \theta(X) - \theta([X, JY]) \\ &= iX \cdot \theta(Y) - JY \cdot \theta(X) - i\theta([X, Y]) \end{aligned}$$

On the other hand, $d\theta$ is of type $(2,0)$, as θ is holomorphic, so that $d\theta(X, JY)$ is equal to $id\theta(X, Y)$. We infer at once that

$$JY \cdot \theta(X) = iY \cdot \theta(X),$$

which proves that the (local) scalar function $\theta(X)$ is holomorphic as claimed. □

2.122 Corollary. *The Lie algebra $\mathfrak{a}(M)$ of infinitesimal automorphisms of the complex structure has a natural structure of a complex Lie algebra, isomorphic with the complex Lie algebra of holomorphic vector fields. The group of automorphisms of the complex structure $\mathfrak{A}(M)$ inherits a natural structure of a complex Lie group.*

Proof. The part of type $(1,0)$ of JX is equal to $iX^{1,0}$ which is holomorphic when $X^{1,0}$ is.

The face that JX is an infinitesimal automorphism of the complex structure when X is, comes also as a direct consequence of the vanishing of the complex torsion tensor N. □

2.123. For brevity, an infinitesimal automorphism of the complex structure will be called a (*real*) *holomorphic vector field.*

Using the operator D^- introduced in 2.52, we have the

2.124 Proposition. *A real vector field X is holomorphic if and only if*

$$D^- X = 0. \tag{2.124}$$

Equivalently, $D\xi$ is of type $(1,1)$ where ξ is the dual 1-form of X.

Proof. Just use the definition (2.119) and the nullity of the torsion of D. □

2.125 Corollary (A. Lichnerowicz [Lic 1]). *On a compact Kähler manifold M of dimension $n = 2m$ we have*

i) *if* $m = 1$ *the Lie algebra of conformal vector fields coincides with the Lie algebra of (real) holomorphic vector fields,*

ii) *if* $m > 1$ *any conformal vector field is Killing (see* 1.80),

iii) *any Killing vector field is a (real) holomorphic vector field,*

iv) *more precisely, a (real) vector field is Killing if and only if it is holomorphic and preserves the volume-form.*

Proof. Use Remark 2.54, Formula 2.56 bis and Proposition 2.124. □

2.126 Corollary. *The Hodge decomposition of the dual* 1-*form* ξ *of any (real) holomorphic vector field* X *may be written*

$$\xi = df + d^c h + \xi_0, \tag{2.126}$$

where f *and* h *are well-defined real functions with zero integral over* M *and* ξ_0 *is the harmonic part of* ξ.

Proof. Apply the dd^c-lemma 2.109 to the 1-form ξ which is dd^c-closed since $d\xi$ is of type $(1, 1)$ as well as $D\xi$. □

2.127 Definition. The real function f appearing in the Hodge decomposition (2.126) is called the *holomorphy potential* (relative to the Kähler metric g) of the (real) holomorphic vector field X.

2.128. (By virtue of Proposition 2.125iv), *the holomorphy potential of* X *vanishes if and only if* X *is a Killing vector field.*

2.129. Observe that the real function h appearing in (2.126) is the holomorphy potential of the (real) holomorphic vector field—JX.

2.130 Remark. If X is a Killing vector field the 1-form $J\xi$ is closed by 2.126 and 2.128.

On the other hand, the Lie derivative of the Kähler form by X is equal to

$$L_X\omega = d(i_X\omega) = dJ\xi, \tag{2.131}$$

so that any Killing vector field is also a *Hamiltonian vector field* for the symplectic structure induced by the Kähler form.

This result can be recovered by using the fact that, on a *compact* Riemannian manifold, each Killing vector field leaves invariant all harmonic forms in particular the Kähler form. (see [Lic 1] p. 130).

2.132. Observe also that X *and* JX *cannot both be Killing vector fields unless* X *is parallel* (by 2.56) since $D\xi$ is then equal to $\frac{1}{2}d\xi$ and ξ is closed.

2.133 Remark. By 2.100 and 2.125iv), we see that two sorts of objects attached to the Kähler metric, namely the Ricci form ρ and the Lie algebra of infinitesimal automorphisms of the Kähler structure (which coincides with the Lie algebra of Killing vector fields of the metric) *do not depend on the whole Kähler structure, but* *only on the weaker structure determined by the complex structure and the volume-form.*

Such a structure may be called a *unimodular complex structure*. In terms of G-structures, it is a $Gl(m, \mathbb{C}) \cap Sl(2m, \mathbb{R})$-structure, *integrable* as a $Gl(m, \mathbb{C})$-structure. *The whole structure is integrable if and only if, in addition, the associated Ricci form vanishes.* By the Calabi-Yau theorem (see 11.15), each unimodular complex structure on a Kählerian manifold is induced by a -non uniquely defined—Kähler metric.

2.134 Proposition (J.L Koszul) ([Kos] p. 567). *Let M be a compact Kähler manifold, ρ its Ricci form. For any two infinitesimal automorphisms of the Kähler structure X and Y, we have:*

$$\rho(X, Y) = -\tfrac{1}{2}\delta(J[X, Y]). \tag{2.134}$$

Proof. Recall first (see 1.58) that the *divergence* δZ of any vector field Z is equal by definition, to $\delta\zeta$, where ζ is the associated 1-form. The divergence δZ depends only on the volume-form μ_g and may be defined by:

$$L_X \mu_g = -(\delta Z) \cdot \mu_g \tag{2.135}$$

(It is the opposite of the usual divergence).

For any two vector fields U and V the divergence of the bracket $[U, V]$ is given by:

$$\delta([U, V]) = U \cdot \delta V - V \cdot \delta U. \tag{2.136}$$

This formula can be easily checked directly, or deduced from the following:

$$[U, V]^\flat = -\delta(U^\flat \wedge V^\flat) + \delta U \cdot V^\flat - \delta V \cdot U^\flat. \tag{2.137}$$

Now to the proof. Let ξ be the 1-form associated to the vector field X. By Proposition 2.140 below, we know that

$$\rho(\xi) = r(J\xi) = \tfrac{1}{2}\Delta J\xi.$$

Since JX is also holomorphic, we get, by (2.33) and 2.126,

$$\delta dJ\xi = [\Lambda, d^c]dJ\xi = -d^c\Lambda dJ\xi = d^c\delta^c J\xi,$$

where $\delta^c J\xi = \delta\xi$ is equal to zero by assumption. Thus $\rho(\xi)$ reduces to

$$\rho(\xi) = \tfrac{1}{2}d\delta J\xi,$$

hence

$$\rho(X, Y) = \tfrac{1}{2}Y \cdot \delta(JX)$$

which is equal, by 2.136, to

$$\rho(X, Y) = \tfrac{1}{2}JX \cdot \delta Y + \tfrac{1}{2}\delta([Y, JX]).$$

Since δY is null and Y is holomorphic, we obtain the result. □

2.138 Proposition. *If the Ricci tensor of a compact Kähler manifold is everywhere negative-definite the group $\mathfrak{A}(M)$ is finite.*

Proof. From (2.53) we infer that

(2.138) $$\langle r(X), X\rangle = \langle D^+X, D^+X\rangle - \langle D^-X, D^-X\rangle$$

for any vector field. By 2.124 we conclude that the Lie algebra $\mathfrak{a}(M)$ is reduced to zero.

Now since the Ricci form ρ is equal, up to factor i, to the curvature of the canonical bundle K, we conclude that K is *ample*. This induces a holomorphic imbedding of M into a complex projective space $\mathbb{C}P^N$ which identifies $\mathfrak{A}(M)$ with a closed complex Lie sub-group of $PGl(N+1, \mathbb{C})$. In particular, the number of connected components of $\mathfrak{A}(M)$ is finite, and so is $\mathfrak{A}(M)$ itself since it is discrete. (see [Kob 4] pp. 82–86 for details). □

2.139 Remark. From 2.138 we infer at once that if the Ricci tensor is zero any holomorphic vector field is parallel.

On the contrary, if r is positive-definite a non-trivial holomorphic vector field cannot be parallel.

Combining (2.53), (2.55) and Proposition 2.124, we obtain the

2.140 Proposition (A. Lichnerowicz). *On a compact Kähler manifold a (real) vector field X is holomorphic if and only if*

(2.140) $$\tfrac{1}{2}\Delta\xi - r(X) = 0$$

where ξ denotes the dual 1-form of X.

Moreover X is (real) holomorphic if and only if the global scalar product $\langle \frac{1}{2}\Delta\xi - r(X), \xi\rangle$ vanishes.

2.141. In view of the Calabi-Futaki theorem (see below 2.160) it is useful to make the above relation 2.140 more explicit by using the Hodge decomposition of the Ricci form

(2.141) $$\rho = \gamma + dd^cF,$$

where γ is the harmonic part of ρ and F a well-defined real function with zero integral over M.

2.142 Definition. The function F above will be called the *Ricci potential* of the Kähler metric.

2.143. If X is a (real) holomorphic vector field, the 1-form $r(X)$ has a Hodge decomposition of the same type (2.126) as the dual 1-form ξ of X. This fact may be deduced at once from 2.140 but can be checked by observing that the exterior derivative $d(r(X))$—which is equal, up to a minus sign, to the Lie derivative of the Ricci form by JX— is of type (1, 1) like ρ since JX is holomorphic. The result is then given by the dd^c-lemma (2.109). More precisely we have the

2.144 Proposition. *For any (real) holomorphic vector field X the 1-form $r(X)$ is equal to*

(2.144) $$r(X) = d(\tilde{f} + X \cdot F) + d^c(\tilde{h} - JX \cdot F),$$

where $\tilde{f}$, $\tilde{h}$ are real functions with zero integral over M determined by

(2.145) $$\Delta\tilde{f} + (dd^c f, \gamma) = 0 \qquad (\text{resp. } \Delta\tilde{h} + (dd^c h, \gamma) = 0)$$

or, equivalently, by

(2.146) $$\delta\delta^c(\tilde{f} \cdot \omega - f \cdot \gamma) = 0 \qquad (\text{resp. } \delta\delta^c(\tilde{h} \cdot \omega - h \cdot \gamma) = 0),$$

where f (resp. h) is the holomorphy potential of X (resp. of $-JX$).

Proof. Using the decomposition 2.141 we get

$$r(X) = -\rho(JX) = -dd^c F(JX) - \gamma(JX).$$

Since JX is holomorphic, like X, the first term of the right member is equal to

(2.147) $$-dd^c F(JX) = -L_{JX} d^c F + d(i_{JX} d^c F) = d(X \cdot F) - d^c(JX \cdot F).$$

As for the second term, we know (applying the argument of 2.143 to γ) that its Hodge decomposition is given by

$$-\gamma(JX) = d\tilde{f} + d^c\tilde{h},$$

where $\tilde{f}$, $\tilde{h}$ are determined -up to an additive constant- by

$$\Delta\tilde{f} = -\delta(\gamma(JX)) \qquad (\text{resp. } \Delta\tilde{h} = -\delta^c(\gamma(JX)).$$

Since γ is co-closed we eventually obtain the relations (2.145). □

2.148 Corollary. *The holomorphy potential f of a real holomorphic vector field X satisfies the equation*

(2.148) $$\tfrac{1}{2}\Delta f - \tilde{f} - X \cdot F = -\int_M (X \cdot F) \cdot \mu_g,$$

where $\tilde{f}$ is determined by (2.145).

2.149 Remark. The Ricci potential F is related to the scalar curvature s by

(2.149) $$\tfrac{1}{2}s = \operatorname{tr}\rho = \tfrac{1}{2}s_0 - \Delta F,$$

where the trace $\frac{1}{2}s_0$ of the harmonic 2-form γ is constant.

In particular, *the scalar curvature is constant if and only if the Ricci potential is zero.*

In that case, Equation (2.148) reduces to

(2.150) $$\tfrac{1}{2}\Delta f - \tilde{f} = 0.$$

Conversely, using the second part of Proposition 2.140, it is easy to check, that, in the case where s is constant, the vector fields $(df)^\#$ and $(d^c f)^\#$ are holomorphic (the latter being an infinitesimal isometry) for any solution f of 2.150.

In other words, each part of the Hodge decomposition of ξ is then the dual 1-form of a (real) holomorphic vector field.

In particular, the harmonic part of ξ is parallel.

We infer at once the following

2.151 Proposition (A. Lichnerowicz [Lic 1]. *On a compact Kähler manifold with constant scalar curvature the complex Lie algebra* $\mathfrak{a}(M)$ *splits into the sum of the (abelian) Lie algebra of parallel vector fields and the (complex) Lie algebra of the holomorphic vector fields orthogonal to the space of harmonic* 1*-forms. The latter Lie algebra is the complexification of the (real) sub-Lie algebra of the infinitesimal isometries of the form* $(d^c f)^{\#}$ *where* f *is any solution of the equation*

$$\Delta^2 f + (dd^c f, \rho) = 0. \tag{2.151}$$

2.152. From Proposition 2.151, it follows easily that the connected isometry group is then a maximal compact, connected subgroup of $\mathfrak{A}(M)$. (see [Lic 1]).

I. The Calabi-Futaki Theorem

2.153 Definition. The *Futaki functional* is the real linear form $\mathscr{F}$ on the space $\mathfrak{a}(M)$ of (real) holomorphic vector fields defined by

$$\mathscr{F}(X) = \int_M (X \cdot F) \cdot \mu_g \qquad X \in \mathfrak{a}(M). \tag{2.153}$$

Alternatively, $\mathscr{F}$ may be defined by

(2.154)

$$\text{i) } \mathscr{F}(X) = \langle df, dF \rangle,$$

$$\text{ii) } \mathscr{F}(X) = \int_M F\delta\xi \cdot \mu_g,$$

$$\text{iii) } \mathscr{F}(X) = -\frac{1}{2}\int_M fs \cdot \mu_g,$$

where f is the holomorphy potential of X.

2.155. In particular, $\mathscr{F}(X)$ is equal to zero when X is Killing. In fact, in this case, the function $X \cdot F$ vanishes identically since X preserves both ρ and γ in (2.141).

2.156. In view of the important Corollary 2.160 we shall compute the first derivative of $\mathscr{F}$ considered as a function from the space $\mathscr{M}^K$ of Kähler metrics on M into the dual space $\mathfrak{a}(M)^*$.

On account of the unhomogeneousness of $\mathscr{F}$ we restrict it to the space $\mathscr{M}_1^K$ of the Kähler metrics of unit total volume. The points of $\mathscr{M}_1^K$ are denoted by the Kähler form and a first variation at ω is given by a 2-form of type (1, 1) satisfying the relation

$$\int_M \operatorname{tr} H \cdot \mu_g = 0. \tag{2.157}$$

The first variations of the volume-form, the Ricci form, the scalar curvature and the Laplace operator in the direction of H are then given by the

2.158 Lemma.

i) $\mu_g'(H) = \operatorname{tr} H \cdot \mu_g$,

ii) $\rho'(H) = -\frac{1}{2} dd^c(\operatorname{tr} H)$,

iii) $\frac{1}{2} s'(H) = \frac{1}{2} \Delta \operatorname{tr} H - (H, \rho)$,

iv) $\Delta'(H)\psi = (H, dd^c\psi)$ *for any real function* ψ.

Proof. i) $\mu_g = \frac{1}{m!}\omega^m$; ii) $\rho_{\alpha\beta} = -\frac{1}{2} dd^c \log \det(g_{\alpha\bar{\beta}})$ (see 2.98), iii) $\frac{1}{2}s = (\rho, \omega)$; iv) $\Delta\psi = -(dd^c\psi, \omega)$. □

2.159 Proposition. *The first derivative of* $\mathscr{F}$ *at any point* ω *of* $\mathscr{M}_1^K$ *in the direction of* H *is given by*

$$\mathscr{F}'(H)(X) = \langle H, f \cdot \gamma - \tilde{f} \cdot \omega \rangle, \tag{2.159}$$

where f *is the holomorphy potential of* X *and* $\tilde{f}$ *is determined by* 2.145.

Proof.

$$\begin{aligned} \mathscr{F}'(H)(X) &= \int_M (X \cdot F'(H)) \cdot \mu_g + \int_M (X \cdot F) \cdot \mu_g'(H) \\ &= \langle f, \Delta(F'(H)) \rangle + \int_M (X \cdot F) \operatorname{tr} H \cdot \mu_g, \end{aligned}$$

where, on account of 2.149 and the above Lemma,

$$\begin{aligned} \Delta(F'(H)) &= (\Delta F)'(H) - \Delta'(H)(F) \\ &= (H, \gamma) - \tfrac{1}{2} \Delta \operatorname{tr} H + \langle \tfrac{1}{2} s\omega - \rho, H \rangle. \end{aligned}$$

Now, use the Equation (2.148) to get the result. □

2.160 Corollary (Calabi-Futaki). *The functional* $\mathscr{F}$ *is constant on each Kähler class.*

Proof. Use the Formula (2.159) remembering that the 2-form $f \cdot \gamma - \tilde{f} \cdot \omega$ is dd^c-coclosed by 2.146. □

2.161 Remark. Corollary 2.160 is a generalization, due to E. Calabi [Cal 7], of a result of A. Futaki [Fut] where the Kähler class is assumed to be a (positive) multiple of the first Chern class. In that case, the harmonic part γ of the Ricci form is a multiple of ω and the 2-form $f \cdot \gamma - \tilde{f} \cdot \omega$ is identically zero. In other words such a metric is critical for $\mathscr{F}$.

Conversely, using the fact that a harmonic form cannot be zero on a non-empty open set of M unless it is identically zero (see [Aro]), we infer easily from (2.159) that *the critical metrics of* $\mathscr{F}$ *are either the Kähler metrics such that the Chern class is a multiple of the Kähler class, or the Kähler metrics relatively to which any (real) holomorphic vector field is an infinitesimal isometry, hence parallel by* 2.132.

2.162 Remark. By Remark 2.149, if a Kähler class $[\omega]$ contains a Kähler metric with constant scalar curvature, then the Futaki functional is zero for any member of ω.

Chapter 3. Relativity

A. Introduction

3.1. Interest in Einstein manifolds developed initially from the interpretation of the Einstein condition $r = 0$ as the field equation for a gravitational field in the absence of matter. This equation was formulated by Einstein in 1915. A brief history of the development of Einstein's field equation through quotes from early papers can be found in [Mi-Th-Wh] (pp. 431–434).

3.2. In this chapter we begin by discussing the physical interpretations of the various concepts in 4-dimensional Lorentz geometry. Then we discuss briefly the physical and geometric content of the Einstein equation. This is followed by a discussion of the Petrov classification of the curvature tensor of an Einstein-Lorentz 4-manifold and a fairly extensive discussion of the Schwarzschild model for the gravitational field outside a spherically symmetric star. The chapter concludes with brief discussions of the Taub-NUT metric and of the singularity theorems which assert that, under certain "physically reasonable" assumptions, spacetime cannot be geodesically complete.

B. Physical Interpretations

3.3. The spacetime of *special relativity* is Minkowski space $\mathbb{R}^4$; that is, it is $\mathbb{R}^4$ provided with a Lorentz inner product g whose signature we take to be $(+\,-\,-\,-)$. A choice of g-orthonormal basis $\{e_0, e_1, e_2, e_3\}$ for $\mathbb{R}^4$ endows $\mathbb{R}^4$ with a coordinate system (x_0, x_1, x_2, x_3) which may be viewed as a preferred coordinate system for a particular observer who perceives himself as "at rest." The coordinate x_0 describes time, as measured by this observer, and the 3-planes $x_0 = \text{constant}$ describe space, as viewed by this observer. This observer's history is described by a curve $\alpha(t) = (t, a, b, c)$: at each time t, according to his clock, he remains at the point (a, b, c) in his space.

A particle with nonzero rest mass which is in motion relative to our "at rest" observer will be described in the observer's coordinate system by the parametrized curve $\beta(t) = (t, x_1(t), x_2(t), x_3(t))$. The speed of this particle, as measured by the observer, is

$$\left[\left(\frac{dx_1}{dt}\right)^2 + \left(\frac{dx_2}{dt}\right)^2 + \left(\frac{dx_3}{dt}\right)^2\right]^{1/2}.$$

The principle that the speed of the particle relative to the observer must be less than the speed of light, which we take to be 1, says that

$$\left(\frac{dx_1}{dt}\right)^2 + \left(\frac{dx_2}{dt}\right)^2 + \left(\frac{dx_3}{dt}\right)^2 < 1$$

or, equivalently, since $x_0(t) = t$, that

$$\left(\frac{dx_0}{dt}\right)^2 - \left(\frac{dx_1}{dt}\right)^2 - \left(\frac{dx_2}{dt}\right)^2 - \left(\frac{dx_3}{dt}\right)^2 > 0.$$

This condition, which expresses the fact that the particle is travelling relative to our observer at a speed less than the speed of light, is independent of the parametrization of β. We can, if we wish, take t to be arc length τ along β rather than taking $t = x_0$. Then τ measures *proper time* along β; that is, τ describes time as perceived by the particle. Notice that the condition that the particle, whose motion is described by β, is travelling at a speed less than the speed of light relative to our observer is actually independent of the observer, it being described invariantly by the inequality $g(\dot{\beta}, \dot{\beta}) > 0$.

3.4. The spacetime of *general relativity* is a 4-dimensional smooth manifold M endowed with a metric of Lorentz signature $(+ - - -)$. A tangent vector v at a point of M is called *timelike* if $g(v, v) > 0$, *spacelike* if $g(v, v) < 0$, and *null* if $g(v, v) = 0$. The motion of a particle with nonzero rest mass in M is described by a *timelike curve* β; that is, by a curve β such that $g(\dot{\beta}, \dot{\beta}) > 0$. Usually, β is parameterized by a parameter t such that $g(\dot{\beta}(t), \dot{\beta}(t)) = m^2$, where m is the rest mass of the particle. When $m = 1$, we say that β describes the motion of a *test particle*. A massless particle, such as a photon, will travel along a *null curve*; that is, along a curve β with $\dot{\beta}(t)$ a null vector for all t.

3.5. Test particles can be viewed as *observers* in spacetime. A unit timelike vector v at $p \in M$, which may be viewed as the 4-*velocity* of a test particle as it passes through p, is called an *infinitesimal observer* at p. Associated with each infinitesimal observer is a *rest space* $v^\perp$, the subspace of M_p consisting of all vectors orthogonal to v. If β describes the motion of a particle of rest mass m ($g(\dot{\beta}, \dot{\beta}) = m^2$) and v is an infinitesimal observer at $\beta(t)$, then $\dot{\beta}(t) = av + w$ where $a \in \mathbb{R}$ and $w \in v^\perp$. The vector w is the 3-*momentum* of the particle, as measured by v, and the vector w/a is the 3-*velocity* of the particle, as measured by v.

A more detailed discussion of these concepts can be found in [Sa-Wu] (Chapters 1 and 2).

C. The Einstein Field Equation

3.6. In general relativity, the Lorentz metric g is viewed as a gravitational potential. As such, it must be related, by a field equation, to the mass/energy distribution that generates the gravitational field. The field equation proposed by Einstein is

$$r - \tfrac{1}{2}sg = T$$

where T is the stress/energy tensor.

The *stress/energy tensor* T is a symmetric 2-covariant tensor field on M. For each infinitesimal observer $v \in M_p$, $T(v, v)$ represents the *energy density* at p, as measured by v, of the distribution of mass/energy in spacetime. The covector $T(v, .)$ represents the *momentum density*, and the symmetric bilinear form $T|_{v^\perp}$ the *stress tensor*, both as measured by v, of the mass/energy at p. Notice that, since $g|_{v^\perp}$ is negative definite, this stress tensor can be diagonalized relative to g. The diagonal entries are the *principal pressures* at p of the mass/energy distribution, as seen by v.

For more details on the physical content of the stress/energy tensor T, see [Sa-Wu] (Chapter 3) or [Mi-Th-Wh] (Chapter 5).

3.7. The Einstein equation can be motivated in a variety of ways. We mention here just three of these.

(i) The stress/energy tensor is a 2-covariant tensor field so we must look for a 2-covariant tensor equation. The tensor $r - \frac{1}{2}sg$ that appears on the left hand side of Einstein's equation is the simplest 2-covariant tensor that can be constructed from the metric coefficients g_{ij} and their first two derivatives, is linear in the second order partial derivatives, and is in addition divergence free. The divergence free condition is a conservation law required by physical considerations.

(ii) The equation $r - \frac{1}{2}sg = 0$ that describes the gravitational field of a vacuum may be obtained as the Euler equation of the variational problem associated with the functional $\int s_g \mu_g$ (cf. Chapter 4).

(iii) In the appropriate "classical limit," the Einstein equation reduces to the Poisson equation describing Newtonian gravity.

For details, see [Fra], [Mi-Th-Wh] and [Web].

3.8 Remark. By a "solution" of Einstein's equation we mean a metric g *and* a stress/energy tensor T which describes a specified type of matter. These tensors g and T must be related by Einstein's equation. In the case where $T = 0$ (vacuum) we simply need to find g satisfying $r - \frac{1}{2}sg = 0$. In other cases, we must find both g and T. One approach is to specify Cauchy data for g and T on a spacelike hypersurface and view the Einstein equation, together with the equation $\operatorname{div} T = 0$, as an evolution equation. See, for example, [Ha-El] (Chapter 7).

3.9 Remark. If we apply the trace to both sides of Einstein's equation we see that

$$s = -\operatorname{trace} \tilde{T},$$

where $\tilde{T}$ is defined by $g(\tilde{T}(v), w) = T(v, w)$ for all $v, w \in M_p$, $p \in M$. Therefore, if $T = 0$ then Einstein's equation

$$r - \tfrac{1}{2}sg = T$$

becomes simply $r = 0$. Thus Ricci flat spacetimes describe vacuum solutions of Einstein's equation. If the Einstein equation is modified as

$$r - \tfrac{1}{2}sg + \lambda g = T,$$

where λ is a constant (the "cosmological constant"), then vacuum spacetimes will be described by geometries satisfying $r = cg$ for some c. Although this modified equation was popular in the early history of general relativity, it is no longer regarded as consistent with observations.

D. Tidal Stresses

3.10. The rate at which nearby freely falling test particles appear to accelerate toward or away from one another is measured by sectional curvature. Suppose we have a freely falling test particle, described by a timelike geodesic γ. A nearby (infinitesimally close) freely falling particle, which is initially in the direction w from $\gamma(w \perp \dot{\gamma}(0)$, $g(w, w) = -1)$ and is initially travelling parallel to γ, will be described by the Jacobi field W along γ with $W(0) = w$ and $W'(0) = 0$. The spatial separation between these particles is measured by the function $f = [-g(V, V)]^{1/2}$. If we calculate the first two derivatives of f and use Jacobi's equation, we find that $f'(0) = 0$ and $f''(0) = -K(\dot{\gamma}(0) \wedge w)$. Hence we see that, in the presence of positive sectional curvature, nearby freely falling particles tend to approach one another, and, in the presence of negative sectional curvature, they tend to separate.

Warning. The relationship between the sign of curvature and the spreading of timelike geodesics varies in the literature, depending on the sign conventions adopted by each author. Many authors use a metric of signature $(-+++)$; that is, they take as metric tensor the negative of the tensor g we use here. This choice will result in a change in the sign of sectional curvature, but will not change the geometry of the geodesics, since the eigenvalues of $W \to R(W, \dot{\gamma})\dot{\gamma}$ are unchanged.

3.11. Given any infinitesimal observer v, the symmetric bilinear form $g(R(v, .)v, .)$ on the rest space $v^\perp$ is the *tidal stress tensor* as measured by v. Since $g|_{v^\perp}$ is negative definite, this bilinear form can be diagonalized relative to g. The diagonal values are the *principal tidal stresses*, as measured by v. Notice that these principal stresses are the stationary values of the function $w \mapsto -K(v \wedge w)$ defined on the unit sphere in $v^\perp$. The number $-\tfrac{1}{3}r(v, v)$, which is equal to one third the sum of the principal tidal stresses as seen by v, is the *mean tidal stress*, as seen by v.

For more details, and a discussion of related ideas, see [Tho 2].

3.12. One consequence of Einstein's equation is that gravitation forces mean tidal stress to be negative (or, at least, nonpositive). For, since trace $\tilde{T} = -s$, we can rewrite Einstein's equation as

$$r = T - \tfrac{1}{2}(\text{trace } \tilde{T})g.$$

Hence the mean tidal stress, as seen by an infinitesimal observer v, is nonpositive provided that

$$r(v, v) = T(v, v) - \tfrac{1}{2}\operatorname{trace} \tilde{T} \geqslant 0.$$

The condition that $T(v, v) - \frac{1}{2}\operatorname{trace} \tilde{T} \geqslant 0$ for every infinitesimal observer v is called the *strong energy condition*. This condition is satisfied by almost every reasonable physical model. It says simply that the sum of the energy density and the three principal pressures must be nonnegative, for every infinitesimal observer v.

In any spacetime satisfying the strong energy condition, mean tidal stress is everywhere nonpositive and hence, *on the average*, nearby freely falling particles tend to approach one another.

E. Normal Forms for Curvature

3.13. Because of the indefiniteness of the metric, the normal form theory for the curvature tensor of a Lorentz manifold is more delicate than the corresponding theory for Riemannian metrics. In the Riemannian case, for example, the definiteness of the metric implies the existence of an orthonormal basis for $\bigwedge^2 M_p$ consisting of eigenvectors of $R: \bigwedge^2 M_p \to \bigwedge^2 M_p$. In the Lorentz case, such orthonormal bases do not necessarily exist.

3.14. A normal form for the curvature tensor is most meaningful geometrically if it arises from a choice of basis for M_p rather than one for $\bigwedge^2 M_p$. To derive such a normal form for Lorentz manifolds of dimension 4, it is convenient to use the Hodge star operator to make $\bigwedge^2 M_p$ into a complex vector space.

3.15. Recall that, given an orientation ω on M_p ($\omega \in \bigwedge^4 M_p$ with $g(\omega, \omega) = -1$) we can define $*: \bigwedge^2 M_p \to \bigwedge^2 M_p$ by $\alpha \wedge \beta = g(*\alpha, \beta)\omega$. Given an orthonormal basis $\{e_0, e_1, e_2, e_3\}$ for $\bigwedge^2 M_p$, with $g(e_0, e_0) > 0$ and $e_0 \wedge e_1 \wedge e_2 \wedge e_3 = \omega$, we can calculate the matrix $[*]$ for $*$ with respect to the orthonormal basis $\{e_0 \wedge e_1, e_0 \wedge e_2, e_0 \wedge e_3, e_2 \wedge e_3, e_3 \wedge e_1, e_1 \wedge e_2\}$ for $\bigwedge^2 M_p$. We find that

$$[*] = \begin{pmatrix} 0 & -I \\ I & 0 \end{pmatrix},$$

where I is the 3×3 identity matrix. Thus $*$ acts on $\bigwedge^2 M_p$ like a complex multiplication. We can make $\bigwedge^2 M_p$ into a complex 3-dimensional vector space by defining $i\alpha = *\alpha$ for each $\alpha \in \bigwedge^2 M_p$.

3.16. The symmetric bilinear form h on $\bigwedge^2 M_p$ defined by

$$h(\alpha, \beta) = g(\alpha, \beta) - ig(*\alpha, \beta)$$

is a nondegenerate complex inner product on the complex vector space $\bigwedge^2 M_p$. For each g-orthonormal basis $\{e_0, e_1, e_2, e_3\}$ for M_p we can check that $\{e_2 \wedge e_3, e_3 \wedge e_1, e_1 \wedge e_2\}$ is an h-orthonormal basis for $\bigwedge^2 M_p$.

3.17. Conversely, given any h-orthonormal basis $\{\alpha_1,\alpha_2,\alpha_3\}$ for $\bigwedge^2 M_p$ we can find a g-orthonormal basis $\{e_0,e_1,e_2,e_3\}$ for M_p such that $e_1 \wedge e_2 = \alpha_3$, $e_3 \wedge e_1 = \alpha_2$, and $e_2 \wedge e_3 = \pm\alpha_1$. Indeed, since

$$\alpha_i \wedge \alpha_i = g(*\alpha_i,\alpha_i)\omega = -\mathfrak{Im}(h(\alpha_i,\alpha_i))\omega = 0,$$

we see that each α_i is decomposible and hence represents a 2-plane $P_i \subset M_p$. Since

$$g(\alpha_i,\alpha_i) = \mathfrak{Re}(h(\alpha_i,\alpha_i)) = 1,$$

each of these 2-planes is spacelike ($g(v,v) < 0$ for all $v \in P_i$). Since

$$\alpha_i \wedge \alpha_j = g(*\alpha_i,\alpha_j)\omega = -(\mathfrak{Im}\, h(\alpha_i,\alpha_j))\omega = 0,$$

we see that these 2-planes must pairwise intersect. Choose unit vectors

$e_1 \in P_2 \cap P_3$,

$e_2 \in P_3$ such that $g(e_1,e_2) = 0$ and $e_1 \wedge e_2 = \alpha_3$,

$e_3 \in P_2$ such that $g(e_1,e_3) = 0$ and $e_3 \wedge e_1 = \alpha_2$, and

$e_0 \in M_p$ such that $g(e_0,e_i) = 0$ for $1 \leqslant i \leqslant 3$ and $e_0 \wedge e_1 \wedge e_2 \wedge e_3 = \omega$.

Then $\alpha_1 = \sum_{j<k} a_{jk} e_j \wedge e_k$ for $a_{jk} \in \mathbb{R}$, and the h-orthonormality of $\{\alpha_1,\alpha_2,\alpha_3\}$ implies that, in fact, $\alpha_1 = \pm e_2 \wedge e_3$. □

3.18. For manifolds of dimension 4, the Hodge star may be used to describe the $O(g)$-invariant subspaces of the space of curvature tensors at each point (cf. 1.128). Recall that $R \in \mathscr{U} \Leftrightarrow R$ has constant curvature, $R \in \mathscr{Z} \Leftrightarrow R$ is purely Ricci and is trace free, and $R \in \mathscr{W} \Leftrightarrow R$ has zero Ricci tensor. We can interpret these conditions in terms of matrices as follows.

3.19. Relative to an orthonormal basis for $\bigwedge^2 M_p$ as in 3.15, the matrix $[Q]$ of the quadratic form $Q(\alpha,\beta) = g(R\alpha,\beta)$ and the matrix $[R]$ of the endomorphism $R\colon \bigwedge^2 M_p \to \bigwedge^2 M_p$ are of the form

$$[Q] = \begin{pmatrix} A & B^t \\ B & C \end{pmatrix} \quad\text{and}\quad [R] = \begin{pmatrix} -A & -B^t \\ B & C \end{pmatrix},$$

where A, B, and C are 3×3 matrices with A and C symmetric. The kernel of the Ricci contraction consists of those R with trace $A = 0$, $C = -A$, and $B = B^t$. Thus $R \in \mathscr{W} \Leftrightarrow *R = R*$ and trace $R = 0$. The map $r \to g \circledA r$ sends the space of symmetric covariant 2-tensors isomorphically onto $\mathscr{W}^\perp$: if $R = g \circledA r$ then $C = A$ and $B = -B^t$. Thus $R \in \mathscr{Z} \Leftrightarrow *R = -R*$, and $R \in \mathscr{U} \Leftrightarrow R$ is a scalar multiple of the identity.

3.20. In particular, we see that M is Einstein ($R \in \mathscr{U} + \mathscr{W}$) if and only $R* = *R$. In terms of the complex structure on M_p, $p \in M$ (cf. 3.15), this says that *M is Einstein if and only if $R\colon \bigwedge^2 M_p \to \bigwedge^2 M_p$ is complex linear*. We can use this fact to derive normal forms for curvature tensors of Einstein-Lorentz manifolds.

3.21. So, when M is Einstein, R is a complex linear endomorphism of $\bigwedge^2 M_p$. Let

$\{\beta_1, \beta_2, \beta_3\}$ be a basis for $\bigwedge^2 M_p$ which casts the matrix $[R]$ for R into Jordan normal form. There are three possible cases.

3.22. Case I. $[R] = \begin{pmatrix} z_1 & 0 & 0 \\ 0 & z_2 & 0 \\ 0 & 0 & z_3 \end{pmatrix}$ for some $z_1, z_2, z_3 \in \mathbb{C}$. In this case, a standard argument (compute both sides of $h(R\beta_i, \beta_j) = h(\beta_i, R\beta_j)$ shows that $h_{ij} = h(\beta_i, \beta_j) = 0$ whenever $z_i \neq z_j$; that is, the eigenvectors are h-orthogonal whenever the eigenvalues are distinct. When the eigenvalues are not distinct, the *eigenspaces* are h-orthogonal and we can orthogonalize $\{\beta_1, \beta_2, \beta_3\}$ using Gram-Schmidt. Normalizing, we may as well assume that $\{\beta_1, \beta_2, \beta_3\}$ is h-orthonormal. The matrix $[h] = (h_{ij})$ for the quadratic form h is then

$$[h] = \begin{pmatrix} 1 & 0 & 0 \\ 0 & 1 & 0 \\ 0 & 0 & 1 \end{pmatrix}.$$

3.23. Case II. $[R] = \begin{pmatrix} z_1 & 0 & 0 \\ 0 & z_2 & 1 \\ 0 & 0 & z_2 \end{pmatrix}$ for some $z_1, z_2 \in \mathbb{C}$. In this case, the standard argument shows that $h_{12} = h_{22} = 0$ and, if $z_1 \neq z_2$, $h_{13} = 0$. If $z_1 = z_2$ and $h_{13} \neq 0$ then we can replace β_1 by $\beta_1 + a\beta_2$ for an appropriate $a \in \mathbb{C}$ to get $h_{13} = 0$. Finally, we can replace $\{\beta_1, \beta_2, \beta_3\}$ by $\{a\beta_1, b\beta_2, b\beta_3 + c\beta_2\}$ where a, b, and c are chosen to make $h_{11} = h_{23} = 1$ and $h_{33} = 0$. The matrix for h relative to this basis, which we shall still call $\{\beta_1, \beta_2, \beta_3\}$, is then

$$[h] = \begin{pmatrix} 1 & 0 & 0 \\ 0 & 0 & 1 \\ 0 & 1 & 0 \end{pmatrix}.$$

3.24. Case III. $[R] = \begin{pmatrix} z & 1 & 0 \\ 0 & z & 1 \\ 0 & 0 & z \end{pmatrix}$ for some $z \in \mathbb{C}$. In this case, the standard argument shows that $h_{11} = h_{12} = 0$ and $h_{13} = h_{22}$. We can replace $\{\beta_1, \beta_2, \beta_3\}$ by $\{a\beta_1, a\beta_2 + \beta_1, a\beta_3 + b\beta_2 + c\beta_1\}$ where a, b, and c are chosen to make $h_{13} = 1$ and $h_{23} = h_{33} = 0$. The matrix for h relative to this $\{\beta_1, \beta_2, \beta_3\}$ is then

$$[h] = \begin{pmatrix} 0 & 0 & 1 \\ 0 & 1 & 0 \\ 1 & 0 & 0 \end{pmatrix}.$$

3.25. To convert these normal forms into normal forms for R relative to a g-orthonormal basis for M_p, we must first describe each one relative to an h-orthonormal basis $\{\alpha_1, \alpha_2, \alpha_3\}$ for $\bigwedge^2 M_p$ and then apply 3.17. For Type I, $\{\beta_1, \beta_2, \beta_3\}$ is h-orthonormal so we may take $\alpha_j = \beta_j$ $(1 \leqslant j \leqslant 3)$. For Type II we may take $\alpha_1 = \beta_1$, $\alpha_2 = \frac{1}{2}\beta_2 + \beta_3$, and $\alpha_3 = i(-\frac{1}{2}\beta_2 + \beta_3)$. For Type III we may take $\alpha_1 = \frac{1}{2}\beta_1 + \beta_3$, $\alpha_2 = \beta_2$, and $\alpha_3 = i(-\frac{1}{2}\beta_1 + \beta_3)$. The matrix for R relative to the h-orthonormal basis $\{\alpha_1, \alpha_2, \alpha_3\}$ is then of one of the following three types:

$$\text{(I)} \begin{pmatrix} z_1 & 0 & 0 \\ 0 & z_2 & 0 \\ 0 & 0 & z_3 \end{pmatrix}, \quad \text{(II)} \begin{pmatrix} z_1 & 0 & 0 \\ 0 & z_2 + 1 & i \\ 0 & i & z_2 - 1 \end{pmatrix}, \quad \text{or (III)} \begin{pmatrix} z & 1 & 0 \\ 1 & z & i \\ 0 & i & z \end{pmatrix}.$$

3.26. If we now construct from $\{\alpha_1, \alpha_2, \alpha_3\}$ a g-orthonormal basis $\{e_0, e_1, e_2, e_3\}$ for M_p as in 3.17 and compute the matrix $[R]$ for R relative to the basis $\{e_0 \wedge e_1, e_0 \wedge e_2, e_0 \wedge e_3, e_2 \wedge e_3, e_3 \wedge e_1, e_1 \wedge e_2\}$ for $\bigwedge^2 M_p$, recalling that $*R = R*$ and that $ie_j \wedge e_k = *e_j \wedge e_k$ for each j and k, we find that

$$[R] = \begin{pmatrix} A & -B \\ B & A \end{pmatrix},$$

where $A + iB$ is the matrix for R relative to the basis $\{\alpha_1, \alpha_2, \alpha_3\}$, as above. (For Type III it may be necessary to replace e_2 by $-e_2$ in order to achieve this formula.) For each of the three types, the matrices A and B are as follows, where $z_j = \lambda_j + i\mu_j$:

$$\text{Type I} \quad A = \begin{pmatrix} \lambda_1 & 0 & 0 \\ 0 & \lambda_2 & 0 \\ 0 & 0 & \lambda_3 \end{pmatrix}, \qquad B = \begin{pmatrix} \mu_1 & 0 & 0 \\ 0 & \mu_2 & 0 \\ 0 & 0 & \mu_3 \end{pmatrix},$$

$$\text{Type II} \quad A = \begin{pmatrix} \lambda_1 & 0 & 0 \\ 0 & \lambda_2 + 1 & 0 \\ 0 & 0 & \lambda_2 - 1 \end{pmatrix}, \qquad B = \begin{pmatrix} \mu_1 & 0 & 0 \\ 0 & \mu_2 & 1 \\ 0 & 1 & \mu_2 \end{pmatrix},$$

$$\text{Type III} \quad A = \begin{pmatrix} \lambda & 1 & 0 \\ 1 & \lambda & 0 \\ 0 & 0 & \lambda \end{pmatrix}, \qquad B = \begin{pmatrix} 0 & 0 & 0 \\ 0 & 0 & 1 \\ 0 & 1 & 0 \end{pmatrix}.$$

3.27. These normal forms were found by Petrov in 1954 and are described in his book [Pet]. They have been used extensively in the study of gravitational radiation (see, e.g., [Pir]). The derivation presented here is from [Tho 1]. Notice that, because of the algebraic Bianchi identity 1.24, we have $\mu_1 + \mu_2 + \mu_3 = 0$ in Type I, $\mu_1 + 2\mu_2 = 0$ in Type II, and $\mu = 0$ in Type III.

F. The Schwarzschild Metric

3.28. The most useful solution of the empty space Einstein equation $r = 0$ is the one found by Schwarzschild in 1916, shortly after Einstein formulated general relativity theory. This solution models the gravitational field outside an isolated, static, spherically symmetric star.

3.29. A spacetime M is *static* if there exists on M a nowhere zero timelike Killing vector field X (see 1.80) such that the distribution of 3-planes orthogonal to X is integrable. The integral manifolds of this distribution will be spacelike hypersurfaces which are, locally, isometric to one another. Locally, these hypersurfaces are $t =$ constant surfaces for some time function t.

3.30. A spacetime M is *spherically symmetric* if there is an isometric action of the special orthogonal group $SO(3)$ on M each of whose orbits is either a spacelike 2-surface or a single point. The 2-dimensional orbits will necessarily have constant positive curvature.

3.31. The staticity and spherical symmetry conditions will both be satisfied on the product manifold $M = \mathbb{R} \times I \times S^2$, where I is an open interval, with any warped product metric

$$g = F^2(\rho)\,dt^2 - d\rho^2 - G^2(\rho)\,d\sigma^2$$

where t and ρ are the standard coordinates on $\mathbb{R}$ and I, and $d\sigma^2$ is the standard (constant curvature 1) metric on S^2. The coordinate function t is a global time function. The group $SO(3)$ acts isometrically on M with orbits the 2-spheres $(t, \rho) =$ constant. The 2-sphere, $t = t_0, \rho = \rho_0$, is isometric, in the induced metric $G^2(\rho_0)\,d\sigma^2$, to a Euclidean 2-sphere of radius $G(\rho_0)$.

We shall find, almost without computation, the Ricci tensor for each such metric, and then shall find the conditions on the functions F and G which will yield zero Ricci curvature.

3.32. Notice first that, for each $t \in \mathbb{R}$, the hypersurface $\{t\} \times I \times S^2$ is totally geodesic since it is the fixed point set of an isometric reflection (time reversal). Since any vector field of constant length normal to a totally geodesic hypersurface is parallel along that hypersurface, it follows that $R(y, z)x = 0$ whenever x is tangent to $\mathbb{R}$ and y, z are perpendicular to x. This implies that $\dfrac{\partial}{\partial t}$ is an eigenvector of the Ricci tensor r.

3.33. Similarly, the hypersurface $\mathbb{R} \times I \times S^1$ is totally geodesic in M for each great circle S^1 in S^2 (it is the fixed point set of a reflection of S^2) and so $R(y, z)x = 0$ whenever x is tangent to S^2 and y, z are perpendicular to x. Hence, each vector tangent to the S^2 factor is also an eigenvector of r.

3.34. It follows that the tangent spaces to the factors $\mathbb{R}$ and S^2, and hence also to I by orthogonality, are eigenspaces of r. Thus *the Ricci tensor is diagonal relative to the product structure on* M and we need only compute the diagonal terms, which are sums of sectional curvatures.

3.35. The relevant sectional curvatures can be obtained as follows.

First note that the curve $\gamma_{(t,p)}(\rho) = (t, \rho, p)$ is a geodesic in M, for each $t \in \mathbb{R}$ and $p \in S^2$, since it is parametrized by arc length and its trajectory is a connected component of the fixed point set of the isometry group generated by reflection of $\mathbb{R}$ about t and reflections of S^2 about great circles through p. The vector field $\frac{\partial}{\partial t}$ is a Jacobi field along $\gamma_{(t,p)}$ and hence

$$K\left(\frac{\partial}{\partial t} \wedge \frac{\partial}{\partial \rho}\right) = \left\langle D^2_{\partial/\partial\rho} \frac{\partial}{\partial t}, \frac{\partial}{\partial t} \right\rangle \Big/ \left\| \frac{\partial}{\partial t} \wedge \frac{\partial}{\partial \rho} \right\|^2 = F''/F. \tag{3.35a}$$

Similarly, for each unit vector x tangent to S^2, $\partial/\partial\theta$ is a Jacobi field along γ, where θ is the angular coordinate on the great circle S^1 in S^2 generated by x, and hence

$$K\left(\frac{\partial}{\partial \rho} \wedge x\right) = \left\langle D^2_{\partial/\partial\rho} \frac{\partial}{\partial \theta}, \frac{\partial}{\partial \theta} \right\rangle \Big/ \left\| \frac{\partial}{\partial \theta} \wedge \frac{\partial}{\partial \rho} \right\|^2 = G''/G. \tag{3.35b}$$

In other words, we have computed the curvature of 2-dimensional totally geodesic subspaces in Gaussian coordinates.

We can obtain the rest of the relevant sectional curvatures by observing that on each ρ = constant hypersurface $\mathbb{R} \times \{\rho\} \times S^2$ the induced metric is the product metric

$$g_\rho = F^2(\rho)\, dt^2 - G^2(\rho)\, d\sigma^2$$

and the second fundamental form of this hypersurface with respect to the unit normal vector field $\frac{\partial}{\partial \rho}$ is given by $\left(\frac{D}{\partial \rho} g_\rho\right)(x, y) = -2g_\rho(S \cdot x, y)$, hence

$$S = \begin{pmatrix} -FF' & 0 \\ 0 & -GG'h \end{pmatrix},$$

where $h = d\sigma^2$. Hence, by the Gauss equation,

$$K\left(\frac{\partial}{\partial t} \wedge x\right) = \frac{F'}{F} \cdot \frac{G'}{G}, \tag{3.35c}$$

whenever x is tangent to S^2, and

$$K(x \wedge y) = -\frac{1}{G^2} + \left(\frac{G'}{G}\right)^2, \tag{3.35d}$$

whenever both x and y are tangent to S^2.

3.36. It follows that the eigenvalues of the Ricci tensor are

$$r_t = \frac{F''}{F} + 2\frac{F'}{F} \cdot \frac{G'}{G}, \tag{3.36a}$$

$$r_\rho = \frac{F''}{F} + 2\frac{G''}{G}, \tag{3.36b}$$

and

$$r_x = \frac{G''}{G} + \frac{F'}{F} \cdot \frac{G'}{G} - \frac{1}{G^2} + \left(\frac{G'}{G}\right)^2 \quad \text{(repeated)}. \tag{3.36c}$$

3.37. We are now ready to solve the Einstein equation $r = 0$ for the warped product metric on $\mathbb{R} \times I \times S^2$.

The condition

$$0 = r_t = \frac{1}{FG^2}(F'G^2)' \qquad \text{(see 3.36a)}$$

says that $F'G^2$ is constant. Set

$$m = F'G^2. \tag{3.37a}$$

The condition

$$0 = r_\rho - r_t = 2\frac{F}{G}\left(\frac{G'}{F}\right)' \qquad \text{(see 3.36a and 3.36b)}$$

says that G'/F is also constant. By rescaling t we may assume that $G'/F = 1$. It then follows that

$$G' = F, \tag{3.37b}$$

and that

$$G'' = F' = m/G^2 \qquad \text{(see 3.37a).} \tag{3.37c}$$

Using these facts, we see from (3.36c) that

$$r_x = \frac{1}{G^2}\left(\frac{2m}{G} - 1 + F^2\right). \tag{3.37d}$$

Hence the condition $r_x = 0$ is satisfied if and only if

$$F = \left(1 - \frac{2m}{G}\right)^{1/2}$$

Thus the metric

$$g = F^2(\rho)\,dt^2 - d\rho^2 - G^2(\rho)\,d\sigma^2 \tag{3.37e}$$

on $\mathbb{R} \times I \times S^2$ satisfies Einstein's empty space equation $r = 0$ provided that

$$F = \frac{dG}{d\rho} = \left(1 - \frac{2m}{G}\right)^{1/2} \tag{3.37f}$$

for some constant m.

3.38. This metric can be cast into a more familiar form by using $\tilde{\rho} = G(\rho)$ as coordinate in place of ρ. This leads to the metric in the form found by Schwarzschild:

$$g = \left(1 - \frac{2m}{\tilde{\rho}}\right)dt^2 - \left(1 - \frac{2m}{\tilde{\rho}}\right)^{-1} d\tilde{\rho}^2 - \tilde{\rho}^2\,d\sigma^2.$$

By taking the domain of $\tilde{\rho}$ as large as possible, we see that this formula defines a static spherically symmetric metric on $\mathbb{R} \times (2m, \infty) \times S^2$. *The function $\tilde{\rho} = G(\rho)$ is usually interpreted as distance from the center of the star* (although neither travel time measurements of reflected light signals nor trigonometric distance measurements would confirm that increments in $\tilde{\rho}$ measure radial distances).

3.39 Remark. Using the equations relating F and G that where established in 3.37 together with the curvature formulas in 3.35 and the vanishing of many curvature components obtained in 3.32, 3.33 we find an eigenbasis and the eigenvalues of the curvature operator:

(3.39a) $\hat{R}$ has the double eigenvalue $-2m \cdot G(\rho)^{-3}$ on span$\left\{\frac{\partial}{\partial t} \wedge \frac{\partial}{\partial \rho}, x \wedge y\right\}$

and

(3.39b) $\hat{R}$ has the fourfold eigenvalue $m \cdot G(\rho)^{-3}$ on span$\left\{\frac{\partial}{\partial t} \wedge x, \frac{\partial}{\partial \rho} \wedge y\right\}$, where x and y are tangent to S^2.

Such a decomposable eigenbasis is useful for computations. Note also that $\hat{R}$ is, pointwise, proportional to the Weyl-tensor of $S^2 \times S^2$.

G. Planetary Orbits

3.40. The Schwarzschild solution is designed to model the gravitational field outside a star. The constant m may be interpreted as the mass of the star by studying the periods of planetary orbits.

3.41. Consider a test planet orbiting the star. Its motion will be described by a timelike geodesic γ in M. This geodesic will necessarily lie in the totally geodesic submanifold $\mathbb{R} \times I \times S^1$ where S^1 is the great circle in S^2 generated by the S^2-component of $\dot{\gamma}(s)$, for any given s. Hence γ is of the form

$$\gamma(s) = (t(s), \rho(s), \theta(s)),$$

where θ is the angular coordinate on S^1.

3.42. The orbit is said to be *circular* if $\rho = a$ for some constant a. Then

(3.42a)
$$\dot{\gamma} = \dot{t}\frac{\partial}{\partial t} + \dot{\theta}\frac{\partial}{\partial \theta},$$

and

(3.42b)
$$D_{\dot{\gamma}}\dot{\gamma} = \ddot{t}\frac{\partial}{\partial t} + (\dot{t}^2 FF' - \dot{\theta}^2 GG')|_{\rho=a}\frac{\partial}{\partial \rho} + \ddot{\theta}\frac{\partial}{\partial \theta},$$

so the geodesic equation $D_{\dot{\gamma}}\dot{\gamma} = 0$ says that

(3.42c)
$$\ddot{t} = \ddot{\theta} = 0 \quad \text{and} \quad \dot{\theta}^2 = \frac{FF'}{GG'}(a)\dot{t}^2.$$

Thus $\dot{t}$ and $\dot{\theta}$ are constant and, since $F' = m/G^2$ and $G' = F$ (cf. (3.37c) and (3.37b),

(3.42d)
$$\dot{\theta}^2 = \frac{m}{G^3(a)}\dot{t}^2.$$

If the parameter s measures proper time along γ then we also have

$$F^2(a)\dot{t}^2 - G^2(a)\dot{\theta}^2 = 1.$$

Since $F = (1 - 2m/G)^{1/2}$, we find that

(3.42e)
$$\dot{t}^2 = \left(1 - \frac{3m}{G(a)}\right)^{-1},$$

and so *the test planet will trace out a circular orbit if and only if*

(3.42f)
$$t(s) = \pm\left(1 - \frac{3m}{G(a)}\right)^{-1/2} s + c_1,$$

$$\theta(s) = \pm\left[\frac{m}{G^2(a)(G(a) - 3m)}\right]^{1/2} s + c_2$$

for some $c_1, c_2 \in \mathbb{R}$. Usually the first $\pm$ sign is taken to be $+$ so that $\dot{t} > 0$.

3.43. We can apply this information to compute the period of the orbit. The proper time required for one revolution is $T_P = |2\pi/\dot{\theta}|$. The cosmic time required is $T_C = |\dot{t}|T_P = |2\pi\dot{t}/\dot{\theta}|$. Since G is interpreted as distance from the center of the star and $\dot{t} = \left(1 - \dfrac{3m}{G(a)}\right)^{-1/2} \to 1$ as $G(a) \to \infty$ we see that, for large circular orbits,

$$T_P^2 \approx T_C^2 = 4\pi^2\dot{t}^2/\dot{\theta}^2 = \frac{4\pi^2}{m}G^3(a).$$

This is just Kepler's third law for a planet orbiting a star of mass m.

3.44 Remark. The above blending of Newtonian concepts (distance from a star, circular orbits) into a relativistic model was possible here only because of the special nature of the model. The existence of a preferred cosmic time t and of a standard space section $I \times S^2$ were essential for this discussion. Note however that, for very large G, cosmic time and standard space can be recovered (to a good approximation) from Newtonian measurements, since curvature decays as m/G^3 and hence the model differs only slightly, when G is large, from the Newtonian model. Thus the Schwarschild model provides a continuous change, as G decreases, from a nearly Newtonian model to a model that is so strongly non-Newtonian that it defies the imagination (cf. 3.45). Our own world is somewhere between these extremes, since time comparison measurements on the earth and in our planetary system easily detect small deviations from the Newtonian model.

3.45 Remark. Note that circular planetary orbits are possible only when $1 - \dfrac{3m}{G(a)} = \dot{t}^{-2} > 0$; that is, only when $G(a) > 3m$. As $G(a)$ approaches $3m$, the proper time

$$T_P = \left|\frac{2\pi}{\dot{\theta}}\right| = 2\pi\left[\frac{G^3(a)}{m}\left(1 - \frac{3m}{G(a)}\right)\right]^{1/2}$$

required for one revolution approaches zero. When $G(a) = 3m$, the circular orbit $\gamma(s) = (t(s), G^{-1}(3m), \theta(s))$ is geodesic when $\ddot{t} = \ddot{\theta} = 0$ and $\dot{t}^2 = 27m^2\dot{\theta}^2$. These orbits are null geodesics. They are interpreted as photon orbits. The "sphere" $G^{-1}(3m) \times S^2 \subset I \times S^2$ is called the *photon sphere* since each photon emitted tangent to this sphere will always remain in it.

3.46. An integral curve of the Killing vector field $\dfrac{\partial}{\partial t}$ (normalized) is interpreted as describing an observer who is "cosmically at rest" (but *not* freely falling!). Observers cosmically at rest on the photon sphere will be able to see into their own pasts. In fact, all past directed null geodesics which are at a point p tangent to the photon sphere focus at a point p' that lies on the past trajectory of the cosmic observer at p (see Figure 3.46). Moreover, observers that are cosmically at rest *outside* the photon spheres will, theoretically, also be able to look into their own pasts simply by aiming their telescopes close to the photon sphere (Figure 3.46). However, for stars of radius $> 3m$, such as our sun, or the earth, the empty space solution is not valid at $r = 3m$ and so this phenomenon is not present.

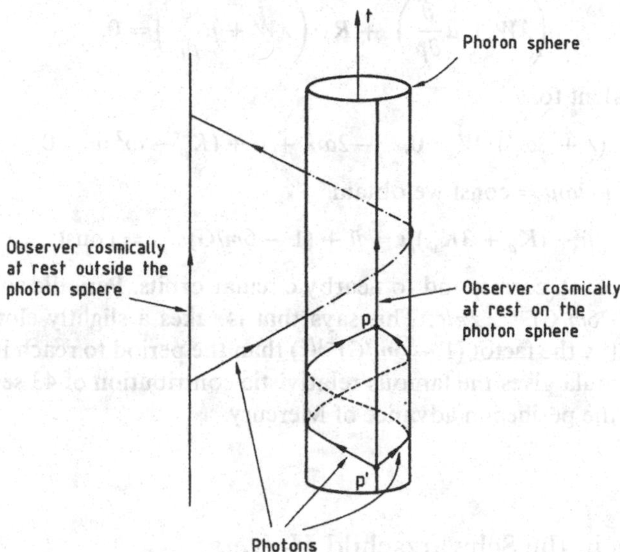

Fig. 3.46

H. Perihelion Procession

3.47. The analysis of circular planetary orbits can be extended to almost circular orbits in order to study perihelion procession.

Let $\gamma(s) = (t(s), \rho(s), \theta(s))$ describe a circular planetary orbit in M, as in 3.42. We may assume that $\rho(s) = a, t(s) = bs$, and $\theta(s) = cs$ where $a \in \mathbb{R}$ is such that $G(a) > 3m$, $b = (1 - 3m/G(a))^{-1/2}$, and $c = [m/G^2(a)(G(a) - 3m)]^{1/2}$. $2\pi/c$ is the period measured in proper time. We shall consider nearby orbits which lie in the totally geodesic hypersurface $\mathbb{R} \times I \times S^1$ containing γ. Such orbits are described infinitesimally by Jacobi fields. We abbreviate the coefficient operator $X \to R(X, \dot{\gamma})\dot{\gamma}$ by $R_{\dot{\gamma}} \cdot X$ and compute with 3.39 its eigendata in $\dot{\gamma}^{\perp}$ $\left(\text{put } W = b \cdot \frac{F}{G}\frac{\partial}{\partial\Theta} + c \cdot \frac{G}{F}\frac{\partial}{\partial t}\right)$.

$$R_{\dot{\gamma}} \cdot \frac{\partial}{\partial\rho} = -(mG^{-3} + c^2) \cdot \frac{\partial}{\partial\rho} =: K_\rho \cdot \frac{\partial}{\partial\rho}$$

$$R_{\dot{\gamma}} \cdot W = mG^{-3} \cdot W =: K_w \cdot W.$$

Of course these unit eigenvectors are not parallel but $\omega := \left\langle \frac{D}{\partial\rho}\left(\frac{\partial}{\partial\rho}\right), W \right\rangle = (1 - 3m/G)bc = (1 - 3m/G)^{1/2}c = K_w^{1/2}$.

Notice that the Newtonian picture predicts $\omega = c$. Since a parallel vector-field along a worldline is physically realized by the angular momentum of a spinning body we conclude from $\omega = c/b$: If the axis of a spinning planet (on a circular orbit) is in the orbital plane then the axis slowly rotates forward relative to its Newtonian position (Focker precession). Next we write the Jacobi equation as

$$\left(\lambda W + \mu\frac{\partial}{\partial\rho}\right)^{\cdot\cdot} + R_{\dot{\gamma}} \cdot \left(\lambda W + \mu\frac{\partial}{\partial\rho}\right) = 0,$$

which is equivalent to

$$(\ddot{\lambda} + 2\omega\dot{\mu}) \cdot W = 0, \quad -2\omega\dot{\lambda} + \ddot{\mu} + (K_\rho - \omega^2)\mu = 0$$

Substituting $\dot{\lambda} + 2\omega\mu = \text{const}$ we obtain

$$\ddot{\mu} + (K_\rho + 3K_w)\mu = \ddot{\mu} + (1 - 6m/G)c^2\mu = \text{const}.$$

Solutions $\mu = \text{const.}$ correspond to nearby circular orbits. Periodic solutions have the period $(1 - 6m/G)^{-1/2} \cdot 2\pi/c$. This says that is takes a slightly elongated orbit slightly longer (by the factor $(1 - 6m/G)^{-1/2}$) than the period to reach its perihelion again. This formula gives the famous relativistic contribution of 43 seconds of arc per century to the perihelion advance of Mercury.

I. Geodesics in the Schwarzschild Universe

3.48. The computations of the previous paragraphs can be extended to obtain qualitative information about general geodesics in the Schwarzschild universe M. For example, if $\gamma(s) = (t(s), \rho(s), \theta(s))$ is any geodesic in the totally geodesic hypersurface $\mathbb{R} \times I \times S^1 \subset M$ then $\mu = G^2\dot{\theta}$ and $\varepsilon = F^2\dot{t}$ are both constant along γ. Indeed, $G^2\dot{\theta} = -\left\langle \dot{\gamma}, \frac{\partial}{\partial\theta} \right\rangle$ and $F^2\dot{t} = \left\langle \dot{\gamma}, \frac{\partial}{\partial t} \right\rangle$, and both of these inner products are constant since γ is a geodesic and $\dfrac{\partial}{\partial\theta}$ and $\dfrac{\partial}{\partial t}$ are Killing vector fields. When γ is a causal geodesic (timelike or null), μ and ε can be interpreted as angular momentum and total energy, respectively, of the particle whose motion is described by γ.

3.49. We shall study the case when γ is a null geodesic in more detail. Then we have

$$0 = \langle\dot\gamma,\dot\gamma\rangle = F^2\dot t^2 - \dot\rho^2 - G^2\dot\theta^2$$

so

$$\dot\rho^2 = F^2\dot t^2 - G^2\dot\theta^2 = \frac{\varepsilon^2}{F^2} - \frac{\mu^2}{G^2} = \frac{\varepsilon^2}{1-\dfrac{2m}{G}} - \frac{\mu^2}{G^2}.$$

Since $\dot\rho^2$ is always nonnegative we must have

$$\left(1-\frac{2m}{G}\right)\Big/G^2 \leqslant \varepsilon^2/\mu^2$$

along each null geodesic. The function $\left(1-\dfrac{2m}{G}\right)\Big/G^2$ attains its maximum value $1/27m^2$ when $G = 3m$ so there are three possible cases.

3.50. *Case I*: $\mu^2 > 27m^2\varepsilon^2$. In this case the geodesic must stay away from the hypersurface $G = 3m$. The function G attains a minimum along each maximally extended null geodesic of this type which lies in the region $G > 3m$. These geodesics represent photons which reach a point of closest approach to the photon sphere and then escape to infinity.

3.51. *Case II*: $\mu^2 = 27m^2\varepsilon^2$. This case includes the null geodesics which circle around in the photon sphere (cf. 3.45), as well as null geodesics which spiral inward or outward, asymptotic to the photon sphere.

3.52. *Case III*: $\mu^2 < 27m^2\varepsilon^2$. In this case $\dfrac{d}{ds}(G\circ\rho)$ is bounded away from zero along γ, since

$$\left[\frac{d}{ds}(G\circ\rho)\right]^2 = [(G'\circ\rho)\dot\rho]^2 = \left(1-\frac{2m}{G\circ\rho}\right)\left(\frac{\varepsilon^2}{1-\dfrac{2m}{G\circ\rho}} - \frac{\mu^2}{(G\circ\rho)^2}\right)$$

$$= \varepsilon^2 - \left[\left(1-\frac{2m}{G\circ\rho}\right)\Big/(G\circ\rho)^2\right]\mu^2$$

$$\geqslant \varepsilon^2 - \frac{1}{27m^2}\mu^2 > 0,$$

and so G is monotone along γ. Along maximal null geodesics of this type, the function G approaches $2m$ as the affine parameter s approaches some finite value. These geodesics are incomplete. They describe photons travelling into, or out of, the star.

J. Bending of Light

3.53. Null geodesics with $\mu^2 > 27m^2\varepsilon^2$ (Type I) describe the paths of photons which pass by the star. To an observer viewing light whose path passes close to the star, it appears that the light path is "bent" by the gravitational field of the star (see Figure 3.53). We can estimate the deflection angle as follows.

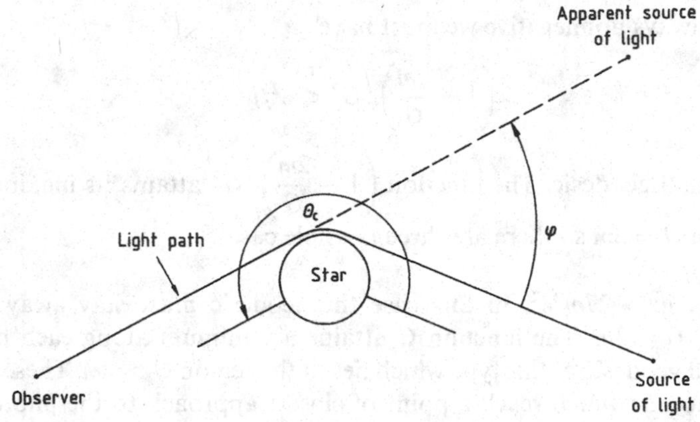

Fig. 3.53

3.54. Assume that both the observer and the light source are very far from the star, i.e. in the almost Minkowskian part of the Schwarzschild geometry. The deflection angle φ is the angle between the directions to the star as computed in the Schwarzschild geometry and in its asymptotic Minkowski space. It is approximately equal to $\theta_c - \pi$, where $\theta_c = \int_{-\infty}^{\infty} \dot{\theta}\, ds$ is the central angle subtended by the light path. To estimate θ_c we note that $\theta(s) = \int_{-\infty} \dot{\theta}\, ds$ is monotone, since $\dot{\theta} = \mu/G^2$ has constant sign, and hence θ may be used as parameter along γ. Then (see the computation in 3.52)

$$\left[\frac{d}{d\theta}(G \circ \rho)\right]^2 = \left[\frac{d}{ds}(G \circ \rho)/\dot{\theta}\right]^2$$

$$= \left\{\varepsilon^2 - \left[\left(1 - \frac{2m}{G \circ \rho}\right)\Big/ G \circ \rho\right)^2\right]\mu^2\right\}(G \circ \rho)^4/\mu^2.$$

If we let $v = m/G \circ \rho$, this equation becomes

$$\left(\frac{dv}{d\theta}\right)^2 = \left(\frac{m\varepsilon}{\mu}\right)^2 - v^2 + 2v^3.$$

Letting θ_0 denote the value of θ when $G \circ \rho$ is minimum (i.e., when v is maximum) and letting $v_0 = v(\theta_0)$, we have

$$0 = \left(\frac{m\varepsilon}{\mu}\right)^2 - v_0^2 + 2v_0^3$$

and hence $\left(\dfrac{dv}{d\theta}\right)^2$ can be rewritten as

$$\left(\frac{dv}{d\theta}\right)^2 = (v_0^2 - v^2) - 2(v_0^3 - v^3).$$

Thus

$$\begin{aligned}\theta_c &= \int_{-\infty}^{\infty} \dot{\theta}\, ds = 2\int_0^{v_0} \frac{dv}{dv/d\theta} \\ &= 2\int_0^{v_0} \frac{1}{(v_0^2 - v^2)^{1/2}}\left(1 - 2\frac{v_0^3 - v^3}{v_0^2 - v^2}\right)^{-1/2} dv \\ &> 2\int_0^{v_0} \frac{1}{(v_0^2 - v^2)^{1/2}}\left(1 + \frac{v_0^3 - v^3}{v_0^2 - v^2}\right) dv = \pi + 4v_0.\end{aligned}$$

We conclude that the deflection angle φ is at least $4v_0 = 4m/G_0$, where G_0 is the minimum value of G along γ. For light passing by the sun, this value is in good agreement with observations. We also mention that the travel time of such closely passing light rays (when measured by observers further away from the star) is measurably longer than a Newtonian explanation predicts (Shapiro effect). The light rays spend more coordinate time in the steeper light cones near the star than observers further away would expect (if they are not relativistically educated).

K. The Kruskal Extension

3.55. The Schwarzschild model can be extended so that causal geodesics which reach the boundary $G = 2m$ in finite proper time do not terminate there. The most popular way of doing this is to imbed the Schwarzschild model in $U \times S^2$, where U is an appropriate open set in $\mathbb{R}^2$, with a metric of the form

$$g = H^2(u, v)(dv^2 - du^2) - J^2(u, v)\, d\sigma^2.$$

Here, as above, $d\sigma^2$ is the standard metric on S^2.

3.56. If we define J implicitly from the equation

$$\left(\frac{J}{2m} - 1\right) e^{J/2m} = u^2 - v^2,$$

where $m \in \mathbb{R}$, and set

$$H = (32m^3/J)e^{-J/2m},$$

then the above formula for g defines a Ricci flat metric on $U \times S^2$, where

$$U = \{(u, v) \in \mathbb{R}^2 | v^2 - u^2 < 1\}.$$

This spacetime is known as the *Kruskal model*. The map $f \times$ identity: $\mathbb{R} \times I \times S^2 \to U \times S^2$, where $f: \mathbb{R} \times I \to U$ is defined by the equations

$$u = \left(\frac{G(\rho)}{2m} - 1\right)^{1/2} e^{-G(\rho)/4m} \cosh(t/4m)$$

$$v = \left(\frac{G(\rho)}{2m} - 1\right)^{1/2} e^{-G(\rho)/4m} \sinh(t/4m),$$

is an isometric imbedding of the Schwarzschild model into the Kruskal model. (How to find such extensions is explained in [Stp]). Notice that under this imbedding, the function J on $U \times S^2$ pulls back to the function G on $\mathbb{R} \times I \times S^2$.

This imbedding maps the Schwarzschild model onto $U_1 \times S^2$, where $U_1 = \{(u, v) \in U | u > |v|\}$. Typical level curves of the Schwarzschild coordinate functions t and G are shown in Figure 3.56.

3.57. The region $U_2 \times S^2$, where $U_2 = \{(u, v) \in U | v > |u|\}$, is the Schwarzschild *black hole*. If a star has "shrunk" to the point where its "radius" J_0 is less than $2m$ then the Kruskal metric, in the region $J_0 < J < 2m$, is interpreted as describing the gravitational field between the star and the *horizon* $J = 2m$. Notice that, since all

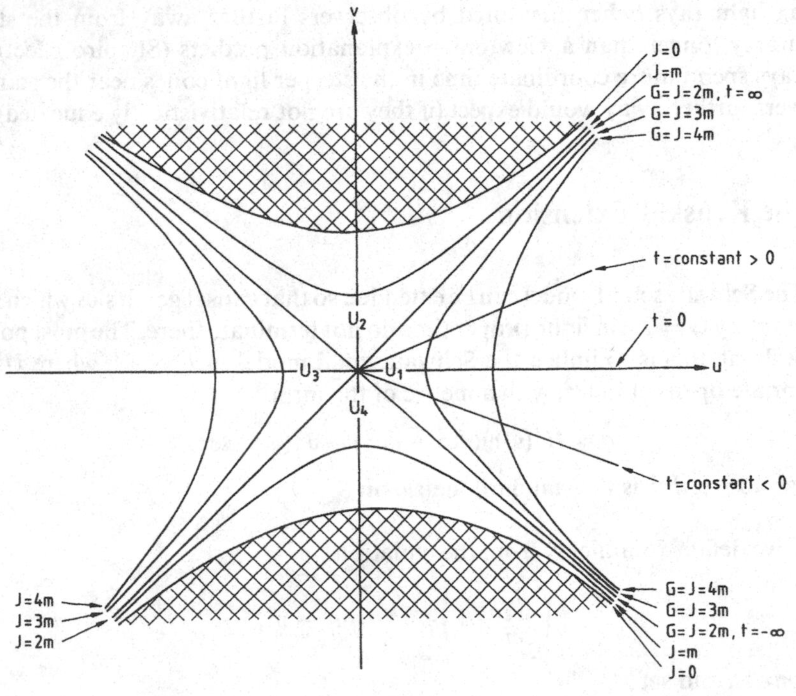

Fig. 3.56

the light cones in the figure are of slope ± 1, all causal curves along which the Kruskal time v is increasing and which enter the black hole must forever remain there, and in fact must terminate at $J = 0$. In particular, any star which collapses inside its Schwarzschild radius $J = 2m$ is doomed to continue collapsing to $J = 0$. The proper time required for geodesics to arrive at $J = 0$ is finite so the Kruskal model is geodesically incomplete. This model, however, cannot be extended further because, for example, $\|R\|^2 = \operatorname{trace} \hat{R} \circ \hat{R}$ is unbounded near $J = 0$.

3.58. The region $U_3 \times S^2$ is isometric to the Schwarzschild model $U_1 \times S^2$ under the isometry $(u, v, p) \to (-u, v, p)$ of $U \times S^2$. The region $U_4 \times S^2$ is isometric to $U_2 \times S^2$ under the (time-reversing) isometry $(u, v, p) \to (u, -v, p)$ of $U \times S^2$. $U_4 \times S^2$ is interpreted as a "white hole": causal curves which leave the region $U_4 \times S^2$ can never return to it.

3.59 Remark. It can be shown (Birkhoff's theorem) that the Kruskal model contains, at least locally, every spherically symmetric solution of Einstein's empty space equation $r = 0$: about each point in every such solution there is a neighborhood which can be mapped isometrically into $U \times S^2$ with the Kruskal metric, for some choice of m. For a proof, see [Ha-El].

L. How Completeness May Fail

3.60. Lorentz manifolds which allow as much physical interpretation as the Schwarzschild geometry are rare in the literature. The vacuum solutions of the Einstein equations (i.e. the Ricci flat Lorentz manifolds) which were found by physicists can however be used to demonstrate how far apart in geometric behaviour Riemannian and Lorentz manifolds are. The Taub-NUT metric, found 1951 by Taub and extended 1963 by Newman, Unti and Tamburino, is a case in point. It is a Ricci flat metric on $\mathbb{R} \times S^3$. The isometry group $U(2)$ has spacelike S^3-orbits. The induced metrics on these spheres are Berger metrics, i.e. they are up to a constant factor isometric to distance spheres in $\mathbb{C}P^2$ or its dual. The Ricci flat metric is obtained by a canonical variation of the standard metric on the fibre bundle $\mathbb{R} \times S^3 \to S^2$ ($S^3 \to S^2$ being the Hopf fibration) as explained in 9.67. The strange behaviour of the geodesics in Taub NUT space occurs mainly in the 2-dimensional totally geodesic fibres. We therefore refer the reader for more details about the proper Taub-NUT space to [Ha-El] pp. 174–178, and present, also following [Ha-El], a simpler flat 2-dimensional example with similar geodesic behaviour which was found in 1967 by Misner. (The fibre metric of Taub-NUT space is not flat).

Consider the discrete isometry group G of the Lorentz plane generated by

$$\begin{pmatrix} \tilde{x} \\ \tilde{t} \end{pmatrix} \to \begin{pmatrix} \cosh \pi & \sinh \pi \\ \sinh \pi & \cosh \pi \end{pmatrix} \cdot \begin{pmatrix} \tilde{x} \\ \tilde{t} \end{pmatrix}, \qquad d\tilde{s} = d\tilde{x} - d\tilde{t}^2.$$

It acts properly discontinuously on the (homogenous but incomplete, see 7.113) half spaces $\tilde{x} + \tilde{t} > 0$ and $-\tilde{x} + \tilde{t} > 0$, but not on the whole Lorentz plane. The quotient is in each case $S^1 \times \mathbb{R}$. After the transformation (valid only on the intersection of the two half planes)

$$t := 1/4(\tilde{t}^2 - \tilde{x}^2), \qquad \psi := 2\,\mathrm{arc\,tanh}(\tilde{x}/\tilde{t})$$

the metric is $ds^2 = -\frac{1}{t}dt^2 + t\,d\psi^2$ on $S^1 \times \mathbb{R}_+$.

The transformations $\qquad \psi_\pm = \psi \mp \log t$

change the metric to

$$ds_+^2 = 2\,d\psi_+\,dt + t\cdot(d\psi_+)^2, \quad \text{valid on } S^1 \times \mathbb{R}$$
$$ds_-^2 = -2\,d\psi_-\,dt + t\cdot(d\psi_-)^2, \quad \text{valid on } S^1 \times \mathbb{R}.$$

The first cylinder then is isometric to the quotient of the half plane $\tilde{x} + \tilde{t} > 0$, the second is isometric to the quotient of the other half plane. Therefore we have two analytic extensions of the Lorentz manifold $(S^1 \times \mathbb{R}_+, ds^2)$ to the Lorentz manifolds $(S^1 \times \mathbb{R}, ds_+^2)$, $(S^1 \times \mathbb{R}, ds_-^2)$, and there is no Hausdorff Lorentz manifold into which both of these extensions can be simultaneously made. Also both these extensions are null geodesically incomplete—although one will not immediately realize this by looking at a picture (Figure 3.60). One family of null geodesics consists of the cylinder generators $\psi_\pm =$ constant with t as affine parameter. They are complete. The circle $t = 0$ is seemingly a closed null geodesic. The other null geodesics spiral around the cylinder, complete as $|t| \to \infty$. In the other direction they wind around infinitely often, seemingly asymptotic to the limit geodesic $t = 0$. However if one checks the parametrization things become very different to what the Riemann trained eye just read out of the picture.

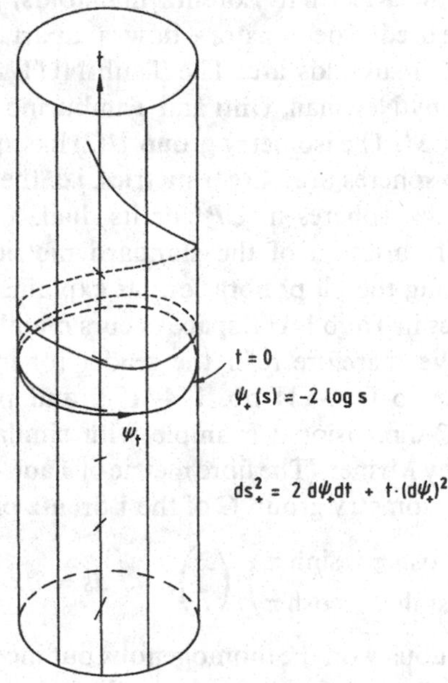

Fig. 3.60. Flat Incomplete Lorentz Manifold (Misner 1967)

For example

$$\psi_+(s) = -2\log s, \qquad t = 0$$

is an affine parametrization of the "closed" geodesic $t = 0$, obviously incomplete as $s \to 0$. For each new turn the affine parameter decreases by less and the infinitely many turns around the cylinder are accomplished with finite affine parameter on the null geodesic. It is true that this incompleteness can more easily be spotted in the above description as a half plane quotient, but probably the reader will agree that a Riemannian geometer would not really expect the described phenomena when first looking at the simple flat metric without singularities:

$$2\,d\psi_+\,dt + t(d\psi_+)^2 \quad \text{on } S_1 \times \mathbb{R}.$$

M. Singularity Theorems

3.61. The Kruskal extension of the Schwarzschild model was discovered in 1960. It sparked a period of intense research in general relativity in which physicists began to use global differential geometric techniques to study spacetime singularities. It was found that, under certain reasonable physical assumptions, singularities in gravitational models are to be expected. We close this chapter by stating one of the most famous singularity theorems, proved by Penrose in 1965.

Theorem. *Let M be any spacetime with the following properties*:

(i) *the Ricci tensor r is everywhere positive semi-definite,*

(ii) *every maximal causal geodesic in M encounters at some point some nonzero tidal stress,*

(iii) *M contains no closed timelike curves,* and

(iv) *M contains a closed trapped surface; that is, a compact spacelike 2-surface with the property that both families of null geodesics normal to S are converging along S. Then M must be geodesically incomplete.*

Condition (i) says essentially that energy density should be everywhere non-negative (cf. 3.12). Condition (iv) is a statement that gravitational collapse is occurring somewhere in M. In the Kruskal model, the 2-sphere $J^{-1}(a) \cap u^{-1}(b)$ is a closed trapped surface, for each $a < 2m$ and $b \in \mathbb{R}$.

A proof of this theorem and an extensive discussion of related theorems can be found in [Ha-El].

3.62 Remark. We have confined our attention in this chapter almost exclusively to vacuum models of spacetime. We have done this because (i) the vacuum metrics are Einstein metrics and this book is about Einstein spaces, and (ii) it seemed preferable to discuss one important example in some detail rather than to discuss several superficially. But there is a vast literature on general relativity, much of which contains information that Riemannian geometers would find fascinating. The books [Be-Eh], [Ha-El], [Mi-Th-Wh], [ONe 3], [Pen 2], [Sa-Wu], [Stp], and [Web] are especially recommended.

Chapter 4. Riemannian Functionals

A. Introduction

4.1. Since the work of Lagrange, it is well known that the equations of classical mechanics can be obtained as solutions of a variational problem, using a suitable functional, called the action on the configuration space (cf. [Ab-Ma], [Arn]). Hilbert proved ([Hil]) that the equations of general relativity can be recovered from the action $g \mapsto \int_M s_g \mu_g$ (total scalar curvature). His paper contains prophetic ideas about the role played by the diffeomorphism group, which he already considered as a "gauge group".

Then a natural way to prove the existence of Einstein metrics appears: one can try to show that the total scalar curvature—suitably normalized, cf. 4.16—has critical points. Alas, it turns out to be somewhat naive.

The use of direct variational methods, often so successful in non-linear analysis (cf. [Nir 4] for instance), becomes here quite difficult. Indeed, since there exist compact manifolds which do not carry any Einstein metric (e.g. $S^1 \times S^2$ and $S^1 \times S^3$, see Chapter 6), analysis alone will not give any result in dimensions 3 and 4. Although such examples are not known in dimension greater than 4, a reasonable guess is the following: the proof of the existence of an Einstein metric by a direct variational method should require some geometric assumption on the manifold. Hamilton's theorem (cf. [Ham 2] and Chapter 5) provides such a result, but it is special to dimension 3.

When restricting the total scalar curvature to a pointwise conformal class of metrics, a larger critical set is obtained, i.e. metrics in that class with constant scalar curvature (cf. 4.25). The existence of such critical points is the famous Yamabe problem. It is generically solved in dimension $\geqslant 6$, but not completely.[1])The main difficulty comes from the failure of the Palais-Smale condition (cf. [Nir 4], II.1)

Paragraph *B* is concerned with basic but crucial properties of Riemannian functionals. Most of them are consequences of the invariance under the diffeomorphism group. Hilbert's theorem (its Riemannian version in fact, which amounts to the same) is proved in *C*, and the Yamabe problem is discussed in *D*.

Combining the results of Yamabe and T. Aubin with those of Kazdan-Warner on the scalar curvature, and also with topological obstructions of Chapter 6, it is

---

[1]) A complete solution has been announced by R. Schoen.

possible to completely describe the functions—including constants!—which are the scalar curvature of some metric on a given compact manifold. This is done in *D* and *E*. We gave more details than elsewhere in the book: these results are indeed a nice piece of analysis for geometers, without being for them as painful as Yamabe's or Calabi's conjecture!

Paragraph F is devoted to the study of the set $\mathfrak{S}$ of metrics of prescribed volume and constant scalar curvature. Under a generic condition, this space is an infinite dimensional manifold, and the total scalar curvature restricted to this manifold has no critical points other than the Einstein metrics. Furthermore, $\mathfrak{S}$ permits a geometric interpretation of the second derivative of the total scalar curvature. This is done in *G*.

In H, we give a summary about quadratic functionals. The reader can consult [Ber 4], [Mut 1, 2] and [Pat] for more details. The results are rather disappointing, but the questions raised seem to us of interest, and not only because they are far from being settled.

B. Basic Properties of Riemannian Functionals

4.2. Let $\mathscr{M}_M$, or just $\mathscr{M}$ denote the set of Riemannian metrics on a compact manifold M. In the compact open C^∞ topology $\mathscr{M}$ is an open convex cone. If $\mathfrak{D}$ denotes the diffeomorphism group of M, there is a natural right action of $\mathbb{R}^* \times \mathfrak{D}$ on $\mathscr{M}$, given by

$$(t, \varphi)\cdot g = t^2 \varphi^* g$$

Clearly, two metrics in the same orbit have the same geometric properties. Some people may feel that the "volume" or the "length of a curve" is a geometric data, but these are changed by the scale factor t. Isometry classes of metrics are described by the quotient $\mathscr{M}/\mathfrak{D}$, which is often called the space of Riemannian *structures*. Classes of (globally) homothetic metrics are given by $\mathscr{M}_1/\mathfrak{D}$, where

(4.3) $$\mathscr{M}_1 = \left\{ g \in \mathscr{M}, \int_M \mu_g = 1 \right\}$$

The main property of the $\mathfrak{D}$-action is the slice theorem of D. Ebin (cf. [Ebi 2] and 12.22). Here, we will only use its infinitesimal version. It roughly says that, from an infinitesimal point of view, $\mathscr{M}/\mathfrak{D}$ and $\mathscr{M}_1/\mathfrak{D}$ actually behave like manifolds. The tangent space to the $\mathfrak{D}$-orbit of a given metric g is just the image of the first-order differential operator

(4.4) $$\delta^*\colon \Omega^1 M \mapsto \mathscr{S}^2 M, \quad \text{defined by}$$

$$\delta^* \xi = \tfrac{1}{2} L_{\xi^\sharp} g \quad \text{(cf. 1.59)}$$

Since the symbol $\sigma_\xi(\delta^*)(x) = t \circ x$ is injective, from Appendix 32 the space $T_g\mathscr{M} = \mathscr{S}^2 M$ admits the following direct sum decomposition, which is orthogonal for the global scalar product

(4.5) $$\mathscr{S}^2 M = \operatorname{Im} \delta_g^* \oplus \operatorname{Ker} \delta_g.$$

Here, the divergence δ_g is the formal adjoint of δ_g^* (see 1.59). Clearly, $T_g\mathscr{M}_1 = \{h \in \mathscr{S}^2 M; \langle h, g\rangle_g = 0\}$ contains $\operatorname{Im}\delta_g^*$, so that the decomposition (4.5) can be refined into

(4.5′) $$T_g\mathscr{M}_1 = \operatorname{Im}\delta_g^* \oplus (\operatorname{Ker}\delta_g \cap T_g\mathscr{M}_1).$$

Here, we have normalized by prescribing the total volume of M to be 1. A further normalization is possible: J. Moser has proved that if two positive C^∞ densities on a compact manifold M have the same integral, they are conjugated by a diffeomorphism of M. So, if μ is a positive density on M with total volume $\int_M \mu = 1$, any metric in $\mathscr{M}_1$ is $\mathfrak{D}$-equivalent to a metric of the subset

(4.6) $$\mathscr{N}_\mu = \{g \in \mathscr{M}; \mu_g = \mu\} \text{ of } \mathscr{M}_1.$$

(cf. [Msr] or [Ham 1], p. 203).

4.7 Definition. A real valued function $\mathbf{F}$ on $\mathscr{M}$, such that $\mathbf{F}(\varphi^* g) = \mathbf{F}(g)$ for every diffeomorphism φ and every g in $\mathscr{M}$ is called a *Riemannian functional.*

4.8 Comments. i). Since φ is by definition an isometry between (M, g) and $(M, \varphi^* g)$, this means that the function F only depends on Riemannian geometric data, and can be viewed as a function on the quotient space $\mathscr{M}/\mathfrak{D}$.

ii) To put some emphasis, we give a trivial counter-example. Fix a point p in M, and take the evaluation of the scalar curvature at p, i.e. the map $g \mapsto s_g(p)$. Indeed, the data of p is not Riemannian.

iii) In [Pal 3], R.S. Palais studies *natural Lagrangians* for Riemannian metrics. They are defined as maps from $\mathscr{M}_M$ to $\mathscr{C}M$, which are $\mathfrak{D}$-invariant and only depend on some k-th order jet of the metric. All the functionals we are going to study are of the type $g \mapsto \int_M L(g)\mu_g$, where L is a natural Lagrangian.

4.9 Definition. A Riemannian functional $\mathbf{F}$ is *differentiable* (resp. differentiable at order r) if there is an integer k such that $\mathbf{F}$ is differentiable (resp. differentiable at order r) on $\mathscr{M}$ equipped with some L_k^2-norm (recall that all of them are equivalent).

Twice in this chapter (4.36 and 4.44), we will use the inverse function theorem (in the Banach or ILH category). So we cannot content ourselves with the notion of directional derivative. However, in all the cases we will consider, differentiability is easy to check: the functional $g \mapsto \int_M P(R_g)v_g$, where P is a polynomial, is of course differentiable. Furthermore, using integrations by parts, it is easy to see in this case that $\left.\frac{d}{dt}\right|_{t=0} \mathbf{F}(g + th) = \mathbf{F}'_g \cdot h = \langle a, h\rangle_g$, where the symmetric two-tensor a is a polynomial with respect to R_g and its first two covariant derivatives (see explicit examples below). This suggests the following definition

4.10 Definition. A differentiable Riemannian functional **F** has a *gradient* at g if there exists a in $\mathscr{S}^2M$ such that, for every h in $\mathscr{S}^2M$,

$$\mathbf{F}'_g \cdot h = \langle a, h\rangle_g.$$

Denote the gradient by grad **F**. If **F** has a gradient everywhere, the map $g \mapsto \operatorname{grad} \mathbf{F}_g$ defines a vector field on $\mathscr{M}$. This vector field satisfies an integrability condition (cf. [Ble]), which in fact goes back to Hilbert (cf. [Hil], p. 403).

4.11 Proposition. *If* $\operatorname{grad} \mathbf{F}_g$ *is the gradient at* g *of a Riemannian functional, then*

$$\delta_g(\operatorname{grad} \mathbf{F}_g) = 0.$$

Proof. The level surface of **F** through g contains the orbit of g under the diffeomorphism group, so the gradient of **F** is L^2-orthogonal to the tangent orbit space, i.e. to $\operatorname{Im} \delta_g^*$. □

4.12 Remark. A similar formula holds if we take L_r^2 scalar products and use the associated gradient and divergence, but it will not provide further information.

The invariance under the diffeomorphism group also has consequences for metrics which admit sufficiently many isometries. The following proposition gives a conceptual proof of former results about special functionals (cf. [Mut 2], [Pat]).

4.13 Proposition (D. Bleecker, [Ble]). *If* (M, g) *is an isotropy irreducible homogeneous space with its canonical Riemannian metric* (cf. 7.39), *then for every Riemannian functional* **F** *admitting a gradient, there exists a constant* λ *such that* $\operatorname{grad} \mathbf{F}_g = \lambda g$.

Proof. We have $\varphi^*(\operatorname{grad} \mathbf{F}_g) = \operatorname{grad} \mathbf{F}_g$ for any isometry φ. Specializing to isometries lying in the isotropy group at a point p, and using the irreductibility of the isotropy representation, we get $\operatorname{grad} \mathbf{F}_g(p) = \lambda(p) g(p)$. Now homogeneity implies that $\lambda(p)$ is a constant. □

4.14 Corollary. *For any Riemannian functional restricted to* $\mathscr{M}_1$, *isotropy irreducible homogeneous metrics are critical points.*

4.15 Remark. Among spaces to which the proposition and the corollary apply, one finds the irreducible Riemannian symmetric spaces (cf. Chapter 7).

C. The Total Scalar Curvature: First Order Properties

4.16 Definition. The *total scalar curvature* of a compact Riemannian manifold (M, g) is $\mathbf{S}(g) = \int_M s_g \mu_g$.

The functional $g \mapsto \mathbf{S}(g)$ is homogeneous of degree $n/2$. So we will either *restrict* it to $\mathscr{M}_1$, or *normalize* it by considering $\mathbf{S}/(\mathrm{vol})^{(n-2)/n}$, which is homogeneous.

From now on, in the body of the proofs, the index g will be omitted if there is no ambiguity.

4.17 Proposition. *For every compact Riemannian manifold* (M, g), *the functional* $\mathbf{S}$ *is differentiable and*

$$\mathbf{S}'_g \cdot h = \langle (s_g/2)g - r_g, h\rangle_g$$

therefore grad $\mathbf{S}_g = (s_g/2)g - r_g$.

Proof. Recall that $s'_g \cdot h = \Delta(\operatorname{tr}(h)) + \delta(\delta h) - (h, r)$ (cf. (1.174e)), so that $\mathbf{S}'_g \cdot h = \int_M ((s'_g \cdot h) + (s_g/2)\operatorname{tr}(h))\mu_g = \langle (s_g/2)g - r_g, h\rangle$, using Stokes' formula. □

4.18 Remark. For $n = 2$, 4.16 gives $\mathbf{S}'_g = 0$. Therefore $\mathbf{S}(g)$ does not depend on the metric and is a differential invariant. This is not surprising, since $(1/4\pi)s_g\mu_g$ is simply the Gauss-Bonnet integrand in dimension 2.

4.19 Corollary. i) *For every Riemannian manifold* (M, g), $\delta_g r_g + (1/2)\,ds_g = 0$;

ii) *if* dim $M > 2$ *and if there is a function* f *such that* $r_g = fg$, *then* f *is a constant and* (M, g) *is Einstein.*

Proof. First recall that $\delta(fg) = -df$ (cf. 1.59). In the compact case, i) follows from 4.11 applied to 4.17. For the general case, note that the formula is local since we can vary g only on a compact set. Now, if $r = fg$, taking the divergence δ and the trace of both sides, we get $(1/2)\,ds = df$ and $s = nf$. Then $ds = df = 0$. □

4.20 Remarks. i) Formula 4.19 is usually proved by using the differential Bianchi identity (cf. 1.94). In fact, both Bianchi identities can be viewed as consequences of the invariance under the diffeomorphism group. This approach, already implicit in Hilbert's paper [Hil], has been developed by J. Kazdan ([Kaz 2]).

ii) The condition to be divergence free is viewed by physicists as a conservation principle. This explains the role played by $r_g - (s_g/2)g$ in the Einstein field equation (see 3.7 and [Ha-El]).

We come to the result of Hilbert.

From now on, we assume dim $M > 2$.

4.21 Theorem. *For a compact Riemannian manifold* (M, g) *of volume* 1, *the following properties are equivalent*:

i) (M, g) *is Einstein;*

ii) g *is a critical point of* $\mathbf{S}/\mathrm{vol}^{\frac{n-2}{n}}$;

iii) g *is a critical point of* $\mathbf{S}$ *restricted to* $\mathscr{M}_1$;

iv) *a volume element* μ *being given,* g *is a critical point of* $\mathbf{S}$ *restricted to the set* $\mathscr{N}_\mu$ *of metrics for which* $\mu_g = \mu$.

Proof. It is clear that ii) and iii) are equivalent, because $\mathbf{S}$ is homogeneous of degree $n/2 - 1$. Recall that

$$T_g\mathcal{M}_1 = \{h \in \mathcal{S}^2M), \langle h, g\rangle = 0\} \text{ and}$$
$$T_g\mathcal{N}_\mu = \{h \in \mathcal{S}^2M), \operatorname{tr}_g h = 0\}$$

Then g is a critical point of $\mathbf{S} \restriction \mathcal{M}_1$ (resp. $\mathbf{S} \restriction \mathcal{N}_\mu$) if and only if the orthogonal projection of $\operatorname{grad} \mathbf{S}_g = (s/2)g - r$ onto $T_g\mathcal{M}_1$ (resp. $T_g\mathcal{N}_\mu$) is zero. In both cases, this means that there is a function f such that $r = fg$, and the result follows from 4.18. □

4.22 Remark. Observe that $\operatorname{grad} \mathbf{S}_g \restriction \mathcal{N}_\mu = (s/n)g - r = z$.

Theorem 4.21 also admits the following finite dimensional version. Let G a compact Lie group acting transitively on M, and denote by $\mathcal{M}_1^G$ the set of G-invariant Riemannian metrics with total volume 1. Then $\mathcal{M}_1^G$ is a finite dimensional manifold. For any g in $\mathcal{M}_1^G$, the scalar curvature s_g is constant and equal to $\mathbf{S}_g$.

4.23 Theorem (Palais, [Pal 3] and [Pal 12]). *A metric g in $\mathcal{M}_1^G$ is an Einstein metric if and only if it is a critical point of $\mathbf{S} \restriction \mathcal{M}_1^G$.*

Proof. This follows by a direct application of the symmetric criticality principle (cf. [Pal 2] and [Pal 4]). We have a manifold V (here $\mathcal{M}_1$), a compact Lie group G acting G-invariant on V, and a smooth function F (here $\mathbf{S}$) on V. Let $V^G = \{v \in V, \gamma \cdot v = v$ for any γ in $G\}$. This principle says that a critical point of $F \restriction V^G$ is a critical point of F. See [Pal 2] and [Pal 4] for a complete discussion in a general setting. □

4.24 Remark. In [Jen 2], G. Jensen proved 4.23 by direct computation for the case $M = G$, and discussed the existence of non trivial critical points. For further results, see also Chapter 9 and D'Atri-Ziller ([DA-Zi]).

One can ask whether some other geometric properties admit such a variational characterisation. The following proposition is a starting point to attack the famous Yamabe problem: given a metric g on a compact manifold, does there exist a constant scalar curvature metric that is pointwise conformal to g?

4.25 Proposition. *For a compact Riemannian manifold (M, g) the following statements are equivalent:*

i) *the scalar curvature is constant;*

ii) *g is a critical point of $\mathbf{S}$ restricted to the set $\operatorname{Conf}_0(g)$ of metrics pointwise conformal to g and having the same total volume.*

Proof. ii) is true if and only if $\mathbf{S}'_g \cdot fg = 0$ for every f whose integral is zero. But this means that s is orthogonal to every such f. Now, the function $f = s_g - (1/(\operatorname{vol}(g)) \int_M s_g \mu_g$ is orthogonal to s and to the constants, therefore to itself, thus s is constant. □

D. Existence of Metrics with Constant Scalar Curvature

4.26. The preceding proposition makes it natural to get such metrics by minimizing the total scalar curvature in a conformal class. Then it will be more convenient to view **S** as defined on a space of functions. Taking the normalization into account, we set, for any positive C^∞ function f,

$$Y_g(f) = \mathbf{S}(fg)/(\mathrm{vol}(M, fg))^{(n-2)/n} \tag{4.27}$$

Setting $f = f_1^{4/(n-2)}$, recall that (cf. 1.161)

$$s_{fg} = f_1^{-(n+2)/(n-2)}\left(4\frac{n-1}{n-2}\Delta_g f_1 + s_g f_1\right).$$

An alternative expression for Y_g is

$$Y_g(f) = \frac{\displaystyle\int_M \left(4\frac{n-1}{n-2}|df_1|_g^2 + s_g f_1^2\right)\mu_g}{\left(\displaystyle\int_M f_1^{2n/(n-2)}\mu_g\right)^{(n-2)/n}}.$$

4.28 Definition. For a metric g on M, *the Yamabe invariant* $\mu(g)$ is the infimum of the total scalar curvature, normalized with respect to the total volume, in the conformal class of g.

The formula above shows that

$$\mu(g) = \inf_{f\in\mathscr{C}M,\, f>0} \frac{\displaystyle\int_M \left(4\frac{n-1}{n-2}|df|_g^2 + s_g f^2\right)\mu_g}{\left(\displaystyle\int_M f^{2n/(n-2)}\mu_g\right)^{(n-2)/n}}.$$

The corresponding Euler-Lagrange equation is

$$4\frac{n-1}{n-2}\Delta_g f + s_g f = \mu(g) f^{(n+2)/(n-2)}.$$

Its non-trivial solutions provide metrics with constant scalar curvature (for $g_1 = f^{4/(n-2)}g$). This fact was already known to us from 4.25. Now, as a consequence of successive work by Yamabe ([Yam 2]), N.S. Trudinger ([Tru 1]) and T. Aubin ([Aub 2], [Aub 3], see also the book [Aub 6]), the infimum $\mu(g)$ is known to be achieved under fairly general assumptions.

4.30 Theorem (cf. [Aub 2]). i) *For any compact Riemannian manifold* (M, g), $\mu(g) \leqslant \mu(S^n, \mathrm{can})$. *If* $\mu(g) < \mu(S^n, \mathrm{can})$ *in the conformal class of g there exists a metric with constant scalar curvature* $\mu(g)$ *and total volume* 1.

ii) *A sufficient condition for having* $\mu(g) < \mu(S^n, \mathrm{can})$ *is that* $n \geqslant 6$ *and that* (M, g) *is not conformally flat.*

We refer to [Aub 2] for the proof. The main point is i). From the Sobolev inclusion lemma (cf. App. 8), there are continuous injections $L_1^2 \hookrightarrow L^q \left(2 \leqslant q \leqslant \frac{2n}{n-2}\right)$, which are compact if $q < \frac{2n}{n-2}$. One approach is, using compactness, to solve first an auxiliary variational problem, where in 4.29 $\frac{2n}{n-2}$ is replaced by a smaller exponent q, and then to prove suitable convergence properties when q converges to $\frac{2n}{n-2}$. For this last point, the determination of the best constant in the Sobolev inequality ([Aub 3]) is needed.

The theorem above does not give a complete answer to the Yamabe conjecture.[1]) However, it permits us to prove the existence, on a given compact manifold, of "many" metrics with constant scalar curvature. The following property will be useful.

4.31 Proposition (L. Bérard Bergery, [BéBer 4]). *If the space of metrics is equipped with the C^2-topology, the map $g \mapsto \mu(g)$ is continuous.*

Proof (sketch). Clearly, μ is upper-semicontinuous. For the lower-continuity, we use the following trick: using Moser's lemma (in fact a version with parameters as in [Ebi 1], p. 38), we may restrict μ to the space of metrics which define the same volume element. We leave the details (which of course involve Hölder's inequality) to the reader. □

4.32 Theorem. i) *Every compact connected manifold M carries a metric with constant negative scalar curvature.*

ii) *If M carries a metric g such that $s_g \geqslant 0$ and $s_g \not\equiv 0$, then there exists on M a metric with constant positive scalar curvature and a metric with zero scalar curvature.*

Proof. i) follows from 4.31, if we show that there exists a metric whose total scalar curvature is strictly negative. Such an existence theorem has been obtained by A. Avez and T. Aubin (cf. [Avz 1], [Aub 1], [Eli], however, there is a simpler argument due to L. Bérard Bergery (cf. [BéBer 4]).

We choose an open disk D^n in M. Let $S^p \times D^q$ be inside D^n such that $p + q = n$, $p \geqslant 1$ and $q \geqslant 2$. Take a positive function f on D^q, depending only on the distance to the origin and identically 1 near the boundary of D^q. If g_2 and g_3 denote the canonical metrics on S^p and D^q, the formula

$$g_1 = f^{-(2p/n-1)}(f^2 g_2 + g_3)$$

defines a metric on $S^p \times D^q$ which extends to a metric g on M. Now

$$\int_M s_g \mu_g = \int_{M - S^p \times D^q} s_g \mu_g + \int_{S^p \times D^q} s_g \mu_g .$$

[1]) R. Schoen has announced a complete affirmative answer to the Yamabe conjecture, cf. [Sce]. On the other hand, the Yamabe problem has no solution on many complete non-compact manifolds, see [Jin].

The first term does not depend on f, and for the second one, we have, by a direct computation,

$$\int_{S^p\times D^q} s_{g_1}\mu_{g_1} = \mathrm{Vol}(S^p, g_2)\int_{D^q} f^{(p/(n-1)-2)}\left(s_{g_2} - \frac{p(n-1-p)}{n-1}|df|^2\right)\mu_{g_3}.$$

Since $p \leqslant n-2$, the coefficient of $|df|^2$ is negative. Now we may choose f such that it remains bounded (say between 1 and 2) and $\int |df|^2 \mu_{g_3}$ becomes as large as we want in order to turn $\mathrm{S}(g) = \int_M s_g \mu_g$ negative.

ii) Let g be a metric such that $s_g \geqslant 0$. Then the elliptic self-adjoint operator $4\frac{n-1}{n-2}\Delta_g f + s_g f$ is positive. Further, it is strictly positive if $s_g \not\equiv 0$. Its first eigenvalue, *denoted* by $\lambda(g)$, is then strictly positive (cf. Appendix 37). Therefore, for any f in $\mathscr{C}M$,

$$\int_M \left(4\frac{n-1}{n-2}|df|_g^2 + s_g f^2\right) v_g \geqslant \lambda(g)\|f\|_2^2.$$

Using Hölder inequality, this proves that $\mu(g) > 0$.

Now, using i) and the continuity of μ, we can find a metric g_1 such that $\mu(g_1) = 0$, and, applying Aubin's theorem 4.30 to g_1, we obtain a metric with zero scalar curvature. If $\mu(g) < \mu(S^n, \mathrm{can})$ the same argument gives also a metric with positive constant scalar curvature; and if ever $\mu(g) = \mu(S^n, \mathrm{can})$, using a continuity argument again, we can diminish $\mu(g)$! □

4.33 Remark. In their works about functions which can be realized as scalar curvatures within a given conformal class, J. Kazdan and F. Warner use the functional $g \to \lambda(g)$ in the very same way as we used μ in the proof of Theorem 4.32. above (cf. [Ka-Wa 4] §3). Further, since the first eigenvalue of a positive elliptic operator is simple, standard perturbation theory (cf. [Kat]) shows that $g \to \lambda(g)$ is differentiable with respect to g. Then Kazdan and Warner prove that the critical points of λ restricted to $\mathscr{M}_1$ are indeed Einstein metrics. The reader may check this using Formula (1.184).

E. The Image of the Scalar Curvature Map

4.34. We are now in a position to answer the following question: given a compact manifold, which functions are scalar curvatures of Riemannian metrics? We note that

i) there exist manifolds which do not carry any metric of positive scalar curvature, e.g., K3 surfaces (Lichnerowicz, [Lic 4]) and tori (Gromov-Lawson, [Gr-La 1], or Schoen-Yau, [Sc-Ya]). Furthermore, there exist manifolds which do not carry any metric with non negative scalar curvature either (cf. [Ka-Wa 4]); the simplest example is provided by connected sums of tori, cf. [Gr-La 1] and [Sc-Ya] again.

ii) just above, we have given some existence theorems for metrics with constant scalar curvature.

Now, results of J. Kazdan and F. Warner essentially say that if we know the constant scalar curvature functions, we know all the functions which are scalar curvatures for some metric. Combining i) and ii), they have the following classification theorem (cf. [Ka-Wa 3]), with an improvement by L. Bérard Bergery (cf. [BéBer 2]).

4.35 Theorem. *Compact manifolds of dimension $n \geqslant 3$ can be divided into three classes:*

(A) *Any (C^∞) function on M is the scalar curvature of some (C^∞) metric;*

(B) *A function on M is the scalar curvature of some metric if and only if it is either identically zero or strictly negative somewhere; furthermore, any metric with vanishing scalar curvature is Ricci-flat;*

(C) *A function on M is a scalar curvature if and only if it is strictly negative somewhere.*

Proof. It consists in first proving the local surjectivity of the map $g \mapsto s_g$ (4.36), and then in using, in a suitable way, the trivial fact that if f belongs to the image of $g \mapsto s_g$, so does $f \circ \varphi$ for any diffeomorphism φ.

From the formula for s_g in local coordinates, we see that the scalar curvature map $g \mapsto s_g$ defines a quasi-linear differential operator of second order. The Sobolev embedding theorem (cf. Appendix 8) shows that, for $p > \dim M$, this operator can be extended to a map from metrics of the class L_2^p into $L^p(M)$. In this context, we can use the implicit function theorem for Banach spaces.

4.36 Proposition. *If, at a metric g_0, the linearized map s'_{g_0} of the scalar curvature map is surjective, and if $p > \dim M$, then the L_2^p extension of $g \mapsto s_g$ is locally surjective, i.e. there exists $\varepsilon > 0$ such that, if f is in $L^p(M)$ and $\| f - s_{g_0} \| < \varepsilon$, there is an L_2^p metric such that $f = s_g$; furthermore, if f is C^∞, so is g.*

Proof. Define a map F from a neighbourhood of zero in $L_4^p(M)$ into $L^p(M)$ by setting $F(f) = s_{g_0} + s'^{*}_{g_0} f$, where $s'^{*}_{g_0}$ is the formal adjoint of s'_{g_0} (see Appendix 16). We apply the inverse function theorem to F, whose differential at zero is $s'_{g_0} \circ s'^{*}_{g_0}$; the last part is an application of a non-linear regularity theorem (see Appendix 41).

From Appendix 32*a*, s'_g is surjective if and only if s'^{*}_g is injective. The next proposition shows that this assumption is generically satisfied.

4.37 Proposition (J.-P. Bourguignon, [Bou 1], VIII.8). *If* $\operatorname{Ker} s'^{*}_g \neq 0$, *then either* (M, g) *is Ricci-flat and* $\operatorname{Ker} s'^{*}_g = \mathbb{R} \cdot 1$, *or the scalar curvature is a strictly positive constant and $s_g/(n-1)$ is an eigenvalue of the Laplacian.*

Proof. Take the divergence of $s'^{*}_g \cdot f$ (cf. 1.21). The Ricci identity computed in (1.187) yields $\delta(Ddf) = d\Delta f - r(\Delta f, .)$, so using 4.19 we get $\delta(s'^{*}_g \cdot f) = (1/2) f ds = 0$. Now, using an argument of elliptic theory (cf. [Aro] and [Hör 3]), f cannot be zero on an open set without being zero everywhere. This proves that s is a constant. Now, $\operatorname{tr}(s'^{*}_g \cdot f) = (n-1)\Delta f - sf = 0$. If $s = 0$, then f is a constant, and returning to $s'^{*}_g \cdot f$, we see that it is possible if and only if $r_g = 0$. □

See [Koo] or [Laf 3] for further study of the equation $s_g'^* \cdot f = 0$. It is not difficult to see that the preceding proposition does not give the best possible result. Indeed, for the standard product metric on $S^2 \times S^3$, $s/4$ is an eigenvalue of the Laplacian, but $\operatorname{Ker} s_g'^* = 0$. Anyway, the condition that $s_g/(n-1)$ does not belong to the spectrum will appear again as an important property in the next paragraph.

4.38 Lemma (see [Ka-Wa 2] for the proof). *If* $\dim M \geqslant 2$ *and if* $f \in C^0(M)$, *then an* L^p-*function belongs to the* L^p-*closure of the set* $\{f \circ \varphi, \varphi \in \mathfrak{D}\}$ *if and only if* $\inf f \leqslant f_1(x) \leqslant \sup f$ *almost everywhere.*

4.39 Lemma. *Let* $f_0 \in \mathscr{C}M$ *be the scalar curvature of some metric. If* $f_1 \in \mathscr{C}M$ *is such that, for a positive real number* λ, *the inclusion* $\lambda(\operatorname{Im}(f_0)) \subset \operatorname{Im}(f_1)$ *holds, then* f_1 *is also the scalar curvature of some metric.*

Proof of the lemma. After a homothetic change of metrics, we may suppose that $\lambda = 1$. If $f_0 = s_{g_0}$ and $\operatorname{Ker} s_{g_0}'^* = 0$, apply the inverse function theorem 4.36 to g_0, and take φ in $\mathfrak{D}$ such that $\| f - f_1 \circ \varphi \|_p < \varepsilon$. If $\operatorname{Ker} s_{g_0}'^* \neq 0$ (which by 4.37 implies that s_{g_0} is constant), we can find a metric g_2 such that $\operatorname{Im}(s_{g_2}) \subset \operatorname{Im}(f_1)$ still holds (take g_2 close enough to g_0), but s_{g_2} is no longer a constant. Then apply the preceding argument to g_2.

4.40. *End of the proof of Theorem* 4.35. If M carries a metric whose scalar curvature is non-negative everywhere, but not identically zero, from 4.32 ii), there are also on M metrics with constant positive, constant negative and zero scalar curvatures. Using Lemma 4.39 with those constant functions as functions f_0, we get that any function is a scalar curvature. The other cases follow just in the same way. That $s_g = 0$ implies $r_g = 0$ in case (B) will be proved in 4.49. □

4.41 Remarks. i) If $\dim M = 2$, using the same technique and only replacing Theorem 4.32 by the conformal representation theorem, one gets the following converse to the Gauss-Bonnet theorem ([Ka-Wa 1]): if $\chi(M) > 0$ (resp. $\chi(M) = 0$, resp. $\chi(M) < 0$) a function f on M is the gaussian curvature of some metric if and only if f is positive somewhere (resp. is identically zero or takes both signs, resp. is negative somewhere).

ii) The use of Sobolev spaces L_k^p (and not only L_k^2 as in most discussions of linear elliptic theory) was crucial: some L^p-norm is needed for the approximation lemma 4.38, which is trivially false for the uniform norm. Then the use of the inverse function theorem in 4.36 makes it necessary to take $p > \dim M$.

F. The Manifold of Metrics with Constant Scalar Curvature[1])

4.42. Since the manifold $\mathscr{M}$ of Riemannian metrics and its submanifolds are Fréchet manifolds modelled on $\mathscr{S}^2M$, the usual inverse function theorem cannot be applied to them. This difficulty is taken care of by the technique of ILH-manifolds, developed

[1]) This section may be skipped on first reading.

by H. Omori ([Omo]). Recall that thanks to Sobolev's lemma (cf. Appendix 8), for any vector bundle E on M, the space $C^\infty(E)$ is the inverse limit $\lim_{s\to+\infty}^{\leftarrow} L_s^2(E)$. In fact, going from Hilbert manifolds modelled on L_s^2 (to which the inverse function theorem can be applied directly) to ILH-manifolds modelled on $C^\infty(E)$ is very similar to going from L_s^2 to C^∞ in elliptic theory. To avoid technicalities, we will not give complete proofs, refering to the work of N. Koiso ([Koi 1] and [Koi 2]) for details.

Since we already know that any compact manifold M carries "many" metrics with constant scalar curvature, we will introduce the set

(4.43) $$\mathfrak{S} = \{g \in \mathscr{M}, \operatorname{vol}(g) = 1 \text{ and } s_g \text{ constant}\}$$

Two technical descriptions of $\mathfrak{S}$ will be useful. On the one hand, it is clear that $g \in \mathfrak{S}$ if and only if $s_g = \int_M s_g \mu_g$. On the other hand, the constancy of s_g is equivalent to $\Delta_g s_g = 0$. Thus the infinitesimal deformations which preserve this condition up to first order are those which lie in the kernel of the fourth order differential operator

(4.43)′ $$\alpha_g \cdot h = \Delta_g(s_g' \cdot h),$$

since $\Delta_g'(h) \cdot f = 0$ when f is constant.

The following theorem of N. Koiso (cf. [Koi 1]) shows that, under a generic condition, $\mathfrak{S}$ is an infinite dimensional manifold, and provides a local version of Yamabe's theorem.

4.44 Theorem. *Let $g_0 \in \mathfrak{S}$ such that $s_{g_0}/(n-1)$ is not a positive eigenvalue of the Laplacian Δ_{g_0}. Then, in a neighbourhood of g_0, $\mathfrak{S}$ is an ILH-submanifold of $\mathscr{M}_1$ such that*

$$T_g\mathfrak{S} = \operatorname{Ker}\alpha_{g_0} \cap \{h \in \mathscr{S}^2M; \langle h, g_0\rangle_{g_0} = 0\}$$

Further, the map $(f,g) \mapsto fg$ from $\mathscr{C}M \times \mathfrak{S}$ into $\mathscr{M}$ is a local ILH-diffeomorphism from a neighbourhood of $(1, g_0)$ to a neighbourhood of g_0.

Sketch of the proof. We first show that the map $g \mapsto S(g) = s_g - \int_M s_g \mu_{g_0}$ from $\mathscr{M}_1$ to $\mathscr{C}_{g_0}M = \{f \in \mathscr{C}M, \langle f, 1\rangle_{g_0} = 0\}$ is an ILH-submersion. In fact, its L_p^2-extension is a submersion in the Hilbertian category for p large enough ($p > (n/2) + 2$, see Appendix 8): the surjectivity comes from the assumption on the spectrum (take $h = fg$). From elliptic theory, we see that

$$\operatorname{Ker} S'(g_0) = \operatorname{Ker}\alpha_{g_0} \cap \{h, \langle h, g_0\rangle_{g_0} = 0\}$$

is a direct factor. Indeed, since the symbol of α_g^* is surjective we have the orthogonal decomposition

(4.45) $$T_{g_0}\mathscr{M}_1 = \{h, \langle h, g_0\rangle_{g_0} = 0\} = (\operatorname{Ker}\alpha_{g_0} \cap T_{g_0}\mathscr{M}_1) \cap \operatorname{Im}\alpha_{g_0}^*.$$

The second part follows from the ILH inverse function theorem, after proving the decomposition

$$\mathscr{S}^2M = T_g\mathfrak{S} \oplus \mathscr{C}M \cdot g_0,$$

under the assumption $s_{g_0}/(n-1) \notin Sp^+(\Delta_{g_0})$. This decomposition, which is no longer orthogonal, is obtained by using elliptic theory again. □

4.46 Remarks. i) It is possible that the assumption $s_{g_0}/(n-1) \notin Sp^+(\Delta_{g_0})$ might be relaxed in the first part of this theorem, but it is crucial in the second part. If (M, g) is a manifold with constant scalar curvature which admits a conformal non isometric vector field X, then $s_g/(n-1) \in Sp^+\Delta_g$. By pulling back g by the flow of X, we get a one parameter family of metrics which are pointwise conformal to g and have the same scalar curvature s_g.

ii) We claim that the theorem can be applied to any Einstein manifold other than the standard sphere. Recall Lichnerowicz's inequality for the first eigenvalue of the Laplacian: for any compact Riemannian manifold, $\lambda_1 \geqslant (n/(n-1))r_{\min}$ (cf. [Lic 1], p. 135, [Be-Ga-Ma] pp. 179–181 or [Gal 3]). Now, if g is an Einstein metric and if $s_g/(n-1)$ belongs to the spectrum, the equality is achieved, and by Obata's theorem (cf. [Be-Ga-Ma] ibidem, or [Oba]), $(M \cdot g)$ is isometric to the standard sphere.

When restricting the total scalar curvature functional to $\mathfrak{S}$, are there other critical points, besides the Einstein metrics? Here is a partial answer.

4.47 Proposition. *Let g be a metric on M with constant scalar curvature such that $s_g/(n-1) \notin Sp^+\Delta_g$. If g is a critical point of the total scalar curvature restricted to $\mathfrak{S}$, then (M, g) is Einstein.*

Proof. For h in $T_g\mathcal{M}_1$, we have $\mathbf{S}'_g \cdot h = \langle h, r - (s/n)g\rangle = \langle h, z\rangle$. By 4.45, if $\mathbf{S}'_g \restriction T_g\mathfrak{S} = 0$, there is a function f such that $z = \alpha^* f = s'^*(\Delta f)$. Taking the trace, we get $0 = (n-1)\Delta^2 f - s\Delta f$, therefore $z = 0$.

4.48 Remark. Even if $s_g/(n-1) \in Sp^+\Delta_g$, the space $\operatorname{Ker}\alpha_g \cap T_g\mathcal{M}_1$ can be viewed as the formal tangent space to $\mathfrak{S}$. So $\mathbf{S} \restriction \mathfrak{S}$ will be critical at g if and only if there is a function f with vanishing mean value such that

$$z_g = D_g df + (\Delta_g f)g - fr_g.$$

Einstein metrics are recovered when $f = 0$. But the existence of a non zero solution is a very strong condition. The only known case is that of the standard sphere. (As seen in 4.46, this equation is equivalent to Obata's equation $Ddf = -(s/(n-1))fg$ when g is Einstein.) It can be conjectured that this is the only possible case.

Proposition 4.47 has the following important consequence for manifolds of type (B) of Theorem 4.35. It is due to J.-P. Bourguignon ([Bou 1]), but our proof is new.

4.49 Corollary. *Let M be a compact manifold which does not carry any metric with nonnegative and not identically zero scalar curvature. Then any metric with zero scalar curvature is Ricci-flat.*

Proof. Zero is a maximum of $\mathbf{S} \restriction \mathfrak{S}$ and therefore a critical value. □

4.50 Examples (cf. 6.D). i) On a K3 surface, every metric with zero scalar curvature is Ricci-flat.

ii) On a torus, every metric with zero scalar curvature is *flat*.

4.51 Remark. This property has been extended by J. Kazdan to the non-compact case (cf. [Kaz 3]).

G. Back to the Total Scalar Curvature: Second Order Properties

Recall that the notion of second derivative is meaningful for functions defined on an open set U of a vector space E: take the first derivative of the first derivative, viewed as a function on $U \times E$ (cf. [Die] or [Ham 1]. This definition cannot be extended to manifolds, unless the extra data of a connection—just made for that, cf. [Die], XVII 18. or [Pal 1] for instance—is given. But the second derivative does have an intrinsic meaning at a critical point, as can be seen by direct computation, or in the jet bundle setting.

The following property will be useful in 4.60 and 4.64.

4.52 Lemma. *Let* $\mathbf{F}$ *be a Riemannian functional which is differentiable at order two. Then, for any* g *in* $\mathscr{M}$ *and any vector field* X *in* $\mathscr{T}M$, *we have*

$$\mathbf{F}''(g)(\delta^* X, .) = 0.$$

The same statement holds for $\mathbf{F} \restriction \mathscr{M}_1$, *provided* g *is a critical point of* $\mathbf{F} \restriction \mathscr{M}_1$.

Proof. This property is just the $\mathfrak{D}$-invariance at the second order. □

4.53. To compute $\mathbf{S}''$, let us take a one parameter deformation $g(t) = g + th + (t^2/2)k$ of g, which, up to order two, lies in $\mathscr{M}_1$. We already know that the first order condition is given by $\int_M (\operatorname{tr}_g h)\mu_g = 0$. Since

$$\left.\frac{d^2}{dt^2}\mu_{g(t)}\right|_{t=0} = \left.(1/2)\frac{d}{dt}\operatorname{tr}_{g(t)}(h + tk)\mu_{g(t)}\right|_{t=0}$$

$$= (1/2)[\operatorname{tr}_g k + (1/2)(\operatorname{tr}_g h)^2 - |h|_g^2]\mu_g$$

the second order condition is given by $\int_M (\operatorname{tr}_g k)\mu_g = \int_M (|h|_g^2 - (1/2)\operatorname{tr}_g^2 h)\mu_g$. Now, we have

$$\left.\frac{d^2}{dt^2}\mathbf{S}_{g(t)}\right|_{t=0} = \frac{d}{dt}\langle -r_{g(t)} + (1/2)s_{g(t)}g(t), h + tk\rangle \tag{4.54}$$

$$= \langle -r'\cdot h + (1/2)(s'\cdot h)g + (1/2)sh, h\rangle + \langle -r + (1/2)sg, k\rangle$$

$$+ (1/2)\langle -r + (s/2)\, g, (\operatorname{tr}_g h)h\rangle + 2\langle r, h \circ h\rangle - \langle sh, h\rangle.$$

4.55 Proposition (cf. [Ber 4], p. 290 and [Mut 1]). *If* g *is an Einstein metric of total volume* 1, *the second derivative of* $\mathbf{S} \restriction \mathscr{M}_1$ *at* g *is*

$$\mathbf{S}_g''(h, h) = \langle h, -(1/2)D^*Dh + \delta^*\delta h + \delta(\delta h)g + (1/2)(\Delta \operatorname{tr}_g h)g$$

$$-(s/n)(\operatorname{tr}_g h)g + \mathring{R}_g h\rangle$$

Proof. When g is Einstein, the last four terms of (4.54) give $((1/n) - (1/2))s\|h\|^2$, and we have also from (1.180a)

$$r' \cdot h = (1/2)D^*Dh - \delta^*\delta h - (1/2)D_g d(\operatorname{tr}_g h) + (s/n)h - \mathring{R}h$$

and
$$s' \cdot h = \Delta(\operatorname{tr}_g h) + \delta(\delta h) - (s/n)\operatorname{tr}_g h. \qquad \square$$

The following remark will suggest to us how to interpret this complicated looking operator.

4.56 Remark. When $\mathbf{S}$ is restricted to a conformal class, from 4.25 the second derivative of a metric with constant scalar curvature makes sense. Indeed, using 4.53 again, we get $\mathbf{S}''_g(fg, fg) = ((n-2)/2)\langle (n-1)\Delta_g f - s_g f, f\rangle_g$. This formula has the following consequences:

i) The index at a critical point is always finite. Taking for instance $(S^1, dt^2) \times (S^{n-1}, \text{can})$, with a long S^1 as first factor, it is possible to get an arbitrarily large index.

ii) As expected from 4.44, g is a degenerate critical point if and only if $s_g/(n-1) \in Sp^+ \Delta_g$, and the null-space is then the eigenspace associated with $s_g/(n-1)$.

iii) If further g is Einstein, and is different from the standard metric on the sphere, then, using 4.46, we see that g is a non-degenerate critical point and that its index is zero. This property can be viewed as the infinitesimal version of a result of M. Obata ([Oba]), which states that an Einstein metric (other than the standard metric on the sphere) is the unique metric with constant scalar curvature in its conformal class with normalized volume.

Let us now come back to the study of $\mathbf{S} \restriction \mathscr{M}_1$. We already know (cf. 4.52) that the null-space of $\mathbf{S}''_g$ contains $\operatorname{Im} \delta_g^*$, but we must eliminate these trivial deformations. Since the preceding Remark 4.56 applies to Einstein metrics, the $\mathbf{S}''_g(fg, fg)$ are already known. So we must look at the orthogonal space to the space $\{\delta_g^*\xi + fg\}$, i.e. $\delta_g^{-1}(0) \cap \operatorname{tr}_g^{-1}(0)$. The following lemma shows that this space is the geometrically interesting part.

4.57 Lemma. i) *For any compact Riemannian manifold (M, g), the algebraic sum $\operatorname{Im} \delta_g^* + \mathscr{S}M \cdot g$ is closed in $\mathscr{S}^2 M$, and we have the orthogonal decomposition*

$$\mathscr{S}^2 M = (\operatorname{Im} \delta_g^* + \mathscr{C}M \cdot g) \oplus (\delta_g^{-1}(0) \cap \operatorname{tr}_g^{-1}(0)).$$

Both factors are infinite dimensional.

ii) *Further, if (M, g) is Einstein, but not the standard sphere, this decomposition can be refined into*

$$\mathscr{S}^2 M = \mathscr{C}M \cdot g \oplus \operatorname{Im} \delta_g^{*\perp} \oplus (\delta_g^{-1}(0) \cap \operatorname{tr}_g^{-1}(0)),$$

and the third component coincides with the tangent space to the manifold $\mathfrak{S}$ of metrics with constant scalar curvature and volume 1.

Proof. Using a device due to Hörmander ([Hör 2], p. 135), we can apply elliptic theory to the operator $(X, f) \to \delta_g^* X + fg$ from $\mathscr{T}M \times \mathscr{C}M$ into $\mathscr{S}^2 M$. As an operator of bidegree (1,0), it is overdetermined elliptic, and the decomposition in i) follows from Appendix 34.

When (M, g) is an Einstein manifold other than the standard sphere, there are no conformal non isometric vector fields (cf. 4.46), so the refinement is straightforward, since $\operatorname{Im} \delta_g^*$ and $\mathscr{C}M \cdot g$ are closed.

Now, recall (4.44) that $T_g\mathfrak{S} = \operatorname{Ker} \alpha_g \cap T_g\mathscr{M}_1$. Since $\operatorname{Im} \delta_g^* \subset \alpha_g^{-1}(0)$ (to prescribe constant scalar curvature is a geometric condition), we can take the trace on $T_g\mathfrak{S}$ of the decomposition $\mathscr{S}^2M = \operatorname{Im} \delta_g^* \oplus \operatorname{Ker} \delta_g$:

$$T_g\mathfrak{S} = \operatorname{Im} \delta_g^* \oplus (\delta_g^{-1}(0) \cap \alpha_g^{-1}(0) \cap T_g\mathscr{M}_1).$$

Since g is Einstein,

$$\alpha_g h = \Delta_g(\Delta_g(\operatorname{tr}_g h) + \delta_g(\delta_g h) - (s/n)\operatorname{tr}_g h).$$

If $\delta_g h = 0$ and $\alpha_g h = 0$, then $\Delta_g(\operatorname{tr}_g h) - (s/n)\operatorname{tr}_g h$ is a constant. This constant is zero if moreover h lies in $T_g\mathscr{M}_1$. Using Lichnerowicz' lower bound for the first eigenvalue (cf. 4.46), we conclude that $\operatorname{tr}_g h$ must be zero. □

4.58 Remarks. i) A by-product of this proof is, in the Einsteinian case, the decomposition of $\mathscr{S}^2M$ which we used to prove the local Yamabe theorem 4.42.

ii) For the standard sphere, we have again $\delta_g^{-1}(0) \cap \alpha_g^{-1}(0) \cap T_g\mathscr{M}_1 = \alpha_g^{-1}(0) \cap \operatorname{tr}_g^{-1}(0)$. However, the geometric interpretation fails.

The results above are summarized in the following theorem, due to M. Berger and N. Koiso (cf. [Ber 4] and [Koi 3]). We shall set

$$\mathscr{C}_gM = \{f \in \mathscr{C}M, \langle f, 1\rangle_g = 0\}. \tag{4.59}$$

4.60 Theorem. *Let (M, g) be a compact Einstein manifold other than the standard sphere. Then the decomposition*

$$T_g\mathscr{M}_1 = \operatorname{Im} \delta_g^* \oplus \mathscr{C}_gMg \oplus \delta_g^{-1}(0) \cap \operatorname{tr}_g^{-1}(0)$$

is orthogonal with respect to $\mathbf{S}_g''$. Furthermore

i) the first factor is contained in the null-space of $\mathbf{S}_g''$;

ii) for f in $\mathscr{C}_gM$ and h in $\delta_g^{-1}(0) \cap \operatorname{tr}_g^{-1}(0)$,

$$\mathbf{S}_g''(fg + h, fg + h) = \tfrac{1}{2}[(n-2)\langle (n-1)\Delta_g f - s_g f, f\rangle_g - \langle D^*Dh - 2\mathring{R}h, h\rangle_g]$$

In particular, the restriction of $\mathbf{S}_g''$ to the second factor is positive definite; the nullity and the coindex of the restriction of $\mathbf{S}_g''$ to the third factor are finite, so that $\mathbf{S}_g'' \restriction T_g\mathfrak{S}$ has a finite coindex.

Proof. Only a few points have not been proved. That $\mathbf{S}_g''(fg, h) = 0$ if $\delta_g h = 0$ and $\operatorname{tr}_g h = 0$ follows from 4.54 by repeated integrations by parts; the computation of $\mathbf{S}_g''(h, h)$ is straightforward. Moreover, D^*D is a self-adjoint positive operator, which leaves the space $\delta_g^{-1}(0) \cap \operatorname{tr}_g^{-1}(0)$ invariant, since for any Einstein metric $\delta\Delta_L = D^*D\delta + (s/n)\delta$ (cf. [Lic 2]). The claims about finiteness follow from Appendix 32. □

4.61 Remarks. i) The same results are true for the standard sphere if, in the second factor, $\mathscr{C}_g M$ is replaced by the orthogonal space to the first order spherical harmonics.

ii) As a consequence of a result of Bourguignon-Ebin-Marsden, ([Bo-Eb-Ma]), the space $\delta_g^{-1}(0) \cap \mathrm{tr}_g^{-1}(0)$ is always *infinite dimensional.*

From the very definition of the second derivative, we get the following corollary (see Chapter 12 for a more direct argument... but which amounts to the same thing).

4.62 Corollary. *The infinitesimal Einstein deformations which preserve the total volume and are orthogonal to the orbit under the diffeomorphism group are solutions of the system*

$$\delta_g h = 0, \qquad \mathrm{tr}_g h = 0, \qquad D^*Dh - 2\mathring{R}h = 0. \qquad \square$$

Since all the geometric information is contained in $T_g\mathfrak{S}$, we introduce the following definition, which follows the general terminology of calculus of variations.

4.63 Definition. An Einstein metric g on M is *stable* if $\mathbf{S}_g'' \restriction \delta_g^{-1}(0) \cap \mathrm{tr}_g^{-1}(0)$ is strongly negative, i.e., if there exists $\lambda > 0$ such that $\langle D^*Dh - 2\mathring{R}h, h\rangle \geqslant \lambda \|h\|^2$ for any h in $\delta_g^{-1}(0) \cap \mathrm{tr}_g^{-1}(0) = T_g\mathfrak{S}$.

In particular, stable Einstein metrics are infinitesimally rigid; therefore, by an important theorem of N. Koiso (see 12.49), they are not deformable. For this reason, the discussion of stability will be postponed until Section H of Chapter 12.

We only, quote a few examples.

—the standard sphere;

—compact Einstein manifolds with strictly negative sectional curvature;

—the standard complex projective space.

On the other hand, the fact that Jensen spheres are not stable appears clearly on Figure 9.72.

4.64 Remark. In [Laf 2], J. Lafontaine studies the total mass of the Ricci curvature, i.e. the functional $g \mapsto \mathbf{V}_r(g) = \int_M \mu_r$. Clearly, $\mathbf{V}_r$ is homogeneous of degree zero. Einstein non Ricci flat metrics are critical and the formulae for $\mathbf{S}$ and $\mathbf{V}_r$ are worth comparing. Indeed, setting $r = \varepsilon g$ and using the Lichnerowicz Laplacian Δ_L, we have, for any f in $\mathscr{C}M$ and h in $\delta_g^{-1}(0) \cap tr_g^{-1}(0)$

$$\mathbf{S}_g''(fg + h, fg + h) = (1/2)[(n-2)\langle (n-1)\mathbf{D}_g f - n\varepsilon f, f\rangle - \langle \mathbf{D}_L h - 2\varepsilon h, h\rangle]$$

and

$$(\mathbf{V}_r)_g''(fg + h, fg + h) = (1/8)|\varepsilon|^{(n-4)/2}[(n-2)\langle (n-1)\mathbf{D}_g^2 f - n\varepsilon \mathbf{D}_g f, f\rangle \\ - \langle \mathbf{D}_L^2 h - 2\varepsilon \mathbf{D}_L h, h\rangle].$$

Using this formula, we see that $\mathbf{V}_r''$ enjoys the same properties as the properties of $\mathbf{S}''$ which were described in Theorem 60 above.

H. Quadratic Functionals

We now come to the case of integrands which are polynomials of degree 2 in the curvature. Since we are looking for Riemannian functionals, these polynomials must be invariant under the natural action of the group $O(n)$ on the space of curvature tensors (cf. [Gil], [Pal 3]). A basis for such a space of polynomials is given by s^2, $|r|^2$, $|R|^2$. Another important basis if $|S|^2$, $|Z|^2$, $|W|^2$ (cf. 1.114) We set

(4.65)
$$\mathbf{Ss}(g) = \int_M s_g^2 \mu_g, \quad \mathbf{Sr}(g) = \int_M |r_g|^2 \mu_g,$$

$$\mathbf{SR}(g) = \int_M |R_g|^2 \mu_g, \quad \mathbf{SZ}(g) = \int_M |Z_g|^2 \mu_g, \quad \mathbf{SW}(g) = \int_M |W_g|^2 \mu_g$$

These functionals are homogeneous of degree $(n/2) - 2$.

4.66 Proposition. *The functionals* $\mathbf{Ss}$ *and* $\mathbf{Sr}$ *are differentiable with gradients*

$$(\operatorname{grad}\mathbf{Ss})_g = 2D_g ds + 2(\Delta_g s_g)g + (1/2)s_g^2 g - 2s_g r_g$$

$$(\operatorname{grad}\mathbf{Sr})_g = D_g^* D_g r_g + D_g ds_g + (1/2)(\Delta_g s_g)g + (1/2)|r_g|^2 g - 2\mathring{R}_g r_g.$$

Proof. We have $\mathbf{Ss}'_g \cdot h = (1/2)\int_M s^2 \operatorname{tr}(h)\mu_g + 2\int_M ss'(h)\mu_g = (1/2)\langle s^2 g, h\rangle + 2\langle s, s'(h)\rangle = \langle (1/2)s^2 g + 2s'^*(s), h\rangle$. In the same way, $\mathbf{Sr}'_g \cdot h = (1/2)\int_M |r|^2 \operatorname{tr}(h)\mu_g + 2\langle r, r'(h)\rangle - 2\langle r \circ r, h\rangle$. (As in 4.53, we must take into account the variation of the scalar product). Recall that $r'(h) = (1/2)\Delta_L h - \delta^*\delta h - (1/2)Dd\operatorname{tr}(h)$, where $\Delta_L h = D^*Dh + h\circ r + r\circ h - 2\mathring{R}h$ (cf. (1.180b), so that $r'^*(h) = (1/2)\Delta_L h - \delta^*\delta h - (1/2)(\delta(\delta h))g$. We have $\operatorname{grad}\mathbf{Sr}_g = (1/2)|r|^2 g - 2r\circ r + 2r'^*(r)$, and the result follows using 4.19. □

4.67 Corollary. i) *Einstein metrics are critical points of* $\mathbf{Ss}\restriction \mathscr{M}_1$ and $\mathbf{Sr}\restriction \mathscr{M}_1$;

ii) *if* $\dim M = 4$, *a metric* g *is critical for* $\mathbf{Ss}$ *if and only if either it is an Einstein metric or has zero scalar curvature;*

iii) *if* $\dim M = 4$ *and if* g *is critical for* $\mathbf{Sr}\restriction \mathscr{M}_1$, *the scalar curvature is constant.*

Proof. The first part is elementary. If $n = 4$. then $\operatorname{tr}(\operatorname{grad}\mathbf{Sr}_g) = 2\Delta s$ and $\operatorname{tr}(\operatorname{grad}\mathbf{Ss}_g) = 6\Delta s$. □

4.68 Remark. For any polynomial with n arguments, consider the functional $\mathbf{F}_P(g) = \int_M P(s_1, s_2, \ldots s_n)\mu_g$, where the s_i are the fundamental symmetric polynomials in the eigenvalues of r_g. Using similar computations, it can be proved (cf. [Laf 2]) that any Einstein metric is a critical point of $\mathbf{F}_P\restriction \mathscr{M}_1$. There may be other critical points, analogous to those in 4.74 below.

Using suitable differential operators, the formula giving $\operatorname{grad}\mathbf{SR}$ can be made quite simple. We introduce the differential operator $d^D\colon \mathscr{S}^2 M \to C^\infty(\Lambda^2 M \otimes T^*M)$ by setting

(4.69) $\qquad d_g^D h(x,y,z) = D_x h(y,z) - D_y h(x,z)$ (compare with 1.12).

This means that we take the exterior differential of the 2-form h viewed as a one form with values in the tangent bundle. See [Bou 9] for developments of this point of view. The formal adjoint of d^D will be *denoted* by δ^D.

4.70 Proposition. *The functional* **SR** *is differentiable with gradient*

$$\operatorname{grad} \mathbf{SR}_g = 2\delta_g^D d_g^D r_g - 2\check{R}_g + (1/2)|R_g|^2 g.$$

(Compare with [Ber 4], p. 292. Recall that $\check{R}_{ab} = R_{ajkl}R_b^{jkl}$) (1.131).

Proof. To avoid misunderstandings, we use covariant tensors only. Then $SR'_g \cdot h = 2\int_M (R, R'(h))_g \mu_g + (1/2)\int_M |R|^2 \operatorname{tr}(h)\mu_g - 4\int_M (\check{R}, h)_g \mu_g$. If $C(h)$ is the covariant 3-tensorfield associated with the variation of the connection, recall that $R'(h)(x,y,z,t) = D_y C(h)(x,z,t) - D_x C(h)(y,z,t) + R(x,y,z,h^\sharp(t))$ (cf. 1.177). So we get

$$\begin{aligned}\int_M (R, R'(h))v_g &= -2\langle R, DC(h)\rangle + \langle \check{R}, h\rangle \\ &= -2\langle D^*R, C(h)\rangle + \langle \check{R}, h\rangle.\end{aligned}$$

Since $2C(h)(y,z,t) = D_y h(z,t) + D_z h(y,t) - D_t h(y,z)$ (see 1.177 again), only the skew-symmetric part in z, t has to be considered. Therefore

$$2\langle D^*R, C(h)\rangle = \langle D^*R(y,z,t), d^D h(z,t,y)\rangle$$

Now, the differential Bianchi identity yields $D^*R(y,z,t) = -d^D r(z,t,y)$, and we finally get $-2\langle D^*R, C(h)\rangle = \langle \delta^D d^D r, h\rangle$. □

4.71 Remark. Checking that $\delta^D d^D r = 2D^*Dr + 2r \circ r - 2\mathring{R}r + Dds$, we recover Formula 6.1 of [Ber 4].

4.72 Corollary. i) *An Einstein metric (and more generally a metric with parallel Ricci tensor) is critical for* $\mathbf{SR} \restriction \mathscr{M}_1$ *if and only if* $\check{R} = (1/n)|R|^2 g$;

ii) *if* $\dim M = 4$, *Einstein metrics are critical for* **SR** *and so are conformally flat metrics with zero scalar curvature; moreover, every critical metric for SR has constant scalar curvature.*

Proof. i) follows from formula 4.66 above. For ii), check that, for any curvature tensor in dimension 4, the algebraic identity

(4.72) $$\check{R} - (1/4)|R|^2 g = (s/3)z + 2\mathring{W}z$$

holds, where $z = r - \dfrac{s}{n}g$ (compare 1.118). From 16.3, $W = s = 0$ implies $d^D r = 0$. Then $\operatorname{grad} SR = 0$ in that case also. For the last claim, observe that $\operatorname{tr}(\operatorname{grad} SR) = 2\Delta s$ in dimension 4. □

4.73 Example. Recall that from 4.13 isotropy irreducible homogeneous spaces, and in particular irreducible symmetric spaces, are critical for **SR**.

4.74 Counter-example. There are non Einstein metrics which are critical for **S***r* and **S***R*: take the product of two space-forms of equal dimensions and whose sectional curvature are opposite. This is not surprising, since when performing squares we loose information about the signs of the eigenvalues of *r*.

4.75 Counter-examples. In dimension larger than 4, there are plenty of Einstein manifolds which are not critical for $\mathbf{S}R \restriction \mathscr{M}_1$, e.g. the Jensen spheres (cf. 9.82) or the product of a flat manifold with a Ricci-flat non flat one—known to exist from Chapter 11.

Now, let us study **S***Z* and **S***W*.

4.76. It is easy to check that **S***Z* enjoys the same properties as **S***r*. The functional **S***W* has been studied a long time ago by relativists. It can be proved that it is differentiable and has a gradient along the same lines as just above. In dimension 4, its gradient is called the *Bach tensor*, and is *denoted* by B. Using 4.66 and 4.70, together with Formula (4.72), we get

(4.77) $$B = -2\delta^D D^* W - 2\mathring{W} r \quad \text{(cf. [Bac]);}$$

Conformally flat metrics are of course critical for **S***W*, but there are many more.

4.78 Proposition. *Let M be a compact four-dimensional manifold. The two following classes of metrics are critical points of* **S***W*:

i) *metrics which are locally conformal to an Einstein metric;*

ii) *half-conformally flat metrics (i.e., such that W^+ or W^- is zero), if M is orientable.*

Proof. i) If g is an Einstein metric, the differential Bianchi identity yields $D^*W = 0$ (cf. 16.A). The property follows, since the nullity of the Bach tensor is a local property which is invariant under conformal changes of the metric.

ii) can also be proved by a direct computation, but a direct global argument will be given in 4.82. □

See the work [Der 3] of A. Derzinski for further results in this direction.

4.79 Remark. We do not know any examples of half conformally flat compact manifolds which are not conformally Einstein.

4.80. The vector space of $SO(4)$-invariant quadratic forms on the space of four dimensional curvature tensors admits $|S|^2$, $|Z|^2$, $|W^+|^2$, $|W^-|^2$ as a basis (cf. 1.126).

If M is a given compact oriented four dimensional manifolds, there are two linear relations between the corresponding functionals on the space of Riemannian metrics. Namely, the Gauss-Bonnet formula says that $\chi(M) = (1/8\pi^2)\int_M(|R|^2 - 4|r|^2 + s^2)\mu_g$ (or $(1/8\pi^2)\int_M(|S|^2 - |Z|^2 + |W|^2)\mu_g$) and the signature formula that $\sigma(M) = (1/12\pi^2)\int_M(|W^+|^2 - |W^-|^2)\mu_g$ (cf. 6.31, 6.34). Using 4.66 and 4.70, together with the algebraic identity $\check{R} - (1/4)|R|^2 g = (s/3)z + 2\mathring{W}z$, it is very easy to check that in dimension 4 the functional **S***R* − 4**S***r* + **S***s* has vanishing derivative.

This fact was first pointed out by Lanczos ([Lan 2]) *before* the discovery by Chern of the generalized Gauss-Bonnet formula. But Lanczos concluded that this Lagrangian was "inactive in the formation of the field, and thus may be omitted". The points of view of mathematicians and physicists are indeed different!

Conversely, starting from the Gauss-Bonnet and the signature formulae, we get a direct proof of 4.72 ii) and 4.78 ii), together with the following additional information (compare with [Bo-La])

4.82 Proposition. i) *The Einstein metrics give an absolute minimum for the functional* **S***R*; *this minimum is* $8\pi^2\chi(M)$;

ii) *conformally flat metrics with zero scalar curvature give an absolute minimum for* **S***R*; *this minimum is* $-8\pi^2\chi(M)$;

iii) *half-conformally flat metrics give an absolute minimum for* **S***W*; *this minimum is* $12\pi^2|\sigma(M)|$.

This result gives topological obstructions to the existence of such metrics on a four dimensional compact manifold (compare with 6.35)

There is some analogy between **S***R* (or **S***W*) and the Yang-Mills functional (compare with [At-Hi-Si], [Bo-La] and Chapter 13). In both cases, one studies the square norm of the curvature of a given bundle. But the differences are important. On the one hand, the Yang-Mills functional makes sense on any vector-bundle equipped with a fixed metrics, and the variable is a metric connection. On the other hand, **S***R* and **S***W* make sense on $\Lambda^2 TM$, and as the metric g varies, the connection and the norm on the bundle both change. Turning back to the formula

$$\operatorname{grad} \mathbf{S}R_g = 2\delta_g^D d_g^D r_g - 2\check{R}_g + (1/2)|R_g|^2 g \qquad \text{(cf. 4.70)},$$

we see that the first term is perfectly analogous to the gradient of the Yang-Mills functional. The two extra terms take the variation of the metric along the fibres into account.

In higher dimensions, the situation is completely different. Indeed, take any compact manifold M, with $\dim M \geqslant 5$. Then M. Gromov pointed out ([Gro 4]) that M carries Riemannian metrics of volume 1 with **S***R* as small as you like.

Quadratic functionals are of high interest in the Kähler framework. There, one restricts to metrics in a given cohomology class, which produces more relations among the functionals: all of them can be expressed in terms of only one. The gradient of the restricted functional has a pleasant geometric interpretation, see 11. E.

Chapter 5. Ricci Curvature as a Partial Differential Equation

Throughout this chapter, we shall investigate the consequences of viewing the expression for the Ricci tensor in terms of the metric as a partial differential operator. In other words, given a metric g, its Ricci curvature

$$\text{Ric}(g) = r \tag{5.1}$$

is computed locally in terms of the first and second partial derivatives of g. We will think of r as prescribed and wish to investigate the properties of the metric. Some natural questions that arise are:

(i) Given a Ricci candidate r, is there a metric g satisfying (5.1)?

(ii) What conditions on r (on a compact manifold, say) insure uniqueness of such a metric (up to a constant multiple)?

(iii) How does the smoothness of r influence the smoothness of g?

We also will examine the analog of (5.1) for Einstein metrics:

$$\text{Ric}(g) = \lambda g \tag{5.2}$$

(for some constant λ) in the context of differential equations and ask similar questions.

A. Pointwise (Infinitesimal) Solvability

We cannot hope to get anywhere if we cannot even solve (5.1) at one point. Let the point be the origin of $\mathbb{R}^n$ ($n \geqslant 3$), and let a symmetric $n \times n$ matrix r_{ij} be given. We will find a metric g defined on a neighborhood of the origin in $\mathbb{R}^n$ such that

$$\text{Ric}(g)|_{x=0} = r_{ij}.$$

To do this, we write down the formula for $\text{Ric}(g)$ in local coordinates. It is

$$\text{Ric}(g)_{ij} = \frac{1}{2}\sum_{s,t} g^{st}\left\{\frac{\partial^2 g_{is}}{\partial x^j \partial x^t} + \frac{\partial^2 g_{js}}{\partial x^i \partial x^t} - \frac{\partial^2 g_{ij}}{\partial x^s \partial x^t} - \frac{\partial^2 g_{st}}{\partial x^i \partial x^j}\right\} + Q(g, \partial g), \tag{5.3}$$

where g^{st} is the inverse of g_{st}, Q is a function of g and its first derivatives, and is homogeneous of degree 2 in the first derivatives of g. Seek g in the simple form $g_{ij} = [1 + \varphi(x)]\delta_{ij}$, where $\varphi(x)$ is a homogeneous quadratic polynomial. Using (5.3), one easily finds that if

$$g = \sum_i \left\{ 1 + \frac{1}{n-2} \sum_{s=1}^n \sum_{t=1}^n \left(\frac{\sum_{k=1}^n r_{kk}}{2n-2} \delta_{st} - r_{st} \right) x^s x^t \right\} dx^i \otimes dx^i,$$

then $\mathrm{Ric}(g)|_{x=0} = r$.

Of course, to solve (5.2) at one point is trivial, since the constant curvature metrics are examples of solutions of (5.2).

B. From Pointwise to Local Solvability: Obstructions

One immediately encounters difficulties in trying to prolong a pointwise solution, even to match the 1-jet of a prescribed r at one point. The central issue is the fact that (5.1) is invariant under the action of the group of diffeomorphisms. Indeed, if φ_t is a family of diffeomorphisms for $t \in \mathbb{R}$ near zero with $\varphi_t(p) = p$ and $\varphi_0 =$ identity, and if $\mathrm{Riem}(g)$ is the Riemann curvature tensor of g, then the invariance of

$$\varphi_t^*(\mathrm{Riem}(g)) = \mathrm{Riem}(\varphi_t^*(g)) \tag{5.4}$$

yields the standard Bianchi identity as integrability conditions. These are obtained by taking the derivative of both sides of (5.4) at $t = 0$, and observing that the resulting identity must hold for any value of $\left.\dfrac{d\varphi_t}{dt}\right|_{t=0}$ (see [Kaz 2] for details of this approach to the Bianchi identities). In particular, for the Ricci curvature this gives the integrability condition $\delta r + \frac{1}{2} ds = 0$. In coordinates this is

$$0 = \mathrm{Bian}(g, r)_k = \sum_{i,j} g^{ij}(r_{ik;j} - \tfrac{1}{2} r_{ij;k}), \quad k = 1, 2, \ldots, n, \tag{5.5}$$

where a semicolon denotes covariant differentiation. In even more detail, (5.5) is:

$$0 = \mathrm{Bian}(g, r)_k = \sum_{i,j} g^{ij} \left[\left(\frac{\partial r_{ik}}{\partial x^j} - \frac{1}{2} \frac{\partial r_{ij}}{\partial x^k} \right) - \sum_l r_k^l \left(\frac{\partial g_{il}}{\partial x^j} - \frac{1}{2} \frac{\partial g_{ij}}{\partial x^l} \right) \right] \tag{5.6}$$

for $k = 1, \ldots, n$. We call (5.5) an integrability condition, even though it contains the unknown metric g as well as r. (Alternately, one could say that since (5.1) contains the same number of equations as unknowns, namely $\binom{n+1}{2}$, and since the Bianchi identity supplies n additional equations for g to satisfy, that the combined system (5.1)–(5.5) is overdetermined.) To see this more vividly, following DeTurck [DeT 1] we will see that one *cannot* always solve (5.1), since for certain Ricci candidates r, there will be no metric g for which the Bianchi identity (5.6) is satisfied.

5.7 Example. In a local coordinate chart $(x^1, \ldots, x^n)$, for any $q_{ij}(x_2, \ldots, x_n)$, there is no Riemannian metric with Ricci curvature

$$r = x^1 \, dx^1 \otimes dx^1 + \sum_{i,j=2}^n q_{ij}(x^2, \ldots, x^n) \, dx^i \otimes dx^j$$

in any neighborhood of any point on the hyperplane $x^1 = 0$. Indeed, if r is the Ricci curvature of a metric g, then the Bianchi identity (5.5) must be satisfied. In particular, letting $k = 1$ in (5.6) we obtain

$$\tfrac{1}{2}g^{11} = 0$$

on the hyperplane $x^1 = 0$, which is impossible for any definite metric g.

If in this example one replaces x^1 by $(x^1)^p$, then the pth derivative of the Bianchi identity is not satisfied at the origin, although (depending on the q_{ij}) it may be possible to satisfy all derivatives of lower order than p by a suitable choice of metric. The next example, also due to DeTurck, shows that it is possible to have non-existence at an *isolated* point.

5.8 Example. There is no Riemannian metric with Ricci curvature

$$r = \sum_{i,j} x^i x^j \, dx^i \otimes dx^j \tag{5.9}$$

in any neighborhood of the origin; however, r is the Ricci curvature of a Riemannian metric in $\mathbb{R}^n \backslash \{0\}$.

To prove the non-existence assertion, let $\text{Bian}(g, r)_1$ be the term in (5.6) with $k = 1$. Then with r as in (5.9) we obtain from (5.6):

$$0 = \left.\frac{\partial \, \text{Bian}(g, r)_1}{\partial x^1}\right|_{x=0} = \sum_i g^{ii},$$

which is an obvious contradiction since g is definite.

Existence of a metric with Ricci curvature (5.9) except at the origin is most easily found using spherical coordinates with $\rho^2 = (x^1)^2 + \cdots + (x^n)^2$. In these coordinates our Ricci candidate is simply

$$r = \rho^2 (d\rho)^2. \tag{5.10}$$

By a straightforward computation in a coordinate chart $(y^1, \ldots, y^n)$, for any smooth positive function $h(y^1)$, if

$$g = (h(y^1))^{2/(n-2)}[h'(y^1)^2 (dy^1)^2 + (dy^2)^2 + \cdots + (dy^n)^2], \tag{5.11}$$

where $h' = \dfrac{dh}{dy^1}$, then

$$\text{Ric}(g) = \alpha^2 \frac{h'(y^1)^2}{h(y^1)^2} (dy^1)^2,$$

where $\alpha^2 = \dfrac{n-1}{n-2}$. To compare with (5.10), we let $\rho = y^1$, let y_j, $j \geqslant 2$ be the angular variables and let $\alpha h'(\rho) = \rho h(\rho)$, so that

$$h(\rho) = \exp \frac{\rho^2}{2\alpha},$$

and we obtain the metric from formula (5.11). Note that the metric degenerates at the origin, since the coefficient of $(d\rho)^2$ becomes zero there.

5.12 Example. There is no metric of any signature with Ricci curvature

$$r = \sum_{i,j} \left(\frac{1}{2} x^i + \frac{1}{2} x^j + \delta_{ij} \sum_k x^k \right) dx^i \otimes dx^j$$

near the origin. In this case, the Bianchi identity (5.6) is $\sum_m g^{mk} = 0$ for $k = 1, 2, \ldots, n$, which contradicts the nonsingularity of g.

The moral of these examples is that the Bianchi identity is a genuine obstruction to solving the Ricci equations, even locally in the special case where the Ricci candidate r is a linear polynomial. Note that in all of these examples the Ricci candidate was singular at the points in question (i.e., r was not an invertible map from T_pM to T_p^*M at these points). From an intuitive viewpoint, one can see that this case might be troublesome since if $\mathrm{Ric}(g) = r$, then

$$\mathrm{Ric}(\varphi^*(g)) = \varphi^*(\mathrm{Ric}(g)) = \varphi^*(r) \tag{5.13}$$

for any diffeomorphism φ. Since g is always nonsingular, if r is singular then the orbit under the action of the diffeomorphism group is "smaller" than that of g, so one suspects difficulties. We thus turn our attention to the non-singular case.

C. Local Solvability of $\mathrm{Ric}(g) = r$ for Nonsingular r

5.14 Theorem (DeTurck [DeT 2]. *Let $r \in C^{m,\alpha}$ ($m > 2$, $0 < \alpha < 1$) be a given symmetric* (2, 0) *tensor field in a neighborhood of a point p on a manifold of dimension $\geqslant 3$. If r is invertible at p, then there exists a $C^{m,\alpha}$ Riemannian metric g such that* $\mathrm{Ric}(g) = r$ *in a neighborhood of p. Moreover, if r is also real analytic then there exists a real analytic metric of any desired signature that satisfies* $\mathrm{Ric}(g) = r$.

It is plausible that the second part of this theorem also holds without assuming real analyticity. In fact, an even stronger result (in the $C^{m,\alpha}$ situation) has been proved in the Lorentz case by DeTurck [DeT 3], who showed that one can solve the Cauchy (= initial value) problem for the Ricci equation for small time and find a metric with Lorentz signature under the assumptions that r is nonsingular and that the initial data satisfy certain necessary compatibility conditions.

The first proof of Theorem 5.14 was given in [DeT 2] and uses the local theory of elliptic systems of partial differential equations and a specialized iteration scheme. Subsequently, DeTurck [DeT 5] found the more straightforward proof that we present below.

The idea of the proof is to apply the local solvability theorem for elliptic equations (Theorem 45 in the Appendix), but some special device must be used since the equation $\mathrm{Ric}(g) = r$ is *not* elliptic. There are several ways to see this non-ellipticity:

First, let $\mathrm{Ric}'(g)$ be the linearization of the (quasi-linear) operator Ric at the metric g (Appendix, Equation 24), and use coordinates in which $g_{ij} = \delta_{ij}$ at the point p we are considering. From the explicit formula (5.3), we find (following Appendix, Section F) that the principal symbol $\sigma_\xi(\mathrm{Ric}'(g))$ at $\xi = (\xi_1, \ldots, \xi_n) \in T_p^*M$ is

$$\textbf{(5.15)}\qquad [\sigma_\xi(\mathrm{Ric}'(g))h]_{ij} = -\frac{1}{2}\left[h_{ij}|\xi|^2 + \sum_s (h_{ss}\xi_i\xi_j - h_{is}\xi_j\xi_s - h_{js}\xi_i\xi_s) \right].$$

For a fixed $\xi \neq 0$, we have $\sigma_\xi(\mathrm{Ric}'(g))$ is a linear map from symmetric matrices to symmetric matrices. But as one easily verifies, say by letting $\xi = (1, 0, 0, \ldots, 0)$, the kernel of this mapping consists precisely of those matrices h of the form $h_{ij} = v_i\xi_j + v_j\xi_i$ for covectors v, i.e., $h = \xi \otimes v + v \otimes \xi$. Since the symbol mapping is not an isomorphism, one concludes that the Ricci system is not elliptic.

A second way to see this non-ellipticity is to let X be a vector field with corresponding flow φ_t on the manifold. Then taking the derivative of (5.13) at $t = 0$ we obtain

$$L_X \mathrm{Ric}(g) = \mathrm{Ric}'(g)(L_X g).$$

The right side contains third derivatives of X, while the left only contains first derivatives. Thus there is some cancellation of terms on the right. Since the symbol of Ric$'$ only uses the highest order derivatives, this shows that tensors in the image of $L_X g$ are in the kernel of $\sigma_\xi(\mathrm{Ric}'(g))$. In greater detail, $L_X g = \delta^* X$ so

$$\sigma_\xi(\mathrm{Ric}'(g)) \circ \sigma_\xi(\delta^*) = 0.$$

But the image of $\sigma_\xi(\delta^*)$ is exactly the tensors of the form $h = \xi \otimes v + v \otimes \xi$. The point of the detailed calculation (5.15) is that these tensors h which arise from the action of the group of diffeomorphisms are the *only* obstructions to the ellipticity of the operator $g \to \mathrm{Ric}(g)$. This is crucial in the study of Ric(g) as a partial differential operator.

As another alternative, one could deduce the non-ellipticity of the Ricci system from the local solvability theorem in the Appendix as follows: We know that Ric$(g) = r$ is always solvable at one point, so if it were elliptic it would always be locally solvable. This contradicts the examples in Section B above.

Recall that the examples of non-solvability given in Section B depend in an essential way on the Bianchi identity (5.5), and that the Bianchi identity is a consequence of the invariance of the Ricci operator under the diffeomorphism group. DeTurck's idea is to use the diffeomorphism group to help, rather than hinder, solvability by replacing (5.1) by the following equation:

$$\mathrm{Ric}(g) = \varphi^*(r)$$

and trying to solve this equation for *both* the unknown metric g *and* the diffeomorphism φ. This equation involves second-order derivatives of g, but only first derivatives of φ, so the definition of ellipticity in the Appendix does not apply, but if one uses the more general definition of Douglis-Nirenberg ellipticity, then this equation turns out to be underdetermined elliptic if the candidate r is invertible (see [DeT 5] for this approach). Alternatively, one could require φ to have some special form—a form that involves first derivatives of something else. Perhaps the simplest scheme is the following, where the diffeomorphism φ also depends on the desired (unknown) metric.

We are ultimately trying to solve Ric$(g) = r$, so let g_0 be a metric such that Ric$(g_0) = r$ at p. We then seek h (small) so that

(5.16) $$\mathrm{Ric}(g_0 + h) = (\exp(v(h)))^* r,$$

where $\exp(v(h))$ is the diffeomorphism that maps $x \in M$ to $\exp_x(v(h)(x))$, and where v is the vector field

$$v^\alpha(h) = \sum_{\beta,s,t} (r^{-1})^{\alpha\beta} (g_0 + h)^{st} \left(h_{s\beta;t} - \frac{1}{2} h_{st;\beta} \right).$$

In this last expression, the semicolons represent covariant differentiation with respect to the g_0 metric. The essential feature of this choice is that the principal symbol of the right-hand side of (5.16) is

$$\frac{1}{2} \sum_s (h_{is}\xi_j\xi_s + h_{js}\xi_i\xi_s - h_{ss}\xi_i\xi_j).$$

When we compare this with (5.15) we see that the principal symbol of (5.16) is $-\frac{1}{2}h_{ij}|\xi|^2$—clearly elliptic in the standard sense. We can now apply the local solvability theorem (Appendix (45)) to prove the $C^{k,\alpha}$ part of Theorem 5.14. (Note that $\exp(v(h))$ will be a diffeomorphism in some open set about p since we can require $h = \nabla h = 0$ at p.) The analytic part of the theorem also follows, since, with g_0 of any signature and nondegenerate, the system (5.16) always has noncharacteristic directions, so that the Cauchy-Kovalevski theorem applies. □

One can also seek a connection (not necessarily coming from a metric) with prescribed Ricci curvature. This is less difficult because there are many more unknowns than equations, so the problem is underdetermined. Gasqui [Gas 1, 2] gave a proof of local existence in the analytic and $C^{m,\alpha}$ cases which was then simplified by DeTurck [DeT 1] who observed that this problem is actually underdetermined elliptic. Despite one's first thought, the symmetry of the Ricci tensor is unrelated to the symmetry of the connection.

D. Local Construction of Einstein Metrics

We now turn to the local existence question for the Einstein metrics equation (5.2). Gasqui [Gas 3] has shown that if the 2-jet of a metric g satisfies $\mathrm{Ric}(g) = \lambda g$ at one point, then it is possible to extend g to an Einstein metric in a neighborhood of the point. More precisely, Gasqui has proved the following.

5.17 Theorem (Gasqui [Gas 3]). *Let R_0 be a given algebraic curvature tensor* (see 1.G), *so $R_0 = R^i{}_{jkl}$ is a given tensor that satisfies the first Bianchi identity at the origin in $\mathbb{R}^n$ $(n \geq 3)$; also let g_0 be a nonsingular metric such that at the origin* $\mathrm{Riem}(g_0) = R_0$ *and* $\mathrm{Ric}(g_0) = \lambda g_0$ *for some constant λ. Then there is a real analytic metric g defined in a neighborhood of the origin such that*

i) $g = g_0$ *and* $\mathrm{Riem}(g) = R_0$ *at the origin*

ii) $\mathrm{Ric}(g) = \lambda g$ *in some neighborhood of the origin.*

Of course, if g_0 is Riemannian, then so is g.

This result is proved using Cartan-Kähler theory. Since the Bianchi identity is trivial for Einstein metrics, one can avoid many of the technicalities of the proof of the corresponding assertion in Theorem 5.14.

We also remark that DeTurck, in [DeT 3] has discussed and proved local existence for the Cauchy problem for (5.2) in the Lorentz case.

E. Regularity of Metrics with Smooth Ricci Tensors

We now turn to the questions: "If $\mathrm{Ric}(g) = r$ and r is smooth, can one conclude that the Riemannian metric g is smooth?" This is again a local question, and its answer involves the regularity theory for elliptic equations (see the Appendix). We follow DeTurck-Kazdan [De-Ka] in our exposition. An obvious example shows that some care is needed.

5.18 Example. Let g be the standard flat metric and let φ be a diffeomorphism of class C^3. Then $\mathrm{Ric}(\varphi^*(g)) = 0$, which is real analytic, but the metric $\varphi^*(g)$ is only of class C^2. By taking products such as $M^n = T^2 \times N^{n-2}$ (where T^2 is the flat torus) with φ smooth on N, one obtains more complicated examples.

In order to resolve the regularity question we first introduce natural local coordinates in which a metric is as smooth as possible. As was pointed out in [Sa-Sh] and [De-Ka], the appropriate coordinates for optimal regularity are *harmonic coordinates*, in which each coordinate function is harmonic (note that, contrary to naive prejudices, geodesic normal coordinates are *not* optimal since one can lose two derivatives; see Example 2.3 in [De-Ka]). Harmonic coordinates were first used by Einstein [Ein] in a special situation, and subsequently by Lanczos [Lan 1], who made the useful observation that they simplify the formula for Ricci curvature. For two-dimensional manifolds, the classical isothermal coordinates are harmonic.

To prove that harmonic coordinates exist in a neighborhood of a point p, one begins with some coordinate chart $(x^1, \ldots, x^n)$ and seeks harmonic functions $u^1(x), \ldots, u^n(x)$ with $u^j(p) = 0$ and the vectors $\operatorname{grad} u^j(p)$ orthonormal. By the local solvability theorem (Appendix, Theorem 45) this can always be done. Moreover, by the regularity theory for elliptic equations (Appendix, Theorem 41), if the metric is in the Holder class $C^{k,\alpha}$ (or C^ω) in the original coordinates, then the harmonic coordinates u^j are of class $C^{k+1,\alpha}$ (or C^ω). Armed with these facts, the next proposition—which asserts that a metric has optimal regularity in harmonic coordinates—is elementary.

5.19 Proposition. *If a given metric g is of class $C^{k,\alpha}$ $1 \leqslant k \leqslant \infty$ (or C^ω) in some coordinate chart then any tensor $T \in C^{k,\alpha}$ (or C^ω) in this chart is also of class $C^{k,\alpha}$ (or C^ω) in harmonic coordinates. In particular, $g \in C^{k,\alpha}$ (or C^ω) in harmonic coordinates.*

Proof. By the above facts, the (isometric) map from the given coordinates to harmonic coordinates is of class $C^{k+1,\alpha}$. Transforming the tensor $T \in C^{k,\alpha}$ to these harmonic coordinates involves at most the first derivatives of the map. Thus $T \in C^{k,\alpha}$ in harmonic coordinates. □

We can now answer the local question, "if Ric(g) is smooth, is g smooth?" In order to control the group of diffeomorphisms, and hence avoid the difficulties of Example 5.18 above, we can either use harmonic coordinates, or else assume that Ric(g) is invertible at the point in question. Some illuminating examples are presented after the statement of the theorem.

5.20 Theorem (DeTurck-Kazdan [De-Ka]). *Let $g \in C^2$ be a Riemannian metric in some neighborhood of a point p.*

(a) *If* Ric(g) *is invertible at p and if in local coordinates* $\mathrm{Ric}(g) \in C^{k,\alpha}$ *(or C^ω) near p, then in these coordinates $g \in C^{k,\alpha}$ (or C^ω).*

(b) *If in harmonic coordinates* $\mathrm{Ric}(g) \in C^{k,\alpha}$, $k > 0$ *(or C^ω) near p, then in these coordinates $g \in C^{k+2,\alpha}$ (or C^ω).*

5.21 Remark. (i) As we saw in Example 5.18, part (a) is false without the assumption Ric(g) is invertible at p. In part (a) one is tempted to try to prove that, in fact, $g \in C^{k+2,\alpha}$. This must fail, since if g is an Einstein metric, say the canonical metric on S^n, and if $\varphi: S^n \to S^n$ is a $C^{k+1,\alpha}$ diffeomorphism, then $g_1 = \varphi^*(g) \in C^{k,\alpha}$ and for some constant $c \neq 0$, $\mathrm{Ric}(g_1) = cg_1 \in C^{k,\alpha}$ as well. Thus, in this situation, the metric is no smoother than its Ricci tensor (but see Theorem 5.26 below).

(ii) Remark 4.10 in [De-Ka] contains an example showing that part (b) is false if one uses geodesic normal coordinates instead of harmonic coordinates (except for the C^∞ and the real analytic cases).

(iii) Note that the smoothness of the whole curvature tensor is reflected in Ric(g). Thus Theorem 5.20 also immediately shows how regularity of Riem(g) implies regularity of the metric.

Proof of Theorem 5.20. (a) Since Ric(g) is invertible at p, then g must satisfy the following basic equation, which was introduced by DeTurck in [DeT 2]:

$$\mathrm{Ric}(g) + \delta^*(r^{-1}\,\mathrm{Bian}(g,r)) = r.$$

This equation is clearly satisfied, since $\mathrm{Bian}(g,r) \equiv 0$; more important, as DeTurck observed, it is elliptic. Note that, since Bian(g,r) involves the first derivatives of r, this equation involves the second derivatives of r. Thus, if $r = \mathrm{Ric}(g)$ is in $C^{k,\alpha}$, then the coefficients of this equation are in $C^{k-2,\alpha}$. Elliptic regularity (Appendix, Theorem 41) then yields the result.

(b) In the coordinate chart $(x^1, \ldots, x^n)$ let Γ^k_{ij} be the Christoffel symbols and let $\Gamma^k = \sum_{i,j} g^{ij}\Gamma^k_{ij}$. Then, by a computation, $\Delta x^k = \Gamma^k$ and

$$\mathrm{Ric}(g)_{ij} = -\frac{1}{2}\sum_{r,s} g^{rs}\frac{\partial^2 g_{ij}}{\partial x^r \partial x^s} + \frac{1}{2}\sum_r \left(g_{ri}\frac{\partial \Gamma^r}{\partial x^j} + g_{rj}\frac{\partial \Gamma^r}{\partial x^i}\right) + \cdots$$

where the dots indicate lower-order terms involving at most one derivative of the metric. In particular, in harmonic coordinates, $0 = \Delta x^k = \Gamma^k$, so that

$$\mathrm{Ric}(g)_{ij} = -\frac{1}{2}\sum_{r,s} g^{rs}\frac{\partial^2 g_{ij}}{\partial x^r \partial x^s} + \text{lower order terms} \tag{5.22}$$

Thus, the differential equation $\mathrm{Ric}(g) = r$ is elliptic (Appendix, Section G) in harmonic coordinates, so we can again apply elliptic regularity. □

Combining the results of Theorem 5.20 and Proposition 5.19, we also obtain the following:

5.23 Corollary. (a) *If in some coordinate chart the Riemannian metric* $g \in C^{k,\alpha}$ $(k \geqslant 2)$ *and* $\mathrm{Ric}(g) \in C^{l,\alpha}$ $(l \geqslant k)$, *then* $g \in C^{k+2,\alpha}$ *in harmonic coordinates.*

(b) *If in addition to the hypotheses of part* (a), $\mathrm{Ric}(g)$ *is invertible, then* $g \in C^{l+2,\alpha}$ *in harmonic coordinates.*

5.24 Remark. It is natural to ask the regularity question about the map from metrics to connections, i.e., can one infer regularity properties of a metric from those of its connection coefficients? DeTurck and Kazdan [De-Ka] have shown that if a connection Γ comes from a metric g and if the connection is of class $C^{k,\alpha}$, then the metric g is of class $C^{k+1,\alpha}$. This holds for nonsingular metrics of any signature, not just Riemannian ones. The proof is just the observation that the first-order differential operator $g \to \Gamma$ is overdetermined elliptic, so one can immediately apply the theorems on regularity of solutions of elliptic equations.

F. Analyticity of Einstein Metrics and Applications

An Einstein metric satisfies

$$\mathrm{Ric}(g) = cg, \tag{5.25}$$

where c is some constant (if $\dim M \geqslant 3$). As we saw in Remark 5.21, by an isometry one can change the differentiability of the metric. From Proposition 5.19 it is clear that to obtain optimal smoothness, one should use harmonic coordinates. The results of this section, while not surprising, are very satisfactory.

5.26 Theorem (DeTurck-Kazdan [De-Ka]). *Let* (M, g) *be a connected* (*Riemannian*) *Einstein manifold of class* C^1 *with* $\dim M \geqslant 3$. *Then* g *is real analytic in harmonic and geodesic normal coordinates. That is, an atlas for* M *can be found with real analytic transition functions, so that* g *is analytic in each coordinate chart.*

Proof. From Formula (5.22), the Equation (5.25) is quasi-linear elliptic in harmonic coordinates. Also, Equation (5.25) is a real analytic function of all of its variables. Therefore, all of its solutions are real analytic in these coordinates. (Appendix, Theorem 41). But then the metrics g are also real analytic in geodesic normal coordinates (indeed, if a metric is real analytic in some coordinate chart, then it is always real analytic in geodesic normal coordinates). Real analyticity of the transition functions follows from the fact that they are harmonic (in the change from one harmonic coordinate chart to another) and that the metric is analytic. □

An immediate corollary concerns local isometric embedding:

5.27 Corollary. *A* (*Riemannian*) *Einstein manifold* (M, g) *of class* C^2 *with* $\dim M = n \geqslant 3$ *is locally isometrically embeddable in* $\mathbb{R}^{n(n+1)/2}$.

Proof. By Theorem 5.26 we may choose local coordinates in which g is real analytic. The assertion now follows from the Cartan-Janet theorem [Jan], [Car 7] (see also [Spi], Vol. 5), which states that any real analytic metric can be locally isometrically embedded in $\mathbb{R}^{n(n+1)/2}$. □

It is not known if for Einstein manifolds the dimension $n(n+1)/2$ can be lowered, except for the case $n = 3$. In dimension three, Einstein metrics necessarily have constant curvature, so a three-dimensional Einstein metric is locally that of the 3-sphere, flat $\mathbb{R}^3$, or hyperbolic 3-space. One has obvious local isometric embeddings of $S^3 \to \mathbb{R}^4$, and of $H^3 \to \mathbb{R}^5$ (as a generalization of the pseudosphere, see [Spi, p. 241]). Note that H^3 cannot be locally isometrically embedded in $\mathbb{R}^4$ (see [Spi], p. 195).

The second application of Theorem 5.26 concerns the unique continuation of Einstein metrics.

5.28 Corollary. *Let M be a connected, simply connected manifold and let g_1 and g_2 be Einstein metrics on M, with respect to which M is complete. If $g_1 = g_2$ (as tensor fields) on some open set, then up to a diffeomorphism $g_1 = g_2$ on all of M. In other words, there is a diffeomorphism $f: M \to M$ such that $g_1 = f^*(g_2)$.*

Proof. Since by Theorem 5.26 the metrics are analytic, this is a consequence of a result of Myers ([Mye], Theorem 3, see also [Ko-No 1], Corollary 6.4, p. 256). Note that it is essential to require the hypotheses of completeness (consider diffeomorphic images of two open subsets of $\mathbb{R}^n$ having different volumes) and simple connectivity (consider two flat tori with different volumes). □

5.29 Remark. Throughout the past two sections, one may have been confused that the Ricci equation (5.1) and the Einstein equation (5.25) are elliptic in certain coordinates, but not in others (whereas, "the definition of elliptic is invariant under changes of coordinates"). This is resolved by observing that in changing coordinates for these equations, one not only changes the independent variables, but also one is making a change in the dependent variables as well, since the metric g is a tensor and hence is transformed, too. The statement that "ellipticity is invariant" refers only to changes of independent variables.

G. Einstein Metrics on Three-Manifolds

We now focus our attention on some global questions. First, we will consider the existence of global solutions of (5.2). In this context, we turn to the following result.

5.30 Theorem (Hamilton [Ham 2]). *Let M be a connected, compact (without boundary), smooth, three-dimensional manifold, and assume that M admits a metric g such that* Ric(g) *is everywhere positive definite. Then M also admits a metric with constant positive sectional curvature.*

5.31 Remark. The conclusion of the theorem implies of course that M is the sphere S^3 or a quotient of S^3 by a discrete group. In a sense, then, this theorem is a Riemannian geometry version of the Poincaré conjecture. Note also that by a theorem of Aubin [Aub 1], it is enough to assume that $\mathrm{Ric}(g) \geqslant 0$ with strict inequality at one point (however, the manifold $S^2 \times S^1$, whose fundamental group is infinite, does not admit any metric with positive Ricci curvature, so it is not enough to assume just that $\mathrm{Ric}(g) \geqslant 0$).

It is natural to ask if there is an analogue of 5.30 for negative Ricci curvature. A counter-example is $S^2 \times S^1$, which Gao-Yau have shown has a strictly negative Ricci curvature metric; of course this cannot have a negative sectional curvature metric since then using the exponential map we see that its universal cover would have to be $\mathbb{R}^3$ while the universal cover of $S^2 \times S^1$ is $S^2 \times \mathbb{R}$.

5.32. The basic idea of the proof is to show that there is a family of metrics g_t for $t \in [0, \infty)$ such that g_0 is the given positive Ricci-curvature metric and $\lim_{t\to\infty} g_t$ is the desired constant-sectional curvature metric. The family g_t is obtained as an "integral curve" of a suitable defined vector field on the (Frechet) space of C^∞ Riemannian metrics. The most natural vector field to try is the gradient of some energy functional. Following Hilbert (see Chapter 4), the first such functional that springs to mind is the total scalar curvature $\mathbf{S}(g) = \int_M s_g \mu_g$. This leads to the following equation for the flow:

$$\frac{\partial}{\partial t} g = \frac{2}{n} s_g g - 2r_g.$$

Unfortunately, this equation is ill-behaved: it does not in general have solutions (even for a short time) because it implies a backward heat equation for the scalar curvature (examples of initial data for which *no* local solutions exist are known for this equation). To circumvent this problem, Hamilton considers instead the equation

$$\frac{\partial}{\partial t} g = \frac{2}{n} \rho g - 2r_g, \tag{5.33}$$

where ρ is the average of the scalar curvature:

$$\rho = \int_M s_g \mu_g \bigg/ \int_M \mu_g.$$

Equation (5.33) always has solutions, at least short-time, on any compact manifold of any dimension for any initial value of the metric. Hamilton showed that, for a compact three-manifold whose initial metric has positive Ricci curvature, the equation (5.33) has a solution for all time, that the Ricci curvature of g_t is positive for all t, and that, as $t \to \infty$ the metric g_t converges to a metric of constant positive curvature.

The complete proof of these statements is too technical and involved to be given in detail here, the estimates required to prove convergence of g_t involve intense local calculations. We will, however, indulge in a few calculations in an attempt to impart the spirit of the proof.

The first obstacle is the proof of short-time existence. Unlike many variational problems (e.g. the harmonic mapping problem), for which the gradient flow leads to parabolic equations, the Equation (5.33) used in this problem is not parabolic. The reason is the same as that for Equation (5.1) not being elliptic: the invariance of (5.33) under the action of the diffeomorphism group of M leads to the same degeneracies we have encountered before. Hamilton [Ham 2] overcomes this problem by using a powerful technique from analysis: the Nash-Moser implicit-function theorem. In fact, he proves a general theorem about "weakly parabolic" systems that can be applied. Recently, DeTurck [DeT 4] has outlined a proof of this part of Hamilton's theorem that is in the spirit of Theorem 5.14 above. It has the feature that the main analysis tool required is the classical parabolic existence and uniqueness theorem ([La-So-Ur], Theorem 5.2, p. 320, Theorem 9.1, pp. 341–342, [Ham], pp. 120–121).

We turn now to the long-time existence and convergence questions for (5.33). To simplify the calculations, we note with Hamilton that there is a change of variables that transforms (5.33) into

(5.34)
$$\frac{\partial}{\partial t} g = -2\,\mathrm{Ric}(g).$$

Namely, beginning with (5.34), choose $\psi(t)$ so that $\tilde{g} = \psi g$ has $\int_M \mu_{\tilde{g}} = 1$ for all t. Then set $\tilde{t} = \int \psi(t)\,dt$. An elementary calculation yields:

$$\frac{\partial}{\partial \tilde{t}} \tilde{g} = \frac{2}{n} \tilde{\rho} \tilde{g} - 2\,\mathrm{Ric}(\tilde{g}).$$

Consequently, we can consider the technically simpler (5.34) for the purpose of deriving estimates on the curvature of the solution, and then rescale them to apply to (5.33). In particular, the solution of (5.34) "blows up" (becomes singular) after some finite time T. An important step in proving existence for (5.33) on $[0, \infty]$ is showing that $\int_0^T \psi(t)\,dt = \infty$.

We now show that, if g_0 has positive definite Ricci curvature, then so does g_t for all $t > 0$. Since this property is scale-invariant, there is no loss of generality in working with (5.34). The first step is to derive an equation that expresses the evolution of the Ricci tensor (see also Chapter 1). From here on we only consider the three dimensional case, $n = 3$.

5.35 Lemma. *If g satisfies* (5.34), *and if* $r = \mathrm{Ric}(g)$, *then*

$$\frac{\partial}{\partial t} r_{ij} = \sum_{p,q} (g^{pq} r_{ij;pq}) - 6 s_{ij} + 3 s r_{ij} - (s^2 - 2v) g_{ij},$$

and

$$\frac{\partial s}{\partial t} = -\Delta s + 2|r|^2$$

where

$$V_{ij} = \sum_{p,q} g^{pq} r_{ip} r_{qj},$$

$$s = \sum_{p,q} g^{pq} r_{pq} = \textit{Scalar curvature of } g,$$

$$v = \sum_{p,q} g^{pq} V_{pq} = \text{trace}(V) = |r|^2.$$

The proof of the lemma is a straightforward calculation with the well-known formula for the first variation of the Ricci tensor (see 1.183), given that the first variation of the metric is equal to $-2r_{ij}$. (The above expression is, in fact, the Lichnerowicz Laplacian of $-r_{ij}$, compare 1.143) One also uses strongly the fact that the sectional curvature is completely determined by the Ricci tensor in three dimensions.

The proof that the Ricci curvature remains positive, as well as the proofs of the other critical estimates in this work, relies on the maximum principle for tensors [Ham 2], Theorem 9.1 (see also [Ham 4] for a clarification).

5.36 Lemma. *Suppose that on $0 \leqslant t \leqslant T$, a symmetric tensor M_{ij} satisfies*

$$\frac{\partial}{\partial t} M_{ij} = -\Delta M_{ij} + \sum_k z^k M_{ij;k} + N_{ij},$$

where the tensor field $N_{ij}(M, g)$ depends smoothly on M and g and satisfies $\sum_{i,j} N_{ij} v^i v^j \geqslant 0$ whenever $\sum_j M_{ij} v^j = 0$, z^k is any vector field, and the Laplacian is taken with respect to g (both z and g may depend on t). If $M \geqslant 0$ at $t = 0$, then it remains so for $0 \leqslant t \leqslant T$.

Armed with Lemmas 5.35 and 5.36, we can prove the following lemma. Its reasoning is typical of the proofs of many of the estimates in this work.

5.37 Lemma. *Assume that g_t satisfies (5.34) for $0 < t < T$. If the scalar curvature $s_{g_0} > 0$ then $s_{g_t} > 0$ for $0 \leqslant t < T$; if $\text{Ric}(g_0) > 0$ then $\text{Ric}(g_t) > 0$ for $0 \leqslant t < T$.*

Proof. From Lemma 5.35, $s(x, t) = s_{g_t}$ satisfies

$$\frac{\partial s}{\partial t} \geqslant -\Delta s.$$

Thus, by the maximum principle, if $t > 0$ then $s(x, t) > s(x, 0) > 0$. Consequently, $s_{g_t} > 0$. (This proof does not work if one attempts to use it to show that $s_{g_0} < 0$ implies $s_{g_t} < 0$, and shows how this heat equation method seems to prefer the positive curvature case.)

The assertion for $\text{Ric}(g_t)$ is proved similarly, only we need Lemma 5.36 with $z^k = 0$ to replace the elementary maximum principle. Let

$$N_{ij} = -6s_{ij} + 3sr_{ij} - (s^2 - 2v)g_{ij}.$$

Note that if the eigenvalues of r are λ, μ, and ν, then the eigenvalues of N are $(\mu - \nu)^2 - 2\lambda^2 + \lambda(\mu + \nu)$ and permutations of this. If, say, $\lambda = 0$, then the corresponding eigenvalue of N is $(\mu - \nu)^2 \geqslant 0$. $\square$

The next important estimate is used to show that the ratio of the eigenvalues of the Ricci tensor of the solution of (5.34) remains bounded for all time. This estimate is proved by applying the maximum principle to the evolution equation for the tensor r_{ij}/s.

5.38 Lemma. *If $s > 0$ and $r_{ij} > \varepsilon s g_{ij}$ for some $\varepsilon > 0$ at $t = 0$, then both conditions continue to hold as long as the solution exists.*

To bound the ratios of the eigenvalues of r_{ij}, one simply combines Lemma 5.38 with the trivial estimate $r \leqslant sg$. Thus, the Ricci curvature (and hence the sectional curvature) can be controlled in terms of the scalar curvature alone.

We now examine the conditions under which the solution of (5.34) will "blow up" at time T. Blow-up cannot occur unless the scalar curvature of g_t approaches infinity. To see this, we note that if u remains bounded, then so do the components of the sectional curvature tensor. But a calculation shows that the norm of each covariant derivative of the curvature, $|\partial^k R|$, satisfies an evolution equation of the form:

$$\frac{\partial}{\partial t}|\partial^k R|^2 = -\Delta|\partial^k R|^2 - 2|\partial^{k+1}R|^2 + \sum_{i+j=k} (\partial^i R|\partial^j R|\partial^k R).$$

Here, $(.|.|.)$ is some linear combination of the contractions of its factors. An interpolation inequality (i.e., Hölder's inequality plus integration by parts) yields:

$$\frac{d}{dt}\int_M |\partial^k R|^2 \leqslant C \int_M |\partial^k R|^2$$

for all k, where C depends on k and $\max|s|$. Thus, if $\max|s|$ is bounded, then so is $\int|\partial^k R|^2$, and so is $\int|\partial^k R|^p$ (by interpolation) and $\max|\partial^l R|$ (by the Sobolev inequality). Using all of this, we see that the form of the evolution equation (5.34) for g then implies that g_t is well-behaved as t approaches T. So we must have $\lim_{t\to T}\max s = \infty$.

The next step to show is that the *minimum* of s also becomes infinite as t approaches T. This will allow us to conclude that for g_t (the solution of (5.34)), its volume becomes infinite as $t \to T$ in such a way that $\int_0^T \psi(t)\,dt$ is infinite (recall the definition of ψ as the scale factor in the transition from (5.33) to (5.34)). This will imply that the solution of (5.33) exists for all time.

To show that $\min s \to \infty$, we estimate the gradient of s, $|Ds|$, as follows:

(5.39) $$|Ds| \leqslant \tfrac{1}{2}\eta^2 s^{3/2} + C(\eta)$$

for any $\eta > 0$. This is another maximum-principle estimate, and is obtained from the evolution equation for the quantity

$$\frac{|Ds|^2}{s} - \eta s^2$$

(these calculations and estimates are the most difficult and cleverest in the paper). Given (5.39) and the fact that $\max s \to \infty$ as $t \to T$, we can find a $T_1 = T_1(\eta)$ such that $C(\eta) < \frac{1}{2}\eta^2(\max s)^{3/2}$ for $T_1 < t < T$. Thus,

$$|\partial u| \leqslant \eta^2(\max s)^{3/2} \quad \text{for } t > T_1.$$

From this last estimate, it is immediate that $s(x) \geqslant (1-\eta)\max s$ at any point x whose geodesic distance from the point at which s achieves its maximum is less than $\frac{1}{\eta}(\max s)^{-1/2}$. But, by Myers' theorem (diameter of M is bounded above), every point of M is within this distance for sufficiently small η. Hence, $\min s \geqslant (1-\eta)\max s$, so $\frac{\max s}{\min s} \to 1$ as $t \to T$.

To complete the proof of Theorem 5.30, we must demonstrate that the solution of (5.33) actually converges to a limit metric, and that the metric has constant sectional curvature. As the reasoning above would indicate, the key step is to show that the eigenvalues of the Ricci curvature become more and more pinched as $t \to \infty$ (for the solution of (5.33)). The (5.34) version of this statement is to show that the quantity

$$\frac{(\lambda-\mu)^2 + (\lambda-\nu)^2 + (\mu-\nu)^2}{(\lambda+\mu+\nu)^2},$$

where λ, μ and ν are the eigenvalues of r, approaches zero as $t \to T$. This is yet another maximum principle estimate, since the above quantity is (in the language of Lemma 5.35),

$$\frac{3v - s^2}{s^2}.$$

The pinching is derived by using the maximum principle on the evolution equation for

$$3\frac{v}{s^{2-\delta}} - s^\delta$$

with $\delta > 0$ carefully chosen, to show that the latter quantity remains bounded as $t \to T$ (recall we have $s \to \infty$).

Given these estimates, rescaling shows that the eigenvalues of the Ricci tensor of the solution of (5.33) must approach each other and become constant as $t \to \infty$; hence the limit metric has constant sectional curvature. This completes our outline of the proof of Theorem 5.30. □

5.40 Remark. The obvious question to ask at this point is whether this result can be generalized to higher dimensions. Positive curvature seems to be "preferred" by the evolution process, since it always increases scalar curvature. The big stumbling block is the fact that the sectional curvature tensor is not completely determined by the Ricci tensor in dimension $\geqslant 4$. This changes the character of many of the auxiliary evolution equations that are derived in the course of the proof, and it is not at all clear how to recover most of the estimates. On the other hand, there are no delicate analytic facts that use the three dimensionality (e.g., no delicate Sobolev inequality estimates or Palais-Smale conditions), and some generalizations can indeed be found, e.g., in [Mag], [Ham 4], see also [Bou 12].

H. A Uniqueness Theorem for Ricci Curvature

We now turn to the second question posed at the beginning of this chapter—when does the Ricci tensor *uniquely* determine the metric? Of course, since $\mathrm{Ric}(cg) = \mathrm{Ric}(g)$ for any positive constant c, one can only have uniqueness up to scaling (and up to independent scaling the factors of a product metric). To save words, we will say that the metric is uniquely determined in this case. A recent theorem of Hamilton, extended by DeTurck and Koiso [De-Ko] shows that the standard metrics on certain symmetric spaces are indeed determined uniquely by their Ricci tensors.

5.41 Theorem. *Let M be an irreducible, Riemannian symmetric space of compact type, and let g_0 be the standard (symmetric) metric on M, scaled so that $\mathrm{Ric}(g_0) = g_0$. If there is another Riemannian metric g_1, on M such that $\mathrm{Ric}(g_1) = g_0$, then $g_1 = cg_0$ for some positive constant c. A priori, g_1 need not respect any homogeneous structure on M associated with g_0.*

The theorem is due to Hamilton [Ham 3] for the spheres S^n, and to DeTurck-Koiso for other manifolds.

This theorem is a corollary of the following more general result which begins with the observation that if two metrics g and $\bar{g}$ induce the same connection, then $\mathrm{Ric}(g) = \mathrm{Ric}(\bar{g})$. Combined with deRham's decomposition theorem for complete simply connected Riemannian manifolds (see 10.43), the strongest statement one can hope to make is that the Ricci curvature uniquely determines the Levi-Civita connection—unless one makes special assumptions on the manifold, such as those of Theorem 5.41.

5.42 Theorem [De-Ko]. *Let $(M, \bar{g})$ be a compact Einstein manifold with $\mathrm{Ric}(\bar{g}) = \bar{g}$ and with non-negative sectional curvature. If g is another Riemannian metric on M with $\mathrm{Ric}(g) = \bar{g}$, then g and $\bar{g}$ have the same Levi-Civita connection.*

The key step is the formula for $\Delta_g(g^{ik}\bar{g}_{ik})$, under the assumption that the identity map $id\colon (M, g) \to (M, \bar{g})$ is a harmonic map, in which case $g^{ik}\bar{g}_{ik}$ is precisely the harmonic mapping energy density. In our situation where $0 < \mathrm{Ric}(g) = \bar{g}$, then $g^{ik}\bar{g}_{ik} = \mathrm{Scal}(g) = s$ and the formula for $\Delta_g(g^{ik}\bar{g}_{ik})$ implies

$$-\Delta s \geqslant 2g^{ik}g^{jl}(\bar{g}_{ij}\bar{g}_{kl} - \bar{R}_{ijkl}), \tag{5.43}$$

with equality holding only if g and $\bar{g}$ have the same Levi-Civita connection. Here $\bar{R}_{ijkl}$ is the curvature tensor of $\bar{g}$ and the fact that $id\colon (M, g) \to (M, \bar{g})$ is a harmonic map is precisely the differential Bianchi identity for Ricci curvature 1.94.

The proof of Theorem 5.42 is completed by showing that under the stated assumptions the right side of (5.43) is non-negative, since then $-\Delta s \geqslant 0$ and hence s is a constant—so one actually has equality in (5.43). Therefore g and $\bar{g}$ have the same Levi-Civita connection. □

The point of this section is that in certain circumstances the metric—or at least the connection—is uniquely determined by the Ricci curvature. There are many situations where there is essential non-uniqueness. One class of examples is the 19-dimensional family of non-cohomologous Ricci-flat metrics on the K3 surface which are obtained using the solution of Calabi's problem (see 12.J and K) and some other explicit examples of non-cohomologous Kähler metrics with the same Ricci tensor constructed by Calabi (see [Cal 5]). In all of these cases, the non-uniqueness is finite dimensional, and leads one to ask if the non-uniqueness is always finite dimensional.

I. Global Non-Existence

Earlier in this chapter we presented some local existence and non-existence results for the equation

$$\mathrm{Ric}(g) = r.$$

Following DeTurck-Koiso, we will now give some situations where a solution does exist locally but *not* globally.

5.44 Theorem [De-Ko]. *Let $(M, \bar{g})$ be a compact Riemannian manifold. There is a constant $c_0(\bar{g})$ such that if $c > c_0(\bar{g})$ there is no Riemannian metric g such that* $\mathrm{Ric}(g) = c\bar{g}$. *In particular, if either the sectional curvature $K_{\bar{g}} \leqslant 1/(n-1)$ or $\bar{g}$ is Einstein with* $\mathrm{Ric}(\bar{g}) = \bar{g}$, *then $c_0(\bar{g}) = 1$.*

To prove this, one shows that under these assumptions the right side of (5.43) is strictly positive, so $-\Delta s > 0$, which is impossible since at a maximum of s we have $-\Delta s \leqslant 0$. □

It is helpful to compare this with Bochner's result which asserts that if $\mathrm{Ric}(g) = r > 0$, then the first Betti number of M must be zero. Thus for most M a solution will not exist. Here we have the more subtle situation that a solution of $\mathrm{Ric}(g) = cr$ may exist for small $c > 0$ but not for large c.

Chapter 6. Einstein Manifolds and Topology

A. Introduction

Which compact manifolds do admit an Einstein metric? Except in dimension 2 (see Section B of this chapter), a complete answer to this question seems out of reach today. At least, in dimensions 3 and 4, we can single out a few manifolds which definitely *do not* admit any Einstein metric.

6.1. In dimension 3, an Einstein manifold has necessarily constant sectional curvature, and so, its universal covering is diffeomorphic either to $\mathbb{R}^3$ or to S^3. In particular, $S^2 \times S^1$ has no Einstein metric. Due to W. Thurston's deep results, the gap between necessary and sufficient conditions for a 3-manifold to be Einstein has considerably shrunk recently. It appears that only a few exceptions are ruled out. Manifolds which admit negative Einstein metrics are by far the general case. In Section C, we list topological restrictions for Einstein 3-manifolds, and quote, without proof, some of W. Thurston's "hyperbolization theorems."

6.2. In dimension 4, a topological obstruction to the existence of an Einstein metric arises from the special form that the integral formulas for the signature and Euler characteristic take. From successive works of M. Berger, J. Thorpe and N. Hitchin, we know that, on a compact Einstein manifold M of dimension 4, the Euler characteristic $\chi(M)$ and the signature $\tau(M)$ satisfy the inequality

$$|\tau(M)| \leqslant \tfrac{2}{3}\chi(M).$$

The same property is at the origin of M. Gromov's inequality

$$\chi(M) \geqslant \frac{1}{2592\pi^2}\|M\|,$$

for a compact Einstein 4-manifold M. Here, $\|M\|$ denotes Gromov's simplicial volume (see Section D).

6.3. In dimensions greater than 4, we do not know of any topological restriction for a manifold to be Einstein. It may very well be that any manifold with dimension greater than 4 admits a negative Einstein metric—or, that most manifolds do.

If one requires that the Einstein constant be positive, then two types of results show up.

6.4. The first type arises from the interaction between the Ricci curvature and the fundamental group. If a complete Riemannian manifold satisfies

$$r \geqslant kg$$

for a positive constant k, then (S.B. Myers) M and its universal cover are compact, so the fundamental group $\pi_1(M)$ is finite. If a compact Riemannian manifold (M,g) has nonnegative Ricci curvature, then (J. Cheeger and D. Gromoll) its universal cover is isometric to a product $\overline{M} \times \mathbb{R}^q$, where $\overline{M}$ is compact, so, for some finite subgroup $F \subset \pi_1(M)$, there is a subgroup $\mathbb{Z}^q$ of finite index in $\pi_1(M)/F$. These and earlier results are collected in Section E.

6.5. The second type of results deals with the scalar curvature. Using the Dirac operator, A. Lichnerowicz proved in 1963 that, on a $4m$-dimensional compact spin manifold, the $\hat{A}$-genus is an obstruction to the existence of metrics with positive scalar curvature. This result, which was subsequently extended by N. Hitchin, M. Gromov and H.B. Lawson, lies at the heart of the problem of existence of metrics with positive scalar curvature. It provides examples of simply connected manifolds which do not admit positive Einstein metrics, and which, of course, are not covered by 6.2 to 6.4. A quick account of this theory has been included in section F.

Finally, we sketch in the last Section G a proof of the Cheeger-Gromoll theorem quoted in 6.4 above.

B. Existence of Einstein Metrics in Dimension 2

6.6. According to Proposition 1.120, a metric on a 2-dimensional manifold is Einstein if and only if it is locally isometric to one of the following three model spaces: S^2, $\mathbb{R}^2$ or H^2 endowed with their canonical metrics. This allows one to construct Einstein manifolds by gluing together manifolds with boundary. This gives an elementary proof of the following Theorem.

6.7 Theorem. *Any 2-dimensional manifold admits a complete metric with constant curvature.*

6.8. Let us first detail the process of gluing together manifolds with constant curvature -1 and geodesic boundary.

For a positive number ε, let T_ε denote the ε-neighborhood of a geodesic L in hyperbolic plane H^2. Let $T^+_{\varepsilon,l}$ (resp., $T^-_{\varepsilon,l}$) denote the quotient of T_ε by the unique orientation preserving (resp., reversing) motion of H^2 which, restricted to L, is a translation by l. Let M_1, M_2 be 2-manifolds whose interiors have constant curvature, and let L_1, L_2 be components of their boundaries which are geodesics with common length l. Then, for ε small enough, any isometry φ of L_i onto the closed geodesic L^+ in $T^+_{\varepsilon,l}$ extends to an isometry of the ε-neighborhood of L_i in M_i onto either half of $T^+_{\varepsilon,l}$. One obtains a homeomorphism of a neighborhood of $L_1 = L_2$ in the gluing $M_1 \bigcup_\varphi M_2$ onto $T^+_{\varepsilon,l}$, which, restricted to M_i, is an isometry. Thus we have constructed an Einstein metric on $M_1 \bigcup_\varphi M_2$. In the same way, any two-fold isometric covering $L_1 \to L^-$ extends to an isometry of an ε-neighborhood of L_1 in M_1 onto $T^-_{\varepsilon, l/2}$, so we can construct an Einstein metric on $M_1 \bmod \varphi$, where φ is translation by $l/2$ in L_1.

The next step is to cut any 2-manifold into simple pieces. This can be done by a purely topological argument.

6.9. Let M be a (possibly open) 2-manifold. Consider pairs (c, U) where c is a simple closed smooth curve in M, and U a regular neighborhood of c. Let us choose a maximal set of pairs (c, U) such that

(i) c does not bound a disk in M,

(ii) if $(c, U) \neq (c', U')$, then the closures $\bar{U}, \bar{U}'$ are disjoint and c, c' are not isotopic in M.

Let us cut M along the curves c into pieces which are (possible open) 2-manifolds with boundary (more precisely, this amounts to deleting the regular neighborhoods U). There might be only one piece, for example if there are no curves c or only a unique one-sided curve. Let P be some piece. We obtain a new manifold without boundary $\hat{P}$ by gluing a disk to each boundary component of P. One readily sees that a) $\hat{P}$ has genus zero, i.e., any simple closed smooth curve in $\hat{P}$ bounds a disk in $\hat{P}$; b) $\hat{P}$ has at most one end, i.e., for any compact K in $\hat{P}$, the complement $\hat{P} - K$ has at most one unbounded connected component. This implies that $\hat{P}$ is diffeomorphic either to S^2 or to $\mathbb{R}^2$. Thus P is diffeomorphic either to the complement of a finite collection of disjoint disks in S^2, or to the complement of a locally finite collection of disks in $\mathbb{R}^2$. Then one easily sees that the number of disks is at most 3 in the first case and 1 in the second.

6.10. Apart from the cases $P = M = S^2$ or $\mathbb{R}^2$, we obtain pieces of one of the following types:

—disk $= S^2 -$ disk;

—annulus $= S^2 -$ 2 disks;

—"pair of pants" $= S^2 -$ 3 disks;

—punctured disk $= \mathbb{R}^2 -$ disk.

The disk may occur only if its boundary doubly covers a one-sided curve in M. In this case, there is only one piece and $M = \mathbb{R}P^2$.

The annulus can occur only if its two boundary components cover the same curve in M. Again, there is only one piece, and M is a torus or a Klein bottle.

The punctured disk occurs either alone, with its boundary doubly covering a one-sided curve in M—in which case M is a Möbius strip—or glued to another punctured disk—in which case M is an open cylinder—or glued to a pair of pants.

6.11. The manifolds obtained by gluing one, two or three punctured disks to the boundary of a pair of pants will be called an *infinite*, a *doubly infinite* or a *triply infinite pair of pants.*

Proposition. *Each of the* 4 *types of pairs of pants admits complete metrics of constant curvature* -1 *with geodesic boundary components of any prescribed length.*

Proof. These metrics are obtained by gluing together two identical right-angled hyperbolic hexagons, along three sides. The lengths of the other three sides may be chosen arbitrarily-including zero, in order to cover the case of (—, doubly, triply) infinite pairs of pants. This is a pleasant exercise in elementary geometry. □

6.12. End of proof. The manifolds S^2 and $\mathbb{R}P^2$ admit metrics with constant positive curvature. The manifolds $\mathbb{R}^2$, torus, Klein bottle, Möbius strip, cylinder, admit complete metrics with curvature zero. If M is not diffeomorphic to one of these three exceptions, then, in the decomposition above, all pieces belong to the above types of pairs of pants. Let all of them be equipped with metrics of curvature -1 such that all boundary components have the same length $e^{\sqrt{2}}$. Then, to any choice of isometries between corresponding boundary circles, there corresponds uniquely, along the process 6.8, an Einstein structure on the initial surface M. □

6.13. This construction is due to W.Thurston (see [Thu 1], [Fa-La-Po], [Bsr]). Pushed further, it gives one of the best description of the Teichmüller space of a surface, together with important applications such as a classification of diffeomorphisms of surfaces. It is already clear from the few lines above, that, except for the triply infinite pairs of pants, the Riemannian structure with constant curvature -1, when it exists, is not unique. For more on this point, see 12.B.

6.14 Remark. The sign of the constant curvature (i.e. $1, 0, -1$) is determined by the sign of the Euler characteristic (assume $-\infty$ is negative). In general, it is -1. The $+1$ and 0 cases occur only for a short list of exceptions.

C. The 3-Dimensional Case

6.15. A Riemannian metric on a 3-dimensional manifold is Einstein if and only if it has constant sectional curvature (Proposition 1.120). In contrast with the 2-dimensional case, every 3-manifold does not admit a metric with constant sectional curvature. We now list a few necessary conditions.

6.16. If a 3-manifold M admits a metric with constant sectional curvature, then its universal covering is diffeomorphic, either to the 3-sphere S^3, or to $\mathbb{R}^3$. In particular, $\pi_2(M) = 0$. More precisely, any embedded 2-sphere S^2 in M bounds an embedded 3-ball B^3 in M. One says that M is *prime* (see [Rol]). Thus clearly $S^2 \times S^1$ has no metric with constant sectional curvature. Neither does any manifold M which admits a non trivial connected sum decomposition $M = N \# P$ (where N and P are not diffeomorphic to S^3).

6.17. Since compact manifolds admitting flat metrics form a restricted family (there are ten of them), and manifolds admitting metrics with constant curvature $+1$ are somehow well understood (see [Wol 1]), let us now restrict our attention to negative curvature. Let M be compact and admit constant sectional curvature -1. Then the fundamental group $\pi_1(M)$ is isomorphic to a discrete, cocompact subgroup Γ of motions of hyperbolic 3-space H^3 (see 1.37). Any element of Γ is hyperbolic, i.e., globally preserves a geodesic of H^3. Any two commuting elements preserve a common geodesic, thus correspond to iterates of the same closed geodesic in M, i.e., are powers of the same element of Γ. As a consequence, Γ contains no subgroup isomorphic to $\mathbb{Z} \oplus \mathbb{Z}$. One then says that M is *homotopically atoroidal.*

6.18. More generally, the fundamental group of a compact manifold with sectional curvature -1 cannot contain solvable subgroups, only cyclic subgroups. (In fact, the relevant notion is that of an *amenable* group, see [Grf]). Indeed, any non-cyclic amenable discrete group of motion of H^3 contains a non-hyperbolic element.

6.19. These simple properties are sufficient to rule out a number of candidates. Let us examine two simple ways of constructing 3-manifolds, circle bundles over surfaces and manifolds fibring over the circle:

—circle bundles over S^2 depend on an integer $k \in \mathbb{Z} = \pi_1(\mathrm{Diff}(S^1))$. For positive k, the fibre has order k in $\pi_1(M_k)$, and $M_k \to S^2$ is covered by the Hopf fibration $M_1 = S^3 \to S^2$. These manifolds admit constant curvature $+1$ metrics. For $k \leqslant 0$, $\pi_1(M_k) = \mathbb{Z}$ and M_k does not admit a constant curvature metric:

—if Σ is a surface with $\chi(\Sigma) \leqslant 0$, then a circle bundle M over Σ is not homotopically atoroidal. Indeed, the fibre is a non trivial central element in a subgroup of finite index of $\pi_1(M)$. Thus M does not admit curvature -1, nor $+1$. However, among the ten flat 3-manifolds, two fiber over the Klein bottle K, the product $S^1 \times K$ and an other manifold F. Only T^3 and $S^1 \times K$ fiber over T^2.

6.20. Any 3-manifold M which fibers over the circle with fibre a surface Σ is obtained by identifying the boundary components of $\Sigma \times [0, 1]$ via a diffeomorphism φ of Σ. The construction only depends on the isotopy class of φ, but still, it produces a huge variety of examples.

—Sphere or $\mathbb{R}P^2$ bundles over the circle are finitely covered by $S^2 \times S^1$, and do not admit constant curvature metrics.

—If Σ is a surface with $\chi(\Sigma) \leqslant 0$, then isotopy classes of diffeomorphisms of Σ are in 1-1 correspondance with outer automorphisms of $\pi_1(\Sigma)$. For example, if Σ is the 2-torus, a torus bundle M_φ is determined by a matrix φ in $Gl(2, \mathbb{Z})$. The fundamental group $\pi_1(M_\varphi)$ is the extension of $\mathbb{Z}^2$ by φ, hence solvable, so M_φ never admits a constant curvature -1 metric. However, M_φ is flat if and only if φ is periodic. Note that M_φ is homotopically atoroidal if and only if φ has two distinct real eigenvalues.

—All compact flat 3-manifolds but one (the manifold F in 6.19) have $b_1(M, \mathbb{R}) \neq 0$. These manifolds fiber over the circle. In seven cases, the fiber is a torus. In the remaining two cases, it is a Klein bottle. Other Klein bottle bundles over the circle do not admit constant sectional curvature metrics.

—Assume $\chi(\Sigma) < 0$. If φ is periodic, then M_φ is covered by $\Sigma \times S^1$ and hence not homotopically atoroidal. In fact, M_φ is homotopically atoroidal iff no iterate of

φ preserves an isotopy class of curves. In this case, φ is called *irreducible* or *pseudo-Anosov* (see [Fa-La-Po]). It is known (see 6.24) that M_φ admits a metric with constant sectional curvature -1.

The examples we have just quoted are non-representative. Indeed, W. Thurston's recent results tend to show that most prime 3-manifolds admit metrics with constant sectional curvature -1. Let us begin with a pleasant sufficient condition, though far from necessary.

6.21 Definition. A two-sided surface N embedded in a 3-manifold M is called *incompressible* if any simple closed curve in N which bounds a disk in M with interior disjoint from N also bounds a disk in N. A compact 3-manifold is called *Haken* if it is prime and contains a two-sided incompressible surface which is not a sphere.

6.22. Among the examples of 6.19 and 6.20, all are Haken but the ones which involve spheres or real projective spaces.

6.23 Theorem (W. Thurston [Thu 3]). *A compact orientable Haken manifold admits a metric with constant sectional curvature* -1 *if and only if it is homotopically atoroidal.*

6.24. The preceding Theorem applies to manifolds M_φ fibering over the circle, with φ an irreducible diffeomorphism of the (orientable) fiber. It produces rather unexpected examples of hyperbolic manifolds. Indeed, the cyclic covering $\hat{M} \to M$ induced by the universal covering of the base circle has infinite volume, but the volume of balls in it grows linearly, a behaviour which cannot happen in dimension 2. The first examples of this phenomenon have been discovered by T. Jørgensen [Jør]. See [Sul] for an exposition of the proof of Theorem 6.23 in the special case of manifolds fibring over the circle.

In [Thu 2], W. Thurston writes that Theorem 6.23 is not fully satisfactory. Indeed, the Haken condition is a technical help for the proof, and probably not a concept profoundly linked with constant curvature. We now explain another method used by W. Thurston to produce even more compact 3-manifolds with constant curvature.

6.25. A *link* in a compact 3-manifold is a finite disjoint union of simple closed curves. The process of *Dehn surgery* along a link L amounts to removing a regular neighborhood of L, and gluing it back in by some new identification. The manifold obtained in this way depends on a discrete parameter: Indeed, an element of $\mathrm{Diff}(T^2)/\mathrm{Diff}(D^2 \times S^1)$ is determined by a pair of mutually prime integers.

6.26 Theorem (W. Thurston [Thu 1]). *Let L be a link in a compact manifold M. Assume that the complement $M - L$ admits a complete metric with constant sectional curvature -1 and finite volume. Then, for all but a finite number of choices of the parameter, the manifolds obtained by Dehn surgery along L admit a metric with constant sectional curvature -1.*

6.27. Every compact 3-manifold is obtained from the 3-sphere S^3 by Dehn surgery along some link L (see [Rol]). Furthermore, L may be chosen so that its complement admits a complete metric with constant curvature -1. Indeed, another important theorem of W. Thurston characterizes the knots whose complement has this property (see [Thu 2]). Thus, in some sense, almost every compact 3-manifold admits a constant sectional curvature -1 metric.

6.28. However, the question of which compact manifolds admit constant curvature is still unsettled. For example, any compact Riemannian manifold with strictly negative sectional curvature satisfies properties 6.17 and 6.18 (see for example [Gro 6]). It is still unknown whether such a manifold admits a metric with *constant* negative sectional curvature. The corresponding question for the class of manifolds admitting negative Ricci curvature has received a negative answer recently. Indeed, this class is stable under connected sums, and connected sum with products $\Sigma \times S^1$ or lens spaces, see [Gao]. Note, in contrast, that a compact 3-manifold with positive Ricci curvature admits constant sectional curvature, this is R. Hamilton's Theorem 5.30.

6.29. In 0.4, we mentioned that a motivation for studying Einstein metrics was the search for especially nice Riemannian metrics on any given compact manifold. In [Thu 2], W. Thurston proposes the following reformulation of the problem for 3-manifolds.

—The nicest metric on a 3-manifold (if there is one) is a locally homogeneous structure. In 3 dimensions, there are eight simply connected homogeneous spaces G/H where G is a maximal group of isometries of G/H (and which admit finite volume quotients). These spaces are described in detail in [*Sco*]. Of course, among these, we find the canonical sphere, euclidean space and hyperbolic space.

—Not every 3-manifold M is locally homogeneous, so one must admit to cut M into pieces. The first step consists in decomposing M into a connected sum of "smallest" manifolds. This can be done in a canonical way. The pieces obtained are called *the prime summands* of M. As we have already seen, all prime manifolds are not locally homogeneous, so a further decomposition is necessary. One cuts M along non trivially imbedded 2-tori, obtaining a family of open manifolds. One will consider that the original program is fulfilled if each piece admits a complete locally homogeneous structure.

—In fact, this modified program is now solved, except in the case of compact, homotopically atoroidal manifolds. Indeed, if M is not homotopically atoroidal, it admits nontrivially embedded submanifolds which are circle bundles—or, more generally, Seifert bundles. There is a maximal submanifold with these properties, unique up to isotopy, which is called *the characteristic variety* of M (see [Joh]). According to another theorem of W. Thurston ([Thu 2], Theorem 2.3), the complement of the characteristic variety admits a complete metric with curvature -1, except for two exceptions which admit complete flat metrics. It remains to further decompose a Seifert bundle into locally homogeneous pieces, in a unique way. Again, hyperbolic geometry serves to model most three manifolds, the other 7 geometries appearing only for a small (but infinite) number of exceptions.

D. The 4-Dimensional Case

6.30. For a differentiable manifold M of dimension n, an important topological invariant, the *Euler characteristic* $\chi(M)$, can be defined in the following way.

Let $b_i = \dim H^i(M, \mathbb{R})$ be *the Betti numbers* of M, then we have

$$\chi(M) = \sum_{i=0}^{n} (-1)^i b_i.$$

For example, $\chi(S^{2m}) = 2$, $\chi(S^{2m+1}) = 0$, $\chi(\mathbb{C}P^m) = m$.

When M is a compact oriented Riemannian manifold of dimension $2m$, the generalised Gauss-Bonnet theorem (cf. [Ko-No 2]) provides a formula for $\chi(M)$ as an integral of a polynomial of degree m in the curvature.

6.31. The 4-dimensional case is very simple (cf. [Bes 2], Exposé n° X). One gets

$$\chi(M) = \frac{1}{8\pi^2} \int_M (\|U\|^2 - \|Z\|^2 + \|W\|^2)\mu_g.$$

where U, Z, W are the irreducible components of the curvature tensor R of M (cf. 1.114).

6.32. If (M, g) is an Einstein manifold, then $Z \equiv 0$. Therefore $\chi(M)$ is positive and can be zero only if (M, g) is flat. This theorem is due to M. Berger (cf. [Ber 2]). It follows that $S^1 \times S^3$ and $T^4 \# T^4$ admit no Einstein metrics.

6.33. Let us suppose now that M is a compact oriented $4k$-dimensional manifold. Then, the cup product defines a symmetric bilinear form B on $H^{2k}(M, \mathbb{Z})$ as follows

$$B(\alpha, \beta) = \langle \alpha \cup \beta, [M] \rangle,$$

where $\alpha, \beta \in H^{2k}(M, \mathbb{Z})$ and where $[M]$ is the fundamental homology class of M.

6.34. The signature of the bilinear form B is called the *signature* of M and is denoted by $\tau(M)$. This number is a topological invariant, and in the 4-dimensional case we have (cf. [Bes 2]).

$$\tau(M) = \frac{1}{12\pi^2} \int_M (\|W^+\|^2 - \|W^-\|^2)\mu_g,$$

where W^+ and W^- are the irreducible components of the Weyl tensor W under the action of the special orthogonal group (cf. 1.126).

If we combine the formulas for $\chi(M)$ and $\tau(M)$ for the given orientation and also for the opposite orientation (in which case τ becomes $-\tau$) we obtain

6.35 Theorem (J. Thorpe cf. [Tho], [Hit 1]). *Let M be a compact oriented Einstein manifold of dimension* 4. *Then the Euler characteristic $\chi(M)$ and the signature $\tau(M)$ satisfy the inequality*

$$\chi(M) \geqslant \tfrac{3}{2}|\tau(M)|.$$

6.36. The equality case $\chi(M) = \frac{3}{2}|\tau(M)|$ in Theorem 6.35 has been studied by N. Hitchin in [Hit 1]. Suppose for example that we have $\frac{2}{3}\chi(M) = -\tau(M)$ (we will see that, if the equality is achieved, M is in fact complex and has a natural orientation).

For the integrands of χ and τ we then have

$$\|U\|^2 + \|W^+\|^2 + \|W^-\|^2 = \|W^-\|^2 - \|W^+\|^2,$$

hence $U = W^+ = 0$ and $R = W^-$.

Let us suppose that R is non zero. The bundle of 2-forms on M, whose fibre can be identified with $\mathfrak{so}(4)$, can be decomposed under the action of the Hodge star operator into the direct sum

$$\wedge^2 M = \wedge^+ M \oplus \wedge^- M.$$

The fibre of $\wedge^+ M$ or of $\wedge^- M$ will be identified with $\mathfrak{su}(2)$. Let $\mathscr{R}$ be the curvature of the bundle $\wedge^2 M$. It is a 2-form with values in the endomorphisms of $\wedge^2 M$. For X, Y in $C^\infty(TM)$ and for ω in $\Omega^2(M)$ we have

$$\mathscr{R}(X, Y)\omega = [R(X, Y), \omega].$$

Denote by $\mathscr{R}^+$ and $\mathscr{R}^-$ the respective curvatures of the bundles $\wedge^+ M$ and $\wedge^- M$ for the natural induced connection.

Then $\mathscr{R} = \mathscr{R}^+ + \mathscr{R}^-$ and, for ω in $\Omega^+(M)$, we have

$$\mathscr{R}(X, Y)\omega = \mathscr{R}^+(X, Y)\omega = [R(X, Y), \omega].$$

Since $R = W^-$ maps $\wedge^- M$ to itself, we have $\mathscr{R}^+ = 0$. Now the Euler characteristic is positive (since $R \not\equiv 0$), so the first Betti number $b_1(M)$ must vanish. Indeed, if $b_1(M)$ is not zero, there exists a harmonic 1-form on M. Since the Ricci curvature vanishes, it follows from Bochner's theorem that it must be parallel. We will then obtain a non-vanishing vector field on M which contradicts the fact that $\chi(M)$ is positive.

Let us suppose that M is simply connected. Then the bundle $\wedge^+ M$ is trivial and the holonomy group can be reduced to $SU(2)$. The manifold M is Kähler and its Ricci curvature is identically zero. Then its first Chern class vanishes automatically also. The manifold must therefore be a K3 surface, by definition (see 12.104).

If M is not a K3 surface, a corollary (cf. [Ch-Gr 1]) of Theorem 6.65, implies that $\pi_1(M)$ must be finite since $\chi(M)$ is positive. The universal cover of M is then a finite covering, therefore the first Betti number and the first Chern class vanishes. The universal cover of M is in fact a K3 surface. □

6.37 Theorem (N. Hitchin cf. [Hit 1]). *Let M be a compact oriented 4-dimensional Einstein manifold. If the Euler characteristic $\chi(M)$ and the signature $\tau(M)$ satisfy*

$$|\tau(M)| = \tfrac{2}{3}\chi(M),$$

then the Ricci curvature vanishes, and M is either flat or its universal cover is a K3 *surface. In that case, M is either a* K3 *surface itself ($\pi_1(M) = \{1\}$), or an Enriques surface ($\pi_1(M) = \mathbb{Z}_2$) or the quotient of an Enriques surface by a free antiholomorphic involution ($\pi_1(M) = \mathbb{Z}_2 \times \mathbb{Z}_2$).*

6.38. Theorem 6.35 allows us to exhibit examples of simply connected manifolds which admit no Einstein metric. If $M = (\mathbb{C}P^2)^{\# p}$, connected sum of p copies of $\mathbb{C}P^2$ with the natural orientation, then

$$\chi(M) = p + 2, \qquad \tau(M) = p.$$

It follows that as soon as $p \geqslant 4$, $(\mathbb{C}P^2)^{\# p}$ admits no Einstein metric. The preceding theorem shows the importance of K3 surfaces and justifies the presentation of the following result.

6.39 Definition. A $4k$-dimensional Riemannian manifold (M,g) is called *hyperkählerian* if it admits three almost complex structures I, J, K such that $I^2 = J^2 = K^2 = -Id$, $IJ = -JI = K$ and if moreover g is Hermitian for I, J, and K. One can show that (M,g) is hyperkählerian if and only if $\mathrm{Hol}(g) \subset Sp(k)$ (cf. 10.33, 14.3).

6.40 Theorem. *On a differentiable manifold M which is homotopy equivalent to a* K3 *surface, given a Riemannian metric g, the following properties are equivalent:*
i) *the scalar curvature of g is non-negative;*
ii) *(M,g) is hyperkählerian;*
iii) *the Ricci curvature r of g is zero.*

Any K3 surface M is spin because $w_2(M) \equiv c_1(M) \bmod 2$ hence is 0. Moreover $\hat{A}(M) = \dfrac{\tau(M)}{16} = -1$ and by Lichnérowicz theorem (cf. § F) M cannot admit a metric with positive scalar curvature. Thus M is in category **B** of Theorem 4.35. In particular, any metric on M with nonnegative scalar curvature is Ricci flat, and i) implies iii).

We saw (6.36) that on M the bundle of positive 2-forms $\Lambda^+ M$ is trivial. Hence we can find three parallel sections I, J, K of $\Lambda^+ M$ such that $I^2 = J^2 = K^2$ and $IJ = -JI = K$ and M is hyperkählerian. Finally ii) implies iii) (cf. 10.67) because non-zero Ricci curvature is an obstruction for a Riemannian manifold to have $SU\left(\dfrac{n}{2}\right)$ as holonomy group and $SU(2) = Sp(1)$. □

6.41 Remarks. One can show (cf. [Pol 1]) that the inequality in Theorem 6.35 is not characteristic of Einstein manifolds. This inequality still holds under various pinching conditions on the curvature.

The method just used to prove Theorem 6.35 is in fact completely algebraic. One evaluates the integrands of the two characteristic numbers at each point x of M and then compares them. One can show that in dimension at least 6, there exist (cf. [Ger], [Bo-Po]) algebraic counter-examples to any generalisation of the inequalities satisfied by the 4-dimensional integrands of characteristic numbers.

If, in the assumptions of Theorem 6.35, one includes the fact that the sectional curvature has a fixed sign, the conclusion may be improved in

$$|\tau(M)| \leqslant (\tfrac{2}{3})^{3/2}\chi(M), \qquad \text{see [Hit 1].}$$

6.42. Another consequence of Formula (6.31) is the following. Let (M,g) be a negative Einstein manifold of dimension 4, normalized so that $r = -(n-1)g$. Then $|U|^2 = \dfrac{n(n-1)}{2} = 6$ and so $\chi(M) \geqslant \dfrac{3}{4\pi^2}\operatorname{Vol}(M,g)$.

But there are compact manifolds for which metrics with $r \geqslant -(n-1)g$ cannot have arbitrarily small volume. Indeed, there is an obstruction known as Gromov's simplicial volume.

6.43. The simplicial volume is a topological invariant of oriented compact manifolds. Let M have dimension n. The fundamental class $[M]$ is a singular homology class in $H_n(M, \mathbb{R})$, i.e., an equivalence class of singular cycles, that is, linear combinations of simplices

$$c = \sum \lambda_i \sigma_i \quad \text{with } \lambda_i \in \mathbb{R} \quad \text{and} \quad \partial c = 0.$$

By definition, the *simplicial volume* $\|M\|$ of M is the infimum of all sums

$$\sum |\lambda_i|$$

over all cycles $c = \sum \lambda_i \sigma_i$ in the class $[M]$.

6.44. Clearly, if there is a map of degree d from M onto N, then

$$\|M\| \geqslant |d|\,\|N\|.$$

Since S^n has self maps of any degree, then $\|S^n\| = 0$ for all $n \geqslant 1$. In fact, M. Gromov has shown that $\|M\| = 0$ for all simply connected manifolds M. More generally ([Gro 2]) vanishing of simplicial volume depends only on the fundamental group.

6.45. On the other hand, there are manifolds for which $\|M\| \neq 0$. This is the case as soon as M admits a metric with negative sectional curvature (Thurston's Theorem, [Gro 2], p. 10 and 20). In case M is a surface with $\chi(M) < 0$, there is an elementary proof that $\|M\| = -2\chi(M)$.

Further examples are obtained by taking connected sums, since, if $\dim M = \dim N \geqslant 3$,

$$\|M \# N\| = \|M\| + \|N\|,$$

or products, since

$$\|M \times N\| \geqslant C\|M\|\,\|N\|,$$

where C is a constant depending only on $\dim(M \times N)$. In dimension 4, one may take $C = 8$.

6.46. The main result concerning simplicial volumes is M. Gromov's Main Inequality: *if* (M,g) *is compact with* $r \geqslant -(n-1)g$, *then* $\operatorname{Vol}(M,g) \geqslant C'\|M\|$, where C' is a (non sharp) constant depending only on the dimension. In dimension 4, one may take $C' = \frac{1}{1944}$ ([Gro 2], p. 12).

Using 6.42, we obtain

6.47 Theorem (M. Gromov [Gro 2], p. 87). *Let M be a 4-dimensional compact manifold. If M admits an Einstein metric, then*

$$\|M\| \leqslant 2592\pi^2\chi(M).$$

6.48 Example. Let Σ be a closed surface of genus γ. Delete p open balls in the product $\Sigma \times \Sigma$. Let M be the double of this manifold with boundary. Then $\chi(M) = 8(\gamma - 1)^2 - 2p$, and $\tau(M) = 0$, whereas, since M has a map of degree 2 onto $\Sigma \times \Sigma$, we have

$$\|M\| \geqslant 2\|\Sigma \times \Sigma\| \geqslant 32(\gamma - 1)^2.$$

For a suitable choice of γ and p, we have $0 < 2592\pi^2\chi(M) < \|M\|$, so M do not admit any Einstein metric, a conclusion which does not follow from Theorem 6.35.

6.49 Remark. Combined with Formula 6.34, Theorem 6.47 has the following improvement. For $\theta \in [0, \frac{3}{2}]$, let $\theta' = 2^{-3}3^{-9}\pi^{-2}(\frac{9}{4} - \theta^2)$. *If M is a 4-dimensional compact Einstein manifold, then, for any $\theta \in [0, \frac{3}{2}]$,*

$$\theta|\tau(M)| + \theta'\|M\| \leqslant \chi(M).$$

E. Ricci Curvature and the Fundamental Group

6.50. Various restrictions on the fundamental group follow from positive Ricci curvature. On the contrary, there are no known topological obstructions to the existence of metrics with negative Ricci curvature r, even non-compact complete ones. When r is negative, Bochner's method gives, in fact, only Riemannian informations.

The next theorems deal with Riemannian manifolds with non-negative Ricci curvature. It follows from a theorem of T. Aubin (cf. [Aub]) that any compact Riemannian manifold with non-negative Ricci curvature, which is positive at a point, admits a metric with positive Ricci curvature. As a result, the cases $r > 0$ and $r \geqslant 0$ are not too far from each other. This is an illustration of rigidity phenomena as explained in [Ch-Eb].

The oldest known result on manifolds with $r > 0$ is due to S.B. Myers (1935). It was obtained by using appropriately second variation formulas (see also [Ch-Eb]).

6.51 Theorem (cf. [Mye]). *Let (M, g) be a complete Riemannian manifold whose Ricci curvature satisfies $r \geqslant k^2 g$ with $k > 0$ a constant. Then M is compact with diameter $d(M) \leqslant \dfrac{\pi}{k}$.*

6.52 Corollary. *If (M, g) is a compact Riemannian manifold with positive Ricci curvature, then its fundamental group is finite.* This follows from the fact that the Riemannian universal cover $\tilde{M}$ of M and M have same curvature. So $\tilde{M}$ has to be compact. □

6.53 Example. The torus $T^m = \mathbb{R}^m/\mathbb{Z}^m$ is a compact manifold with fundamental group $\mathbb{Z}^m$. Therefore, on the torus, there does not exist any metric with positive Ricci curvature, but of course there are flat metrics.

6.54. This corollary cannot be extended to complete non-compact manifolds. L. Bérard Bergery and P. Nabonnand have given examples of manifolds with positive Ricci curvature, in dimension at least four, whose fundamental group is $\mathbb{Z}$ (cf. [Nab]). On $]0, \infty[\times S^{n-1} \times \mathbb{R}$, let us consider the following metric:

$$ds^2 = dr^2 + h^2(r) g_0 + f^2(r)\, dt^2,$$

where g_0 is the standard metric on S^{n-1}, f the restriction to $\mathbb{R}^+$ of an even positive $\mathscr{C}^\infty$ function and h the restriction to $\mathbb{R}^+$ of a positive odd function such that $h'(0) = 1$. Then it can be shown that $g = dr^2 + h^2(r) g_0$ is a complete metric on $\mathbb{R}^n$ and, in this way, one gets a complete metric on $\mathbb{R}^{n+1}$, the warped product $(M, g) = (\mathbb{R}^n, g) \times_f \mathbb{R}$ (cf. 7.11). The identity component of the isometry group of M contains a subgroup isomorphic to $\mathbb{Z}$. Hence $\mathbb{R}^{n+1}/\mathbb{Z}$ is a differentiable manifold with positive Ricci curvature and fundamental group isomorphic to $\mathbb{Z}$.

In dimension 3 the situation is different. In fact R. Schoen and S.T. Yau prove the following strong result:

6.55 Theorem (cf. [Sc-Ya 4]). *A complete non-compact manifold with dimension 3 and positive Ricci curvature is diffeomorphic to* $\mathbb{R}^3$.

In the case of a compact manifold (M, g), S. Bochner proved (1946) the following result on the first cohomology group $H^1(M, \mathbb{R})$.

6.56 Theorem (cf. [Boc 1]). *Let (M, g) be a compact Riemannian manifold. If the Ricci curvature of M is non-negative, then* $\dim H^1(M, \mathbb{R}) = b_1 \leqslant \dim M$. *Moreover, the universal cover $\tilde{M}$ of M is the product $\mathbb{R}^{b_1} \times M$.*

If the Ricci curvature of M is positive, then $H^1(M, \mathbb{R}) = \{0\}$.

This theorem is based on the study of harmonic 1-forms and uses the Weitzenböck formula (cf. Chapter 1)

$$(d\delta + \delta d)\alpha = D^* D\alpha + r(\alpha^\#),$$

which expresses the Laplacian of a 1-form α in terms of its rough Laplacian and Ricci curvature. (Here D denotes the Levi-Civita connection and $\alpha^\#$ the vector field associated with α by the metric). Since the operator D^*D is non-negative if $r > 0$, then no harmonic form can exist. If $r \geqslant 0$, any harmonic form is parallel. But $\dim H^1(M, \mathbb{R}) =$ dimension of harmonic forms $= b_1$ by Hodge-de Rham theorem. □

6.57. In the case of manifolds with positive Ricci curvature this result is weaker than Myers theorem since, if $\pi_1(M)$ is finite, then automatically $H^1(M, \mathbb{R}) = \{0\}$.

A generalisation of this result has been obtained by S. Gallot (cf. [Gal 4]) and by M. Gromov (cf. [Gro 1]).

6.58 Theorem. *Let M be a compact connected n-dimensional Riemannian manifold. Denote by k a lower bound of the Ricci curvature, and by d the diameter of M. Then, there exists a positive number ε depending only on n and d such that if $k \geqslant -\varepsilon \cdot \mathrm{Vol}(M)^{2/n}$, then $b_1(M) \leqslant n$.*

6.59 Example. On the torus $T^m = \mathbb{R}^m/\mathbb{Z}^m$, we have $b_1 = m$.

6.60. The following theorem, due to J.W. Milnor, refines Bochner theorem by giving informations of an algebraic nature on the fundamental group.

Let M be a manifold and G be a subgroup of the fundamental group generated by a finite number of generators $\mathscr{H} = \{h_1, \ldots, h_p\}$. To each positive integer s, we can associate the number $\gamma(s)$ of reduced words of length at most s built using the h_i's and their inverses. The map $\gamma_{\mathscr{H}}: s \mapsto \gamma_{\mathscr{H}}(s)$ is called *the growth function* of the group G associated with the generating system $\mathscr{H}$.

6.61 Theorem (cf. [Mil 1]). *Let (M, g) be a complete Riemannian manifold with non-negative Ricci curvature. Then the growth function $\gamma_{\mathscr{H}}(s)$ of any finitely generated subgroup of the fundamental group of M satisfies*

$$\gamma_{\mathscr{H}}(s) \leqslant ks^n$$

where k is a constant and n the dimension of M.

The proof is based on the following comparison result due to R.L. Bishop (cf. [Bis]) (see also 0.64).

Let x_0 be a point of the universal cover $\tilde{M}$ of M and $v(r)$ the volume of the ball of radius r centered at x_0. Since the Ricci curvature is non negative, we have $v(r) \leqslant \omega_n r^n$, where ω_n is the volume of the unit ball in $\mathbb{R}^n$.

The generators $h_1, \ldots, h_p$ can be interpreted as deck transformations on $\tilde{M}$. If $W = \mathrm{Max}_{i=1,\ldots,p}\, d(x_0, h_i(x_0))$, we notice that the ball with center x_0 and radius Ws contains at least $\gamma(s)$ distinct points of the form $h(x_0)$, $h \in \pi_1(M)$.

For ε small enough, the sets $B(x_0, \varepsilon)$ and $h(B(x_0, \varepsilon))$. $h \in \pi_1(M)$, $h \neq Id$, are disjoint. Then, the ball $B(x_0, Ws + \varepsilon)$ contains at least $\gamma(s)$ disjoint sets of the form $h(B(x_0, \varepsilon))$. Hence

$$\gamma_{\mathscr{H}}(s) v(\varepsilon) \leqslant v(Ws + \varepsilon)$$

from which the result follows by using Bishop's inequality. □

6.62 Remark. On a non-compact manifold one does not know if the existence of a complete metric with non-negative Ricci curvature implies that the fundamental group is finitely generated.

6.63 Example (cf. [Mil 1]). Let G be the nilpotent group of matrices of the form

$$\begin{pmatrix} 1 & a & b \\ 0 & 1 & c \\ 0 & 0 & 1 \end{pmatrix}$$

and H be the subgroup of G consisting of integral matrices. Then G/K is a T^2-bundle over S^1. One can show that the growth of $\pi_1(G/K)$ is polynomial of order 4. Hence G/K does not admit any metric with non-negative Ricci curvature.

The structure of manifolds with non-negative Ricci curvature has been clarified by J. Cheeger and D. Gromoll (cf. [Ch-Gr]). See also [Es-He] for a more recent proof. Recall the following notions from 1.68.

6.64 Definition. In a complete Riemannian manifold we call a *line* (respectively a *ray*) any geodesic γ minimizing the distance between any two of its points, which is defined for all real values of the parameter (respectively for $t \in [0, +\infty[$).

6.65 Theorem (J. Cheeger and D. Gromoll [Ch-Gr 1]). *Let (M,g) be a connected complete Riemannian manifold with non-negative Ricci curvature. Then (M,g) is a Riemannian product $(\overline{M} \times \mathbb{R}^q, \overline{g} \times g_0)$, where g_0 is the canonical (flat) metric of $\mathbb{R}^q$ and $(\overline{M},\overline{g})$ is a complete Riemannian manifold with non-negative Ricci curvature and* without any line.

6.66 Remarks. (a) A topological consequence of the last assertion is that $\overline{M}$ has at most one end.

(b) A proof of this fundamental theorem will be sketched in the last Section G of the chapter (see 6.78 and following). This proof shows more precisely that the Busemann function (see 6.79) of any ray on M is subharmonic.

(c) If one assumes that (M,g) has non-negative *sectional* curvature, this theorem is due to V.A. Toponogov [Top]. In the particular case of dimension 2, the classification of complete (non-compact) manifolds with non-negative Gaussian curvature is due to S. Cohn-Vossen [Coh].

(d) Notice that if the dimension of $\overline{M}$ is 0 (respectively 1), then $\overline{M}$ is a point (respectively a circle S^1) and (M,g) is flat.

(e) In the special case where (M,g) is *Ricci-flat*, obviously $(\overline{M},\overline{g})$ is also Ricci-flat. In this case, if the dimension of $\overline{M}$ is $\leqslant 3$, then (M,g) is flat.

(f) In the special case where (M,g) is homogeneous, obviously $(\overline{M},\overline{g})$ is also homogeneous, but it is furthermore compact, since any non-compact homogeneous Riemannian manifold admits at least one line (exercise!).

Finally, if we assume furthermore that M is compact, the preceding theorem gives many informations on the fundamental group of M and its coverings. We refer to [Ch-Gr 2] for the proofs of the following results.

6.67 Corollary (J. Cheeger and D. Gromoll [Ch-Gr 1, 2]). *Let (M,g) be a* compact *connected Riemannian manifold with non-negative Ricci curvature. Then*

(a) *there exists a finite normal subgroup F of $\pi_1(M)$ such that $\pi_1(M)/_F$ is an extension of some $\mathbb{Z}^q$ by a finite group;*

(b) *the universal Riemannian covering $(\tilde{M},\tilde{g})$ of (M,g) is isometric to a Riemannian product $(\overline{M} \times \mathbb{R}^q, \overline{g} \times g_0)$, where g_0 is the canonical (flat) metric of $\mathbb{R}^q$ and $(\overline{M},\overline{g})$ is a* compact simply connected *Riemannian manifold with non-negative Ricci curvature;*

(c) *there exists a finite quotient $(\hat{M},\hat{g})$ of the above $(\overline{M},\overline{g})$ such that a finite covering of (M,g) is both diffeomorphic to $\hat{M} \times T^q$, locally isometric to $(\hat{M} \times T^q, \hat{g} \times g_1)$ and fibred over (T^q, g_1), where (T^q, g_1) is some flat torus;*

(d) *if the isometry group of the above* $(\hat{M}, \hat{g})$ *is finite* (*this happens in particular when* (M,g) *is Ricci-flat*), *then the preceding finite covering of* (M,g) *in* (c) *is globally isometric to* $(\hat{M} \times T^q, \hat{g} \times g_1)$.

In order to illustrate what may happen in assertion (c) above, we only quote the following example from [Ch-Gr 2]. The Riemannian product $S^2 \times \mathbb{R}$ (for the canonical metrics) admits free actions of $\mathbb{Z}$ of the following form: a generator acts on S^2 through a rotation with angle α where $\alpha/2\pi$ is *irrational* and on $\mathbb{R}$ by a translation. Then $M = (S^2 \times \mathbb{R})/_{\mathbb{Z}}$ is diffeomorphic to $S^2 \times S^1$, fibred over S^1 and locally isometric to a Riemannian product $S^2 \times S^1$, but no covering of M is isometric to $S^2 \times S^1$ with a product metric.

The argument to get 6.67 from 6.65 is the following; let (M,g) be a compact Riemannian manifold such that its universal covering $(\tilde{M}, \tilde{g})$ is non compact. Then $(\tilde{M}, \tilde{g})$ admits at least one line (exercise!).

F. Scalar Curvature and the Spinorial Obstruction

The results in the preceding section provide examples of non simply connected manifolds with dimension $\geqslant 5$ which do not admit any positive Einstein metric. There are also simply connected examples, which have been found when studying a weaker condition: existence of a metric with merely positive scalar cuvature.

The essential argument is contained in the use of Weitzenböck formulas for Dirac operators.

6.68. Let (M,g) be a spin manifold, i.e., a manifold whose second Stiefel-Whitney class vanishes. We denote by $Cl(T_xM)$ the Clifford algebra attached to $(T_xM, g(x))$ and we set $Cl(n) = Cl(\mathbb{R}^n)$. Let us suppose that $\dim M = n$. Then a $Cl(n)$-bundle E on M is a vector bundle whose fibre is a representation space of $Cl(n)$. Such a bundle is naturally equipped with a fibre metric and with a connection which is associated with the Riemannian connection D of M. We will denote it later by D. We then define *the Dirac operator* in the following way. Let φ be a section of E and (e_i) be a local orthonormal basis of vector fields on M. We set

$$\mathscr{D}_\varphi = \sum_{i=1}^{n} e_i \cdot D_{e_i}\varphi$$

where $\cdot$ denotes Clifford multiplication.

The principal symbol of $\mathscr{D}$ at a point $\alpha \in T^*M$ is Clifford multiplication by α. Hence, the principal symbol of $\mathscr{D}^2$ is $-\|\alpha\|^2 Id$ (see Appendix 15).

The operators $\mathscr{D}$ and $\mathscr{D}^2$ are obviously elliptic; in addition both are self-adjoint.

6.69 Examples. If E is the bundle of Clifford algebras over M, then $\mathscr{D}$ is just $d + \delta$.

When E is the spinor bundle ΣM on M, we have (cf. 1.150)

$$\mathscr{D}^2 = D^*D + \tfrac{1}{4}s, \tag{6.70}$$

where s is the scalar curvature of M.

If, moreover, M is compact without boundary, the space $\mathscr{H}$ of *harmonic spinors* is defined as $\mathscr{H} = \mathrm{Ker}(\mathscr{D}) = \mathrm{Ker}(\mathscr{D}^2)$.

In even dimensions, ΣM can be decomposed as the direct sum of two bundles $\Sigma^+ M$ and $\Sigma^- M$, respectively called *the bundle of positive spinors* and *the bundle of negative spinors*, whose spaces of sections are mapped onto one another by the Dirac operator. Set $\mathscr{D}^+ = \mathscr{D}_{|\Sigma^+ M}$. Recall that, by definition, Index $(\mathscr{D}^+) = \dim \mathrm{Ker}\, \mathscr{D}^+ - \dim \mathrm{Coker}\, \mathscr{D}^+$. Applying the "Bochner method" to the Weitzenböck formula (6.70) yields

6.71 Theorem (A. Lichnerowicz [Lic 4]). *Let (M, g) be a compact spin manifold. If the scalar curvature s of (M, g) is non-negative and not identically zero, then all harmonic spinors vanish. In particular, if the dimension n of M is even, the index of $\mathscr{D}^+$ must vanish.*

If $s \equiv 0$, then all harmonic spinors are parallel.

6.72. Using the Atiyah-Singer Index Theorem [At-Si 1], which links the index of $\mathscr{D}^+$ with a characteristic number of M, the *$\hat{A}$-genus* $\hat{A}(M)$, A. Lichnerowicz concluded that, *if a compact connected spin manifold with dimension $n = 4m$ admits a metric with positive scalar curvature, then $\hat{A}(M) = 0$.*

For example, in dimension 4, $\hat{A}(M) = \frac{1}{16}\tau(M)$. Since a K3 surface is spin and has signature -16 (see 12.108), we see that a K3 surface admits no metric with positive scalar curvature (in particular, no positive Einstein metric). On the other hand, complex projective plane $\mathbb{C}P^2$ has signature 1 and is not spin. It admits metrics with positive scalar curvature, showing that the assumption that M be spin is essential in Theorem 6.71.

6.73. Using a refined version of the Index Theorem ([At-Si 3]), N. Hitchin has been able to improve Lichnerowicz' conclusion. The natural framework for Index Theorems is K-theory. The index of an elliptic operator naturally lives in some K-theory group. The $\hat{A}$-genus can be extended to live in K-theory too. This has been done by J. Milnor in [Mil 2]. Precisely, there is a surjective homomorphism α from the spin cobordism ring Ω_*^{spin} onto $KO^{-*}(\text{point})$ such that $\alpha(M) = \hat{A}(M)$ if $n = 8m$.

Now the Index Theorem implies

$$\begin{aligned} \dim\ker \mathscr{D}^+ - \dim\ker \mathscr{D}^- &= \hat{A}(M) && \text{if } n = 4m \\ \dim\ker \mathscr{D} &= \alpha(M) \bmod 2 && \text{if } n = 8m+1 \\ \dim\ker \mathscr{D}^+ &= \alpha(M) \bmod 2 && \text{if } n = 8m+2. \end{aligned}$$

6.74 Corollary (N. Hitchin [Hit 1]). *Let M be compact and spin. If M admits a metric with positive scalar curvature, then $\alpha(M) = 0$.*

6.75 Example ([Hit 1]). Certain exotic spheres of dimension $n = 8m + 1$ or $8m + 2$ are spin manifolds for which $\alpha(M) \neq 0$. They consequently do not admit any metric with positive scalar curvature (in particular, no positive Einstein metric).

6.76. The beauty of N. Hitchin's result is that the necessary condition $\alpha(M) = 0$ for admitting a metric with positive scalar curvature is probably also sufficient, if M is simply connected.

Let $\mathfrak{P}$ be the class of compact manifolds which admit a metric with positive scalar curvature. It should be noted that condition $\alpha = 0$ is invariant under cobordism. In some sense, condition $\mathfrak{P}$ is invariant under cobordism too. The precise statement (due to M. Gromov and H.B. Lawson [Gr-La 2]) is as follows. If M has dimension >5, is *simply connected* and spin (resp. not spin) and M is spin- (resp. oriented-) cobordant to a manifold in $\mathfrak{P}$, then M is in $\mathfrak{P}$. In fact, thanks to results by S. Smale, the question reduces to showing that $\mathfrak{P}$ is stable under codimension $\geqslant 3$ surgeries (this explicit construction had also been discovered by R. Schoen and S.T. Yau [Sc-Ya 3]).

Since the generators of the oriented cobordism ring have explicit realizations, which can easily be checked to be in $\mathfrak{P}$, one concludes that *any* simply connected *non-spin compact manifold with dimension* >5 *admits a metric with positive scalar curvature.* Spin-cobordism is more delicate, but simply-connected spin cobordism and oriented cobordism rings only differ by torsion. Moreover, there are spin generators for $\Omega_*^{SO} \otimes \mathbb{Q}$ (quaternionic projective spaces), which are in $\mathfrak{P}$, so M. Gromov and H.B. Lawson conclude that the ideal defined by $\mathfrak{P}$ in $\Omega_{*>5}^{\mathrm{spin}}$ has finite index in $\ker \alpha$. This means that, *if M is compact*, simply-connected *and spin, with dimension* >5, *and if* $\alpha(M) = 0$, *then some connected sum* $M \# \cdots \# M$ *admits a metric with positive scalar curvature.*

6.77. New obstructions to the existence of a metric with positive scalar curvature appear for non simply connected manifolds. Since they do not provide us with more examples of manifolds with no positive Einstein metrics, we will not discuss them here. Obstructions appear for rather large groups ([Sc-Ya 3, 4], [Gr-La 1, 3]), some do not depend very much on the group (see [Crr]). Notice that the case of a finite fundamental group does not reduce to the simply connected case. Indeed, there exist manifolds M with a compact universal cover M carrying positive scalar curvature, whereas $\alpha(M) \neq 0$, see [BéBer 4] and [Rbg 1, 2].

G. A Proof of the Cheeger-Gromoll Theorem on Complete Manifolds with Non-Negative Ricci Curvature

6.78. As we have seen in § E above, the three main general theorems concerning the topology of manifolds with non-negative Ricci curvature are

(a) Myers' Theorem 6.51, whose proof is sketched in 1.92;

(b) Bochner' Theorem 6.56, whose proof uses Weitzenböck formula 1.155;

(c) Cheeger-Gromoll' Theorem 6.65.

Up to now, no proof of the Cheeger-Gromoll theorem is available in a book. Recently, J.H. Eschenburg and E. Heintze succeeded in giving in [Es-He] a completely "elementary" proof, which does not involves in particular the regularity theory of solutions of elliptic equations (as did the original proof). We will sketch below a simplification of Cheeger-Gromoll's original proof, which is also suggested in [Es-He]. For the convenience of the reader, we recall first the result to be proved, under a different, but equivalent formulation.

6.79 Theorem (J. Cheeger and D. Gromoll [Ch-Gr 1], J.H. Eschenburg and E. Heintze [Es-He]). *Let (M, g) be a complete connected Riemannian manifold with non-negative Ricci curvature. Assume that (M, g) admits a* line. *Then (M, g) is a Riemannian product $(\overline{M} \times \mathbb{R}, \overline{g} \times dt^2)$, where $(\overline{M}, \overline{g})$ is a complete connected Riemannian manifold with non-negative Ricci curvature and dt^2 is the canonical metric on $\mathbb{R}$.*

Proof. A key ingredient of the proof is the *Busemann function* of a ray. Let γ: $[0, +\infty[\to M$ be any ray. The functions f_t (for t in $[0, +\infty[$) defined by

$$f_t(x) = t - d(x, \gamma(t))$$

form a family of equicontinuous functions on M, bounded by $d(x, \gamma(0))$ (with the bound attained on γ) and non-decreasing in t. Hence the limit

$$b = \lim_{t \to +\infty} f_t$$

exists and is a continuous function on M, bounded by $d(x, \gamma(0))$, with the bound attained on γ. We call b the *Busemann function* of γ. Notice that b is not necessarily even C^1. In order to deal with this non-smoothness of b, we introduce C^∞ *support functions* (this tool was intensively used in [Wu 4]). □

6.80 Definition. A function h is a C^∞ *support function* of the continuous function f *at* x_0 in M if:

(a) h is defined and C^∞ in a neighborhood W of x_0,
(b) $h(x_0) = f(x_0)$, and
(c) $h(x) \leqslant f(x)$ for any x in W.

The following construction gives support functions for the Busemann function b of a ray γ, with additional properties when (M, g) has non-negative Ricci curvature. For any x_0 in M and any $t \geqslant 0$, let X_t be the initial vector (with norm 1) of a minimizing geodesic connecting x_0 and $\gamma(t)$. Let X be an limit point of the X_t in $T_{x_0}M$ when t goes to $+\infty$. Now Lemma 1.67 shows that the geodesic c_X generated by X is also a ray. We set

$$b_{X,t}(x) = b(x_0) + t - d(x, c_X(t))$$

and we have

6.81 Lemma. *The functions $b_{X,t}$ are C^∞ support functions for the Busemann function b of the ray, at x_0 of M.*

Moreover, if (M, g) has non-negative Ricci curvature, we have

$$\Delta(b_{X,t})(x_0) \leqslant \frac{n-1}{t}$$

where n is the dimension of M.

Proof. Since c_X is a ray, $c_X(t)$ does not belong to the cut locus of x_0 and conversely x_0 does not belong to the cut-locus of $c_X(t)$, so $b_{X,t}$ is C^∞ in a neighbourhood W of x_0, since the distance function is C^∞ away from the cut-locus. Now $b_{X,t}(x_0) = b(x_0)$ since $d(x_0, c_X(t)) = t$ (c_X is minimizing).

Then

$$\begin{aligned} b_{X,t}(x) - b(x) &= b(x_0) + t - d(x, c_X(t)) - b(x) \\ &= \lim_{u \to +\infty} (u - d(x_0, \gamma(u)) + t - d(x, c_X(t)) - u + d(x, \gamma(u))). \end{aligned}$$

Let γ_u be the minimizing geodesic from x_0 to $\gamma(u)$ generated by X_u (see Figure 6.81). Then, if $u > t$, we get

$$d(x_0, \gamma(u)) = t + d(\gamma_u(t), \gamma(u)), \quad \text{hence}$$

$$\begin{aligned} b_{X,t}(x) - b(x) &= \lim_{u \to +\infty} (d(x, \gamma(u)) - d(\gamma_u(t), \gamma(u)) - d(x, c_X(t))) \\ &\leqslant \lim_{u \to +\infty} (d(\gamma(u), c_X(t)) - d(\gamma_u(t), \gamma(u))) \\ &\leqslant \lim_{u \to +\infty} d(c_X(t), \gamma_u(t)). \end{aligned}$$

But there exists a subsequence of u_n such that $\lim_{n \to +\infty} u_n = +\infty$ and $\lim_{n \to +\infty} X_{u_n} = X$. This gives $b_{X,t}(x) - b(x) \leqslant 0$.

For the last assertion, we have

$$\Delta b_{X,t}(x) = -\Delta d(x, c_X(t)),$$

and the result follows from the following lemma. □

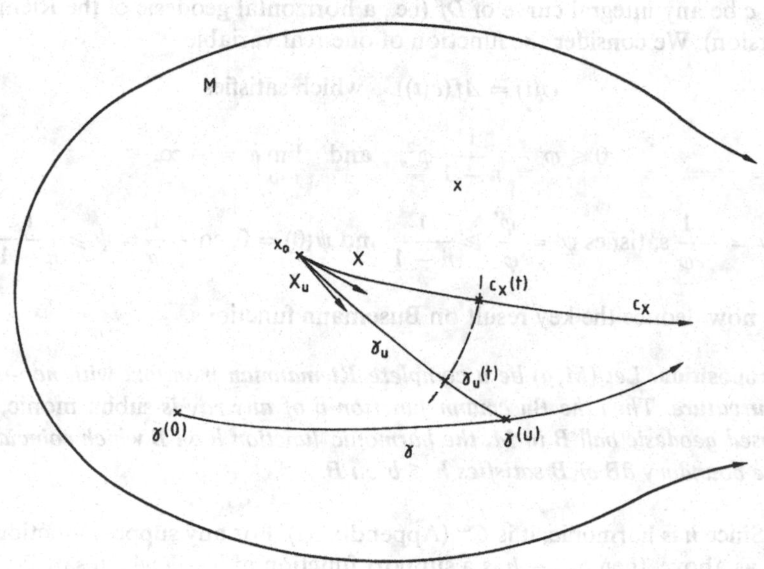

Fig. 6.81

6.82 Lemma. (a) *Let x_0 be a point in a Riemannian manifold (M,g) and C_{x_0} the cut-locus of x_0. Then the function $f(x) = d(x_0, x)$ is C^∞ on $W = M \backslash (\{x_0\} \cup C_{x_0})$ and satisfies $\|Df\| = 1$ and $\Delta f(x_0) = -\infty$.*

(b) *Moreover, if (M,g) has non-negative Ricci curvature, we have*

$$\Delta f \geqslant -\frac{n-1}{f} \quad \text{on } W.$$

Proof. Part (a) is standard and follows in particular from Gauss Lemma 1.46. We only prove Part (b). This may be checked through a direct elementary calculation. But we prefer to note that the condition $\|Df\| = 1$ implies that f is a Riemannian submersion from W to $\mathbb{R}$ (with its standard metric dt^2). So we may apply O'Neill's formulas for submersions, which are detailed in Chapter 9 below. Since the base $\mathbb{R}$ is 1-dimensional, the tensor A of the submersion vanishes and the tensor T satisfies

$$T_U V = -Ddf(U,V)Df \quad \text{for "vertical" vectors } U, V \text{ (i.e., } Uf = Vf = 0)$$

and

$$N = (\Delta f)Df.$$

Since $\|T\|^2 = \dfrac{1}{n-1}\|N\|^2 + \|T^0\|^2 \leqslant \dfrac{1}{n-1}\|N\|^2$, we get

$$\|Ddf\|^2 \leqslant \frac{1}{n-1}(\Delta f)^2 \quad \text{(notice that } Hf(Df, U) = 0 \text{ for any } U).$$

From (9.36c), we deduce easily

$$0 \leqslant r(Df, Df) = Df(\Delta f) - \|Ddf\|^2 \leqslant Df(\Delta f) - \frac{1}{n-1}(\Delta f)^2.$$

Let c be any integral curve of Df (i.e., a horizontal geodesic of the Riemannian submersion). We consider the function of one real variable

$$\varphi(t) = \Delta f(c(t)), \quad \text{which satisfies}$$

$$0 \leqslant \varphi' - \frac{1}{n-1}\varphi^2, \quad \text{and} \quad \lim_{t \to 0} \varphi = -\infty.$$

Then $\psi = -\dfrac{1}{\varphi}$ satisfies $\psi' = \dfrac{\varphi'}{\varphi^2} \geqslant \dfrac{1}{n-1}$ and $\psi(0) = 0$, so $-\dfrac{1}{\varphi} = \psi \geqslant \dfrac{t}{n-1}$. □

We now deduce the key result on Busemann functions.

6.83 Proposition. *Let (M,g) be a complete Riemannian manifold with non-negative Ricci curvature. Then the Busemann function b of any ray is* subharmonic, *i.e. for any closed geodesic ball B in M, the harmonic function h on B which coincides with b on the boundary ∂B of B satisfies $h \leqslant b$ on B.*

Proof. Since h is harmonic, it is C^∞ (Appendix 31). For any support function $b_{x,t}$ of b at x_0 as above, then $b_{x,t} - h$ is a support function of $b - h$ which satisfies

$$\Delta(b_{x,t} - h) = \Delta b_{x,t} \leqslant \frac{n-1}{t}.$$

Since $b - h = 0$ on ∂B, the desired conclusion $h \leqslant b$ follows immediately from the following generalization, by E. Calabi, of the maximum principle of E. Hopf. □

6.84 Proposition (E. Calabi [Cal 8], E. Hopf [Hop]). *Let (M,g) be a connected Riemannian manifold and f a continuous function on M. Assume that, for any x on M and any $\varepsilon > 0$, there exists a C^∞ support function $f_{x,\varepsilon}$ of f at x such that $\Delta f_{x,\varepsilon}(x) \leqslant \varepsilon$. Then f attains no maximum unless it is constant.*

Proof. Assume on the contrary that there exists a point x in M where f is maximum and such that, for any neighbourhood W of x, then f is not constant on W. We may choose a closed geodesic ball B centered at x such that f does not coincides with $f(x)$ on the whole boundary ∂B of B. We set $\partial' B = \{y \in \partial B \text{ such that } f(x) = f(y)\} \neq \partial B$. Then we may choose a C^∞ function φ on M such that $\varphi(x) = 0$, φ is negative on $\partial' B$, φ is positive on at least one point of $\partial B \setminus \partial' B$ and the gradient $D\varphi$ is never 0 on B (see Figure 6.84). Now let $h_\alpha = e^{\alpha\varphi} - 1$ for any positive α. Then h_α is a C^∞ function on M, $h_\alpha(x) = 0$, h_α is negative on $\partial' B$ and $\Delta h_\alpha = (-\alpha^2 \|D\varphi\|^2 + \alpha\Delta\varphi)e^{\alpha\varphi}$. Since B is compact, there exists α such that Δh_α is negative on the whole of B. We fix this α. Now there exists $\eta > 0$ such that $(f + \eta h_\alpha)(y) < f(x)$ for any y on ∂B. Hence $f + \eta h_\alpha$ attains its maximum ($\geqslant f(x)$) at some point z inside the interior B^0 of B. For any $\varepsilon > 0$, the function $f_{z,\varepsilon} + \eta h_\alpha$ is a C^∞ support function of $f + \eta h_\alpha$ at z and attains also its maximum at z. So $\Delta(f_{z,\varepsilon} + \eta h_\alpha)(z) \geqslant 0$. Since $\Delta h_\alpha(z)$ is negative and $\Delta f_{z,\varepsilon}(z) \leqslant \varepsilon$ for any $\varepsilon > 0$, we get a contradiction. □

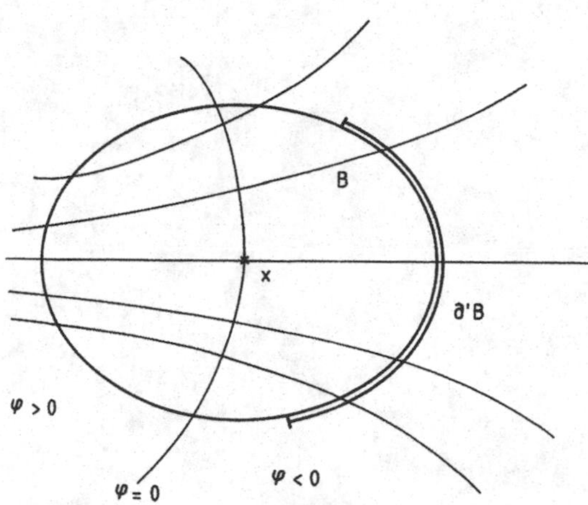

Fig. 6.84

6.85 *End of the Proof of Theorem* 6.79. Let γ be a *line* on M. It defines two rays $\gamma_+(t) = \gamma(t)$ and $\gamma_-(t) = \gamma(-t)$. We denote by b_+ and b_- the corresponding Busemann functions, which are subharmonic continuous functions on M.

Now $(b_+ + b_-)(x) = \lim_{t\to+\infty}(2t - d(x,\gamma(t)) - d(x,\gamma(-t))) \leqslant 0$, with $b_+ + b_- = 0$ on γ. Applying Proposition 6.84 (with the support functions given above), we conclude that $b_+ = -b_-$ on M. So b_+ is both subharmonic and superharmonic, hence harmonic and C^∞. Moreover, the support functions $b^+_{x,t}$ and $b^-_{Y,t}$ of b_+ and b_- at x are C^∞ functions with a gradient of norm 1 at x and satisfy $b^+_{x,t} \leqslant b_+ = -b_- \leqslant -b^-_{Y,t}$ with $b^+_{x,t}(x) = -b^-_{Y,t}(x)$. All that implies that the gradient of b_+ also has norm 1.

And the theorem follows from the last

6.86 Lemma. *Let (M,g) be a connected complete Riemannian manifold with non-negative Ricci curvature and f a C^∞ function on M with $\|Df\| = 1$.*

Then M is a Riemannian product $(\overline{M} \times \mathbb{R}, \bar{g} \times dt^2)$, and f is the projection onto the second factor.

Proof. Here again, f is a Riemannian submersion onto $(\mathbb{R}, dt^2)$ as in the proof of Lemma 6.82. With the same notation $\varphi = \Delta f \circ c$ as in the lemma, we let ζ be any integral of $\dfrac{\varphi}{n-1}$ and $\xi = e^{-\zeta}$. We get immediately

$$\xi'' = \left(-\frac{\varphi'}{n-1} + \frac{\varphi^2}{n-1}\right)\xi \leqslant 0.$$

But ξ is positive on $\mathbb{R}$, hence constant, so $\varphi = 0$, which gives $\Delta f = 0$, then $\|Ddf\|^2 = 0$, $T = 0$ and the Riemannian submersion is indeed a Riemannian product since $\mathbb{R}$ is simply connected. □

Chapter 7. Homogeneous Riemannian Manifolds

A. Introduction

7.1. In this chapter, we sketch the general theory of homogeneous Riemannian manifolds and we use it to give some examples of (homogeneous) Einstein manifolds. Up to now, the general classification of homogeneous Einstein manifolds is not known even in the compact case. In particular, the following question is still an open problem.

7.2 Problem. Classify the compact simply connected homogeneous manifolds $M = G/K$ (with compact G and K) which admit a G-invariant Einstein metric.

See below 7.54 for a detailed discussion. Notice that there are examples (due to M. Wang and W. Ziller) of compact simply connected G/K admitting no G-invariant Einstein metric (see 9.95).

7.3. However, many examples of homogeneous Einstein manifolds are known and we will give some of them here. After the general definitions of homogeneous Riemannian manifolds (§ B) and the computation of their curvatures (§ C), we give the simplest examples of homogeneous Einstein manifolds, namely spaces with constant curvature and projective spaces (§ D). Also, we obtain many more examples (without any computation) among isotropy irreducible spaces; and we give (without proof) J.A. Wolf's classification of strongly isotropy irreducible spaces.

In this chapter, we also recall some general results on homogeneous Einstein manifolds (§ E), deduced from more general results on Ricci curvature, and prove (in § E) the following rough classification.

7.4 Theorem. *Let (M, g) be a homogeneous Einstein manifold with (constant) scalar curvature s;*

a) *if $s > 0$, then M is compact with finite fundamental group;*
b) *if $s = 0$, then M is flat;*
c) *if $s < 0$, then M is non compact.*

Concerning 7.4c), we mention the following

7.5. ***Question*** (D.V. Alekseevskii). Given a non-compact homogeneous Einstein manifold $M = G/K$, is K a maximal compact subgroup of G?

7.6. Later on, we give other examples. In § F, we give E. Cartan's theory of Riemannian symmetric spaces (including a sketch of their classification); and finally we give in § G (without proof) the recent classification by M. Wang and W. Ziller of Einstein metrics among the standard homogeneous metrics. All these examples are summarized in Tables 1–8 (§ H). Notice that more examples will be given later in Chapters 9 and 15.

7.7. In the last § I, we consider the case of homogeneous Lorentzian manifolds. We do not give the general theory, and we only show the main differences from the Riemannian case. We illustrate this with a few examples, some Einstein and some not.

B. Homogeneous Riemannian Manifolds

7.8 Definition. A Riemannian manifold (M, g) is *homogeneous* if the group of isometries $I(M, g)$ acts transitively (i.e., for each x and y in M, there exists an isometry f such that $f(x) = y$).

Following F. Klein's Erlangen Programm [Kle] in spirit, we will use a more precise definition emphasizing the group acting on M. We recall that there may be more than one Lie group acting transitively on a given homogeneous space, so that a manifold may appear as a homogeneous space under different groups (see a few examples below in 7.13).

7.9 Definition. A Riemannian manifold (M, g) is *G-homogeneous* (or *homogeneous under the Lie group* G) if G is a closed subgroup of $I(M, g)$ which acts transitively on M.

7.10 Remarks. i) We note that G need not be all of $I(M, g)$; for example, the isometry group of the canonical Euclidean space is the group of orientation preserving and orientation reversing motions, but the subgroup of translations also acts transitively on Euclidean space.

ii) We use the condition that G is closed only to simplify the exposition. Indeed, if G is any Lie group acting effectively and transitively on M, and if G leaves invariant some Riemannian metric on M, then there exists a unique subgroup $\bar{G}$ of Diff(M) such that, for any G-invariant Riemannian metric g on M, the group $\bar{G}$ is the closure of G in $I(M, g)$.

7.11 Isotropy Subgroup. Since G is a closed subgroup of $I(M, g)$, the *isotropy subgroup* $K = \{f \in G; f(x) = x\}$ at the point x in M is a compact subgroup of $I_x(M, g)$.

Notice that M is compact if and only if G is compact.

Since an isometry f is determined by giving only the image $f(x)$ of a point x and the corresponding tangent map $T_x f$, the *linear isotropy representation* $\chi(f) = T_x f$ of K in $Gl(T_x M)$ is faithfull (i.e., injective).

7.12 A Note on Effectivity. With our definition 7.9, the group G must act *effectively* on G/K, i.e., K contains no non-trivial normal subgroup of G.

In general, given a Lie group G and a compact subgroup K, then G does not act necessarily effectively on G/K (and correspondingly, the linear isotropy representation is not necessarily faithfull). Let C be the maximal normal subgroup of G contained in K. Then $G' = G/C$ acts on G/K with isotropy subgroup $K' = K/C$ and G' acts effectively on $G/K = G'/K'$.

Usually, in explicit examples, we will *not* give the precise group G' which acts effectively, since it is always possible to determine it by explicit computations. Here is an example: the complex projective space $\mathbb{C}P^n$ is usually defined as the homogeneous space $SU(n+1)/S(U(1)U(n))$ but $G = SU(n+1)$ does not act effectively on $\mathbb{C}P^n$; here $C = \mathbb{Z}_{n+1}$ (the center of $SU(n+1)$) and $G' = SU(n+1)/\mathbb{Z}_{n+1} = PSU(n+1)$.

In all the examples given in both this chapter and the following one, the action of G will be *almost effective* on G/K, i.e., K contains no *non-discrete* normal subgroup of G. In this case, the Lie algebras of G and G' (respectively K and K') are the same. In the tables (§ H), we will give the Lie algebras of G and K (see 7.101).

7.13 Examples. *Transitive actions on spheres.* The canonical sphere S^n may be viewed as the homogeneous manifold $SO(n+1)/SO(n)$, but there also exist other compact connected Lie groups acting effectively and transitively on some sphere. They have been classified by D. Montgomery and H. Samelson ([Mo-Sa 1]) and A. Borel ([Bor 1 and 2]). They gave the following list

G	$SO(n)$	$U(n)$	$SU(n)$	$Sp(n)Sp(1)$	$Sp(n)U(1)$	$Sp(n)$
Sphere	S^{n-1}	S^{2n-1}		S^{4n-1}		
K	$SO(n-1)$	$U(n-1)$	$SU(n-1)$	$Sp(n-1)Sp(1)$	$Sp(n-1)U(1)$	$Sp(n-1)$
m	0	1	1	1	2	6

G_2	Spin(7)	Spin(9)
S^6	S^7	S^{15}
$SU(3)$	G_2	Spin(7)
0	0	1

In this table, the action of G on the corresponding sphere S^p is obtained by considering some special linear representation ρ of G in $\mathbb{R}^{p+1}$, such that G acts transitively on the unit sphere of $\mathbb{R}^{p+1}$. For example, we have the canonical representation of $SO(n)$ in $\mathbb{R}^n$, $U(n)$ in $\mathbb{C}^n$ $(=\mathbb{R}^{2n})$ and $Sp(n)Sp(1)$ on $\mathbb{H}^n$ $(=\mathbb{R}^{4n})$ (for this last group, see [Hel 1] and Chapters 10 and 14). Also K denotes the isotropy subgroup and m indicates the dimension of the space of G-invariant Riemannian metrics *up to homotheties* (i.e., isometries and multiplication by a positive constant).

We will study which of these G-invariant Riemannian metrics are Einstein in a next chapter, see 9.86.

7.14 Remark. Notice that we have many natural inclusions between those groups, namely $SU(n) \subset U(n) \subset SO(2n)$, $Sp(n) \subset SU(2n)$, $Sp(n)U(1) \subset U(2n)$, $Sp(n) \subset Sp(n)U(1) \subset Sp(n)Sp(1) \subset SO(4n)$, $G_2 \subset SO(7)$, $SU(4) \subset \mathrm{Spin}(7) \subset SO(8)$ and $\mathrm{Spin}(9) \subset SO(16)$.

For an inclusion such as $G \subset G'$, the same sphere appears as a homogeneous manifold under G and G', but usually there are some G-invariant Riemannian metrics which are not G'-invariant. This is not true only in the following three cases: $G_2 \subset SO(7)$, $\mathrm{Spin}(7) \subset SO(8)$ and $SU(n) \subset U(n) (n \geqslant 2)$; in particular G_2, $\mathrm{Spin}(7)$, or $SU(n)$ are *not* the full group of isometries of any Riemannian metric on the corresponding sphere.

We recall that any $SO(n)$-invariant metric on S^{n-1} is (proportional to) the canonical one, so the full group of isometries is $O(n)$. Thus even $SO(n)$ is not the full group of isometries of any Riemannian metric on S^{n-1}.

7.15 Examples. *Projective spaces.* We recall that the projective spaces are homogeneous manifolds, namely,

$$\mathbb{R}P^n = SO(n+1)/O(n),$$
$$\mathbb{C}P^n = SU(n+1)/S(U(1)U(n)),$$
$$\mathbb{H}P^n = Sp(n+1)/Sp(n)Sp(1),$$
$$\mathbb{C}aP^2 = F_4/\mathrm{Spin}(9).$$

All the groups in 7.13 also act transitively on $\mathbb{R}P^n$. In addition $Sp(n)$ acts transitively on $\mathbb{C}P^{2n-1}$ (through the identification $\mathbb{H}^n = \mathbb{C}^{2n}$) with isotropy subgroup $Sp(n-1)U(1)$.

This gives the complete list of compact connected Lie groups acting effectively and transitively on projective spaces (A.L. Oniščik [Oni]). For the Einstein metrics among these, see 9.86.

7.16 Examples. *Tori.* On the other hand, notice that the only compact connected Lie group acting effectively and transitively on the torus T^n is the torus itself (D. Montgomery and H. Samelson [Mo-Sa 2]).

7.17 Some Non-compact Examples. The hyperbolic space is the homogeneous manifold $H^n = SO_0(n,1)/SO(n)$, where $SO_0(n,1)$ is the connected component of the identity in the orthogonal group $O(n,1)$ of the canonical quadratic form with signature $(n,1)$ on $\mathbb{R}^{n+1}$ (compare 1.36).

There are also some complex, quaternionic and Cayleyan analogues:

$$\mathbb{C}H^n = SU(n,1)/S(U(n)U(1))$$
$$\mathbb{H}H^n = Sp(n,1)/Sp(n)Sp(1)$$
$$\mathbb{C}aH^2 = F_4^{-20}/\mathrm{Spin}\,9$$

(we use for exceptional simple Lie groups, the same notations as in [Hel 1]). Notice that there are many different closed subgroups of $SO_0(n,1)$ (resp. $SU(n+1)$, $Sp(n,1)$, F_4^{-20}) acting transitively on H^n (resp. $\mathbb{C}H^n$, $\mathbb{H}H^n$, $\mathbb{C}aH^2$). In particular, there exists a solvable Lie subgroup of $SO_0(n,1)$ (resp. $SU(n,1)$, $Sp(n,1)$, F_4^{-20}) acting *simply* transitively on H^n (resp. $\mathbb{C}H^n$, $\mathbb{H}H^n$, $\mathbb{C}aH^2$); this follows for example from Iwasawa's decomposition $G = KAN$ (see [Hel 1]).

7.18. In the following, we fix once and for all a G-homogeneous Riemannian manifold (M,g) (always connected) and a point x in M. Then $K = G \cap I_x(M,g)$ is a compact subgroup of G and M is diffeomorphic to the quotient manifold G/K (see for example [Cha], pp. 109–111 or [Hel 1], p. 113). In particular, M is compact if and only if G is compact. We note that G acts (nearly) effectively on M, that is, K contains no (non-discrete) invariant subgroup of G. We recall that, since the isometries of M may be used to extend geodesics as far as we wish (see Figure 7.19), we have the following well-known result.

7.19 Theorem. *A homogeneous Riemannian manifold is complete.*

For a detailed proof, see for example [Ko-No 1], p. 176 or the legend of Figure 7.19 □

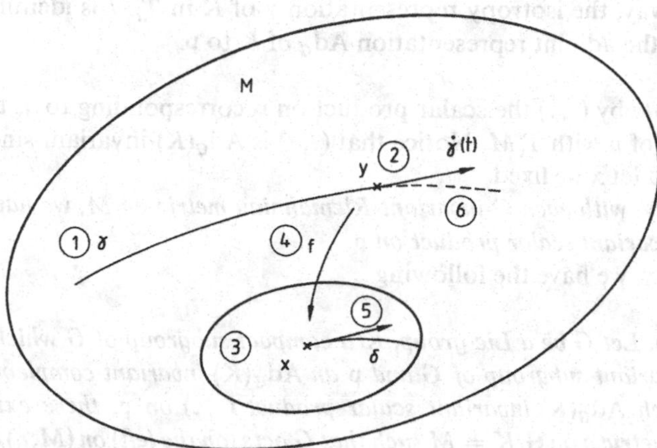

Fig. 7.19. ① γ a geodesic, $y \in \gamma$ and ② $\dot{\gamma}(t)$ tangent to γ at $y = \gamma(t)$, ③ a strictly convex ball $B(x,r)$ in x, ④ an isometry f with $f(y) = x$, ⑤ $T_y f(\dot{\gamma}(t))$ generates a geodesic δ from x with length at least r, ⑥ the geodesic $f^{-1} \circ \delta$ extends γ of a length r.

C. Curvature

7.20. We *denote by* $\mathfrak{g}$ the Lie algebra of G. For each X in $\mathfrak{g}$ we *denote* by $\operatorname{Exp}(tX)$ the one-parameter subgroup of G generated by X. The action of $\operatorname{Exp}(tX)$ on M turns it into a one-parameter group φ_t of diffeomorphisms of M, defined by

$$\varphi_t(y) = \operatorname{Exp}(tX)y.$$

From now on we will *identify* X in g with the vector field on M generated by φ_t. In doing so, we identify $\mathfrak{g}$ with the set of those Killing vector fields of (M, g) which generate one-parameter subgroups of G (we recall that G is not necessarily the whole isometry group). There is one subtle point in this identification

7.21 Warning. *Let* $[\ ,\ ]$ *be the Lie bracket of vector fields in* M *and* $[\ ,\ ]_{\mathfrak{g}}$ *the Lie algebra bracket of* $\mathfrak{g}$ *(we will use this convention in all the chapter). Then, using the identification given above, we have for* X *and* Y *in* $\mathfrak{g}$

$$[X, Y]_{\mathfrak{g}} = -[X, Y].$$

See [Wal] or [Ko-No 2], p. 469 for comments, and references for the rest of this paragraph.

7.22. The Lie subalgebra $\mathfrak{k}$ of $\mathfrak{g}$ generated by K is then identified with the subalgebra of those Killing vector fields in $\mathfrak{g}$ which vanish at x. Let Ad_G be the adjoint representation of G in $\mathfrak{g}$. Since K is compact, $\mathrm{Ad}_G(K)$ is a compact subgroup of the linear group of $\mathfrak{g}$. By a well known averaging procedure, there exists an $\mathrm{Ad}_G(K)$-invariant subspace $\mathfrak{p}$ of $\mathfrak{g}$ such that $\mathfrak{g}$ is the direct sum $\mathfrak{k} \oplus \mathfrak{p}$.

We *choose* once and for all such a $\mathfrak{p}$ (which is not necessarily unique). Then we identify $\mathfrak{p}$ with T_xM by taking the value of the corresponding Killing vector field at x. In this way, the isotropy representation χ of K in T_xM is identified with the restriction of the adjoint representation Ad_G of K to $\mathfrak{p}$.

7.23. We *denote* by $(.,.)$ the scalar product on $\mathfrak{p}$ corresponding to g_x through that identification of $\mathfrak{p}$ with T_xM. Notice that $(.,.)$ is $\mathrm{Ad}_G(K)$-invariant since K acts by isometries and let x be fixed.

In this way, *with each G-invariant Riemannian metric on M, we have associated an* $\mathrm{Ad}_G(K)$*-invariant scalar product on* $\mathfrak{p}$.

Conversely, we have the following

7.24 Theorem. *Let G be a Lie group, K a compact subgroup of G which contains no nontrivial invariant subgroup of G and* $\mathfrak{p}$ *an* $\mathrm{Ad}_G(K)$*-invariant complement to* $\mathfrak{k}$ *in* $\mathfrak{g}$*. Then, for each* $\mathrm{Ad}_G(K)$*-invariant scalar product* $(.,.)$ *on* $\mathfrak{p}$*, there exists a unique Riemannian metric g on* $G/K = M$ *such that G acts (on the left) on* (M, g) *by isometries and the scalar product induced by g on* $\mathfrak{p}$ *(as in 7.23) coïncides with the given* $(.,.)$.

For a detailed proof, see [Ko-No 2], p. 200. The idea is to define g_y on T_yM as f^*g_x for some f in G with $f(y) = x$. □

7.25 Example. Let G be any (connected) Lie group. Then G acts transitively on itself by translations on the left. Hence any scalar product on the Lie algebra $\mathfrak{g}$ of G gives rise by left translations to a G-invariant Riemannian metric on G. Such a metric is usually called a *left-invariant* Riemannian metric. Here $K = \{e\}$, so $\mathfrak{k} = 0$ and $\mathfrak{p} = \mathfrak{g}$.

7.26. In some sense, 7.23 and 7.24 state that a G-homogeneous Riemannian manifold (M, g) is completely determined by an $\mathrm{Ad}_G(K)$-invariant scalar product on $\mathfrak{p}$. A consequence of this fact is that the *curvature* of (M, g) may be computed in terms of the scalar product on $\mathfrak{p}$ and the Lie algebra structure of $\mathfrak{g}$. Notice that, because of the invariance of the curvature under isometries, we need only to know the curvature at the point x.

Moreover, the curvature tensor at x is identified with a tensor on the vector space $\mathfrak{p}$. Since K acts by isometries, the resulting tensor on $\mathfrak{p}$ is in particular $\mathrm{Ad}_G(K)$-invariant.

We first determine the Levi-Civita connection.

7.27 Lemma. *Let X, Y, Z be Killing vector fields on a Riemannian manifold (M, g). Then*

$$2g(D_X Y, Z) = g([X, Y], Z) + g([X, Z], Y) + g(X, [Y, Z]). \tag{7.27}$$

Proof. If X and Z are Killing vector fields, then

$$g([X, Z], X) = g(D_X Z, X) - g(D_Z X, X) = g(D_X X, Z).$$

By the usual "polarization" trick, we get

$$g(D_X Y, Z) + g(D_Y X, Z) = g([X, Z], Y) + g(X, [Y, Z]).$$

Then (7.27) follows from $D_X Y - D_Y X = [X, Y]$. □

7.28 Proposition. *Let (M, g) be G-homogeneous and $\mathfrak{p}$ defined as in 7.22. Let X, Y be Killing vector fields in $\mathfrak{p}$. Then, at the point x, we have*

$$(D_X Y)_x = -\tfrac{1}{2}[X, Y]_{\mathfrak{p}} + U(X, Y) \tag{7.28a}$$

where $U: \mathfrak{p} \times \mathfrak{p} \to \mathfrak{p}$ is defined by

$$2(U(X, Y), Z) = ([Z, X]_{\mathfrak{p}}, Y) + (X, [Z, Y]_{\mathfrak{p}}), \tag{7.28b}$$

for all Z in $\mathfrak{p}$.

Proof. This follows directly from Formula (7.27) and (7.21). (Here and in the following, $[\ ,\ ]_{\mathfrak{p}}$ and $[\ ,\ ]_{\mathfrak{k}}$ are the components on $\mathfrak{p}$ and $\mathfrak{k}$ of $[\ ,\ ]_{\mathfrak{g}}$). □

7.29 Remark. Notice that, if X and Y are Killing vector fields, then $D_X Y$ is not necessarily a Killing vector field on M. As a consequence the Formula (7.28a) is only valid at x. Of course, the value of $D_X Y$ at another point y on M may be obtained by applying some isometry f with $f(x) = y$, but we will not need it here.

7.30 Theorem. *Given two vectors X and Y at a point x of a homogeneous Riemannian manifold M, we have, with the identifications of 7.22, 7.23 and (7.28b), the following formula for the curvature*

$$\begin{aligned} g_x(R(X, Y)X, Y) = {} & -\tfrac{3}{4}|[X, Y]_{\mathfrak{p}}|^2 - \tfrac{1}{2}([X, [X, Y]_{\mathfrak{g}}]_{\mathfrak{p}}, Y) \\ & - \tfrac{1}{2}([Y, [Y, X]_{\mathfrak{g}}]_{\mathfrak{p}}, X) + |U(X, Y)|^2 - (U(X, X), U(Y, Y)). \end{aligned} \tag{7.30}$$

In particular, if $|X| = |Y| = 1$ and $(X, Y) = 0$, then (7.30) gives the sectional curvature of the 2-plane spanned by X and Y.

Proof.

$$\begin{aligned}
g_x(R(X,Y)X,Y) &= (D_{[X,Y]}X,Y) - (D_X D_Y X,Y) + (D_Y D_X X,Y)\\
&= -(D_Y X,[X,Y]) - X(D_Y X,Y) + (D_Y X, D_X Y) + Y(D_X X,Y)\\
&\quad -(D_X X, D_Y Y)\\
&= |D_Y X|^2 - (D_X X, D_Y Y) + Y([X,Y],X)\\
&= \tfrac{1}{4}|[X,Y]_{\mathfrak{p}}|^2 + ([X,Y]_{\mathfrak{p}}, U(X,Y)) + |U(X,Y)|^2\\
&\quad -(U(X,X),U(Y,Y)) + ([Y,[X,Y]],X) + ([X,Y],[Y,X])\\
&= |U(X,Y)|^2 - (U(X,X),U(Y,Y)) + \tfrac{1}{4}|[X,Y]_{\mathfrak{p}}|^2\\
&\quad +\tfrac{1}{2}([[X,Y]_{\mathfrak{p}},X]_{\mathfrak{p}},Y) + \tfrac{1}{2}(X,[[X,Y]_{\mathfrak{p}},Y]_{\mathfrak{p}})\\
&\quad +([Y,[X,Y]_{\mathfrak{k}}]_{\mathfrak{p}},X) + ([Y,[X,Y]_{\mathfrak{p}}]_{\mathfrak{p}},X) - |[X,Y]_{\mathfrak{p}}|^2.
\end{aligned}$$

Using the $\mathrm{Ad}_G(K)$-invariance of $(.,.)$ we have $([Y,[X,Y]_{\mathfrak{k}}]_{\mathfrak{p}},X) = (Y,[[X,Y]_{\mathfrak{k}},X]_{\mathfrak{p}})$ and formula (7.30) follows. □

7.31. In the following, we identify the curvature tensors at x with the corresponding tensors on $\mathfrak{p}$ through the identification of $\mathfrak{p}$ with T_xM. For example, we will write

$$g_x(R(X,Y)X,Y) = (R(X,Y)X,Y).$$

Now let (X_i) be an orthonormal basis of $\mathfrak{p}$ for $(.,.)$. In order to simplify the formula for the Ricci curvature we set $Z = \sum_i U(X_i,X_i)$.

7.32 Lemma. *The vector* $Z = \sum_i U(X_i,X_i)$ *is the unique vector in* $\mathfrak{p}$ *such that, for each vector* X *in* $\mathfrak{p}$,

$$(Z,X) = \mathrm{tr}(\mathrm{ad}\,X) \tag{7.32}$$

(*where we* denote *by* tr *the trace of the linear map* ad X *in End*($\mathfrak{g}$)).

Proof. $[X,\mathfrak{k}] \subset \mathfrak{p}$, so $\mathrm{tr}(\mathrm{ad}\,X) = \sum_i([X,X_i]_{\mathfrak{p}},X_i) = (Z,X)$. □

Notice that $Z = 0$ if and only if G is *unimodular*.
Then a straightforward computation using the basis (X_i) gives

7.33 Corollary. *The Ricci curvature* r *is given, at* x, *by*

$$\begin{aligned}
r(X,X) = &-\frac{1}{2}\sum_j |[X,X_i]_{\mathfrak{p}}|^2 - \frac{1}{2}\sum_i([X,[X,X_i]_{\mathfrak{p}}]_{\mathfrak{p}},X_i)\\
&-\sum_i([X,[X,X_i]_{\mathfrak{k}}]_{\mathfrak{p}},X_i) + \frac{1}{4}\sum_{i,j}([X_i,X_j]_{\mathfrak{p}},X)^2\\
&-([Z,X]_{\mathfrak{p}},X).
\end{aligned} \tag{7.33}$$

We may simplify this formula a little bit by introducing the Killing-Cartan form of $\mathfrak{g}$.

7.34 Definition. The *Killing-Cartan form B* is the symmetric, $\mathrm{Ad}(G)$-invariant, bilinear form on $\mathfrak{g}$ defined by $B(X, Y) = \mathrm{tr}(\mathrm{ad}\, X \circ \mathrm{ad}\, Y)$ for each X, Y in $\mathfrak{g}$.

7.35. We may extend the scalar product $(.,.)$ on $\mathfrak{p}$ into an $\mathrm{Ad}_G(K)$-invariant scalar product (still denoted by $(.,.)$) on all of $\mathfrak{g}$, by setting $(\mathfrak{k}, \mathfrak{p}) = 0$, and by choosing on $\mathfrak{k}$ an auxiliary $\mathrm{Ad}(K)$-invariant scalar product (there always exists one since K is compact).

We choose an orthonormal basis (V_j) of $\mathfrak{k}$ for that scalar product. Together with the orthonormal basis (X_i) on $\mathfrak{p}$ as in 7.31, this gives an orthonormal basis on $\mathfrak{g}$. Then we get, for any X in $\mathfrak{g}$

$$B(X, X) = \sum_i ([X, [X, X_i]_{\mathfrak{g}}]_{\mathfrak{p}}, X_i) + \sum_j ([X, [X, V_j]_{\mathfrak{g}}]_{\mathfrak{k}}, V_j). \tag{7.35}$$

We deduce first

7.36 Lemma. *The restriction of B to* $\mathfrak{k}$ *is negative definite.*

(And in particular $-B \restriction \mathfrak{k}$ is an $\mathrm{Ad}_G(K)$-invariant scalar product on $\mathfrak{k}$).

Proof. We apply (7.35) to X in $\mathfrak{k}$ and we get by $\mathrm{ad}(\mathfrak{k})$-invariance,

$$B(X, X) = -\sum_i |[X, X_i]_{\mathfrak{p}}|^2 - \sum_j |[X, V_j]_{\mathfrak{k}}|^2, \tag{7.36}$$

so $B(X, X) \leqslant 0$; moreover $B(X, X) = 0$ if and only if X is in the center of $\mathfrak{g}$, but this implies $X = 0$ by effectivity (see 7.12). □

Notice that $\mathfrak{p}$ is not necessarily B-orthogonal to $\mathfrak{k}$ and that B is not necessarily definite or of constant sign on $\mathfrak{p}$.

7.37. Now, for X in $\mathfrak{p}$, an easy computation using the $\mathrm{ad}(\mathfrak{k})$-invariance gives

$$\sum_j ([X, [X, V_j]_{\mathfrak{g}}]_{\mathfrak{k}}, V_j) = \sum_i ([X, [X, X_i]_{\mathfrak{k}}]_{\mathfrak{p}}, X_i), \tag{7.37a}$$

so we get

$$B(X, X) = 2\sum_i ([X, [X, X_i]_{\mathfrak{k}}]_{\mathfrak{p}}, X_i) + \sum_i ([X, [X, X_i]_{\mathfrak{p}}]_{\mathfrak{p}}, X_i), \tag{7.37b}$$

and we deduce

7.38 Corollary. *The Ricci curvature r can also be expressed at x by*

$$r(X, X) = -\frac{1}{2}\sum_i |[X, X_i]_{\mathfrak{p}}|^2 - \frac{1}{2}B(X, X) + \frac{1}{4}\sum_{i,j} ([X_i, X_j]_{\mathfrak{p}}, X)^2 - ([Z, X]_{\mathfrak{p}}, X). \tag{7.38}$$

Finally, we obtain the following formula for the scalar curvature.

7.39 Corollary. *The scalar curvature s is given (at x and everywhere) by*

$$s = -\frac{1}{4}\sum_{i,j} |[X_i, X_j]_{\mathfrak{p}}|^2 - \frac{1}{2}\sum_i B(X_i, X_i) - |Z|^2. \tag{7.39}$$

D. Some Examples of Homogeneous Einstein Manifolds

7.40. Since a Riemannian manifold with constant sectional curvature is obviously Einstein, we obtain examples of homogeneous Einstein manifolds by considering the homogeneous Riemannian manifolds with constant sectional curvature such that

$$\left.\begin{aligned} S^n &= SO(n+1)/SO(n) \\ \mathbb{R}^n &= D(n)/SO(n) \\ H^n &= SO_0(n,1)/SO(n) \end{aligned}\right\} \quad \text{with their canonical metrics.}$$

Some quotients of these are still homogeneous, for example $\mathbb{R}P^n = SO(n+1)/O(n)$, T^n (or more generally $T^p \times \mathbb{R}^{n-p}$) and also some other quotients of S^n (see J. Wolf [Wol 4], p. 89 for a complete list). Notice that H^n has no *homogeneous* (non trivial) quotient.

7.41. Many other examples generalizing the preceding are given by symmetric spaces such as the projective spaces

$$\begin{aligned} \mathbb{C}P^n &= SU(n+1)/S(U(1)U(n)), \\ \mathbb{H}P^n &= Sp(n+1)/Sp(n)Sp(1), \\ \mathbb{C}aP^2 &= F_4/\mathrm{Spin}(9). \end{aligned}$$

We recall that the compact rank one symmetric spaces are two-point homogeneous (see for example [Hel 1]) and in particular G is transitive on the unit sphere bundle of M; this gives easily that their metrics are Einstein.

In dimension 4, this exhausts (up to products) the possibilities; a case-by-case check using S. Ishihara's classification of 4-dimensional homogeneous spaces [Ish 1] gives the following.

7.42 Theorem (G. R. Jensen [Jen 1]). *A 4-dimensional homogeneous Einstein manifold is symmetric.*

A. Derdzinski has now a proof avoiding this case-by-case check (personal communication).

Notice that, even in dimension 5, the complete classification of Einstein homogeneous manifold is still an open problem. Now a simple remark yields many examples without computing actually the curvature.

7.43 Definition. A G-homogeneous manifold $M = G/K$ (with compact K) is called an *isotropy-irreducible space* if the linear isotropy representation of K is irreducible.

Notice that, chosing $\mathfrak{p}$ as in 7.22, the condition says that the representation $\mathrm{Ad}_G(K)$ of K on $\mathfrak{p}$ is irreducible.

7.44 Theorem (J.A. Wolf [Wol 3]). *Let $M = G/K$ be a G-homogeneous isotropy irreducible space. Then (up to homotheties) M admits a unique G-invariant Riemannian metric. This Riemannian metric is Einstein.*

Proof. Any G-invariant Riemannian metric on M is associated with some $\mathrm{Ad}_G(K)$-invariant scalar product $(.,.)$ on $\mathfrak{p}$. Since the representation of $\mathrm{Ad}_G(K)$ is irreducible, all $\mathrm{Ad}_G(K)$-invariant scalar products are proportional, so all G-invariant Riemannian metrics on M are proportional.

Moreover any $\mathrm{Ad}_G(K)$-invariant symmetric bilinear form on $\mathfrak{p}$ is proportional to $(.,.)$. In particular the Ricci tensor r is proportional to the metric at x and hence everywhere by homogeneity; therefore the G-invariant Riemannian metric is Einstein. □

7.45 Examples. The main examples of isotropy-irreducible spaces are the *irreducible symmetric spaces*, which we will study in greater details in paragraph F.

In particular, in the *non-compact* case, this exhausts all the possibilities. One may show the following

7.46 Proposition. *A non-compact isotropy irreducible space is symmetric.*

Proof. Let $M = G/K$ be isotropy irreducible and choose $\mathfrak{p}$ and $(.,.)$ as in 7.22. We may extend $(.,.)$ to an $\mathrm{Ad}_G(K)$-invariant scalar product on all of $\mathfrak{g}$ by setting $(.,.)\restriction_{\mathfrak{k}} = -B\restriction_{\mathfrak{k}}$ and $(\mathfrak{k}, \mathfrak{p}) = 0$ (see 7.36).

The eigenspaces E_λ of the decomposition of B in $(.,.)$ are $\mathrm{Ad}_G(K)$-invariant, and $\mathfrak{k} \subset E_{-1}$. By irreducibility, either $\mathfrak{p} \subset E_{-1}$ (and $(.,.) = -B$) or $\mathfrak{p} = E_\lambda$, $\lambda \neq -1$. If M is non compact, G is non compact so B is not negative definite hence $\lambda \geqslant 0$. If $\lambda > 0$, then G is semi-simple and $\mathfrak{k}$ is maximal compact in $\mathfrak{g}$, so M is symmetric (see 7.79 below). If $\lambda = 0$, $\mathfrak{p}$ is an ideal of $\mathfrak{g}$. By irreducibility, $\mathfrak{p}$ must be abelian and M is flat. □

7.47 Further Examples. In the paper [Wol 3], J.A. Wolf classifies the G-homogeneous Riemannian manifolds G/K such that the *connected component of the identity* in K has an irreducible isotropy representation. We will call these spaces *strongly isotropy irreducible*, since this condition is slightly stronger than that of Definition 7.43. Notice that the conditions are obviously the same if K is *connected*; this happens in particular if G is connected and G/K is simply connected.

The proof of Wolf's classification is quite long so we will not give it here; see [Wol 3] or also [Wa-Zi 1] for a more recent and conceptual proof, together with a few corrections. We give the complete list later in Tables 1–6 of § H, where they are part of the more general classification of standard homogeneous manifolds. We give here only some compact examples, where G is a simple classical group (if G is not simple, a strongly isotropy irreductible space is symmetric).

7.48. *Construction of spaces which are homogeneous under a simple classical compact Lie group.*

Let ρ be any representation of a compact Lie group K in some $\mathbb{R}^N$. If the representation ρ has a complex or symplectic structure, or none of them, it gives an embedding of $\rho(K)$ into $G = SU\left(\frac{N}{2}\right)$, or $Sp\left(\frac{N}{4}\right)$, or $SO(N)$, respectively. Of course $\rho(K)$ is a compact subgroup of G and the quotient space $G/\rho(K)$ is G-homogeneous (here G is always nearly effective if $\rho(K) \neq G$).

A first class of examples is given by

7.49 Proposition (J.A. Wolf [Wol 3]). *Let K be any simple compact Lie group with trivial center and ρ its adjoint representation. Then the homogeneous manifold $SO(\dim K)/_{\rho(K)}$ is isotropy irreducible.*

In fact, the Ad_G representation of $\rho(K)$ is $\wedge^2\rho$, which decomposes into the sum of ρ itself (adjoint representation of $\rho(K)$ on itself) and another representation χ, which is precisely the isotropy representation of $\rho(K)$ in $SO(\dim K)/\rho(K)$. Now it is true (but not obvious) that χ is irreducible. A conceptual proof is given in [Wa-Zi 1].

7.50. One gets also some examples by considering some well-known representations. We *denote* by ρ_1 the canonical representations of $SO(n)$ in $\mathbb{R}^n$, of $SU(n)$ in $\mathbb{C}^n$ or of $Sp(n)$ in $\mathbb{H}^n = \mathbb{R}^{4n}$. We denote by S or $\wedge$ the symmetrization or alternation procedures, and we recall that the tensor product of two symplectic representations is real, whereas the tensor product of a real and a symplectic representation is symplectic.

Then, with the construction 7.48, the following examples give an isotropy-irreducible space:

K	ρ	G
$SU(p)SU(q)$	$\rho_1 \otimes \rho_1$	$SU(pq)$
$Sp(1)Sp(n)$	$\rho_1 \otimes \rho_1$	$SO(4n)$
$Sp(1)SO(n)$	$\rho_1 \otimes \rho_1$	$Sp(n)$
$SU(n)$	$\wedge^2\rho_1$	$SU\left(\frac{n(n-1)}{2}\right)$
$SU(n)$	$S^2\rho_1$	$SU\left(\frac{n(n+1)}{2}\right)$
$SO(n)$	$S^2\rho_1 - 1$ [1]	$SO\left(\frac{n(n+1)}{2} - 1\right)$
$Sp(n)$	$\bar{S}^2\rho_1 - 1$ [2]	$SO(2n^2 - n - 1)$

[1]) This is the representation of $SO(n)$ in *trace-free symmetric* 2-tensors

[2]) Here we take the representation of $Sp(n)$ in *trace-free hermitian-quaternionic* matrices

7.51 Remark. There is a general construction, due to C.T.C. Wall (see the end of [Wol 3]), which covers 7.49 and 7.50, together with a few more examples. Let G/K be a compact irreducible symmetric space. Then, with a few exceptions (essentially the real and quaternionic grassmannians), the construction of 7.48, applied to the linear isotropy representation of K, gives a non-symmetric strongly isotropy irreducible space (in the complex or quaternionic case, we take only one part of K).

In fact, Wall's construction gives all non-symmetric strongly isotropy irreducible spaces, with a few exceptions like $SO(7)/G_2$ or those with an exceptional G. A conceptual proof is given in [Wa-Zi 1]. The precise final result is the following

7.52 Theorem (J.A. Wolf [Wol 3]). *A compact simply connected strongly isotropy irreducible space is either an irreducible symmetric space* (*Tables* 1–4 *of* § H) *or is in the families of Table* 5 *or in the finite list of Table* 6 *of* § H.

Added in proof: Recently, A. Gray has informed the author that this classification has been first obtained by O.V. Manturov [Man 1, 2, 3], up to a few errors.

E. General Results on Homogeneous Einstein Manifolds

We give here the implications of general results on Ricci curvature to the particular case of homogeneous Einstein manifolds. We begin with the case of Einstein manifolds with positive scalar curvature.

7.53 Theorem. *Let (M, g) be a homogeneous Einstein manifold with positive scalar curvature. Then M is compact with finite fundamental group. Also G is compact and the maximal semi-simple Lie subgroup of G acts transitively on M.*

Proof. This is a particular case of Myers' theorem 6.51, since (M, g) is complete and the Ricci curvature is a positive constant. The last assertion is an easy exercise in topology. □

The complete classification of compact homogeneous Einstein manifolds with positive scalar curvature is an open problem. In particular, the following question is still unsolved

7.54 Problem. Classify the compact simply connected homogeneous manifolds $M = G/K$ (with compact G and K) which admit a G-invariant Einstein Riemannian metric.

We have seen in 7.44 that the isotropy irreducible G-homogeneous spaces admit exactly one G-invariant Einstein metric (up to homotheties).

On the other hand, we describe below in 9.95 some examples (due to M. Wang and W. Ziller) of compact simply connected G/K which admit no G-invariant Einstein metric. Some of them admit no homogeneous Einstein metric at all (for any G acting transitively).

We also describe in Chapter 9 many examples of G/K admitting two (or more) non-homothetic G-invariant Einstein metrics. This includes for example the spheres $S^{2q+3} = Sp(q+1)/Sp(q)$, see 9.82.

But the following questions are still unsolved

7.55 Problem. Given a compact simply connected homogeneous space $M = G/K$ (with compact G and K), is the set of G-invariant Einstein metrics on M finite (up to homotheties)? Or conversely, may there exist families of (non homothetic) G-invariant Einstein metrics?

Notice that in the non-compact case, there do exist examples of non-trivial families of homogeneous Einstein metrics (with negative scalar curvature), for example some families of bounded homogeneous domains in $\mathbb{C}^n$, $n \geqslant 7$, (with their Bergman metric), see [PiSH] (this was pointed out to the author by J.E. D'Atri).

We consider now the general results for homogeneous Einstein manifolds with negative scalar curvature. We have first

7.56 Theorem. *Let (M, g) be a homogeneous Einstein manifold with negative scalar curvature. Then M and G are non compact.*

Proof. This is an obvious corollary of Bochner's Theorem 1.84. □

We mention the following

7.57 Conjecture (D.V. Alekseevskii [Ale 4]). Let $M = G/K$ be a non-compact homogeneous Einstein manifold. Then K is a maximal compact subgroup of G.

7.58. This seems quite optimistic since the corresponding statement is false under the weaker assumption of negative Ricci curvature. M. Leite and I. Doti Miatello [Le-DM] have recently shown that $Sl(n, \mathbb{R})$ ($n \geqslant 3$) admits a left invariant metric with negative Ricci curvature.

If the conjecture were true, there would be a solvable subgroup of G acting *simply* transitively on M, so we would be left with the classification of solvable Lie group with an Einstein left-invariant metric. Many examples are known, some with a symmetric M (see for example 7.17) and some not (see for example E.D. Deloff's thesis [Del]). We also recall that there are explicit examples of families of non-homothetic left-invariant Einstein metrics. But the complete classification is still an open problem, except for the unimodular case, where we have the following

7.59 Theorem (I. Doti-Miatello [DoM]). *Let G be a unimodular solvable Lie group. Then any Einstein left-invariant metric on G is flat.*

Proof. We use formula (7.38). Here $Z = 0$ and $B = 0$. We consider the derived Lie algebra $\mathfrak{n} = [\mathfrak{g}, \mathfrak{g}]$ of $\mathfrak{g}$. Then $\mathfrak{n}$ is nilpotent, so has a nontrivial center $\mathfrak{c}$. Let (X_i), (resp(Y_j), resp.(Z_k)) be an orthonormal basis of $\mathfrak{c}$ (resp. the orthogonal of $\mathfrak{c}$ in $\mathfrak{n}$ and the orthogonal of $\mathfrak{n}$ in $\mathfrak{g}$). Then $[X_i, Y_j] = 0$ and $[X_i, Z_k] \in \mathfrak{c}$. In particular

$$\textbf{(7.59)} \qquad r(X_i, X_i) = -\frac{1}{2}\sum_{k,i'}([X_i, Z_k], X_{i'})^2$$

$$+\frac{1}{2}\sum_{i',k}([X_{i'}, Z_k], X_i)^2$$

$$+\frac{1}{2}\sum_{j,k}([Y_j, Z_k], X_i)^2$$

$$+\frac{1}{4}\sum_{j,j'}([Y_j, Y_{j'}], X_i)^2 + \frac{1}{4}\sum_{k,k'}([Z_k, Z_{k'}], X_i)^2.$$

Consequently $\sum_i r(X_i, X_i) \geqslant 0$. Since G is non compact, it can be Einstein only if $r \equiv 0$ and the last assertion follows from Theorem 7.61 given below. □

7.60 Remarks. (i) In [DoM], there is a more general result on the Ricci curvature.
(ii) The nilpotent case is due to E. Heintze [Hei].

Finally, the Ricci-flat case is simpler, and we have the complete classification.

7.61 Theorem (D.V. Alekseevskii and B.N. Kimelfeld [Al-Ki 1]). *A homogeneous Ricci-flat manifold is flat, and hence the product of a torus by a Euclidean space.*

Proof (after [BéBer 1]). Using the Cheeger-Gromoll theorem 6.65, we see that the universal covering $(\tilde{M}, \tilde{g})$ is a Riemannian product of a Euclidean space and a compact simply connected homogeneous Riemannian manifold N. In the case above, N is Ricci-flat, so by Bochner's theorem 1.84, it is reduced to a point. □

Since there are many different groups acting transitively on the Euclidean space, the classification of flat G-homogeneous manifold is quite different. We only mention the result (see for example [BéBer 1]).

7.62 Theorem. *A G-homogeneous Riemannian manifold (M, g) is flat if and only if it satisfies the following three conditions*
(i) *the maximal invariant connected nilpotent subgroup G_1 of G is abelian,*
(ii) *the isotropy subgroup K contains a maximal connected semi-simple subgroup of G,*
(iii) *let $\mathfrak{g}_1$ be the Lie subalgebra of G_1 and choose $\mathfrak{p}$ containing $\mathfrak{g}_1$, then the restriction of* $(.,.)$ *to* $\mathfrak{g}_1$ *is* $\mathrm{Ad}(G)$*-invariant.*

F. Symmetric Spaces

7.63 Definition. A Riemannian manifold (M, g) is called *symmetric* (or a Riemannian symmetric space) if for each x in M there exists an isometry f_x of (M, g) such that $f_x(x) = x$ and $T_x(f_x) = -Id_{T_xM}$.

The isometry f_x (necessarily unique if M is connected) is called the *symmetry around* x.

7.64. The study of Riemannian symmetric spaces began with the celebrated work of E. Cartan ([Car 6]) who classified them completely. There is now a large literature on them (see for example [Hel 1], [Ko-No 2], Chapter 9 or [Los]). Here we do not give a complete discussion of them. We only recall a few properties and apply them to the computation of their Ricci curvature.

7.65 Proposition. *A Riemannian symmetric space is homogeneous.*

Sketch of Proof. First, M is complete since any geodesic segment may be extended at each of its ends via the symmetries (see Figure 7.65a). Then for any two points y and z in M, the symmetry around the middle point of any geodesic segment from y to z (there exists at least one by completeness) interchanges y and z, (see Figure 7.65b) so the isometry group acts transitively. □

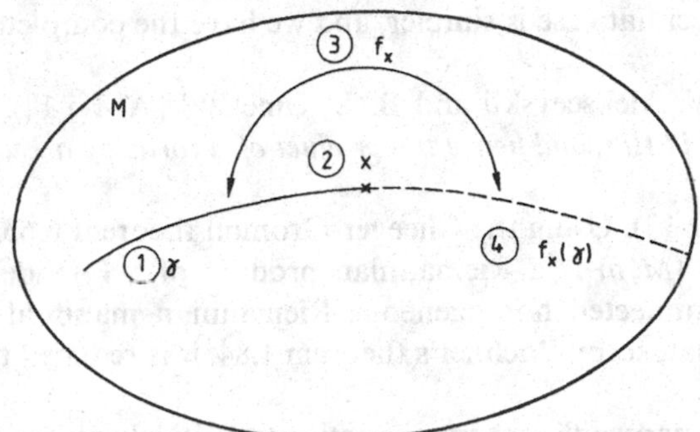

Fig. 7.65a. ① a geodesic γ, ② x "near the end", ③ apply f_x, ④ $f_x(\gamma)$ is a geodesic which extends γ

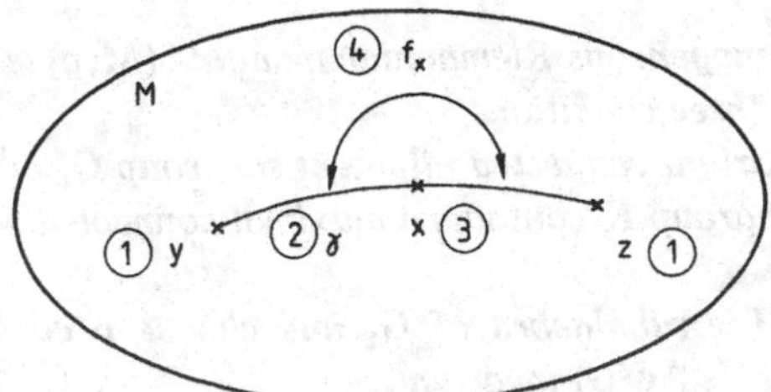

Fig. 7.65b. ① y and z on M, ② γ any geodesic form y to z, ③ x the mid-point of γ, ④ f_x maps y to z

7.66. We *denote* by G the connected component of the identity in the group of isometries of (M, g) and by K the isotropy subgroup at some point x (fixed once and for all). Then the symmetry f_x around x belongs to K and generates an involutive automorphism σ of G by

$$\sigma(f) = f_x \circ f \circ f_x^{-1}.$$

We *denote* by $G^\sigma = \{f \in G \colon \sigma(f) = f\}$ and *denote* by G_0^σ the connected component of the identity in G^σ. Then

7.67 Theorem (E. Cartan). (i) *Given any (connected) symmetric space M and any x in M, then the corresponding involutive automorphism σ of G is such that $G^\sigma \supset K \supset G_0^\sigma$.*

(ii) *Conversely, if G is a Lie group, σ an involutive automorphism of G and K a compact subgroup of G with $G^\sigma \supset K \supset G_0^\sigma$, then any G-invariant Riemannian metric on G/K is symmetric.*

Proof. We prove only (i). We refer to [Hel 1] for (ii). For each f in K, we have $f(x) = x$ and $\sigma(f)(x) = f_x \circ f \circ f_x^{-1}(x) = x$. Moreover $T_x(\sigma(f)) = T_x(f_x \circ f \circ f_x^{-1}) = T_x(f_x) \circ T_x f \circ T_x(f_x^{-1}) = T_x f$, so $f = \sigma(f)$.

Let f_t be a one-parameter subgroup of G_0^σ. We have $\sigma(f_t) = f_t$ so $f_x \circ f_t = f_t \circ f_x$. In particular $f_x(f_t(x)) = f_t(x)$, so $f_t(x)$ is a fixed point of f_x. But $f_0(x) = x$ and x is an *isolated* fixed point of f_x; hence $f_t(x) = x$ for each t and f_t belongs to K. □

7.68. Let $\mathfrak{g}$ be the Lie algebra of G, $\mathfrak{k}$ the subalgebra corresponding to K and $\sigma = \mathrm{Ad}_G(f_x)$ the automorphism of $\mathfrak{g}$ induced by σ on G. We choose $\mathfrak{p} = \{X \in \mathfrak{g} \colon \sigma(X) = -X\}$. Then $\mathfrak{p}$ is $\mathrm{Ad}_G(K)$-invariant. On the other hand $\sigma(X) = X$ for each X in $\mathfrak{k}$, which is the common Lie algebra of G^σ, K and G_0^σ. Then we have

7.69 Fundamental Lemma. *Given an involutive automorphism σ of $\mathfrak{g}$ with root spaces $\mathfrak{k}$ for* the eigenvalue 1 *and $\mathfrak{p}$ for* the eigenvalue -1, *we have*

$$[\mathfrak{k}, \mathfrak{k}] \subset \mathfrak{k},\ [\mathfrak{k}, \mathfrak{p}] \subset \mathfrak{p} \text{ and } [\mathfrak{p}, \mathfrak{p}] \subset \mathfrak{k}.$$

Proof. We only prove the last inclusion: let X and Y be in $\mathfrak{p}$.

Then $\sigma(X) = -X$, $\sigma(Y) = -Y$, so

$$[X, Y]_{\mathfrak{g}} = [\sigma(X), \sigma(Y)]_{\mathfrak{g}} = \sigma([X, Y]_{\mathfrak{g}}).$$

Therefore $[X, Y]_{\mathfrak{g}}$ belongs to $\mathfrak{k}$. □

7.70. Conversely, a Riemannian homogeneous space such that $[\mathfrak{p}, \mathfrak{p}] \subset \mathfrak{k}$ is not necessarily symmetric, however it is *locally symmetric* (see definition in 10.76) and its universal cover is symmetric. Such are the lens spaces $S^{2n+1}/\mathbb{Z}_p$ ($p \geqslant 3$).

7.71. The computation of the curvature of symmetric spaces is easy for this choice of $\mathfrak{p}$, since U vanishes (i.e., symmetric spaces are naturally reductive see below 7.84). Using $\mathrm{ad}(\mathfrak{k})$-invariance of the scalar product, and the fact that for X in $\mathfrak{p}$, $\mathrm{ad}\,X$ interchanges $\mathfrak{k}$ and $\mathfrak{p}$, we get easily the following

7.72 Proposition. *The curvature of a Riemannian symmetric space satisfies for X, Y in $\mathfrak{p}$*

(7.72a) $$(R(X, Y)X, Y) = -((\mathrm{ad}\,X \circ \mathrm{ad}\,X)Y, Y)$$

(7.72b) $$r(X, Y) = -\mathrm{tr}((\mathrm{ad}\,X \circ \mathrm{ad}\,Y) \restriction \mathfrak{p}).$$

Using σ, we easily see that $B(\mathfrak{k}, \mathfrak{p}) = 0$ and we deduce

7.73 Theorem. *The Ricci curvature of a Riemannian symmetric space satisfies:*

$$r = -\tfrac{1}{2}B \upharpoonright \mathfrak{p}.$$

Proof. Since, if $X \in \mathfrak{p}$, $\operatorname{ad} X$ interchanges $\mathfrak{k}$ and $\mathfrak{p}$, we have

$$\operatorname{tr}((\operatorname{ad} X \circ \operatorname{ad} Y) \upharpoonright p) = \operatorname{tr}((\operatorname{ad} X \circ \operatorname{ad} Y) \upharpoonright k) = \tfrac{1}{2}B(X, Y).$$

(Notice that R and r depend only on $\mathfrak{g}$ and $\mathfrak{p}$). □

7.74 Corollary. *Let (M, g) be a Riemannian symmetric space. Then g is Einstein if and only if the restriction of B to $\mathfrak{p}$ is*

(i) *either identically zero,*
(ii) *or definite and proportional to the scalar product $(.,.)$ on $\mathfrak{p}$.*

Of course the scalar curvature will be positive (zero, negative) if B is negative definite (zero, positive definite).

7.75. Obviously, the assumption in Corollary 7.74 is satisfied if the symmetric space is assumed to be isotropy-irreducible, i.e., if the adjoint representation $\operatorname{Ad}_G(K)$ of K in $\mathfrak{p}$ is irreducible. In this case, we say for short that the symmetric space is *irreducible.* Then an important step in the classification program for symmetric spaces is the following

7.76 Theorem. *A simply-connected Riemannian symmetric space is the Riemannian product of a Euclidean space and a finite number of* irreducible *Riemannian symmetric spaces.*

Sketch of Proof. Here we give only the main ideas of this proof (the reader will not find it difficult to fill in the technical details; or he may refer to [Hel 1] for a complete proof).

Let $\mathfrak{p} = \bigoplus_{i=0}^{r} \mathfrak{p}_i$ be the unique direct sum decomposition of $\mathfrak{p}$ such that

(i) $B \upharpoonright \mathfrak{p}_i = \lambda_i(.,.) \upharpoonright \mathfrak{p}_i$,
(ii) $B(\mathfrak{p}_i, \mathfrak{p}_j) = (\mathfrak{p}_i, \mathfrak{p}_j) = 0$ if $i \neq j$,
(iii) each $\mathfrak{p}_i$ is $\operatorname{Ad}_G(K)$-invariant,
(iv) if $i \neq 0$, the representation of $\operatorname{Ad}_G(K)$ in $\mathfrak{p}_i$ is irreducible,
(v) $\lambda_0 = 0$ (possibly, $\mathfrak{p}_0 = 0$ or $r = 0$).

We recall that $B \upharpoonright \mathfrak{k}$ is negative definite. Using the $\operatorname{Ad}(G)$-invariance of B, it follows easily that, if $i \neq j$, $[\mathfrak{p}_i, \mathfrak{p}_j] = 0$ and that $[\mathfrak{p}_0, \mathfrak{p}_0] = 0$. For each $i \neq 0$ we let $\mathfrak{k}_i = [\mathfrak{p}_i, \mathfrak{p}_i] \subset k$. Notice that $B(\mathfrak{k}_i, \mathfrak{k}_j) = 0$. Moreover $\mathfrak{k}_i$ is an ideal of $\mathfrak{k}$. We define $\mathfrak{k}_0$ to be the unique ideal such that $\mathfrak{k} = \bigoplus_{i=0}^{r} \mathfrak{k}_i$. Now $\mathfrak{g}_i = \mathfrak{k}_i \oplus \mathfrak{p}_i$ is an ideal of $\mathfrak{g}$ and $[\mathfrak{g}_i, \mathfrak{g}_j] = 0$ if $i \neq j$. Hence $\mathfrak{g} = \bigoplus_{i=0}^{r} \mathfrak{g}_i$ is a decomposition of $\mathfrak{g}$ as a direct sum of ideals.

We denote by G_i the subgroup of G generated by $\mathfrak{g}_i$ and $K_i = G_i \cap K$. The restriction $(.,.) \upharpoonright \mathfrak{p}_i$ defines a G_i-invariant Riemannian metric g_i on $G_i/K_i = M_i$, and of course (M_i, g_i) is a symmetric space. Using de Rham's theorem 10.43 (we recall that M is simply connected), we may show that (M, g) is the Riemannian product

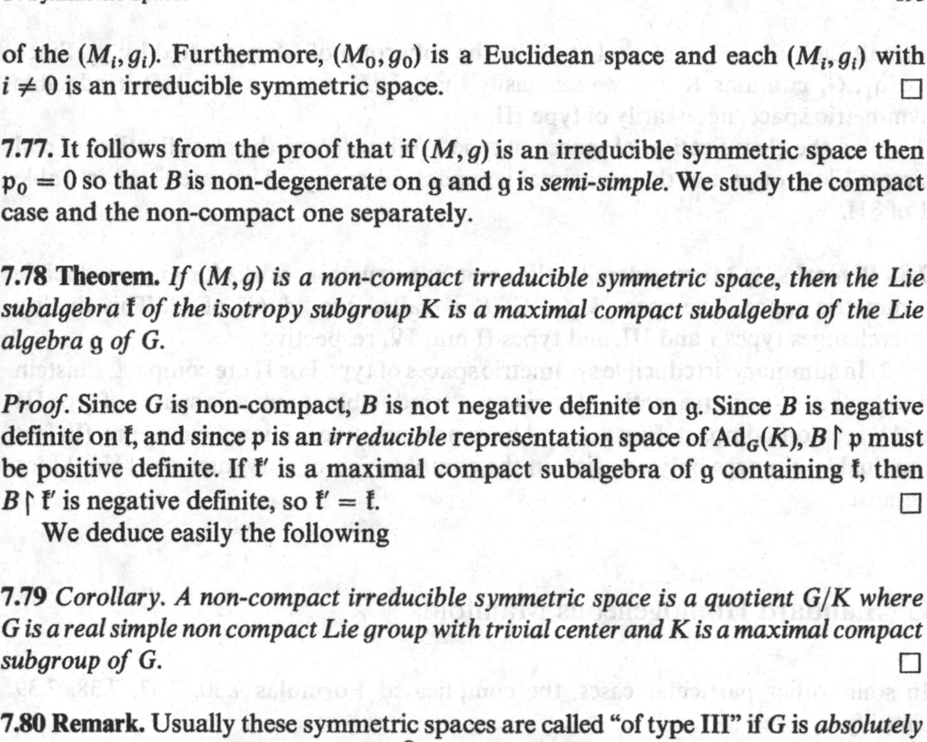

of the (M_i, g_i). Furthermore, (M_0, g_0) is a Euclidean space and each (M_i, g_i) with $i \neq 0$ is an irreducible symmetric space. □

7.77. It follows from the proof that if (M, g) is an irreducible symmetric space then $\mathfrak{p}_0 = 0$ so that B is non-degenerate on $\mathfrak{g}$ and $\mathfrak{g}$ is *semi-simple*. We study the compact case and the non-compact one separately.

7.78 Theorem. *If (M, g) is a non-compact irreducible symmetric space, then the Lie subalgebra $\mathfrak{k}$ of the isotropy subgroup K is a maximal compact subalgebra of the Lie algebra $\mathfrak{g}$ of G.*

Proof. Since G is non-compact, B is not negative definite on $\mathfrak{g}$. Since B is negative definite on $\mathfrak{k}$, and since $\mathfrak{p}$ is an *irreducible* representation space of $\mathrm{Ad}_G(K)$, $B \upharpoonright \mathfrak{p}$ must be positive definite. If $\mathfrak{k}'$ is a maximal compact subalgebra of $\mathfrak{g}$ containing $\mathfrak{k}$, then $B \upharpoonright \mathfrak{k}'$ is negative definite, so $\mathfrak{k}' = \mathfrak{k}$. □

We deduce easily the following

7.79 *Corollary. A non-compact irreducible symmetric space is a quotient G/K where G is a real simple non compact Lie group with trivial center and K is a maximal compact subgroup of G.* □

7.80 Remark. Usually these symmetric spaces are called "of type III" if G is *absolutely* simple (i.e., if the complexification $\mathfrak{g}^{\mathbb{C}} = \mathfrak{g} \otimes \mathbb{C}$ is simple as a complex Lie algebra), and "of type IV" if not, in which case G is a simple complex Lie group. The lists of these symmetric spaces are given in Tables 3 and 4 of § H.

7.81 Theorem. *If (M, g) is a compact simply-connected irreducible symmetric space, then either G is simple or there exists a real simple compact simply-connected Lie group H with center Z such that $G = (H \times H)/Z$ and $K = H/Z$, where Z and H are imbedded in $H \times H$ via the diagonal embedding $h \mapsto (h, h)$.*

Remark. These last symmetric spaces $H \times H/H$ are called "of type II". See the list in Table 2 of § H.

Proof. The automorphism σ of 7.68 interchanges the simple ideals $\mathfrak{g}_i$ of $\mathfrak{g}$ (here $\mathfrak{g}$ is semi-simple, hence a direct sum of simple ideals). If $\sigma(\mathfrak{g}_i) = \mathfrak{g}_i$, then $\mathfrak{p}_i = \{x \in \mathfrak{g}_i | \sigma(x) = -x\}$ is an $\mathrm{Ad}_G(K)$-invariant non zero subspace of $\mathfrak{p}$, and by irreducibility $\mathfrak{p} = \mathfrak{p}_i$. Hence $\mathfrak{g} = \mathfrak{g}_i$ is simple.

If $\sigma(\mathfrak{g}_i) = \mathfrak{g}_j$ with $i \neq j$, then $\mathfrak{p}_i = \{(x, -\sigma(x)) \in \mathfrak{g}_i \oplus \mathfrak{g}_j\}$ is an $\mathrm{Ad}_G(K)$-invariant subspace of $\mathfrak{p}$, thus $\mathfrak{p} = \mathfrak{p}_i$, and $\mathfrak{g} = \mathfrak{g}_i \oplus \mathfrak{g}_j$. We let $\mathfrak{h} = \mathfrak{g}_i$ and identify $\mathfrak{g}_j$ with $\mathfrak{h}$ via σ. Then $\mathfrak{g} = \mathfrak{h} \oplus \mathfrak{h}$ and $\mathfrak{k} = \{(x, \sigma(x)) \in \mathfrak{g}_i \oplus \mathfrak{g}_j\} = \{(x, x) \in \mathfrak{h} \oplus \mathfrak{h}\}$. □

7.82. Now we are left with the case of compact irreducible symmetric spaces such that G is simple. These ones are called "of type I". Another trick shows that once again the classification is a consequence of the classification of real simple Lie groups. It goes as follows; let $\mathfrak{g}^{\mathbb{C}} = \mathfrak{g} \otimes \mathbb{C}$ be the complexification of $\mathfrak{g}$ and $G^{\mathbb{C}}$ the simply-connected complex Lie group generated by $\mathfrak{g}^{\mathbb{C}}$. Now $\mathfrak{g}_1 = \mathfrak{k} \oplus i\mathfrak{p}$ is obviously a (real) Lie subalgebra of $\mathfrak{g}^{\mathbb{C}}$. Let G_1 be the subgroup of $G^{\mathbb{C}}$ generated by $\mathfrak{g}_1$. Since $\mathfrak{k} \subset \mathfrak{g}_1$, G_1 contains K and we see easily that G_1/K is a non compact irreducible symmetric space, necessarily of type III.

Now the classification of spaces of type I follows from the classification of real simple Lie groups and their maximal compact subgroups. They are listed in Table 1 of § H.

7.83 Remarks. 1) More generally, the same construction works for any irreducible symmetric space. The space $M_1 = G_1/K$ is called the "dual" of M. This duality interchanges types I and III, and types II and IV, respectively.

2) In summary, irreducible symmetric spaces of type I or II are compact, Einstein and have non-negative sectional curvature. Irreducible symmetric spaces of type III or IV are non-compact, Einstein and have non-positive sectional curvature. (In fact the duality changes only one sign in the computation of curvature, see [Hel 1] for details).

G. Standard Homogeneous Manifolds

In some other particular cases, the complicated Formulas 7.30, 7.33, 7.38, 7.39 simplify.

7.84 Definition. Let (M, g) be a G-homogeneous Riemannian manifold and $\mathfrak{p}$ as in 7.22. Then g is called *naturally reductive* (with respect to $\mathfrak{p}$) if $U \equiv 0$.

Notice that the notion of natural reductivity *depends* on the choice of the group G *and* on the choice of $\mathfrak{p}$. Here is another characterization of naturally reductive spaces.

7.85 Theorem (B. Kostant [Kos]). *An* $\mathrm{Ad}_G(K)$*-invariant scalar product* $(.,.)$ *on* $\mathfrak{p}$ *defines a naturally reductive G-invariant Riemannian metric on* $M = G/K$ *if and only if on the ideal* $\mathfrak{g}' = \mathfrak{p} + [\mathfrak{p}, \mathfrak{p}]$ *of* $\mathfrak{g}$, *there exists a non-degenerate symmetric bilinear form q such that*

i) $q(\mathfrak{g}' \cap \mathfrak{k}, \mathfrak{p}) = 0$,

ii) $(.,.)$ *is the restriction of q to* $\mathfrak{p}$,

iii) *q is* $\mathrm{Ad}(G')$*-invariant, where* G' *is the subgroup of G with Lie algebra* $\mathfrak{g}'$.

For a proof, see [Kos] or the more recent reference [DA-Zi] p. 4. □

Notice that q is not necessarily positive definite on $\mathfrak{g}'$. Also the group G' acts transitively on M, and the metric is naturally reductive with respect to G' and $\mathfrak{p}$. Conversely, if q is a non-degenerate, symmetric, $\mathrm{Ad}(G)$-invariant bilinear form on $\mathfrak{g}$ with $q \restriction \mathfrak{k}$ non-degenerate and $q \restriction \mathfrak{k}^{\perp}$ positive definite (where $\mathfrak{k}^{\perp}$ is the q-orthogonal complement of $\mathfrak{k}$), then $q \restriction \mathfrak{k}^{\perp}$ is a naturally reductive metric on G/K and $G' = G$. Hence for most purposes it suffices to assume $G' = G$.

A particular case has been studied especially in the compact case.

7.86 Definition. A G-homogeneous Riemannian manifold (M, g) is called *normal* if there exists an $\mathrm{Ad}(G)$-invariant scalar product q on $\mathfrak{g}$ such that, if $\mathfrak{p}$ is the orthogonal complement of $\mathfrak{k}$ in $\mathfrak{g}$ for q, then g is associated with the restriction of q to $\mathfrak{p}$.

7.87 Proposition. *The sectional curvature of a naturally reductive G-homogeneous Riemannian manifold is given by*

(7.87a) $$(R(X,Y)X,Y) = ([[X,Y]_{\mathfrak{k}},X]_{\mathfrak{p}},Y) + \tfrac{1}{4}|[X,Y]_{\mathfrak{p}}|^2,$$

or equivalently by

(7.87b) $$(R(X,Y)X,Y) = q([X,Y]_{\mathfrak{k}},[X,Y]_{\mathfrak{k}}) + \tfrac{1}{4}|[X,Y]_{\mathfrak{p}}|^2.$$

Notice that the sectional curvature of a *normal* G-homogeneous Riemannian manifold is always non-negative (and identically zero if and only if $G = T^n$ and $K = \{1\}$). This is not necessarily true for the naturally reductive case (since q is not necessarily positive on $\mathfrak{k}$).

It is useful to introduce the Casimir operator of the isotropy representation.

7.88 Definition. Let q be a non-degenerate symmetric $\mathrm{ad}(\mathfrak{k})$-invariant bilinear form on a Lie algebra $\mathfrak{k}$. The *Casimir operator* of an N-dimensional representation ρ of $\mathfrak{k}$ is the element

$$C_{\rho,q} = -\sum_j \rho(U_j)\circ\rho(V_j)$$

of $\mathfrak{gl}(\mathbb{R}^N)$, where (U_j) and (V_j) are dual bases of $\mathfrak{k}$ (i.e., $q(U_i, V_j) = \delta_{ij}$).

Notice that $C_{\rho,q}$ commutes with any element of $\rho(\mathfrak{k})$.

7.89 Proposition. *The Ricci curvature of a naturally reductive G-homogeneous Riemannian manifold satisfies*

(7.89a) $$r(X,X) = -\frac{1}{4}B(X,X) + \frac{1}{2}\sum_i q([X,X_i]_{\mathfrak{k}},[X,X_i]_{\mathfrak{k}})$$

and, if $\mathfrak{g}' = \mathfrak{g}$, (compare 7.85),

(7.89b) $$r(X,X) = -\tfrac{1}{4}B(X,X) + \tfrac{1}{2}(C_{\chi,q}X,X),$$

where χ is the isotropy representation of K in $\mathfrak{gl}(\mathfrak{p})$.

Proof. (7.89a) follows from (7.87b), (7.37b) and the $\mathrm{Ad}(G')$-invariance of q.

(7.89b) follows from

$$\begin{aligned}(C_{\chi,q}X,Y) &= -\sum_j([U_j,[V_j,X]_{\mathfrak{p}}]_{\mathfrak{p}},Y)\\ &= -\sum_j q([Y,[X,V_j]_{\mathfrak{p}}]_{\mathfrak{k}},U_j)\\ &= -\sum_j([Y,[X,X_i]_{\mathfrak{k}}]_{\mathfrak{p}},X_i)\\ &= \sum_i q([Y,X_i]_{\mathfrak{k}},[X,X_i]_{\mathfrak{k}}).\end{aligned}$$ □

We specialize to a still more particular case.

7.90 Definition. A G-homogeneous Riemannian manifold is called *standard* if it is normal with $q = -B$.

Notice that G must be compact and semisimple in order that B be negative definite. If G is a simple compact Lie group, any naturally reductive G-homogeneous Riemannian manifold is standard (up to a scaling factor). It follows easily from the proof of 7.44

7.91 Proposition. *Let $M = G/K$ be a compact isotropy irreducible space. Then its (essentially unique) G-invariant Riemannian metric is standard (choosing the appropriate scaling factor).*

Notice that not all standard homogeneous metrics are Einstein, unlike the isotropy irreducible spaces. For example, the standard metric of the $SU(n)$-homogeneous space $S^{2n-1} = SU(n)/SU(n-1)$ is not Einstein. It follows immediately from Formula (7.89b).

7.92 Proposition. *A standard homogeneous Riemannian manifold is Einstein if and only if there is a constant λ such that $C_{\chi,-B} = \lambda Id$. where χ is the isotropy representation of $\mathfrak{k}$ in $\mathfrak{p}$.*

Notice that $C_{\chi,-B}$ is a (possibly different) multiple of Id when restricted to every irreducible factor of χ. Hence 7.92 gives a strong condition if χ is not irreducible. In particular, there cannot be any element in $\mathfrak{p}$ on which χ acts trivially.

7.93 Proposition. *If a standard Riemannian homogeneous manifold G/K is Einstein with $r = \lambda g$, then the constant λ satisfies $\frac{1}{4} \leqslant \lambda \leqslant \frac{1}{2}$* and

(i) $\lambda = \frac{1}{2}$ *if and only if G/K is locally symmetric,*

(ii) $\lambda = \frac{1}{4}$ *if and only if $K = \{1\}$.*

Proof. Since $(C_{\chi,-B}X, X)$ is $\geqslant 0$ and identically 0 if and only if $\chi = id$, we have $\lambda \geqslant \frac{1}{4}$ and $\lambda = \frac{1}{4}$ if and only if $K = \{1\}$.

Using (7.35) and the Ad(G)-invariance of B, we have

$$\frac{1}{2}(C_{\chi,q}X, X) = -\frac{1}{4}B(X, X) - \sum_i([X, X_i]_{\mathfrak{p}}, [X, X_i]_{\mathfrak{p}})$$

hence $\lambda \leqslant \frac{1}{2}$ and $\lambda = \frac{1}{2}$ if and only if $[\mathfrak{p}, \mathfrak{p}] \subset \mathfrak{k}$. □

7.94 Remarks. (i) One can also find a more explicit formula for the Einstein constant λ of a standard homogeneous Einstein metric. If $\mathfrak{k}$ is the direct sum of its center $\mathfrak{k}_0$ and simple ideals $\mathfrak{k}_1, \ldots, \mathfrak{k}_r$ with $s_i = \dim \mathfrak{k}_i$, if the Killing-Cartan forms $B_{\mathfrak{k}_i}$ of the ideals $\mathfrak{k}_i$ satisfy $B_{\mathfrak{k}_i} = c_i B$, then

$$\lambda = \frac{1}{4} + \frac{1}{2n}\sum_{i=1}^{r} s_i(1 - c_i). \tag{7.94}$$

(ii) Formula (7.89b), Propositions 7.92 and 7.93 were first observed in [Wa-Zi 2]. In this paper, M. Wang and W. Ziller classify completely the simply connected standard homogeneous Riemannian manifolds, with G simple, which are Einstein. They also exhibit some non-trivial examples where G is not simple and where the standard G-homogeneous metric is Einstein. If G is not simple, there arise many naturally reductive G-homogeneous Einstein metrics (see Chapter 9), most of which are not standard. We will not give here the proof of their classification in the case where G is simple, but we will describe the result below (Theorem 7.99). We begin with some examples.

7.95 Examples. The same construction as in 7.49 gives the following class of examples. Let K be any compact semi-simple Lie group with trivial center. Its adjoint representation embeds K in $SO(\dim K)$, and the standard Riemannian metric on $SO(\dim K)/K$ is Einstein.

7.96. We use the same notations as in 7.48 and 7.50. We obtain (non-strongly isotropy irreducible) standard homogeneous Einstein metrics in the following cases

K	ρ	G
$SO(n)\dots SO(n)$ (k times, $k \geqslant 3$)	$\rho_1 \otimes id \otimes \cdots \otimes id + id \otimes \rho_1 \otimes \cdots \otimes id + \cdots + id \otimes id \otimes \cdots \otimes \rho_1$	$SO(kn)$
$S(U(n)\dots U(n))$ (k times, $k \geqslant 3$)	$\rho_1 \otimes id \otimes \cdots \otimes id + id \otimes \rho_1 \otimes \cdots \otimes id + \cdots + id \otimes id \otimes \cdots \otimes \rho_1$	$SU(kn)$
$Sp(n)\dots Sp(n)$ (k times, $k \geqslant 3$)	$\rho_1 \otimes id \otimes \cdots \otimes id + id \otimes \rho_1 \otimes \cdots \otimes id + \cdots + id \otimes id \otimes \cdots \otimes \rho_1$	$Sp(kn)$
$SO(n)SO(n)$	$\rho_1 \otimes \rho_1$	$SO(n^2)$
$Sp(n)Sp(n)$	$\rho_1 \otimes \rho_1$	$SO(4n^2)$
$Sp(n)U(2n-1)$	$\rho_1 \otimes id + id \otimes \rho_1$	$Sp(3n-1)$
$SO(n)U(n+1)$	$\rho_1 \otimes id + id \otimes \rho_1$	$SO(3n+2)$
$S(U(l)U(m_1)U(m_2))$ [1]	$\rho_1 \otimes id \otimes id + id \otimes \rho_1 \otimes \rho_1$	$SU(l + m_1 m_2)$

[1] if $m_1^2 + m_2^2 + 1 = l m_1 m_2$ ($m_1 \geqslant m_2 \geqslant 2$).

7.97. As in 7.51, the following construction generalizes the previous examples. Let G_i/K_i, $i = 1, \dots, r$ be a sequence of compact irreducible symmetric spaces with $\dim(G_i/K_i) = n_i$ and with isotropy representation π_i. Then the standard homogeneous metric on $SO(n_i)/\pi_i(K_i)$ is Einstein if and only if K_i is simple or $G_i/K_i = SO(2n)/SO(n)SO(n)$ or $Sp(2n)/Sp(n)Sp(n)$. (This follows from irreducibility if K_i is simple and from 7.96 n° 4, 5 above). Assuming that this is satisfied for each i, we consider the homogeneous space $SO(N)/\rho(K)$ where $N = \sum_i n_i$, $K = K_1 \dots K_r$ and

$$\rho = \pi_1 \otimes id \otimes \cdots \otimes id + id \otimes \pi_2 \otimes \cdots \otimes id + id \otimes id \otimes \cdots \otimes \pi_r.$$

Then the standard homogeneous metric on $SO(N)/\rho(K)$ is Einstein if and only if $n_i/\dim K_i$ is independent of i.

We obtain the examples in 7.95 if all G_i/K_i are compact irreducible symmetric spaces of type II.

Notice that the condition that $n_i/\dim K_i$ is independent of i is in particular satisfied if the same symmetric space is repeated (but also in some other cases). The first case of 7.96 is of this type with $G_i/K_i = SO(n+1)/SO(n)$.

7.98. An isolated example is $M = \mathrm{Spin}(8)/G_2$ for the inclusion $G_2 \subset \mathrm{Spin}(7) \subset \mathrm{Spin}(8)$ where the first inclusion is that of 7.13 with quotient $S^7 = \mathrm{Spin}(7)/G_2$ and the second one is the classical one with quotient S^7 also.

Topologically, M is diffeomorphic to $S^7 \times S^7$ (but not isomorphic as a homogeneous space). Its standard Spin(8)-invariant metric is Einstein (here the isotropy representation χ of G_2 is twice the unique 7-dimensional irreducible representation which gives the inclusion $G_2 \subset \mathrm{Spin}(7)$) but is *not* isometric to a product metric.

7.99 Theorem (M. Wang-W. Ziller [Wa-Zi 2]). *A compact simply-connected standard G-homogeneous Riemannian manifold with G simple is Einstein if and only if it is listed in table* 1, 2 (*with* $G = H$) 5, 6, 7 *or* 8 *of* § H *below.*

H. Tables

7.100 Introduction. In the following tables 1–8, we list some G-homogeneous Riemannian manifolds, namely:

(a) in Tables 1, 2, 3, 4, the irreducible symmetric spaces (of type I, II, III or IV respectively);

(b) in Tables 5, 6, the other strongly isotropy irreducible spaces;

(c) in Tables 7, 8, the other spaces (with G simple) whose standard Riemannian metric is Einstein.

In fact, we will not really list all those spaces, because it is not so easy to determine explicitely all the corresponding G and K's. Hence

7.101. *We will only give the Lie algebra* $\mathfrak{g}$ *of G and the Lie subalgebra* $\mathfrak{k}$ *of K*

This is sufficient to determine M up to a finite covering. For more details, see [Hel 1], [Wol 3] or [Wa-Zi 2].

Notice that it is not sufficient in general to give the type of $\mathfrak{k}$ as a Lie algebra in order to determine its embedding in $\mathfrak{g}$, (except in some particular cases, such as symmetric spaces). Hence at some places in the tables, we also indicate how we embed $\mathfrak{k}$ in $\mathfrak{g}$, often by refering to some place in the chapter where such a construction is performed.

Because of low-dimensional isomorphisms between classical groups, the same space may appear in various ways. To avoid that, we choose some set of conditions on dimensions. We recall the following isomorphisms between classical Lie algebras:

(1) $\mathfrak{so}(2, \mathbb{C}) = \mathbb{C}$; $\mathfrak{so}(2) = \mathfrak{so}(1, 1) = \mathfrak{so}(1, \mathbb{H}) = \mathbb{R}$;

(2) $\mathfrak{sl}(2, \mathbb{C}) = \mathfrak{so}(3, 1) = \mathfrak{so}(3, \mathbb{C}) = \mathfrak{sp}(1, \mathbb{C})$; $\mathfrak{su}(2) = \mathfrak{sl}(1, \mathbb{H}) = \mathfrak{so}(3) = \mathfrak{sp}(1)$;
$\mathfrak{su}(1, 1) = \mathfrak{sl}(2, \mathbb{R}) = \mathfrak{so}(2, 1) = \mathfrak{sp}(1, \mathbb{R})$;

(3) $\mathfrak{so}(4, \mathbb{C}) = \mathfrak{sl}(2, \mathbb{C}) \oplus \mathfrak{sl}(2, \mathbb{C})$; $\mathfrak{so}(4) = \mathfrak{su}(2) \oplus \mathfrak{su}(2)$;
$\mathfrak{so}(2, 2) = \mathfrak{sl}(2, \mathbb{R}) \oplus \mathfrak{sl}(2, \mathbb{R})$; $\mathfrak{so}(2, \mathbb{H}) = \mathfrak{su}(2) \oplus \mathfrak{sl}(2, \mathbb{R})$;

(4) $\mathfrak{so}(5, \mathbb{C}) = \mathfrak{sp}(2, \mathbb{C})$; $\mathfrak{so}(5) = \mathfrak{sp}(2)$; $\mathfrak{so}(4, 1) = \mathfrak{sp}(1, 1)$; $\mathfrak{so}(3, 2) = \mathfrak{sp}(2, \mathbb{R})$;

(5) $\mathfrak{sl}(4,\mathbb{R}) = \mathfrak{so}(6,\mathbb{C})$; $\mathfrak{su}(4) = \mathfrak{so}(6)$; $\mathfrak{su}(3,1) = \mathfrak{so}(3,\mathbb{H})$; $\mathfrak{su}(2,2) = \mathfrak{so}(4,2)$;
$\mathfrak{sl}(4,\mathbb{R}) = \mathfrak{so}(3,3)$; $\mathfrak{sl}(2,\mathbb{H}) = \mathfrak{so}(5,1)$;
(6) $\mathfrak{so}(4,\mathbb{H}) = \mathfrak{so}(6,2)$.

Finally, notice that there may exist different notations for the same Lie algebra. For example $\mathfrak{sl}(n,\mathbb{H})$ is often denoted by $\mathfrak{su}^*(2n)$ and $\mathfrak{so}(n,\mathbb{H})$ by $\mathfrak{so}^*(2n)$. Concerning exceptional simple Lie algebras, E_6 is the *compact* Lie algebra, E_6^δ the non-compact real Lie algebra with character δ, whose complexified Lie algebra is $E_6^{\mathbb{C}}$; and similarly for E_7, E_8, F_4, G_2 (see [Hel 1] for details).

7.102 Table 1. (Compact) irreducible symmetric spaces of type I

$\mathfrak{g}$	$\mathfrak{k}$	dim M	Condition	$\mathfrak{g}$	$\mathfrak{k}$	dim M
$\mathfrak{su}(p+q)$	$\mathfrak{su}(p) \oplus \mathfrak{su}(q) \oplus \mathbb{R}$	$2pq$	$1 \leqslant p \leqslant q$	E_6	$\mathfrak{su}(6) \oplus \mathfrak{su}(2)$	40
$\mathfrak{su}(n)$	$\mathfrak{so}(n)$	$\frac{(n-1)(n+2)}{2}$	$3 \leqslant n$	E_6	$\mathfrak{so}(10) \oplus \mathbb{R}$	32
$\mathfrak{su}(2n)$	$\mathfrak{sp}(n)$	$(n-1)(2n+1)$	$2 \leqslant n$	E_6	$\mathfrak{sp}(4)$	42
$\mathfrak{so}(2n)$	$\mathfrak{su}(n) \oplus \mathbb{R}$	$n(n-1)$	$5 \leqslant n$	E_6	F_4	26
$\mathfrak{so}(p+q)$	$\mathfrak{so}(p) \oplus \mathfrak{so}(q)$	pq	$1 \leqslant p \leqslant q$ $7 \leqslant p+q$	E_7	$\mathfrak{su}(8)$	70
$\mathfrak{sp}(n)$	$\mathfrak{su}(n) \oplus \mathbb{R}$	$n(n+1)$	$2 \leqslant n$	E_7	$\mathfrak{so}(12) \oplus \mathfrak{su}(2)$	64
$\mathfrak{sp}(p+q)$	$\mathfrak{sp}(p) \oplus \mathfrak{sp}(q)$	$4pq$	$1 \leqslant p \leqslant q$	E_7	$E_6 \oplus \mathbb{R}$	54
G_2	$\mathfrak{su}(2) \oplus \mathfrak{su}(2)$	8		E_8	$\mathfrak{so}(16)$	128
F_4	$\mathfrak{so}(9)$	16		E_8	$E_7 \oplus \mathfrak{su}(2)$	112
F_4	$\mathfrak{sp}(3) \oplus \mathfrak{su}(2)$	28				

7.103 Table 2. (Compact) irreducible symmetric spaces of type II

$\mathfrak{k}$	dim M	Condition	$\mathfrak{k}$	dim M
$\mathfrak{su}(n)$	n^2-1	$2 \leqslant n$	F_4	52
$\mathfrak{so}(n)$	$\frac{n(n-1)}{2}$	$7 \leqslant n$	E_6	78
$\mathfrak{sp}(n)$	$n(2n+1)$	$2 \leqslant n$	E_7	133
G_2	14		E_8	248

Notes

(i) Here $\mathfrak{g} = \mathfrak{k} \oplus \mathfrak{k}$, and the embedding $\mathfrak{k} \subset \mathfrak{g}$ is the diagonal one.

(ii) M is a (connected) simple compact Lie group.

7.104 Table 3. (Non-compact) irreducible symmetric spaces of type III

$\mathfrak{g}$	$\mathfrak{k}$	dim M	Condition	$\mathfrak{g}$	$\mathfrak{k}$	dim M
$\mathfrak{su}(p,q)$	$\mathfrak{su}(p) \oplus \mathfrak{su}(q) \oplus \mathbb{R}$	$2pq$	$1 \leqslant p \leqslant q$	E_6^2	$\mathfrak{su}(6) \oplus \mathfrak{su}(2)$	40
$\mathfrak{sl}(n,\mathbb{R})$	$\mathfrak{so}(n)$	$\frac{(n-1)(n+2)}{2}$	$3 \leqslant n$	E_6^{-14}	$\mathfrak{so}(10) \oplus \mathbb{R}$	32
$\mathfrak{sl}(n,\mathbb{H})$	$\mathfrak{sp}(n)$	$(n-1)(2n+1)$	$2 \leqslant n$	E_6^6	$\mathfrak{sp}(4)$	42
$\mathfrak{so}(n,\mathbb{H})$	$\mathfrak{su}(n) \oplus \mathbb{R}$	$n(n-1)$	$5 \leqslant n$	E_6^{-26}	F_4	26
$\mathfrak{so}(p,q)$	$\mathfrak{so}(p) \oplus \mathfrak{so}(q)$	pq	$1 \leqslant p \leqslant q$ $7 \leqslant p+q$	E_7^7	$\mathfrak{su}(8)$	70
$\mathfrak{sp}(n,\mathbb{R})$	$\mathfrak{su}(n) \oplus \mathbb{R}$	$n(n+1)$	$2 \leqslant n$	E_7^{-5}	$\mathfrak{so}(12) \oplus \mathfrak{su}(2)$	64
$\mathfrak{sp}(p,q)$	$\mathfrak{sp}(p) \oplus \mathfrak{sp}(q)$	$4pq$	$1 \leqslant p \leqslant q$	E_7^{-25}	$E_6 \oplus \mathbb{R}$	54
G_2^2	$\mathfrak{su}(2) \oplus \mathfrak{su}(2)$	8		E_8^8	$\mathfrak{so}(16)$	128
F_4^{-20}	$\mathfrak{so}(9)$	16		E_8^{-24}	$E_7 \oplus \mathfrak{su}(2)$	112
F_4^4	$\mathfrak{sp}(3) \oplus \mathfrak{su}(2)$	28				

Notes

(i) Here $M = G/K$ where G is a non-compact absolutely simple real Lie group and K is a maximal compact subgroup.

(ii) The duality defined in 7.83 interchanges spaces of type I and III in the same order.

7.105 Table 4. (Non-compact) irreducible symmetric spaces of type IV

$\mathfrak{g}$	$\mathfrak{k}$	dim M	Condition	$\mathfrak{g}$	$\mathfrak{k}$	dim M
$\mathfrak{sl}(n,\mathbb{C})$	$\mathfrak{su}(n)$	n^2-1	$2 \leqslant n$	$F_4^{\mathbb{C}}$	F_4	52
$\mathfrak{so}(n,\mathbb{C})$	$\mathfrak{so}(n)$	$\frac{n(n-1)}{2}$	$7 \leqslant n$	$E_6^{\mathbb{C}}$	E_6	78
$\mathfrak{sp}(n,\mathbb{C})$	$\mathfrak{sp}(n)$	$n(2n+1)$	$2 \leqslant n$	$E_7^{\mathbb{C}}$	E_7	133
$G_2^{\mathbb{C}}$	G_2	14		$E_8^{\mathbb{C}}$	E_8	248

Notes

(i) Here $M = G/K$ where G is a simple complex Lie group and K a maximal compact subgroup.

(ii) The duality defined in 7.83 interchanges spaces of type II and IV in the same order.

7.106 Table 5. (Compact) non-symmetric strongly isotropy irreducible spaces
(a) *Infinite families*

$\mathfrak{g}$	$\mathfrak{k}$	Condition	$\mathfrak{g}$	$\mathfrak{k}$	Condition
$\mathfrak{su}\left(\frac{n(n-1)}{2}\right)$	$\mathfrak{su}(n)$	$5 \leqslant n$	$\mathfrak{so}\left(\frac{(n-1)(n+2)}{2}\right)$	$\mathfrak{so}(n)$	$5 \leqslant n$
$\mathfrak{su}\left(\frac{n(n+1)}{2}\right)$	$\mathfrak{su}(n)$	$3 \leqslant n$	$\mathfrak{so}((n-1)(2n+1))$	$\mathfrak{sp}(n)$	$3 \leqslant n$
$\mathfrak{su}(pq)$	$\mathfrak{su}(p) \oplus \mathfrak{su}(q)$	$2 \leqslant p \leqslant q$ $p+q \neq 4$	$\mathfrak{so}(n(2n+1))$	$\mathfrak{sp}(n)$	$2 \leqslant n$ (*)
$\mathfrak{so}(n^2-1)$	$\mathfrak{su}(n)$	$3 \leqslant n$ (*)	$\mathfrak{so}(4n)$	$\mathfrak{sp}(1) \oplus \mathfrak{sp}(n)$	$2 \leqslant n$
$\mathfrak{so}\left(\frac{n(n-1)}{2}\right)$	$\mathfrak{so}(n)$	$7 \leqslant n$ (*)	$\mathfrak{sp}(n)$	$\mathfrak{sp}(1) \oplus \mathfrak{so}(n)$	$3 \leqslant n$

Note: These spaces are constructed in 7.50 (or in 7.49 for those with a (*)).

7.107 Table 6. (Compact) non-symmetric strongly isotropy irreducible spaces
(b) *Exceptions*

$\mathfrak{g}$	$\mathfrak{k}$	Note	$\mathfrak{g}$	$\mathfrak{k}$	Note	$\mathfrak{g}$	$\mathfrak{k}$	Note
$\mathfrak{su}(16)$	$\mathfrak{so}(10)$	1	$\mathfrak{so}(133)$	E_7	3	E_6	$\mathfrak{su}(3)$	4
$\mathfrak{su}(27)$	E_6	1	$\mathfrak{so}(248)$	E_8	3	E_6	$3\mathfrak{su}(3)$	4, 5
$\mathfrak{so}(7)$	G_2	2	$\mathfrak{sp}(2)$	$\mathfrak{su}(2)$	1	E_6	G_2	4
$\mathfrak{so}(14)$	G_2	3	$\mathfrak{sp}(7)$	$\mathfrak{su}(6)$	1	E_6	$G_2 \oplus \mathfrak{su}(3)$	4
$\mathfrak{so}(16)$	$\mathfrak{so}(9)$	1	$\mathfrak{sp}(10)$	$\mathfrak{so}(12)$	1	E_7	$\mathfrak{su}(3)$	4
$\mathfrak{so}(26)$	F_4	1	$\mathfrak{sp}(16)$	$\mathfrak{sp}(3)$	1	E_7	$\mathfrak{su}(6) \oplus \mathfrak{su}(3)$	4
$\mathfrak{so}(42)$	$\mathfrak{sp}(4)$	1	$\mathfrak{sp}(28)$	E_7	1	E_7	$G_2 \oplus \mathfrak{sp}(3)$	4
$\mathfrak{so}(52)$	F_4	3	G_2	$\mathfrak{su}(3)$	4	E_7	$F_4 \oplus \mathfrak{su}(2)$	4
$\mathfrak{so}(70)$	$\mathfrak{su}(8)$	1	G_2	$\mathfrak{so}(3)$	4	E_8	$\mathfrak{su}(9)$	4
$\mathfrak{so}(78)$	E_6	3	F_4	$\mathfrak{su}(3) \oplus \mathfrak{su}(3)$	4	E_8	$F_4 \oplus G_2$	4
$\mathfrak{so}(128)$	$\mathfrak{so}(16)$	1	F_4	$G_2 \oplus \mathfrak{su}(2)$	4	E_8	$E_6 \oplus \mathfrak{su}(3)$	4

Notes
(1) Defined in 7.51.
(2) Defined in 7.13
(3) Defined in 7.49.
(4) Here $\mathfrak{k}$ is a maximal subalgebra of $\mathfrak{g}$ (and this is sufficient, in those cases, to characterize the embedding, see [Dyn 2]).
(5) $p\mathfrak{k}$ means $\mathfrak{k} \oplus \mathfrak{k} \oplus \cdots \oplus \mathfrak{k}$ (p times).

7.108 Table 7. (Compact) non-strongly isotropy irreducible G-homogeneous manifolds with G simple, whose standard metric is Einstein

(a) *Infinite families*

$\mathfrak{g}$	$\mathfrak{k}$	Condition	$\mathfrak{g}$	$\mathfrak{k}$	Condition
$\mathfrak{su}(kn)$	$k\mathfrak{su}(n) \oplus (k-1)\mathbb{R}$	$3 \leqslant k$ $1 \leqslant n$ (1)	$\mathfrak{so}(n^2)$	$\mathfrak{so}(n) \oplus \mathfrak{so}(n)$	$3 \leqslant n$
$\mathfrak{su}(l+pq)$	$\mathfrak{su}(l) \oplus \mathfrak{su}(p)$ $\oplus\, \mathfrak{su}(q) \oplus 2\mathbb{R}$	$2 \leqslant p \leqslant q$, $lpq = p^2 + q^2 + 1$	$\mathfrak{so}(4n^2)$	$\mathfrak{sp}(n) \oplus \mathfrak{sp}(n)$	$2 \leqslant n$
$\mathfrak{so}(N)$	$\rho(\mathfrak{k})$	See note (2)	$\mathfrak{sp}(kn)$	$k\mathfrak{sp}(n)$	$3 \leqslant k$ $1 \leqslant n$ (1)
$\mathfrak{so}(3n+2)$	$\mathfrak{su}(n+1) \oplus \mathfrak{so}(n)$ $\oplus\, \mathbb{R}$	$3 \leqslant n$	$\mathfrak{sp}(3n-1)$	$\mathfrak{su}(2n-1)$ $\oplus\, \mathfrak{sp}(n) \oplus \mathbb{R}$	$1 \leqslant n$

Notes

(1) $k \cdot \mathfrak{h}$ denotes $\mathfrak{h} \otimes \cdots \otimes \mathfrak{h}$ (k times)

(2) Here $\mathfrak{k} = \mathfrak{k}_1 \oplus \cdots \oplus \mathfrak{k}_r$ $(r > 1)$, and there exists compact irreducible symmetric spaces G_i/K_i such that $(\mathfrak{g}_i, \mathfrak{k}_i)$ satisfies
either $\mathfrak{k}_i$ simple
or $(\mathfrak{g}_i, \mathfrak{k}_i) = (\mathfrak{so}(2n), \mathfrak{so}(n) \oplus \mathfrak{so}(n))$
or $(\mathfrak{g}_i, \mathfrak{k}_i) = (\mathfrak{sp}(2n), \mathfrak{sp}(n) \oplus \mathfrak{sp}(n))$
and $\dfrac{\dim G_i}{\dim K_i}$ independant of i; and ρ is the representation $\pi_1 \otimes id \otimes \cdots \otimes id + id \otimes \pi_2 \otimes \cdots \otimes id + \cdots + id \otimes id \otimes \cdots \otimes \pi_r$ of $\mathfrak{k}$ in $\mathbb{R}^N$ with $N = \sum_i (\dim G_i - \dim K_i)$, where π_i is the isotropy representation of $\mathfrak{k}_i$ in G_i/K_i (see 7.97).

7.109 Table 8. (Compact) non-strongly isotropy irreducible G-homogeneous manifolds with G simple whose standard metric is Einstein

(b) *Exceptions*

$\mathfrak{g}$	$\mathfrak{k}$	ρ (see note 1) or $\mathfrak{h}$ (see note 2)	$\mathfrak{g}$	$\mathfrak{k}$	$\mathfrak{h}$
$\mathfrak{so}(8)$	G_2	$\rho_1 \oplus id$ (1)	E_8	$\mathbb{R}^8$	Cartan subalgebra
$\mathfrak{so}(26)$	$\mathfrak{sp}(1) \oplus \mathfrak{sp}(5)$ $\oplus\, \mathfrak{so}(6)$	$(\rho_1 \otimes \rho_1 \otimes id)$ $\oplus\, (id \otimes id \otimes \rho_1)$ (1)	E_8	$8\,\mathfrak{su}(2)$	$\mathfrak{so}(16)$ (5)
F_4	$\mathfrak{so}(8)$	$\mathfrak{so}(9)$	E_8	$4\mathfrak{su}(2)$	$E_6 \oplus \mathfrak{su}(3)$
E_6	$\mathbb{R}^6$	Cartan subalgebra	E_8	$4\mathfrak{su}(3)$	$E_6 \oplus \mathfrak{su}(3)$
E_6	$3\mathfrak{su}(2)$	$3\mathfrak{su}(3)$ (3)	E_8	$2\mathfrak{su}(3)$	$\mathfrak{su}(9)$ (6)
E_6	$\mathfrak{so}(8) \oplus \mathbb{R}^2$	$\mathfrak{so}(10) \oplus \mathbb{R}$	E_8	$2\mathfrak{su}(5)$	$\mathfrak{k}$ maximal
E_6	$\mathfrak{so}(6) \oplus \mathfrak{su}(2)$	$\mathfrak{su}(6) \oplus \mathfrak{su}(2)$	E_8	$\mathfrak{so}(9)$	$\mathfrak{su}(9)$
E_7	$\mathbb{R}^7$	Cartan subalgebra	E_8	$\mathfrak{so}(9)$	$\mathfrak{so}(16)$ (7)
E_7	$7\mathfrak{su}(2)$	$\mathfrak{so}(12) \oplus \mathfrak{su}(2)$	E_8	$2\mathfrak{so}(8)$	$\mathfrak{so}(16)$
E_7	$\mathfrak{so}(8)$	$\mathfrak{su}(8)$	E_8	$\mathfrak{sp}(2)$	$\mathfrak{k}$ maximal
E_7	$\mathfrak{so}(8) \oplus 3\mathfrak{su}(2)$	$\mathfrak{so}(12) \oplus \mathfrak{su}(2)$ (4)	E_8	$2\mathfrak{sp}(2)$	$\mathfrak{so}(16)$ (8)

7.109 Table 8. (Continued).

Notes

(1) Here ρ is the representation of $\mathfrak{k}$ in $\mathbb{R}^N$ which gives the embedding of $\mathfrak{k}$ in $\mathfrak{so}(N)$; ρ_1 is the standard representation or the unique 7-dimensional irreducible representation of G_2.

(2) Except in the first three cases, we characterize the embedding of $\mathfrak{k}$ in $\mathfrak{g}$ either by an explicit description or by giving a maximal subalgebra $\mathfrak{h}$ of $\mathfrak{g}$ such that $\mathfrak{k} \subset \mathfrak{h} \subset \mathfrak{g}$. Following [Dyn 2], the type of $\mathfrak{h}$ as a Lie algebra is sufficient to characterize its embedding in $\mathfrak{g}$ and the inclusion $\mathfrak{k} \subset \mathfrak{h}$ will always be a standard one (except in the case of the note (7)).
We insist that the space to be considered is G/K corresponding to $(\mathfrak{g}, \mathfrak{k})$.

(3) $3\mathfrak{su}(2) = 3\mathfrak{so}(3) \subset 3\mathfrak{su}(3)$ via $\mathfrak{so}(3) \subset \mathfrak{su}(3)$.

(4) $\mathfrak{so}(8) \oplus 2\mathfrak{su}(2) = \mathfrak{so}(8) \oplus \mathfrak{so}(4) \subset \mathfrak{so}(12)$.

(5) $8\mathfrak{su}(2) = 4\mathfrak{so}(4) \subset \mathfrak{so}(16)$.

(6) $2\mathfrak{su}(3) \subset \mathfrak{su}(9)$ via $\rho_1 \otimes \rho_1$.

(7) $\mathfrak{so}(9) \subset \mathfrak{so}(16)$ via a half-spin representation.

(8) $\mathfrak{sp}(2) \oplus \mathfrak{sp}(2) \subset \mathfrak{so}(16)$ via $\rho_1 \otimes \rho_1$.

I. Remarks on Homogeneous Lorentz Manifolds

7.110. There is a theory of homogeneous Lorentz manifolds (or more generally of homogeneous pseudo-Riemannian manifolds) which parallels the theory of Riemannian ones, but is more difficult. Indeed, in the Lorentz or pseudo-Riemannian case, the isotropy subgroup at a point is not necessarily compact; as a consequence it may happen that there is no $\mathrm{Ad}_G(K)$-invariant complement $\mathfrak{p}$ to $\mathfrak{k}$ in $\mathfrak{g}$. In order to replace such a $\mathfrak{p}$, we have to work in the general case with the quotient vector space $\mathfrak{g}/\mathfrak{k}$ and the action induced on it by the bracket of $\mathfrak{g}$. We will not develop this general theory here (see for example Chapter 10 in [Ko-No 2]). We only indicate a few definitions and examples, and point out the main differences with the Riemannian case, especially for Einstein manifolds.

7.111 Definition. A Lorentz manifold (M, h) is *G-homogeneous* if G is a closed subgroup of $I(M, h)$ which acts transitively on M.

7.112 Examples. i) The Minkowski space is $(\mathbb{R}^n, h_{\mathrm{can}})$, where $h_{\mathrm{can}} = dx_1^2 - dx_2^2 - \cdots - dx_n^2$. It is G-homogeneous for any closed subgroup G of $\mathbb{R}^n \times O(1, n-1)$ which contains the translations. This is the model space for Lorentzian geometry. It is flat (i.e., $R \equiv 0$) and of course simply-connected and complete.

ii) The complete Lorentz manifolds with non-zero constant curvature are discrete quotients of

$$S_1^n = SO_0(1, n)/SO_0(1, n-1),$$

or the universal covering $\tilde{H}_1^n$ of

$$H_1^n = SO(2, n-1)/SO_0(1, n-1).$$

They are often presented as submanifolds,

$$S_1^n \subset \mathbb{R}^{n+1} \quad \text{as} \quad \{x_0^2 - x_1^2 - \cdots - x_n^2 = 1\}$$

with the structure induced from $dx_0^2 - dx_1^2 - \cdots - dx_n^2$ on $\mathbb{R}^{n+1}$ and

$$H_1^n \subset \mathbb{R}^{n+1} \quad \text{as} \quad \{x_0^2 + x_1^2 - x_2^2 - \cdots - x_n^2 = 1\}$$

with the structure induced from $dx_0^2 + dx_1^2 - dx_2^2 - \cdots - dx_n^2$ on $\mathbb{R}^{n+1}$ (see [Wol 4]).

7.113 A Non-complete Example. We consider the 2-dimensional Minkowski space $(\mathbb{R}^2, dx^2 - dy^2)$ and the open submanifold $M = \{(x, y) \in \mathbb{R}^2 | x + y > 0\}$.

Of course $(M, dx^2 - dy^2)$ is a flat non-complete simply connected Lorentz manifold, but it is also *homogeneous*. Indeed, let G be the non-abelian Lie group $\mathbb{R} \times \mathbb{R}$ with product

$$(a, t)(a', t') = (a + a'e^t, t + t').$$

Then G acts *simply transitively* on M by

$$(a, t)(x, y) = (xcht + ysht + a, xsht + ycht - a),$$

so M is homogeneous.

As a consequence Theorem 7.19 *is not true* in the Lorentz case, even in the locally symmetric case (here R is zero, hence parallel).

7.114 Definition. A Lorentz manifold (M, h) (or more generally a pseudo-Riemannian manifold) is *symmetric* if for each x in M there exists an isometry f_x of M such that $f_x(x) = x$ and $T_x(f_x) = -Id_{T_xM}$.

7.115 Remarks. i) One may show that a symmetric Lorentz manifold is homogeneous and complete (the proof is the same as in the Riemannian case).

ii) But example 7.113 shows that a simply connected homogeneous locally symmetric Lorentz manifold is not necessarily symmetric.

iii) On the other hand, a *complete* simply connected locally symmetric Lorentz manifold is symmetric (see [Ko-No 2] Chapter 11).

iv) Notice also that complete means only that any geodesic is indefinitely extendible. We recall that Hopf-Rinow theorem is *not* true for pseudo-Riemannian manifolds. For example, $\tilde{H}_1^n$ is complete (and symmetric) but it is not true that any two points on it may be joined by a geodesic.

7.116. Furthermore the theory of symmetric Lorentz manifolds is somewhat different from that of Riemannian ones. Here one may show (see for example M. Cahen and N. Wallach [Ca-Wa]) that a simply-connected symmetric Lorentz manifold is the product of a symmetric *Riemannian* manifold (with metric $-g$ in our normalization) and one of the following three manifolds

i) $(\mathbb{R}, dt^2)$,

ii) a complete simply connected Lorentz manifold with constant curvature,

iii) one of the manifolds described below in 7.117.

7.117 A Particular Class of Symmetric Examples. Let $(\mathbb{R}^n, (.,.))$ be the canonical Euclidean space and $f: \mathbb{R}^n \to \mathbb{R}^n$ any *symmetric* linear map. We consider the solvable Lie algebra $\mathfrak{g} = \mathbb{R}^{2n+2} = \mathbb{R}^n \times \mathbb{R}^n \times \mathbb{R} \times \mathbb{R}$ with the bracket

$$[(x,y,t,u),(x',y',t',u')] = (uf(y') - u'f(y), ux' - u'x, (x, f(y')) - (x', f(y)), 0).$$

Then $\mathfrak{k} = \{(x,0,0,0)\}$ is an abelian subalgebra and $\mathfrak{p} = \{(0,y,t,u)\}$ is an ad($\mathfrak{k}$)-invariant complement to $\mathfrak{k}$ in $\mathfrak{g}$.

On $\mathfrak{p}$, the symmetric bilinear form defined by

$$q((0,y,t,u),(0,y',t',u')) = tu' + ut' - (y,y')$$

has signature $(1, n+1)$ and is ad($\mathfrak{k}$)-invariant.

Let G be the simply connected solvable Lie group generated by $\mathfrak{g}$ and K the subgroup corresponding to $\mathfrak{k}$. Then K is closed and q induces a G-invariant Lorentz metric h on $M = G/K$ by the same construction as in the Riemannian case (see 7.24). Then

7.118 Theorem (M. Cahen and N. Wallach [Ca-Wa]). *The homogeneous Lorentz manifold* (M,h) *above* (in 7.117) *is symmetric. Moreover its Ricci curvature is zero if ad only if the trace of* f *satisfies* $\operatorname{tr}(f) \equiv 0$, *and its sectional curvature is zero if and only if* $f \equiv 0$.

For the proof, see [Ca-Wa]. Calculations similar to those in § C also work in this case.

7.119 Corollary. *There exist Ricci-flat non-flat symmetric Lorentz manifolds* (*at least for* $n \geqslant 4$).

(Compare with 7.61.)

7.120 Remarks. (i) In the case of dimension 4, i.e. when $n = 2$, then the two conditions $f \neq 0$ and $\operatorname{tr}(f) = 0$ imply that (up to a homothety on M) we may take $f = \begin{pmatrix} 1 & 0 \\ 0 & -1 \end{pmatrix}$. This (essentially unique) 4-dimensional Lorentz manifold has a curvature of type II in Petrov's classification (see 3.23).

(ii) Notice also that for appropriate choices of f, then G/K has compact quotients (see [Ca-Wa]).

(iii) Finally, notice that, if $\operatorname{tr}(f) \neq 0$, then the manifold (M,h) is symmetric irreducible, but not Einstein.

Chapter 8. Compact Homogeneous Kähler Manifolds

0. Introduction

8.1. The only known examples of complex manifolds admitting Kähler-Einstein structures with positive scalar curvature are, so far, compact homogeneous Kähler manifolds.

These manifolds have been explored extensively from various points of view in the 1950's. *Simply-connected ones are exactly the orbits of a compact Lie group in the adjoint representation, all of them admitting a canonical Kähler-Einstein structure, unique in a sense.*

The latter assertion will be proved in full detail (Theorem 8.2), while indications are given to prove the former (Theorem 8.89).

All invariant Kähler structures are then described. Among them we find the canonical symplectic structure of Kirillov-Kostant-Souriau.

Finally, we give a rough classification of the orbits of the classical compact simple groups.

8.2 Theorem. *Let G be a (connected) compact Lie group acting on its Lie algebra $\mathfrak{g}$ by the adjoint representation.*

Each orbit M in $\mathfrak{g}$ under the action of G admits a canonical, G-invariant complex structure and a unique (up to homothety) G-invariant Kähler-Einstein metric (with positive scalar curvature).

Any Kähler-Einstein structure on M is homogeneous under its own group of isometries and is obtained from a G-invariant one by an automorphism of the complex structure.

The existence of complex structures on the orbits of the compact Lie groups was noticed in [Bor 2], § 29. (See [Bor 4], [Ser 1] and [Bo-Hi], § 12).

The existence of an invariant Kähler-Einstein metric is essentially due to J.L. Koszul in [Koz 1], ((5.1) and Th.3).

The equivalence under complex automorphisms of all the Kähler-Einstein structures was proved in [Mat 1], Th.3 by Y. Matsushima.

A. The Orbits of a Compact Lie Group for the Adjoint Representation

8.3. Before proving the theorem, we develop some aspects of the geometry of an orbit of G as an imbedded manifold in a Euclidean space. For that, we choose, once and for all, a *G-invariant scalar product* on $\mathfrak{g}$ denoted *by* $(.,.)$, for instance the opposite of the Killing form. The particular choice for $(.,.)$ is not important in the sequel.

Then we give an explicit construction of a canonical, G-invariant almost-complex structure on each orbit M and show its integrability.

Afterwards, using a formula of J.L. Koszul established in Chapter 2, we compute the (unique) G-invariant Ricci form ρ attached to the complex structure and show that ρ is positive definite, from which the second part of the theorem follows at once.

The third assertion in the theorem follows easily from the peculiar feature of the identity component of the isometry group of a compact Kähler-Einstein manifold with positive scalar curvature and from a basic result of C. Montgomery.

8.4. Since the center of G acts trivially on $\mathfrak{g}$ we may as well assume the center reduced to the identity, so that the Lie algebra $\mathfrak{g}$ is *semi-simple* and G is its *adjoint group*.

8.5. Let $\mathfrak{g} = \sum_{i=1}^{N} \mathfrak{g}_i$ be a decomposition into a direct sum of simple ideals and $G = \prod_{i=1}^{N} G_i$ the corresponding decomposition of G, where G_i is the adjoint group of the simple Lie algebra $\mathfrak{g}_i$. Then each orbit M of G in $\mathfrak{g}$ decomposes into the product of the orbits M_i of G_i in $\mathfrak{g}_i$, which are called *simple components* of M (relative to the above decomposition of $\mathfrak{g}$).

Observe that each G_i acts effectively on M_i if M_i is not reduced to the origin, as it is easily proved, and so does G on M if M is a *complete orbit*, that is if no one of the M_i reduces to the origin.

In the sequel of the chapter, $\mathfrak{g}$ will be a semi-simple Lie algebra, G its adjoint group, M a complete orbit of G in its Lie algebra $\mathfrak{g}$. Generic elements of $\mathfrak{g}$ will be denoted by $U, V, W, \dots$, but a generic element of the orbit M will be denoted by w when considered as a point of the manifold M and by W when viewed as an element of the Lie algebra. In the same way, the bracket operation $\mathrm{ad}(W)$ will be also denoted by ad_w for w belonging to M.

8.6 *Fundamental vector fields.* Each element U of the Lie algebra $\mathfrak{g}$ induces a 1-parameter group of transformations $\varphi_t(U)$ on M defined by:

$$\varphi_t(U)(W) = \mathrm{Ad}(\exp(t \cdot U))(W) \quad \forall w \in M, \forall t \in \mathbb{R}.$$

The associated vector field X_U is then equal to:

$$X_U(w) = [U, W] = -\mathrm{ad}(W)(U) \quad \forall w \in M. \tag{8.6}$$

and is called *the fundamental vector field attached to U.*

8.7 Lemma. *The bracket of two fundamental vector fields X_U and X_V attached to U and V respectively is the fundamental vector field $-X_{[U,V]}$ attached to $-[U \cdot V]$.*

Proof. Consider the natural extension of X_U and X_V to the whole affine space g defined by (8.6) and denoted by the same symbols. Their bracket is equal to:

$$\begin{aligned}[X_U, X_V](w) &= [V, X_U] - [U, X_V] \\ &= [V,[U,W]] - [U,[V,W]] \\ &= [[V,U],W],\end{aligned}$$

whose restriction to M is just $-X_{[U,V]}$ as we claimed. □

8.8 Remark. Let w_0 be a distinguished point of the orbit M, so that M is realized as the quotient G/G_{w_0} of G by the stabilizer G_{w_0} of w_0. It is easily seen that *each right-invariant vector field on G induced by an element U of* g *is projectable and projects onto the fundamental vector field X_U.*

We obtain, in this way, a homomorphism of Lie algebras from the Lie algebra of right-invariant vector fields on G—that is the dual Lie algebra of g—into the Lie algebra of the (smooth) vector fields on M.

Since G acts effectively, this homomorphism is clearly injective.

We see, by 8.6, that *it doesn't depend on any distinguished point on M*.

8.9 The Tangent Bundle of M. Since G acts transitively on M, the values, at any point w of M, of the fundamental vector fields X_U generate the whole tangent space T_wM of M at w.

We infer that *T_wM is the affine sub-space of* g *at w associated with the* (*vector*) *sub-space* Im ad_w, *the image of the endomorphism* ad_w *on* g. For simplicity, we shall write M_w instead of Im ad(W) and the above identification of T_wM with M_w will be denoted by

(8.10) $$T_wM \simeq M_w$$

8.11. In particular, *a* (*smooth*) *vector field on M will be considered as a* (*smooth*) *function of M into* g *taking its value at w in M_w.*

8.12 Warning. It is of most importance, in this context, to distinguish thoroughly *brackets of vector fields* on M and *brackets of elements of* g.

8.13 Definition. The *normal* metric on M is the G-invariant Riemannian metric induced by the Ad_G-invariant scalar product on g.

The G-invariance of the scalar product $(.,.)$ is described, at the infinitesimal level, by the relation

(8.14) $$([W,U],V) + (U,[W,V]) = 0 \quad \forall U,V,W \in \mathfrak{g}.$$

The endomorphism ad_w is thus *skew-symmetric* (relatively to the scalar product) and g decomposes into the orthogonal direct sum

(8.15) $$\mathfrak{g} = \operatorname{Ker} \operatorname{ad}_w \oplus \operatorname{Im} \operatorname{ad}_w,$$
where $\operatorname{Ker} \operatorname{ad}_w$ is the kernel of ad_w.

8.16. In particular, the restriction of ad_w to $\operatorname{Im} \operatorname{ad}_w$ has no kernel. We thus obtain a *skew-symmetric automorphism* of the tangent bundle TM of M, still denoted by ad_w at w.

8.17. *The sub-space* $\operatorname{Ker} \operatorname{ad}_w$ *is the Lie algebra of the stabilizer group* G_w *at* w.
Its associated affine sub-space at w is the fiber at w of the normal bundle of M in $\mathfrak{g}$. We still identify that fiber with the vector sub-space $\operatorname{Ker} \operatorname{ad}_w$ which we'll denote by L_w for simplicity.

8.18. Observe that w belongs to L_w so that we get a distinguished (smooth) section—*the tautological section*—of the normal bundle of M. This section never vanishes on M.

8.19. We recall here the following two results about compact Lie groups (for the proof, see, for example, [Lie], Exposé 23). *Let G be any connected, compact Lie group.*
i) *The maximal tori in G are all conjugate.* (Their common dimension is the *rank* of G).
ii) *The commutator group $C(S)$ of any toral sub-group S of G, is the union of all maximal tori containing S. In particular, $C(S)$ is connected.*

8.20 Proposition. *The stabilizer group G_w of w is connected. It is the commutator group $C(S_w)$ of its connected center S_w.*

Proof. G_w is clearly equal to the commutator group of the closure, in G, of the 1-parameter sub-group $\exp(t \cdot w)$. G_w is thus connected by 8.19 ii).
The Lie algebra $\mathfrak{s}_w$ of the connected center S_w of G_w is the center of L_w. Now, since ad_w has no kernel on M_w, we infer that L_w is exactly the commutator of $\mathfrak{s}_w$ in $\mathfrak{g}$, so that the commutator group of S_w is exactly the connected component of the identity of G_w, that is G_w itself. □

8.21 Proposition. *The disjoint union of the $\mathfrak{s}_w$, when w runs over M, is the total space of a trivial vector sub-bundle of the normal bundle of M in $\mathfrak{g}$.*

Proof. If w_0 is any distinguished point of M, the mapping which associates with any element Z in $\mathfrak{s}_{w_0}$ the point $\operatorname{Ad}(g) \cdot Z$ in $\mathfrak{s}_{\operatorname{ad}(g) \cdot w_0}$ is a well defined isomorphism, independent of the particular choice of g in its class $g \cdot G_{w_0}$. □

8.22 Definitions. An element w of the Lie algebra $\mathfrak{g}$ is *regular* if L_w is Abelian, or, equivalently, if $\mathfrak{s}_w$ coïncides with L_w.
The orbit of a regular element of $\mathfrak{g}$ is a *flag manifold* of G.

8.23. We'll see below that regular elements constitute a dense open subset of $\mathfrak{g}$, so that *the generic orbit is a flag manifold*.

By 8.17 the normal bundle of a flag manifold is trivial. In particular, the Stiefel-Whitney classes and the Pontryagin classes of the flag manifolds are equal to zero.

By 8.19 the flag manifolds are all diffeomorphic to the quotient G/T where T is any maximal torus in G.

B. The Canonical Complex Structure

8.24 The Isotropy Representation. The tangent space $T_w M \simeq M_w$ to the orbit M at w is obviously preserved by ad_{L_w} and Ad_{G_w}. The letter action is the *isotropy representation* at w.

The restriction of the isotropy representation at w to the connected centre S_w of G_w is completely reducible. We get an Ad_{S_w}-decomposition of M_w into an orthogonal direct sum

$$\textbf{(8.25)} \qquad M_w = \sum_{i=1}^{m} E_{w,i}. \qquad \dim M = n = 2m,$$

where each $E_{w,i}$ is a 2-dimensional vector space isomorphic, as a S_w-representation space, to the irreducible 2-dimensional representation Γ_{a_i} of S_w in $\mathbb{R}^2$ defined by the matrix

$$\textbf{(8.26)} \qquad \Gamma_{a_i}(s) = \begin{pmatrix} \cos 2\pi a_i(s) & -\sin 2\pi a_i(s) \\ \sin 2\pi a_i(s) & \cos 2\pi a_i(s) \end{pmatrix} \quad \forall s \in S_w.$$

Here a_i is the *weight* of Γ_{a_i} and can be considered either as a linear form on $\mathfrak{s}_w$ taking integral values on the integral lattice $\exp^{-1}(e)$, or as a(linear) function on S_w taking its values in $\mathbb{R}/\mathbb{Z}$.

8.27. Observe that non of the a_i's is zero. In fact, ad_w has no kernel on M_w so that $a_i(w)$ is non-null for all $i = 1, \ldots, m$.

8.28 Positive roots. Relatively to any orthogonal basis of $E_{w,i}$, the action of S_w is described by the matrix $\Gamma_{a_i(s)}$ or the matrix $\Gamma_{-a_i(s)}$ depending on the orientation.

From the two opposite forms $\pm a_i$, we *choose* the one whose value on W is *positive* and denote it by a_i.

8.29 Definition. The forms a_i are *the* (*positive*) *roots* of S_w in $\mathfrak{g}$.

The roots may be viewed as elements of $\mathfrak{s}_w$ via the scalar product.

Each irreducible component $E_{w,i}$ is then oriented by the (orthonormal) basis relatively to which the action of S_w is described by the matrix Γ_{a_i}.

8.30 Almost-complex Structure on M. An orientation on the 2-dimensional Euclidean space $E_{w,i}$ is just a complex structure. We get, in this way, a canonical complex structure J_w on the whole of $M_w \simeq T_w M$.

From 8.26 we infer easily that the derived action of $\mathfrak{s}_w$ on M_w is described, on each $E_{w,i}$, by

(8.31) $$[Z, X_i] = 2a_i(Z)J_w X_i \quad \forall Z \in \mathfrak{s}_w, \forall X_i \in E_{w,i}, \forall i = 1, \ldots, m.$$

Now, the above decomposition of M_w is not uniquely defined since roots a_i and a_j may be equal for two different indices i and j. Putting together the sub-spaces $E_{w,i}$ attached to a same root α we obtain an Ad_{S_w}-decomposition of M_w into *eigen-subspaces*:

(8.32) $$M_w = \sum_{\alpha \in \Psi} E_{w,\alpha},$$

where Ψ is the set of the (positive) roots of $\mathfrak{s}_w$ in $\mathfrak{g}$, and $E_{w,\alpha}$ the direct sum of the subspaces $E_{w,i}$ such that a_i is equal to α. Each subspace $E_{w,\alpha}$ is determined by S_w alone and its dimension is twice the multiplicity m_α of α. This decomposition is clearly G-invariant when w runs over its orbit M.

On each component $E_{w,\alpha}$ of M_w we get immediately, from 8.31, that the complex operator J_w is defined by:

(8.33) $$J_w X = \frac{1}{2\pi\alpha(w)}[W, X] = \frac{1}{2\pi\alpha(w)} \mathrm{ad}_w X \quad \forall X \in E_{w,\alpha}, \forall w \in M.$$

We infer, at once, that J_w doesn't depend on a particular Ad_{S_w}-decomposition of M_w into irreducible components.

8.34. *We thus obtain an almost-complex structure J on M which, by* 8.33, *is clearly G-invariant.*

8.35 Remark. For each point w of M, the two automorphisms J_w and ad_w of the tangent space $T_w M \simeq M_w$ commute. We have:

$$\mathrm{ad}_w \circ J_w X = J_w \circ \mathrm{ad}_w X = -2\pi\alpha(w)X \quad \forall X \in E_{w,\alpha}, \forall \alpha \in \Psi.$$

In the following, J_w will denote as well the endomorphism of $\mathfrak{g}$ obtained by putting J_w equal to zero on L_w. This endomorphism still commutes with ad_w. It is defined for all w in $\mathfrak{g}$ and can be considered as a (smooth) field of endomorphisms of the affine space $\mathfrak{g}$.

If Π_w denotes the orthogonal projection of $\mathfrak{g}$ onto M_w, we have

(8.36) $$\Pi_w = -J_w^2.$$

8.37 *The complexified tangent bundle of M.*

Let $M_w^{\mathbb{C}}$ be the complexified vector space $M_w \otimes \mathbb{C}$ of M_w and extend the action of S_w and J_w $\mathbb{C}$-linearly to $M_w^{\mathbb{C}}$.

Each (complexified component) $E_{w,i}^{\mathbb{C}}$ of $M_w^{\mathbb{C}}$ decomposes into the direct sum of two complex 1-dimensional eigen spaces $T_{w,i}$ and $\bar{T}_{w,i}$, conjugate to each other, associated with the eigen-values $2\pi i a_i$ and $-2\pi i a_i$ respectively.

Let T_w *(resp.* $\bar{T}_w$) denote the direct sum of the $T_{w,i}$ (resp. $\bar{T}_{w,i}$), so that

(8.38) $$M_w^{\mathbb{C}} = T_w \oplus \bar{T}_w \quad \forall w \in M.$$

This decomposition is clearly G-invariant on M. Now, J_w reduces to i-times (resp. $-i$-times) the identity on T_w (resp. $\overline{T}_w$) so that T_w (resp. $\overline{T}_w$) determines, as w runs over its orbit M, the complex vector bundle of vectors of type $(1,0)$ (resp. $(0,1)$).

8.39 Proposition. *The G-invariant almost-complex structure J is integrable.*

Proof. We have to prove that the complex torsion tensor N is equal to zero (see 2.11). For this, it is sufficient to prove that the vector field $4N(X_U, X_V) = 0$ when X_U and X_V are any two fundamental vector fields (see 8.6).

8.40 Lemma. *For any two fundamental vector fields X_U and X_V, the vector field $4N(X_U, X_V)$ is equal to*

$$4N_w(X_U, X_V) = [[U,V] + J_w[J_wU, V] + J_w[U, J_wV] - [J_wU, J_wV], W] \quad \forall w \in M.$$

Proof. Since fundamental vector fields preserve J, $4N(X_U, X_V)$ reduces to

$$4N(X_U, X_V) = [X_U, X_V] - [JX_U, JX_V].$$

The first bracket is, by lemma 8.7, the fundamental vector field attached to $-[U,V]$.

To evaluate the second bracket we use, as we did in Lemma 8.7, the natural extensions of the vector fields X_U and X_V to the whole affine space $\mathfrak{g}$. We need the following

8.41 Lemma. *The (ordinary) derivative $\dot{J}_w(X)$ of the endomorphism J_w at w, in the direction of any vector X tangent to $\mathfrak{g}$ at w, is defined by*

$$\dot{J}_w(X)(V) = [U, J_wV] - J_w[U,V], \quad \forall V \in \mathfrak{g} \simeq T_w\mathfrak{g}, \forall w \in \mathfrak{g},$$

where U is any element of $\mathfrak{g}$ whose image by Ad_w is equal to $-X$.

Proof. The vector X is tangent, at $w(o) = w$, to the curve

$$w(t) = \mathrm{Ad}(\exp(t \cdot U)) \cdot w$$

By G-invariance of J_w, we get then:

$$J_{w(t)}(V) = \mathrm{Ad}(\exp(t \cdot U)) \circ J_w \circ \mathrm{Ad}(\exp(-t \cdot U))W$$

for every V in $\mathfrak{g}$. Lemma 8.41 follows immediately. □

Proof of Lemma 8.40 (continued). We have

$$[JX_U, JX_V]_w = (J\dot{X}_V)(JX_U)_w - (J\dot{X}_U)(JX_V)_w$$

with

$$(J\dot{X}_V)(JX_U)_w = [JV, JX_U] + [\dot{J}_w(JX_U)(JV), W]$$

and the corresponding formula for the other term. By using Lemma 8.41 we obtain readily

$$[JX_U, JX_V]_w = [[JU, JV] - J[JU, V] - J[U, JV], W],$$

which ends the proof of Lemma 8.40. □

By virtue of Lemma 8.40, the question of integrability of the almost-complex structure on the orbit M is reduced to an algebraic question concerning the Lie algebra $\mathfrak{g}$ and the field of endomorphisms J_w. More precisely, Lemma 8.40 may be reformulated as follows.

8.42. *The almost-complex structure J on M is integrable if and only if $S_w(U, V)$ belongs to L_w for every elements U and V in $\mathfrak{g}$, where we have*

$$4S_w(U, V) = [U, V] + J_w[J_w U, V] + J_w[U, J_w V] - [J_w U, J_w V].$$

For a fixed w in M, we may assume that U and V belong to M_w. It is, in fact, more convenient to deal with complex vectors. We thus assume that U and V are elements of the complexified space $M_w^{\mathbb{C}}$ (see above 8.37), both having a type $(1, 0)$ or $(0, 1)$.

If U and V are of different types, $S_w(U, V)$ is obviously equal to zero.

If U and V both are of type $(1, 0)$, we get

$$2S_w(U, V) = [U, V] + iJ_w[U, V] \quad \forall U, V \in T_w.$$

Consider now two elements U and V in T_w belonging respectively to the eigen-spaces $T_{w,i}$ and $T_{w,j}$. The bracket $[U, V]$ is equal to zero or else is an eigen-vector for ad_w associated with $2\pi i(a_i + a_j)(w)$ in case where the latter is itself an eigen-value. Since $(a_i + a_j)(w)$ is positive, we conclude that $[U, V]$ belongs to T_w, so that $S_w(U, V)$ is equal to zero. This ends the proof of Proposition 8.39. □

C. The G-Invariant Ricci Form

8.43. As we have established above (2.100), the Ricci form of a Kähler metric depends only on the complex structure and the volume-form.

On the other hand, there is one and only one (up to homothety) G-invariant volume-form on each orbit M, the volume-form Ω of the normal metric, so that the Ricci form of *any* Kähler metric relative to the canonical G-invariant complex structure J is a well-defined 2-form ρ which we can determine from J only.

8.44. Now, the fundamental vector fields X_U on M preserve, by definition, each G-invariant geometric object. In particular, they preserve Ω and J, hence *are infinitesimal automorphisms of any G-invariant Kähler structure*, by Proposition 2.125 iv.

Since each vector tangent to M at any point w of M is the value of a fundamental vector field (determined up to an element of L_w), the Ricci form ρ is completely defined by its values on fundamental vector fields. Applying Formula (2.134) of J.L. Koszul, we get

(8.45) $$\rho_w(X, Y) = -\tfrac{1}{2}\delta(J[X, Y])(w),$$

where X and Y are fundamental vector fields attached to U and V respectively.

We already know, by 8.7 and 8.35, that the vector field $J_w[X, Y]$ is equal to $[W, J_w[U, V]]$.

It remains to compute the divergence of that vector field (which is *not*, of course, a fundamental vector field).

8.46 Definition. An *adapted basis* of M_w is a basis $\{X_i, JX_i\}$, $i = 1, \ldots, m$, such that $\{X_i, JX_i\}$ constitutes for each i a positively oriented orthonormal basis of $E_{w,i}$ in an Ad_{S_w}-decomposition of M_w into irreducible sub-spaces (see 8.25).

8.47 Proposition. *At w, the G-invariant Ricci form ρ is*

(8.47) $$\rho_w(X, Y) = \left(\sum_{i=1}^{m} [X_i, JX_i], [U, V]\right),$$

where $\{X_i, JX_i\}$, $i = 1, \ldots, m$ is any adapted basis of M_w and X, Y are fundamental vector fields attached to U and V respectively.

Proof. Consider any adapted basis $\{X_i, JX_i\}$, $i = 1, \ldots, m$ of M_w and extend it by parallel translation along the geodesics (relative to the normal metric) starting from w on M. In this way we obtain a local orthonormal frame in a neighbourhood of w in M, which we denote by $\{e_k\}$, $k = 1, \ldots, 2m$, where the covariant derivative of each e_k is zero at w. Now, since the normal metric is the restriction to M of the Euclidean structure of $\mathfrak{g}$, the covariant derivative of any vector field ξ_w in the direction of e_k is just the projection into $T_wM \simeq M_w$ of the (ordinary) derivative in the direction of e of any extension of ξ_w to $\mathfrak{g}$. Therefore

(8.48) $$\delta\xi_w = -\sum_{k=1}^{2m} ([e_k, \xi](w), e_k) = -\sum_{k=1}^{2m} (\dot{\xi}_w(e_k), e_k)$$

for any vector field ξ.

Applying Lemma 8.41 and Formula (8.48) to the vector field $\xi_w = [W, J_w[U, V]]$, we get

$$\begin{aligned}
\rho_w(X, Y) = -\frac{1}{2}\delta\xi_w &= \frac{1}{2}\sum_{k=1}^{2m} ([\dot{W}(e_k), J_w[U, V]], e_k) \\
&\quad + \frac{1}{2}\sum_{k=1}^{2m} ([W, \dot{J}_w(e_k)([U, V]), e_k) \\
&= \frac{1}{2}\sum_{k=1}^{2m} ([e_k, J_w[U, V]], e_k) \\
&\quad - \frac{1}{2}\sum_{k=1}^{2m} ([W, [\mathrm{ad}_w^{-1}(e_k), J_w[U, V]]], e_k) \\
&\quad + \frac{1}{2}\sum_{k=1}^{2m} ([W, J_w[\mathrm{ad}_w^{-1}(e_k), [U, V]]], e_k).
\end{aligned}$$

The first sum is obviously equal to zero, and so is the second sum because

$\mathrm{ad}_w^{-1}(e_k)$ and $\mathrm{ad}_w(e_k)$ are colinear since each e_k belongs to a subspace $E_{w,i}$ by assumption. As for the third sum, it is equal to

$$-\frac{1}{2}\sum_{k=1}^{2m}([U,V],[\mathrm{ad}_w^{-1}(e_k),J_w\circ\mathrm{ad}_w(e_k)])=\frac{1}{2}\sum_{k=1}^{2m}([e_k,Je_k],[U,V])$$
$$=\sum_{k=1}^{m}([U,V],[X_i,JX_i]),$$

which completes the proof of Proposition 8.47. □

8.49 Corollary. *The sum* $\gamma(w)=\frac{1}{2\pi}\sum_{i=1}^{m}[X_i,JX_i]$ *is independent of the adapted basis.*

It belongs to the center, $\mathfrak{s}_w$*, of* L_w *and is equal to the sum* $\Sigma_{\alpha\in\Psi}m_\alpha\cdot\alpha$ *of the (positive) roots of* S_w *in* $\mathfrak{g}$ *with their multiplicities.*

Proof. Since $\mathfrak{g}$ is semi-simple, $[\mathfrak{g},\mathfrak{g}]$ is equal to $\mathfrak{g}$: that proves the first assertion in view of 8.47. On the other hand, the vector field X vanishes at w whenever U belongs to L_w. We infer from (8.47)

$$([\gamma(w),U],V)=0\quad\forall U\in L_w,\forall V\in\mathfrak{g},$$

which proves that $[\gamma(w),U]$ is zero for every U in L_w, or, equivalently, that $\gamma(w)$ belongs to the centre of L_w.

Finally, consider any element Z in $\mathfrak{s}_w$. We have, for every $i=1,\ldots,m$

$$(Z,[X_i,JX_i])=(JX_i,[Z,X_i])=2\pi a_i(Z),$$

hence

$$\left(\sum_{i=1}^{m}[X_i,JX_i],Z\right)=2\pi\sum_{i=1}^{m}a_i(Z)=2\pi\sum_{\alpha\in\Psi}m_\alpha\cdot\alpha(Z),$$

which ends the proof. □

8.50. Before proving that the Ricci form is positive definite we have to enter a little further the theory of roots in a semi-simple algebra.

As we saw above (8.19), the stabilizer group G_w of any point w of its orbit M is the union of all maximal tori containing the torus S_w. Let T be one of those maximal tori and consider the restriction of the adjoint representation to T which is still completely reducible. We obtain a decomposition

(8.51)
$$\mathfrak{g}=H\oplus\sum_{\pm\alpha_T\in R_G}E_{\pm\alpha_T},$$

where H is the Lie algebra of T and each $E_{\pm\alpha_T}$ is a 2-dimensional irreductible representation space for T, isomorphic to one of the representation $\Gamma_{\pm\alpha_T}$, where, this time, the weights $\pm\alpha_T$ are elements of H^*, or H via the scalar product. The set R_G of the weights $\pm\alpha_T$ forms a system of roots in the abstract sense (see, for example [Ser 2], Chapter V). In particular, the pairs $\pm\alpha_T$ are distincts.

The sub-spaces L_w and M_w are preserved by the action of T. Accordingly, the set R_G is divided into the subsets R_{G_w} and D of the weights of Ad_T relative to

L_w and M_w respectively. As long as no orientation has been choosen on the 2-dimensional irreductible components $E_{\pm\alpha_T}$ the weights are defined up to sign only.

Now, the decomposition

$$M_w = \sum_{\pm\alpha_T \in D} E_{\pm\alpha_T} \tag{8.52}$$

of M_w is also a decomposition of M_w into irreducible components under the action of the sub-torus S_w of T, and the weights $\pm a_i$ of that action are the restriction of the weights $\pm\alpha_T$ to $\mathfrak{s}_w^*$ (or the orthogonal projection of the $\pm\alpha_T$ onto $\mathfrak{s}_w$ according to the adopted point of view).

8.53 Definition. The (positive) *complementary* roots of T (relatively to w) are the weights of Ad_T whose restrictions to $\mathfrak{s}_w$ are (positive) roots of S_w.

Observe that the set D_+ of the (positive) complementary roots of T is just the subset of elements of R_G whose value at w is positive.

8.54. Let α_T any element of D_+ and any orthonormal basis $\{X_{\alpha_T}, JX_{\alpha_T}\}$ in $E_{\pm\alpha_T}$ (which we may write E_{α_T}). Clearly the bracket $[X_{\alpha_T}, JX_{\alpha_T}]$ commutes with any element of H, hence belongs to H and, in fact, is equal to $2\pi\alpha_T$. From Corollary 8.49 we infer that *the sum of the (positive) complementary roots (relatively to w) of any maximal torus containing S_w actually belongs to $\mathfrak{s}_w$ and is equal to $\gamma(w)$.*

8.55 Lemma. *The scalar product of the sum of the elements of D_+ with any element of D_+ is positive, greater than or equal to the square of the norm of that element.*

Proof ([Koz 1], p. 574, Lemma 2, or [Bo-Hi] 14-8). Let α_T be any distinguished element in D_+ and consider β_T in D_+ whose scalar product with α_T is negative. We recall the two following facts from the general theory of roots in a semi-simple Lie algebra:

(8.56)
- i) The number $k = -2\dfrac{(\alpha_T, \beta_T)}{(\alpha_T, \alpha_T)}$ is a (positive) integer.
- ii) $\beta_T + l\cdot\alpha_T, l = 0, 1, \ldots, k$ are weights of Ad_T and, in fact, belong to D_+ since $(\beta_T + l\cdot\alpha_T)(w)$ is positive for l positive.

The sum $\sum_{l=0}^{k} (\beta_T + l\cdot\alpha_T)$ is equal to $(k+1)\beta_T + \dfrac{k(k+1)}{2}\alpha_T$ whose scalar product with α_T is equal to zero.

Now D_+ is clearly the disjoint union of strings $\{\beta_T + l\cdot\alpha_T\}, l = 0, \ldots, k$ as above, where (β_T, α_T) is negative, and elements whose scalar product with α_T is non-negative, among which we find α_T itself. □

8.57 Proposition. *The G-invariant Ricci form ρ is positive definite.*

Proof. By 8.47, $\rho_w(X, Y)$ is equal to;

$$\rho_w(X, Y) = 2\pi(\gamma(w), [U, V]) \quad \forall X, Y \in M_w \simeq T_wM,$$

where U, V are any elements in $\mathfrak{g}$ whose images by ad_w are equal to X, Y, and where $\gamma(w)$ belongs to $\mathfrak{s}_w$. We see immediately that $\rho_w(X, Y)$ is zero whenever X and Y belong to distinct irreducible components $E_{w,i}$ and $E_{w,j}$ of M_w. On the other hand, we have:

$$\rho_w(X, JX_i) = \frac{(\gamma(w), a_i)}{2\pi(a_i(w))^2}|X_i|^2 \qquad i = 1, \ldots, m$$

for any element X_i in $E_{w,i}$.

Let T be any maximal torus of G containing S_w. Then a_i is the orthogonal projection of a (positive) complementary root α_T in D_+ while $\gamma(w)$ is equal, by 8.54 to the sum of the elements of D_+. By Lemma 8.55 above, the scalar product $(\gamma(w), a_i)$, which is equal to $(\gamma(w), \alpha_T)$, is positive.

8.59 Corollary. *Each orbit M admits a unique (up to homothety) G-invariant Kähler-Einstein metric, compatible with the canonical complex structure, with positive scalar curvature.*

Proof. ρ is the Ricci form of *any* G-invariant Kähler metric on M compatible with the canonical complex structure. Since ρ is positive definite it is itself the Kähler form of a Kähler metric g_0 which is clearly Kähler-Einstein with scalar curvature equal to the (real) dimension n of M. Any G-invariant Kähler-Einstein metric is homothetic to g_0 since its Kähler form has to be a multiple of ρ by a (positive) constant factor. □

8.60 Proposition. *Any Kähler-Einstein structure on M, compatible with the canonical complex structure, is the image, by an automorphism of the complex structure, of a G-invariant Kähler-Einstein structure.*

Proof ([Mat 3], Theorem 3). Let g_1 be any Kähler-Einstein metric on M, compatible with the canonical complex structure J. By Corollary 8.59 there exists a (unique) G-invariant Kähler-Einstein metric g_0 whose total volume is equal to that of g_1. Let G_0 and G_1 be the connected groups of isometries of g_0 and g_1, respectively. It must be observed that G may be a proper sub-group of G_0, but, in any case, the latter is still a compact Lie group without centre (see below 8.88i), so that the connected group of complex automorphisms $\mathfrak{A}(M)$ of M is a semi-simple complex Lie group of which both G_0 and G_1 are compact real forms (by Theorem 11.52), hence conjugate each other by an element φ of $\mathfrak{A}(M)$:

(8.60) $$G_1 = \mathrm{Ad}(\varphi) \cdot G_0$$

Let's now recall the following result of D. Mongomery.

8.61. *Each maximal compact Lie sub-group of a Lie group of homeomorphisms acting transitively on a compact, simply-connected space acts itself transitively.* ([Mon]).

As a compact real form of a semi-simple complex Lie group, G_1 is a maximal compact sub-group of $\mathfrak{A}(M)$. On the other hand, M is simply-connected by Theorem

11.26. It follows that G_1 acts transitively on M. We infer that the volume-form Ω_1 of G_1 is equal to $\varphi^*\Omega_0$, both being G_1-invariant and having same total volume. Since φ preserves, by definition, the complex structure on M, it follows that the Ricci form ρ_1 of g_1 is the image $\varphi^*\rho_0$ of the Ricci form ρ_0 of g_0. This completes the proof. In the course of the proof, we also proved.

8.62 Corollary. *Any Kähler-Einstein metric on M, compatible with the canonical complex structure, is homogeneous under its own connected group of isometries.*

8.63. Theorem 8.2 is completely proved. □

D. The Symplectic Structure of Kirillov-Kostant-Souriau

8.64. So far, we have constructed on each orbit M of any compact Lie group G a G-invariant canonical complex structure and a G-invariant Kähler-Einstein structure described by its Kähler form ρ, given by 8.58, where $\gamma(w)$ can be viewed as a "constant", that is G-invariant section of the trivial bundle $\mathfrak{s}$ over M. Consider now the 2-form F on M defined by:

$$F_w(X, Y) = (w, [U, V]) \quad \forall X, Y \in T_wM \simeq M_w, \tag{8.65}$$

where, as above, U and V are any elements of $\mathfrak{g}$ whose image by ad_w are X and Y respectively. Here, w is viewed as the tautological section of $\mathfrak{s}$ (see 8.18).

8.66 Proposition. *The 2-form F is the Kähler-form of a G-invariant Kähler structure compatible with the canonical complex structure of M.*

Proof. The 2-form F is clearly G-invariant. Proposition 8.66 is an immediate consequence of the following three lemmas.

8.67 Lemma. *The exterior differential of a G-invariant 2-form B is given by*

$$dB(X, Y, Z) = B([X, Y], Z) + B([Y, Z], X) + B([Z, X], Y) \tag{8.67}$$

for any fundamental vector field X, Y, Z on M.

Proof. Since B is G-invariant, we have, for any three fundamental vector fields X, Y, Z:

$$X \cdot B(Y, Z) = B([X, Y], Z) + B(Y, [X, Z]).$$

On the other hand, we have, for any vector fields X, Y, Z

$$\begin{aligned} dB(X, Y, Z) &= X \cdot B(Y, Z) + Y \cdot B(Z, X) + Z \cdot B(X, Y) \\ &\quad - B([X, Y], Z) - B([Y, Z], X) - B([Z, X], Y). \end{aligned}$$

Lemma 8.67 follows at once. □

8.68 Lemma. *F is closed.*

Proof. Let X, Y, Z three fundamental vector fields on M attached respectively to the elements U, V, T of $\mathfrak{g}$. By Lemma 8.67 we get, at the point w of M:

$$dF_w(X,Y,Z) = F_w([X,Y],Z) + F_w([Y,Z],X) + F_w([Z,X],Y)$$
$$= (w, [[U,V],T] + [[V,T],U] + [[T,U],V])$$

which is zero by the Jacobi identity. □

8.69 Lemma. *F is a 2-form of type $(1,1)$ and is positive definite.*

Proof. Consider once more an Ad_{S_w}-decomposition of M_w into irreducible components for a point w of M. Writing $F_w(X,Y)$ as $([W,U],V)$ in (8.65) we see immediately, as in 8.57, that $F_w(X,Y)$ is zero whenever X and Y belong to distinct sub-spaces $E_{w,i}$, and that $F_w(X_i, JX_i)$ is equal to $(2\pi a_i(w))^{-1} \cdot |X_i|^2$ for any vector X_i in $E_{w,i}$. Since $a_i(w)$ is positive for all $i = 1, \ldots, m$, we conclude as in Proposition 8.57. □

8.70. If we forget the complex structure of M, the 2-form F determines a *symplectic structure* on M, which is just the *symplectic structure of Kirilov-Kostant-Souriau* (see, for instance, [Kir] § 15). We shall refer to it as the *canonical symplectic structure* of M.

Observe that F is *not*, in general, the Kähler form of the G-invariant Kähler-Einstein structure (compare 8.58 and 8.65).

8.71 Remark. The canonical symplectic 2-form F is the bilinear form associated, via the normal metric, to the automorphism ad_w^{-1} of the tangent bundle of M

(8.71) $$F(X,Y) = (\mathrm{ad}_w^{-1} X, Y) \quad \forall X, Y \in M_w \simeq T_w M, \forall w \in M$$

In contrast, the bilinear forms associated to ad_w or J_w are non-degenerate and even positive definite 2-forms of type $(1,1)$ but are not closed in general. In particular *the normal metric is Hermitian but non Kähler in general* (see Proposition 8.86 below).

E. The Invariant Kähler Metrics on the Orbits

The two G-invariant Kähler forms ρ and F that we have considered so far are both of the form

(8.72) $$B_w(X,Y) = (\sigma(w), [U,V]) \quad \forall X, Y \in T_w M \simeq M_w, \forall w \in M,$$

where U and V are, as usual, elements of $\mathfrak{g}$ whose images by ad_w are X and Y respectively and σ is a G-invariant section of the fiber bundle $\mathfrak{s}$ over M. (see 8.58 and 8.65 respectively).

8.73 Lemma. *For any non-zero G-invariant section σ of $\mathfrak{s}$, the 2-form B defined by 8.72 is a non-zero, G-invariant, closed 2-form of type (1, 1). It is positive definite if and only if the scalar product $(\alpha, \sigma(w))$ is positive for every w in M and every positive root α of S_w.*

Proof. B is clearly G-invariant. One proves that it is closed in the same manner as for F in Lemma 8.68. Again, as in Lemma 8.69, for any decomposition under Ad_{S_w} of M_w into irreducible sub-spaces we get

(8.74)
$$B(X, Y) = 0 \quad \text{whenever } X, Y \text{ belong to distinct irreducible components } E_{w,i} \text{ and } E_{w,j},$$
$$B(X_i, JX_i) = \frac{(\sigma(w), a_i)}{2\pi(a_i, w)^2}|X_i|^2 \quad \forall X_i \in E_{w,i}\ \forall i = 1, \dots, m.$$

The Lemma follows at once. □

8.75 Definition. The 2-form B defined by 8.72 is the image of the G-invariant section σ by *transgression*.

8.76 Proposition. *Each closed, G-invariant 2-form B is the image by transgression of a (unique) G-invariant section of $\mathfrak{s}$.*

Proof. Since B is G-invariant, we get, in particular:

$$B_w(\mathrm{Ad}(b)\cdot X, \mathrm{Ad}(b)\cdot Y) = B_w(X, Y) \quad \forall X, Y \in M_w, \forall w \in M, \forall b \in G_w,$$

which implies, at the infinitesimal level

$$B_w([Z, X], Y) + B_w(X, [Z, Y]) = 0 \quad \forall X, Y \in M_w, \forall w \in M, \forall Z \in L_w. \tag{8.77}$$

In the following we consider B as well as a $\mathbb{C}$-bilinear 2-form on the complexified tangent bundle.

Let T be any maximal torus containing S_w—hence contained in G_w—, $\mathfrak{t}$ its Lie algebra, and consider the corresponding Ad_T-decomposition of $M_w^{\mathbb{C}}$ into 1-dimensional eigen-spaces

$$M_w^{\mathbb{C}} = \sum_{\alpha_T \in D_+} \mathfrak{g}_{\alpha_T} + \sum_{\alpha_T \in D_+} \mathfrak{g}_{-\alpha_T}, \tag{8.78}$$

where $\mathfrak{g}_{\alpha_T}$ (resp. $\mathfrak{g}_{-\alpha_T}$) is the eigen-space of $M_w^{\mathbb{C}}$ associated with the eigen-value $2\pi i\alpha_T$ (resp. $-2\pi i\alpha_T$), and each pair $\mathfrak{g}_{\alpha_T}$, $\mathfrak{g}_{-\alpha_T}$ is conjugate.

From 8.77 applied to any Z in $\mathfrak{t}$, we find easily

(8.79)
i) $B_w(X, \bar{Y}) = 0$ whenever X and Y are eigen vector (of type (1, 0)) attached to distinct (positive) roots α_T and β_T in D_+.

ii) $B_w(X, \bar{X}) = i\psi(\alpha_T)|X|^2$ for any eigen vector X attached to the (positive) root α_T in D_+.

where $\psi(\alpha_T)$ depends on α_T only.

Since B is a closed G-invariant 2-form we get from 8.67 the following relation at w

(8.80) $0 = B_w(\mathrm{ad}_w[U,V], \mathrm{ad}_w Z) + B_w(\mathrm{ad}_w[V,Z], \mathrm{ad}_w U)$
$$+ B_w(\mathrm{ad}_w[Z,U], \mathrm{ad}_w V) \qquad \forall w \in M, \forall U, V, Z \in \mathfrak{g}^{\mathbb{C}}.$$

Let α_T, β_T be any pair in D_+ such that the sum $\alpha_T + \beta_T$ is still a weight of Ad_T, hence belongs itself to D_+. Let U, V, $\bar{Z}$ be unitary eigen vectors in $M_w^{\mathbb{C}}$ attached respectively to the (positive) roots α_T, β_T and $\alpha_T + \beta_T$ (U and V are of type $(1,0)$, Z is of type $(0,1)$). Since the eigen spaces are $\mathbb{C}$-1-dimensional, we have:

$$[U,V] = \lambda \cdot \bar{Z} \qquad [V,Z] = \mu \cdot \bar{U} \qquad [Z,U] = \nu \cdot \bar{V},$$

where λ, μ, ν are non-zero complex numbers. The latter are all equal to $\lambda (\neq 0)$. This follows at once from

$$([U,V],Z) + (V,[U,Z]) = 0 = ([V,Z],U) + (Z,[V,U]),$$

where $(.,.)$ denotes the $\mathbb{C}$-bilinear extension to $\mathfrak{g}^{\mathbb{C}}$ of the scalar product.

Applying 8.77 to the triplet (U,V,Z) we obtain:

$$0 = ((\alpha_T + \beta_T)(w))^2 \cdot B(\bar{Z}, Z) + (\alpha_T(w))^2 \cdot B(\bar{U}, U) + (\beta_T(w))^2 \cdot B(\bar{V}, V)$$

for every unitary element U, V, $\bar{Z}$ belonging respectively to $\mathfrak{g}_{\alpha_T}$, $\mathfrak{g}_{\beta_T}$, $\mathfrak{g}_{(\alpha_T+\beta_T)}$ and for each pair α_T, β_T of elements of D_+ such that $\alpha_T + \beta_T$ is itself in D_+ (such a pair will be called a *closed* pair).

It follows that:

$$((\alpha_T + \beta_T)(w))^2 \psi(\alpha_T + \beta_T) = (\alpha_T(w))^2 \psi(\alpha_T) + (\beta_T(w))^2 \psi(\beta_T)$$

for all closed pairs (α_T, β_T) in D_+.

Since the (positive) roots generate $\mathfrak{t}$, we get a (unique) element $\sigma(w)$ in $\mathfrak{t}$ such that

(8.81) $$\psi(\alpha_T) = \frac{(\alpha_T, \sigma(w))}{2\pi(\alpha_T(w))^2} \quad \forall \alpha_T \in D_+.$$

Now, the maximal tori containing S_w are obtained from each other by conjugation by an element b of G_w, which permutes the eigen spaces $\mathfrak{g}_{\alpha_T}$ such that the restrictions of α_T to $\mathfrak{s}_w$ are the same. On the other hand, $\sigma(w)$ belongs to the intersection of the various sub-algebras $\mathfrak{t}$, that is to $\mathfrak{s}_w$.

The whole construction is clearly G-invariant. We obtain in this way a G-invariant section σ of $\mathfrak{s}$, of which B is the image by transgression. □

8.82 Corollary. *The closed, G-invariant 2-forms on M are of type* $(1,1)$ *relatively to the canonical complex structure.*

In view of Lemma 8.73, we infer immediately from Proposition 8.76 the

8.83 Proposition. *The set of the G-invariant Kähler metrics on M, compatible with the canonical complex structure, is in* 1.1 *correspondance, via the homomorphism of trangression, with the set of G-invariant sections of the fiber bundle* $\mathfrak{s}$ *over M whose scalar product with any positive root is positive.*

8.84 Remark. Let w_0 be a distinguished point of M, and p the natural projection from G onto $M \simeq G/G_{w_0}$. The G-invariant sections σ of $\mathfrak{s}$ are in 1.1 correspondance with the elements σ_0 of $\mathfrak{s}_{w_0}$. Each σ_0 in $\mathfrak{s}_{w_0}$ induces a bi-invariant 1-form on the fibre G_{w_0} of the above fibration p. It is well-known and easy to establish that the space of bi-invariant 1-forms on G_{w_0} is identified with the first De Rham cohomology space $H^1_{DR}(G_{w_0}; \mathbb{R})$. On the other hand, harmonic forms on M (for any G-invariant Riemannian metric) are G-invariant, so that *the trangression induces an isomorphism of $H^1_{DR}(G_{w_0}; \mathbb{R})$ onto $H^2_{DR}(M; \mathbb{R})$.*

As an immediate corollary of Proposition 8.83 we get the

8.85 Proposition. *If the rank of the fiber bundle $\mathfrak{s}$ M is one, M admits only one G-invariant Kähler structure, up to homothety, which is Kähler-Einstein. In particular, the canonical symplectic structure coincides with the G-invariant Kähler-Einstein structure.*

8.86 Remark. Consider the orbits for which the rank of the vector bundle $\mathfrak{s}$ is one or, equivalently, the orbits of elements w of $\mathfrak{g}$ such that the centre of their commutator in $\mathfrak{g}$ reduces to the line generated by w itself.

Orbits of this type exist only when $\mathfrak{g}$ is simple. Among them we find all the *irreducible Kähler symmetric spaces*, for which *the isotropy representation is irreducible* (see 7.45, 7.75, 10 Table 2). This implies that all the G-invariant metrics coincide, up to homothety. In particular, the *normal metric is then Kähler-Einstein.*

Conversely, from the proof of 8.76 (see 8.79 and 8.81) we infer easily that the normal metric is Kähler if and only if the (positive) roots a_i of S_w coincide, for any w in the orbit—equivalently, $a_i(w)$ is the same for all the (positive) roots a_i—, and this implies clearly that the normal metric is symmetric. Indeed, the normal metric on an orbit M is symmetric if and only if the bracket $[M_w, M_w]$ is contained in L_w for any w in M.

F. Compact Homogeneous Kähler Manifolds

8.87 Definition. A *compact homogeneous Kähler manifold* is a compact Kähler manifold on which the (connected) group of isometries acts transitively.

8.88. We recall without proof the two following basic informations about *simply-connected*, compact homogeneous Kähler manifolds

i) *The (connected) group of isometries has no center.* (see [Mat 1]. [Bo-Re]).

ii) *The isotropy group B of any point is the commutator $C(S)$ of a toral sub-group S of G, where S may be chosen to be the connected center of B. (In particular B is connected)* (see [Mat 1], [Bor 4], [Ser 1], [Lic 5]).

We infer the

8.89 Theorem. *Every compact, simply-connected homogeneous Kähler manifold is isomorphic, as a homogeneous complex manifold, to an orbit of the adjoint representation of its connected group of isometries, endowed with its canonical complex structure.*

Proof. Let M be a simply-connected, compact homogeneous Kähler manifold, G its (connected) group of isometries, x_0 a distinguished point of M, B_0 the stabilizer sub-group of x_0 in G, S_0 the connected center of B_0, so that B_0 is the centralizer $C(S_0)$ of S_0 in virtue of 8.88 ii).

Let T be any maximal torus of G containing S_0 (hence contained in B_0), and let $\mathfrak{g}$, $\mathfrak{b}_0$, $\mathfrak{s}_0$, $\mathfrak{t}$ be the Lie algebras of G, B_0, S_0 and T respectively.

We endow $\mathfrak{g}$ with a G-invariant scalar product and consider the orthogonal complement $\mathfrak{m}_0$ of $\mathfrak{b}_0$ in $\mathfrak{g}$. The sub-space $\mathfrak{m}_0$ is clearly Ad_{B_0}-invariant and is identified with the isotropy representation of the tangent space at x_0 to M.

Now, the G-invariant complex structure of M determines an Ad_{B_0}-invariant complex structure J on $\mathfrak{m}_0$.

Let

$$\mathfrak{g} = \mathfrak{t} \oplus \sum_{\pm\alpha_T \in R_G} E_{\pm\alpha_T} = \mathfrak{t} \oplus \sum_{\pm\beta_T \in R_{B_0}} E_{\pm\beta_T} \oplus \sum_{\pm\alpha_T \in D} E_{\pm\alpha_T} \tag{8.90}$$

be the decomposition of $\mathfrak{g}$ into irreducible components.

Consider now the *primitive* roots in R_G, that is the positive roots (relative to the above ordering) which are not the sum of two positive roots.

8.91. *Primitive roots form a basis of* $\mathfrak{t}$ *and each positive root is a sum of primitive roots* (see [Ser 2]).

Among the primitive roots in R_G we get all the primitive roots of R_B which are still primitive in R_G since D_+ is closed.

We obtain in this way, a basis of $\mathfrak{t}$

$$\{{}^1\beta_T, \ldots, {}^{r-s}\beta_T, {}^1\alpha_T, \ldots, {}^s\alpha_T\} \quad r = \mathrm{rank}(G)$$
$$s = \dim(S_0),$$

where $\{{}^1\beta_T, \ldots, {}^{r-s}\beta_T\}$ is the set of the primitive roots in R_{B_0} and $\{{}^1\alpha_T, \ldots, {}^s\alpha_T\}$ the set of the primitive roots in D.

Let

$$\{{}^1\beta_T^\sharp, \ldots, {}^{r-s}\beta_T^\sharp, {}^1\alpha_T^\sharp, \ldots, {}^s\alpha_T^\sharp\}$$

be the *dual* basis for the normal metric and let

$$w = \sum_{i=1}^{s} w_i \cdot {}^i\alpha_T^\sharp \tag{8.92}$$

be any combination of the ${}^i\alpha_T^\sharp$ with *positive* coefficients w_i under Ad_T (we refer the reader to 8.50–8.57 for the notation). Here $\mathfrak{t} \oplus \sum_{\pm\beta_T \in R_{B_0}} E_{\pm\beta_T}$ is the Lie algebra $\mathfrak{b}_0$, while $\sum_{\pm\alpha_T \in D} E_{\pm\alpha_T}$ is a decomposition of $\mathfrak{m}_0$ under Ad_T.

The endomorphism J of M_0 can be extended to the whole of $\mathfrak{g}$ by putting it equal to zero on $\mathfrak{b}_0$. We obtain an Ad_{B_0}-invariant endomorphism of $\mathfrak{g}$, which we denote also by J, whose kernel is exactly $\mathfrak{b}_0$, and which preserves the sub-spaces $E_{\pm\alpha_T}$, for each $\pm\alpha_T$ in D, inducing in each of them a preferred orientation. We denote by α_T the weight inducing the same orientation: α_T is said to be a (*positive*) *root of the complex structure J*.

We refer the reader to [Bo-Hi] for the two following facts.

8.93. *The integrability of J implies that the set D_+ of the (positive) roots of J is closed,* in the sense that the sum of two elements of D_+ belongs itself to D_+ whenever it is a weight of Ad_T. ([Bo-Hi], Lemma 12-4).

8.94. *We can then find an ordering on the whole set of weights R_G of* Ad_T *such that the roots in D_+ are still positive relatively to this ordering.* ([Bo-Hi], Proposition 13-7).

The commutator of w in $\mathfrak{g}$ is clearly $\mathfrak{b}_0$. We infer that the stabilizer of w in G, which is connected as we saw already in 8.19, is B_0.

On the other hand, the scalar product of w with any (positive) root of J is positive. Thus M is isomorphic to the orbit of w endowed with its canonical complex structure. □

In view of the preceding theorem, Theorem 8.2 may be reformulated as follows:

8.95 Theorem (Y. Matsushima [Mat 3]). *Every compact, simply connected, homogeneous Kähler manifold admits a unique (up to homothety) invariant Kähler-Einstein structure (under the connected group of isometries of the initial Kähler structure).*

Each Kähler-Einstein structure is deduced from the former by an automorphism of the complex structure.

8.96 Remark. By virtue of (8.60), we obtain on every compact, simply-connected, homogeneous Kähler manifold M a family, parametrized by an open subset of the connected group $\mathfrak{A}(M)$ of automorphisms of the complex structure of M, of distinct —but isometric—Kähler-Einstein metrics compatible with that complex structure (compare 11.46).

In the non-simply-connected case we quote, without proof, the following basic result

8.97 Theorem (Y. Matsushima [Mat 3], Borel-Remmert [Bo-Re]). *Each compact homogeneous Kähler manifold is the Kähler product of a flat complex torus and a simply-connected compact homogeneous Kähler manifold.*

8.98 Corollary. *A compact homogeneous Kähler manifold admits a Kähler-Einstein structure if and only if it is a complex torus or is simply-connected.*

8.99 Remark. More generally, Borel and Remmert show in [Bo-Re] that each compact *Kählerian* (i.e., admitting a Kähler metric) homogeneous complex manifold is the product, as a homogeneous complex manifold, of a complex torus and compact, simply-connected, Kählerian homogeneous complex manifold.

Moreover, the latter is a *rational algebraic manifold.* ([Got]).

In fact, it can be proved [Bo-Re] that each compact Kählerian homogeneous complex manifold can be equipped with a Kähler structure for which it becomes a homogeneous Kähler manifold.

The fact that M is projective-algebraic when simply-connected can be also deduced from the existence of a Kähler-Einstein metric with positive scalar curvature (see 11.11).

G. The Space of Orbits

Let G be, as usual, a compact (connected) Lie group with trivial center, T a fixed maximal torus of G, $\mathfrak{g}$ and $\mathfrak{t}$ the Lie algebras of G and T, respectively. (The Lie algebra of a maximal torus is called a *Cartan sub-algebra* of $\mathfrak{g}$).

We first prove

8.100 Proposition (R. Bott). *Each orbit M for the adjoint representation of G in $\mathfrak{g}$ intersects H in a finite non-empty set of points.*

Proof (R. Bott in [Bot 2]). Let Z be any regular element in $\mathfrak{t}$ so that $\mathfrak{t}$ is the commutator of Z in $\mathfrak{g}$ and let f_Z the function defined by

$$f_Z(w) = (Z, W) \quad \forall w \in M.$$

At any critical point w_0 of f_Z on M and for any vector X tangent to M at w_0, we get

$$0 = df_Z(w_0)(X) = (Z, X) \quad \forall X \in T_{w_0}M \simeq M_{w_0}.$$

It follows that

$$0 = (Z, [W_0, U]) = +([Z, W_0], U) \quad \forall U \in \mathfrak{g},$$

which implies that $[Z, W_0]$ is equal to zero, or equivalently, that W_0 belongs to $\mathfrak{t}$.

The subalgebra H is contained in L_{w_0} (recall the notation of 8.17), so that M intersects H orthogonally at w_0. The intersection set then is discrete hence finite.

8.101 Remark. If w_0 is a point in the intersection of the orbit M with $\mathfrak{t}$, the other points of the intersection are obtained from w_0 by the action of any element g of G preserving globally the maximal torus T, that is by any element of the normaliser N_T of T in G. The *Weyl group* $W(G)$, of G relative to T is, by definition, the group of automorphisms of T induced by the elements of N_T, or, equivalently, the quotient N_T/T. It is a finite group (see [Ada] p. 29). We have shown that *the intersection $M \cap \mathfrak{t}$ is just the orbit of any of its element w_0 by the Weyl group $W(G)$.*

Consider again the set R_G of weights of Ad_T in $\mathfrak{g}$, viewed as elements of $\mathfrak{t}$ via the G-invariant scalar product. With any element Z of $\mathfrak{t}$ we associate an ordering on R_G by saying that a weight in R_G is positive whenever its scalar product with Z—which is non-null by assumption—is positive.

Then, $\mathfrak{t}$ is spanned, as a real vector space, by the primitive roots $\{{}^1\alpha_T, \ldots, {}^r\alpha_T\}$ (see the proof of Theorem 8.95).

8.102. Definition. The *closed* (resp. *open*) *positive Weyl chamber* $\bar{C}$ (resp. C) is the set of those elements of H whose scalar product with any positive root—equivalently, with any primitive root—is non-negative (resp. positive).

The *walls* of the closed Weyl chamber $\bar{C}$ are the hyperplanes ${}^iH$ orthogonal to the primitive roots ${}^i\alpha_T$, $i = 1, \ldots, r$.

We recall the known result (see [Ada], p. 110).

8.103. *The orbit, under the Weyl group $W(G)$, of any element of* $\mathfrak{t}$ *has one and only one element in the closed positive Weyl chamber* $\bar{C}$.

We infer immediately the

8.104 Proposition. *Each orbit M in* $\mathfrak{g}$ *is the orbit of one (unique) element w_0 in the closed Weyl chamber $\bar{C}$ in* $\mathfrak{t}$.

8.105 Remark. M is a flag manifold of G if and only if its representative w_0 in $\bar{C}$ belongs to C. Flag manifolds then appear as being the generic orbits. The others are the orbits of elements w_0 belonging to one or several walls of the closed Weyl chamber. It follows readily from the proof of Proposition 8.100 that *the dimension of the torus S_{w_0} is exactly equal to the rank r of G minus the number of walls ${}^t\!H$ containing w_0.*

In a sense, flag manifolds generate the orbits. More precisely we have the

8.106. Proposition. *Each flag manifold of G admits a holomorphic fibration over each orbit, with fiber a flag manifold.*

Proof. Consider the closed Weyl chamber $\bar{C}$ as above and the orbits $\tilde{M}$ and M of two elements z_0 and w_0 belonging to C and $\bar{C}$ respectively.

The tangent space $T_{z_0}\tilde{M}$ to $\tilde{M}$ at z_0 is canonically identified with the orthogonal complement of $\mathfrak{t}$ in $\mathfrak{g}$, since z_0 is regular, while the tangent space $T_{w_0}M$ to M at w_0 is identified with the sub-space of the preceding orthogonal to the kernel of ad_{w_0}. In the notations of 8.51, we have

$$T_{z_0}\tilde{M} \simeq \sum_{\alpha_T \in R_G^+} E_{\alpha_T} \qquad T_{w_0}M \simeq \sum_{\alpha_T \in D^+} E_{\alpha_T},$$

where R_G^+ is the set of the positive roots (relative to the fixed ordering of the weights of Ad_T in $\mathfrak{t}$) and D^+ the set of the (positive) complementary roots relatively to the sub-group G_{w_0}, where G_{w_0} is the stabilizer of w_0 in G.

We get a G-equivariant mapping Φ from $\tilde{M}$ onto M by putting:

$$\Phi(\mathrm{Ad}(g)\cdot z_0) = \mathrm{Ad}(g)\cdot w_0 \quad \forall g \in G. \tag{8.107}$$

Φ is well defined since the stabilizer sub-group of z_0, which is equal to the maximal torus T, is contained in G_{w_0}. Φ is clearly surjective and smooth.

The tangent mapping $d\Phi$ of Φ at z_0 is defined by:

$$d\Phi_{z_0}(X) = \mathrm{ad}_{w_0}\cdot \mathrm{ad}_{z_0}^{-1}(X) \quad \forall X \in T_{z_0}\tilde{M} \simeq \tilde{M}_{z_0} \tag{8.108}$$

which reduces to the product by $\alpha_T(w_0)/\alpha_T(z_0)$ on each irreductible component E_{α_T}.

It follows at once that Φ has maximal rank and commutes with the complex operators of $\tilde{M}$ and M. Since Φ is G-equivariant we conclude that Φ is a (locally trivial) holomorphic fibration.

The fiber over w_0 is itself a compact homogeneous Kähler manifold, diffeomorphic to G_{w_0}/T, hence a flag manifold of G_{w_0} by Theorem 8.89. This completes the proof. □

8.109 Corollary. *The flag manifolds of a compact Lie group G are all isomorphic as complex homogeneous manifolds.*

8.110. More generally, we show, by the same argument, that the orbits of two elements w_1 and w_2 having the *same* stabilizer in G are isomorphic as complex homogeneous manifolds.

H. Examples

8.111 Example. Orbits of $SU(N)$.

The Lie algebra $\mathfrak{su}(N)$ is the Lie algebra of the trace zero skew-hermitian matrices of order N. A Cartan sub-algebra $\mathfrak{t}$ is the (Abelian) Lie algebra of the diagonal matrices in $\mathfrak{su}(N)$. We put

$$(\lambda_1,\ldots,\lambda_N) = \begin{pmatrix} i\lambda_1 & & & \\ & i\lambda_2 & 0 & \\ & 0 & \ddots & \\ & & & i\lambda_N \end{pmatrix}, \quad \sum_{i=1}^{N} \lambda_i = 0,$$

and hence also by λ_i the linear form which associate with the N-tuple $(\lambda_1,\ldots,\lambda_N)$ its ith component.

The weights of Ad_T—where T is the maximal torus corresponding to $\mathfrak{t}$, that is, the torus of the diagonal matrices in $SU(N)$—are the forms $(\lambda_i - \lambda_j)$, $i \neq j$. For the canonical ordering, the positive roots are the $\dfrac{N(N-1)}{2}$ forms $(\lambda_i - \lambda_j)$, $i < j$, and the primitive roots the $(N-1)$ forms $(\lambda_i - \lambda_{i+1})$, $i = 1,\ldots,N-1$. (see [Ada]).

The (closed) positive Weyl chamber $\bar{C}$ is then the set of the N-tuples $(\lambda_1,\ldots,\lambda_N)$ such that

$$\lambda_1 \geqslant \cdots \geqslant \lambda_N.$$

By 8.104, each orbit of $\mathrm{Ad}_{SU(N)}$ is the orbit of an element of $\bar{C}$. Any element of $\bar{C}$ may be expressed as:

$$w = (\underbrace{\mu_1,\ldots,\mu_1}_{p_1 \text{ times}}, \underbrace{\mu_2,\ldots,\mu_2}_{p_2 \text{ times}},\ldots,\underbrace{\mu_q,\ldots,\mu_q}_{p_q \text{ times}})$$

with

$$\mu_1 > \mu_2 \cdots > \mu_q$$

and

$$1 \leqslant p_i \leqslant N \quad \sum_{i=1}^{q} p_i = N \quad \forall i = 1,\ldots,q.$$

It is easily seen that the commutator of w doesn't depend on the values of the

μ_i but only on the ordered sequence $(p_1, \ldots, p_q)$, and is equal to the sub-group $S(U(p_1) \times \cdots \times U(p_q))$ of $SU(N)$.

Any orbit of $SU(N)$ is thus naturally isomorphic, as $SU(N)$-homogeneous manifold, to one of the quotient spaces

$$\frac{SU(N)}{S(U(p_1) \times \cdots \times U(p_q))} \simeq \frac{U(N)}{U(p_1) \times \cdots \times U(p_q)} = M^{SU(N)}_{(p_1, \ldots, p_q)},$$

where $(p_1, \ldots, p_q)$ is any (ordered) partition of N by positive integers.

By 8.110, $M^{SU(N)}_{(p_1, \ldots, p_q)}$ has a well-defined structure of a complex homogeneous manifold.

8.112. By contrast, orbits associated with the same partition of N, but differently ordered may have quite different complex structures (see, for an example, [Bo-Hi]).

From a geometric viewpoint, the manifold $M^{SU(N)}_{(p_1, \ldots, p_q)}$ is the manifold of the (partial) flags of type $(p_1, \ldots, p_q)$ in $\mathbb{C}^N$, that is, the (ordered) q-tuple $(P_1, \ldots, P_q)$ of mutually orthogonal (for the natural Hermitian structure) complex sub-spaces of $\mathbb{C}^N$, where each P_i has (complex) dimension equal to p_i, $i = 1, \ldots, q$.

Flag manifolds of $SU(N)$ are thus diffeomorphic to the manifold of flags in $\mathbb{C}^N$ in the usual sense.

The dimension of the center of $S(U(p_1) \times \cdots \times U(p_q))$ is equal to $(q-1)$.

The orbits for which that dimension is equal to 1 (see 8.86) are thus diffeomorphic to

$$\frac{SU(N)}{S(U(p) \times U(q))} \simeq \frac{U(N)}{U(p) \times U(q)} \qquad p + q = N.$$

These manifolds are the Grassmann manifolds of complex p-subspaces of $\mathbb{C}^N$. Among them we find the complex projective space $\mathbb{C}P^{N-1}$ endowed with its canonical complex structure for $p = 1$, or with the conjugate complex structure for $p = N - 1$ (the two complex structures are equivalent).

It is easily checked and well-known that the normal metric of those manifolds is Kähler for their canonical complex structure, and symmetric (see 8.86).

8.113 Example. Orbits of $SO(2N)$.

The Lie algebra $\mathfrak{su}(2N)$ is the Lie algebra of the skew-symmetric real matrices of (even) order $2N$. A Cartan sub-algebra $\mathfrak{t}$ is the (Abelian) Lie algebra of the matrices of the form:

$$(\lambda_1, \ldots, \lambda_n) = \begin{pmatrix} \boxed{\begin{matrix} 0 & -\lambda_1 \\ \lambda_1 & 0 \end{matrix}} & & & 0 \\ & \boxed{\begin{matrix} 0 & -\lambda_2 \\ \lambda_2 & 0 \end{matrix}} & & \\ & 0 & \ddots & \\ & & & \boxed{\begin{matrix} 0 & -\lambda_n \\ \lambda_n & 0 \end{matrix}} \end{pmatrix}.$$

The weights of Ad_T—where T is the corresponding maximal torus—are the 1-forms $(\lambda_i - \lambda_j)$, $\pm(\lambda_i + \lambda_j)$, $i \neq j$. For the canonical ordering the (positive) roots are the 1-forms $(\lambda_i - \lambda_j)$, $(\lambda_i + \lambda_j)$, $i < j$ and the primitive roots are the N forms $(\lambda_i - \lambda_{i+1})$, $i = 1, \ldots, N-1$ and $(\lambda_{N-1} + \lambda_N)$ (see [Ada], Example 4.19, p. 87).

The (closed) positive Weyl chamber $\bar{C}$ is then the set of the N-tuples $(\lambda_1, \ldots, \lambda_N)$ such that

$$\lambda_1 \geqslant \cdots \geqslant \lambda_{N-1} \geqslant |\lambda_N|.$$

Any element of $\bar{C}$ may be expressed as:

$$(\underbrace{\mu_1, \ldots, \mu_1}_{p_2}, \underbrace{\mu_2, \ldots, \mu_2}_{p_2}, \ldots, \underbrace{\mu_q, \ldots, \mu_q}_{p_q}, \underbrace{0, 0, \ldots, 0}_{l})$$

with

$$\mu_1 > \mu_2 \cdots > \mu_q > 0$$

and

$$1 \leqslant p_i \leqslant N \qquad 0 \leqslant l \leqslant N \qquad \sum_{i=1}^{q} p_i + l = N$$

or as

$$(\underbrace{\mu_1, \ldots, \mu_1}_{p_1}, \underbrace{\mu_2, \ldots, \mu_2}_{p_2}, \ldots, \underbrace{\mu_q, \ldots, \mu_q, -\mu_q}_{p_q})$$

with

$$\mu_1 > \mu_2 > \cdots > \mu_q > 0$$

and

$$1 \leqslant p_i \leqslant N \qquad \sum_{i=1}^{q} p_i = N.$$

Again the stabilizer of the elements of one or the other type depends only on the (ordered) sequence $(p_1, \ldots, p_q, l)$ or $(p_1, \ldots, p_q)$.

We get, in the first case, the sub-group

$$U(p_1) \times \cdots \times U(p_q) \times SO(2l),$$

and in the second case, the sub-group

$$U(p_1) \times \cdots \times U(p_{q-1}) \times \tilde{U}(p_q),$$

where, each $U(p_i)$ is the unitary group of order p_i canonically imbedded in $SO(2p_i)$, imbedded itself in $SO(2N)$ in a diagonal way at the position determined by the position of $U(p_i)$ in the product, while $\tilde{U}(p_q)$ is the unitary sub-group of $SO(2p_q)$ associated with the complex structure $\tilde{J}$ of $\mathbb{R}^{2p_q}$ determined by

$$\tilde{J}e_1 = e_2, \ldots, \tilde{J}e_{2p_q-3} = e_{2p_q-2}, \tilde{J}e_{2p_q-1} = -e_{2p_q},$$

where $(e_1, e_2, \ldots, e_{2p_q-1}, e_{2p_q})$ is the natural basis of $\mathbb{R}^{2p_q}$.

8.114. From a geometric point of view, we get three sorts of manifolds.

—The first manifold is $M^{SO(2N)}_{(p_1,\ldots,p_q,l)}$, (where l is ant non-zero integer not greater than N, and $(p_1,\ldots,p_q)$ any ordered partition of $(N-l)$), of (partial) complex flags of type $(p_1,\ldots,p_q)$ in $\mathbb{R}^{2N}$, that is of ordered q-tuples $(p_1,\ldots,p_q)$ of mutually orthogonal sub-spaces P_i, each of them being of dimension $2p_i$ and endowed with a complex structure compatible with the canonical Euclidian metric.

—The second manifold is $M^{SO(2N)}_{(p_1,\ldots,p_q)}$—where $(p_1,\ldots,p_q)$ is any ordered partition of N—of positive complex flags of type $(p_1,\ldots,p_q)$, that is flags defined as above, where, in addition, the total complex structure of $\mathbb{R}^{2N}$ determined by the flag induced the canonical orientation of $\mathbb{R}^{2N}$.

—The third manifold is $\tilde{M}^{SO(2N)}_{(p_1,\ldots,p_q)}$ of negative complex flags of type $(p_1,\ldots,p_q)$ —where $(p_1,\ldots,p_q)$ is any ordered partition of N—defined as above, except that the orientation induced by the total complex structure on $\mathbb{R}^{2N}$ is the opposite to the canonical one.

The orbits whose stabilizer sub-groups have a 1-dimensional center are the manifolds $M^{SO(2N)}_{(N-l,l)}$, $M^{SO(2N)}_{(N)}$ and $\tilde{M}^{SO(2N)}_{(N)}$. The two latter are conjugate, by an element of $O(2N)$—not belonging to $SO(2N)$—which exchange their canonical complex structure. They are, respectively, the manifolds of complex structures of $\mathbb{R}^{2N}$, compatible with the canonical Euclidean structure and inducing the canonical orientation (resp. the opposite to the canonical orientation), while $M^{SO(2N)}_{(N-l,l)}$ is the manifold of the (proper) sub-spaces of $\mathbb{R}^{2N}$ of dimension $2(N-l)$ endowed with a complex structure compatible with the induced Euclidean structure.

It can be easily checked that the normal metric is Kähler and symmetric for $M^{SO(2N)}_{(N)}$ and $\tilde{M}^{SO(2N)}_{(N)}$, both diffeomorphic to $\dfrac{SO(2N)}{U(N)}$, while it is neither Kähler nor symmetric for $M^{SO(2N)}_{(N-l,l)}$, except for $l = N-1$, where $M^{SO(2N)}_{(1,N-1)}$ is diffeomorphic to $\dfrac{SO(2N)}{U(1)\times SO(2N-2)} \simeq \dfrac{SO(2N)}{SO(2)\times SO(2N-2)}$ (see 8.86). The latter may be interpreted, as well, as the manifold of oriented 2-planes of $\mathbb{R}^{2N}$.

8.115 Example. Orbits of $SO(2N+1)$.

The rank of $SO(2N+1)$ is equal to N, just as $SO(2N)$. The Lie algebra $\mathfrak{so}(2N+1)$ is the Lie algebra of the skew-symmetric real matrices of (odd) order $2N+1$. A Cartan sub-algebra $\mathfrak{t}$ is the Abelian Lie algebra of the matrices of the form

$$(\lambda_1,\ldots,\lambda_n) = \begin{pmatrix} \begin{matrix} 0 & -\lambda_1 \\ \lambda_1 & 0 \end{matrix} & & & & \\ & \begin{matrix} 0 & -\lambda_2 \\ \lambda_2 & 0 \end{matrix} & & 0 & \\ & 0 & \ddots & & \\ & & & \begin{matrix} 0 & -\lambda_n \\ \lambda_n & 0 \end{matrix} & \\ & & & & 0 \end{pmatrix}.$$

The weights of Ad_T—where T is the corresponding maximal torus—are the 1-forms $(\lambda_1-\lambda_j)\ i\neq j$, $\pm(\lambda_i+\lambda_j)\ i\neq j$, $\pm\lambda_i$. For the canonical ordering, the (positive) roots are the 1-forms $(\lambda_i - \lambda_j)$ $i < j$, $(\lambda_i + \lambda_j)$, $i \neq j$, λ_i and the primitive roots are the N 1-forms $(\lambda_i - \lambda_{i+1})$ $i = 1, \ldots, N-1$ and λ_N (see [Ada], Example 4.20, p. 89).

The (closed) positive Weyl chamber $\bar{C}$ is then the set of the N-tuples $(\lambda_1, \ldots, \lambda_N)$ such that

$$\lambda_1 \geqslant \lambda_2 \geqslant \cdots \geqslant \lambda_N \geqslant 0.$$

As in the preceding example, we see easily that the stabilizer sub-groups of the elements of $\bar{C}$ are equal to:

$$U(p_1) \times U(p_2) \times \cdots \times U(p_q) \times SO(2l+1)$$

with:

$$0 \leqslant l \leqslant N \qquad \sum_{i=1}^{q} p_i = N - l.$$

The corresponding orbit $M^{SO(2N+1)}_{(p_1,\ldots,p_q)}$ is the manifolds of (partial) complex flags of type $(p_1, \ldots, p_q)$ defined as in Example 8.113.

Among these sub-groups, the ones whose center is one-dimensional are the sub-groups

$$U(N-l) \times SO(2l+1).$$

The corresponding orbits are the manifolds of the $(2N-2l)$-dimensional sub-spaces of $\mathbb{R}^{2N+1}$ equipped with a complex structure compatible with the canonical Euclidean structure. Again, using 8.86, we check easily that the normal metric of $M^{SO(2N+1)}_{(N-1)}$ is neither Kähler nor symmetric except for $l = N-1$. The latter orbit is diffeomorphic to

$$M^{SO(2N+1)}_{(1)} \simeq \frac{SO(2N+1)}{U(1) \times SO(2N-1)} \simeq \frac{SO(2N+1)}{SO(2) \times SO(2N-1)},$$

which is still the manifold of oriented 2-planes of $\mathbb{R}^{2N+1}$.

8.116 Example. Orbits of $Sp(N)$.

The Lie algebra $\mathfrak{sp}(N)$ is the Lie algebra of skew-Hermitian quaternionic matrices of order N. A Cartan sub-algebra $\mathfrak{t}$ is the (Abelian) Lie algebra of the matrices of the form

$$(\lambda_1, \ldots, \lambda_N) = \begin{pmatrix} i\lambda_1 & & 0 \\ & \ddots & \\ 0 & & i\lambda_N \end{pmatrix},$$

where the λ_i's are *real* numbers. The weights of Ad_T—where T is the corresponding maximal torus of $Sp(N)$—are the 1-forms $(\lambda_i - \lambda_j)$ $i \neq j$, $\pm(\lambda_i + \lambda_j)$ $i \neq j$ and $\pm 2\lambda_i$. For the canonical ordering, the (positive) roots are $(\lambda_i - \lambda_j)$ $i < j$, $(\lambda_i + \lambda_j)$ $i \neq j$, $2\lambda_i$ and the primitive roots are $(\lambda_i - \lambda_{i+1})$ $i = 1, \ldots, N-1$ and $2\lambda_N$. (see [Ada], Example 4.18, p. 85).

The (closed) positive Weyl chamber $\bar{C}$ is then the set of N-tuples $(\lambda_1, \ldots, \lambda_N)$ such that

$$\lambda_1 \geqslant \lambda_2 \geqslant \cdots \lambda_{N-1} \geqslant \lambda_N \geqslant 0.$$

The stabilizer sub-groups of the elements of $\bar{C}$ are then equal to one of the following

$$U(p_1) \times \cdots \times U(p_q) \times Sp(l),$$

where l is any integer between 0 and N, and $(p_1, \ldots, p_q)$ any (ordered) partition of $(N - l)$.

In order to give a geometric interpretation to the corresponding orbits $M^{Sp(N)}_{(p_1,\ldots,p_q)}$ it is convenient to consider $Sp(N)$ as the group of (complex) automorphisms of $\mathbb{C}^{2N}$ preserving both the canonical Hermitian scalar product and the canonical complex symplectic structure (see [Chv], I. § VI to VIII).

It is then easily seen that $M^{Sp(N)}_{(p_1,\ldots,p_q)}$ is isomorphic, as a $Sp(N)$-homogeneous space, to the manifold of isotropic flags of type $(p_1, \ldots, p_q)$, that is of q-tupler $(P_1, \ldots, P_q)$ of mutually orthogonal (for the Hermitian product) complex sub-spaces P_i of $\mathbb{C}^{2N}$ each of (complex) dimension p_i and completely isotropic relatively to the complex structure.

Since any complex 1-dimensional sub-space of $\mathbb{C}^{2N}$ is isotropic we recover among the orbits of $Sp(N)$ the complex projective space $\mathbb{C}P^{2N-1}$, isomorphic to

$$\frac{Sp(N)}{U(1) \times Sp(N-1)}$$

(see 7.15).

More generally, the stabilizer sub-groups whose center is 1-dimensional are the sub-groups

$$U(N - l) \times Sp(l).$$

Among the corresponding orbits $M^{Sp(N)}_{(N-l)}$, which are the manifolds of totally isotropic complex $(N - l)$-subspaces of $\mathbb{C}^{2N}$, the only ones for which the normal metric is Kähler (hence Kähler symmetric) are $M^{Sp(N)}_{(1)}$, that is, $\mathbb{C}P^{2N-1}$, already seen, and $M^{Sp(N)}_{(N)}$, isomorphic to:

$$\frac{Sp(N)}{U(N)},$$

which is the manifold of totally isotropic complex N-subspaces of $\mathbb{C}^{2N}$. (Apply 8.86).

Chapter 9. Riemannian Submersions

A. Introduction

9.1. The notion of *Riemannian immersion* (see 1.70) has been intensively studied since the very beginning of Riemannian geometry. Indeed the first Riemannian manifolds to be studied were surfaces imbedded in $\mathbb{R}^3$. As a consequence, the differential geometry of Riemannian immersions is well known and available in many textbooks (see for example [Ko-No 1, 2], [Spi]).

On the contrary, the "dual" notion of *Riemannian submersion* appears to have been studied only very recently and its (local) differential geometry was first exposed in 1968 (B. O'Neill [ONe 1] A. Gray [Gra 6], see also earlier works [Her], [Kob 9], [Nag] among others).

As a consequence, this theory is not yet available in textbook form, so we will give here a detailed presentation of it.

9.2. The first five paragraphs (B to F) are devoted to the general theory, first from the "local" viewpoint; after the basic definitions (§ B) we introduce O'Neill's invariants A and T of a submersion (§ C). We use them to compute the curvature of the total space in terms of curvature of the base and the fibre of the submersion (§ D). Then we consider the global theory, under the assumption that the total space is complete (§ E). In particular we get in 9.42

9.3 Theorem (R. Hermann [Her]). *A Riemannian submersion whose total space is complete is a locally trivial fibre bundle.*

We study in § F the particular case of Riemannian submersions with totally geodesic fibres, which give a very interesting family of examples.

9.4 Theorem (R. Hermann [Her]). *In a Riemannian submersion with totally geodesic fibres, the fibres are all isometric and the holonomy group of the fibration is included in the isometry group of the corresponding fibre, and hence is a Lie group.*

Also, the formulas for curvature can be simplified. We study how they vary when we change the scale of the base and the fibre (§ G).

9.5. In § H, we give applications of the theory of Riemannian submersions to the study of homogeneous Einstein manifolds. With a simple technical lemma (due to L. Bérard Bergery), we get many examples of homogeneous Einstein manifolds. In particular, in 9.86, we get all homogeneous Einstein metrics on spheres and projective spaces (a classification due to W. Ziller [Zil 2]). To be complete, we give in § I some homogeneous Einstein manifolds which are not obtained by the preceding lemma.

9.6. In § J, we study the case of warped products.

Finally in § K, by combining the techniques of § F, G, J, we sketch the construction of the first known examples of non-homogeneous Einstein manifolds with positive scalar curvature (the 4-dimensional case is due to D. Page [Pag] and generalizations in all even dimensions are due to L. Bérard Bergery [BéBer 3]). See also the Addendum for further non-homogeneous examples.

B. Riemannian Submersions

9.7. Let (M, g) and $(B, \check{g})$ be two Riemannian manifolds and $\pi\colon M \to B$ a (smooth) submersion. For each x in M, with $b = \pi(x)$, we *denote* by

$F_b = \pi^{-1}(b) = F_x$,
$\mathscr{V}_x$, the tangent subspace to F_b in T_xM (identified with $T_x(F_b)$),
$\mathscr{H}_x$, the orthogonal complement to $\mathscr{V}_x$ in T_xM,
$\mathscr{V}$, (resp $\mathscr{H}$) the corresponding distributions of subspaces (and also the orthogonal projections onto them),
$\hat{g}_b$, the restriction of g to F_b.

We call M the *total space*,
B the *base* (or base space),
$F_b = F_x$ the *fibre* (at b or through x),
$\mathscr{V}_x$ the *vertical subspace* at x,
$\mathscr{V}$ the *vertical distribution* (or projection),
$\mathscr{H}_x$ the *horizontal subspace* at x,
$\mathscr{H}$ the *horizontal distribution* (or projection),
a vector field U *vertical* if $U_x \in \mathscr{V}_x$, $\forall x \in M$,
a vector field X *horizontal* if $X_x \in \mathscr{H}_x$, $\forall x \in M$.

Throughout this chapter, $U, V, W, U', \ldots$ will always be vertical vector fields and $X, Y, Z, X', \ldots$ horizontal vector fields.

Notice that the tangent map

$$\pi_*\colon T_xM \to T_bB$$

has $\mathscr{V}_x$ as kernel and hence induces a *linear isomorphism* from $\mathscr{H}_x$ to T_bB.

9.8 Definition (B. O'Neill [ONe 1]). The map π (more precisely the data $\{(M, g), (B, \check{g}), \pi\}$) is a *Riemannian submersion* if π_* induces an isometry from $(\mathscr{H}_x, g_x \restriction \mathscr{H}_x)$ to $(T_bB, \check{g}_b)$ for every x in M, with $b = \pi(x)$.

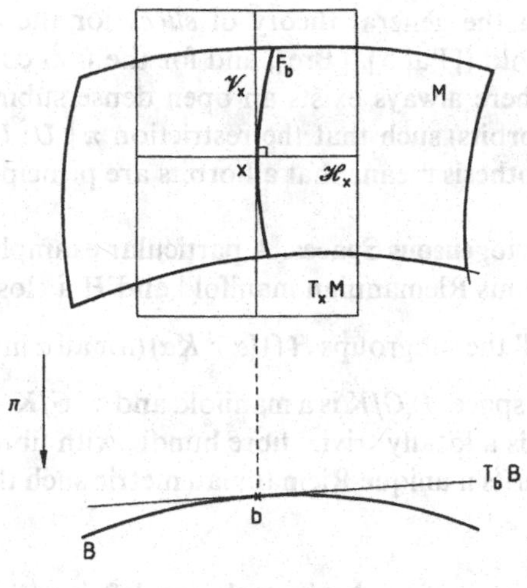

Fig. 9.8

We give below various types of examples.

9.9 Products. Let $(B, \check{g})$ and $(F, \hat{g})$ be two Riemannian manifolds and $(M, g) = (B \times F, \check{g} + \hat{g})$ their Riemannian product (see 1.37). Then the canonical projections p_1 and p_2 on the factors B and F are Riemannian submersions.

9.10 Products with Varying Metrics on the Fibres. More generally, let $(B, \check{g})$ be a Riemannian manifold and $\hat{g}_b$ be a smooth family of Riemannian metrics on F, *indexed by B*. On the product $M = B \times F$, we consider the Riemannian metric $g = \check{g} + \hat{g}_b$ whose value at $(b, x) \in B \times F$ is precisely

$$g(b, x) = p_1^*(\check{g}(b)) + p_2^*(\hat{g}_b(x)).$$

Then the *first* projection $p_1 \colon M \to B$ is a Riemannian submersion.

Notice that $\mathscr{V}_{(b,x)}$ is identified with T_bB and $\mathscr{H}_{(b,u)}$ with T_xF. In particular, in these examples, $\mathscr{H}$ is an *integrable distribution* (this is not the case in general, see 9.24).

9.11 Warped Products. A special case of the preceding example is well-known. Take $\hat{g}_b = f(b)\hat{g}$ where $\hat{g}$ is some fixed Riemannian metric on F and f a *positive function* on B.

Then the resulting Riemannian manifold $(B \times F, \check{g} + f\hat{g})$ is called the *warped product* of $(B, \check{g})$ and $(F, \hat{g})$ by f, and denoted by $B \times_f F$ (see for example [Bi-ON]).

9.12 Quotients by an Isometric Action. Let (M, g) be a Riemannian manifold and G a closed subgroup of the isometry group of (M, g).

(9.12) *Assume* that the projection π from M to the quotient space M/G is a *smooth submersion*. Then there exists one and only one Riemannian metric $\check{g}$ on $B = M/G$ such that π is a Riemannian submersion.

(We recall that, using the general theory of *slices* for the action of a group of isometries on a manifold ([Pal 5], [Bre], and for the non compact case [Yau 6]), one may show that there always exists an open dense submanifold U of M (the union of "principal" orbits) such that the restriction $\pi \restriction U: U \to U/G$ is a smooth submersion. Our hypothesis means that all orbits are principal, i.e., $U = M$.)

9.13 Fibrations of Homogeneous Spaces. A particular example appears when $M = G/K$ is a G-homogeneous Riemannian manifold and H a closed subgroup of G.

(9.13) *Assume* that all the subgroups $H \cap (\alpha^{-1} K \alpha)$ (for all α in G) are *conjugate* in H.

Then the double coset space $H\backslash G/K$ is a manifold and $\pi\colon G/K \to H\backslash G/K$ is a smooth submersion (in fact it is a locally trivial fibre bundle with fibre $H/H \cap K$).

Here, $H\backslash G/K$ inherits a unique Riemannian metric such that π is a Riemannian submersion.

9.14 Remarks. a) *Warning.* If we begin with any left invariant metric on G (and $K = \{e\}$), this construction does *not* necessarily give a G-homogeneous Riemannian metric on $H\backslash G$.

b) *A special case*: The above assumption (9.13) is satisfied in particular if H is a closed *invariant* subgroup of G. Let G_1 be the quotient group G/H and K_1 the image of K in the quotient space $G \to G/H$. Then we get a Riemannian submersion from G/K to a G_1-homogeneous Riemannian manifold G_1/K_1, with fibre $H/H \cap K$ (see for example [Bé Ber 5]).

9.15 A General Construction. Given a smooth submersion $\pi\colon M \to B$, a distribution $\mathscr{H}$ on M which is supplementary to the distribution $\mathscr{V} = \operatorname{Ker}(\pi_*)$, a Riemannian metric $\check{g}$ on B and finally a smooth family $\hat{g}_b$ of Riemannian metrics on the fibres F_b of π, then there exists one and only one Riemannian metric g on M such that π is a Riemannian submersion from (M, g) to $(B, \check{g})$ with horizontal distribution $\mathscr{H}$ and induced metric $\hat{g}_b$ on the fibres F_b (exercise !).

C. The Invariants A and T

9.16. We denote by D the Levi-Civita connection of g and $\hat{D}$ the collection of all Levi-Civita connections of the Riemannian metrics $\hat{g}_b$ of the fibres F_b, for all b in B. Notice that $\hat{D}_U V$ is a well-defined *vertical* vector field on M (we recall that U, V are always vertical vector fields), more precisely

(9.16) $$\hat{D}_U V = \mathscr{V} D_U V.$$

Furthermore, each fibre F_b is a (closed) submanifold of M, hence it admits a second fundamental form (see 1.71), which gives the first tensorial invariant of the Riemannian submersion. In order to simplify the notation, we embed all theses forms and some of their adjoints (for all b) into one tensor field on M.

9.17 Definition (B. O'Neill [ONe 1]). We *denote* by T the (2, 1) tensor field on M whose values on vector fields E_1, E_2 are given by

(9.17) $$T_{E_1}E_2 = \mathscr{H}D_{\mathscr{V}E_1}\mathscr{V}E_2 + \mathscr{V}D_{\mathscr{V}E_1}\mathscr{H}E_2.$$

An easy computation (using 1.5) shows that T is indeed a tensor field.

9.18. The following properties of T are easy consequences of the definition:

(9.18a) $T_X U = T_X Y = 0.$

(9.18b) $T_U V = \mathscr{H}D_U V$ and $T_U X = \mathscr{V}D_U X.$

(9.18c) $T_U V = T_V U.$

(9.18d) T_U is alternating, in particular $(T_U V, X) = -(T_U X, V).$

(For the last time, we recall that U, V, W are vertical vector fields and X, Y, Z horizontal vector fields).

Notice that $T_U V$ is the second fundamental form of each fibre, hence is known to be symmetric (9.18c). Also (9.18d) derives $T_U X$ from g and $T_U V$.

9.19. Of course, the vertical distribution is integrable, since $\mathscr{V}$ is the tangential distribution defined by the fibres. But the horizontal distribution $\mathscr{H}$ is not necessarily integrable and the natural "obstruction" to integrability of $\mathscr{H}$ gives the second tensorial invariant of the Riemannian submersion. Here also, we embed it (and its adjoint) into a tensor field on M.

9.20 Definition (B. O'Neill [ONe 1]). We *denote* by A the (2, 1) tensor field on M whose values on vector fields E_1, E_2 are given by

(9.20) $$A_{E_1}E_2 = \mathscr{H}D_{\mathscr{H}E_1}\mathscr{V}E_2 + \mathscr{V}D_{\mathscr{H}E_1}\mathscr{H}E_2.$$

Here also, an easy computation (using 1.5) shows that A is a tensor field.

9.21. The following properties of A are easy consequences of the definition.

(9.21a) $A_U X = A_U V = 0.$

(9.21b) $A_X U = \mathscr{H}D_X U$ and $A_X Y = \mathscr{V}D_X Y.$

(9.21c) $A_X Y = -A_Y X$ (see below 9.24).

(9.21d) A_X is alternating, in particular $(A_X Y, U) = -(A_X U, Y).$

Using (9.21d), we may obtain $A_X U$ from $A_X Y$ and g. We will see below in 9.24 that $A_X Y$ is related to the obstruction to integrability of $\mathscr{H}$, but before showing this, we need a useful auxiliary notion.

9.22 Basic Vector Fields. We call a vector field E on M *projectable* if there exists a vector field $\check{E}$ on B such that $\pi_*(E_x) = \check{E}_b$ for each x in M, and we say that E and $\check{E}$ are π-related.

Obviously, if E_1 and E_2 are π-related to $\check{E}_1$ and $\check{E}_2$, then $[E_1, E_2]$ is π-related to $[\check{E}_1, \check{E}_2]$. Since a vector field is vertical if and only if it is π-related to 0, we deduce that if E is projectable, then $[E, U]$ is *vertical*.

9.23 Definition. A vector field X on M is called *basic* if it is projectable and horizontal.

Notice that, for every vector field $\check{X}$ on B, there exists one and only one *basic* vector field X on M which is π-related to $\check{X}$ (if π surjective). Moreover, if X and Y are *basic*, then $\mathscr{H}[X, Y]$ is precisely the basic vector field π-related to $[\check{X}, \check{Y}]$ and $\mathscr{H} D_X Y$ is the basic vector field π-related to $\check{D}_{\check{X}} \check{Y}$ (where $\check{D}$ is the Levi-Civita connection of B). Observe that $g(X, Y) = \check{g}(\check{X}, \check{Y})$ is constant on the fibres if X and Y are basic. Finally, if X is a basic vector field and U a vertical vector field, then $[X, U]$ is vertical.

9.24 Proposition (B. O'Neill [ONe 1]). *For every horizontal vector fields X and Y,*

$$A_X Y = \tfrac{1}{2}\mathscr{V}[X, Y]. \tag{9.24}$$

In this sense, $A_X Y$ measures the obstruction to integrability of the distribution $\mathscr{H}$.

Proof. It is sufficient to prove $A_X X = 0$, since

$$\mathscr{V}[X, Y] = \mathscr{V} D_X Y - \mathscr{V} D_Y X = A_X Y - A_Y X.$$

We may take X to be basic (A is a tensor). Then for any (vertical) U,

$$(U, A_X X) = (U, D_X X) = -(D_X U, X)$$

Since $[X, U]$ is vertical, $(D_X U, X) = (D_U X, X)$, so $(U, A_X X) = \frac{1}{2} U(X, X)$, and this is zero since (X, X) is constant along the fibres. □

9.25 Definition (Summary of formulas for connections). If U, V are vertical vector fields and X, Y horizontal vector fields, then

$$D_U V = \hat{D}_U V + T_U V \tag{9.25a}$$

$$D_U X = T_U X + \mathscr{H} D_U X \tag{9.25b}$$

$$D_X U = \mathscr{V} D_X U + A_X U \tag{9.25c}$$

$$D_X Y = A_X Y + \mathscr{H} D_X Y \tag{9.25d}$$

are the decomposition into vertical and horizontal components of the Levi-Civita connection.

9.26 *The vanishing of T and A.* Since T is related to the second fundamental form, it is identically zero if and only if each fibre is totally geodesic. We will study this particular case in greater details in § F.

Similarly, since A is related to the obstruction to integrability of $\mathscr{H}$, it is identically zero if and only if $\mathscr{H}$ is integrable. Integrating $\mathscr{H}$, we easily see that, if $A \equiv 0$, then at least *locally*, M is isometric to $B \times F$ with a Riemannian metric $\check{g} + \hat{g}_b$

as in example 9.10 (and π is p_1). Finally, if A and T vanish identically, then M is (at least locally) a Riemannian product $B \times F$ as in example 9.9. Indeed the orthogonal decomposition $TM = \mathscr{V} \oplus \mathscr{H}$ is parallel and we may apply de Rham's holonomy theorem (see 10.43).

D. O'Neill's Formulas for Curvature

9.27. We *denote* by R the curvature tensor of g, by $\hat{R}$ the collection of all curvature tensors of the Riemannian metrics $\hat{g}_b$ on the fibres F_b and by $\check{R}(X, Y)Z$ the horizontal vector field such that

$$\pi_*(\check{R}(X, Y)Z) = \check{R}(\pi_* X, \pi_* Y)\pi_* Z$$

at each point x of M, where $\check{R}$ is the curvature tensor of $\check{g}$ on B.

Then a straightforward computation that we will *not* reproduce here (see [ONe 1]) gives

9.28 Theorem (B. O'Neill [ONe 1], A. Gray [Gra 6]).

(9.28a) $(R(U, V)W, W') = (\hat{R}(U, V)W, W') - (T_U W, T_V W') + (T_V W, T_U W')$.

(9.28b) $(R(U, V)W, X) = ((D_V T)_U W, X) - ((D_U T)_V W, X)$.

(9.28c) $$(R(X, U)Y, V) = ((D_X T)_U V, Y) - (T_U X, T_V Y) + ((D_U A)_X Y, V) + (A_X U, A_Y V).$$

(9.28d) $$(R(U, V)X, Y) = ((D_U A)_X Y, V) - ((D_V A)_X Y, U) + (A_X U, A_Y V) - (A_X V, A_Y U) - (T_U X, T_V Y) + (T_V X, T_U Y).$$

(9.28e) $$(R(X, Y)Z, U) = ((D_Z A)_X Y, U) + (A_X Y, T_U Z) - (A_Y Z, T_U X) - (A_Z X, T_U Y).$$

(9.28f) $$(R(X, Y)Z, Z') = (\check{R}(X, Y)Z, Z') - 2(A_X Y, A_Z Z') + (A_Y Z, A_X Z') - (A_X Z, A_Y Z').$$

We deduce in particular the following sectional curvatures (K, $\check{K}$, $\hat{K}$ are the sectional curvatures of the Riemannian metrics g, $\check{g}$ and $\hat{g}_b$ respectively).

9.29 Corollary. Let $|U \wedge V| = 1$, $|X| = |U| = 1$, $|X \wedge Y| = 1$. Then

(9.29a) $$K(U, V) = \hat{K}(U, V) + |T_U V|^2 - (T_U U, T_V V).$$

(9.29b) $$K(X, U) = ((D_X T)_U U, X) - |T_U X|^2 + |A_X U|^2.$$

(9.29c) $$K(X,Y) = \check{K}(\check{X}, \check{Y}) - 3|A_X Y|^2.$$

9.30 Remarks. (a) Formula (9.29c) implies immediately that, if M has positive (respectively nonnegative) sectional curvature, then B also has positive (respectively nonnegative) sectional curvature. This fact has been used for many applications, see for example [Che], [Gr-Me], [ESC], [Rig].

(b) In the special case where the fibres are totally geodesic (i.e., $T \equiv 0$), then the "mixed" sectional curvature $K(X, U)$ is always nonnegative; moreover $K(X, U)$ is positive for any nonzero X and U if and only if the bundle is "fat" in the sense of Weinstein [Wns 1] (see also [BéBer 6] and [De-Ri]).

9.31 Remarks. The covariant derivatives of A and T appear effectively in the formulas for R. There are many relations between the various components (on $\mathscr{V}$ and $\mathscr{H}$) of A, T, DA and DT. We list them below without proof. But notice that $((D_U A)_X Y, V)$, $((D_X A)_Y Z, V)$, $((D_X T)_U V, Y)$ and $((D_U T)_V W, X)$ cannot be recovered from A and T (at x).

9.32 Relations Between A, T, DA and DT.

$(D_{E_1} T)_{E_2}$ and $(D_{E_1} A)_{E_2}$ are alternating;

$((D_E T)_U V, X)$ is symmetric in U, V; $((D_E A)_X Y, U)$ is alternating in X, Y

$$(D_X T)_Y = -T_{A_X Y}; \; (D_U T)_X = -T_{T_U X}; \quad (D_U A)_V = -A_{T_U V}; \; (D_X A)_U = -A_{A_X U};$$

$$((D_X T)_U V, W) = (A_X V, T_U W) - (A_X W, T_U V); \qquad ((D_U A)_X V, W) = (T_U V, A_X W) - (T_U W, A_X V);$$

$$((D_X T)_U Y, Z) = (A_X Y, T_U Z) - (A_X Z, T_U Y); \qquad ((D_U A)_X Y, Z) = (A_X Z, T_U Y) - (A_X Y, T_U Z);$$

$$((D_U T)_V W, W') = (T_U W, T_V W') - (T_U W', T_V W); \qquad ((D_X A)_Y U, V) = (A_X U, A_Y V) - (A_X V, A_Y U);$$

$$((D_U T)_V X, Y) = (T_U X, T_V Y) - (T_V X, T_U Y); \qquad ((D_X A)_Y Z, Z') = (A_X Z, A_Y Z') - (A_X Z', A_Y Z);$$

$$((D_X A)_Y Z, U) + ((D_Y A)_Z X, U) + ((D_Z A)_X Y, U) = (A_X Y, T_U Z) + (A_Y Z, T_U X) + (A_Z X, T_U Y);$$

$$((D_U A)_X Y, V) + ((D_V A)_X Y, U) = ((D_Y T)_U V, X) - ((D_X T)_U V, Y).$$

For the proof, see [ONe 1] or [Gra 6] (the last relation appears only in [Gra 6], with wrong signs). □

9.33 Ricci Curvature: Preparatory Material. Since the Ricci curvature is our main object in this book, we introduce some special notation for it.

We *denote* by $(X_i)_{i \in I}$ a (local) orthonormal basis of $\mathscr{H}_x$,

$(U_j)_{j \in J}$ a (local) orthonormal basis of $\mathscr{V}_x$,

(9.33a) $$(A_X, A_Y) = \sum_i (A_X X_i, A_Y X_i) = \sum_j (A_X U_j, A_Y U_j),$$

(9.33b) $$(A_X, T_U) = \sum_i (A_X X_i, T_U X_i) = \sum_j (A_X U_j, T_U U_j),$$

(9.33c) $$(AU, AV) = \sum_i (A_{X_i} U, A_{X_i} V),$$

(9.33d) $$(TX, TY) = \sum_j (T_{U_j} X, T_{U_j} Y),$$

and for any *tensor field E* on *M*

(9.33e) $$\check{\delta} E = -\sum_i (D_{X_i} E)_{X_i},$$

(9.33f) $$\hat{\delta} E = -\sum_j (D_{U_j} E)_{U_j},$$

(9.33g) $$\delta E = \check{\delta} E + \hat{\delta} E.$$

Finally we define the symmetric 2-tensor $\tilde{\delta} T$ by

(9.33h) $$(\tilde{\delta} T)(U, V) = \sum_i ((D_{X_i} T)_U V, X_i).$$

9.34 Mean Curvature Vector. The mean curvature vector (defined in 1.73) along each fibre gives the horizontal vector field

(9.34) $$N = \sum_j T_{U_j} U_j.$$

Notice that N is identically zero if and only if each fibre is a *minimal submanifold* of M. This is also equivalent (in the case of submersions) to the fact that the Riemannian submersion is a *harmonic map* (see [Ee-Sa]).

Exercise. Prove the following formulas:

$$\hat{\delta} A = A_N; \quad \hat{\delta} T = 0; \quad \sum_j ((D_E T)_{U_j} U_j, X) = (D_E N, X);$$

$$\check{\delta} N = -\sum_j (\tilde{\delta} T)(U_j, U_j); \quad \frac{1}{2}((D_Y N, X) - (D_X N, Y)) = \sum_j ((D_{U_j} A)_X Y, U_j).$$

9.35 Definition. We say that $\mathscr{H}$ satisfies the *Yang-Mills condition* if, for any vertical vector U and any horizontal vector X, we have:

(9.35) $$(\check{\delta} A(X), U) - (A_X, T_U) = 0.$$

Notice that this condition depends only on $\mathscr{H}$ and $\check{g}$ and *not* on the family of metrics $\hat{g}_b$ on the fibres (indeed, the tensor $\mathscr{V}(D_X A)_Y Z - T_{A_Y Z} X$ for horizontal vectors X, Y, Z, depends only on $\mathscr{H}$ and $\check{D}$).

This notion was introduced by Yang and Mills in physics and is now intensely studied, both in mathematical physics and in pure mathematics (see for example [At-Hi-Si] and [Don]). It is also important for Einstein Riemannian submersions, as we will see below in 9.73 and following.

Now a straightforward computation gives

9.36 Proposition. *Let $r, \check{r}, \hat{r}$ be the Ricci curvatures of the Riemannian metrics $g, \check{g}$ and $\hat{g}_b$, respectively. Then*

(9.36a) $r(U,V) = \hat{r}(U,V) - (N, T_U V) + (AU, AV) + (\tilde{\delta}T)(U,V),$

(9.36b) $r(X,U) = ((\mathring{\delta}T)U, X) + (D_U N, X) - ((\check{\delta}A)X, U) - 2(A_X, T_U),$

(9.36c) $r(X,Y) = \check{r}(X,Y) - 2(A_X, A_Y) - (TX, TY) + \frac{1}{2}((D_X N, Y) + (D_Y N, X)),$

where $\check{r}$ is the (horizontal) symmetric 2-form on M such that

$$\check{r}(X,Y) = \check{r}(\pi_* X, \pi_* Y).$$

For some applications to the construction of manifolds with positive Ricci curvature, see [Che], [Nas], [Poo 2], [BéBer 6].

9.37 Corollary. *Let $s, \check{s}, \hat{s}$ be the scalar curvatures (at x or b) of the Riemannian metrics $g, \check{g}$ and $\hat{g}_b$ respectively. Then*

(9.37) $$s = \check{s} \circ \pi + \hat{s} - |A|^2 - |T|^2 - |N|^2 - 2\check{\delta}N.$$

Here we have written

$$|A|^2 = \sum_i (A_{X_i}, A_{X_i}) = \sum_j (AU_j, AU_j),$$

$$|T|^2 = \sum_i (TX_i, TX_i) = \sum_j (T_{U_j}, T_{U_j}).$$

For an application of this formula to the scalar curvature of homogeneous spaces, see [BéBer 1].

E. Completeness and Connections

9.38. We consider now the *global* structure of Riemannian submersions. We assume that M is *connected* and π onto, so B is also connected, but the fibres are not necessarily connected.

In this paragraph, a *path* is a piece-wise smooth map from some interval with everywhere nonzero derivative. A path in M is called *horizontal* if its derivative is everywhere horizontal. Given a path γ in B, a *horizontal lift* of γ is any horizontal path c in M such that $\pi \circ c = \gamma$. (see Figure 9.38).

9.39 Definition (C. Ehresmann [Ehr 1]). Let $\pi: M \to B$ be a submersion and $\mathscr{H}$ a distribution on M which is supplementary to the vertical distribution $\mathscr{V} = \mathrm{Ker}(\pi_*)$.

We say that $\mathscr{H}$ *is* (Ehresmann-)*complete* or that $\mathscr{H}$ *is a* (Ehresmann-)*connection* if, for any path γ in B with starting point $b \in B$, and any x in $F_b = \pi^{-1}(b)$, there exists a horizontal lift c of γ starting from x.

9.40 Theorem (C. Ehresmann [Ehr 1]). *If $\mathscr{H}$ is complete, the submersion π is a locally trivial fibration.*

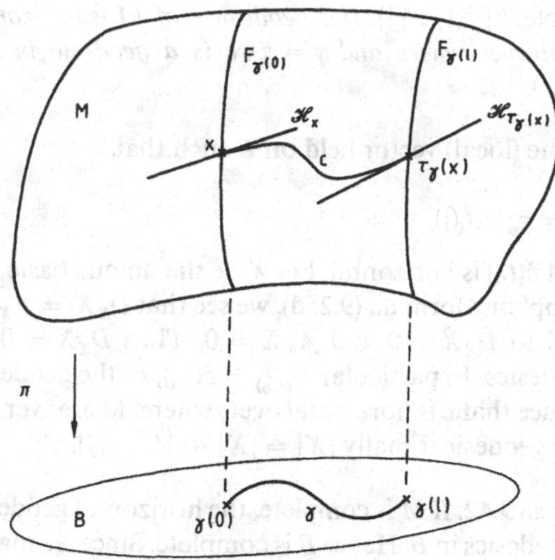

Fig. 9.38

Sketch of the proof. For any path $\gamma\colon [0,l] \to B$ and any x in $F_{\gamma(0)}$, we *denote* by $\tau_\gamma(x)$ the endpoint of the unique horizontal lift c of γ starting from x (i.e., $c(0) = x$, $c(l) = \tau_\gamma(x)$). We easily see that τ_γ is a diffeomorphism from $F_{\gamma(0)}$ to $F_{\gamma(l)}$ (see figure 9.38).

We obtain a local trivialization of the fibration by taking a covering of B by strictly convex open balls U_i (for some Riemannian metric) and in each U_i, by considering the horizontal lifts of the geodesics starting from the central point. □

9.41 Definition. The *holonomy group* G_b of the connection $\mathscr{H}$ at the point b of B is the group of all diffeomorphisms τ_σ of F_b corresponding to *closed* paths σ in B starting at b.

Notice that for any path γ, the groups $G_{\gamma(0)}$ and $G_{\gamma(l)}$ are conjugated by τ_γ.

9.42 Theorem (R. Hermann [Her]). *Let $\pi\colon M \to B$ be a Riemannian submersion. If (M,g) is* complete, *then $\mathscr{H}$ and B are complete and π is a locally trivial fibration.*

In order to prove this theorem, we need some preliminary remarks on basic vector fields and geodesics.

9.43. Let X be a basic vector field on M which is π-related to $\check{X}$ on B. We denote by φ_t (respectively $\check{\varphi}_t$) the 1-parameter (semi-) group of (local) diffeomorphisms generated by X (respectively $\check{X}$). Integrating $\pi_*(X) = \check{X}$, we get

$$\pi \circ \varphi_t = \check{\varphi}_t \circ \pi \text{ for each } t.$$

In particular, φ_t sends any fibre F_b into the fibre $F_{\check{\varphi}_t(b)}$. More precisely, the path $c(t) = \varphi_t(x)$ (whose derivative is X) is exactly the horizontal lift (starting from x) of the path $\gamma(t) = \check{\varphi}_t(\pi(x))$ (whose derivative is $\check{X}$). In the particular case of geodesics, we get

9.44 Lemma (B. O'Neill [ONe 1]). *If a geodesic c of M is horizontal at some point, then it is horizontal everywhere and $\gamma = \pi \circ c$ is a geodesic in B (with the same arc-length parameter).*

Proof. Let $\check{X}$ be some (local) vector field on B such that

(i) $\check{D}_{\check{X}}\check{X} = 0$

(ii) $\check{X}(\pi \circ c(t_0)) = \pi_*(\dot{c}(t_0))$,

where t_0 is such that $\dot{c}(t_0)$ is horizontal. Let X be the unique basic vector field which is π-related to $\check{X}$. Applying formula (9.25d), we see that $D_X X = A_X X + \mathscr{H} D_X X$. But $\mathscr{H} D_X X$ is π-related to $\check{D}_{\check{X}}\check{X} = 0$ and $A_X X = 0$. Thus $D_X X = 0$ and the integral curves of X are geodesics. In particular $X_{c(t_0)} = \dot{c}(t_0)$, so the geodesic c is an integral curve of X. We deduce that c is horizontal everywhere. Moreover $\pi \circ c$ is an integral curve of $\check{X}$, hence a geodesic. Finally $|X| = |\check{X}|$ so $|\dot{c}| = |\dot{\gamma}|$. □

9.45 *Proof of Theorem* 9.42. If M is complete, the horizontal geodesics are obviously horizontal lifts of geodesics in B. Hence B is complete. Since we may lift any geodesic of B, by a limiting process (that we will not detail here), we may lift any path of B from any starting point. Therefore $\mathscr{H}$ is complete and we apply 9.40. □

9.46 Remark. If M is complete, any fibre F_b is complete for the induced Riemannian metric $\hat{g}_b$, since F_b is a closed submanifold of M.

Unfortunately, it may happen that $(B, \check{g})$, $\mathscr{H}$ and all the $(F_b, \hat{g}_b)$ be complete, but M is not complete. If we want M to be complete, we need additional conditions. One such sufficient condition is that, for any point b in B, there exists a neighborhood W and a constant K such that $\tau_\gamma^* \hat{g}_{\gamma(1)} \geqslant K \hat{g}_b$ for any curve γ in W starting at b. Such a condition is satisfied for example in the case of warped products.

9.47 Structural Groups. Very often, a locally trivial fibre bundle $\pi: M \to B$ is described by a family of local trivializations, i.e., a covering of B by open sets $(U_i)_{i \in I}$ and associated diffeomorphisms

$$\varphi_i: \pi^{-1}(U_i) \to U_i \times F$$

satisfying $p_1 \circ \varphi_i = \pi \restriction \pi^{-1}(U_i)$, where p_1 is the projection onto the first factor.

When $U_i \cap U_j \neq \varnothing$, the diffeomorphism

$$\varphi_{ij} = \varphi_i \circ \varphi_j^{-1} \restriction [(U_i \cup U_j) \times F]: (U_i \cap U_j) \times F \to (U_i \cap U_j) \times F$$

satisfies $\varphi_{ij}(x, y) = (x, \psi_{ij}(x)(y))$ where $\psi_{ij}(x)$ is some diffeomorphism of F.

We say that a subgroup G of the group of diffeomorphisms of F is a *structural group* of π (for the given trivialization) if $\psi_{ij}(x)$ belongs to G for any i, j and x in $U_i \cap U_j$.

Notice that we may always enlarge G, or conjugate it in the group of diffeomorphisms of F (see [Nom 3] for some details).

Then, from the proof of Theorem 9.40, it follows

9.48 Proposition (J.A. Wolf [Wol 1]). *Let $\mathscr{H}$ be a Ehresmann-connection for $\pi: M \to B$, with holonomy group G_b at b in B. Then G_b is a structural group of π.*

Consequently, any subgroup G of Diff(F) such that G_b is conjugated to a subgroup of G is a structural group of π. So we set

9.49 Definition. Let $\mathscr{H}$ be a Ehresmann-connection for $\pi: M \to B$ and F_b the fibre at b in B. Let G be a subgroup of Diff(F_b).

We say that $\mathscr{H}$ is a *G-connection* if G_b is conjugate in Diff(F_b) to a subgroup of G.

This notion of a G-connection is a generalization of the notions of principal, associated or linear connections. We indicate briefly the links between these various definitions.

9.50 Example 1: Products. If there exists on $\pi: M \to B$ a connection $\mathscr{H}$ with structural group $G = \{Id_F\}$ (i.e., G is reduced to the identity of F), then M is (diffeomorphic to) the product $B \times F$ and π is the projection onto the first factor. Moreover, in the canonical decomposition $TM = TB \times TF$, then $\mathscr{H}$ coincides with TB, hence $\mathscr{H}$ is integrable and A vanishes. So we are in the situation described in 9.10. (Notice that the induced metrics on the fibres $F_b = \{b\} \times F$ may vary with b.)

9.51 Example 2: Linear Connections. We recall that a vector bunde $\pi: M \to B$ is a locally trivial fibre bundle whose fibre is a vector space F and whose structural group is $Gl(F)$.

Let $\mathscr{H}$ be a $Gl(F)$-connection. Let X be a tangent vector of B at b, and s a section of π. For any curve γ tangent to X at b (with $\gamma(0) = b$), we consider the path γ_t restriction of γ to $[0, t]$. Then

(9.51) $$\nabla_X s = \frac{d}{dt}(\tau_{\gamma_t}^{-1}(s \circ \gamma(t)))\bigg|_{t=0}$$

does not depend on γ and defines a covariant derivative ∇ on π.

Conversely, given a covariant derivative ∇ on π and a path γ in B, we may consider the sections s of π which are parallel along γ (i.e., $\nabla_{\dot\gamma} s = 0$). We define a distribution $\mathscr{H}$ by taking as horizontal vectors at any point x in M the tangent vectors to parallel sections (along a curve γ down in B) through that point.

9.52. *Thus a $Gl(F)$-connection on a vector bundle is equivalent to a covariant derivative.*

Notice that if the vector bundle is Euclidean, complex or Hermitian (corresponding to structural groups $O(F)$, $Gl(F, \mathbb{C})$ or $U(F)$), then $\mathscr{H}$ is an $O(F)$-connection (respectively a $Gl(F, \mathbb{C})$-connection or an $U(F)$-connection) if and only if it corresponds to a covariant derivative which is metric (respectively complex or Hermitian).

9.53 *Curvature.* Observe that, given a submersion $\pi: M \to B$ and a distribution $\mathscr{H}$ which is supplementary to $\mathrm{Ker}(\pi_*) = \mathscr{V}$, we may always define the obstruction tensor field to integrability

$$A: \mathscr{H} \times \mathscr{H} \to \mathscr{V} \quad \text{through}$$

(9.53a) $$A_X Y = \tfrac{1}{2}\mathscr{V}[X, Y]$$

(for vector fields X, Y in $\mathscr{H}$).

When $\mathscr{H}$ is a $Gl(F)$-connection on a vector bundle, A is related to the curvature of the associated covariant derivative by the formula

(9.53b) $$(A_X Y)_x = -\tfrac{1}{2} R^{\nabla}(\pi_*(X), \pi_*(Y))_x$$

for any x in M, and X, Y in $\mathscr{H}_x$ (as usual, we identify $\mathscr{V}_x T_x F_{\pi(x)}$ with $F_{\pi(x)}$).

9.54 Example 3: Principal Connections. We recall that a principal bundle is a locally trivial fibre bundle $p: P \to B$ which is the quotient map of the *free* action of a Lie group G on P (on the right by convention). In our setting a principal bundle is a locally trivial fibre bundle whose fibre is a Lie group G and whose structural group is G acting on itself *on the left*.

Now a *principal connection* is a 1-form θ on P with values in the Lie algebra $\mathfrak{g}$ of G such that:

(a) for any α in G, we have

(9.54a) $$\alpha^*\theta = \mathrm{Ad}(\alpha^{-1}) \circ \theta,$$

(b) for any V in $\mathfrak{g}$, we have at each point of P

(9.54b) $$\theta(\hat{V}) = V$$

where the (vertical) vector field $\hat{V}$ generates the one-parameter group of diffeomorphisms of P given by the action of the subgroup $\mathrm{Exp}(tV)$ of G.

Notice that, when V varies in $\mathfrak{g}$, the corresponding $\hat{V}$'s generate $\mathscr{V}$. We deduce that $\mathscr{H} = \operatorname{Ker}\theta$ is a G-connection on P and that conversely, any G-connection $\mathscr{H}$ gives rise to a unique principal connection θ with $\mathscr{H} = \operatorname{Ker}\theta$. In particular, a distribution $\mathscr{H}$ is a G-connection if and only if it is G-invariant. For more details, see [Ko-No 1].

Finally the tensor field A of $\mathscr{H}$ is related to the curvature Ω of θ through

(9.54c) $$(A_X Y)_x = -\tfrac{1}{2}(\theta_x \restriction \mathscr{V}_x)^{-1}(\Omega_x(X, Y)).$$

(We recall that θ induces an isomorphism from $\mathscr{V}_x$ to $\mathfrak{g}$).

9.55 Example 4: Associated Connections. Let $p: P \to B$ be a principal G-bundle and F any manifold on which G acts (on the left). Then G acts *freely* on $P \times F$ (on the right) by

$$P \times F \times G \to P \times F$$
$$(x, y, \alpha) \to (x\alpha, \alpha^{-1}y).$$

The quotient manifold $M = (P \times F)/G$ is usually denoted by $P \times_G F$. The map $(x, y) \mapsto p(x)$ from $P \times F$ to B induces a map $\pi: M \to B$ and we obtain a locally trivial fibre bundle, called the *associated* bundle with fibre F to the principal bundle $p: P \to B$. Notice that any local trivialization of p induces a local trivialization of π with structural group G.

Now any principal connection θ on P induces a G-connection $\mathscr{H}$ on M by taking $\mathscr{H}$ to be the image of $\operatorname{Ker}\theta \times TF$ under the projection $P \times F \to P \times_G F$. Conversely any G-connection $\mathscr{H}$ on M comes from a unique principal connection θ on P (see [Ko-No 1] or [Nom 3]).

F. Riemannian Submersions with Totally Geodesic Fibres

We recall that the fibres of a Riemannian submersion are totally geodesic if and only if $T \equiv 0$ (see 9.26).

9.56 Theorem (R. Hermann [Her]). *Let $\pi\colon M \to B$ be a Riemannian submersion with totally geodesic fibres. Then, for any path $\gamma\colon [0, l] \to B$, the diffeomorphism τ_γ is* isometric (*where it is defined*).

Moreover, if $\mathscr{H}$ is complete, *τ_γ is an* isometry *from $F_{\gamma(0)}$ to $F_{\gamma(l)}$ and for all $b \in B$ the holonomy group G_b is a subgroup of the isometry group of $(F_b, \hat{g}_b)$ and hence a* Lie group.

Proof. If X is basic and U, V vertical,

$$(D_U X, V) = -(X, T_U V) = 0.$$

We denote by $\hat{g}$ the collection of all $\hat{g}_b$'s on M. Its Lie derivative with respect to X satisfies

$$(\mathscr{L}_X \hat{g})(U, V) = (D_U X, V) + (D_V X, U) = 0.$$

So $\hat{g}$ is parallel for basic vector fields. Now τ_γ is (part of) a local one-parameter group of diffeomorphisms associated with a basic vector field (see 9.40), thus τ_γ preserves $\hat{g}$ and the theorem follows. □

9.57 Remark. There exist locally trivial fibre bundles whose structural group cannot be reduced to a Lie group, for any family of local trivializations (this was noticed by J.P. Serre). For example, any principal G-bundle over S^3 is trivial if G is a Lie group (since $[S^3, BG] = \pi_2(G) = 0$, a result due to E. Cartan [Car 11]), whereas S^3 does admit some non trivial sphere bundles (see for example [An-Bu-Ka]).

For these bundles, the total space admits no Riemannian metric such that the projection becomes a Riemannian submersion with totally geodesic fibres.

9.58 Corollary. *Let $\pi\colon M \to B$ be a Riemannian submersion with totally geodesic fibres and X, Y two* basic *vector fields. Then the vertical vector field $A_X Y$ is a Killing vector field along each fibre. More precisely, along the fibre F_b, $A_X Y$ is tangent to the action of the holonomy group G_b.*

Proof. This follows from the formula $A_X Y = \frac{1}{2}\mathscr{V}[X, Y]$. □

Notice that the proof of Theorem 9.56 is in fact an equivalence so we have

9.59 Theorem (J. Vilms [Vil]). *Let G be a Lie group, $p\colon P \to B$ a principal G-bundle, F any manifold on which G acts. Let $\pi\colon M \to B$ be the associated bundle with fibre F* (i.e., *$M = P \times_G F$*).

Given a Riemannian metric $\check{g}$ on B, a G-invariant Riemannian metric $\hat{g}$ on F and a principal connection θ on P, there exists one and only one Riemannian metric $\hat{g}$ on M such that π is a Riemannian submersion from (M, g) to $(B, \check{g})$ with totally geodesic fibres isometric to $(F, \hat{g})$ and (complete) horizontal distribution associated with θ.

Moreover, if $(B, \check{g})$ and $(F, \hat{g})$ are complete, then (M, g) is complete.

9.60 Example. Let $\pi: M \to B$ be a Euclidean vector bundle with fibre the Euclidean space F, and ∇ a metric connection for π. Given any Riemannian metric $\check{g}$ on B and any $O(F)$-invariant Riemannian metric $\hat{g}$ on F (for example the canonical metric induced by the Euclidean structure), there exists one and only one Riemannian metric on M as in Theorem 9.59.

We will see many other (homogeneous) examples in Paragraph H.

We now study the Ricci curvature and the Einstein condition. From Formulas 9.36, we get immediately the following

9.61 Proposition. *Let $\pi: M \to B$ be a Riemannian submersion with totally geodesic fibres. Then, (M, g) is Einstein if and only if, for some constant λ, $\check{r}$, $\hat{r}$ and A satisfy,*

(9.61a) $\check{\delta} A = 0$ (*i.e. $\mathscr{H}$ is a Yang-Mills connection*),

(9.61b) $\hat{r}(U, V) + (AU, AV) = \lambda(U, V)$ *for each vertical vectors U and V,*

(9.61c) $\check{r}(X, Y) - 2(A_X, A_Y) = \lambda(X, Y)$ *for each horizontal vectors X and Y.*

Unfortunately, proving the existence of at least one Yang-Mills connection on a given fibre bundle is a difficult problem, and the other two conditions are even more restrictive. For example we deduce immediately

9.62 Corollary. *Let $\pi: M \to B$ be a Riemannian submersion with totally geodesic fibres. If (M, g) is Einstein, then $\hat{s}$ and $|A|^2$ are constant on M, and $\check{s}$ is constant on B.*

Proof. The traces of formulas (9.61b) and (9.61c) give

(9.62a) $$\hat{s} + |A|^2 = \lambda \dim F$$

(9.62b) $$\check{s} \circ \pi - 2|A|^2 = \lambda \dim B.$$

Since $\check{s} \circ \pi$ is constant on each fibre, then $|A|^2$ and $\hat{s}$ are also constant on each fibre. Now any two fibres are isometric so $\hat{s}$ is constant on all of M, and so are $|A|^2$, and $\check{s}$ on B. □

Note that Formulas (9.61) do *not* imply that $\hat{g}$ and $\check{g}$ be necessarily Einstein (see below 9.93).

9.63 A Particular Case. As an illustration, we consider the particular case of a principal G-bundle, with G compact and simple. We take $\hat{g}$ to be the standard Riemannian metric, which is bi-invariant and Einstein (see 7.75). Now Formula (9.54c) easily shows that for U and V in the Lie algebra of G, then $(A\hat{U}, A\hat{V})$ is constant on each fibre and moreover Ad(G)-invariant with respect to U and V, so in particular, for any vertical vector fields W and W', we have

$$(AW, AW') = \mu(W, W'), \quad \text{with} \quad \mu = \frac{|A|^2}{\dim G}.$$

Hence we get the following result, which is the translation in our notations of Palais' treatment of the Kaluza-Klein equations (see [Pal 4]).

9.64 Corollary. *Let $p\colon P \to B$ be a principal G-bundle (with a compact simple Lie group G). Assume that p is a Riemannian submersion with totally geodesic fibres isometric to the standard Riemannian metric on G. Then (P, g) is Einstein if and only if*

(9.64a) *$\mathscr{H}$ is a Yang-Mills connection and $\check{s}$ is constant*

and

(9.64b)
$$\check{r}(X, Y) - 2(A_X, A_Y) = \frac{\hat{s} + |A|^2}{\dim G}(X, Y)$$

for any horizontal vector fields X and Y.

Notice that, because of the G-invariance, Equation (9.64b) is in fact an equation on B.

9.65 S^1-Bundles. Another interesting case is that of a principal S^1-bundle $p\colon P \to B$. Here it is known that the curvature Ω of a principal connection θ is the pull-back to P of a closed 2-form ω on B.

Assume that p is a Riemannian submersion with totally geodesic fibres *of length* 2π. The principal connection associated to θ via 9.54 is such that $\theta(\hat{U}) = 1 \in \mathbb{R}$, where $\hat{U}$ is the unit vertical vector field generating the action of S^1 on P. Then formula (9.54c) becomes

(9.65)
$$A_X Y = -\tfrac{1}{2}\omega(\check{X}, \check{Y})\hat{U}$$

for any horizontal vectors X and Y.

It easily follows that $\check{\delta} A = 0$ if and only if ω is *coclosed* (i.e., $\delta\omega = 0$) on B. Hence Proposition 9.61 becomes

9.66 Corollary. *Let $p\colon P \to B$ be a principal S^1-bundle and a Riemannian submersion with totally geodesic fibres of length 2π. Let ω be the 2-form on B whose pull-back on P is the curvature of the principal connection associated to $\mathscr{H}$. Then (P, g) is Einstein if and only if*

(9.66a) *ω is closed, coclosed and has constant norm,*

(9.66b)
$$\check{r}(X, Y) - \frac{1}{2}(\omega_X, \omega_Y) = \frac{|\omega|^2}{4}(X, Y),$$

for any vectors X, Y on B, where we denote by

$$(\omega_X, \omega_Y) = \sum_i \omega(X, X_i)\omega(Y, X_i)$$

for some orthonormal basis (X_i) on B.

We recall that if ω is closed and coclosed, it is harmonic and that the converse is true if B is compact. We do not know (so far) all the solution of (9.66b). We will give some examples below in 9.76, 9.77 and Add.A.

G. The Canonical Variation

We obtain a very interesting deformation of the metric g of the total space of a Riemannian submersion by changing the relative "scales" of B and F. More precisely

9.67 Definition. Let $\pi: M \to B$ be a Riemannian submersion. We *define* the *canonical variation* g_t of the Riemannian metric g on M by setting

$$g_t \restriction \mathscr{V} = tg \restriction \mathscr{V},$$

$$g_t \restriction \mathscr{H} = g \restriction \mathscr{H},$$

$$g_t(\mathscr{V}, \mathscr{H}) = 0.$$

9.68 Properties. (a) *For every $t > 0$, $\pi: (M, g_t) \to (B, \check{g})$ is a Riemannian submersion with the same horizontal distribution $\mathscr{H}$ for all t, and on each fibre F_b, $(\hat{g}_t)_b = t\hat{g}_b$.*

In particular, vertical, horizontal and basic vector fields are the same for all g_t's.

(b) *If g has totally geodesic fibres, then g_t also has totally geodesic fibres for all $t > 0$.*

This follows in particular from the following formulas, whose proofs are straightforward.

9.69 Lemma. *We denote by $A^t, T^t, \ldots$ the tensorial invariants for g_t and $A, T, \ldots$ those for g $(=g_1)$. Then*

(9.69a) $A^t_X Y = A_X Y$ *and* $A^t_X U = tA_X U$;

(9.69b) $T^t_U V = tT_U V$ *and* $T^t_U X = T_U X$;

(9.69c) $[((D_X A)_Y Z, U)]^t = t((D_X A)_Y Z, U)$;

(9.69d) $[((D_U A)_X Y, V)]^t = t((D_U A)_X Y, V) + (t - t^2)\{(A_X U, A_Y \check{V}) - (A_X V, A_Y U)\}$;

(9.69e) $[((D_X T)_U V, Y)]^t = t((D_X T)_U V, Y)$;

(9.69f) $[((D_U T)_V W, X)]^t = t((D_U T)_V W, X) + (t - t^2)(T_V W, A_X U)$;

(9.69g) $[\check{\delta} A]^t = \check{\delta} A$;

(9.69h) $[((\hat{\delta} T)U, X)]^t = ((\hat{\delta} T)U, X) + (1 - t)(T_U, A_X)$;

(9.69i) $(\tilde{\delta} T)^t = t(\tilde{\delta} T)$;

(9.69j) $N^t = N$.

In particular, the curvature of g_t may be expressed by using only $\check{R}$, $\hat{R}$ and the invariants A, T, DA and DT *for* g.

[Notice that (9.69d) and (9.69f) suggest that $((D_U A)_X Y, V) + (A_X U, A_Y V) -$

$(A_X V, A_Y U)$ and $((D_U T)_V W, X) + (T_V W, A_X U)$ are more interesting tensors than $((D_U A)_X Y, U)$ and $((D_U T)_V W, X)$ alone.]

Here, we give only the Ricci curvature in the case where the fibres are totally geodesic.

9.70 Proposition. *Let $\pi: M \to B$ be a Riemannian submersion with totally geodesic fibres. Then the Ricci curvature r_t and the scalar curvature s_t of the canonical variation g_t are described as follows*

(9.70a) $r_t(U, V) = \hat{r}(U, V) + t^2(AU, AV)$ *for any vertical vectors U and V,*

(9.70b) $r_t(X, Y) = \check{r}(X, Y) - 2t(A_X, A_Y)$ *for any horizontal vectors X and Y,*

(9.70c) $r_t(X, U) = t((\check{\delta} A)(X), U)$,

(9.70d) $s_t = \dfrac{1}{t}\hat{s} + \check{s} \circ \pi - t|A|^2.$

9.71 An Auxiliary Function. We recall that (at least in the compact case), Einstein metrics are critical points of some Riemannian functionals, one of which is the (volume-normalized) total scalar curvature as described in Chap. 4. Hence any Einstein metric among the g_t's must be a critical point for the restriction of that functional to the canonical variation g_t. Using Corollary 9.62, we may assume that $\hat{s}$, $\check{s}$ and $|A|^2$ are *constant* and $|A| \neq 0$. We denote by φ the function on $\mathbb{R}_+$ defined by

$$\varphi(t) = \frac{s_t \operatorname{vol}(M, g_t)^{2/n}}{\operatorname{vol}(M, g)^{2/n}} = t^{(\dim F)/n}\left(\frac{1}{t}\hat{s} + \check{s} - t|A|^2\right). \tag{9.71}$$

Hence any Einstein metric among the g_t must satisfy $\varphi'(t) = 0$. Now the variations of φ are easy to describe and it admits 0, 1 or 2 critical points on $\mathbb{R}_+$. More precisely, an elementary computation gives

9.72 Proposition. *The function φ of* (9.71) *admits on $\mathbb{R}_+$*

(a) *2 critical points if $\hat{s} > 0$, $\check{s} > 0$ and*
$d = (\check{s} \dim F)^2 - 4\hat{s}|A|^2 \dim B(\dim B + 2 \dim F) > 0$;

(b) *1 critical point if*
either $\hat{s} > 0$, $\check{s} > 0$ and $d = 0$,
or $\hat{s} = 0$ and $\check{s} > 0$,
or $\hat{s} < 0$;

(c) *no critical point in the other cases.*

Moreover, at a critical point t we have
$s_t > 0$ *if $\hat{s} \geqslant 0$ or if $\hat{s} < 0$ and $\check{s} > 0$ and $\check{s}^2 + 4\hat{s}|A|^2 > 0$;*
$s_t = 0$ *if $\hat{s} < 0$, $\check{s} > 0$ and $\check{s} + 4\hat{s}|A|^2 = 0$;*
$s_t < 0$ *if $\hat{s} < 0$ and $\check{s} \leqslant 0$ or if $\hat{s} < 0$ and $\check{s} > 0$ and $\check{s}^2 + 4\hat{s}|A|^2 < 0$.*

Hence the graph of φ is one of the following:

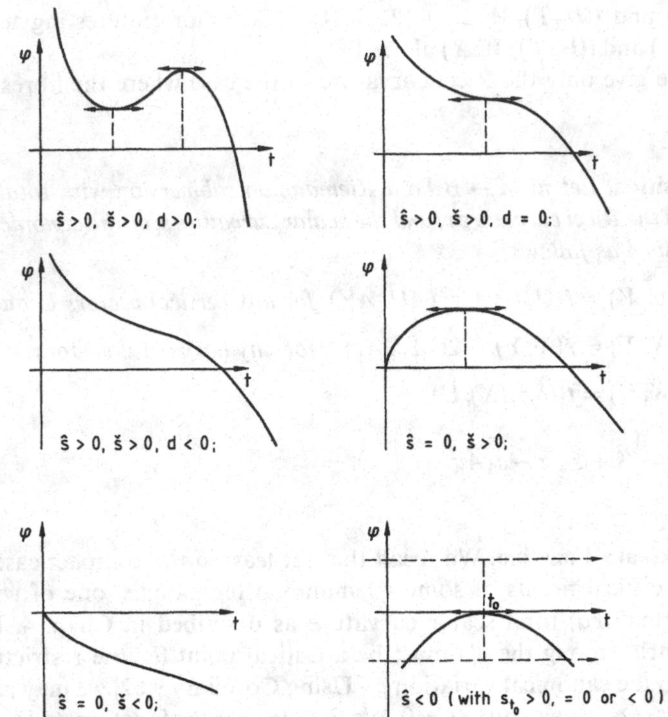

Fig. 9.72

Of course, the critical points of φ do not necessarily give an Einstein metric g_t. But we may characterize the case where *both* give Einstein metrics in case (a) of 9.72. Formulas 9.70 give immediately the following

9.73 Theorem (L. Bérard Bergery, unpublished). *Let $\pi: M \to B$ be a Riemannian submersion with totally geodesic fibres. Then there are* two *different Einstein metrics in the canonical variation if and only if π satisfies the following conditions:*

(9.73a) *$\mathscr{H}$ is a (non-integrable) Yang-Mills connection;*

(9.73b) *$\hat{g}$ is Einstein with positive scalar curvature;*

(9.73c) *$\check{g}$ is Einstein with positive scalar curvature;*

(9.73d) *$|A|^2$ is constant, and moreover*

$$(AU, AV) = \mu(U, V) \quad \textit{for any vertical vectors } U \textit{ and } V$$

$$(A_X, A_Y) = \nu(X, Y) \quad \textit{for any horizontal vectors } X \textit{ and } Y$$

(9.73e) $$(\check{\lambda})^2 - 3\hat{\lambda}(\mu + 2\nu) > 0.$$

(Here $\hat{r} = \hat{\lambda}\hat{g}$ and $\check{r} = \check{\lambda}\check{g}$; note that $|A|^2 = \mu \dim F = \nu \dim B$).

We will use Theorem 9.73 in the next paragraph to give new examples of homogeneous Einstein manifolds, starting from one Riemannian submersion with Einstein total space. We reformulate Theorem 9.73 for that purpose.

9.74 Main Technical Lemma. *Let π: $M \to B$ be a Riemannian submersion with totally geodesic fibres. Assume that g, $\hat{g}$ and $\check{g}$ are Einstein and $A \not\equiv 0$. Then there exists a (unique) positive $t \neq 1$ such that g_t is also Einstein if and only if $0 < \hat{\lambda} \neq \frac{1}{2}\check{\lambda}$.*

Proof. Formulas (9.61) give first that $\mathscr{H}$ is Yang-Mills, so $r_t(X, U) = 0$ for any t. Then

$$(AU, AV) = (\lambda - \hat{\lambda})(U, V) \quad \text{and} \quad 2(A_X, A_Y) = (\check{\lambda} - \lambda)(X, Y),$$

so $\hat{\lambda} < \lambda < \check{\lambda}$ and (9.73a to d) are satisfied. Finally g_t is Einstein if t satisfies $(\check{\lambda} - \hat{\lambda})t^2 - \check{\lambda}t + \hat{\lambda} = 0$. One solution is $t = 1$ and the other $t_0 = \dfrac{\hat{\lambda}}{\check{\lambda} - \hat{\lambda}}$ is positive and $\neq 1$ if and only if $0 < \hat{\lambda} \neq \frac{1}{2}\check{\lambda}$. □

9.75 S^1-Bundles with Compact Einstein Base. In a similar way, even if we do not know the general solution of (9.66b), we may characterize the solution in the special case where the base B is a compact Einstein manifold. Up to a 2-fold covering, we may suppose that p: $P \to B$ is a principal S^1-bundle.

We recall that principal S^1-bundles on a compact connected manifold B are classified by the cohomology group $H^2(B, \mathbb{Z})$. Moreover, for any principal connection θ on P, the cohomology class $[\omega] \in H^2(B, \mathbb{R})$ of the associated closed 2-form ω (see 9.65) satisfies

$$[\omega] = 2\pi\alpha^{\mathbb{R}}, \tag{9.75}$$

where α classifies p in $H^2(B, \mathbb{Z})$ and $\alpha^{\mathbb{R}}$ is the image of α by the universal "change of coefficients" morphism $H^2(B, \mathbb{Z}) \to H^2(B, \mathbb{R})$.

Conversely, for any closed 2-form ω in the cohomology class $2\pi\alpha^{\mathbb{R}}$, there exists a principal connection θ on P whose curvature is $\Omega = p^*\omega$ (see for example [Kob 8]).

9.76 Theorem. *Let $(B, \check{g})$ be a compact Einstein manifold (with $\check{r} = \check{\lambda}\check{g}$), and p: $P \to B$ be a principal S^1-bundle on B, classified by α in $H^2(B, \mathbb{Z})$.*

Then P admits a (unique) S^1-invariant Einstein Riemannian metric g such that p is a Riemannian submersion with totally geodesic fibres if and only if we have either (a) *$\check{\lambda} = 0$, $\alpha^{\mathbb{R}} = 0$ and a finite covering of P is the Riemannian product $B \times S^1$, or* (b) *$\check{\lambda} > 0$ and there exists on B an* almost Kähler structure *$(J, \check{g}, \omega')$ such that $[\omega']$ is a multiple of $\alpha^{\mathbb{R}}$.*

[We recall that an *almost Kähler structure* is an almost Hermitian structure such that ω is closed; it follows automatically that ω is also coclosed, see [Gr-He].]

Proof of (b). With $\check{r} = \check{\lambda}\check{g}$, Equation (9.66b) becomes $(\omega_X, \omega_Y) = \left(2\check{\lambda} - \dfrac{|\omega|^2}{2}\right)(X, Y)$.

If we define J by $\omega(x, y) = \check{g}(Jx, y)$, an elementary calculation shows that a multiple of J becomes an almost complex structure whose associated Kähler form ω' is a multiple of ω.

Conversely, given an almost Kähler structure $(J, \check{g}, \omega)$ such that $[\omega]$ is a multiple of $\alpha^{\mathbb{R}}$, then up to a factor we may take $[\omega] = 2\pi\alpha^{\mathbb{R}}$. Then there exists a principal connection θ on P with curvature $p^*\omega$. Since ω is coclosed, the corresponding connection $\mathscr{H}$ is Yang-Mills. By letting the metric on B be equal to $c\check{g}$ for some constant c, one easily shows that one can satisfy (9.66b) for an appropriate choice of c. □

9.77 Examples. Condition (b) in Theorem 9.76 is satisfied in particular if $(B, \check{g}, J, \omega)$ is a compact Kähler-Einstein manifold with positive scalar curvature and α any rational multiple of $c_1(B)$; indeed $[\omega] = 2\pi c_1(B)^{\mathbb{R}}$. Theorem 9.76 in that case is due to S. Kobayashi [Kob 9]. The (small) generalization given here is due to L. Bérard Bergery. It is conjectured that any compact almost-Kähler Einstein manifold is in fact Kähler (see for example [Gol]).

We give below two examples.

(a) Let $B = \mathbb{C}P^n$ with its canonical (Fubini-Study) Kähler-Einstein metric, and let α be the positive generator of $H^2(B, \mathbb{Z})$; then $P = S^{2n+1}$ with its standard metric and p is the Hopf fibration (see below 9.81);

(b) Let $B = G^2_{n+1}$ be the Grassmannian of oriented 2-planes in $\mathbb{R}^{2n+1}$, viewed as the complex quadric with its symmetric Kähler-Einstein metric. Here, for $\alpha = \dfrac{c_1(B)}{n}$, P is the unit tangent bundle T_1S^n of S^n and admits a homogeneous Einstein metric (Stiefel manifold). Notice that for $n = 3$ or 7, T_1S^3 (respectively T_1S^7) is diffeomorphic to $S^2 \times S^3$ (respectively to $S^6 \times S^7$), but is not a product either as a homogeneous manifold nor as an Einstein manifold.

Theorem 9.76 also applies to non-homogeneous bases, see Add.A, Add.C.

9.78 Remark. Let $(B, \check{g}, J, \omega)$ be an almost Kähler manifold with $\check{g}$ Einstein but $\check{\lambda} < 0$, and $p: P \to B$ be a principal bundle over B, classified by α in $H^2(B, \mathbb{Z})$ with $[\omega] = \mu\alpha^{\mathbb{R}}$ (for some μ). Then there exists on P a unique S^1-invariant Einstein *Lorentz* metric h such that, for the connection $\mathscr{H}$ associated to ω, we have

(a) $h(\mathscr{V}, \mathscr{H}) = 0$,

(b) $h \restriction \mathscr{H}$ is positive definite (and $h \restriction \mathscr{V}$ negative definite),

(c) $p_*(x)$ induces an isometry from $h \restriction \mathscr{H}_x$ to $\check{g}$ on $T_{p(x)}B$, for any x in P.

In a similar way, we may use 9.71 to 9.73 to get Einstein pseudo-Riemannian metrics, with signature $(\dim B, \dim F)$, by looking at negative critical points of φ.

H. Applications to Homogeneous Einstein Manifolds

9.79. A simple construction gives many explicit examples of Riemannian submersions with totally geodesic fibres between homogeneous manifolds. Let G be a Lie group and H, K two compacts subgroups of G with $K \subset H$. Then we have the natural fibration

$$\pi\colon G/K \to G/H$$
$$\alpha K \to \alpha H$$

with fibre H/K (in eK) and structural group H. More precisely, π is the associated bundle with fibre H/K to the H-principal bundle $p\colon G \to G/H$.

Let $\mathfrak{g}$ be the Lie algebra of G and $\mathfrak{h} \supset \mathfrak{k}$ the corresponding subalgebras for H and K. We choose once and for all an $\mathrm{Ad}_G(H)$-invariant complement $\mathfrak{m}$ to $\mathfrak{h}$ in $\mathfrak{g}$, and an $\mathrm{Ad}_G(K)$-invariant complement $\mathfrak{p}$ to $\mathfrak{k}$ in $\mathfrak{h}$. Then $\mathfrak{p} \oplus \mathfrak{m}$ is an $\mathrm{Ad}_G(K)$-invariant complement to $\mathfrak{k}$ in $\mathfrak{g}$. An $\mathrm{ad}_G(H)$-invariant scalar product $(.,.)$ on $\mathfrak{m}$ defines a G-invariant Riemannian metric $\check{g}$ on G/H, and an $\mathrm{Ad}_G(K)$-invariant scalar product on $\mathfrak{p}$ defines a H-invariant Riemannian metric $\hat{g}$ on H/K. Finally, the orthogonal direct sum for these scalar products on $\mathfrak{p} \oplus \mathfrak{m}$ defines a G-invariant Riemannian metric g on G/K. (Notice that there may exist G-invariant Riemannian metrics on G/K which are not obtained by this construction). Then we have

9.80 Theorem (L. Bérard Bergery [BéBer 5]). *The map π is a Riemannian submersion from $(G/K, g)$ to $(G/H, \check{g})$ with totally geodesic fibres isometric to $(H/K, \hat{g})$.*

Proof. Since the submersion π is G-equivariant, it is sufficient to look at π_* in eK. With the identification of 7.22, it is obvious that π is a Riemannian submersion. Now, let V, W be Killing vector fields in p (at the point eK, the vectors V, W are tangent to the fibre H/K). From 7.28, we get at eK

$$D_V W = -\tfrac{1}{2}[V, W]_{\mathfrak{p}\oplus\mathfrak{m}} + U(V, W).$$

Now $[V, W]_{\mathfrak{p}\oplus\mathfrak{m}}$ is in fact in $\mathfrak{p}$ and we claim that $U(V, W)$ is also in $\mathfrak{p}$ (at eK); indeed if X is in $\mathfrak{m}$,

$$(U(V, W), X) = \tfrac{1}{2}\{([X, V]_{\mathfrak{p}\oplus\mathfrak{m}}, W) + (V, [X, W]_{\mathfrak{p}\oplus\mathfrak{m}})\}.$$

But $[X, V]$ belongs to $\mathfrak{m}$, so $U(V, W)_{\mathfrak{m}} = 0$.
Finally, $D_V W$ belongs to $\mathfrak{p}$, so $T \equiv 0$. □

9.81 Examples 1: Hopf Fibrations. Let $G = SU(m+1)$, $H = S(U(1)U(m))$ and $K = SU(m)$. From 7.13–15, we see that the corresponding fibration is

$$\pi\colon S^{2m+1} \to \mathbb{C}P^m,$$

which is called the *Hopf fibration.*

Notice that the standard $SU(m+1)$-invariant metric on S^{2m+1} is *not* Einstein. We choose on S^{2m+1} the canonical metric g with constant curvature 1 (which is $SU(m+1)$-invariant); then π is a Riemannian submersion with totally geodesic fibres onto the symmetric metric $\check{g}$ on $\mathbb{C}P^m$ (with curvature ranging between 1 and 4). Notice that the fibre $S^1 = H/K$ is a great circle of S^{2m+1}, and hence has length 2π.

Here $\hat{r} = 0$, so g is the unique Einstein Riemannian metric in the family g_t. With the notation of 9.72–74, we have $\lambda = 2m$, $\hat{\lambda} = 0$, $\check{\lambda} = 2m + 2$, so $|A|^2 = 2m$ and

$$\varphi(t) = 2mt^{\frac{1}{2m+1}}(2m + 2 - t). \tag{9.81}$$

It is known that these metrics g_t may be realized as "distance spheres" $S(x_0, r) =$

$\{x + d(x, x_0) = r\}$ in the complex projective space $\mathbb{C}P^{m+1}$ (for $t < 1$) or its symmetric dual, the complex hyperbolic space $\mathbb{C}H^{m+1}$ (for $t > 1$) (see [Bo-Ka]).

We consider now the generalizations to the quaternionic and Cayleyan cases.

9.82 Example 2. Let $G = Sp(q+1)$, $H = Sp(1) \cdot Sp(q)$, $K = Sp(q)$. We obtain π: $S^{4q+3} \to \mathbb{H}P^q$ with fibre S^3. We consider g with constant curvature 1 (so $\hat{g}$ has also constant curvature 1) and $\check{g}$ symmetric with curvature ranging between 1 and 4.

Here $\lambda = 4q + 2$, $\hat{\lambda} = 2$, $\check{\lambda} = 4q + 8$, so $|A|^2 = 12q$ and

$$\varphi(t) = 2t^{-\frac{4q}{4q+3}}(3 + 8q(q+2)t - 6qt^2) \tag{9.82}$$

For $t_0 = \dfrac{1}{2q+3}$, the $Sp(q+1)$-invariant Riemannian metric g_{t_0} on S^{4q+3} is *Einstein* (and *not* isometric to g).

9.83 Example 3. Let $G = Sp(q+1)$, $H = Sp(1) \cdot Sp(q)$, $K = U(1)Sp(q)$. We obtain π: $\mathbb{C}P^{2q+1} \to \mathbb{H}P^q$ with fibre S^2. If we take g to be symmetric with curvature ranging between 1 and 4, then also $\check{g}$ is symmetric with curvature ranging between 1 and 4, ad $\hat{g}$ has constant curvature 4.

Here $\lambda = 4q + 4$, $\hat{\lambda} = 4$, $\check{\lambda} = 4q + 8$, so $|A|^2 = 8q$ and

$$\varphi(t) = 8t^{-\frac{2q}{2q+1}}(1 + 2q(q+2)t - qt^2), \tag{9.83}$$

For $t_0 = \dfrac{1}{q+1}$, the $Sp(q+1)$-invariant Riemannian metric on $\mathbb{C}P^{2q+1}$ is Einstein (and is *not* isometric to g).

9.84 Example 4. Let $G = \text{Spin}(9)$, $H = \text{Spin}(8)$, $K = \text{Spin}(7)$ where the inclusion $H \subset G$ is the usual one, but the inclusion $K \subset H$ is the usual one *followed by a* (non-trivial) *triality automorphism* (we recall that such an automorphism of Spin(8) is *outer* and is not the restriction to Spin(8) of any automorphism of Spin(9), so the resulting Spin(7) is different from the usual one in Spin(9), see for example [Bor 2]).

We obtain π: $S^{15} \to S^8$ with fibre S^7. With appropriate choices, we get g, $\hat{g}$ and $\check{g}$ with constant curvature (1 for g and $\hat{g}$, 4 for $\check{g}$).

Here $\lambda = 14$, $\hat{\lambda} = 6$, $\check{\lambda} = 28$, so $|A|^2 = 56$ and

$$\varphi(t) = 14t^{-8/15}(3 + 16t - 4t^2) \tag{9.84}$$

For $t_0 = \frac{3}{11}$, the Spin(9)-invariant Riemannian metric g_{t_0} on S^{15} is Einstein (and is not isometric to g or to that of 9.82 for $q = 3$).

9.85 Remarks. All these examples are called *generalized Hopf fibrations*. As in 9.81, the spheres of 9.82 and 9.84 may be described as distance spheres in appropriate symmetric spaces (see [Wns 2] and [Bo-Ka]). The non-canonical Einstein metrics were found by G. Jensen [Jen 3] for S^{4q+3} in 9.82, W. Ziller [Zil 2] for $\mathbb{C}P^{2q+1}$ in 9.83 and J.P. Bourguignon and H. Karcher [Bo-Ka] for S^{15} in 9.84.

Notice that some of these examples may also appear in different forms with $G = U(m+1)$ or $Sp(1)Sp(q+1)$. In particular, the full (connected) isometry group of g_{t_0} is $U(m+1)$ for S^{2m+1}, $Sp(1)Sp(q+1)$ for S^{4q+3} and $U(1)Sp(q+1)$ for $\mathbb{C}P^{2q+1}$. A careful analysis of all these cases, together with the classification of transitive actions of a Lie group on spheres and projective spaces (see 7.13 and 7.15) shows that if $(G/K, g)$ is a homogeneous sphere or projective space, either G/K is isotropy-irreducible, or g is a metric in the families g_t above, or we are in the case

$$G = Sp(q+1), \qquad H = Sp(1)Sp(q), \qquad K = Sp(q)$$

and the fibres the $Sp(1)$-homogeneous S^3 (non necessarily with constant curvature). Then an explicit computation of the Ricci curvature in this case gives

9.86 Theorem (W. Ziller [Zil 2]). *The homogeneous Einstein Riemannian metrics on spheres and projective spaces are (homothetic to) the canonical ones or the metrics g_{t_0} described above. Hence (up to a scaling factor), S^{15} has 3 homogeneous Einstein Riemannian metrics, $\mathbb{C}P^{2q+1}$ and S^{4q+3} $(q \neq 3)$ have 2 homogeneous Einstein Riemannian metrics and the other ones ($S^{2m}, S^{4q+1}, \mathbb{C}P^{2m}, \mathbb{H}P^q, \mathbb{C}aP^2$) have only one homogeneous Einstein Riemannian metric.*

9.87. More generally, the theory of standard Einstein homogeneous manifold, combined with the techniques of Paragraphs F and G, will give a large number of new examples of Einstein homogeneous manifolds. We begin with some new examples of Einstein metrics on some $M = G/K$ which have already such a metric.

Starting with the assumptions of 9.79, we suppose furthermore that

(a) G is compact and semi-simple.

(b) the standard G-invariant metrics g on G/K and $\check{g}$ on G/H are Einstein.

(c) the H-invariant Riemannian metric $\hat{g}$ on H/K induced by $(-B_G) \upharpoonright \mathfrak{h}$ is also Einstein (where B_G is the Killing form of G).

Then, if $0 < \hat{\lambda} \neq \frac{1}{2}\check{\lambda}$, the Main Technical Lemma 9.74 implies that there also exists a *non-standard* G-invariant Einstein Riemannian metric on G/K.

Of course, $\hat{\lambda} > 0$ as soon as H/K is not a torus. Notice that (c) is satisfied if we have instead

(c′) $B_H = cB_G$, or more generally $B_{H'} = cB_G$ if H is not effective on H/K and H' is the normal subgroup of H acting effectively on H/K. (In particular, (c′) is satisfied e.g. if H (or H') is simple).

Indeed, $\hat{g}$ is a multiple of the standard metric on H/K and 7.92 implies that H/K is Einstein since G/K is Einstein and the isotropy representation of H/K is contained in the isotropy representation of G/K.

9.88. Many such examples can be obtained by choosing G/K from tables 7 in 7.108, or 8 in 7.109 and finding a maximal subgroup H in G with $K \subset H \subset G$. If we replace (c) by (c′)—and assume G is simple—a complete list of such examples can be found in [Wa-Zi 2]. There are several cases where $\hat{\lambda} = \frac{1}{2}\check{\lambda}$. The following special case is particularly interesting.

9.89. Let $K = \{1\}$, $g = -B_G$ and assume that $\mathfrak{h}$ is not an ideal in $\mathfrak{g}$. Then it was shown in [DA-Zi] that the left-invariant metric g_t on G is Einstein for a (unique) $t \neq 1$ if and only if the standard metric on G/H is Einstein and $B_H = cB_G$. Using the tables in 7.102–109, one can easily give a complete list of all such examples in the case where G is simple, see [Wa-Zi 2]. A large family are the examples in 7.95, which all satisfy $B_H = cB_G$. In [DA-Zi] it was shown that g_t is Einstein for a unique $t < 1$ and that the metric g_t is normal homogeneous with respect to the group $G \times H$ acting transitively on G (where H acts by *right* translations).

In the special case where G/H is irreducible symmetric, then $t < 1$ follows immediately, since $\check{\lambda} = \frac{1}{2}$ and $\hat{\lambda} = c\frac{1}{4}$ (see 7.93) and $c < 1$ since $\mathfrak{h}$ is not an ideal in $\mathfrak{g}$. In this special case, the result is due to G. Jensen [Jen 3]. In the general case, the proof of $t < 1$ requires a closer examination of the constants c. As an example, the symmetric space $H \times H/H$ (of type II) gives rise to a left invariant Einstein metric on $H \times H$ which is not isometric to a product metric (this includes the case $S^3 \times S^3$ if $H = SU(2) = S^3$).

9.90. Corollary 9.70 may also be used directly to construct many new homogeneous Einstein metrics. Let $K \subset H \subset G$ as in 9.79 and assume that g, $\check{g}$ and $\hat{g}$ are given by $-B_G$, and that $\check{g}$ and $\hat{g}$ are Einstein.

Then, using the biinvariance of B, it follows easily that

(a) $\check{\delta} A = 0$,

(b) $(AU, AV) = -\frac{1}{4}B_G(U, V) + \frac{1}{4}B_H(U, V)$

(c) $(A_X, A_Y) = \frac{1}{4}(C_{\chi^*,q}X, Y) - \frac{1}{4}(C_{(\chi^* \upharpoonright \mathfrak{k}),q}X, Y)$,

where χ^* is the isotropy representation of G/H and q the restriction of $-B_G$.

Thus $(AU, AV) = \mu(U, V)$ if and only if $B_H \upharpoonright \mathfrak{p} = c(B_G \upharpoonright \mathfrak{p})$ and then $\mu = \dfrac{1-c}{4}$. Similarly $(A_X, A_Y) = \nu(X, Y)$ if and only if $C_{(\chi^* \upharpoonright \mathfrak{k}),q} = a \cdot Id$ and then

$$\nu = \mu \frac{\dim F}{\dim B} = \frac{(1-c)\dim F}{4 \dim B}.$$

Notice that $B_H \upharpoonright \mathfrak{p} = c(B_G \upharpoonright \mathfrak{p})$ is in particular satisfied if $B_H = cB_G$ (e.g., if H is simple) or more generally if $B_{H'} = cB_G$ (where H' is the normal subgroup of H acting effectively on G/K).

Hence, if the standard homogeneous metrics on G/H and H/K are Einstein and $C_{(\chi^* \upharpoonright \mathfrak{k}),q} = aId$, and $B_H = cB_G$ (or $B_{H'} = cB_G$) and if

$$\begin{aligned} d &= \check{\lambda}^{*2} - 4\hat{\lambda}(\mu + 2\nu) \\ &= \left(\frac{1}{4} + \frac{1-c}{2\dim B}\dim H\right)^2 - c(1-c)\left(1 + 2\frac{\dim F}{\dim B}\right)\left(\frac{1}{4} + \frac{1}{2\dim F}\sum_i s_i(1-c_i)\right) \\ &\geqslant 0, \end{aligned}$$

then 9.70 implies there exist one (if $d = 0$) or two (if $d > 0$) Einstein metrics g_t on G/K (one of which could be standard). Again we do not know in which cases the extra condition is satisfied, but there are examples with $d < 0$ or $d = 0$ as well as $d > 0$, see 9.95.

9.91. Assume furthermore that G/H is isotropy irreducible, that $\chi^* \upharpoonright \mathfrak{k}$ is also irreducible, and that the standard homogeneous metric on H/K is Einstein. For example, let $G/H = SO(p+q)/SO(p)SO(q)$ and $K = K_1 \cdot SO(q)$ where $K_1 \subset SO(p)$ is such that the standard metric of $SO(p)/K_1$ is Einstein. Then if $d = \check{\lambda}^2 - 4\hat{\lambda}(\mu + 2\nu) > 0$, there exist two Einstein metrics on G/K.

An interesting example is obtained as follows. For the symmetric space $SO(2n)/U(n)$, the subgroup $SO(2n-1) \subset SO(2n)$ acts also transitively on $SO(2n)/U(n)$ with isotropy subgroup $U(n-1)$. Then we get the fibration

$$SO(2n-2)/U(n-1) \to SO(2n-1)/U(n-1) \to SO(2n-1)/SO(2n-2),$$

where the fibre and the base are symmetric spaces and the total space is diffeomorphic to a symmetric space. Since χ^* is the standard representation of $SO(2n-2)$, $\chi^* \restriction \mathfrak{k}$ is irreducible. Here we have $\check{\lambda} = \frac{1}{2}, c = \frac{2n-4}{2n-3}, \hat{\lambda} = c \cdot \frac{1}{2} = \frac{n-2}{2n-3}$, $\mu + 2\nu = \frac{n-1}{4(2n-3)}$ and hence

$$d = \check{\lambda}^2 - 4\hat{\lambda}(\mu + 2\nu) = \frac{1}{4} - \frac{(n-2)(n-1)}{(2n-3)^2} > 0.$$

One of these Einstein metrics g_t is the symmetric metric on $SO(2n)/U(n)$ and the other one is a new Einstein metric see [Wa-Zi 3].

9.92. In the following special case, one always has $d > 0$. Assume that G/H is isotropy irreducible, K is a non-trivial normal subgroup in H with $H = K \cdot K'$ and K' non-abelian with $B_{K'} = c' B_G$. Then there exist a $t < 1$ and a $t > 1$ such that g_t is Einstein.

Indeed, we have $c_{\chi^* \restriction \mathfrak{k}} = a Id$ and $H/K = K'$ is standard Einstein. If $\psi(t) = \hat{\lambda} - \check{\lambda} t + (\mu + 2\nu)t^2$, then g_t is Einstein if and only if $\psi(t) = 0$. Now $\psi(0) = \hat{\lambda} > 0$ and $\psi(1) = \mu + 2\nu - \check{\lambda} + \hat{\lambda} = \frac{1-c'}{4}\left(1 + 2\frac{\dim K'}{\dim B}\right) - \left(\frac{1}{4} + \frac{1}{2\dim B}[(1-c)\dim K + (1-c')\dim K']\right) + \frac{c'}{4} = -\frac{(1-c)\dim K}{2\dim B} < 0$. Hence there exists t_1 and t_2 with $0 < t_1 < 1 < t_2$ and with $\psi(t_1) = \psi(t_2) = 0$.

In the case where G/H is an irreducible symmetric space, this result is due to G. Jensen [Jen 3] (and in the particular case $G/H = SO(p+q)/SO(p)SO(q)$ even earlier to Sagle [Sag]) and in the general case, it is due to D'Atri and Ziller [DA-Zi]. In [DA-Zi], it was also shown that the metrics g_t are all naturally reductive with respect to the transitive action of $G \times K'$ on G/K.

9.93. We can also obtain many Einstein metrics from the submersion $H/K \to G/K \to G/H$ in the case where G/H is Einstein, but H/K is only a product of Einstein metrics with different scaling factors t_i on each factor of H/K. The Einstein condition then becomes a system of coupled quadratic equations in the $t_{i's}$. In [DA-Zi], all Einstein metrics that one can obtain by this method were found in the case where G/H is isotropy irreducible and K is normal in H, with $H = K \cdot K'$. Since $H/K = K'$, then K' has to be a product of at least two simple Lie groups and hence H has to be a product of at least 3 (respectively 2 if $K = \{1\}$) simple Lie groups. Hence most examples obtained in this way are left invariant metrics on Lie groups (if $K = \{1\}$). A typical example is $G/H = SO(p+q)/SO(p)SO(q)$. By scaling the Killing form B on $SO(p+q)$ with t_1 and t_2 on $SO(p)$ and $SO(q)$, we obtain four Einstein metrics on $SO(p+q)$, one of which is (minus) the Killing form ($t_1 = t_2 = 1$) and the others satisfy $t_1 < 1, t_2 < 1; t_1 > 1, t_2 < 1; t_1 < 1, t_2 > 1$, respectively (if $p = q$, the last two are isometric). On the other hand, for $G/H = Sp(p+q)/Sp(p)Sp(q)$, two of these four solutions are complex if $|p-q|$ is small. Again, all these examples are naturally reductive with respect to the transitive group $G \times K'$.

9.94 *Examples of compact simply-connected G-homogeneous manifolds admitting no G-invariant Einstein metric.* We consider as before a compact semi-simple Lie group G and two compact subgroups $K \subset H$. With the notations of 9.79, we assume furthermore that the two representations of K (by $\mathrm{Ad}_G(K)$) in $\mathfrak{p}$ and $\mathfrak{m}$ are irreducible and non-isomorphic. We deduce that every G-invariant metric on G/K is (up to homothety) in the canonical variation g_t of the Riemannian submersion (with totally geodesic fibres) $\pi: G/K \to G/H$ corresponding to the standard metric, as in 9.80. Since G/H and H/K are obviously isotropy-irreducible they are Einstein manifolds (7.44), and moreover, the conditions (9.73a) to d)) of Theorem 9.73 are satisfied. We have already seen many examples where the quantity $d = (\check{\lambda})^2 - 4\hat{\lambda}(\mu + 2\nu)$ of (9.73e) is positive. But there also exist many examples where d is negative, which, in this case, implies that G/K has no G-invariant Einstein metric. M. Wang and W. Ziller [Wa-Zi 3] give a number of such examples, one of which we describe now.

9.95 Proposition. *Let $G = SO(2n)$, $H = U(n)$ and $K = U(1) \times SO(n)$ where $U(1)$ is the center of $U(n)$ and the embeddings $U(n) \subset SO(2n)$ and $SO(n) \subset SU(n)$ are the canonical ones.*

Then, if $n \geqslant 4$, $SO(2n)/U(1) \times SO(n)$ admits no $SO(2n)$-invariant Einstein metric.

Proof. Notice that G/H and H/K are irreducible symmetric spaces. Hence, for the standard metric, $\check{\lambda} = \frac{1}{2}$ and $\hat{\lambda} = c\frac{1}{2}$, where $B_{SU(n)} = cB_{SO(n)}$. An easy computation gives $c = \frac{n}{2(n-1)}$. As in 9.90, we get

$$d = \check{\lambda}^2 - 4\hat{\lambda}(\mu + 2\nu) = \frac{1}{4} - \frac{(1-c)c}{2}\left(1 + \frac{2\dim F}{\dim B}\right)$$

$$= \frac{1}{4} - \frac{(n-2)(n+1)}{4(n-1)^2}.$$

which is negative when $n \geqslant 4$. □

9.96 Remarks. a) In the case $n = 3$ of the above proposition, we get $d = 0$, and $SO(6)/U(1) \times SO(3)$ is a non-isotropy-irreducible homogeneous space with only one $SO(6)$-invariant Einstein metric (up to homotheties). There are also many examples of this kind.

b) We know that a manifold may admit more than one Lie group acting transitively. But M. Wang and W. Ziller also exhibit a compact simply-connected homogeneous manifold without any homogeneous Einstein metric. We state the result without proof.

9.97 Proposition. *Let $G = SU(4)$, $H = Sp(2)$ (with the standard embedding $Sp(2) \subset SU(4)$) and we consider the (essentially unique) embedding of $K = SU(2)$ as a maximal subgroup of $Sp(2)$. Then the corresponding manifold $G/K = SU(4)/SU(2)$ admits no homogeneous Einstein metric* (*for* any *group acting transitively on it*).

I. Further Examples of Homogeneous Einstein Manifolds

9.98. The methods of paragraph H do not cover all the examples of compact homogeneous Einstein manifolds. Indeed most homogeneous Riemannian manifolds are not *total spaces* of any non trivial Riemannian submersion with totally geodesic fibres (notice that G/K is always the *base* of the submersion $G \to G/K$).

We describe below homogeneous Einstein manifolds which are obtained by different methods.

9.99. Let χ be the isotropy representation of G/K and assume that $\mathfrak{p} = \mathfrak{p}_1 \oplus \cdots \oplus \mathfrak{p}_r$ and $\chi = \chi_1 \oplus \cdots \oplus \chi_r$, where the χ_i's are irreducible *non-equivalent* representations (in $\mathfrak{p}_i$). Then the most general G-invariant Riemannian metric on G/K is given in this case by

$$(.,.) = t_1(-B_G) \restriction \mathfrak{p}_1 + \cdots + t_r(-B_G) \restriction \mathfrak{p}_r.$$

On each factor, $\mathfrak{p}_i$, the corresponding Ricci tensor is a multiple of $(.,.)$, and hence the Einstein condition will consist of r polynomial equations in the r unknowns $t_1, \ldots, t_r$.

In order to illustrate this condition, let us assume that $r = 2$. We may normalize by setting $t_1 = 1$ and $t_2 = t$. Using (7.30) and

$$U(X, Y) = \frac{1-t}{2}[X, Y]_1 + \frac{1-t}{2t}[X, Y]_2$$

(where $[X, Y]_1$ and $[X, Y]_2$ are the components of $[X, Y]$ in $\mathfrak{p}_1$ and $\mathfrak{p}_2$), a direct computation shows that the metric g_t is Einstein if and only if

$$\begin{aligned} (9.99) \qquad & -t^3\left\{\sum_i\left(\frac{1}{2}|[X, X_i]_2|^2 + \frac{1}{4}|[Y, X_i]_1|^2\right)\right\} \\ & + t^2\left\{B_G(C_{\chi_1, -B_G}X, X) + \sum_i\left(\frac{1}{4}|[X, X_i]_1|^2 + |[X, X_i]_2|^2\right)\right\} \\ & - t\left\{B_G(C_{\chi_2, -B_G}Y, Y) + \sum_j\left(|[Y, Y_j]_1|^2 + \frac{1}{4}|[Y, Y_j]_2|^2\right)\right\} \\ & + \sum_j\left(\frac{1}{2}|[Y, Y_j]_1|^2 + \frac{1}{4}|[X, Y_j]_2|^2\right) = 0, \end{aligned}$$

for any X in $\mathfrak{p}_1$ and Y in $\mathfrak{p}_2$ with $B_G(X, X) = B_G(Y, Y) = -1$, where (X_i) (respectively (Y_j)) is an orthonormal basis of $\mathfrak{p}_1$ (respectively $\mathfrak{p}_2$) for $-B_G$ and $| \ |$ is the norm associated to $-B_G$.

Notice that if $[\mathfrak{p}_1, \mathfrak{p}_1] \subset \mathfrak{p}_1 \oplus \mathfrak{k}$ (or if $[\mathfrak{p}_2, \mathfrak{p}_2] \subset \mathfrak{p}_2 \oplus \mathfrak{k}$), then the coefficient of t^3 (or the constant coefficient) vanishes and the equation becomes quadratic. Of course, the metric in this case is again a submersion metric, as in 9.90 with $\mathfrak{h} = \mathfrak{p}_1 \oplus \mathfrak{k}$ (or $\mathfrak{p}_2 \oplus \mathfrak{k}$) and we have seen examples with two, one or no G-invariant Einstein metric.

In the other cases, the coefficient of t^3 is negative and the constant coefficient is positive, hence equation (9.99) always admits a positive solution and G/K has at least one G-invariant Einstein metric. (This was pointed out to the author by M. Wang and W. Ziller.) Notice that all solutions of (9.99) are positive. The author does not know when there are three (or two) solutions.

9.100. Finally, we give a family of examples where $\mathrm{Ad}_G(K)$ has four irreducible components.

Let $G = SU(3)$ and $K = SO(2)$ where K is given by the diagonal matrices in $SU(3)$ with entries $e^{2\pi i k\theta}$, $e^{2\pi i l\theta}$, $e^{-2\pi i(k+l)\theta}$, where k and l are fixed relatively prime integers. Then $M_{k,l} = G/K$ is a 7-manifold with $H^4(M_{k,l}, \mathbb{Z})$ a cyclic group of order $|k^2 + l^2 + kl|$, and hence there exist infinitely many homotopy types among the $M_{k,l}$. Then M. Wang showed the following result (that we will not prove in details)

9.101 Theorem (M. Wang [Wan]). *For every k, l, there exists an $SU(3)$-invariant Einstein metric on $M_{k,l} = SU(3)/SO(2)$ and hence there are infinitely many homotopy types of Einstein metrics in dimension* 7.

To illustrate what needs to be done for the proof of this theorem, let χ be the isotropy representation of G/K. Then $\mathfrak{p}$ has a decomposition $\mathfrak{p} = \mathfrak{p}_0 \oplus \mathfrak{p}_1 \oplus \mathfrak{p}_2 \oplus \mathfrak{p}_3$ with $\dim \mathfrak{p}_0 = 1$ and $\dim \mathfrak{p}_i = 2$ $(i = 1, 2, 3)$, and χ acts irreducibly on each $\mathfrak{p}_i$. Hence a G-invariant metric on M_k is of the form

$$(.,.) = \lambda(-B)\restriction_{\mathfrak{p}_0} + \frac{1}{x}(-B)\restriction_{\mathfrak{p}_1} + \frac{1}{y}(-B)\restriction_{\mathfrak{p}_2} + \frac{1}{z}(-B)\restriction_{\mathfrak{p}_3}.$$

If we normalize $\lambda = 8\pi^2(k^2 + l^2 + kl)^2$, then a direct computation using (7.38) shows that $(.,.)$ is Einstein if and only if

$$\text{(9.101)} \qquad \begin{cases} 3m^2x^2 + 3l^2y^2 + 2k^2z^2 = 6x + 6y - \dfrac{2xy}{z} \\[2ex] 2m^2x^2 + 3l^2y^2 + 3k^2z^2 = 6y + 6z - \dfrac{2yz}{x} \\[2ex] 3m^2x^2 + 2l^2y^2 + 3k^2z^2 = 6x + 6z - \dfrac{2xz}{y} \end{cases}$$

(where $m = -(k + l)$).

Then a topological fixed point argument shows that for every k, l, m, these equations have real solutions with $x, y, z > 0$.

If $k = l = 1$, we can obtain an Einstein metric on M by using 9.90 with $G/H = SU(3)/S(U(1) \times U(2))$ and $K = SO(2)$ the center of H. One can show that for all other values of k and l these Einstein metrics cannot be the total space of a non-trivial Riemannian submersion with totally geodesic fibres and Einstein base. Notice that, because of the use of the topological fixed point argument, we do not get *explicit* solutions.

In a similar way, another "non-effective" method gives the following (recent) result, that we quote without proof.

9.102 Theorem (M. Wang and W. Ziller [Wa-Zi 3]). *Let G be a compact connected Lie group and K a compact subgroup whose Lie algebra* $\mathfrak{k}$ is maximal *in the Lie algebra* $\mathfrak{g}$ *of G. Then the homogeneous space G/K admits a G-invariant Riemannian Einstein metric.*

Indeed, they show that, under the hypothesis above, the "normalized total scalar curvature" (i.e. the scalar curvature when the metric is scaled in such a way that the volume of G/K is 1) is bounded from above and a *proper* function on the space of G-invariant Riemannian metrics on G/K. Then they consider a metric which gives the maximum of that function, and it is Einstein. Notice that there may exist other Einstein metrics, which do not necessarily realize the maximum.

J. Warped Products

9.103. Up to that point, we used only Riemannian submersions with totally geodesic fibres to construct new explicit examples of Einstein metrics. We may also consider for that purpose other special cases among Riemannian submersions. The simplest one is that of warped products (defined in 9.11), but unfortunately, nobody yet has been able to find new examples of *compact* Einstein manifolds in that way. Nevertheless warped products do give new examples of complete Einstein manifolds and the Einstein equations are quite interesting, so we give them below.

In order to illustrate clearly the special properties of warped products among Riemannian submersions, we introduce the "trace-free" part T^0 of T, defined by

$$\text{(9.103)} \qquad \begin{cases} T^0_U V = T_U V - \dfrac{(U,V)}{p} N \\ T^0_U X = T_U X + \dfrac{(N,X)}{p} U \\ T^0_X = 0 \end{cases}$$

where $p = \dim F$.

We deduce easily

9.104 Proposition. *Let* $M = B \times_{f^2} F$ *be the warped product of* $(B, \check{g})$ *and* $(F, \hat{g}_0)$ *by* *the (positive) function f^2 on B as in 9.11. Then the projection $\pi: M \to B$ onto the first factor is a Riemannian submersion whose fibre at b of B is $(F, f(b)^2\hat{g}_0)$.*

Moreover the tensorial invariants of π satisfy

(9.104a) $$A = 0,$$

(9.104b) $$T^0 = 0,$$

(9.104c) $$N \text{ is basic.}$$

And more precisely, N is π-related to the vector field $-\frac{p}{f}\check{D}f$ on B, where $\check{D}f$ is the gradient of f for $\check{g}$.

Conversely, the conditions (9.104a, b, c) *characterize* locally *warped products among Riemannian submersions.*

9.105. As a corollary, all tensors appearing in curvature formulas for a warped product may be expressed by using f and its derivatives (besides $\hat{R}$ and $\check{R}$). We obtain in particular, for any vector E, any horizontal vectors X, Y and any vertical vectors U, V,

(9.105a) $$((D_E T)_U V, X) = \frac{1}{p}(U, V)(D_E N, X),$$

(9.105b) $$(\hat{\delta}T(U), X) = (D_U N, X) = 0,$$

(9.105c) $$(\check{\delta}T)(U, V) = -\frac{1}{p}(U, V)\check{\delta}N,$$

(9.105d) $$(D_X N, Y) = p\left[-\frac{1}{f}\check{D}df(\check{X}, \check{Y}) + \frac{1}{f^2}df(\check{X})df(\check{Y})\right] \circ \pi,$$

(9.105e) $$\check{\delta}N = p\left[\frac{\check{\Delta}f}{f} + \frac{|df|^2}{f^2}\right] \circ \pi,$$

where $\check{X} = \pi_*(X)$, $\check{Y} = \pi_*(Y)$ and $\check{D}df$ (respectively $\check{\Delta}f$) is the Hessian (respectively the Laplacian) of f for $\check{g}$.

Below, we only give the formulas for the Ricci curvature.

9.106 Proposition. *The Ricci curvature r of the warped product $M = B \times_{f^2} F$ satisfies*

(9.106a) $$r(U, V) = \hat{r}_0(U, V) + (U, V)\left[\left(\frac{\check{\Delta}f}{f} - (p-1)\frac{|df|^2}{f^2}\right) \circ \pi\right],$$

(9.106b) $$r(U, V) = 0,$$

(9.106c) $$r(X, Y) = \check{r}(\check{X}, \check{Y}) - \frac{p}{f}\check{D}df(\check{X}, \check{Y}),$$

for any vertical vectors U, V and any horizontal vectors X, Y, with $\check{X} = \pi_*(X)$ and $\check{Y} = \pi_*(Y)$, where $\hat{r}_0$ is the Ricci curvature of $(F, \hat{g}_0)$ and $\check{r}$ the Ricci curvature of $(B, \check{g})$.

Hence the Einstein equations become in this case

9.107 Corollary. *The warped product $M = B \times_{f^2} F$ is Einstein (with $r = \lambda g$) if and only if $\hat{g}_0$, $\check{g}$ and f satisfy*

(9.107a) $(F, \hat{g}_0)$ *is Einstein (with $\hat{r}_0 = \hat{\lambda}_0 \hat{g}_0$),*

(9.107b) $$\frac{\check{\Delta} f}{f} - (p-1)\frac{|df|^2}{f^2} + \frac{\hat{\lambda}_0}{f^2} = \lambda,$$

(9.107c) $$\check{r} - \frac{p}{f}\check{D}df = \lambda \check{g}.$$

9.108. Obviously, (9.107a) gives a condition on $(F, \hat{g}_0)$ alone, whereas (9.107b) and (9.107c) are two differential equations for f on $(B, \check{g})$. It is not known in full generality when these two equations do have a positive solution f on a given Riemannian manifold (even locally). Of course, (9.107b) and (9.107c) admit a (positive) *constant* solution if and only if $\check{g}$ is Einstein with the same sign as $\hat{g}_0$ for the scalar curvature (the equations become $\check{\lambda} = f^2 \hat{\lambda}_0$). In this case, M is an ordinary Riemannian product.

On the other hand, if we are interested in *non-constant* solutions, we may replace the two equations (9.107b) and (9.107c) by the unique equation

(9.108) $$\check{r} - \frac{p}{f}\check{D}df = \frac{1}{2}\left(\check{s} + 2p\frac{\check{\Delta} f}{f} - p(p-1)\frac{|df|^2}{f^2} + p\frac{\hat{\lambda}_0}{f^2} - (p+q-2)\lambda\right)\check{g}$$

where $q = \dim B$ (this was communicated to the author by L. Bérard Bergery).

Indeed, the trace-free part of (9.108) and (9.107c) are the same. Then we recover (9.107b) and the trace of (9.107c) by considering appropriate linear combinations of the trace of (9.108) and the equation obtained by factoring out df in the *divergence* of (9.108). (We recall that $\check{r} - \frac{1}{2}\check{s}\check{g}$ is divergence-free, see 1.94.)

Below, we study Equation (9.108) in the special cases $q = \dim B \leqslant 2$.

9.109 Warped Products with a 1-dimensional Basis. Here we assume $q = \dim B = 1$ (and $p > 1$). Then we may parametrize B by arc-length t with respect to $\check{g}$ and view f as a function of t. Then (9.108) becomes the ordinary (well-known) differential equation

(9.109) $$f'^2 + \frac{\lambda}{p} f^2 = \frac{\hat{\lambda}_0}{p-1}.$$

The behaviour of the solutions depends on the signs of λ and $\hat{\lambda}_0$. Up to homotheties on g, $\hat{g}_0$ or $\check{g}$ (this last one reduces to some affine change on the parameter t), we are reduced to the following set of solutions (besides the constant case):

$\hat{\lambda}_0$	$-(p-1)$	0	$p-1$	$p-1$	$p-1$
λ	$-p$	$-p$	$-p$	0	p
f	$\operatorname{ch} t$	e^t	$\operatorname{sh} t$	t	$\sin t$

We deduce in particular

9.110 Theorem. *Let M be a warped product $B \times_{f^2} F$ with $q = \dim B = 1$ and $p = \dim F > 1$. If M is a complete Einstein manifold, then either M is a Ricci flat Riemannian product, or $B = \mathbb{R}$, F is Einstein with non positive scalar curvature and M has negative scalar curvature.*

Proof. If M is complete, then B is complete too and f has to be defined on the whole of $\mathbb{R}$. Moreover, if B is S^1, f has to be periodic. Then the only possibilities are the products (with $\lambda = \hat{\lambda}_0 = 0$) and the first two solutions ($f = cht$ or e^t) corresponding to the cases where the fibre has negative or zero scalar curvature. □

9.111. Now warped products may appear also as open (dense) submanifolds of (complete) manifolds. An easy example is that of "polar coordinates" in the Euclidean space.

Let $M = \mathbb{R}^n$ with its canonical flat metric, and $U = \mathbb{R}^n - \{0\}$. Then, with the induced metric, U is a warped product $]0, +\infty[\times_{t^2} S^{n-1}$ where $\hat{g}_0$ is the canonical metric on S^{n-1} (the parameter t is the distance to 0 in $\mathbb{R}^n$). This is exactly the meaning of the well-known formula

$$g = dt^2 + t^2\hat{g}_0$$

which expresses the Euclidean metric of $\mathbb{R}^n$ in polar coordinates.

It is well-known that this generalizes immediately to all (symmetric) spaces with constant curvature (in every case, we endow S^{n-1} with its canonical metric). Let $M = S^n$ (with its canonical metric). Then for a given point x (and its antipodal point $-x$), and the induced metric, $U = S^n - \{x, -x\}$ is the warped product $]0, \pi[\times_{\sin^2 t} S^{n-1}$. Similarly if $M = \mathbb{R}P^n$ (with its canonical metric) and if C_x is the cut-locus of x (i.e., the conjugate hyperplane), then $U = \mathbb{R}P^n - (\{x\} \cup C_x)$ is the warped product $\left]0, \frac{\pi}{2}\right[\times_{\sin^2 t} S^{n-1}$.

Finally, if $M = H^n$ (hyperbolic space), then $U = H^n - \{x\}$ is the warped product

$$]0, +\infty[\times_{sh^2 t} S^{n-1}.$$

Notice that this involves the last three solutions of (9.109) and that t is always the distance to some point.

9.112. We may also consider the distance to some hypersurface. This gives some new examples:

(a) Let α be a non-trivial isometry of a complete Ricci-flat manifold $(F, \hat{g}_0)$. On the Riemannian product $\mathbb{R} \times F$, the group $\mathbb{Z}$ of isometries generated by $\bar{\alpha}(t, x) = (t + l, \alpha(x))$ (where l is some positive constant), acts freely and the Riemannian quotient (M, g) of $\mathbb{R} \times F$ by $\mathbb{Z}$ is complete and Ricci-flat. Moreover the projection of $\mathbb{R} \times F$ onto the first factor induces a Riemannian submersion $\pi \colon M \to S^1 = \mathbb{R}/_{l\mathbb{Z}}$, which is *not* a warped product, but which is a product when restricted to any proper connected open subset of S^1.

(b) Similarly, let α and β be two isometries acting freely on a complete Ricci-flat Riemannian manifold $(F, \hat{g}_0)$, satisfying $\alpha^2 = \beta^2 = Id_F$. On the Riemannian product $\mathbb{R} \times F$, the group G of isometries generated by $\bar{\alpha}(t, x) = (-t, \alpha(x))$ and $\bar{\beta}(t, x) = (l - t, \beta(x))$ (where l is some positive constant), acts freely and the Riemannian quotient (M, g) of $\mathbb{R} \times F$ by G is Ricci-flat and complete. Then the open dense subset U of M, projection of $(\mathbb{R} - l\mathbb{Z}) \times F$, is isometric to the product $]0, l[\times F$.

(c) Let α be an isometry acting freely on a complete Einstein manifold $(F, \hat{g}_0)$ with negative scalar curvature, and satisfying $\alpha^2 = Id_F$. By 9.110, the warped product $\mathbb{R} \times_{ch^2 t} F$ is Einstein, and the map $\overline{\alpha}(t, x) = (-t, \alpha(x))$ is obviously an isometry acting freely on it, and satisfies $\overline{\alpha}^2 = Id$. Hence the Riemannian quotient $M = (\mathbb{R} \times_{ch^2 t} F)/\{Id, \overline{\alpha}\}$ is complete and Einstein (with negative scalar curvature). Notice that M is a 1-dimensional (non-trivial) vector bundle over the quotient $F/\{Id, \alpha\}$ and that the open complement U of the zero section in M is isometric to the warped product $]0, +\infty[\times_{ch^2 t} F$.

(d) Finally, we may consider the hyperbolic space H^n as the warped product $\mathbb{R} \times_{e^{2t}} \mathbb{R}^{n-1}$. (This is a special case of 9.110 where $\mathbb{R}^{n-1}$ is endowed with the canonical Euclidean metric). Let α be any isometry of H^n which commutes with the projection pr_1 onto the first factor and induces a translation of length $l \neq 0$ on $\mathbb{R}$. Then the group $\mathbb{Z} = \{\alpha^n, n \in \mathbb{Z}\}$ acts freely on H^n and the Riemannian quotient (M, g) of H^n by $\mathbb{Z}$ is complete and Einstein (with sectional curvature -1). Moreover pr_1 induces a Riemannian submersion $\pi\colon M \to S^1 = \mathbb{R}/_{l\mathbb{Z}}$ which is not a warped product, but which becomes the warped product $]0, l[\times_{e^{2t}} \mathbb{R}^{n-1}$ when restricted to $\pi^{-1}(S^1 - \{0\})$.

Now we may state the following

9.113 Theorem. *Let (M, g) be a complete Einstein manifold containing an open dense subset U which is (for the induced metric) a warped product on a 1-dimensional basis with* complete *fibre F. Then*

either (M, g) is a warped product as in Theorem 9.110,

or (M, g) is one of the preceding examples in 9.111 *and* 9.112.

Proof. We only sketch the proof of this theorem, which seems to be (more or less) well-known. For more details, we refer the reader to the forthcoming work of R.S. Palais, C.L. Terng, and A. Derdzinski, (or to [BéBer 3] where only the special case of "cohomogeneity one" manifolds is considered).

We take U as large as possible. If the base B is complete, then M is a warped product and Theorem 9.110 applies. If not, B is an open interval $]0, l[$ (with eventually $l = +\infty$) and $U =]0, l[\times F$. We easily see that the warping function f has a finite limit at the boundaries at finite distance in B, and that M is then the quotient of $[0, l] \times F$ or $[0, +\infty[\times F$ by appropriate identifications on the boundary. (One way to see that is to look at horizontal geodesics in U.) If $f(0) = 0$, then the fiber $\{t\} \times F$ in U shrinks to one point m in M when t goes to 0. Since, for t small enough, $\{t\} \times F$ is precisely the set of points at distance t from m, we see that F has to be a sphere. Then we use the following well-known result (that we give without proof)

9.114 Lemma. *If we identify $\{x \in \mathbb{R}^n, 0 < |x| < \varepsilon\}$ with $]0, \varepsilon[\times S^{n-1}$ in polar coordinates, the C^∞ Riemannian metric $dt^2 + \varphi(t)^2 \hat{g}_0$ (where t is the parameter on $]0, \varepsilon[$ and $\hat{g}_0$ a metric on S^{n-1}) extends to a C^∞ Riemannian metric on $\{x \in \mathbb{R}^n, |x| < \varepsilon\}$ if and only if $\hat{g}_0$ is λg_{can} where g_{can} is the canonical metric on S^{n-1} and λ some positive constant, and $\frac{1}{\lambda}\varphi$ is the restriction on $]0, \varepsilon[$ of a C^∞ odd function on $]-\varepsilon, \varepsilon[$ with $\frac{1}{\lambda}\varphi'(0) = 1$.*

(See for example [Ka-Wa 5] for references and a generalization).

This gives the various cases of 9.111, according to the behaviour of f in l.

If now f has a non-zero limit at 0 (and also eventually at l), the identification on the boundary of $[0, l] \times F$ or $[0, +\infty[\times F$ can be: *either* an isometry of order two of $\{0\} \times F$ with itself acting freely (and similarly at l if l is finite), and then f has to be the restriction of a C^∞ even function near 0 (and similarly at l); *or* an isometry of $\{0\} \times F$ with $\{l\} \times F$ and then f has to be the restriction of a C^∞ function on R satisfying $f(0)f(t + l) = f(l)f(t)$.

Now a careful check gives the various cases of 9.112. Notice that in the last case (d), we need the following well-known result (which we quote without proof). □

9.115 Lemma. *The complete Riemannian manifolds (M, g) and $(M, \mu g)$ are isometric for some constant $\mu \neq 1$ if and only if (M, g) is isometric to the canonical Euclidean space.*

9.116 Warped Products with a 2-dimensional Basis. In the special case of a warped product $B \times_{f^2} F$ over a 2-dimensional basis, we have (with the notation of 9.104 and following) $\check{r} = \frac{\check{s}}{2}\check{g}$ (and $q = 2$), hence Equation (9.108) simplifies to

$$\text{(9.116)} \qquad \check{D}\,df = -\frac{1}{2}\left(2\check{\Delta} f - (p-1)\frac{|df|^2}{f} + \frac{\hat{\lambda}_0}{f} - \lambda f\right)\check{g}.$$

It is well-known that such an equation admits non-constant solutions only in the special case of a metric $\check{g}$ admitting locally a non-zero Killing field. More precisely, we have (exercise!)

9.117 Lemma. *On a 2-dimensional manifold $(B, \check{g})$ the equation $\check{D}df = \varphi\check{g}$ admits a non-constant solution f if and only if, locally at points where $df \neq 0$, there exist local coordinates (t, u) such that f is a function of t alone and $\check{g} = dt^2 + f'(t)^2\, du^2$ $\left(\text{where } f' = \dfrac{df}{dt}\right)$; then $\varphi = f''$.*

(In the n-dimensional case, this is due to H.W. Brinkmann [Bri].)

With the notations of the lemma, Equation (9.116) becomes (at least locally on B) an ordinary differential equation for f in the variable t

$$\text{(9.117a)} \qquad 2f'' + (p-1)\frac{f'^2}{f} - \frac{\hat{\lambda}_0}{f} + \lambda f = 0.$$

This is easy to integrate once, and gives

$$\text{(9.117b)} \qquad f'^2 = \frac{\hat{\lambda}_0}{p-1} - \frac{\lambda}{p+1}f^2 + cf^{1-p}$$

(where c is some constant).

9.118. We describe below some solutions of equation (9.117b) which give complete Einstein metrics on the (differentiable) product manifold $M = \mathbb{R}^2 \times F$. Here $(F, \hat{g}_0)$ is a complete p-dimensional Einstein manifold with $\hat{r}_0 = \hat{\lambda}_0\hat{g}_0$. On $\mathbb{R}^2$, we consider:

either polar coordinates (t, θ) and a metric $\check{g} = dt^2 + h(t)^2 d\theta^2$, where h is an odd function on $\mathbb{R}$ satisfying $h(t) > 0$ if $t > 0$ and $h'(0) = 1$;
or Cartesian coordinates (u, v) and a metric $\check{g} = du^2 + h(u)^2 dv^2$, where h is a positive function on $\mathbb{R}$.

Finally, we consider on M the warped product metric $\check{g} + f^2\hat{g}_0$, where f is a function of t (or u) alone. (Hence we consider f as a function of one variable.)

(a) Let $\hat{\lambda}_0 = p - 1$ and f be the unique function on $[0, +\infty[$ satisfying $f(0) = 1$, $f' \geqslant 0$ and $f'^2 = 1 - f^{1-p}$.

Then the warped product metric $g = dt^2 + \dfrac{4f'(t)^2}{(p-1)^2} d\theta^2 + f(t)^2\hat{g}_0$ on $M = \mathbb{R}^2 \times F$ is complete and Ricci-flat.

This example is known as the (Riemannian) Schwarschild metric (at least when $F = S^2$). Indeed, if we set $f = r$ and express t as a function of r, we get $g = \dfrac{dr^2}{1 - r^{1-p}} + \dfrac{4(1 - r^{1-p})}{(p-1)^2} d\theta^2 + r^2\hat{g}_0$, which (if $p = 2$) coincides, up to a sign, with the Schwarschild metric in Lorentz geometry (compare with 3.28).

(b) Let $\hat{\lambda}_0 = 0$. Then $g = du^2 + e^{2u} dv^2 + e^{2u}\hat{g}_0$ is a complete Ricci-flat metric on $\mathbb{R}^2 \times F$. Notice that g is a warped product metric both over the base $(\mathbb{R}, du^2)$ and the base $(\mathbb{R}^2, du^2 + e^{2u} dv^2)$.

(c) Let $\hat{\lambda}_0 = 1 - p$ and f be the unique function on $\mathbb{R}$ satisfying $f(0) = 1$, $f' > 0$ and $f'^2 = -1 + f^2 + 2\dfrac{(p-1)^{p-1}}{(p+1)^{p+1}} f^{1-p}$. Then the warped product metric $g = du^2 + f'(u)^2 dv^2 + f(u)^2\hat{g}_0$ on $M = \mathbb{R}^2 \times F$ is complete and Einstein (with $r = -(p+1)g$).

In the examples (b) and (c), notice that the translating along the coordinate v are isometries for g. In particular we get a family of Einstein metrics on $\mathbb{R} \times S^1 \times F$ by taking the quotient of M by the group $\mathbb{Z}$ of isometries generated by one non-trivial translation. Moreover, if F admits free isometries of order 2, we may take further quotients.

(d) Let a be any real number such that $a > 0$ if $\hat{\lambda}_0 \geqslant 0$ and $a > \sqrt{\dfrac{-\hat{\lambda}_0}{p+1}}$ if $\hat{\lambda}_0 < 0$. Let b be such that $\dfrac{1}{b} = \dfrac{p+1}{2}a + \dfrac{\hat{\lambda}_0}{2a}$, and let f be the unique function on $[0, +\infty[$ such that $f(0) = a$, $f' \geqslant 0$ and $f'^2 = \dfrac{\hat{\lambda}_0}{p-1} + f^2 - \left(a^{p+1} + \dfrac{\hat{\lambda}_0}{p-1} a^{p-1}\right) f^{1-p}$.

Then the warped product metric $g = dt^2 + b^2 f'(t)^2 d\theta^2 + f(t)^2\hat{g}_0$ on $M = \mathbb{R}^2 \times F$ is complete and Einstein with $r = -(p+1)g$.

Notice that, if a varies, the corresponding g's form a one-parameter family of non-isometric Einstein metrics (with the same constant).

In the special case $\hat{\lambda}_0 = 1 - p$ and $a = 1$, we get $f = cht$ and $\check{g}$ on $\mathbb{R}^2$ is the metric $dt^2 + sh^2 t\, d\theta^2$ with constant curvature -1.

All these examples are described in [BéBer 3] (where they were studied in the special case of an homogeneous fibre F).

Now we state (without proof) the following

9.119 Theorem. *Let M be a complete Einstein manifold which is a warped product* *over a 2-dimensional basis. Then either M is a Riemannian product or M is one of the examples in* 9.118.

We refer to the forthcoming work of R.S. Palais, C.L. Terng and A. Derdzinski for this theorem and also for the more general case where only an open dense subset of M is a warped product over a 2-dimensional basis. Unfortunately, in that way we do not obtain any new example of *compact* Einstein manifold.

9.120. We sketch only one example. Let M be $\mathbb{R}^n$, S^n or H^n (with their canonical metric) and C a complete geodesic in M. Then $U = M - C$ is isometric to the warped product $N \times_{f^2} S^{n-2}$, with the canonical metric on S^{n-2}, where N is one connected component of $\mathbb{R}^2$, S^2 or H^2 (respectively) minus a complete geodesic and f is t, $\sin t$ or sht (respectively) where t is the distance to that geodesic.

9.121. As a final remark, we want to emphasize that warped products appear quite naturally as open dense subsets of complete manifolds in the following situation. Let (M, g) be a complete Riemannian manifold and G a closed (connected) subgroup of the group of isometries of M, such that the principal orbits of the action of G on M are *isotropy-irreducible* G-homogeneous spaces F. (Notice that this implies that non-principal orbits can be only $\mathbb{Z}_2$-quotients of F, or perhaps points in the special case where F is a sphere.) We saw in 9.12 that the union U of all principal orbits is an open dense subset of M and the quotient map $U \to U/_G$ becomes a Riemannian submersion for the restriction of g to U and the quotient metric on $U/_G$. Now, from the isotropy-irreducibility of the principal orbits follows easily the following

9.122 Proposition. *The Riemannian submersion $U \to U/G$ is, at least locally on U/G, a warped product.*

From Proposition 9.122, it follows that the preceding studies give a classification of complete Einstein manifolds whose isometry group has isotropy-irreducible principal orbits of codimension one or two. Roughly speaking, such manifolds are either homogeneous or one of the above examples (with F homogeneous). One easily sees that this gives also a classification of complete Einstein manifolds whose isometry group has codimension one principal orbits which are products of $\mathbb{R}$ or S^1 with an isotropy irreducible space. (See [BéBer 3] for a detailed study of the codimension one case).

K. Examples of Non-Homogeneous Compact Einstein Manifolds with Positive Scalar Curvature

9.123. Until recently, the only known examples of compact Einstein manifolds with positive scalar curvature were homogeneous. The first non-homogeneous example was an Einstein metric on the connected sum $\mathbb{C}P^2 \# \overline{\mathbb{C}P}^2$ of two copies of $\mathbb{C}P^2$ with opposite orientation. This was discovered by D. Page (in [Pag]), as a Riemannian dual of the Taub-NUT metric of Lorentz geometry (through a computation in local coordinates). Then, L. Bérard Bergery has given (in [BéBer 3]) a construction which gives a family of examples in any even dimension and generalizes Page's example. We describe below L. Bérard Bergery's construction and slightly generalize it. The somewhat technical parts of the proof are only sketched. (We refer to [BéBer 3] for more details.)

9.124. Let $(B, \check{g}, J, \omega)$ be a compact almost-Kähler Einstein manifold with positive scalar curvature as in 9.76. We let $n-2$ be the dimension of B (with n even) and we normalize $\check{g}$ such that the Ricci curvature satisfies $\check{r} = n\check{g}$ (this is the case for example for the canonical complex projective space with sectional curvature between 1 and 4). As in 9.76, we assume furthermore that the cohomology class $[\omega]$ of ω in $H^2(B, \mathbb{R})$ is proportional to the real image $\alpha^{\mathbb{R}}$ of an indivisible integral cohomology class α in $H^2(B, \mathbb{Z})$, and we define the positive number q to be such that $n[\omega] = 2\pi q \alpha^{\mathbb{R}}$.

Among the known examples of such manifolds are the compact simply-connected Hermitian symmetric spaces. Here q is the unique positive *integer* such that $c_1(B) = q\alpha$ in $H^2(B, \mathbb{Z})$, with α indivisible. This q is computed in [Bo-Hi] in the irreducible case. We get the following table:

G/K	q	G/K	q
$SU(p+p')/S(U(p)U(p'))$	$p+p'$	$SO(2p)/U(p)$	$2p-2$
$Sp(p)/U(p)$	$p+1$	$SO(p+2)/SO(p)SO(2)$	p
$E_6/\mathrm{Spin}(10)SO(2)$	12	$E_7/E_6SO(2)$	18

Notice that in the Kähler case, S. Kobayashi and T. Ochiai have shown in [Ko-Oc] that q is always smaller than $n/2$ with equality only if B is the canonical complex projective space $\mathbb{C}P^{(n-2)/2}$.

We denote by $P(s)$ the (total space of the) S^1-bundle over B classified by $s\alpha$ in $H^2(B, \mathbb{Z})$, where s is any positive integer (notice that $P(s)$ is nothing but $P(1)/\mathbb{Z}_s$). Finally, we denote by $M(s)$ the S^2-bundle over B associated with $P(s)$ for the usual action of $S^1 = SO(2)$ over S^2 (by rotations around N-S axis). Then we get the following

9.125 Theorem (L. Bérard Bergery [BéBer 3]). *With the above notation, if* $1 \leqslant s < q$, *then* $M(s)$ *admits an Einstein metric with positive scalar curvature.*

Proof. We only give the main outline of the proof and we refer to [BéBer 3] for the missing technical details.

We may consider $M(s)$ as the quotient manifold of the product $[0, l] \times P(s)$ under the identifications on the two boundaries given by the projection $p\colon P(s) \to B$. (Notice that the fibre is S^1, so we get indeed a manifold without boundary.) On $[0, l] \times P(s)$, we consider the Riemannian metric (singular at the boundaries) $dt^2 + g(f(t), h(t))$, where t is the variable along $[0, l]$, f and h are two functions on $[0, l]$, and $g(a, b)$ is the two-parameter family of metrics on $P(s)$ such that p is a Riemannian submersion of $(P(s), g(a, b))$ over $(B, b\check{g})$ with totally geodesic fibre S^1 of length $2\pi a$, and horizontal distribution associated to the principal connection with curvature $p^*(\omega)$. (See the end of 9.75.)

We first remark that $dt^2 + g(f(t), h(t))$ induces a (non-singular) Riemannian metric on $M(s)$ if and only if the functions f and h satisfy:

(a) f is positive on $]0, l[$, $f'(0) = 1 = -f'(l)$ and f is "odd" at 0 and l, i.e., f is the restriction of an odd function on $\mathbb{R}$ satisfying $f(l+t) = -f(l-t)$;

(b) h is positive on $[0, l]$ and "even" at 0 and l, i.e., h is the restriction of an even function such that $h(l+t) = h(l-t)$. This is similar to the characterization of metrics in polar coordinates (Lemma 9.114).

It is easy to compute the Ricci curvature of $g(a, b)$ using (9.65) and to compute the Ricci curvature of $dt^2 + g(f(t), h(t))$, using the fact that the projection onto the first factor is a Riemannian submersion with $A = 0$. We only give the final result. The Riemannian metric $dt^2 + g(f(t), h(t))$ is Einstein (with constant λ) if and only if f and h satisfy the following three ordinary differential equations:

(1) $$-\frac{f''}{f} - (n-2)\frac{h''}{h} = \lambda,$$

(2) $$-\frac{f''}{f} - (n-2)\frac{f'h'}{fh} + \frac{(n-2)n^2s^2f^2}{4q^2h^4} = \lambda,$$

(3) $$-\frac{h''}{h} - \frac{f'h'}{fh} - (n-3)\frac{h'^2}{h^2} + \frac{n}{h^2} - \frac{n^2s^2f^2}{2q^2h^4} = \lambda.$$

Two linear combinations of these equations are especially interesting. Considering (2) − (1) and (2) − (1) + $(n-2)$(3) and dividing by $(n-2)$, we get:

(4) $$\frac{h''}{h} - \frac{f'h'}{fh} + \frac{n^2s^2f^2}{4q^2h^4} = 0 \quad \text{and}$$

(5) $$-2\frac{f'h'}{fh} - (n-3)\frac{h'^2}{h^2} - \frac{n^2s^2f^2}{4q^2h^4} + \frac{n}{h^2} = \lambda.$$

Integrating (4) gives $f = \dfrac{|2qhh'|}{ns\sqrt{1-ah^2}}$, where a is some constant. In order to get a solution satisfying (a) and (b), we must take $a > 0$ and, up to a homothetical change of g, we may choose $a = 1$. We introduce the (increasing) function φ such that $f = \dfrac{2q\varphi'}{ns}$ and $h = \sqrt{1-\varphi^2}$. Then (5) becomes

(6) $$\frac{2\varphi\varphi''}{1-\varphi^2} - (n-3)\frac{\varphi^2\varphi'^2}{(1-\varphi^2)^2} - \frac{\varphi'^2}{(1-\varphi^2)^2} + \frac{n}{1-\varphi^2} = \lambda.$$

Integrating once gives

(7) $$\varphi'^2 = 1 - \varphi^2 + \frac{((n-1)-\lambda)P(\varphi) + b\varphi}{(1-\varphi^2)^{(n/2)-1}},$$

where b is some constant and P is the even polynomial such that $P(0) = 1$ and $P''(\varphi) = n(1-\varphi^2)^{(n/2)-1}$. Here we must take $b = 0$ to get a solution.

Now there exist two positive numbers λ and x, unique, such that $(1-x^2)^{n/2} + ((n-1)-\lambda)P(x) = 0$ and $-\frac{nsx}{q} + n = \lambda(1-x^2)$. Then we take for φ the unique solution of (7) (with $b = 0$) such that $\varphi(0) = -x$, and for l the integral $2\int_0^x \sqrt{1-y^2 + \frac{((n-1)-\lambda)P(y)}{(1-y^2)^{(n/2)-1}}}\,dy$. We have $\varphi(l) = x$ and it follows that f and h satisfy (a) and (b). □

9.126 Remarks. (a) In the special case $n = 4$, the only possibility is $B = S^2$, hence $q = 2$ and we obtain only one example (for $s = 1$). This is precisely D. Page's example.

(b) The manifold $M(s)$ admits always an almost-complex structure, which is complex when B is Kähler. The Einstein metric of Theorem 9.125 is Hermitian, but *not* Kähler (there are other metrics which are Kähler on $M(s)$ if B is Kähler, but they cannot be Einstein).

(c) If B is G-homogeneous (as in Chapter 8), then G acts on $M(s)$ by isometries and the principal orbits have codimension one, but the Einstein metric on $M(s)$ is never homogeneous.

(d) The antipodal map on S^2 commutes with the action of $SO(2)$, and induces a free isometry (of order two) on $M(s)$. Hence $M(s)$ has a Riemannian $\mathbb{Z}_2$-quotient. In the special case of D. Page's example, we get an Einstein metric on the connected sum of $\mathbb{C}P^2$ with $\mathbb{R}P^4$.

(e) A closer look at equations (1), (2) and (3) above shows that there is no other solution satisfying (a) and (b), in particular there is no solution if $s \geqslant q$. In the special case where B is S^2, then $M(s)$ is diffeomorphic to $S^2 \times S^2$ if s is even and $\mathbb{C}P^2 \# \overline{\mathbb{C}P}^2$ if s is odd. But the equations give only one solution for $\mathbb{C}P^2 \# \overline{\mathbb{C}P}^2$ (the product metric on $S^2 \times S^2$ corresponds to $M(0)$ in some sense). In the general case, we may deduce from that a characterization of these examples. We will not do it here, and we consider only the following special case (that we quote without proof, but all ingredients to obtain it have already been given)

9.127 Theorem (L. Bérard Bergery [BéBer 3], A. Derdzinski [Der 3]). *Let M be a compact 4-dimensional Einstein manifold. Assume that the isometry group G of M satisfies one of the following conditions*

either (a) *the dimension of G is at least 4,*

or (b) $G = T^3$,

or (c) $G = SO(3)$ *and the principal orbits are S^2 or $\mathbb{R}P^2$.*

Then M is either locally symmetric or D. Page's example or its $\mathbb{Z}_2$-quotient (defined in 9.126(d)).

((a) is due to L. Bérard Bergery, and (c) to A. Derdzinski; (b) was noticed by both of them, but was probably already known).

Notice that, when $\dim(G) = 3$, the only case left is that of $G = SU(2)$ or $SO(3)$ with a 3-dimensional principal orbit.

9.128. *Some non-compact complete examples of Einstein manifolds.* The above equations (1), (2), (3) give many other Einstein metrics, which are complete for appropriate choices of boundary conditions.

Let $E(s)$ be the 2-dimensional vector bundle on B, associated to $P(s)$ for the ordinary 2-dimensional representation of $S^1 = SO(2)$. We may write $E(s)$ as the quotient of $[0, +\infty[\times P(s)$ by the fibration $\{0\} \times P(s) \to B$. Then a metric (singular at the boundary) $dt^2 + g(f(t), h(t))$ on $[0, +\infty[\times P(s)$ gives a complete (non-singular) Riemannian metric on $E(s)$ if and only if h is an even positive function and f an odd function, positive on $t > 0$, and satisfying $f'(0) = 1$. Now (1), (2), (3) have a family of solutions (with $b \neq 0$ in (7)) with these boundary conditions, providing a one-parameter family of (non-homothetic) complete Einstein metrics (with $\lambda < 0$) on $E(s)$. Notice that $E(s)$, being a complex line bundle over B, has a natural almost-complex structure (which is complex if B is) and these metrics are almost-Hermitian, but not almost-Kähler (except in some special cases, see proposition below).

Of course, $dt^2 + g(f(t), h(t))$ is a metric on $P(s) \times \mathbb{R}$, which is complete if f and h are positive functions on the whole of $\mathbb{R}$. But Equations (1), (2) (3) have *no* such solution, hence we do not get any complete Einstein metric on $P(s) \times \mathbb{R}$ in this way.

Finally, we may also consider the case where B is still a compact Einstein almost-Kähler manifold (with $n[\omega] = 2\pi q\alpha^{\mathbb{R}}$, with α an indivisible integral class), but where $\check{\lambda}$ is non-positive. For some equations very similar to the preceding ones, the same construction gives complete Einstein metrics on $E(s)$ and $P(s) \times \mathbb{R}$ (but *no* Einstein metric on $M(s)$). We only summarize the results below, and we refer to [BéBer 3] for more details (and the corresponding Lorentz case).

9.129 Theorem (L. Bérard Bergery [BéBer 3]). *Let $(B, \check{g}, J, \omega)$ be a compact Einstein (with $\check{r} = \check{\lambda}\check{g}$) almost Kähler manifold such that the cohomology class of ω satisfies $n[\omega] = 2\pi q\alpha^{\mathbb{R}}$, where α is an integral indivisible class (we normalize by setting $\check{\lambda} = \pm n$ or $q = 1$ if $\check{\lambda} = 0$). Let $P(s)$ be the S^1-principal bundle classified by $s\alpha$ and $E(s)$ the associated complex line bundle. Then the above construction gives a complete Einstein metric on $E(s)$ and $P(s) \times \mathbb{R}$ only in the following cases:*

(a) *for any $\check{\lambda}$ and any $s \geqslant 1$, $E(s)$ admits a one-parameter family of (non-homothetic) complete Einstein almost-Hermitian metrics (with $\lambda < 0$);*

(b) *if $\check{\lambda} \leqslant 0$, for any $s \geqslant 1$, $P(s) \times \mathbb{R}$ admits a one-parameter family of (non-homothetic) complete Einstein almost-Hermitian metrics (with $\lambda < 0$);*

(c) *if $\check{\lambda} > 0$, $E(s)$ admits furthermore:*
for $s < q$, a complete Ricci-flat almost-Hermitian metric;
for $s = q$, a complete Ricci-flat almost-Kähler metric;
for $s > q$, a complete Einstein almost-Kähler metric (with $\lambda < 0$);

(d) *if $\check{\lambda} \leqslant 0$, $E(s)$ and $P(s) \times \mathbb{R}$ admit furthermore a complete Einstein almost-Kähler metric (with $\lambda < 0$).*
Moreover, if B is complex, these almost-Hermitian (respectively almost-Kähler) metrics are Hermitian (respectively Kähler).

In the special case where B is the complex projective space $\mathbb{C}P^{(n/2)-1}$ with its canonical metric, then $P(1)$ is the sphere S^{n-1} and we may also consider the boundary condition given at $t = 0$ by odd functions f and h with $f'(0) = h'(0) = 1$. In the compact case, this gives only the canonical metrics on S^n, $\mathbb{R}P^n$ or $\mathbb{C}P^{n/2}$ for appropriate boundary conditions at $t = l$. But we do obtain also a family of solutions on $[0, +\infty[$, which gives

9.130 Theorem (L. Bérard Bergery [BéBer 3]). *There exists a one-parameter family of (non-homothetic) $U(n/2)$-invariant complete Einstein Hermitian metrics on $\mathbb{R}^n$ $(=\mathbb{C}^{n/2})$, one of which is Ricci-flat and not flat, and the others have negative scalar curvature.*

Chapter 10. Holonomy Groups

A. Introduction

10.1. This chapter on holonomy groups is included in the present book, devoted to Einstein manifolds, for the following reason: a corollary of the main classification Theorem 10.90 states that, in some suitable context, a Riemannian manifold is automatically Einstein, and moreover sometimes Ricci flat: see Section 10.111. The way to this result, through holonomy groups, being quite long and intricated, we thought it worthwhile to devote a complete chapter to holonomy groups. Another motivation is the special case of manifolds with holonomy representation equal to $SU(n)$ or to $Sp(1)\cdot Sp(n)$, see Sections 10.28 and 10.32 and Chapters 14 and 15.

In reading this chapter the reader is advised to take a look at Table 1 and to the key Sections 10.15, 19, 38, 43, 51, 58, 79, 85, 92, 96, 108, 111, 114, 117.

10.2. The notion of holonomy group seems to have appeared for the first time explicitly in Elie Cartan's articles [Car 5] and [Car 8]. In [Car 8] Cartan gave a systematic description of manifolds of dimension 2 and 3 endowed with an affine (more than linear) connection whose affine holonomy group is a given affine group. Cartan made a systematic use of holonomy groups in his classification of symmetric spaces in [Car 6] (note that after finishing his classification with the holonomy method, Cartan recognized that his list coincided with that of real simple Lie algebras then immediately after he proved that fact directly: see section 10.71). With that exception holonomy groups seem to have been completely ignored by differential geometers up to 1950; then came the basic work of Borel, Chevalley, Lichnerowicz, Nijenhuis, Ambrose-Singer: [Bo-Li], [Nij 1], [Nij 2], [Am-Si].

In 1955 appeared the main classification Theorem 10.90 and 10.92, which was afterwards proved in a more direct way by J. Simons. Then again holonomy groups were forgotten; finally, starting in 1968 with Aleksevskii, followed by Brown-Gray in 1972 and Calabi in 1979 and finally in 1982 and 1985 (using in an essential way Yau's proof of the Calabi conjecture) most problems on holonomy groups were solved (see [Ber 1], [Sim], [Br-Gr], [Ch-Gr 1], [Ale 1], [Cal 5], [Bea 2], [Bry 1, 2]). See also Sections 10.95, 10.115, and Add.B, and Chapter 14 for the remaining open problems.

The non simply connected case is almost completely open (see part I of this chapter). However, it is of less concern for Einstein manifolds.

Please do not mix the present notion of holonomy with that of holonomy which appears in foliations: see for this [Hae] for example. They have nothing in common, at least today, but the name. Concerning the ethymology in our case, it comes from the greek. "ὅλος" means "entire, totality" and "νομος" means "law, rule". Then holonomy means "exact law" really and (compare with the word "isotropy group") the group of holonomy means the group which measures the defects of holonomy, the failure of the parallel transport to be the identity.

10.3. The notion of holonomy group can be defined in a much more general context than that of Riemannian manifolds (which concerns us here). In fact since it is based only on the notion of parallel transport it can be defined for every principal or vector bundle equipped with a connection. For example for any linear connection on the tangent bundle to a manifold; this connection need not have zero torsion for example. But in such a context there cannot exist any theorem like the main classification Theorem 10.92. For in fact a result of Hano and Ozeki (see [Ha-Oz]) asserts the following: as soon as the tangent bundle of the manifold M of dimension n can be reduced (this reduction needs only a topological condition) to some given closed connected subgroup G of the general linear group $GL(n, \mathbb{R})$ then there exists some connection on the tangent bundle to M whose holonomy group is precisely equal to G. See for example in [Mic], 1144 and [Kob 7] recent applications of the notion of holonomy group. Remark also that the various notions of Sections 10.15, 19, 21, 24, 57, 58 are valid for any linear connection.

10.4. On the other hand the connection, whose existence is asserted by [Ha-Oz] and whose holonomy group is equal to G, will have torsion. For in fact the property for a linear connection to have zero-torsion will impose some restrictions on its holonomy group: see [Ber 1] for this question. However these results seem not to have been pursued further.

10.5. If now (M, g) is a Riemannian manifold, its canonical linear connection has zero torsion and moreover leaves the Riemannian metric g invariant; this implies extremely strong restrictions on its holonomy group: through intermediate results such as Borel-Lichnerowicz closedness, the de Rham decomposition theorem, the final result the main classification Theorem 10.90. As said above it is a corollary of that theorem that will justify the introduction of holonomy groups in the present book. The underlying reason why holonomy groups of Riemannian manifolds are so few can be seen in the conjunction of the Bianchi identities and the fact that the Lie algebra of the holonomy group is generated by the various curvature tensors of (M, g): this is the Ambrose-Singer Theorem 10.58.

10.6. We saw in Section 10.3 that holonomy groups appear merely as subgroups of linear groups; that is why, when needed by the context, we shall prefer the name "holonomy representation"; a label which seems to be never used in the literature.[1])

[1]) One reason might be the following fact: except in the case of symmetric spaces (and of low dimensions where special isomorphisms occur for classical groups) abstract holonomy groups of Riemannian manifolds are in one-to-one correspondence with their linear holonomy representation: see Section 10.93.

Now when studying holonomy groups there are *two natural questions*: A. What are the possible holonomy representations for a Riemannian manifold? B. Compute the holonomy representation of a given Riemannian manifold.

10.7. For a more detailed treatment than ours we refer the reader to the only three systematic existing references: [Lic 3], [Wak], [Ko-No 1], [Ko-No 2]. Reference [Wak] is the more complete, and includes an excellent bibliography; but it is quite hard to find. In [Ko-No 1] the reader is advised to consult both Notes 1 and 12 and the bibliography of these Notes.

The reference [No-Da] presents an very interesting point of view which is not in our purpose.

Recent references, interested in various points of view on holonomy groups, are [Mar], [Poo 1] (see p. 162), [Mic] (see p. 1144), [Gra 1].

B. Definitions

We consider here only C^∞, connected Riemannian manifolds, but not necessarily complete; they are always equipped with the canonical connection (cf. 1.39).

10.8. Let p be a point in M and λ be a C^1-piecewise loop based at p. Let us denote by $\tau(\lambda)$ the parallel transport along λ. Then $\tau(\lambda)$ is an element of the orthogonal group $O(T_p)$. The inverse loop λ^{-1} and the composed loop $\lambda \cup \mu$ of two such loops obey the relations $\tau(\lambda^{-1}) = (\tau(\lambda))^{-1}$ and $\tau(\lambda \cup \mu) = \tau(\lambda) \circ \tau(\mu)$. This relations permit the following:

10.9 Definitions. We call the *holonomy group of* (M,g) *at* p, (or sometimes *the holonomy representation of* (M,g) *at* p) and *denote* it by $\mathrm{Hol}(p)$, the subgroup of the orthogonal group $O(T_p)$ generated by the set of all $\tau(\lambda)$ where λ runs through the set of C^1-piecewise loops of M based at p. If one consider only loops which are homotopic to the identity, one gets a subgroup called the *restricted holonomy group* (or *representation*) *of* (M,g) *at* p, and *denoted* by $\mathrm{Hol}^0(p)$.

10.10. Note that $\mathrm{Hol}^0(p)$ is always contained in the special orthogonal group $SO(T_pM)$: Section 10.24. In Section 10.51 we shall see that $\mathrm{Hol}^0(p)$ indeed is a closed subgroup of $SO(T_pM)$, but this fact is highly non trivial; moreover it is false in general for linear connections (see [Ha-Oz] for an example).

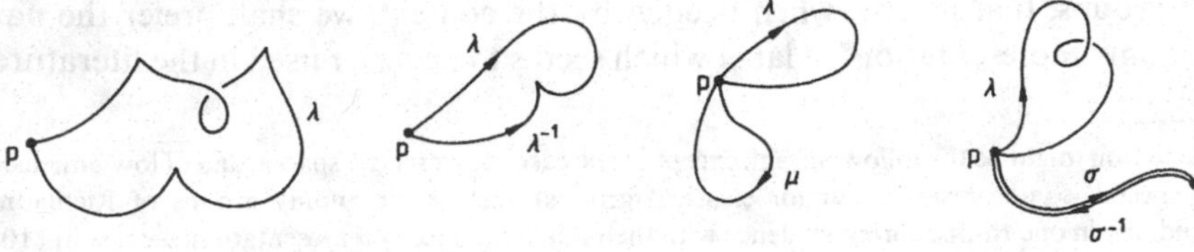

10.11. Let us change the point p to the point q of M; let us fix some C^1-piecewise curve σ from p to q; then $\mathrm{Hol}(q)=\tau(\sigma)\mathrm{Hol}(p)(\tau(\sigma))^{-1}$. And the same for the restricted groups: $\mathrm{Hol}^0(q)=\tau(\sigma)\mathrm{Hol}^0(p)\tau(\sigma)^{-1}$. As a consequence the holonomy representations at various points of M are in fact all isomorphic. This enables us to talk more simply of the holonomy groups (or representations) of the Riemannian manifold (M,g), the full one and the restricted one. *Notations* are $\mathrm{Hol}(g)$, $\mathrm{Hol}^0(g)$.

10.12. If the fundamental group of M is zero: $\pi_1(M)=0$, then of course $\mathrm{Hol}^0(g)=\mathrm{Hol}(g)$. Note that such an equality can happen for non-simply connected manifolds, flat tori for example, see Section 10.25. As an exercise, the reader may prove (and in more than one way) that the holonomy representation of canonical spheres (S^n, can) is the full special orthogonal group $SO(n)$. In Section 10.54 we will see that is indeed the case for a generic Riemannian manifold.

10.13 An Example. Let $M=\mathbb{R}^2\times S^1$ the manifold product of the euclidean plane $\mathbb{R}^2$ by the circle S^1; it can be realized by identifying $\mathbb{R}^2\times\{0\}$ with $\mathbb{R}^2\times\{1\}$ in the manifold with boundary $\mathbb{R}^2\times[0,1]$. If the identification is, more precisely, the one obtained with $f(\alpha)$ the linear rotation of angle α of $\mathbb{R}^2$, one then obtains on M a flat Riemannian metric denoted by $(M,g(\alpha))$. It is a complete, flat manifold; then $\mathrm{Hol}^0(g(\alpha))=0$, this being valid for any flat manifold, as can be seen in Section 10.25. But $\mathrm{Hol}(g(\alpha))$ in $O(2)$ is the subgroup generated by $f(\alpha)$, namely a finite subgroup if α is a rational multiple of π, otherwise an everywhere dense (hence non closed) subgroup abstractly isomorphic to $\mathbb{Z}$. As a consequence one sees that $\mathrm{Hol}(g)$ can be non compact for complete manifolds; the situation for compact manifolds is an open problem today, see Section 10.115.

10.14. A naive philosophy would be that from the structure of the holonomy representation one could read the structure of (M,g). In other words: classify Riemannian manifolds according to their holonomy representation. The classification theorem shows that there are very few possible holonomy representations; hence such a classification would be very coarse. A more effective philosophy is that $\mathrm{Hol}^0(g)$ teaches us quite a lot about structure of (M,g) when the dimension of $\mathrm{Hol}^0(g)$ is quite small compared to that of $O(n)$, namely $\dfrac{n(n+1)}{2}$; see details in Section 10.97.

By the very definition of $\mathrm{Hol}^0(g)$, there exists a canonical homomorphism $\pi_1(M)\to\mathrm{Hol}(g)/\mathrm{Hol}^0(g)$. From this one deduces the

10.15 Proposition. $\mathrm{Hol}^0(g)$ *is a normal subgroup of* $\mathrm{Hol}(g)$. *Moreover the quotient* $\mathrm{Hol}(g)/\mathrm{Hol}^0(g)$ *is countable.*

In fact the fundamental group of a manifold is countable. We know more: $\mathrm{Hol}^0(g)$ is indeed the connected component of the identity element of $\mathrm{Hol}(g)$: this is harder and will be seen in Section 10.48.

10.16. Finally if $(\tilde{M},\tilde{g})$ denotes the universal Riemannian covering of (M,g), then $\mathrm{Hol}(\tilde{g})=\mathrm{Hol}^0(\tilde{g})=\mathrm{Hol}^0(g)$.

C. Covariant Derivative Vanishing Versus Holonomy Invariance. Examples

10.17. On a Riemannian manifold (M, g) let us consider a tensor field α (cf. 1.4). Let us suppose first that α is invariant by parallel transport, i.e. for every p and q in M and every C^1-piecewise path λ from p to q one has $\tau^*(\lambda)(\alpha(p)) = \alpha(q)$, where $\tau^*(\lambda)$ denotes the tensorial extension of the parallel transport $\tau(\lambda)$ along λ. Then, by the Definition 10.9 the tensor $\alpha(p)$ at T_pM is invariant by the tensorial extension of the holonomy representation $\mathrm{Hol}(p) \subset O(T_pM)$.

Conversely given any tensor on T_pM, where p is a fixed point of M, if α_0 is invariant under the tensorial extension of $\mathrm{Hol}(p)$ one can built up on (M, g) a tensor field α such that $\alpha(p) = \alpha_0$ and which is invariant by parallel transport, this being done in the following evident manner: define $\alpha(q)$ by the above formula for some C^1-piecewise path from p to q. One has only to be careful in proving the differentiability of α.

10.18. The above two properties are equivalent to a third one namely: α has vanishing covariant derivative: $D\alpha = 0$. This can be seen using the formula $(D\alpha)(X_1, \ldots, X_s; X) = D_X(\alpha(X_1, \ldots, X_s)) - \sum_{i=1}^{s} \alpha(X_1, \ldots, D_X X_i, \ldots, X_s)$. For let us choose along a given path λ vector fields $X_1, \ldots, X_s$ which are parallel and take $X = \lambda'$ the velocity vector of λ. Then $D_X X_i = 0$ for every $i = 1, \ldots, s$ and $D\alpha = 0$ is equivalent to $(D_X\alpha)(X_1, \ldots, X_s) = 0$, i.e. the value $\alpha(X_1, \ldots, X_s)$ is constant along λ. And conversely since having the choice of path we can always take one with $\lambda' =$ any given value $X(p)$. We have then proved the

10.19 Fundamental Principle. *Let (M, g) be a Riemannian manifold. Let us give a fixed type (r, s) for tensors on M. Then the following three properties are equivalent:*

(i) *there exists on (M, g) a tensor field of type (r, s) which is invariant by parallel transport;*

(ii) *there exists on (M, g) a tensor field α of type (r, s) which has zero covariant derivative: $D\alpha = 0$;*

(iii) *there exists p in M and on T_p a tensor α_0 of type (r, s) which is invariant by the tensorial extension of type (r, s) of the holonomy representation* $\mathrm{Hol}(p)$.

10.20 Remarks. Such a result holds good for linear connections. In case one only knows the invariance under Hol^0 one will get of course only the corresponding local result. Such a principle can be extended to geometric objects more general than tensor fields. The following particular case will be useful in the sequel:

10.21 Proposition. *Let k be an integer between* 1 *and* $n - 1$. *There is an equivalence between the following:*

(i) *there exists on (M, g) a field of k-dimensional tangent subspaces (a distribution as one calls it sometimes in the statement of Frobenius' theorem) which is invariant by parallel transport;*

(ii) *the holonomy representation* $\mathrm{Hol}(g)$ *leaves invariant a subspace of dimension k.*

Moreover such a distribution is necessarily involutive (in the sense of Frobenius' theorem).

The equivalence is obvious. The involutive property is proven as follows: if X, Y are two vector fields belonging to the distribution under question $\mathscr{V}$, then $D_X Y$ belongs again to $\mathscr{V}$ by the invariance under parallel transport, and the same for $D_Y X$. Finally the bracket $[X, Y] = D_X Y - D_Y X$ still belongs to $\mathscr{V}$. □

10.22 Remark. Of course the preceeding proposition can be formulated with the concept of exterior differential forms: (i) and (ii) are equivalent to the existence of an exterior differential form α, which has to be non-zero, decomposable and to satisfy $D\alpha = 0$. For let us generate the distribution $\mathscr{V}$ by $X_1, \ldots, X_k$ (locally). Define α by $\alpha(X_1, \ldots, X_k) = 1$ and $\alpha(X_{i_1}, \ldots, X_{i_k}) = 0$ for any other choice in any basis extending $X_1, \ldots, X_k$. Then the result will follow from the formula in Section 10.18.

10.23 Philosophy. Finding on a given Riemannian manifold (M, g) the geometric objects whose covariant derivative vanishes is equivalent to the algebraic problem of finding the invariants of the holonomy representation $\mathrm{Hol}(g)$. We now will give quite numerous examples of such objects since we are going to meet them again in the sequel.

10.24 Example 1: Orientable Manifolds. The manifold (M, g) is orientable if and only if $\mathrm{Hol}(g)$ is included in $SO(n)$. The obvious proof is left to the reader. But one can also use Principle 10.19 if one remarks that $SO(n)$ leaves invariant an exterior form of maximum degree.

10.25 Example 2: Flat Manifolds. One knows that a Riemannian manifold is flat if and only if it admits, locally, a basis of parallel vector fields. Hence the Fundamental Principle 10.19 asserts that (M, g) is flat if and only if $\mathrm{Hol}^0(g)$ is reduced to the identity. Note that this result does not extend to $\mathrm{Hol}(g)$; for example $\mathrm{Hol}(g) = \mathbb{Z}_2$ for a flat Klein bottle as a consequence of Section 10.24. However Bieberbach's theorem (see for example [Wol 4], p. 100) implies that, for a flat compact manifold, $\mathrm{Hol}(g)$ is always finite and in particular compact (compare with the example of Section 10.13).

10.26 Example 3: "Kählerian" Manifolds. Let (M, g) be a Riemannian manifold with dimension $n = 2m$. Let us say that (M, g) is "Kählerian" if it admits an exterior 2-form ω, which is of maximal rank, and whose covariant derivative vanishes, i.e., $D\omega = 0$. The Fundamental Principle 10.19 asserts that this is equivalent to the fact that the holonomy representation $\mathrm{Hol}(g)$ in $O(n = 2m)$ leaves invariant an exterior 2-form of maximal rank. This can be written as $\mathrm{Hol}(g) \subset U(m)$, where $U(m)$ denotes the unitary group of $\mathbb{C}^m$. The complex structure J_p on a tangent space T_pM here comes from the exterior form together with the Euclidean structure. These J_p form an almost complex structure on M which is integrable, see 2.11. It follows from the Newlander-Nirenberg Theorem that M admits the structure of a complex manifold, for which the metric g is hermitian and Kähler (see 2.28). Thus "Kählerian" Riemannian manifolds are no more than Kähler manifolds.

10.27. As an exercise, the reader could prove that $\mathrm{Hol}(g) = U(m)$ for the canonical Kähler metric on the complex projective space $\mathbb{C}P^m$. Later on in Section 10.55 we shall see that in fact $\mathrm{Hol} = U(m)$ for a generic Kähler manifold.

10.28 Special Kähler Manifolds. On a Kähler manifold (M, g) of complex dimension m a *complex volume form* θ is by definition an exterior form of type $(m, 0)$ (cf. 2.6) with $\theta \neq 0$ and $D\theta = 0$. We remark that on the standard space $\mathbb{C}^m$ the standard form $z_1^* \wedge \cdots \wedge z_m^*$ is invariant under some element g of the unitary group $U(m)$ if and only if $\det g = 1$ that is to say if $g \in SU(m)$ where $SU(m)$ denotes the special unitary group: $SU(m) = \{g \in U(m) \colon \det g = 1\}$. We now have the:

10.29 Proposition (Iwamoto [Iwa], Lichnerowicz [Lic 3], p. 261). *Let us consider for a Kähler manifold (M, g) the conditions:*

(i) *(M, g) Ricci flat: $r = 0$*

(ii) *there is on (M, g) a complex volume form*

(iii) $\mathrm{Hol}(g) \subset SU(m)$.

Then the conditions (ii) *and* (iii) *are equivalent and imply condition* (i). *Conversely condition* (i) *implies* $\mathrm{Hol}^0(g) \subset SU(m)$ *and, if M is moreover simply connected, implies* (ii) *and* (iii).

10.30 Definition. *A Kähler manifold (M, g) will be said to be* special *if it satisfies condition* (i), *i.e. if it is Ricci flat.*

The equivalence of (ii) and (iii) follows immediately from the fundamental principle 10.19. The best way to enter (i) in the picture is to remark that the Levi-Civita connections extends to the *canonical complex line bundle* $\bigwedge^m (T'M)^*$, whose curvature is the Ricci form ρ introduced in 2.44. Condition (ii) implies that the canonical line bundle is flat for the above connection. In particular $\rho = 0$ and $r = 0$. This argument is given in detail in 2.F. □

10.31 Example 5: Hyperkählerian Manifolds. These manifolds are those which admit more than one Kähler structure for a given metric, for more details see Chapter 14. We here only sketch their properties. They can be defined as manifolds (M, g) of dimension $4m$ admitting three almost complex structures I, J, K which are of zero covariant derivative, which are hermitian i.e. respect to Riemannian metric g, and obey the quaternionic relations $I^2 = J^2 = K^2 = -Id$, $IJ = JI = -K$. By the very definition of the symplectic orthogonal group $Sp(m) \subset O(4m)$, and in view of the Principle 10.19, one sees that (M, g) is hyperkählerian if and only if $\mathrm{Hol}(g) \subset Sp(m)$. Note that when $m = 1$ one has $Sp(1) = SU(2)$ and our manifolds are those of the preceeding example. We remark also that a Kählerian manifold (M, g) being given, then it is hyperkählerian if and only if it carries a parallel exterior 2-form of type $(2, 0)$ and of maximal rank. Note also the simple fact: $Sp(m) \subset SU(2m)$. This implies that hyperkählerian manifolds are automatically Ricci flat by Proposition 10.29.

Concerning examples, see Chapters 14 and 15. Recall only that Calabi was among the first in 1979 to find examples of hyperkählerian manifolds, complete but not compact. Compact ones were finally obtained by Beauville ([Bea 2]) using Yau's proof of Calabi's conjecture.

10.32 Example 6. Quaternion-Kähler Manifolds. These manifolds are studied at length in Chapter 14; for the reader's convenience we give here a brief account of the theory. To be hyperkählerian is a strong condition for a manifold, for example the quaternionic projective space $\mathbb{H}P^n$ is not hyperkählerian. But it obeys the following. We consider, as in the preceeding example, a Riemannian manifold (M, g) of dimension $n = 4m$ with a Riemannian metric g which is invariant under three almost-complex structures I, J, K obeying the quaternions axioms. But we relax the condition that I, J, K have covariant derivative vanishing by asking only that locally there exists on (M, g) three 1-forms α, β, γ such that

$$\begin{aligned} D_X I &= \quad\quad\quad\;\; \gamma(X)\cdot J - \beta(X)\cdot K \\ D_X J &= -\gamma(X)\cdot I \quad\quad\quad\quad\;\; \alpha(X)\cdot K \\ D_X K &= \quad \beta(X)\cdot I - \alpha(X)\cdot J \end{aligned}$$

for every X, i.e. I, J, K generate a rank 3 subbundle in the vector bundle of endomorphisms, this subbundle being of zero covariant derivative globally. One proves then that such a manifold, called quaternionic, has a holonomy representation $\mathrm{Hol}(g)$ contained in $Sp(1)\cdot Sp(m) \subset O(4m)$, and conversely. Note that if one denotes by ξ, η, ζ the Kähler forms associated to the hermitian structures (M, g, I), (M, g, J), (M, g, K) then in general no one has vanishing covariant derivative but the sum $\xi^2 + \eta^2 + \zeta^2 = \theta$ has zero covariant derivative. This is obvious from the formulas above. One proves that such a quaternionic manifold is automatically Einstein. This is one of our motivations for including holonomy groups in the present book! See also Sections 10.96 and 109. For examples of quaternion-Kähler manifolds, besides $\mathbb{H}P^n$, see Chapter 14, where the 4-form θ is also introduced.

D. Riemannian Products Versus Holonomy

10.33 Riemannian Products. Let (M_1, g_1), (M_2, g_2) be two Riemannian manifolds; then the product manifold $M_1 \times M_2$ carries naturally a product Riemannian metric, *denoted by* $g_1 \times g_2$, and defined by the condition that both projections are Riemannian submersions, see 1.37 and 9.8. Equivalently on the tangent space to $M_1 \times M_2$ at (p_1, p_2) one defines $g_1 \times g_2$ to be the product Euclidean structure on $T_{(p_1,p_2)}(M_1 \times M_2) = T_{p_1}M_1 \times T_{p_2}M_2$. Note that both distributions defined by the subspaces tangent to M_1 and M_2 respectively are involutive; their leaves are nothing but the manifolds $\{p_1\} \times M_2$ and $M_1 \times \{p_2\}$; moreover they are totally geodesic submanifolds. A Riemannian manifold (resp. locally) isometric to a Riemannian product will be called (resp. *locally*) *reducible.*

10.34. These two distributions are invariant by the holonomy representation $\mathrm{Hol}(g_1 \times g_2)$ in the sense of Section 10.21. If moreover both M_1 and M_2 are orientable, this can be interpreted by saying that $M_1 \times M_2$ carries two decomposable exterior forms, non zero, of degree $n_1 = \dim M_1$ and $n_2 = \dim M_2$, which have zero covariant derivative: this comes from Section 10.21. Hence the Principle 10.19 implies that the representation $\mathrm{Hol}(g_1 \times g_2)$ in $O(n_1 + n_2)$ leaves invariant two complementary and orthogonal vector subspaces, of dimensions d_1 and d_2 respectively. We write this as $\mathrm{Hol}(g_1 \times g_2) \subset O(n_1) \times O(n_2) \subset O(n_1 + n_2)$.

10.35. In fact the "reduction" of $\mathrm{Hol}(g_1 \times g_2)$ is even more striking. Namely we have the formula

$$\text{(10.36)} \qquad \mathrm{Hol}(g_1 \times g_2) = \mathrm{Hol}(g_1) \times \mathrm{Hol}(g_2),$$

where $\mathrm{Hol}(g_1) \times \mathrm{Hol}(g_2)$ denotes the direct sum representation of the representations $\mathrm{Hol}(g_1)$ and $\mathrm{Hol}(g_2)$, namely $\mathrm{Hol}(g_1) \times \mathrm{Hol}(g_2)$ is the set of all elements of the form $\tau \times \sigma$ in $O(n_1) \times O(n_2)$ where independently τ runs through $\mathrm{Hol}(g_1)$ and σ through $\mathrm{Hol}(g_2)$. This implies that τ in $O(n_1)$ for example induces the identity on $O(n_2)$. To prove formula 36 one has only to remark that a loop completely contained in a leaf like $M_1 \times \{p_2\}$ yields a parallel transport in $O(n_1 + n_2)$ which is the identity on the $O(n_2)$ part.

10.37 Example. Be careful that a flat torus is not in general a Riemannian product (globally); for example the torus $\mathbb{R}^2/\Lambda$ where Λ is the lattice in $\mathbb{R}^2$ generated by $u = (0, 1)$ and $v = (x, y)$ will never be a product if $x \in]0, 1/2]$ and $x^2 + y^2 \geqslant 1$.

A quite surprising fact is the converse of Formula (10.36). More precisely:

10.38 Theorem. *Let (M, g) be a Riemannian manifold, not necessarily complete and p be some point in M. Denote by $T_0 \subset T_pM$ the subspace of T_pM which is acted on trivially by* $\mathrm{Hol}(p)$, *i.e. every vector in T_0 should be fixed under every element of* $\mathrm{Hol}(p)$; *let $T_0^\perp$ denote the orthogonal complement of T_0 in T_pM. Since* $\mathrm{Hol}(p)$ *is orthogonal its action on T_0 can be split into irreducible representations acting respectively on the orthogonal decomposition $T_0^\perp = T_1 \overset{\perp}{\oplus} T_2 \overset{\perp}{\oplus} \cdots \overset{\perp}{\oplus} T_k$. Then (M, g) is locally a Riemannian product $g_0 \times g_1 \times g_2 \times \cdots \times g_k$ where g_0 is flat. Moreover* $\mathrm{Hol}(p)$ *is a direct sum representation* $\mathrm{Hol}(p) = A_1 \times A_2 \times \cdots \times A_k$ *where every A_i is a subgroup of $O(T_i)$ which acts irreducibly on T_i and trivially on T_0 and every T_j with $j \neq i$.*

10.39 Remark. What is surprising is this: as soon as a holonomy representation is reducible it becomes a direct sum representation. For example the subgroup $O(m) \cdot O(m) \subset O(2m)$ consisting of the set of all $\tau \times \tau$ where τ runs through $O(m)$ will never appear as a holonomy representation. A deep reason for this can be seen in Bianchi identities together with the fact that holonomy is generated by curvature tensors, see more on this in Section 10.63.

Proof of Theorem 10.38. A. Let us show first that the reducibility of $\mathrm{Hol}(p)$ implies that g is locally a Riemannian product. By Proposition 10.21 there exists on M two involutive distributions $\mathscr{V}$, $\mathscr{W}$ where $\mathscr{W}$ is the orthogonal distribution $\mathscr{W} = \mathscr{V}^\perp$ of $\mathscr{V}$. Moreover, by Frobenius' theorem, they are integrable and we have now on (M, g) two orthogonal foliations. We claim now that every leaf is a totally geodesic submanifold. In fact the tangent spaces to a given leaf are invariant under parallel translation (Proposition 10.21). In particular a geodesic in (M, g) whose initial velocity vector is tangent to this leaf will stay for ever in this leaf. The fact that (M, g) is a Riemannian product follows now immediately. □

B. Using 10.16, noting that we are seeking only a local statement, and by an easy induction, we are left with proving that if M is simply connected and if $\mathrm{Hol}(p)$ is reducible then $\mathrm{Hol}(p)$ is globally a direct sum representation. This follows from the following:

10.40 Lasso Lemma ([Lic 3], p. 51, [Ko-No 1], p. 184, [Wak], p. 58). *Let M be a manifold and $\mathscr{U}$ a given open covering of M. Then any loop in M, if it is homotopic to zero, is a product of lassos (see figures) whose noose is always contained in some open set of $\mathscr{U}$.*

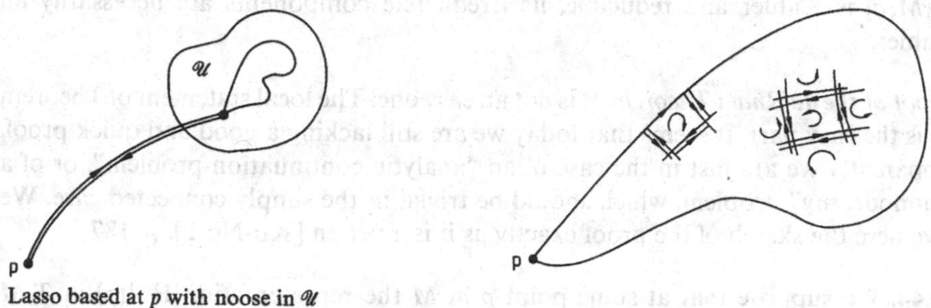

Lasso based at p with noose in $\mathscr{U}$

Fig. 10.40

One starts with a compactness argument in order to involve only a finite number of elements of $\mathscr{U}$. Then Figure 10.40 shows how to decompose the loop in lassos: all the sides disappear two by two except those which give the boundary of the homotopy, which is exactly the given loop.

10.41 Corollary. *For every Riemannian manifold (M, g) the restricted holonomy $\mathrm{Hol}^0(g)$ admits a decomposition as a direct sum $\mathrm{Hol}^0(g) = A_1 \times A_2 \times \cdots \times A_k \subset O(n_1) \times O(n_2) \times \cdots \times O(n_k) \subset O(n_0 + n_1 + n_2 + \cdots + n_k)$ where every orthogonal piece A_i $(i = 1, \ldots, k)$ is irreducible.*

10.42 Remark. There is a delicate point here when (M, g) is not complete. This point is that the A_i are not a priori (restricted) holonomy groups of some Riemannian manifolds. When (M, g) is complete, this is certainly the case because A_i is $\mathrm{Hol}^0(g_i)$ where g_i is the Riemannian metric involved in the decomposition of the universal covering $(\tilde{M}, \tilde{g})$ as Riemannian product by virtue of the de Rham decomposition Theorem 10.43 below. Contrariwise if (M, g) is not complete as in Figure 10.42 below

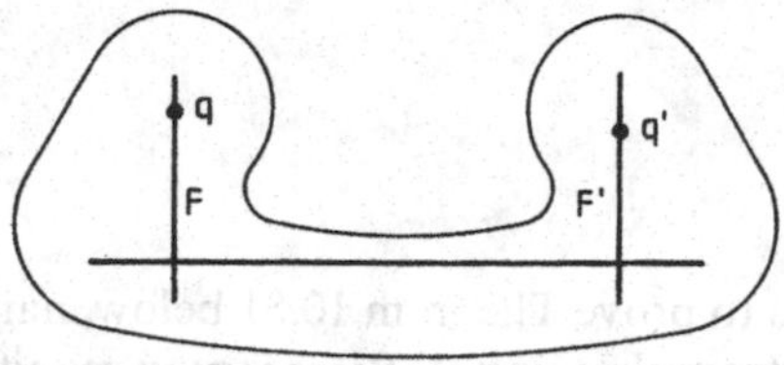

Fig. 10.42

then the two vertical leaves F, F' will in general have at q, q' different Riemannian metrics. The total holonomy group Hol(g) will have to include Hol(q) and Hol(q') of F and F'. We will see however in Section 10.108 that every A_i is the holonomy group of some Riemannian manifold, but the proof requires more knowledge than we have now.

10.43 Theorem (de Rham, [Rha]). *If a Riemannian manifold is complete, simply connected and if its holonomy representation is reducible, then (M, g) is a Riemannian product.*

Note. For an extension of the de Rham Theorem to more general Riemannian connections, see Note 12 of [Ko-No 2] and 10.J. Note also that, after [Lic 3], p. 264, if (M, g) is Kähler and reducible, its irreducible components are necessarily all Kähler.

Proof of the de Rham Theorem. It is not an easy one. The local statement of Theorem 38 is the easy part. It seems that today we are still lacking a good and quick proof. Apparently we are just in the case of an "analytic continuation problem", or of a "monodromy" problem, which should be trivial in the simply connected case. We give here the sketch of the proof exactly as it is given in [Ko-No 1], p. 187.

10.44. We suppose that at some point p in M the representation Hol(p) in T_pM leaves invariant the orthogonal direct sum $T_pM = T_1 \overset{\perp}{\oplus} T_2$. Denote by M_1, M_2 respectively the maximal leaves of M for the corresponding distributions. They are totally geodesic and complete submanifolds of M. We want to prove that (M, g) is the Riemannian product $(M_1, g|M_1) \times (M_2, g|M_2)$.

For any curve $z\colon [0,1] \to M$ with $z(0) = p$ we can define its projection z_1 on M_1 as follows. Let Z be the development of z in the affine tangent space T_pM. Let Z_1 be the Euclidean projection of Z onto T_1. Then we can develop back Z_1 onto z_1 in $(M_1, g|M_1)$. We can now associate to the point $Z(1)$ in M the point $z_1(1)$ in M_1. The simply connectedness shows that this point $z_1(1)$ depends in fact only on the point $z(1)$ and not on the curve z joining p to z. Then we can define a projection $p_1\colon M \to M_1$ and similarly $p_2\colon M \to M_2$. It is now easy to check that the map $(p_1, p_2)\colon M \to M_1 \times M_2$ is a local isometry. This achieves the proof. □

10.45 Moral. One sees now that in order to answer questions A and B of Section 10.6 we need only to do it for Riemannian manifolds for which Hol^0 is irreducible, this being valid at least when one deals with simply connected and complete manifolds. The preceeding assertion will follow directly from Sections 10.41 and 43. For the non complete case see Section 10.107.

E. Structure I

In this section our aim is to prove Theorem 10.51 below, namely that the restricted holonomy group of any (complete or not) Riemannian manifold is always compact. This result does not hold for non-Riemannian manifolds, see [Ha-Oz]. The above result is not obvious and its proof needs intermediate steps. Recall that Riemannian manifolds in this section are not necessarily complete.

10.46 Proposition. Hol^0 *is always an arcwise connected subgroup of* $O(n)$.

Let λ be a loop which is homotopic to zero. An easy result in approximation theory yields that our homotopy can be realized by intermediate loops λ_t which are all C^1-piecewise. Since parallel transport is given by solutions of a differential equation, the classical theorem of continuity with respect to parameter of solutions of such a differential equation will show that the map $t \mapsto \tau(\lambda_t)$ with values in $O(n)$ is continuous in t. Then $\tau(\lambda)$ belongs to the arcwise connected component of the identity element of $O(n)$. □

10.47 Corollary. Hol^0 *is a Lie subgroup of* $O(n)$.

The precise meaning of this statement is that Hol^0 is the image, under the exponential map, of a Lie subalgebra of the Lie algebra of $O(n)$. Of course such an image need not be closed in general. The proof of Corollary 10.47 is the consequence of a very delicate theorem of Yamabe to the effect that any arcwise connected subgroup of a Lie group is indeed a Lie subgroup. For this proof see [Yam 1]. But note the following: in the above proof of Proposition 46 one could get easily not only a continuous arcwise connectedness but in fact a C^1-piecewise connectedness in the manifold $O(n)$. In this case the proof of Yamabe's theorem is much easier, see such a proof in [Ko-No 1], p. 275. The key idea is here to obtain the bracket of the left invariant vector fields as a limit of suitable parallelograms. □

10.48 Corollary. Hol^0 *is the connected component of the identity in* Hol.

We saw in 10.15 that $\mathrm{Hol}/\mathrm{Hol}^0$ is countable. Since we are dealing with Lie groups this implies that $\dim \mathrm{Hol}^0 = \dim \mathrm{Hol}$. But Hol^0 is connected. This achieves the proof. □

10.49 Lemma. *Let G be a Lie subgroup of $O(n)$. If moreover G is connected and acts irreducibly, then G is closed in $O(n)$ hence compact.*

Let $\mathfrak{g}$ be the Lie algebra of G. Since $\mathfrak{g}$ is a subalgebra of the Lie algebra $\mathfrak{o}(n)$ of $O(n)$ the Killing form $(A, B) \to -\mathrm{trace}(A \cdot B)$ of $\mathfrak{o}(n)$ is still positive definite on $\mathfrak{g}$. In particular $\mathfrak{g}$ admits the decomposition $\mathfrak{g} = S + Z$ where the decomposition is orthogonal with respect to this Killing form and Z is the center of $\mathfrak{g}$. Moreover the center is abelian and its complement S is a semi-simple Lie algebra. The proof now shows separately that the images by the exponential map of both S and Z are closed. For the semi-simple part S this is quite classical, see for example [Ko-No 1], p. 279. For the center Z one shows that its dimension is necessarily equal to 0 or 1, by using an algebraic lemma on skewsymmetric endomorphisms which commute with an irreducible subgroup of $O(n)$, see [Ko-No 1], p. 277 or [Wak], p. 17. Now, when the dimension is equal to 1 the image has to be a circle, hence closed. For otherwise its closure would be of dimension at least 2. But this was precisely excluded above. □

10.50 Note. The above Result 10.49 is still valid for subgroups of the full linear group $GL(n, \mathbb{R})$. The proof is longer, see [Wak], p. 17. And do not forget the counter-example of [Ha-Oz], see Section 10.10.

10.51 Theorem (Borel-Lichnerowicz: [Bo-Li]). *For any Riemannian manifold* (M, g) *the restricted holonomy group* $\mathrm{Hol}^0(g)$ *is closed in* $O(d)$, *and in particular is compact.*

This results directly from 10.48 and 10.49. □

We saw in 10.13 that the full holonomy group Hol is not compact in general. Note also that Theorem 10.51 is valid only for Riemannian manifolds. There are counterexamples even for linear connections with zero torsion: [Ha-Oz].

To study the structure of holonomy groups we need first to learn various things on symmetric spaces, and also to see that holonomy groups are generated by curvature tensors.

F. Holonomy and Curvature

10.52 How to Compute Hol^0? The theoretical answer is: with the curvature. In fact let p be a point in M and $x, y \in T_pM$. Extend x, y by two commuting vector fields X, Y on M, $[X, Y] = 0$. Look at the parallel transport along the loop λ_t made up by the parallelogram built up with X and Y, based at p and with side lengths all equal to t. Denote by $R(x, y)$ the curvature endomorphism of T_pM given by the curvature tensor at p and by x and y. A classical formula (which sometimes is viewed as the definition of curvature) states that the bracket in the manifold TM of the horizontal lifts X^H, Y^H of X, Y respectively is precisely equal to $[X^H, Y^H](p) = -R(x, y)$.

But now the definition of parallel transport and the formula giving the geometric interpretation of the bracket of vector fields implies that

$$[X^H, Y^H](p) = \frac{d}{dt}(\tau(\lambda_t))(0).$$

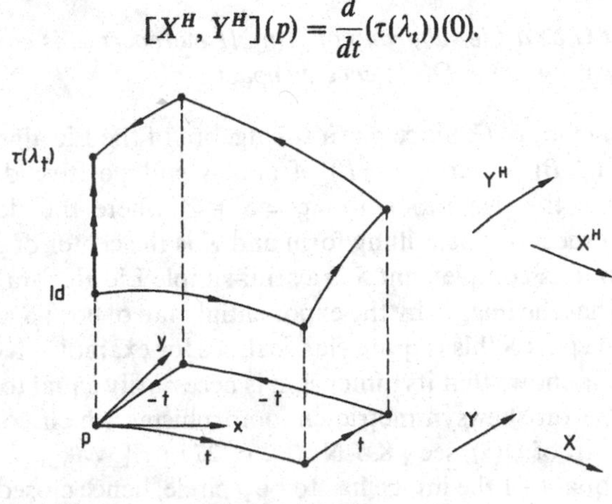

Fig. 10.52

10.53. In conclusion the Lie algebra $\mathfrak{hol}(p)$ of $\mathrm{Hol}(p) \subset O(T_pM)$ certainly contains all the endomorphisms $R(x, y)$ for x, y running through T_pM.

10.54 Corollary. *For a generic Riemannian manifold* $\mathrm{Hol}^0 = SO(n)$.

In fact if (M, g) is generic and if $\{x_i\}$ is an orthonormal generic basis of some T_pM, then the $n(n-1)$ endomorphisms $R(x_i, x_j)$ will be linearly independent in the space of antisymmetric endomorphisms (that is to say the Lie algebra of $O(n)$). □

10.55. We do not want to be more precise concerning genericity; instead let us give some examples:

(i) any small perturbation of the canonical metric of the sphere will have holonomy group equal to $SO(n)$.

(ii) in the same spirit the holonomy group of a generic Kähler manifold (for example a small deformation of $(\mathbb{C}P^m, \mathrm{can})$) is equal to $U(m)$.

(iii) the philosophy of normal coordinates (cf. [Be-Ga-Ma], p. 91 for example) implies in particular that, for any given algebraic object which satisfies the curvature tensor identities, there is a Riemannian manifold (M, g) whose curvature tensor at some point p is the given tensor.

10.56. Let now p, q be two points in M and choose any C^1-piecewise curve from p to q, say λ. Then exactly the same reasoning as above implies that

$$(\tau(\lambda))^{-1} \circ R(\tau(\lambda)(x), \tau(\lambda)(y)) \circ \tau(\lambda) \in \mathfrak{hol}(p) \tag{10.57}$$

for every x, y in T_pM. One needs only to work with lassos whose nooses are small parallelograms based at the point q.

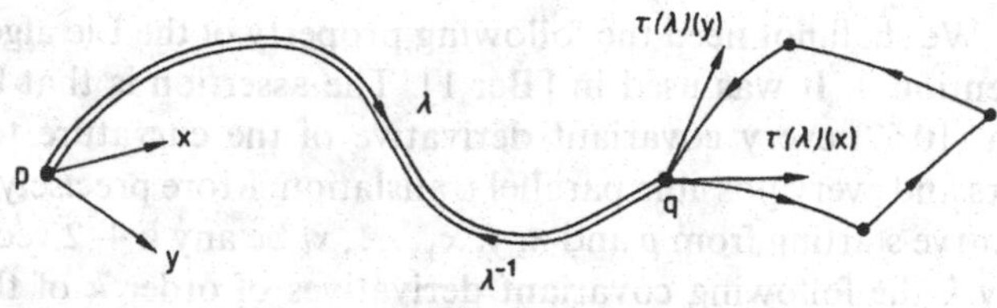

Fig. 10.57

In conclusion one can say that the Lie algebra $\mathfrak{hol}$ contains all the curvature endomorphisms, at every point and for every pair of vectors, but everything has to be pulled back to a fixed point and this by all possible parallel transportations. The basic result is that $\mathfrak{hol}$ contains no more than those:

10.58 Theorem (Ambrose-Singer, [Am-Si], [Ko-No 1], p. 89). *The Lie algebra* $\mathfrak{hol}(p)$ *is exactly the subalgebra of* $\mathfrak{so}(T_pM)$ *generated by the elements* $(\tau(\lambda))^{-1} \circ R(\tau(\lambda)(x), \tau(\lambda)(y)) \circ \tau(\lambda)$, *where* x, y *run through* T_pM *and* λ *runs through all* C^1-*piecewise paths starting from* p.

The proof goes as follows. We work inside the principal bundle P over M built up as follows. Let us fix some orthonormal frame f_0 at some point p in M. Consider all the parallel-transported frames f from f_0 along all curves starting from p? The set P of these frames is clearly a principal bundle over M with $\mathrm{Hol}(p)$ as structural group. Let us denote by $\mathfrak{g}$ the Lie algebra of $\mathrm{Hol}(p)$ and by $\mathfrak{g}'$ the subalgebra of the Lie algebra of $O(n)$ built up as follows. Given p in M, vectors x and y in T_pM and an orthonormal frame f of P at p, the curvature endomorphism $R(x, y)$, together with f, generates an element in the Lie algebra of $O(n)$. Let us denote by $\mathfrak{g}'$ the Lie subalgebra of that of $O(n)$ generated by all such elements when p runs through M, x, y through T_pM and f through the frames of P at p.

The Formula (10.57) tells us exactly that $\mathfrak{g}'$ is a subalgebra of $\mathfrak{g}$ and that the condition $\mathfrak{g}' = \mathfrak{g}$ is the theorem we want to prove.

Consider now the distribution D of P defined as follows. At every $f \in P$, the linear space $D(f)$ is the direct sum of the horizontal subspace of P at f and of the linear vertical subspace defined by $\mathfrak{g}'$ at f. One checks that the distribution D is integrable. Then consider the maximal leaf $L(f_0)$ through f_0. By the very construction of P and the definition of D, we have $L(f_0) = P$, hence $\mathfrak{g}' = \mathfrak{g}$. □

10.59 Remark. There are more intuitive proofs of the Ambrose-Singer theorem. Consider a loop based at p which is homotopic to zero and fill it up by a smooth surface. Then replace the finite decomposition of the Lasso Lemma 40 by a double integral on the surface; this will yield the parallel translation along the loop as a double integral of elements of the desired form. This technique is due to Nijenhuis: see [Nij 1].

Note. The Ambrose-Singer theorem is valid for any connection without modification of the proof.

10.60 Remark. We shall not need the following property of the Lie algebra $\mathfrak{hol}$; but we wish to mention it. It was used in [Ber 1]. The assertion is that $\mathfrak{hol}$ contains, as in Formula (10.57) every covariant derivative of the curvature tensor for all possible vectors and every possible parallel translation. More precisely, let λ be any C^1-piecewise curve starting from p and $x, y, x_1, \ldots, x_k$ be any $k+2$ vectors in T_pM. Then for every k the following covariant derivatives of order k of the curvature tensor R obey:

$$(\tau(\lambda))^{-1} D_{\tau(\lambda)(x_1)} \cdots D_{\tau(\lambda)(x_k)} R(\tau(\lambda)(x), \tau(\lambda)(y)) \circ \tau(\lambda) \in \mathfrak{hol}(p).$$

One will find proofs in [Ko-No 1], p. 101 and 152 or in [Wak], p. 107. A geometric idea, extending that of Figure 10.57 is to take curves as in the Figure 10.60; this yields $D_{x_1} R(x, y)$ for example.

10.61 Note. Besides its esthetic appeal Theorem 10.58 has some theoretical importance, but it is not too effective for computing Hol. In fact this theorem looks quite redundant since it contains the whole holonomy in the generators! We will come back in Section 10.101 to that apparently vicious circle. For the moment we note first that Theorem 10.58 will be used in Sections 10.66 and 10.95 below. Note also that Theorem 10.58 can be used to prove what Remark 10.60 makes conceivable,

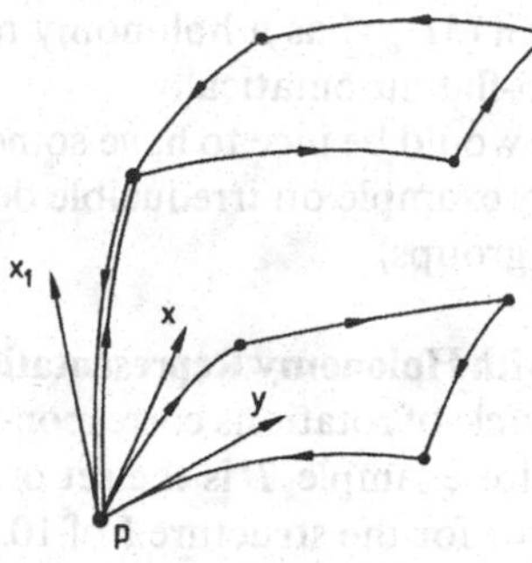

Fig. 10.60

namely: if (M, g) is a real analytic Riemannian manifold then $\mathfrak{hol}(p)$ is generated by the set of all covariant derivatives of the curvature tensor evaluated at p. For the proof see [Ko-No 1], Sections II.10 and III.9.

10.62 Examples. We give a few applications of the trivial part of Theorem 10.58 namely Formula (10.57). Some of these applications are not really needed but we hope they will help the reader to get some intuition about holonomy groups. Other applications are more basic, some of them explain in particular the link with Einstein manifolds. Recall that this link will be summarized in Corollary 10.111.

10.63 Example 1: About Remark 10.39. We return here to the surprising fact that the reducibility of holonomy implies that the holonomy representation is a direct sum. Assume some space T_1 is invariant under Hol(p) and write $T_pM = T_1 \overset{\perp}{\oplus} T_2$. Let us denote by $a, b, \ldots$ (resp. $l, m, \ldots$) indices for elements in T_1 (resp. T_2). For the curvature tensor (at p for example) we then have $R_{xyal} = 0$ for every indices x, y. In particular $R_{bmal} = 0$; and $R_{ablm} = 0$ follows now from the algebraic Bianchi identity 1.24. But $R_{ablm} = 0$ says exactly that the T_1-part of the holonomy acts trivially on T_2.

10.64 Example 2: Manifolds with Holonomy Representation $G_2 \subset O(7)$. One can define G_2 as the subgroup of $O(7)$ whose Lie algebra is made up by all matrices which are antisymetric and obey all relations

$$A_{i+1,i+3} + A_{i+2,i+6} + A_{i+4,i+5} = 0,$$

where indices are computed in the group $\mathbb{Z}/7\mathbb{Z}$. Then Bianchi identities and Formula (10.57) imply, after some computations, that the Ricci curvature of our manifold is necessarily zero (see for example Bonan [Bon 1] for details). In conclusion if Hol $\subset G_2$ for a 7-dimensional Riemannian manifold then this manifold is Ricci-flat.

10.65 Example 3: Manifold with Holonomy Representation Spin(7) $\subset O(8)$. One can define Spin(7) as the subgroup of $O(8)$ whose Lie algebra consists of antisymmetric matrices which obey all relations of the type

$$A_{i,7} + A_{i+1,i+3} + A_{i+2,i+6} + A_{i+4,i+7} = 0 \quad \text{(mod. 8)}.$$

As above one sees again that if (M, g) has a holonomy representation contained in $\mathrm{Spin}(7) \subset O(8)$ then it is Ricci-flat automatically.

Both for G_2 and Spin(7) it would be nice to have some non-computational proof of the Ricci flatness (based for example on irreducible decompositions of curvature tensors with respect to these groups).

10.66 Example 4: Manifolds with Holonomy Representation Contained in $T \cdot Sp(m) \subset O(4m)(m \geq 2)$. Here T is the circle of rotations corresponding to one almost-complex structure (see Section 10.32) for example T is the set of multiplication by complex numbers of modulus equal to 1 for the structure I of 10.32. Then Formulas (10.57), Bianchi identities and some computations imply that the curvature tensor part corresponding to the Lie algebra of T is always equal to zero. Theorem 10.58 then yields the following fact: if the holonomy representation of some Riemannian manifold is contained in $T \cdot Sp(m)$, then it is necessarily contained in fact in $Sp(m)$ in particular (M, g) has to be hyperkählerian.

10.67 Example 5: Special Kähler Manifolds. We want here to make Proposition 10.29 plausible with help of Formula 10.57. We have only to note that the Lie algebra of $SU(m)$ is that consisting of complex skewhermitian matrices with complex trace equal to 0. Then Formula 10.57 implies, if $\mathrm{Hol}^0(g) \subset SU(n)$ that $\mathrm{trace}_{\mathbb{C}} R(x, y) = 0$ for every x, y. But the symmetry properties of the curvature tensor of kähler manifolds (see 2.42) imply that $\mathrm{trace}_{\mathbb{C}} R = -\rho$. □

10.68 How to Actually Compute the Holonomy Group? In general the best way, if any, is to apply the main classification Theorem 10.92. Then, to decide in its list, one applies Table 1, looking at the covariant constant forms that (M, g) has to carry. Now if an exterior form α is such that $D\alpha = 0$, then first it is closed, because (see Remark 1.19c) or [Gr-Kl-Me], p. 42) its exterior differential $d\alpha$ is equal to the alternated covariant derivative: $d\alpha = \mathrm{Alt}(D\alpha)$. Moreover the Hodge dual $*\alpha$ of α will again have vanishing covariant derivative, hence again closed. In conclusion any exterior form whose covariant derivative vanishes is *harmonic*. By Hodge's theorem and de Rham theorem we know now something about the topology of the manifold M. See in [Mic], p. 1144 and [Kob 7] some applications of this technique.

G. Symmetric Spaces; Their Holonomy

10.69 Introduction and some History. We sketch in this part basic facts about symmetric spaces: most of these facts will be needed for the sequel (see among others the statement of the main classification Theorem 10.92). For more details and proofs on symmetric spaces, good references are [Hel], [Ko-No 2], Chapter XI. See also [Los]. See also in this book 7.F.

10.70. The notion of symmetric space was introduced in 1925 by Elie Cartan in the following context. Cartan was trying to classify Riemannian manifolds whose curvature tensor has vanishing covariant derivative: $DR = 0$. His colleagues of that time were concerned with this problem. Cartan attacked the problem by applying 10.19 and 10.57: the holonomy representation of such a manifold should leave the curvature tensor invariant. One then is led to an algebraic problem about curvature tensors and orthogonal representations. Cartan completely solved the problem; his computations were quite lengthy, especially for the exceptional Lie groups. He was applying in particular his own classification of irreducible linear representations of Lie groups.

10.71. To his astonishment, he found the following. The totality of the Riemannian manifolds with $DR = 0$ and which moreover are irreducible, complete and simply connected is made up of two distinct series. Each element in one series is made up of a pair of spaces, a compact one and a non-compact one. The first series is made up of pairs $(G, G^{\mathbb{C}})$ where G runs through all compact simple simply connected Lie groups and where $G^{\mathbb{C}}$ denotes the complex group associated to G. The second series consists of pairs $(G/H, G^*/H)$ where G runs through all simply connected, simple and non-compact Lie groups. The subgroup H is necessarily a maximal compact connected subgroup of G (unique up to conjugacy) and G^* is the compact form of G. In view of this striking fact he succeeded in proving the existence of a direct link between manifolds with $DR = 0$ and real forms of simple Lie groups (whose classification was achieved by him in [Car 3]. Then the classification of manifolds with $DR = 0$ is reduced to the more classical (and cheaper by far) classification of real forms of simple Lie algebras. For a modern proof of this classification "à la Dynkin" see [Hel], second edition.

The name "symmetric space" comes in fact from a third equivalent definition of the spaces under study: namely they are the Riemannian manifolds for which, at least locally, the geodesic symmetry around any point is an isometry, see 7.63. In short the above Cartan results can be summarized in the following:

10.72 Theorem (Elie Cartan). *For a Riemannian manifold (not necessarily complete) the following conditions are equivalent:*

(i) $DR = 0$;

(ii) *the geodesic symmetry s_p around any point p (which is defined only locally) is an isometry.*

For a complete Riemannian manifold (M, g) the two following conditions are equivalent:

(iii) *for every point p in M the geodesic symmetry around p is well defined and is an isometry;*

(iv) *the manifold M is a homogeneous space G/H where G is a connected Lie group, H a compact subgroup of G, and where there exists an involutive automorphism σ of the group G for which, if S denotes the fixed point set of σ and S_e its connected component of the identity one has $S_e \subset H \subset S$. Moreover the Riemannian metric under consideration on G/H is invariant under G.*

Furthermore: if (M, g) satisfies (iii) *or* (iv), *then it satisfies* (i) *and* (ii). *If (M, g) satisfies* (i) *or* (ii) *and if it is simply connected and complete then it satisfies* (iii) *and* (iv).

Note. The relation between (iv) and the real forms of simple Lie groups is a pure Lie algebra result: see [Hel]. Of course the involution is the one coming from the complex conjugation of the complex form.

10.73. *Keys for the proof.* The link between (i) and (ii) is as follows: assume $DR = 0$. Then Jacobi fields (1.30) along a given geodesic are solutions of a differential equation with constant coefficients, so their norms at a given time t, for fixed initial conditions, are the same. In particular their norm is kept fixed under the geodesic symmetry. This proves that the geodesic symmetry is an isometry by the philosophy of normal coordinates (Cartan-Riemann theorem on exponential map: [Be-Ga-Ma], p. 91. Conversely, assume the geodesic symmetry around p is an isometry. Then it respects any Riemannian invariant, in particular the covariant derivative DR of the curvature tensor. At the point p we then have $DR(p) = -DR(p)$ because DR is of order 5; then $DR(p) = 0$.

10.74. *The link between* (iii) *and* (iv) *is as follows.* Consider the group of all isometries of (M, g). Classically it is a Lie group (Myers-Steenrod Theorem 1.77, see also [Ko-No 2], p. 239). Inside it the subgroup generated by the geodesic symmetries around all points in M will be a subgroup, and a Lie subgroup too. Call it G. We then get M as an homogeneous space G/H where H is the isotropy subgroup of a fixed point of M. The involutive automorphism which we are looking for is simply σ defined by $g \mapsto s_p g s_p$ where s_p denotes the geodesic symmetry around p. Conversely, in the situation of (iv), first decompose the Lie algebra $\mathfrak{g}$ of G as the direct sum $\mathfrak{g} = \mathfrak{h} + \mathfrak{m}$ where $\mathfrak{h}$ is the Lie algebra of H and where $\mathfrak{m}$ is the eigenspace of $\mathfrak{g}$ for the eigenvalue -1 of the action of σ on $\mathfrak{g}$. Pick any Euclidean structure on $\mathfrak{m}$ which is invariant under the (compact) adjoint action of H. Then such a Euclidean structure on C can be transported by the action of G as a Riemannian metric on G/H which clearly obeys (iii).

10.75. *The link between* (i) ∪ (ii) *and* (iii) ∪ (iv). This is the more delicate part. For the following reason: the fact that M is simply connected does not imply straightforwardly that the geodesic symmetries are globally well defined. There are two kinds of proof to get (iii) from (ii) in the simply connected case. The first is that given in [Ko-No 2], p. 223 for example; it consists of showing that the geodesic symmetries are indeed globally defined. This is quite lengthy and achieved through the real analyticity of the structure.

The second is one of Elie Cartan's wonderful contributions, in fact it is the spirit of his first classification. We need to find a pair of groups G, H as in (iv) only from the condition $DR = 0$. This is achieved as follows. Fix a point m in M, denote by $\mathfrak{m}$ the tangent space $T_p M$. Denote by $\mathfrak{h}$ the Lie subalgebra of $\mathfrak{o}(T_p M)$ generated by the curvature endomorphisms $R(x, y)$ where x, y run through $\mathfrak{m} = T_p M$. We have an action of $\mathfrak{h}$ precisely by these curvature endomorphisms; this defines a bracket operation $[r, x]$ between $r \in \mathfrak{h}$ and $x \in \mathfrak{m}$. We have of course the bracket operation $[r, s]$ for r, $s \in \mathfrak{h}$. Finally we define the bracket operation $[x, y]$ for x, $y \in \mathfrak{m}$ by $[x, y] = R(x, y) \in \mathfrak{h}$. Now an easy calculation shows that these brackets define on this direct sum $\mathfrak{g} = \mathfrak{h} + \mathfrak{m}$ the structure of a Lie algebra because $DR = 0$ via the

Ricci commutation formula (see 1.21). Then we are done with the introduction of Lie groups G, H whose Lie algebras are $\mathfrak{g}$, $\mathfrak{h}$ respectively. Note the following: the Ambrose-Singer theorem and the philosophy of part C show that for a manifold with $DR = 0$ its holonomy group $\mathrm{Hol}(p)$ is generated by the set of $R(x, y)$ with x, y running through T_pM; from this one guesses that for a symmetric space its holonomy group will coincide more or less with the isotropy subgroup H. See more on this in section 10.79. □

10.76 Definition. *A Riemannian manifold is said to be* locally symmetric *if it satisfies* (i) *or* (ii) *in* 72; *it is said to be* symmetric *if it satisfies* (iii) *or* (iv) *in* 72. *It is said to* be irreducible symmetric *if it is symmetric and if its holonomy* Hol^0 *is irreducible. Instead of a symmetric Riemannian manifold one often says* symmetric space, *respectively* irreducible symmetric space.

Note. A notion of symmetric space can be defined in a more general setting than that of Riemannian manifold, for example in the affine or in the Lorentzian case. For this see [Ko-No 2], Chapter 11 and Part J below.

10.77 Irreducibility. A symmetric space is decomposed in a unique manner into a Riemannian product of irreducible symmetric spaces; more precisely this is a local statement except in the simply connected case where we have a real Riemannian product. If not we have statements up to discrete coverings. This explains why in what follows we will restrict ourselves to SIMPLY CONNECTED IRREDUCIBLE SYMMETRIC SPACES. That the simply connected case, in the setting of (iv) of Theorem 10.72, certainly when G is simply connected and when $H = S_e$ (for S_e see 10.72). For all these affirmations see [Hel], second edition on 7.F.

10.78 Holonomy of Symmetric Spaces. From what was said in section 10.75 one sees that:

10.79 Proposition. *For an irreducible simply connected symmetric space G/H the holonomy group* Hol *is equal to H, as acting by the adjoint representation.*

To achieve the proof after the considerations of Section 10.75 one needs only to be careful about the following: in general a manifold M can be written as a homogeneous space G/H in more than one way. Example: $S^{2n+1} = O(2n+2)/O(2n+1) = U(n+1)/U(1)\cdot U(n)$, or $S^6 = O(7)/O(6) = G_2/SU(3)$. We need only to prove that such a representation is unique when asking for the conditions of (iv) in Theorem 72. This is simple. □

Thus symmetric spaces are Riemannian manifolds for which the holonomy group is known. They give then important examples of these groups. In fact the classification Theorem 10.92 will say that besides symmetric spaces there are extremely few possibilities for holonomy groups.

Conversely for non-symmetric homogeneous spaces G/H there is in general no way to compute their holonomy. One uses in fact the classification theorem; see [Ko-No 2], Section X.4 for more on this.

10.80 Rank and Transitivity of Holonomy. We still consider only irreducible symmetric spaces and the Lie algebra decomposition $\mathfrak{g} = \mathfrak{h} + \mathfrak{m}$. E. Cartan proved that all abelian subalgebras of $\mathfrak{m}$ which are of maximal dimension are conjugate under the adjoint action of H. This common dimension he called the *rank* of G/H; for the proof see [Hel 1]. This extends the previous simpler result for ordinary Lie algebras ([Hel 1]).

These abelian maximal subalgebras of $\mathfrak{m}$ will be called *Cartan subalgebras* of $G/H = M$. In the simply connected case their images in M under the exponential map are flat tori of M when G is compact and euclidean spaces of M when G is non-compact.

From Cartan's result we first deduce the behavior of geodesics on any symmetric space in the simply connected case. When G is non-compact they are lines (isomorphic to $\mathbb{R}$). When G is compact they behave as geodesics in a flat torus do.

A second conclusion of Cartan's result is the following: the rank of G/H is equal to 1 if and only if the isotropy action of H on the unit tangent sphere at the origin is *transitive*. Note that this implies easily that for a compact simply connected symmetric space of rank equal to 1 all its geodesics are closed: see [Bes 1] for more on this subject.

Let us give an elementary way to look at the transitivity of the isotropy action. The orbit $H(x)$ of some $x \in \mathfrak{m}$ has the bracket $[\mathfrak{h}, x]$ for tangent space. For the Killing form $\langle .,. \rangle$ we then have: $\langle [\mathfrak{h}, x], y \rangle = \langle \mathfrak{h}, [x, y] \rangle$ for every y in $\mathfrak{m}$. In particular if y is orthogonal to the orbit $H(x)$ it follows that $[x, y] = 0$ i.e. x, y are in a Cartan subalgebra. We will meet again this computation in Section 10.99 as the crucial step in Simons's proof of the classification theorem. The property $\langle a, [b, c] \rangle = \langle [a, b], c \rangle$ enjoyed by the Killing form will be replaced by Bianchi's identity for the curvature tensor.

10.81 Symmetric Spaces of Rank Equal to 1. They form an important class among symmetric spaces. In the compact case for example they consist exactly of the spheres and the projective spaces over $\mathbb{C}$, $\mathbb{H}$ and Cayley numbers. They coincide with the class (in the simply connected case) of metric spaces whose isometry group is transitive on pairs of equidistant points. See [Bes 1], [Cha], [Mat], [Bus 2]. They are listed in Tables 2 and 3.

10.82 Note. As foreseen by Cartan in [Car 6], p. 590, symmetric spaces play in geometry a role as important as spaces of constant curvature do. Look for example in Tables 2 and 3 at the column of geometric interpretations. See also the following references: [Los], [Bes 1].

10.83 Note. Every irreducible symmetric space is automatically Einstein by 10.79 and 7.44. Moreover the sign is never 0 by 7.61. It is + in the compact case and − in the non-compact case. In the reducible case such a space will admit an Einstein metric if and only if all its components are of the same compactness type.

10.84 Topology. Irreducible simply connected non-compact symmetric spaces are diffeomorphic to $\mathbb{R}^d$. The compact ones have a topology which is well understood.

For this see [Be-La], p. 25, [Bo-Hi], [Gr-Ha-Va] which also contain information about general homogeneous spaces.

10.85 Classification. Irreducible simply connected spaces have been completely classified by Cartan in [Car 6]. For a more modern and simpler classification the reader can consult for example [Hel] (the second edition!). Note that the classification process concerns a good part of mathematics. We may remark that the first edition of [Hel], a book entirely devoted to symmetric spaces, involved giving that classification but without proof! This gap was filled up in the second edition sixteen years later. The classification first put symmetric spaces (irreducible, simply connected) in four types I, II, III, IV. Then for every type there is a finite list consisting of an infinite series (involving classical groups) and a finite number (involving exceptional groups). The reader will find these lists in the Tables at the end of this chapter. We here give only the general structure of the four possible types.

Type I. Here G is a compact simple simply connected Lie group and H is the connected component of the identity element of some non-trivial involutive automorphism of G. Then $M = G/H$.

Type II. Here M, as a manifold, is a compact simple simply connected Lie group endowed with a Riemannian metric which is at the same time left and right invariant. The representation (G, H, σ) is obtained as follows: $G = M \times M$, the direct product group of M with itself, H is the diagonal subgroup of this direct product (isomorphic to M then) and σ is the symmetry which exchanges the coordinates of an element in $M \times M$.

Type III. Here G is a non-compact simple simply connected Lie group, H is a maximal compact connected subgroup of G and σ is the involution which defines it (see the Note following 10.72).

Type IV. Here G is a complex simple simply connected Lie group, H is a maximal compact connected subgroup and σ is the complex conjugation on G which defines H.

10.86 Duality. The above types go in pairs, I and III together and II and IV together. In the pairing under consideration, first the groups G have the same complexified group, second the subgroups H are the same, more precisely their isotropy representation on the tangent space to G/H are the same. The pairs (I, III) are in bijection with the non-compact simply connected simple Lie groups, the pairs (II, IV) are in bijection with simple compact simply connected Lie groups. Hence the list of pairs (I, III) is much longer than that of pairs (II, IV).

10.87 Rank. For the pairs of (II, IV) the rank is nothing but the rank of the corresponding simple Lie group in disguise. For the really interesting pairs (I, III) the rank is listed in the Tables.

10.88 The Kählerian Case. Cartan proved that a symmetric space is hermitian, and then automatically Kähler, if and only if its isotropy group H contains a center of dimension greater than or equal to 1 (at least in the irreducible case). This center is then automatically of dimension equal to 1. See their list in the Tables and, for the proof, see [Ko-No 2], p. 261.

10.89 Tables. We are using mainly Helgason's book notations: [Hel]. In contradiction with our convention (but only for the case of orthogonal Lie algebras) we are using the notation for the orthogonal groups, ignoring the fact that they are not simply connected. This will not change the nature of the quotient spaces $M = G/H$: spheres, real grassmannians, etc... We thought that the reader would prefer the $O(n)$-notation to the spinor-notation: Spin(n).

H. Structure II

The basic result is the following:

10.90 Theorem (Berger [Ber 1] and Simons [Sim]). *Let (M, g) be a Riemannian manifold and assume* Hol^0 *irreducible. Then either* Hol^0 *is transitive on the unit sphere (namely transitive on $U_pM \subset T_pM$ when one works with* $\mathrm{Hol}^0(p)$*) or (M, g) is a locally symmetric space of rank greater than or equal to* 2 *(and then* Hol^0 *is the isotropy group of such a space).*

10.91 The original demonstration of [Ber 1] was using Elie Cartan's classification of irreducible linear real representations (here orthogonal ones) of real Lie groups: see [Car 1], [Car 2]. Cartan's main point was that such a representation can always be obtained with so-called fundamental representations by mean of tensor product representations (and sometimes suitable reductions of these products). The proof showed first that, with few exceptions, such a tensor product representation cannot occur: for Bianchi identities and Formula (10.57) imply $DR = 0$ as soon as the holonomy representation involves more than one tensor product. One is then left with the fundamental representations, explicitly given in Cartan's work [Car 1, 2]. Then again Bianchi identities and Formula (10.57) showed (after) lengthy computations) that either $DR = 0$ or $R = 0$ with a list of possible exceptions listed in Corollary 10.92 below.

But (up to one exception that of Section 10.66: $T \cdot Sp(n)$) this list is exactly that of Lie groups acting transitively on spheres (this was noted to the author in a letter by A. Borel). For the classification of Lie groups acting transitively on spheres, see [Mo-Sa 1, 2], [Bor 1], [Bor 2] and Section 7.13. Then it was tempting to try a direct proof of the fact that if the holonomy representation is not transitive on the unit sphere, then the Riemannian manifold is locally symmetric. This was achieved by J. Simons only seven years after in [Sim]. To make our text more pleasant to read and because Simons proof is still quite long even if shorter than the original one, we prefer to give immediately the pleasant consequences of Theorem 10.90. For the reducible case, see Section 10.107 and for the symmetric case see Section 10.79. A sketch of Simon's proof will be given in Section 10.98.

10.92 Corollary. *Let (M, g) be a Riemannian manifold of dimension n which is not locally symmetric and whose holonomy representation* Hol^0 *is irreducible. Then its holonomy representation* Hol^0 *is one of the following:*

(I) $\mathrm{Hol}^0 = SO(n)$
(II) $n = 2m$ and $\mathrm{Hol}^0 = U(m)$ $(m \geqslant 2)$
(III) $n = 2m$ and $\mathrm{Hol}^0 = SU(m)$ $(m \geqslant 2)$
(IV) $n = 4m$ and $\mathrm{Hol}^0 = Sp(1) \cdot Sp(m)$ $(m \geqslant 2)$
(V) $n = 4m$ and $\mathrm{Hol}^0 = Sp(m)$ $(m \geqslant 2)$
(VI) $n = 16$ and $\mathrm{Hol}^0 = \mathrm{Spin}(9)$
(VII) $n = 8$ and $\mathrm{Hol}^0 = \mathrm{Spin}(7)$
(VIII) $n = 7$ and $\mathrm{Hol}^0 = G_2$.

In the above list we have not made precise the holonomy *representation* for the following reason: each time a group appears this is when the dimension makes its representation completely determined, up to conjugacy by an orthogonal transformation. These representations have been given explicitely above, in Paragraphs 10.28, 31, 32, 64, 65. There is one exception (that of number (VI)) in which case it is the isotropy representation of the symmetric space $\mathbb{C}a P^2 = F_4/\mathrm{Spin}(9)$, see Paragraph 10.96 (VI) below.

10.93 Remark. It follows from Corollary 10.92 that certain spaces are nearly determined by their holonomy representation. Namely, if (M, g) is a simply connected symmetric space of rank bigger than one, and if (N, h) is simply connected and has the same holonomy representation, then (N, h) is isometric to (M, g) or to its dual. We shall see in 10.96 (VI) that this also applies to $\mathbb{C}a P^2$.

10.94. To prove the corollary one applies the results of [Mo-Sa 1, 2], [Bor 1, 2] which give a complete classification of Lie groups G acting transitively on spheres. The list is the following for the sphere S^{n-1} of dimension $n-1$ (see also 7.13):

(I) $G = SO(n)$
(II) $n = 2m$ and $G = U(m)$ $(m \geqslant 2)$
(III) $n = 2m$ and $G = SU(m)$ $(m \geqslant 2)$
(IV) $n = 4m$ and $G = Sp(1) \cdot Sp(m)$ $(m \geqslant 2)$
(IV bis) $n = 4m$ and $G = T \cdot Sp(m)$ $(m \geqslant 2)$
(V) $n = 4m$ and $G = Sp(m)$ $(m \geqslant 2)$
(VI) $n = 16$ and $G = \mathrm{Spin}(9)$
(VII) $n = 8$ and $G = \mathrm{Spin}(7)$
(VIII) $n = 7$ and $G = G_2$.

10.95. Note that the results above give a classification of the abstract groups G acting on S^{n-1}, they say nothing about this action in itself. But in our case we know that *the action is linear* (even orthogonal) because it is a holonomy representation. But then we already remarked in Section 10.92 that the given dimensions tell us exactly (up to conjugacy by an orthogonal transformation) which linear representation is involved. So Theorem 10.90 will be proved if we can throw out the case (IV bis) above. But that was precisely done in Section 10.66.

10.96 Answers to Question A of 10.6. This question was: what are the possible holonomy representations for Riemannian manifolds? We take now the eight representations of Corollary 10.92 one by one. The symmetric case is discarded in view of 10.79. For each case we consider three questions: does there exist at least one Riemannian manifold with that holonomy representation? does there exist a complete one? does there exists a compact one? The answers are summarized in Table 1.

(I): $SO(n)$. We saw in Section 10.54 that this is the general case (generic).

(II): $U(m)$ $(n = 2m, m \geqslant 2)$. This is the general case of a Kähler manifold, see section 10.55.

(III): $SU(m)$ $(n = 2m, m \geqslant 2)$. In Section 10.55 we saw that $\mathrm{Hol}^0(g) \subset SU(m)$ is the case of special Kähler manifolds but note we want $\mathrm{Hol}^0(g) = SU(m)$ the whole group. Already in 1960 Calabi gave examples of such special Kähler manifolds but his examples were only local, not complete: see [Cal 5] on 15.C. In 1970 Calabi got complete examples (unpublished) which are quite difficult already to exhibit. One had to wait until 1978 and the solution of the Calabi conjecture by Yau (Theorem 11.15) to find compact examples. We know now that on any compact Kähler manifold whose first Chern class is zero there will exist some Kähler metric with $\mathrm{Hol}^0 \subset SU(m)$: apply [Yau 3] or [Ast]. But we want $\mathrm{Hol}^0 = SU(m)$ exactly. To achieve that take as a starting Kähler manifold some algebraic hypersurface of degree $m + 2$ in $\mathbb{C}P^{m+1}$; its first Chern class vanishes: 11. Note first that Hol^0 is irreducible because if not the underlying manifold will be a topological product as simply connected (apply 10.43). But this is forbidden by results on the topology of algebraic objects, see [Bea 2]. So we are left with the list of representations in Corollary 10.92. The only one with which is strictly contained in $SU(m)$ is $Sp(m/2)$. Hence (M, g) would have to be hyperkählerian something which is forbidden here by Sections 10.31, 32, 68 since the second Betti number of our hypersurface cannot be larger than $2 (n \geqslant 3)$ by again classical results: see for all this [Bea 2].

(IV): $Sp(1) \cdot Sp(m)$ $(n = 4m, m \geqslant 2)$. As noted in Section 10.32 $\mathrm{Hol}^0 \subset \mathrm{Sp}(1) \cdot Sp(m)$ for any quaternion-Kähler manifold but here we want moreover $\mathrm{Hol}^0 = Sp(1) \cdot Sp(m)$ exactly. Of course this the case for the quaternionic projective space $\mathbb{H}P^m$ and any m. But are they other ones? We note the non-compact homogeneous examples of Aleksevskii have $\mathrm{Hol}^0 = Sp(1) \cdot Sp(m)$, see [Ale 2]. For more on this question see Chapter 14.

(V): $Sp(m)$ $(n = 4m, m \geqslant 2)$. By Section 10.31 we have that $\mathrm{Hol}^0 \subset Sp(m)$ is equivalent to (M, g) is hyperkählerian. It is not before 1979 that Calabi succeeded in constructing non-trivial (i.e. non-flat) hyperkählerian manifolds in [Cal 5]. They were complete but non-compact manifolds. It is not hard to see that Calabi's examples have $\mathrm{Hol}^0 = Sp(m)$. Compact examples were only gotten in 1981 for $m = 2$. Yau's proof of Calabi's conjecture is essential in the construction. For higher m this was achieved in 1982. For references see [Bea 2].

(VI): Spin(9). In [Br-Gr] Brown and Gray have shown that the condition $\mathrm{Hol}^0 = \mathrm{Spin}(9)$ implies the following: either the manifold is flat or at every point its curvature tensor is equal up to a constant to that of the Cayley plane $(\mathbb{C}aP^2, \mathrm{can})$. This was indicated by Aleksevskii in [Ale 1]. The reason is as indicated in Section 10.97: the codimension of Spin(9) in $O(16)$ is big enough to force the curvature tensor, via Formula 10.57 and Bianchi's identity, to be a given one. Then Brown-Gray show that the pointwise condition can be integrated in the sense that the given manifold is everywhere locally isometric (up to scaling) to one of the two symmetric spaces ($\mathbb{C}aP^2$, can) or its dual (see Table 1. The first case appears when the curvature is positive the other in the negative situation. In conclusion the case of holonomy Spin(9) is extremely special in the theory: this group appears only for the two symmetric spaces above.

(VII) and (VIII): Spin(7) and G_2. R. Bryant [Bry 1] has proved recently that there exist germs of metrics in $\mathbb{R}^8$ and $\mathbb{R}^7$ respectively, with holonomy Spin(7) and G_2. His method consists in writing the condition on Hol_0 as an exterior differential system with a differential form φ as unknown. Indeed, any metric on $\mathbb{R}^7$ with holonomy G_2 admits a unique parallel 3-form φ with unit norm. Conversely, any "non-degenerate" differential 3-form φ on $\mathbb{R}^7$ determines a metric $g(\varphi)$. Then A. Gray [Gra 1] has proved that $\mathrm{Hol}_0(g(\varphi)) \subset G_2$, i.e., $D^{g(\varphi)}\varphi = 0$ if and only if $d\varphi = 0$ and $\delta_{g(\varphi)}\varphi = 0$. Cartan-Kähler theory shows existence of solutions of this system. In fact, solutions depend on 20 arbitrary functions of 6 variables. Furthermore, it can be shown that any Ricci flat curvature tensor at one point can be achieved. A similar argument applies to Spin(7). Further examples can be found in Add.B.

However, whether there exists complete manifolds with $\mathrm{Hol}_0 = \mathrm{Spin}(7)$ or G_2 is still unknown. About this question, another reference is [Mar]. Note that these manifolds will be Ricci flat, according to 10.64, 10.65.

10.97 Philosophy. The preceeding discussion can be summarized as follows:

(i) when the codimension of the holonomy representation is small (like $O(n)$, $U(m)$) nothing can said be about the Riemannian manifold;

(ii) when this codimension is quite large (like in the symmetric case or for Spin(9)) then the manifold is completely determined;

(iii) when this codimension is medium then one has some information on the manifold: namely it is Ricci flat in the case of $SU(m)$, Spin(7), G_2, $Sp(m)$ and it is Einstein in the case of $Sp(1)\cdot Sp(m)$.

10.98 Keys for the Proof of Theorem 10.90 Following Simons [Sim]. One starts with $G = \mathrm{Hol}^0(p)$ irreducible and one assumes that G is not transitive on the unit sphere U_pM. We have to show that (M, g) is symmetric. It is mainly an algebraic argument.

10.99. *First step.* It consists of showing that in (M, g) there are many curvature endomorphisms which are zero: $R(a, b) = 0$ for suitable vectors. We apply Formula (10.57) to the effect that the orbit $\mathrm{Orb}(x)$ of a given unit vector $x \in U_pM$ under the action of G has a tangent space $T_x\mathrm{Orb}(x)$ at x which contains every vector like $R(u, v)x$ whatever the vectors u, v in T_pM are. Let now z be orthogonal to $T_x\mathrm{Orb}(x)$, such non zero z exists because G is not transitive. Then one has $g(R(u, v)x, z) = 0$ for every u, v. The symmetry of the curvature tensor implies now $g(R(x, z)u, v) = 0$ for every u, v i.e. we have

(10.100) $$R(x, z) = 0$$

as an endomorphism of T_pM.

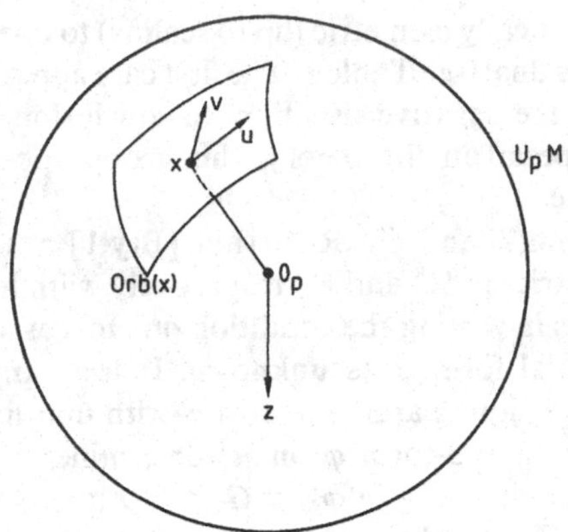

Fig. 10.100

To have these zero curvature endomorphisms is important but we shall have to work much more on them. One reason is the following. Consider a Riemannian product $(M_1 \times M_2, g_1 \times g_2)$. Then $R(x_1, x_2) = 0$ as soon as x_1 is tangent to M_1 and x_2 to M_2. In particular we will have to use in a crucial way the fact that the holonomy representation under consideration is irreducible if we want to prove that "many zero curvature endomorphisms imply symmetry".

10.101. *Second step.* We make again here another heuristic remark which is suggested already in Section 10.61. Let us introduce a notation: V being a euclidean space, let R be a tensor on V which is algebraically like a curvature tensor and let g be some element of the special orthogonal group $SO(V)$. *Define the action* $g(R)$ *of* g *on* R by

(10.102) $$g(R)(x,y) = g^{-1} \circ R(g(x), g(y)) \circ g.$$

Now in the statement of Theorem 10.58 take any two loops with the same end-points p and q. From this one sees that

(10.103) $$g(R)(x,y) \in \mathfrak{g} \quad \text{for any } g \in G$$

where $R = R(p)$ denotes the curvature tensor of (M, g) at p and $G = \mathrm{Hol}^0(p)$ is the holonomy group at p whose Lie algebra is denoted by $\mathfrak{g}$. We are facing now a vicious circle since in the Ambrose-Singer theorem the whole holonomy group is needed together with $R(p)$ to yield its own Lie algebra! This vicious circle is mixture of vice and smartness which stays at the bottom of Simons proof. As a revealing example consider the case of a symmetric manifold. Then by we will have $g(R) = R$ for any g in G. The main idea of Simons is to use Formula 10.100 jointly with the action of G namely the fact that $g(R)(g(x), g(z)) = 0$. We define a *saturated holonomy system* a triple $\{V, R, G\}$ made up of a Euclidean vector space V, a tensor on V enjoying the symmetries of a curvature tensor and a subgroup $G \subset SO(V)$ which has the following property: the Lie algebra $\mathfrak{g}$ of G is generated by the set of the $g(R)(x,y)$ where g runs through G, where x, y run through V and where $g(R)$ is defined by Formula 10.102. A saturated holonomy system will be said to be *symmetric* if $g(R) = R$ for every g in G. This is equivalent to showing that $A(R) = 0$ for every A

in $\mathfrak{g}$ where A denotes the obvious adjoint action of $\mathfrak{g}$ associated to Formula 10.102. Simons first proves that a *saturated holonomy system which is irreducible and not transitive on the unit sphere is symmetric*. This is a purely algebraic statement. We will sketch the proof in steps 3 and 4 below. Then in the fifth step we will pass from this pointwise algebraic symmetry to the manifold (M, g) itself.

10.104. *Third step.* From now on we consider a saturated holonomy system V, R, G which is irreducible and not transitive on the unit sphere. Let $W \subset V$ a vector subspace of V such $R(x, y) = 0$ for every x, y in W and assume that W is maximal under that condition. By Formula (10.100) and by hypothesis we know that there exist such W of dimension bigger than 1. By hypothesis we know that $g(R)(x, y) = 0$ also for every x, y in W. Using Bianchi's identity, some computation and the irreducibility Simons proves the existence of some maximal $Z(W) \supset W$ which is *totally geodesic* i.e. such that $g(R)(x, y)x$ is still in W for every x, y, z in W and every g in G. The subspace $Z(W)$ is built in such a way that $Z(W) = V$ will imply the symmetry of the system. Simons now goes by induction on the dimensions of V and G simultaneously.

10.105. *Fourth step.* By the induction hypothesis and the way the W's are built up one can write the Lie algebra $\mathfrak{g}$ as a direct vector sum $\mathfrak{g} = \sum_i G_i$ with the following property: for every i the dimension of the vector subspace of $\mathfrak{g}$ which is generated by the adjoint actions $A(R)$ when A runs through G_i is smaller than $\dim \mathfrak{g}$. The induction hypothesis and the fact that $Z(W)$ is totally geodesic imply then that $A(R) = 0$ for every A in every G_i. Hence the symmetry of $\{V, R, G\}$.

10.106. *Fifth step.* One needs now to go from an algebraic statement concerning only saturated holonomy systems to a conclusion on Riemannian manifolds. The Ambrose-Singer Theorem 10.58 gives us a saturated holonomy system $\{V = T_pM, R(p), \mathrm{Hol}^0(p)\}$ at every point p in M. Assuming irreducibility and non-transitiviness we know that every $\{T_pM, R(p), \mathrm{Hol}^0(p)\}$ is symmetric. We have to organize these systems in some sense. The situation is similar to that of the case (VI) of Spin(9) in Section 10.96 but more complicated. In the case of Spin(9) one knew that all curvature tensors $R(p)$ were the same up to some constant (depending on the point p). An argument as in Schur's Theorem 1.89 (see [Br-Gr]) shows that this function has to be a constant. In the present case one has to show first that curvature tensors at various points of M are proportional. This is achieved by a theorem of B. Kostant (unpublished) on the structure of the curvature tensor of a symmetric holonomy system. □

10.107 Back to the Reducible Case. We remarked in Section 10.42 that the classification Theorem 10.92 together with Corollary 10.41 does not solve the classification of holonomy groups at least when the Riemannian manifold under consideration is reducible and non-complete. We can finish this problem here now. It is enough to remark that the groups A_i figuring in Corollary 10.41 yield in fact holonomy systems by their very contruction. Then Simons proof goes without modification for the A_i's. These A_i need not in full generality be holonomy groups of some Riemannian manifold but they will have to be by Simons proof taken in the lists of either symmetric spaces either in the list of Corollary 10.92. We have then proved the following result (which to our knowledge is not in the literature):

10.108 Theorem. *The restricted holonomy group* Hol^0 *of any Riemannian manifold (not necessarily complete)* (M, g) *is always a direct sum representation* $A_1 \times \cdots \times A_k$ *where every representation* A_i *is either that of an irreducible symmetric space (they are listed for example in Tables* n°s 2, 3, 4 *or in the list of Corollary* 10.92.

Applications. We now give some applications of the main classification theorem (Corollary 10.92).

10.109. *First application: exterior forms whose covariant derivative is vanishing.* The philosophy is the following: the classification theorem shows that the possible holonomy representations for a Riemannian manifold are very few. Then the philosophy of part C (more precisely the Fundamental Principle 10.19) shows us immediately that there are very few Riemannian manifolds which admit a non trivial exterior form with zero covariant derivative. Note first that we can consider only the case of irreducible manifolds (see Section 10.108). Second we can also assume that our manifold is not a symmetric one: apply 10.79. So we are left with the list of Corollary 10.92. We look now separately at each case. We omit the trivial case of forms of degree 0 (constant functions) and of degree $n = \dim M$ (the volume form which might be only local when M is not orientable, see Section 10.24).

(I): $SO(n)$. No form with vanishing covariant derivative.

(II): $U(m)$ $(n = 2m, m \geqslant 2)$. The group $U(m) \subset O(2m)$ leaves invariant only the standard 2-form and its powers. Then (M, g) can have only its Kähler form (Section 10.26) and the various powers of this Kähler form.

(III): $SU(m)$ $(n = 2m, m \geqslant 2)$. The group $SU(m) \subset O(2m)$ leaves invariant, besides the powers of the standard 2-form, only the complex volume form (see Section 10.28) and its complex conjugate. So on (M, g), our special Kähler manifold with exactly $\mathrm{Hol}^0 = SU(m)$, the only forms with zero covariant derivative could be the Kähler form and its powers, the complex volume form and its conjugate. The volume forms will only exist if $\mathrm{Hol} = SU(m)$.

(IV): $Sp(1) \cdot Sp(m)$ $(n = 4m, m \geqslant 2)$. It is not hard to see that the only exterior forms invariant by $Sp(1) \cdot Sp(m) \subset O(4m)$ are the various powers of the form θ met in Section 10.32. This can be seen either algebraically either with the following trick: in the spirit of Section 10.68 such an invariant form would yield a harmonic form on $\mathbb{H}P^m$. But the Betti numbers of $\mathbb{H}P^m$ are known, they are equal to 1 in dimensions 4, 8, , $4m - 4$. Hence no other forms than the θ^k $(k = 1, \ldots, m-1)$ are permitted. Now a manifold (M, g) with holonomy group $Sp(1) \cdot Sp(m)$ will have as zero covariant derivative exterior forms exactly the θ^k $(k = 1, \ldots, m-1)$.

(V): $Sp(m)$ $(n = 4m, m \geqslant 2)$. The only invariant forms of $\mathbb{R}^{4m}$ under $Sp(m) \subset O(4m)$ are probably the various products and sums of the three Kähler forms ξ, η, ζ of Section 10.32. We are convinced that this simple fact is true. Down on Riemannian manifolds with holonomy group $Sp(m)$, we believe that only the forms belonging to the exterior algebra generated by the three Kähler forms will have zero covariant derivative.

(VI): Spin(9). Since there are only two explicit possible Riemannian manifolds (see Section 10.96 (VI) the situation is not really interesting. We mention however that Spin(9) $\subset O(16)$ leaves invariant exactly one 8-form, as it will follow for example of the topology of $\mathbb{C}aP^2$ (whose Betti numbers are all zero but 1 in dimension 8) by applying Section 10.68.

(VII) and (VIII): Spin(7) and G_2. We mention here that Spin(7) $\subset O(8)$ leaves invariant one form of degree 4 (and no more up to a scalar) and that G_2 leaves in $\mathbb{R}^7$ one 3-form invariant and its dual 4-form and no more. See for example [Bon 1]. This is crucial in R. Bryant's approach, see [Bry 1, 2].

10.110 Corollary. *Assume that a Riemannian manifold (M, g) possesses some exterior form α whose covariant derivative vanishes. Assume that α is not zero and not decomposable. Then (M, g) should fall in one of the following categories: symmetric manifolds, Kähler manifolds, quaternionic manifolds, manifolds with holonomy group G_2 or* Spin(7).

The above corollary follows immediately of the preceeding analysis and from what has been said in Section 10.96 concerning the case of Spin(9) and finally from Remark 10.22. Let us now add the results of Sections 10.29, 31, 32, 64, 65 to the effect that in every case we get an Einstein manifold. We then have proved the even more striking:

10.111 Corollary. *Let (M, g) be a Riemannian manifold admitting an exterior form α whose covariant derivative is vanishing. Assume now that α is not zero, not decomposable and not some power of the Kähler form if (M, g) is Kähler. Then (M, g) is automatically an Einstein manifold.*

This corollary has been used recently in various contexts, see for example [Mic] and [Kob 6].

I. The Non-Simply Connected Case

10.112. Using section 10.15 one obtains the inclusion $\mathrm{Hol}(g) \subset \mathrm{Norm}(\mathrm{Hol}^0(g))$ where Norm denotes the normalizer in $O(n)$ for subgroups of $SO(n)$. Independently of any geometry we should first study the possible algebraic normalizers for the various possible Hol^0. Using Section 10.43 we should first study the normalizer in $O(n = n_0 + n_1 + \cdots + n_k)$ of a direct sum representation of representations $A_1, \ldots,$ A_k of $O(n_1), \ldots, O(n_k)$ for the direct orthogonal decomposition $T_0 \overset{\perp}{\oplus} T_1 \overset{\perp}{\oplus} \cdots \overset{\perp}{\oplus} T_k$, where the representation under consideration acts trivially on the part T_0. We have the easy:

10.113 Lemma. *The quotient of normalizers* $\mathrm{Norm}(A_1 \times \cdots \times A_k)/\mathrm{Norm}_{O(0)}(Id) \times \mathrm{Norm}_{O(1)}(A_1) \times \cdots \times \mathrm{Norm}_{O(k)}(A_k)$ *is finite.*

In conclusion, up to finite quotients, we are led to compute the normalizers of the groups of Corollary 10.92 and those of holonomy representations of symmetric spaces (which coincide with the isotropy representation by Section 10.79). We have the:

10.114 Proposition. *For symmetric spaces the normalizer of the isotropy representation (or the holonomy representation) is always finite. For the "eight maffia" they are respectively*:

$$
\begin{aligned}
&\text{(I): Norm}(SO(n)) = O(n)\\
&\text{(II): Norm}(U(m)) = U(m)\\
&\text{(III): Norm}(SU(m)) = U(m)\\
&\text{(IV): Norm}(Sp(1)\cdot Sp(m)) = Sp(1)\cdot Sp(m)\\
&\text{(V): Norm}(Sp(m)) = Sp(1)\cdot Sp(m)\\
&\text{(VI): Norm}(\mathrm{Spin}(9)) = \mathrm{Spin}(9)\\
&\text{(VII): Norm}(\mathrm{Spin}(7)) = \mathrm{Spin}(7)\\
&\text{(VIII): Norm}(G_2) = G_2.
\end{aligned}
$$

The case of symmetric spaces in treated in [Wol 5], p. 540. For the maffia this is elementary, see for example [Ber 1], pp. 326–327.

10.115. We now turn to the question already raised in Section 10.13:

Is the holonomy group of a compact manifold compact?

Today there is no complete answer to this question. However we will see below that the answer is yes in many cases. But note that our proofs are expensive. They involve the classification theorem, the reduction theorem (somewhat simpler) and the Cheeger-Gromoll reduction theorem for Ricci-flat non compact manifolds. It would be nice to have a direct proof.

10.116. Remark that we can work up to finiteness since Hol^0 is always compact (Section 10.51). Then the conjunction of Lemma 113 and of the list of Proposition 10.114 shows that we can get infinite stuff only when in the decomposition into irreducible factors one or more factors of the three following types occur: the identity Id on $O(n_0)$ if $n_0 \geqslant 2$, $SU(m)$, $Sp(m)$. This is already something. But now we use Sections 10.26 and 10.31 to the effect that a factor like $SU(m)$ and/or $Sp(m)$ would yield some zero Ricci curvature. Looking now at the universal covering of (M, g) and splitting it into a product according to de Rham's Theorem 10.43, an infinite stuff in a Ricci flat manifold will imply a further decomposition into a product, something which is impossible for such a factor like $SU(m)$ or $Sp(m)$ since they are irreducible: the product decomposition follows from the deep theorem of Cheeger and Gromoll [Ch-Gr] see 6.65. Then the only possibility left is to have a factor $Id \subset O(n_0)$ with $n_0 \geqslant 2$, that is to say our Riemannian manifold is locally a product with a flat part of dimension bigger than 1. But then as already remarked in Section 10.25 we can appeal to Bieberbach's theorem: a pure flat manifold always has finite holonomy group. So the only possibility to get into trouble would be to have a mixture of a flat part (in the de Rham decomposition theorem) of dimension greater than 1 and an infinite fundamental group for some part with holonomy group $SU(m)$ or $Sp(m)$. In particular we have proved the following:

10.117 Theorem (Cheeger-Gromoll [Ch-Gr]). *Let (M, g) be a compact Riemannian manifold and let $\tilde{M} = \mathbb{R}^{n_0} \times N$ be the de Rham decomposition of the universal Riemannian covering of $(\tilde{M}, \tilde{g})$. Then* $\mathrm{Hol}(g)$ *is certainly compact as soon as $n_0 \leqslant 1$ or N is compact.*

J. Lorentzian Manifolds

10.118. We consider now Lorentzian manifolds, i.e. 4-dimensional smooth manifolds endowed with a bilinear symmetric form, non degenerate and of signature $(+---)$. For such manifolds we already said in 10.3 that there is a notion of holonomy groups. Then the question now is: what are the possibilities for the holonomy representations of these manifolds?

To our knowledge this question is treated in the literature in a quasi-folkloric way. In particular it is quite difficult to find one's way through it. As a consequence, we will in this section give just some facts concerning the restricted holonomy groups $\mathrm{Hol}^0(g)$ *(denoted by H)* and give few references.

10.119. The key tool here is the extension by Wu of the Theorem 10.43 concerning reducible holonomy representations. Since for subgroups of $SO_0(1,3)$ reducibility does not necessarily imply complete reducibility one has to be careful here. We consider more generally smooth manifolds (M,g) where g is a non-degenerate bilinear symmetric form and its restricted holonomy group $H = \mathrm{Hol}^0(g)$. Assume $H = H_0 \times H_1 \times \cdots \times H_k$ where there is an orthogonal direct sum decomposition

$$T = \sum_{i=0}^{k} T_k$$

of the tangent space of M with the following links: for every i the group H_i acts trivially on T_j for every $j \neq i$ and H_i acts on T_i *weakly irreducibly* (i.e. the only non-degenerate invariant subspaces are $\{0\}$ and T_i). Moreover H_0 consists only in the identity and the restriction of g to T_0 is non-degenerate.

Wu's theorem ([Wu 1, 2, 3]) is now that locally (and globally if (M,g) is simply connected and complete) (M,g) is isometric to the direct product of a flat (M_0,g_0) and of manifolds (M_i,g_i) with weakly irreducible holonomy groups $(i \geqslant 1)$. When one has such a trivial decomposition one will call (M,g) *decomposable.*

10.120. When M is 2-dimensional with g of signature $(+-)$ the group H can only be $SO_0(1,1)$ or $\{1\}$. By the above either $H = SO_0(1,1)$ either (M,g) is flat. Both possibilities occur in fact, and also for Einstein and even symmetric ones.

10.121. For 3-dimensional (M,g) with g of signature $(+--)$ obvious occuring cases are $H = \{1\}$ and $H = SO_0(1,2)$. Also occur the two possible decomposition cases: $H = \{1\} \times SO(2)$ and $H = SO_0(1,1) \times \{1\}$. There are only two possible weakly irreducible subgroups of $SO_0(1,2)$, called A^1 and A^2. Written with the identification $SO_0(1,2) = SL(2,\mathbb{R})/\pm 1$ they are the following set of matrices:

$$A^1 = \left\{\begin{pmatrix} 1 & b \\ 0 & 1 \end{pmatrix} : b \in \mathbb{R}\right\}, \qquad A^2 = \left\{\begin{pmatrix} a & b \\ 0 & 1/a \end{pmatrix} : a \in \mathbb{R}_+^*, b \in \mathbb{R}\right\}.$$

In $SO_0(1,2)$ the subgroup A^1 consists of the elements leaving invariant a given null-vector. And A^2 of those leaving invariant a null-line. Examples of g with $H = A^1$ and $H = A^2$ are:

for A^1: $f(z)\,dx^2 + 2\,dy\,dz$, $f > 0$
for A^2: $f(z)\,dx^2 + 2\,dy\,dz + g(z)\,dz^2$, f and $g > 0$.
There is even a symmetric space with $H = A^1$ namely the Cohen-Wallach one quoted in 7.117.

10.122. We turn finally to the Lorentzian case: g is of signature $(+ - - -)$. Again we have trivially occuring possibilities for H:

$SO_0(1,3)$, $\{1\} \times SO(3)$, $SO_0(1,1) \times SO(2)$, $SO_0(1,2) \times \{1\}$,
$SO_0(1,1) \times \{1\} \times \{1\}$, $\{1\} \times \{1\} \times SO(2)$,
$A^1 \times \{1\} \times \{1\}$, $A^2 \times \{1\} \times \{1\}$, $\{1\}$.

Now there are only four other —a priori—possibilities for H. The first is B^1, namely a one-parameter subgroup of $SO_0(1,1) \times SO(2)$ which is neither $SO_0(1,1) \times \{1\} \times \{1\}$ nor $\{1\} \times \{1\} \times SO(2)$. The three others are B^2, B^3_θ, B^4 which we are going to write with matrices under the identification $SO_0(1,3) = SL(2,\mathbb{C})$. Namely:

$$B^2 = \left\{\begin{pmatrix} 1 & b \\ 0 & 1 \end{pmatrix} : b \in \mathbb{C}\right\},$$

$$B^3_\theta = \left\{\begin{pmatrix} \theta & b \\ 0 & \theta/a \end{pmatrix} : a \in \mathbb{R}^*_+, b \in \mathbb{R}\right\} \quad \text{where } \theta \in \mathbb{C} \text{ is of norm 1},$$

$$B^4 = \left\{\begin{pmatrix} a & b \\ 0 & 1/a \end{pmatrix} : a \in \mathbb{C}^*, b \in \mathbb{R}\right\}.$$

The group B^4 consists of the elements which leave invariant (globally) a given null-plane and B^2 of those which leave invariant moreover a null-vector of this null-plane.

10.123. Concerning the existence of Lorentzian manifolds (eventually Einstein and symmetric) with holonomy group equal to B^1, B^2, B^3_θ, B^4 the situation to our knowledge is as follows. The group B^1 can never occur. The reason is that the surprising 10.38 still works here by Wu's theorem. In [Sch] and [Go-Ke] there are examples with B^2 and B^4, which are Einstein
For B^2 take

$$dx^2 + dy^2 + 2\,du\,dv + \omega(x,y)\,du^2 \quad \text{with } \Delta\omega = 0.$$

For B^4 take

$$dx^2 + dy^2 + 2\,du\,dv + 2\rho\,dx\,du + \left(\omega - \frac{\partial\rho}{\partial x}v\right)dv^2$$

(with some conditions on ω and ρ).

For B^3_θ in [Sch] it is proved that there is no Einstein Lorentz manifold with $H = B^3_\theta$. But the general existence is an open problem as far as we know.

The realization of these various groups by symmetric Lorentz spaces seems to be open, as far as we know, except for B^2 which is the holonomy group of the Cahen-Wallach example quoted in 7.117.

K. Tables

Table 1. The Eight Candidates to a Holonomy Fellowship

$G \subset SO(n)$		$\dim G$	$\text{Norm}_{O(n)} G$	Signification of $\text{Hol}(g) \subset G$	Ricci curvature	Algebra of invariant exterior differential forms	Existence			Existence of non-trivial compact deformations
							Local	Complete	Compact	
(I) $SO(n)$		$\frac{n(n-1)}{2}$	$O(n)$			None	Generic Riemannian manifold			Yes
(II) $U(n)$	$2m$	m^2	$U(m)$	Equivalent to be Kähler		The Kähler form ω	Generic Kähler manifold			Yes
(III) $SU(m)$	$2m$	$m^2 - 1$	$U(m)$	Equivalent to Ricci flat Kähler	Ricci flat	The Kähler form ω, the complex volume form and its dual	Yes	Yes	Yes	Yes
(IV) $Sp(1) \cdot Sp(m)$	$4m$	$m^2 + 2m + 3$	$Sp(1) \cdot Sp(m)$	Equivalent to quaternion-Kähler	Einstein	The 4-form θ	Yes	Some homogeneous by Aleksevskii	Yes but only standards known	?
(V) $Sp(m)$	$4m$	$m^2 + 2m$	$Sp(1) \cdot Sp(m)$	Equivalent to hyperkälherian	Ricci flat	Probably three 2-forms	Yes	Yes	Yes	Yes
(VI) Spin(9)	16	36	Spin(9)	?	Einstein	One 8-form	Necessarily isometric (locally) to (CaP^2, can) or to its non-compact dual			No
(VII) Spin(7)	8	21	Spin(7)	?	Ricci flat	One 4-form	Yes	?	?	?
(VIII) G_2	7	14	G_2	?	Ricci flat	One 3-form and its dual	Yes	?	?	?

Table 2. Irreducible Symmetric Spaces of Type I

Helgason's type	G	H	$\dim G/H$	rank	Isotropy representation[1])	Kähler or not	Geometric realization
A I	$SU(n)$	$SO(n)$	$\frac{(n-1)(n+2)}{2}$	$n-1$	$\Lambda^p SO(n) \underline{\otimes} \Lambda^p SO(n)$ $p = 2$ if n odd $p = 3$ if n even	No	Set of the $\mathbb{R}P^{n-1}$'s in $\mathbb{C}P^{n-1}$ or set of the real structures of $\mathbb{C}^n$ (which leave invariant the complex determinant)
A II	$SU(2n)$	$Sp(n)$	$(n-1)(2n+1)$	$n-1$	$\underline{\Lambda}^2 Sp(n)$	No	Set of quaternionic structures of $\mathbb{C}^{2n}$ compatible with its Hermitian structure or set of the metric compatible fibrations $S^3 \to \mathbb{C}P^{2n-1} \to \mathbb{H}P^{n-1}$
A III	$SU(p+q)$	$S(U(p) \times U(q))$ $p \leqslant q$	$2pq$	$\min(p,q)$	$S(U(p) \otimes U(q))$	Yes	Complex p-Grassman manifold of $\mathbb{C}^{p+q}$ (in particular $\mathbb{C}P^q$ if $p = 1$) or set of the $\mathbb{C}P^{p-1}$'s in $\mathbb{C}P^{p+q-1}$
BD I	$SO(p+q)$	$SO(p) \times SO(q)$	pq	$\min(p,q)$	$SO(p) \otimes SO(q)$	Yes if and only if $p = 2$	real p-Grassman manifold of $\mathbb{R}^{p+q}$ (in particular $\mathbb{R}P^q$ if $p = 1$) or set of the $\mathbb{R}P^{p-1}$'s in $\mathbb{R}P^{p+q-1}$
D III	$SO(2n)$	$U(n)$	$n(n-1)$	$\left[\frac{n}{2}\right]$	$\Lambda^2 U(n)$	Yes	Set of complex structures of $\mathbb{R}^{2n}$ compatible with its Euclidean structures or set of the metric-compatible fibrations $S^1 \to \mathbb{R}P^{2n-1} \to \mathbb{C}P^{n-1}$

C I	$Sp(n)$	$U(n)$	$n(n+1)$	n	$U(n) \underline{\otimes} U(n)$	Yes	Set of the $\mathbb{C}P^{n-1}$'s in $\mathbb{H}P^{n-1}$ or set of the complex structures of $\mathbb{H}^n$
C II	$Sp(p,q)$	$Sp(p) \times Sp(q)$ $p \leqslant q$	$4pq$	$\min(p,q)$	$Sp(p) \otimes Sp(q)$	No	Quaternionic p-Grassman manifold of $\mathbb{H}^{p+q}$ (in particular $\mathbb{H}P^q$ if $p=1$) or set of the $\mathbb{H}P^{p-1}$'s of $\mathbb{H}P^{p+q-1}$
E I	E_6	$Sp(4)$	42	6	$\underline{\wedge}^4 Sp(4)$	No	Antichains of $(\mathbb{C} \otimes \mathbb{C}a)P^2$
E II	E_6	$SU(6) \times SU(2)$	40	4	$\wedge^3 SU(6) \otimes SU(2)$	No	Set of the $(\mathbb{C} \otimes \mathbb{H})P^2$'s in $(\mathbb{C} \otimes \mathbb{C}a)P^2$
E III	E_6	$SO(10) \times SO(2)$	32	2	$\mathrm{Spin}(10) \otimes SO(2)$	Yes	Rosenfeld's elliptic projective plane $(\mathbb{C} \otimes \mathbb{C}a)P^2$
E IV	E_6	F_4	26	2	F_4	No	Set of the $\mathbb{C}aP^2$'s in $(\mathbb{C} \otimes \mathbb{C}a)P^2$
E V	E_7	$SU(8)$	70	7	$\wedge^4 SU(8)$	No	Antichains of $(\mathbb{H} \otimes \mathbb{C}a)P^2$
E VI	E_7	$SO(12) \times SU(2)$	64	4	$\mathrm{Spin}(12) \otimes SU(2)$	No	Rosenfeld's elliptic projective plane $(\mathbb{H} \otimes \mathbb{C}a)P^2$
E VII	E_7	$E_6 \times SO(2)$	54	3	$E_6 \otimes SO(2)$	Yes	Set of the $(\mathbb{C} \otimes \mathbb{C}a)P^2$'s in $(\mathbb{H} \otimes \mathbb{C}a)P^2$
E VIII	E_8	$SO(16)$	128	8	$\mathrm{Spin}(16)$	No	Rosenfeld's elliptic projective plane $(\mathbb{C}a \otimes \mathbb{C}a)P^2$ [2])
E IX	E_8	$E_7 \times SU(2)$	112	4	$\wedge^2 E_7 \otimes SU(2)$	No	Set of the $(\mathbb{H} \otimes \mathbb{C}a)P^2$ in "$(\mathbb{C}a \otimes \mathbb{C}a)P^2$" [2])

Table 2 (continued)

Helgason's type	G	H	$\dim G/H$	rank	Isotropy representation[1])	Kähler or not	Geometric realization
F I	F_4	$SP(3) \times SU(2)$	28	4	$\underline{\Lambda}^3 Sp(3) \otimes SU(2)$	No	Set of the $\mathbb{H}P^2$'s in $\mathbb{C}aP^2$
F II	F_4	$SO(9)$	16	1	Spin(9)	No	Cayley elliptic projective plane $\mathbb{C}aP^2$
G I	G_2	$SU(2) \times SU(2)$	8	2	$\underline{\otimes}^3 SU(2) \otimes SU(2)$	No	Set of the quaternionic subalgebras of $\mathbb{C}a$

[1]) here Λ (resp. $\otimes$) denotes the exterior (resp. tensor) product representation and $\underline{\Lambda}$ (resp. $\underline{\otimes}$) denotes the natural irreducible representation deduced from it

[2]) up to this day an algebraic definition of this projective plane over $\mathbb{C}a \otimes \mathbb{C}a$ seems pending, see [Fre] and [Ros]

Table 3. Irreducible Symmetric Spaces of Type III

Helgason's type	G	H	$\dim G/H$	rank	Isotropy representation[1])	Kähler or not	Geometric interpretation
A I	$SL(n,\mathbb{R})$	$SO(n)$	$\frac{(n-1)(n+2)}{2}$	$n-1$	$\Lambda^p SO(n)\,\underline{\otimes}\,\Lambda^p SO(n)$ $p=2$ if n odd $p=3$ if n even	No	Set of Euclidean structures on $\mathbb{R}^n$ or set of the $\mathbb{R}P^n_{hyp}$'s in $\mathbb{C}P^n_{hyp}$
A II	$SU^*(2n)=$ $SL(n,\mathbb{H})$	$Sp(n)$	$(n-1)(2n+1)$	$n-1$	$\underline{\Lambda}^2 Sp(n)$	No	Set of the $\mathbb{H}P^{n-1}_{hyp}$'s in $\mathbb{C}P^{2n-1}_{hyp}$
A III	$SU(p,q)$	$S(U(p)\times U(q))$ $p\leqslant q$	$2pq$	$\min(p,q)$	$S(U(p)\otimes U(q))$	Yes	Grassman manifold of positive definite $\mathbb{C}^p$'s in $\mathbb{C}^{p,q}$, or set of the $\mathbb{C}P^{p-1}_{hyp}$'s in $\mathbb{C}P^{p+q+1}_{hyp}$ (in particular, complex hyperbolic space $\mathbb{C}P^q_{hyp}$ if $p=1$)
BD I	$SO_0(p,q)$	$SO(p)\times SO(q)$ $p\leqslant q$	pq	$\min(p,q)$	$SO(p)\otimes SO(q)$	Yes if and only if $p=2$	Grassman manifold of positive definite $\mathbb{R}^p$'s in $\mathbb{R}^{p,q}$, or set of the $\mathbb{R}P^{p-1}_{hyp}$'s in $\mathbb{R}P^{p+q+1}_{hyp}$ (in particular, real hyperbolic space $\mathbb{R}P^q_{hyp}$—denoted by H^q in 1.37—if $p=1$)
D III	$SO^*(2n)=$ $SO(n,\mathbb{H})$	$U(n)$	$n(n-1)$	$[n/2]$	$\Lambda^2 U(n)$	Yes	Set of quaternionic quadratic forms on $\mathbb{R}^{2n}$, or set of the $\mathbb{C}P^{n-1}_{hyp}$'s in $\mathbb{R}P^{2n-1}_{hyp}$
C I	$Sp(n,\mathbb{R})$	$U(n)$	$n(n+1)$	n	$U(n)\,\underline{\otimes}\,U(n)$	Yes	Set of Lagrangian subspaces of $\mathbb{R}^{2n}$ or set of the $\mathbb{C}P^n_{hyp}$'s in $\mathbb{H}P^n_{hyp}$
C II	$Sp(p,q)$	$Sp(p)\times Sp(q)$ $p\leqslant q$	$4pq$	$\min(p,q)$	$Sp(p)\otimes Sp(q)$	No	Grassman manifold of positive definite $\mathbb{H}^p$'s in $\mathbb{H}^{p,q}$, or set of the $\mathbb{H}P^{p-1}_{hyp}$'s in $\mathbb{H}P^{p+q+1}_{hyp}$ (in particular, quaternionic hyperbolic space $\mathbb{H}P^q_{hyp}$ if $p=1$)

Table 3 (continued)

Helgason's type	G	H	$\dim G/H$	rank	Isotropy representation[1])	Kähler or not	Geometric interpretation
E I	E_6^6	$Sp(4)$	42	6	$\underline{\wedge}^4 Sp(4)$	No	Anti-chains of $(\mathbb{C}\otimes\mathbb{C}a)P^2_{hyp}$
E II	E_6^2	$SU(6)\times SU(2)$	40	4	$\wedge^3 SU(6)\otimes SU(2)$	No	Set of the $(\mathbb{C}\otimes\mathbb{H})P^2_{hyp}$'s in $(\mathbb{C}\otimes\mathbb{C}a)P^2_{hyp}$
E III	E_6^{-14}	$SO(10)\times SO(2)$	32	2	$\mathrm{Spin}(10)\cdot SO(2)$	Yes	Rosenfeld's hyperbolic projective plane $(\mathbb{C}\otimes\mathbb{C}a)P^2_{hyp}$
E IV	E_6^{-26}	F_4	26	2	F_4	No	Set of the $\mathbb{C}a\,P^2_{hyp}$'s in $(\mathbb{C}\otimes\mathbb{C}a)P^2_{hyp}$
E V	E_7^7	$SU(8)$	70	7	$\wedge^4 SU(8)$	No	Anti-chains of $(\mathbb{H}\otimes\mathbb{C}a)P^2_{hyp}$
E VI	E_7^{-5}	$SO(12)\times SU(2)$	64	4	$\mathrm{Spin}(12)\otimes SU(2)$	No	Rosenfeld's hyperbolic projective plane $(\mathbb{H}\otimes\mathbb{C}a)P^2_{hyp}$
E VII	E_7^{-25}	$E_6\times SO(2)$	54	3	$E_6\otimes SO(2)$	Yes	Set of the $(\mathbb{C}\otimes\mathbb{C}a)P^2_{hyp}$'s in $(\mathbb{H}\otimes\mathbb{C}a)P^2_{hyp}$
E VIII	E_8^8	$SO(16)$	128	8	$\mathrm{Spin}(16)$	No	Rosenfeld's hyperbolic projective plane "$(\mathbb{C}a\otimes\mathbb{C}a)P^2_{hyp}$" [2])
E IX	E_8^{-24}	$E_7\times SU(2)$	112	4	$\wedge^2 E_7\otimes SU(2)$	No	Set of the $(\mathbb{H}\otimes\mathbb{C}a)P^2_{hyp}$'s in "$(\mathbb{C}a\otimes\mathbb{C}a)P^2_{hyp}$" [2])
F I	F_4^4	$Sp(3)\times SU(2)$	28	4	$\underline{\wedge}^3 Sp(3)\otimes SU(2)$	No	Set of the $\mathbb{H}P^2_{hyp}$'s in $\mathbb{C}aP^2_{hyp}$
F II	F_4^{-20}	$SO(9)$	16	1	$\mathrm{Spin}(9)$	No	Hyperbolic Cayley projective plane $\mathbb{C}a\,P^2_{hyp}$
G I	G_2^2	$SU(2)\times SU(2)$	8	2	$\underline{\otimes}^3 SU(2)\otimes SU(2)$	No	Set of the non-division quaternionic sub-algebras of the non-division Cayley algebra

[1]) here $\wedge$ (resp. $\otimes$) denotes the exterior (resp. tensor) product representation and $\underline{\wedge}$ (resp. $\underline{\otimes}$) denotes the natural irreducible representation deduced from it

[2]) up to this day an algebraic definition of this hyperbolic plane over $\mathbb{C}a\otimes\mathbb{C}a$ seems pending, see [Fre] and [Ros]

Table 4. Irreducible Symmetric Spaces of Type II and IV

Type II: $M = G$	*Type* IV: G	*dim* G	*rank* G
$SU(n+1)$	$SL(n+1,\mathbb{C})$	$n(n+2)$	n
$SO(2n+1)$	$SO(2n+1,\mathbb{C})$	$n(2n+1)$	n
$Sp(n)$	$Sp(n,\mathbb{C})$	$n(2n+1)$	n
$SO(2n)$	$SO(2n,\mathbb{C})$	$n(2n-1)$	n
E_6	$E_6^{\mathbb{C}}$	78	6
E_7	$E_7^{\mathbb{C}}$	133	7
E_8	$E_8^{\mathbb{C}}$	248	8
F_4	$F_4^{\mathbb{C}}$	52	4
G_2	$G_2^{\mathbb{C}}$	14	2

Chapter 11. Kähler-Einstein Metrics and the Calabi Conjecture

11.1. Among the known examples of Einstein manifolds, a good many are Kähler. In fact, all compact examples with zero or negative scalar curvature are either Kähler, or locally homogeneous. On a complex manifold, one often gets Kähler-Einstein metrics by specific techniques. One reason is perhaps, in the Kähler case, the relative autonomy of the Ricci tensor with regard to the metric, once the complex structure is given. The Ricci tensor—or, to be precise, the Ricci form—only depends on the volume form. On the other hand, on a compact manifold, the cohomology class of the Ricci form is determined by the complex structure (so that, in particular, the sign of an eventual Kähler-Einstein metric is itself determined by the complex structure). Due to these circumstances, it has been possible to exhibit some existence theorems of Einstein metrics in the Kähler framework (Calabi-Yau and Aubin-Calabi-Yau theorems) which have no counterpart in general Riemannian geometry.

In Section A, we discuss the fundamental necessary condition on the cohomology of a compact complex manifold in order that it admits a Kähler-Einstein metric. It turns out that this condition is also sufficient in the negative and Ricci flat cases. This is a consequence of the existence theorems which are stated in Section B together with their main applications. Some of these show how Einstein's equations may be used to prove results apparently independant of it. Section C contains a few words about the proofs of the existence theorems. Of course, we could not treat the hard analysis in such a short space.

The present proofs definitely fail in the positive case. In fact, there are further obstructions, which are gathered in Section D. Observing that all known obstructions are linked with holomorphic vector fields, E. Calabi and A. Lichnerowicz suggested studying a weaker condition on a Kähler metric, that it be "extremal". E. Calabi indeed showed that certain compact complex manifolds, which do not admit Kähler-Einstein metrics, carry such extremal metrics. A short treatment of this extension of the subject has been included in Section E.

A. Kähler-Einstein Metrics

In this chapter, M will denote a complex manifold (i.e., a manifold endowed with a fixed complex structure), of complex dimension m.

11.2 Definition. A Kähler metric on a complex manifold is said to be *Kähler-Einstein* if it is Einstein as a Riemannian metric. Equivalently, the Ricci form ρ is proportional to the Kähler form

(11.2) $$\rho = \lambda\omega,$$

or else, the Kähler form ω is an eigenform of the Kähler curvature operator (associated with the eigenvalue $\lambda = s/2m$, where m is the complex dimension.)

11.3 Remark. In the Kähler case, Theorem 1.97 has a short proof. Since ρ and ω are both closed, the assumption $\rho_x = \lambda_x\omega_x$ implies that λ is constant if $m \geqslant 2$.

11.4 Definition. A cohomology class in $H^2(M, \mathbb{R})$ is said to be *positive* (resp. *negative*) if it can be represented by a real positive (resp. negative) 2-form of type (1, 1) (see 2.26).

It is easily checked that the two cases above are mutually exclusive: a given cohomology class cannot be both positive and negative. In particular, the sign—if any—of a cohomology class depends only on the given complex structure on M.

11.5 Proposition. *The sign of the (constant) scalar curvature s of a Kähler-Einstein metric—if any—on a given (compact) complex manifold M is determined by the complex structure of M. Moreover, the value of s is then determined by*

(11.5) $$V \cdot s^m = \frac{(4\pi m)^m}{m!} c_1^m$$

where c_1^m denotes the Chern number associated with the m-th power of the first Chern class of M, depending on the complex structure only, and V the total volume.

Proof. By Proposition 2.75 the Ricci form of any Kähler metric on M represents, up to a positive factor, the first Chern class of M, and the first part of the proposition is a consequence of 11.4. Then, (11.5) directly follows from (11.2). □

11.6 Remark. The total volume V of a given Kähler metric depends only on the Kähler class $[\omega]$ of the Kähler form ω of the metric, and so does, consequently, the value of the Einstein constant $\frac{s}{2m}$. Up to a change of scale, the latter may be chosen equal to -1, 0 or $+1$. The three cases are mutually exclusive from each other and are determined by the complex structure of M.

11.7 Remark. As we just saw, a necessary condition for a given (compact) complex manifold to admit a Kähler-Einstein metric is that *its first Chern class have a sign*, negative, zero or positive. It follows that a "generic" (compact) complex manifold admits no Kähler-Einstein metric.

On the other hand, it is by no means obvious whether that necessary condition would be also sufficient. We shall be concerned at length with this question in the present chapter. The answer happens to be "yes" in the negative and the zero case, but "no" in the positive one.

11.8. In the same way as the simplest Einstein manifolds are the Riemannian manifolds with constant sectional curvature, the simplest Kähler-Einstein manifolds are the Kähler manifolds with constant *holomorphic* sectional curvature, that is the (compact factor spaces of the) open unit disk D^m in $\mathbb{C}^m$ endowed with a Bergmann metric in the negative case, the (factor spaces of the) flat complex tori $T^m = \mathbb{C}^m/\Gamma$ in the zero case and the complex projective space $\mathbb{C}P^m$, endowed with a Fubini-Study metric in the positive case (see § D of Chapter 2).

11.9. As we saw in Chapter 8, any compact simply-connected homogeneous Kähler manifold admits a Kähler-Einstein metric with positive scalar curvature (see 8.95). Those are the only known examples of positive Kähler-Einstein metrics.

11.10. It may be difficult, in general, to exhibit the first Chern class of a given complex manifold and to check whether it does have a sign or not. But there are some situations where this computation becomes straightforward by using standard techniques of complex geometry. This is the case when the complex manifold M is a (smooth) complex hypersurface of the complex projective space $\mathbb{C}P^N$, that is the locus of the zeros of a homogeneous (complex) polynomial, of degree d say. We get then

$$c_1(M) = (N + 1 - d)h$$

where h denotes the restriction to M of the (positive) generator of $H^2(\mathbb{C}P^N, \mathbb{Z})$. In particular, $c_1(M)$ has always a sign, positive for low degrees (d less than or equal to the complex dimension N of the ambient projective space), null if d is equal to $N + 1$, negative if d is greater.

By the Calabi-Yau and Aubin-Yau theorems (see below 11.15 and 11.17) all those complex manifolds admit, for $d > N$, a Kähler-Einstein metric.

On the other hand the latter is *never* the induced metric (see [Ko-No 2], Note 26 or [Smy]). More generally, we get an explicit (and straightforward) computation of $c_1(M)$ when M is the intersection of several complex hypersurfaces of $\mathbb{C}P^N$ in general position (see [Hir] Appendix One). Formula (11.10) remains valid where d is understood as the sum of the degrees of the complex hypersurfaces.

11.11. By a celebrated theorem of K. Kodaira, a (compact) complex manifold M whose first Chern class is positive or negative is *projective algebraic*, that is to say can be holomorphically imbedded in a complex projective space $\mathbb{C}P^N$. More precisely, some r-th power of the canonical line bundle (see 2.95) K in the negative case or of the anti-canonical line bundle K^* in the positive case is *very ample*: the canonical mapping which associates with every point of M the hyperplane of holomorphic sections of K^r or $(K^*)^r$ vanishing at this point is a (well defined) holomorphic imbedding of M into the dual projective space associated with the space of holomorphic sections of K^r or $(K^*)^r$.

Moreover, since the canonical line bundle and its dual are complex invariants, the group $\mathfrak{A}(M)$ of complex automorphisms of M is exactly the subgroup of complex automophisms of the ambient projective space leaving M invariant. In particular, $\mathfrak{A}(M)$ is an algebraic group and the number of its connected components is *finite* (see 2.138).

11.12. Although the case where the first Chern class of a complex manifold has no sign may be considered as generic, it may seem difficult to exhibit explicit examples of this kind, outside products.

If M is a complex surface ($m = 2$), we have a simple criterion given by the following obvious observation. *If* $c_1(M)$ is positive, negative or zero, *the Chern number* $c_1^2(M)$ *is non-negative.*

In order to obtain a complex surface with negative $c_1^2(M)$, we use a standard construction in complex geometry: *blowing up* a point on a complex manifold M, consisting in replacing a point p of M by the set of (complex) tangent directions around the point, leaving unchanged the remainder of M. We thus get a new complex manifold $\hat{M}$ and a holomorphic mapping from $\hat{M}$ onto M, biholomorphic over $M - \{p\}$, and having a fiber isomorphic to $\mathbb{C}P^{m-1}$ over p (see, for instance, [Bes 2], Exp. VI § 5 and enclosed references). The fiber over p is called the *exceptional divisor* of the blowing up.

When M is a complex surface, the exceptional divisor induces, by Poincaré duality, an element e of $H^2(\hat{M}, \mathbb{Z})$ whose square, considered as an integer, is equal to -1. The Chern class $c_1(\hat{M})$ is easily shown to be equal to $c_1(M) - e$, while the Chern number $c_1^2(\hat{M})$ is just equal to $c^2(M) - 1$ ([Bes 2] p. 28).

In other words, the blowing up process *decreases* the Chern number c_1^2. In particular, starting from the complex projective plane $\mathbb{C}P^2$—for which c_1^2 is equal to 9—, by blowing up 9 points or more we certainly obtain a complex surface whose first Chern class is indefinte.

11.13. In return, it can be proved ([Hit 3] or [Yau 4]) that the complex surfaces Σ_r obtained from $\mathbb{C}P^2$ by blowing up r distinct points, $0 \leqslant r \leqslant 8$, *do have* a positive first Chern class, whenever those points are in general position, that is: no 3 of them lie on a same line, no 6 of them lie on a same conic, and if $r = 8$, they are all simple points of each cubic curve passing through all eight of them.

Moreover, the manifolds Σ_r are the only (compact) complex surfaces having positive first Chern class, with $\mathbb{C}P^2$ and $\mathbb{C}P^1 \times \mathbb{C}P^1$ (loc. cit.). Consequently, those are the only compact complex surfaces on which the existence of a Kähler-Einstein metric with positive scalar curvature can be expected. On the other hand the complex surfaces Σ_1, Σ_2 and Σ_3 are proved to admit *no* Kähler-Einstein metric, by showing that their connected group of complex automorphisms is not reductive (see, below, 11.54).

Concerning this question, nothing is known for the other complex surfaces Σ_r $4 \leqslant r \leqslant 8$.

11.14 *Note.* For $1 \leqslant r \leqslant 6$, the manifolds Σ_r are known as Del Pezzo surfaces (of degree 9-r). (see [Bea 1] p. 60). Among them, two families only can be realized as a complete intersection (s. 11.10) in a complex projective space: Σ_6 which is a (complex) hypersurfaces of degree 3 in $\mathbb{C}P^3$ and Σ_5 which is the intersection of two quadrics (complex hypersurface of degree 2) of $\mathbb{C}P^4$ (loc. cit.) (recall that $\mathbb{C}P^1 \times \mathbb{C}P^1$ itself is realized as a quadric of $\mathbb{C}P^3$).

B. The Resolution of the Calabi Conjectures and its Consequences

By 2.75, the Ricci form of any Kähler metric on a compact complex manifold M belongs to a fixed real cohomology class, namely to $2\pi c_1(M)$. Conversely, we have the crucial

11.15 Theorem (E. Calabi [Cal 2], S.T. Yau [Yau 3]). *Let M be a compact Kähler manifold, ω its Kähler form, $c_1(M)$ the real first Chern class of M. Any closed (real) 2-form of type (1, 1) belonging to $2\pi c_1(M)$ is the Ricci form of one and only one Kähler metric in the Kähler class of ω.*

11.16. As an immediate consequence, we get the following fundamental facts (recall that a complex manifold is *Kählerian* if it admits a Kähler metric).

i) *The compact Kählerian manifolds with zero (real) first Chern class are exactly the compact complex manifolds admitting a Kähler metric with zero Ricci form* (equivalently, by 10.29, with restricted holonomy group contained in the special unitary group).

ii) *The compact complex manifolds with positive (negative) first Chern class are exactly the compact complex manifolds admitting a Kähler metric with positive (negative) Ricci form.*

As for the question asked in 11.7, a partial answer is given by the

11.17 Theorem (T. Aubin [Aub 5], S.T. Yau [Yau 3]). *Any compact complex manifold with negative first Chern class admits a Kähler-Einstein metric (with negative scalar curvature). This metric is unique up to homothety.*

11.18 Remark. Using 2.138, 11.11 and 11.16 ii) above we obtain that the group of automorphisms $\mathfrak{A}(M)$ of a compact complex manifold M with negative first Chern class is finite. (For a more elementary proof avoiding the use of Theorem 11.15, see [Lic 1] § 13).

Moreover, using the uniqueness part of Theorem 11.17, we see that any transformation in $\mathfrak{A}(M)$ actually is an isometry for the Kähler-Einstein metric.

11.19. As we have seen in 11.13, the statement analogous to Theorem 11.17, when the first Chern class is assumed to be positive, is false. A source of trouble here is the possible existence of a non-trivial Lie algebra $\mathfrak{a}(M)$ of infinitesimal holomorphic transformations. As a matter of fact, the only known obstructions to the existence —not to speak of uniqueness—of Kähler-Einstein metrics on a (compact) complex manifold with positive first Chern class are all related to $\mathfrak{a}(M)$ (see, below, part D of this Chapter for a detailed account of this obstruction). On the other hand, these obstructions are empty when the Lie algebra $\mathfrak{a}(M)$ is reduced to zero and nothing is known in this case.

11.20. Theorem 11.15 and Theorem 11.17 are generally quoted as Calabi-Yau theorem and Aubin-Calabi-Yau theorem respectively (Théorème I and Théorème II$^-$ respectively in [Ast]).

Accordingly, the corresponding Kähler-Einstein metrics are called *Calabi-Yau metrics* (one in each Kähler class, unique up to homothety) when the (real) first Chern class of the manifold vanishes, and *Aubin-Calabi-Yau metric* when the first Chern class is negative (in the latter case the metric is usually normalized as in 11.6, so that the scalar curvature is equal to $-2m$, the Einstein constant is equal to -1, and the metric then is called *the* (unique) Aubin-Calabi-Yau metric of the complex manifold).

A sketch of the proof, following [Ast], will be given in part C of the chapter. The remainder of Part B is devoted to giving a brief survey of the important geometric consequences of the two Theorems 11.15 and 11.17 above.

Theorem 11.15—then not completely proved—was used for the first time by E. Calabi to give a description of the (Kählerian) compact, complex manifolds with vanishing first Chern class, using the natural mapping of the complex manifold into its Albanese torus ([Cal 2]).

In the meantime, a celebrated splitting theorem, due to J. Cheeger and D. Gromoll, concerning Riemannian manifolds with non-negative Ricci curvature see 6.66e)—in addition to the complete proof by S.T. Yau of Theorem 11.15—eventually gives a simple proof of the

11.21 Theorem. *Let M be a compact Kählerian manifold with vanishing real first Chern class. Then M admits as a finite holomorphic covering the product of a complex torus by a simply-connected Kählerian manifold, again with vanishing real first Chern class.*

Sketch of proof (see [Bea 2] for details). By Theorem 11.15, M admits a Ricci-flat Kähler metric and so does its universal covering $\tilde{M}$. In application of the Cheeger-Gromoll theorem quoted above, the De Rham decomposition of $\tilde{M}$ is the (Kähler) product of a flat complex Euclidean space by a *compact* simply-connected Kähler manifold with zero Ricci curvature. The remainder of the proof follows easily from the fact that the group of isometries of a simply-connected Ricci-flat Riemannian metric on a compact manifold is finite (see 1.84). □

11.22 Remark. By 2.139, the connected group of complex automorphisms of a compact Kählerian manifold M with vanishing real first Chern class is a complex torus whose (real) dimension is equal to the first Betti number b_1 of M. The latter coincides with the connected group of isometries of any Calabi-Yau metric on M. Observe that the dimension of the complex torus in Theorem 11.21 is greater than or equal to b_1.

11.23. Theorem 11.21 allows us to restrict the study of the compact Kählerian manifolds with vanishing real first Chern class to the study of simply-connected ones.

Those, in turn, split, relatively to any given Calabi-Yau metric, into the Kähler product of irreducible (simply-connected, compact) Kähler manifolds with holonomy groups either a special unitary group $SU(m_i)$ or a symplectic group $Sp(m_j/2)$, where m_i, m_j is the complex dimension of the corresponding factor (Theorems 10.43 and 10.92).

Observe that these manifolds cannot be symmetric since the first Chern class of any simply-connected, compact homogeneous Kähler manifold is positive—see Chapter 8.

That decomposition doesn't depend on the chosen Calabi-Yau metric in the following sense: the complex manifold M splits into a product $\prod M_i$ of (compact, simply-connected) complex manifolds (of complex dimension m_i) in such a way that for any Ricci-flat Kähler metric g on M there are also Ricci-flat Kähler metrics on the factors M_i so as to get a De Rham splitting of (M, g) into irreducible components with holonomy $SU(m_i)$ or $Sp(m_j/2)$. (This fact is an easy consequence of Theorem 11.15—see [Bea 2] § 5).

Moreover the existence of a Kähler metric with holonomy $SU(m)$ or $Sp(m/2)$ on a compact complex manifold (of complex dimension m) can be seen in the complex structure. For m greater than 2 the two cases are mutually exclusive ([Bea 2] § 3 and 4). The first category contains all the examples given in this chapter, in particular in 11.10, of compact simply-connected complex manifolds with vanishing first Chern class, while examples in the second category, the so-called *symplectic Kähler*—or *hyperkählerian*—manifolds have been discovered only recently for $m > 2$ (see Chapter 14).

11.24. When the complex manifold M has positive first Chern class, we have no information on the group $\mathfrak{A}(M)$, except when M admits, in addition, a Kähler-Einstein metric or, at least, a Kähler metric with constant (positive) scalar curvature. But, we have the

Lemma (S. Bochner). *A compact, complex manifold with positive first Chern class has no non-trivial holomorphic p-form, $p = 1, \ldots, m$.*

Proof. Take any Kähler metric with positive Ricci form and consider the general Bochner formula applied, by $\mathbb{C}$-linearity, to complex exterior forms. Observe now that, when applied to a form of type $(p, 0)$, the zero order term in the Bochner formula (see 1.156) involves only the Ricci curvature and conclude as for the 1-forms. □

11.25 Corollary. *On any compact, complex manifold with positive first Chern class the numbers $h^{p,0}$ and $h^{0,p}$ are equal to zero. In particular, the Euler number $\chi(M, \mathcal{O})$ is equal to* 1.

Proof. 2.104, 2.105 and 2.116. □

As a direct consequence, we get the

11.26 Theorem (S. Kobayashi). *A compact, complex manifold with positive first Chern class is simply connected.*

Proof. By Myers' Theorem 6.52, the universal cover $\tilde{M}$ of our manifold M is compact. Moreover $\tilde{M}$ is naturally endowed with a complex structure for which the covering projection is holomorphic. For any Kähler metric with positive Ricci form on M, the induced metric on $\tilde{M}$ is again a Kähler metric of the same type. In particular, the Euler numbers $\chi(M, \mathcal{O})$ and $\chi(\tilde{M}, \tilde{\mathcal{O}})$ of M and $\tilde{M}$ are both equal to 1. On the other hand, they can both be expressed as integrals, over M and $\tilde{M}$ respectively, of the same universal polynomial in the Kähler curvature (see 2.117). It follows that $\chi(\tilde{M}, \tilde{\mathcal{O}})$ is just equal to k times $\chi(M, \mathcal{O})$ if k is the order of the covering. Thus, k is equal to 1. □

Up to now, we only considered the first Chern class. Considering now the two first Chern classes we get, as an application of Theorem 11.15, the following

11.27 Corollary. *If the first and the second (real) Chern class of a compact Kählerian manifold M vanish, M is (holomorphically) covered by a complex torus and all the subsequent Chern classes vanish too.*

Proof. By Theorem 11.15 M admits a Ricci-flat metric so that the curvature R reduces to its Weyl part W. By (2.80a), W itself vanishes and the metric is flat. Now apply 2.60 and 2.73. □

11.28 Corollary (S.T. Yau [Yau 5]). *Let M be a Kählerian manifold of complex dimension m with negative first Chern class. Then*

$$(-1)^m(2(m+1)c_2c_1^{m-2} - mc_1^m) \geq 0. \tag{11.28}$$

Equality holds if and only if the holomorphic sectional curvature of the Aubin-Calabi-Yau metric is constant (negative) so that M is (holomorphically) covered by the unit ball in $\mathbb{C}^m$.

Proof. Just use the Aubin-Calabi-Yau metric (Theorem 11.17 and Remark 11.20) whose Kähler form ω represents $-2\pi c_1(M)$ and consider formula (2.82a). □

11.29 Remark. When m is equal to 2, inequality (11.28) specializes to

$$3\tau - \chi = c_1^2 - 3c_2 \leq 0 \tag{11.29}$$

involving (oriented) topological invariants only. We thus obtain, in the Kähler-Einstein case, a refinement of the general Thorpe inequality 6.35.

Remark. Assuming the existence of a Kähler-Einstein metric with zero or negative scalar curvature respectively, the conclusion of Corollaries 11.27 and 11.28 become a mere application of the Apte formula 2.80, while inequality (11.29) is due to H. Guggenheimer.

Finally we quote here two results concerning constant holomorphic curvature spaces.

11.30 Theorem. *Any Kählerian manifold homeomorphic to $\mathbb{C}P^m$ is (biholomorphically) isomorphic to it.*

11.31 Theorem (S.T. Yau [Yau 5]). *Let N be a compact complex surface that is covered by the ball in $\mathbb{C}^2$. Then any complex surface M that is oriented homotopically equivalent to N is (biholomorphically) isomorphic to it.*

Theorem 11.30 was proved by K. Kodaira and F. Hirzebruch in case where m is odd and "half-proved" in the opposite case (see [Ko-Hi]). The remaining incertitude then is cancelled out using Theorem 11.17. When m is equal to 2, Theorem 11.30 can be strengthened as follows

11.32 Theorem ([Yau 5]). *Let M be a compact complex surface homotopically equivalent to $\mathbb{C}P^2$. Then M is (biholomorphically) equivalent to $\mathbb{C}P^2$.*

C. A Brief Outline of the Proofs of the Aubin-Calabi-Yau Theorems

We don't give a complete proof of Theorems 11.15 and 11.17. We only describe the general framework and explain a few easy steps. For the proofs of the difficult estimates, the reader should consult the original papers [Cal 2], [Aub 5], [Yau 3], and the books [Ast] and [Aub 6].

11.33 Restating Theorem 11.15. As observed in 2.101, Theorem 11.15 means that, given any Kähler form ω on M, any non-vanishing volume element on M, of the same total volume, is the volume form of a unique Kähler metric cohomologous to ω. In other words, for any smooth function f such that $\int_M e^f \omega^m = \int_M \omega^m$, there exists a unique positive definite $(1,1)$-form $\tilde{\omega} \in [\omega]$ such that $\tilde{\omega}^m = e^f \omega^m$.

The dd^c-Lemma 2.110 allows us to parametrize the real cohomology class $[\omega]$ by smooth real functions φ such that $\int_M \varphi\omega^m = 0$. To deal with partial differential equations, it will be convenient to use Hölder spaces $C^{k+\alpha}$ (see App.A). Let us denote by $\mathscr{H}$ the space of real $C^{2+\alpha}$ functions φ on M such that the 2-form $\omega - \frac{1}{2}dd^c\varphi$ is positive definite, and $\int_M \varphi\omega^m = 0$, and by $\mathscr{H}'$ the space of real C^α functions f on M such that $\int_M e^f \omega^m = \int_M \omega^m$. Then Theorem 11.15 follows from the following claim.

11.34 Claim. *The mapping* Cal: $\mathscr{H} \to \mathscr{H}'$, $\varphi \mapsto \mathrm{Cal}(\varphi) = \log((\omega - \frac{1}{2}dd^c\varphi)^m/\omega^m)$ *is a diffeomorphism onto* $\mathscr{H}'$.

Notice that, whereas Einstein's equations form a system of equations with a vector valued function as unknown, finding a *Kähler* Ricci flat metric on a fixed complex manifold reduces to solving only one equation, with unknown a real function. In local complex coordinates, one has

$$\textbf{(11.34)} \qquad \operatorname{Cal}(\varphi) = \log\det\left(g_{\alpha\bar\beta} - \frac{\partial^2\varphi}{\partial z^\alpha \partial \bar z^\beta}\right) - \log\det(g_{\alpha\bar\beta}).$$

This is an example of a *complex Monge-Ampère equation.*

11.35. Implicitly, we used the fact that, if f is smooth, then any $C^{2+\alpha}$ solution of the equation $\operatorname{Cal}(\varphi) = f$ is smooth. This follows from general properties of Ricci curvature (Theorem 5.20). In our context, there is a straightforward proof (compare Appendix, Remark following Theorem 40). Let g, $\tilde g$ be the metrics corresponding to the forms ω and $\tilde\omega = \omega - \frac{1}{2}dd^c\varphi$. Assume that $\varphi \in C^{k+\alpha}$. Differentiating Equation (11.35), one gets

$$\frac{1}{2}\tilde\Delta\left(\frac{\partial\varphi}{\partial z^\gamma}\right) = \frac{\partial f}{\partial z^\gamma} - \sum_{\alpha,\beta}(\tilde g^{\alpha\beta} - g^{\alpha\bar\beta})\frac{\partial g_{\alpha\bar\beta}}{\partial z^\gamma}$$

where the Laplace operator $\tilde\Delta$ of $\tilde g$ is elliptic with coefficients in $C^{k-2+\alpha}$. Theorem 40 of the Appendix implies that $\dfrac{\partial\varphi}{\partial z^\gamma} \in C^{k+\alpha}$ and so $\varphi \in C^{k+1+\alpha}$. □

11.36. To prove Claim (11.34), one first shows that the mapping Cal is injective. Then, that it is a local diffeomorphism. Finally, and this is the hard part which we will do not detail here, one shows that Cal is proper. This implies that its image is closed, hence Cal is surjective.

11.37 Injectivity. The remarkable fact that the mapping Cal is injective has been discovered by E. Calabi in 1955 ([Cal 2]). His proof goes as follows. Assume that the cohomologous Kähler forms ω_1 and $\omega_2 = \omega_1 - \frac{1}{2}dd^c\varphi$ have the same volume form. Since the exterior algebra is commutative in even degrees, one may write the condition $\omega_1^m = \omega_2^m$ in the form

$$0 = \omega_2^m - \omega_1^m = \left(\sum_{k=0}^{m-1} \omega_1^k \wedge \omega_2^{m-k-1}\right) \wedge (\omega_2 - \omega_1)$$

That is, $\frac{1}{2}dd^c\varphi \wedge \sigma = 0$, where $\sigma = \sum_{k=1}^{m-1} \omega_1^k \wedge \omega_2^{m-k}$ is a closed real $(2m-2)$-form of type $(m-1, m-1)$.

In local coordinates, σ can be written

$$\sigma = \sum_{\alpha,\beta} M^{\alpha\bar\beta}\, dz_1 \wedge \cdots \wedge \widehat{dz_\alpha} \wedge \cdots \wedge d\bar z_1 \wedge \cdots \wedge \widehat{d\bar z_\beta} \wedge \cdots \wedge d\bar z_m.$$

One easily checks that the matrix $(M^{\alpha\bar\beta})$ is positive definite. Thus φ, which satisfies $\displaystyle\sum_{\alpha,\beta} M^{\alpha\bar\beta}\frac{\partial^2\varphi}{\partial z^\alpha \partial \bar z^\beta} = 0$ on a compact manifold must be constant, that is, $\varphi \equiv 0$.

11.38 Local Inversion. From the expression in coordinates, (11.34), a simple calculation gives the differential of the mapping Cal: at a point $\varphi \in \mathscr{H}$, $\mathrm{Cal}'_\varphi = \frac{1}{2}\tilde{\Delta}$, where $\tilde{\Delta}$ denotes the Laplace operator of the Kähler metric $\tilde{\omega} = \omega - \frac{1}{2}dd^c\varphi$. Since $\tilde{\omega}$ is positive definite, the operator $\tilde{\Delta}$ is elliptic. According to Theorem 27 of the Appendix, $\tilde{\Delta}$ is an isomorphism of $T_\varphi \mathscr{H} = \{\varphi\colon \int_M \varphi\omega^m = 0$ onto $\{h\colon \int_M h\tilde{\omega}^m = 0\}$ which is precisely $T_{\mathrm{Cal}(\varphi)}\mathscr{H}'$.

Now the Inverse Function Theorem applies (App. 43) and the mapping Cal is a local diffeomorphism.

11.39 Surjectivity. One shows that the mapping Cal is proper. This amounts to proving an a priori bound on the solution of the equation $\mathrm{Cal}(\varphi) = f$. One has to give and upper bound on the $C^{2,\alpha}$ norm of φ and a lower bound of the Kähler metric $\tilde{\omega} = \omega - \frac{1}{2}dd^c\varphi$ by a fixed Kähler metric depending only on f.

The C^0 estimate, and the inequality const. $\omega \leqslant \tilde{\omega} \leqslant$ const$'$ ω, which is in fact a C^2 estimate on φ, are due to S.T. Yau. The C^2 estimate is obtained by differentiating twice the equation $\mathrm{Cal}(\varphi) = f$. Thanks to an idea going back to A.V. Pogorelov ([Pog 1]), the terms containing third derivatives of φ cancel.

One needs a $C^{2,\alpha}$ estimate. In fact, a C^3 estimate is obtained. The trick consists in estimating norms of the following type, computed in the *unknown* metric:

$$\tilde{g}^{\alpha\bar{\lambda}}\tilde{g}^{\mu\bar{\beta}}\tilde{g}^{\gamma\bar{\nu}}\varphi_{\alpha\bar{\beta}\gamma}\varphi_{\bar{\lambda}\mu\bar{\nu}}.$$

This step of the proof goes back to E. Calabi [Cal 3].

In fact, as is now known, the $C^{2+\alpha}$ estimate follows from the C^2 estimate. It is a general property of equations of Monge-Ampère type, see [Eva] or [Gi-Tr, 2nd edition] Th. 17.14 page 461.

11.40 Restating Theorem 11.17. It will be convenient to discuss simultaneously the existence of Kähler-Einstein metrics with Einstein constant -1 and $+1$. So let $\varepsilon = \pm 1$, and let us assume that the cohomology class $\varepsilon c_1(M)$ is positive. Let ω be a Kähler form in this class. Its Ricci form ρ sits in $c_1(M)$, so there must exist a smooth function f such that $\rho = \varepsilon\omega - \frac{1}{2}dd^c f$. If $\tilde{\omega} = \omega - \frac{1}{2}dd^c\varphi$ is a Kähler form cohomologous to ω, its Ricci form $\tilde{\rho}$ is given by Formula (2.98):

$$\tilde{\rho} - \rho = -\tfrac{1}{2}dd^c \log(\tilde{\omega}^m/\omega^m).$$

Thus the condition that $\tilde{\omega}$ is a Kähler-Einstein metric, $\tilde{\rho} = \varepsilon\tilde{\omega}$, is equivalent to $\log(\tilde{\omega}^m/\omega^m) - \varepsilon\varphi = f +$ constant. Hence Theorem 11.17 is a consequence of the following claim.

11.41 Claim. *If* $\varepsilon = -1$, *the mapping* $\varphi \mapsto \mathrm{Cal}^\varepsilon(\varphi) = \mathrm{Cal}(\varphi) - \varepsilon\varphi$ *is a diffeomorphism of* $\mathscr{V} = \{$functions $\in C^{2+\alpha}$ such that $\omega - \frac{1}{2}dd^c\varphi$ is positive definite$\}$ *onto* $C^\gamma(M)$.

11.42 Injectivity of Cal^-. Assume that two Kähler forms ω_1 and $\omega_2 = \omega_1 - \frac{1}{2}dd^c\varphi$ satisfy

$$\log(\omega_2^m/\omega_1^m) - \varepsilon\varphi = 0.$$

Let $x \in M$ be a point where φ is maximum. Then, at x, $\log(\omega_2^m/\omega_1^m)$ is non negative.

Indeed, choosing coordinates such that $g^1_{\alpha\bar\beta} = \delta_{\alpha\bar\beta}$ and $\partial^2\varphi/\partial z^\alpha \partial \bar z^\beta = \lambda_\alpha \delta_{\alpha\beta}$, all λ_α are non positive, and $\log(\omega_2^m/\omega_1^m) = -\sum \log(1+\lambda_\alpha)$. Thus $-\varepsilon\varphi(x) \leqslant 0$. In the same way, at a minimum y, $-\varepsilon\varphi(y) \geqslant 0$. If $\varepsilon = -1$, one can conclude that $\varphi = 0$.

11.43 A C^0-estimate. The same argument gives a C^0 estimate for the solutions of $\mathrm{Cal}^-(\varphi) = f$. Indeed, if x is a maximum and y a minimum of φ, one gets from the equation

$$\log((\omega - \tfrac{1}{2}dd^c\varphi)^m/\omega^m) - \varepsilon\varphi = f$$

that $-\varepsilon\varphi(x) \leqslant f(x)$ and $-\varepsilon\varphi(y) \geqslant f(y)$. If $\varepsilon = -1$, one can conclude that

$$\mathrm{Sup}\,\varphi - \mathrm{Inf}\,\varphi \leqslant \mathrm{Sup}\,f - \mathrm{Inf}\,f.$$

11.44 Local inversion. The differential of the mapping Cal^ε at a point $\varphi \in V$ is

$$\mathrm{Cal}^{\varepsilon\prime}_\varphi = \tfrac{1}{2}\tilde\Delta - \varepsilon,$$

where $\tilde\Delta$ denotes the Laplace operator of the metric $\omega - \frac{1}{2}dd^c\varphi$. Hence the Inverse Function Theorem applies as soon as 2ε is not an eigenvalue of $\tilde\Delta$. This is certainly the case if $\varepsilon = -1$.

11.45. The estimates needed to finish the proof of Theorem 11.17 are very similar to those evoked in 11.39. However, they are in some sense easier, as the C^0-estimate above indicates.

D. Compact Complex Manifolds with Positive First Chern Class

11.46. The method of proof of Theorem 11.17 fails to handle the case where $c_1(M)$ is positive at each of its steps.

The uniqueness statement in Theorem 11.17 certainly does not generalize. Indeed, complex projective space carries the Fubini-Study metric ω, which is Kähler-Einstein, and admits non isometric automorphisms. If ψ is such an automorphism, the form $\psi^*\omega$ is a Kähler-Einstein metric different from ω. If moreover ψ lies in the identity component of $\mathfrak{A}(\mathbb{C}P^m)$, then one has $\psi^*\omega = \omega - \frac{1}{2}dd^c\varphi$, and equation $\mathrm{Cal}^+(\varphi) = 0$ admits two distinct solutions 0 and φ. This phenomenon occurs for all complex manifolds M with positive first Chern class and a non trivial identity component of $\mathfrak{A}(M)$. See however Add.D.

11.47. The failure of proof of uniqueness of solutions is reflected, at the infinitesimal level, by the failure of local inversion. We have seen that Cal^+ is invertible at $\varphi = 0$ iff $2s/m$ is not an eigenvalue of the Laplacian of ω. We shall see in 11.52 that, if ω is Kähler-Einstein, the kernel of Δ identifies with the space of holomorphic vector fields, i.e., the Lie algebra of $\mathfrak{A}(M)$.

11.48. Finally, the estimates also break down when $c_1(M)$ is positive (this is clear for the C^0 estimate 11.43). We now show that the existence of a positive Kähler-Einstein metric indeed implies more necessary conditions on a compact complex manifold, and particularly, on its automorphism group.

Let us begin with the following

11.49 Theorem (A. Lichnerowicz [Lic 1]). *Let M be a compact complex manifold, of complex dimension m, endowed with a Kähler metric g, whose Ricci tensor satisfies*

$$r \geqslant \frac{k}{2m} g, \tag{11.49}$$

where k is a positive constant. Then the first non zero eigenvalue λ_1 of the Laplace operator Δ (acting on functions) is greater than or equal to k/m.

Combining Formulas (2.53), (2.55) and (2.56 ter) and integrating over M we get, for any 1-form ξ

$$\langle \Delta\xi, \xi\rangle - 2\langle r(\xi), \xi\rangle + \tfrac{1}{2}\langle \delta\xi, \delta\xi\rangle = \langle \delta^*\xi, \delta^*\xi\rangle. \tag{11.50}$$

Let f be any eigenfunction for Δ relative to the eigenvalue λ_1. The 1-form df is an eigenform for Δ (acting on 1-forms) relative to the same eigenvalue, and so is Jdf, since Δ commutes with J. Applying the above relation (11.50) to the 1-form $d^c f$—which is co-closed—we get

$$\begin{aligned}\lambda_1 \langle Jdf, Jdf\rangle &= 2\langle r(Jdf), Jdf\rangle + \langle \delta^*(Jdf), \delta^*(Jdf)\rangle \\ &\geqslant \frac{k}{m}\langle Jdf, Jdf\rangle + \langle \delta^*(Jdf), \delta^*(Jdf)\rangle.\end{aligned}$$

The result follows at once. □

11.51 Remark. It may happen that, under the assumptions of Theorem 11.49, k/m actually belongs to the spectrum of Δ. In that case, the proof above shows that, for any eigenfunction f relative to k/m, the vector field $(Jdf)^\sharp$ is Killing (see 1.81c). Under the stronger assumption that g is Einstein, we get

11.52 Theorem (Y. Matsushima [Mat 3]). *Let M be a compact, complex manifold of (real) dimension $n = 2m$, endowed with a positive Kähler-Einstein metric. The Lie algebra $\mathfrak{i}(M)$ of Killing vector fields is a real form of the (complex) Lie algebra of (real) holomorphic vector fields. If $\mathfrak{i}(M)$ is non trivial, the first non zero eigenvalue of the Laplace operator Δ on functions is equal to s/m, and the corresponding eigenspace (endowed with the Lie algebra structure induced by the Poisson bracket) identifies with the Lie algebra of Killing vector fields.*

This is an immediate consequence of Proposition 2.151. Indeed, since the Ricci form is equal to $\dfrac{s}{m}\omega$, equation (2.151) reduces to

$$\Delta f - \frac{s}{m} f = 0,$$

while the Lie algebra of parallel vector fields reduces to zero since the Ricci tensor is positive definite (1.84). In particular, we obtain an isomorphism between the eigenspace of Δ relative to s/m and the space of Killing vector fields. One easily checks that the bracket of two (globally) Hamiltonian vector fields respectively dual to the 1-forms $d^c f$ and $d^c h$ is itself Hamiltonian and dual to the 1-form $d^c\{f, h\}$, where the *Poisson bracket* $\{f, h\}$ is defined by

$$\{f, h\} = (df, d^c h) = -\Lambda(df \wedge dh). \qquad \square$$

11.53. The Lie algebra $\mathfrak{g}$ of a compact Lie group is the direct sum (as a Lie algebra) of its center and its commutator Lie subalgebra $[\mathfrak{g}, \mathfrak{g}]$, and this also holds for the complexified Lie algebra $\mathfrak{g}^{\mathbb{C}}$.

More generally, a complex Lie algebra having this property is called *reductive* (see [Hoc]). Hence we have proved

11.54 Corollary. *The identity component $\mathfrak{A}^0(M)$ of the automorphism group of a compact complex manifold carrying a Kähler-Einstein metric is reductive.*

11.55 Remark. By Proposition 2.151, the same conclusion holds if we only assume that M admits a Kähler metric with constant scalar curvature. For all known examples of compact Kähler manifolds with constant positive scalar curvature (that is, Add.C and the homogeneous spaces of Chapter 8) the Lie groups $\mathfrak{A}(M)$ and $I(M)$ are semi-simple, that is, the center of $I(M)$ is finite. We do not know whether this fact is general.

11.56 Examples. Let M be the complex surface obtained by blowing up one point p in complex projective plane $\mathbb{C}P^2$ (cf. 11.12). It is easily seen that the automorphism group $\mathfrak{A}(M)$ of M coincides with the isotropy subgroup of p in $\mathfrak{A}(\mathbb{C}P^2) = PGl(3, \mathbb{C})$. The Lie algebra of $\mathfrak{A}(M)$ may be visualized as follows

$$\begin{pmatrix} 0 & \times & \times \\ 0 & \times & \times \\ 0 & \times & \times \end{pmatrix}$$

where each cross stands for an arbitrary complex number. This Lie algebra is not reductive. Indeed, its center is trivial, whereas any commutator has vanishing first row (the commutator subalgebra is isomorphic to $\mathfrak{gl}(2, \mathbb{C})$).

In the same way, the identity component of the automorphism group of the surface obtained by blowing up two points in $\mathbb{C}P^2$ is the subgroup of $PGl(3, \mathbb{C})$ fixing the two points, and its Lie algebra is not reductive either. From Corollary 11.54, we conclude that these two manifolds do not carry any Kähler-Einstein metric.

Another obstruction arises from the Futaki Theorem 2.160 (see also Remark 2.162).

11.57 Theorem (A. Futaki [Fut]). *Let M be a compact complex manifold with positive first Chern class $c_1(M)$. If M admits a Kähler-Einstein metric, then the Futaki functional of any Kähler metric whose Kähler form belongs to $c_1(M)$ is zero.*

11.58. Here is an example due to A. Futaki ([Fut]) of a compact complex manifold with positive first Chern class *and* reductive group of automorphisms (so that obstruction 11.54 doesn't apply), but admitting no Kähler-Einstein metric.

For $i = 1, 2$, let H_i denote the *hyperplane bundle* on the complex projective space $\mathbb{C}P^i$, that is, the holomorphic line bundle dual to the *tautological line bundle* (or *Hopf bundle*) whose fiber at a point of $\mathbb{C}P^i$ is just this point viewed as a complex line in $\mathbb{C}^{i+1}$. Observe that the natural Hermitian metric of $\mathbb{C}^{i+1}$ induces a Hermitian fiber metric on H_i in a natural way. Consider the product $H_1 \times H_2$ as a (holomorphic) vector bundle over the product $\mathbb{C}P^1 \times \mathbb{C}P^2$ and let M be the corresponding projective bundle. M is a compact complex manifold of complex dimension 4.

The projective group $P(Gl(2, \mathbb{C}) \times Gl(3, \mathbb{C}))$ acts, in an obvious way, on M as a group of automorphisms and is proved to be actually isomorphic to $\mathfrak{A}(M)$, which is thus reductive.

11.59. We consider, in particular, the action of $\mathbb{C}^*$ on M induced by the multiplication on the factor H_1 of $H_1 \times H_2$ (alternatively, H_1 and H_2 each determine a section of $\mathbb{C}P^1 \times \mathbb{C}P^2$ in M and $\mathbb{C}^*$ is the subgroup of $\mathfrak{A}(M)$ preserving both sections punctually). We denote by X the corresponding holomorphic vector field.

11.60. In order to get a Kähler form in the first Chern class $c_1(M)$, we proceed as follows. First consider the Kähler forms $\tilde{\omega}_\varepsilon$ on the total space $H_1 \times H_2$-{zero section} determined by

$$\tilde{\omega}_\varepsilon = p^*\eta + \varepsilon\, dd^c \log |v|^2$$

where η is the Kähler form on $\mathbb{C}P^1 \times \mathbb{C}P^2$ induced by the Fubini-Study metric on each factor, p is the projection of $H_1 \times H_2$ onto $\mathbb{C}P^1 \times \mathbb{C}P^2$, $|v|$ is the norm of v in $H_1 \times H_2$ and ε is a positive real number small enough for $\tilde{\omega}_\varepsilon$ to be positive. Clearly $\tilde{\omega}_\varepsilon$ induces a Kähler form ω_ε on M, whose Ricci form is denoted by ρ_ε.

Letting ε tend to zero, we get, as a limit for ρ_ε, a positive form ω belonging to $c_1(M)$ as required. Let ρ be the Ricci form of the Kähler form ω. Both ω and ρ belong to $c_1(M)$ so that ω is equal to the harmonic part of ρ and the Ricci potential is determined by $dd^c F = \rho - \omega$.

Then A. Futaki shows that $\int_M (X \cdot F)\omega^4$ is positive, so that, by 11.57, M admits no Kähler-Einstein metric.

11.61. On a Kähler manifold of dimension m, the isometry group has dimension at most $m^2 + 2m$ ([Kob 4], theorem 5.1). According to Theorem 11.52, this also bounds $\dim_{\mathbb{C}} \mathfrak{A}(M)$ if the metric is Einstein. In fact, more is known.

Theorem (Y. Sakane [Sak]). *Let M be a compact complex manifold with complex dimension $m \geqslant 5$. Assume that M admits a Kähler-Einstein metric. Then, if $\dim_{\mathbb{C}} \mathfrak{A}(M) \geqslant m^2 - 2m + 8$, M is isomorphic either to canonical $\mathbb{C}P^m$, $\mathbb{C}P^1 \times \mathbb{C}P^{m-1}$ or $\mathbb{C}P^2 \times \mathbb{C}P^{m-2}$.*

E. Extremal Metrics

11.62. Let Ω be a fixed Kähler class on a compact complex manifold M of (real) dimension $2m$. An element of Ω will denote as well a Kähler form and the associated Kähler metric. We consider the three functionals

$$\mathbf{S}s(g) = \int_M s^2 \cdot \mu_g \quad \mathbf{S}\rho(g) = \int_M |\rho|^2 \cdot \mu_g \qquad \mathbf{S}R(g) = \int_M |R|^2 \cdot \mu_g$$

for g any Kähler metric belonging to Ω, where s, ρ, R denote as usual the scalar curvature, the Ricci form and the whole curvature tensor respectively. By (2.80a) and (2.81a) each one can be expressed affinely by any one of them—$\mathbf{S}s$ say—with coefficients depending only on the Kähler class Ω and the first two Chern classes of M.

11.63 Definition (E. Calabi, [Cal 1]). A Kähler metric g is *extremal* if it is critical for any one of the three functionals $\mathbf{S}s$, $\mathbf{S}\rho$, $\mathbf{S}R$ above (so that it is critical for the two others as well) in its Kähler class.

A first variation of the Kähler form ω of g in Ω is determined by a 2-form dd^cf, that is by a scalar function f (see 2.114). We have the

11.64 Lemma. *The first derivative of* $\mathbf{S}s$ *in the direction of* f *is equal to*

$$\mathbf{S}s'_g(f) = -4\langle f, \delta\delta D^- ds\rangle.$$

Proof. By Lemma 2.158, we get readily

$$\mathbf{S}s'_g(f) = \langle dd^cf, (2\Delta s + s^2)\omega - 4s\rho\rangle = \langle f, \delta\delta^c[(2\Delta s + s^2)\omega - 4s\rho]\rangle.$$

Now, we have

$$\delta^c[(2\Delta s + s^2)\omega] = -d^c(2\Delta s + s^2) \lrcorner \omega = d(2\Delta s + s^2)$$

while, by (2.32),

$$-4\delta^c(s\rho) = -4s\delta^c\rho + 4d^cs \lrcorner \rho = -2sds + 4d^cs \lrcorner \rho.$$

We get thus

$$\begin{aligned}\delta^c[(2\Delta s + s^2)\omega - 4s\rho] &= 2(\Delta ds + 2d^cs \lrcorner \rho)\\ &= 4\delta D^- ds \qquad \text{(see (2.53) and (2.55)).}\end{aligned}$$

Lemma 11.64 follows at once. □

The Euler equation of the variational problem related to $\mathbf{S}s$—or $\mathbf{S}\rho$, $\mathbf{S}R$ as well—in a given Kähler class is then

$$\delta\delta D^- ds = 0$$

which, on M compact, is clearly equivalent to

$$D^- ds = 0.$$

From Proposition 2.124 we infer immediately the

11.65 Theorem. *A Kähler metric is extremal if and only if the gradient Ds is a (real) holomorphic vector field.*

11.66 Remark. Furthermore, according to Remark 2.128, the vector field JDs is *Killing*.

11.67. Let α be the real number defined by

$$c_1(M) = \frac{\alpha}{2\pi}\Omega + (c_1(M))_0,$$

where $(c_1(M))_0$ denotes the primitive part, relative to the Kähler class Ω, of the first Chern class $c_1(M)$. It is easily seen that the trace of the harmonic part γ of the Ricci form is independent of the choice of the metric in the Kähler class and is equal to $m\alpha$.

Thus, the scalar curvature s is equal to

$$s = 2m\alpha - 2\Delta F,$$

where F denotes the Ricci potential (see 2.142). It follows that the functional Ss is bounded from below by $4m^2\alpha^2 V$ where V is the total volume, $V = \int_M \Omega^2$. Moreover, this bound is reached exactly by metrics with harmonic Ricci form or, equivalently, with constant scalar curvature.

Notice furthermore that, if $c_1(M) = 0$, or if $c_1(M)$ is positive or negative definite and $\Omega = \pm 2\pi c_1(M)$, then a metric with constant scalar curvature in Ω has to be Einstein. (Indeed, the Ricci form ρ is the unique harmonic representative of $2\pi c_1(M)$, so $\rho = 0$, ω or $-\omega$ respectively). Thus we obtain

—*If $c_1(M) = 0$, then, for any Kähler class Ω, $\inf_\Omega \mathrm{S}s = 0$ is attained exactly for the unique Ricci-flat metric in Ω.*

—*If $c_1(M)$ is negative definite and $\Omega = -2\pi c_1(M)$, then $\inf_\Omega \mathrm{S}s = 4m^2 V$ is attained exactly for the unique Kähler-Einstein metric of M.*

11.68. The above features—existence of a positive lower bound, which is sometimes attained, sometimes not (for example, if $\mathfrak{A}^0(M)$ is not reductive), makes the variational problem of minimizing the functional Ss attractive in itself. However, E. Calabi's main motivation when he introduced extremal metrics in 1954 (see also [Lic 6]) was to widen the problem of existence of Kähler-Einstein metrics. Since there are manifolds with positive first Chern class which do not carry any Kähler-Einstein metric, let us ask whether any compact complex manifold with positive first Chern class admits an extremal metric.

11.69. Unfortunately, as observed by E. Calabi himself [Cal 7], the existence of an extremal metric on a compact complex manifold M imposes some restriction on the automorphism group. Indeed, Remark 11.66 implies that, if non discrete, the automorphism group of M must contain a non trivial connected compact subgroup.

Thus, there are manifolds which do not admit any extremal metric.

A simple example ([Lev]) is obtained by blowing up $\mathbb{C}P^1 \times \mathbb{C}P^1$ at the four points $(0, \infty)$, $(0, 1)$, (∞, ∞) and $(1, \infty)$. The identity component of the automorphism group is then the additive group $\mathbb{C}$. For more examples, see [Lev].

11.70. On the other hand, E. Calabi [Cal 6] has shown that there exist extremal Kähler metrics with non constant scalar curvature.

The simplest is given on the complex surface F_1 obtained from the (positive) Hopf bundle H over $\mathbb{C}P^1$ by compactifying each fiber by adding its infinity point. Since the Hopf bundle can be naturally identified, as a complex surface, with $\mathbb{C}P^2$ with a point deleted, F_1 may still be viewed as the complex surface obtained from $\mathbb{C}P^2$ by blowing up this point.

Let S_0 (resp. S_∞) be the 0- (resp. ∞-) section of H. Both are complex curves in F_1 isomorphic to $\mathbb{C}P^1$, and the open set $U = F_1 - (S_0 \cup S_\infty)$ is naturally identified with $\mathbb{C}^2 - \{(0,0)\}$.

The Kähler metric we are looking for will have the unitary group $U(2)$ as a symmetry group, just like D. Page's metric (see 9.125). More precisely, it will be defined on U by a Kähler potential depending only upon the usual norm on $U = \mathbb{C}^2 - \{(0,0)\}$.

Following [Cal 6] we write the Kähler potential as a (real) function u of t, where e^t is the usual square norm in $\mathbb{C}^2$. The Kähler form then is given by

$$\omega = \tfrac{1}{4} dd^c u$$

and the corresponding Kähler metric g is described as follows.

Each point x in $\mathbb{C}^2 - \{(0,0)\}$ determines a complex line l_x and its orthogonal $l_x^\perp$ for the usual Hermitian structure. The tangent space at x of $\mathbb{C}^2 - \{(0,0)\}$ splits as the sum $l_x \oplus l_x^\perp$. Then the g-square norm is equal to u'' and u' on l_x and $l_x^\perp$ respectively, and l_x and $l_x^\perp$ are g-orthogonal. In particular, u' and u'' have to be positive for $-\infty < t < +\infty$. Now we want the metric to be extended to S_0 and S_∞. Consider the case S_0 (the case S_∞ is similar).

The complex curve $S_0 \simeq \mathbb{C}P^1$ is the limit, in a sense, of the spheres Σ_t of points in $\mathbb{C}^2$ with (usual) square norm equal to e^t. Each sphere Σ_t is the sphere S^3 endowed with a "Berger metric" defined by the parameters $u'(t)$ and $u''(t)$.

More precisely, the metric induced on $\mathbb{C}P^1$, through the Hopf fibration, has constant holomorphic sectional curvature equal to $(u'(t))^{-1}$, while the length of the S^1-fibers is $2\,(u''(t))^{1/2}$.

Now it is easy to see that the metric g can be extended to S_0 (resp. to S_∞) if and only if u' and $u''e^t$ (resp. $u''e^{-t}$) both have a positive limit. Moreover, the metric is C^∞ at $t = -\infty$ (resp. at $t = +\infty$) if and only if u' and $u''e^t$ (resp. $u''e^{-t}$) are.

Let a (resp. b) be the limit of u' when $t = -\infty$ (resp. $t = +\infty$). Since u'' is positive, we have $a < b$. Recalling that S_0 and S_∞ generate $H_2(F_1, \mathbb{Z})$ (see [Bes 2], exposé IV), we infer that the pair (a, b) determines the cohomology class of the Kähler form ω.

Recall that the first Chern class of F_1 is given by the pair $(1, 3)$, [Bes 2] p. 121. Let g be any Kähler metric of this type on F_1. On the open set U, the volume-form μ_g is given by

(11.71) $$\mu_g = u'u''e^{-2t}\mu_{g_0}$$

where μ_{g_0} is the volume-form of the usual flat metric (associated with the Kähler potential $u(t) = e^t$). Moreover, for any function w on U depending only on t, we have

(11.72) $$\operatorname{grad}(w) = \frac{2w'}{u''}X$$

where X denotes the tautological vector field on $\mathbb{C}^2 - \{(0,0)\}$ which associates to each point the same, viewed as a tangent vector. Then,

(11.73) $$\Delta w = -4\left(\frac{w'}{u'} + \frac{w''}{u''}\right).$$

Using (11.71), we infer from 2.98 that the Ricci form ρ is equal to

(11.74) $$\rho = \tfrac{1}{2}dd^c v$$

where v is the function of t defined by

(11.75) $$v = 2t - \log u' - \log u''.$$

Thus, by (11.74), the scalar curvature s is equal to

(11.76) $$s = -\Delta v = 4\left(\frac{v'}{u'} + \frac{v''}{u''}\right).$$

On the other hand, the metric g is extremal if and only if grad(s) is a (real) holomorphic vector field. Since X is itself holomorphic on U, we infer immediately from 11.72 that *g is extremal if and only if $\frac{s'}{u''}$ is constant*, or, equivalently,

(11.77) $$s = \alpha\psi + \beta$$

where α, β are real constants and ψ, henceforth, will denote the function u'. The latter equation, after two integrations, yields

(11.78) $$\psi\psi' = -\tfrac{1}{48}\alpha\psi^4 - \tfrac{1}{24}\beta\psi^3 + \psi^2 - \tfrac{1}{4}\gamma\psi - \delta = P(\psi),$$

where γ, δ are real constants.

The constants are determined by the asymptotic conditions above, to which the Kähler potential is submitted to extend to S_0 and S_∞, namely,

the polynomial P vanishes for ψ equal to a and b and is positive for $a < \psi < b$, and

the quotient $\dfrac{P(\psi)}{\psi - a}\left(\text{resp. } \dfrac{P(\psi)}{b - \psi}\right)$ is equal to a (resp. b) for $\psi = a$ (resp. $\psi = b$).

We get

$$\psi\psi' = \frac{(\psi - a)(b - \psi)(2a\psi^2 + (b^2 - a^2 2ab)\psi + 2a^2 b)}{(b - a)(a^2 + b^2 + 4ab)}.$$

We eventually obtain (unfortunately in a non-explicit way!) a (unique) extremal Kähler metric of this type of F_1 in each Kähler class. Observe that the constants α

and β are equal to

$$\alpha = \frac{96a}{(b-a)(a^2+b^2+4ab)} \qquad \beta = \frac{24(b^2-3a^2)}{(b-a)(a^2+b^2+4ab)}.$$

One can check that, for all values of the parameters a and b, the scalar curvatures s of the extremal metrics are everywhere positive. Of course, s cannot be constant since $\psi' = u''$ is positive for $-\infty < t < +\infty$ (see also 11.56).

The construction above can be generalized in a similar way to any projective line bundle over projective space $\mathbb{C}P^m$.

11.79. Recall D. Page and L. Bérard Bergery's construction of Einstein metrics 9.125. It applies to a wide class of $\mathbb{C}P^1$-bundles over $\mathbb{C}P^m$. Thus every manifold in this class carries two kinds of distinguished metrics invariant under the action of $U(m)$: the unique—up to scaling—Einstein metric, and the 1-parameter family—up to scaling—of extremal Kähler metrics, constructed in a very similar way. At least in dimension four, the parallel between these two constructions is a special case of a striking dictionary between a class of extremal Kähler metrics, and a class of Einstein non Kähler metrics, discovered by A. Derdzinski.

11.80. Let us consider local metrics on $\mathbb{R}^4$. Start with an Einstein metric. Can it be made Kähler by a conformal change? Conversely, given a Kähler metric, can it be made Einstein by a conformal change. One first observes that the conformal factor, if any, is completely determined.

Let g be an Einstein metric. Assume that, for some positive function f, the metric fg is Kähler. Then, up to a constant,

$$f = |W^+|^{2/3} \quad \text{or} \quad f = |W^-|^{2/3}$$

Conversely, let g be a Kähler metric. Assume that, for some positive function k, the metric kg is Einstein. Then, up to a constant,

$$k = s^{-2}.$$

11.81 Theorem (A. Derdzinski, [Der 3]). i) *Let (M, g) be an oriented 4-dimensional Einstein manifold. Assume that the self-dual Weyl tensor W^+ does not vanish. Then we have two equivalent properties*

a) *W^+ has, at each point, at most two distinct eigenvalues on $\bigwedge^+$*

b) *up to a double cover, the metric $|W^+|^{2/3}g$ is Kähler (with respect to some complex structure on M, compatible with the given orientation).*

ii) *Conversely, let (M, J, g) be a 4-dimensional Kähler manifold, with non constant scalar curvature s. Then we have two equivalent properties*

a) *g is an extremal Kähler metric, and the function*

$$s^3 - 6s\Delta s - 12|ds|^2$$

is constant

b) *the metric $s^{-2}g$, where defined, is Einstein.*

Part i) is proved in 16.67.

We have seen in 4.78 that metrics conformal to Einstein metrics are critical points of the quadratic functional **S***W*. Thus ii) in a sense follows from the fact that, for a Kähler metric, **S***W* is a combination of **S***s* and the Kähler class.

11.82 Example. Let F_1 denote the manifold obtained by blowing up one point in $\mathbb{C}P^2$, and let g be the Page metric 9.125. Then W^+ does not vanish identically. (There are global reasons, Theorem 13.30, or the following argument: since $\tau(F_1) = 0$, Formula 6.34 would imply that $W^- = 0$ but, according to Kuiper's Theorem 1.171, F_1 which is simply-connected does not admit any conformally flat metric). Since, at each point, there is a non trivial isotropy group of isometries, the operator W^+ has a double eigenvalue, and the conformal metric $|W^+|^{2/3}g$ is Kähler (with respect to some $U(2)$-invariant complex structure, that is, the ordinary complex structure of F_1). Fact 11.81 ii) implies that this new metric is extremal and $U(2)$-invariant, so it is one of the Calabi examples.

Notice that W^- also has a double eigenvalue at each point, and so, the metric $|W^-|^{2/3}g$ is also Kähler with respect to another complex structure, inducing the opposite orientation. This amounts to the fact that F_1 admits antiholomorphic involutions, which commute with the $U(2)$-action, and preserve the Page metric (cf. 9.126d)).

Conversely, since the Page example is unique (cf. 9.126a)), it turns out that, among the one parameter family of Calabi's extremal metrics on F_1, only one metric satisfies the extra condition on scalar curvature.

As a matter of fact, using (11.72), (11.73) and the expressions (11.77), (11.78), we get easily, with the notations of 11.70,

$$s^3 - 6s\Delta s - 12|ds|^2 = 6\alpha(8\alpha\delta - \beta\gamma)\psi^{-1} + \text{constant},$$

which is constant if and only if $(8\alpha\delta - \beta\gamma)$ is equal to zero, or, equivalently, if and only if the numbers a and b are related by

$$(b^2 - 3a^2)^2 = 16a^3b.$$

It is easily checked that the equation $(x^2 - 3)^2 = 16x$ has a unique real solution greater than one, say μ. Observe that the Kähler class of that distinguished extremal metric is *not* the first Chern class (μ is equal to 3.184... while the slope of the first Chern class is 3).

11.83 Remark. All known examples of compact Einstein 4-manifolds either satisfy $W^\pm \equiv 0$, or are conformal to a Kähler metric. Indeed, here is the list of examples:

—locally symmetric spaces. The reducible ones, i.e., locally products of surfaces, are Kähler. In the irreducible case, one checks easily that $DW^+ = 0$ and $W^+ \not\equiv 0$ implies that $W^- \equiv 0$ (see 16.70).

—Kähler-Einstein metrics produced by Theorems 11.15, 11.17, and their quotients by involutions.

—the Page metric and its quotient.

11.84 Remark. Eventual companions of the Page metric in dimension four. Only a finite number of compact 4-manifolds may carry an Einstein metric, which is conformal to an extremal Kähler metric. Indeed, such a manifold is a Kähler complex surface, which carries a holomorphic vector field with zeros. According to [Ca-Ho-Ko], such surfaces are obtained either from $\mathbb{C}P^2$ by blowing up a number k of points, or from a holomorphic $\mathbb{C}P^1$-bundle over $\mathbb{C}P^1$ by blowing up l points. Now the Hitchin-Thorpe inequality 6.35 implies that $k \leqslant 8, l \leqslant 7$, which, according to [Ma-Mo], leaves only 10 distinct topological types. On the other hand, there are plenty of non compact Page-like metrics, see [Bes 2], Exposé XVII and [Der 5].

Chapter 12. The Moduli Space of Einstein Structures

A. Introduction

12.1. This chapter is devoted to studying the set of all Einstein structures on a given compact smooth manifold M. By an *Einstein structure*, we mean an equivalence class of Riemannian metrics. We do not distinguish between an Einstein metric g and equivalent tensor fields $\tilde{g} = c\varphi^*g$, where φ is a diffeomorphism of M, and c a positive constant. In the sequel, the quotient space of Einstein metrics under this relation is called the *Moduli Space* of Einstein structures on M, and *denoted* by $\mathscr{E}(M)$.

12.2. The case of surfaces has been and is the subject of a huge literature. Indeed, the Moduli Space of Einstein structures of an orientable surface M coincides with the moduli space of complex structures, or of conformal structures on M. A few facts about these spaces are gathered in Section B. In general, $\mathscr{E}(M)$ is not a manifold itself, but it admits a connected ramified covering $\mathscr{E}(M)$ called the *Teichmüller Space* of M, with the following properties. The space $\tilde{\mathscr{E}}(M)$ is a smooth manifold; the space $\mathscr{E}(M)$ is the quotient of $\tilde{\mathscr{E}}(M)$ under a discrete group action; the stabilizer of a structure $[g]$—which correspond to a singular point of $\mathscr{E}(M)$—is the discrete part of the isometry group of g. A very similar situation occurs in the case where M is the 3-torus T^3 (then Einstein metrics coincide with flat metrics).

12.3. As suggested by the preceding examples, one might hope that, in general, the Moduli Space is, up to a discrete group action, a finite dimensional manifold. A great part of this chapter is a (partly successful) attempt to prove this.

Unlike the constant sectional curvature condition, the Einstein equation cannot be defined through reference to a local model. In dimensions $n \geqslant 4$, this makes the sheaf theoretic approach of [Cal] and [BB-Bo-La] rather uneasy. Our approach of the Moduli Space is via analysis on the space of metrics, as in Chapter 4.

The Moduli Space is, by definition, a subset of the space of all Riemannian structures on M (see 4.2). In a neighborhood of the class of an Einstein metric g, a chart for this quotient space is given by a *slice* $\mathfrak{S}_g$ at g to the action of the diffeomorphism group on metrics. In this chart, the Moduli Space of Einstein Structures is seen as the subset of Einstein metrics in $\mathfrak{S}_g$. It is defined by a system of nonlinear P.D.E.'s: $E(\tilde{g}) = r_{\tilde{g}} - \frac{1}{n}\left(\int_M s_g \cdot \mu_{\tilde{g}}\right)\tilde{g} = 0.$

12.4. In order to understand these non linear equations, one studies the linearization, $E'(g)$, defined as follows. Given a symmetric 2-tensor field h, write $h = \frac{d}{dt}\tilde{g}(t)\Big|_{t=0}$ and let $\tilde{g}(0) = g$. Then we set

$$E'(g) \cdot h = \frac{d}{dt} E(\tilde{g}(t))\Big|_{t=0} .$$

In Section C, we prove that this linear operator on the tangent space to the slice $\mathfrak{S}_g$ is elliptic. As a consequence, its kernel, the vector space of *infinitesimal Einstein deformations at* g, is finite dimensional. This space may be considered as a "formal" tangent space of the Moduli Space at the point $[g]$.

12.5. The implicit function theorem allows us, in Section F, to exponentiate the space of infinitesimal Einstein deformations into a finite dimensional real analytic submanifold Z of the slice, which contains a neighborhood of $[g]$ in the Moduli Space as a *real analytic subset*. This is the main structure theorem. It reduces the question of the Moduli Space being a manifold (i.e., being all of Z) to a formal problem. The algebra there is still mysterious. N. Koiso found an example where the Moduli Space is a proper subset of Z. In particular, in that case, there exist "formal" tangent vectors which do not integrate into a continuous family of Einstein structures.

12.6. Section G gathers what global properties of the Moduli Space are known. For instance, there exist disconnected Moduli Spaces. Indeed, the Hopf fibrations give rise to two, sometimes three Einstein structures on spheres S^{4n+3}, with different Einstein constants. This raises attractive questions, such as the following. What is the set of (suitably normalized) Einstein constants of a given compact manifold, of all manifolds of a given dimension?

12.7. As a consequence of the structure theorem, proving rigidity of an Einstein structure $[g]$ (i.e., that $[g]$ is an isolated point of the Moduli Space) reduces to establishing a vanishing theorem for solutions of the linear elliptic operator $E'(g)$. Two such theorems are derived in Section H from Weitzenböck formulas (cf. 1.I). In particular, they show that most locally symmetric spaces are rigid. In the same spirit, an estimate on the dimension of the kernel of $E'(g)$ follows from the general method of P. Li ([Li]) and S. Gallot ([Gal 3]). This yields an upper bound on the dimension of the Moduli Space.

12.8. Apart from the case of surfaces, and moduli of flat metrics, all known examples of continuous families of Einstein structures follow from the Aubin-Calabi-Yau Theorems 11.15 and 11.17. In fact, when an Einstein metric g is Kähler with respect to some complex structure J, one may hope to understand completely the operator $E'(g)$. For example, one shows that, if J has negative or vanishing first Chern class, and a universal family of deformations, the Aubin-Calabi-Yau metrics corresponding to neighboring complex structures and Kähler classes fill in a whole neighborhood of $[g]$ in the Moduli Space, which, in this case, is a manifold near $[g]$. (See Section J).

In particular, any infinitesimal Einstein deformation at g integrates into a curve of Kähler-Einstein structures. This very satisfactory local behaviour of the Moduli Space of Kähler-Einstein structures leads one to consider it as the most relevant generalization of Teichmüller theory. (See Chapter 0 and [Koi 4]).

12.9. By investigating the limiting case in the Hitchin-Thorpe inequality 6.35, one shows that any Einstein metric on a manifold which is homotopy equivalent to a flat 4-manifold (resp., to a K3 surface) is flat (resp., is Kähler with respect to some K3 complex structure). As a consequence, for these manifolds (and, unfortunately, only these in dimension $n \geqslant 4$), the Moduli Space is entirely known. The case of flat manifolds, which is elementary, is explained in Section B. The more difficult Moduli Space of the underlying manifold of K3 surfaces has been delayed until Section K.

12.10. In spite of the heavy machinery developed in Sections C to F, only partial results have been obtained. The reader will feel that our present knowledge of the Moduli Space is very poor. Mainly due to the lack of examples, see however Add.A, most basic questions remain unanswered. For example,

—may a Moduli Space be really singular?

—may a Moduli Space have infinitely many connected components?

—are there continuous families of positive Einstein structures?

B. Typical Examples: Surfaces and Flat Manifolds

A two dimensional Riemannian manifold is Einstein if and only if it has constant curvature. The correspondance between metrics of constant curvature, conformal structures, and complex structures on surfaces makes the subject extremely rich. We shall give a very short account of the results which admit generalizations to the case of Einstein or Kähler-Einstein manifolds.

12.11. Only S^2 and $\mathbb{R}P^2$ admit a Riemannian structure with constant positive curvature, which is unique.

12.12. Flat structures exist only on manifolds with vanishing Euler characteristic, i.e., the torus T^2 and the Klein bottle. Flat structures on T^2 are parameterized by the modular domain

$$SO(2)\backslash SL(2, \mathbb{R})/SL(2, \mathbb{Z}).$$

Flat Klein bottles are quotients of flat rectangular tori, and their parameter space is a line (and not half a line, think two seconds about it).

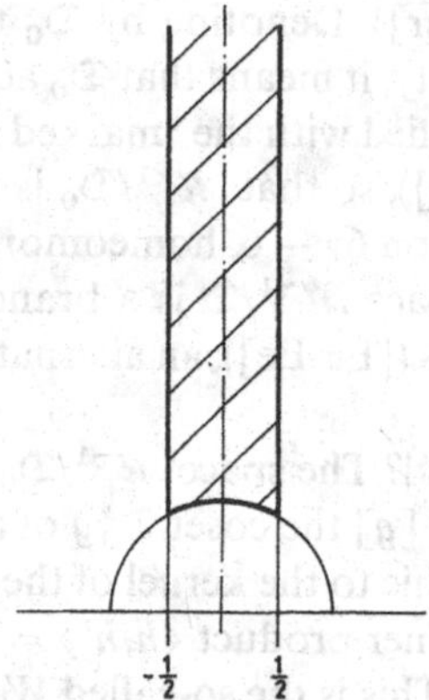

12.13. The negative case, which occurs if and only if the manifold M has negative Euler characteristic, lies much deeper. Indeed, there is a well-known dictionary "1-dimensional complex structure versus 2-dimensional Riemannian structure with constant curvature". The question amounts to the study of complex structures on a compact manifold of genus $\gamma > 1$. Much work has been done on this subject, we only give a few typical references ([Ea-Ee], [Rau]). For the Riemannian geometric point of view, see [BB-Bo-La] and [Bsr]. The theory can be summarized as follows (we shall use the dictionary freely).

12.14. *Local deformations, local theory.* Let us remark first that all germs of complex manifolds are biholomorphic, and that all the germs of Riemannian manifolds with curvature -1 are isometric. Therefore, whichever point of view is taken (complex or Riemannian geometric), sheaf theory will be relevant. We take the metric point of view, which is closer to our concerns. Infinitesimal deformations which preserve constant curvature -1 satisfy the equation $s'_g \cdot h = 0$, where s'_g is the linearized operator of the scalar curvature map $g \mapsto s_g$ (cf. 1.184 and 4.17). As we already saw in 12.3, we have to consider the deformations which are orthogonal to the orbit of g under the diffeomorphism group. Namely, h must satisfy the system of equations

(12.14) $$s'_g \cdot h = 0, \qquad \delta_g \cdot h = 0.$$

It can be proved that such h's are the real parts of the holomorphic quadratic differentials (for the complex structure defined by the metric g). This is not surprising, since an alternative sheaf theoretic argument (cf. [Ea-Ee] for instance) proves that deformations of complex curves are parametrized by holomorphic quadratic differentials. According to the Riemann-Roch theorem, this last space has complex dimension $3\gamma - 3$, so that the space of solutions of the system (12.14) has real dimension $6\gamma - 6$.

12.15 Global Theory. If we denote by $\mathscr{M}^{-1}$ the space of metrics on M with constant curvature -1, the Moduli Space is nothing but $\mathscr{M}^{-1}/\mathfrak{D}$. It is not a manifold, since some metrics of constant curvature on M admit non-trivial isometries (i.e., non trivial biholomorphic maps, in the complex language). Such an isometry cannot be isotopic to the identity (cf. [Bsr]). Denoting by $\mathfrak{D}_0$ the group of diffeomorphisms which are isotopic to the identity, it means that $\mathfrak{D}_0$ acts freely on $\mathscr{M}^{-1}$. The element cosets in $\mathscr{M}^{-1}/\mathfrak{D}_0$ can be identified with the "marked Riemann surfaces" of the classical theory, (cf. [Rau] or [Bsr]), so that $\mathscr{M}^{-1}/\mathfrak{D}_0$ is the famous *Teichmüller* space. It is a manifold of real dimension $6\gamma - 6$, homeomorphic to a ball, and the natural projection onto the moduli space $\mathscr{M}^{-1}/\mathfrak{D}$ is a branched covering. This approach comes from C. Earle and J. Eells ([Ea-Ee]); an alternative purely geometric approach is due to P. Buser [Bsr].

Now, what more can be said? The space $\mathscr{M}^{-1}/\mathfrak{D}_0$ carries a natural Riemannian metric. Indeed, if we denote by $[g]$ the coset $\mathfrak{D}_0^* g$ of a metric g in $\mathscr{M}^{-1}$, the tangent space $T_{[g]}(\mathscr{M}^{-1}/\mathfrak{D}_0)$ is isomorphic to the kernel of the system (12.14). This space can be equipped with the global inner product $\langle h, h' \rangle = \int_M (h, h')\mu_g$, and compatibility relations are straightforward. This is the so-called *Weil-Petersson* metric (cf. [Ahl]). Moreover, if we take the complex point of view, $\mathscr{M}^{-1}/\mathfrak{D}_0$ carries an obvious almost complex structure. It can be proved that these structures are compatible, and that the Weil-Petersen metric is in fact Kähler.

12.16. What about non-orientable surfaces of constant curvature? It is easy to see (cf. [BB-Bo-La]) that the space of infinitesimal deformations has dimension $3\gamma - 3$, if γ is the genus of the orientable covering. Alternatively, by computing the Euler characteristic of the sheaf of germs of Killing vector fields on M, it can be proved that this dimension is $-3\chi(M)$, as soon as M, orientable or not, has negative Euler characteristic.

12.17. We shall now give a brief description of the moduli space of flat structures on tori and their finite quotients. Again, the nature of this space shows what one should expect, in good cases, for the moduli space of Einstein structures. In fact, on flat manifolds of dimension $n \leqslant 4$, Einstein and flat structures coincide. Indeed, it follows from Proposition 1.120 that any three dimensional Einstein manifold has constant sectional curvature. For T^3 and its quotients, this constant can be neither positive (by Myers' Theorem 6.52) nor negative (again, because of the growth of balls, see [Mil 1]); thus it is zero. In four dimensions, there are obstructions to the existence of Einstein metrics; one of them 6.37 implies that any Einstein metric on a manifold whose Euler characteristic vanishes has to be flat. This applies to T^4, so Einstein metrics on T^4 are flat, and the same conclusion holds for other flat 4-manifolds.

12.18 Flat Structures on T^n. Let $\mathfrak{D}'$ denote the group of diffeomorphisms of $M = T^n$ inducing the identity on the fundamental group, let $\mathscr{F}$ denote the quotient space of flat metrics of volume one under $\mathfrak{D}'$. The moduli space $\mathscr{F}$ of flat structures on M is obtained from $\mathscr{F}$ by dividing by the discrete group $\mathfrak{D}/\mathfrak{D}' = Gl(n, \mathbb{Z})$.

Any flat metric on M has an $\mathbb{R}^n/\mathbb{Z}^n$ as identity component of its isometry group. Any two such groups of diffeomorphisms of M are conjugated by a diffeomorphism in $\mathfrak{D}'$. Thus, in order to describe $\mathscr{F}$, one may restrict to metrics left-invariant for a given group structure on M.

Let g, g' be flat metrics, and assume that $g' = \varphi^* g$, with $\varphi \in \mathfrak{D}'$. Then, the lift

$\varphi: \tilde{M} \to \tilde{M}$ to the universal cover is an isometry between euclidean spaces, thus, is affine. Since φ induces the identity on a lattice of M, φ is a translation, and $g = g'$. Thus $\mathscr{F}$ is homeomorphic to the space of left-invariant metrics of volume one on $\mathbb{R}^n/\mathbb{Z}^n$, that is, an open, convex subset of a vector space of dimension $\left(\frac{n(n+1)}{2} - 1\right)$.

12.19 Other Flat 2-, 3- or 4-manifolds. According to Bieberbach's theorem [Bie], M admits a finite covering $T^n \to M$. Thus, if F denotes the group of deck transformations—a crystallographic group—$\mathscr{F}(M)$ identifies to the subset of F-invariant structures in T^n, or to F-invariant scalar products of volume one on $\mathbb{R}^n/\mathbb{Z}^n$. Again, *$\mathscr{F}$ is an open convex set.*

To obtain $\mathscr{F}(M)$, one still must divide by the (affine) action of $\mathfrak{D}(M)/\mathfrak{D}'(M)$, the *normalizer* of F in $Gl(n, \mathbb{Z})$.

Remark. Although it is still unknown, it seems likely that there exists flat manifolds whose Moduli Space reduces to one point.

C. Basic Tools

12.20. By definition, the moduli space of Einstein structures on M is a subset of the space of all Riemannian structures on M, namely, the coset space of $\mathscr{M}_1 = \{$metrics with total volume 1$\}$ under the action of the diffeomorphism group $\mathfrak{D}$, endowed with the quotient topology. In order to describe this set locally, it would be convenient to have a local section of the map $\mathscr{M}_1 \to \mathscr{M}_1/\mathfrak{D}$. This is nearly achieved by D.G. Ebin's Slice Theorem [Ebi 2].

12.21. We recall the following notation from Chapter 4 and 1.59.

The tangent space of $\mathscr{M}_1$ at g, *denoted by* $T_g\mathscr{M}_1$, consists of all symmetric 2-tensor fields h whose total trace (relative to g) is zero.

The tangent space of the orbit $\mathfrak{D}^*g$ at g consists of all Lie derivatives of g with respect to vector fields X. Since, by 1.60, $L_X g = 2\delta_g^* X^\flat$, we *denote it by* $\mathrm{Im}(\delta_g^*)$.

Since the differential operator $\delta_g^*: \Omega^1 M \to T_g\mathscr{M}_1$ is over-determined elliptic, Corollary 34 of the Appendix applies. The image $\mathrm{Im}(\delta_g^*)$ is closed, the orthogonal complement is the kernel of the adjoint operator $\delta: T_g\mathscr{M}_1 \to \Omega^1 M$ (see 1.59). That is,

$$T_g\mathscr{M}_1 = \mathrm{Im}(\delta_g^*) \oplus [T_g\mathscr{M}_1 \cap \mathrm{Ker}\,\delta].$$

12.22. The Slice Theorem [Ebi 2] states that this orthogonal complement exponentiates into a real analytic submanifold $\mathfrak{S}_g$ of $\mathscr{M}_1$, which is a *slice* to the action of $\mathfrak{D}$. It has the following properties:

—$\mathfrak{S}_g$ is invariant under the group $I(M, g)$ of isometries of g.

—If $\varphi \in \mathfrak{D}$ and $\varphi^*\mathfrak{S}_g \cap \mathfrak{S}_g \neq \varnothing$, then $\varphi \in I(M, g)$.

—There is a local cross-section $\chi: \mathfrak{D}/I(M, g) \to \mathfrak{D}$ defined on a neighborhood of the coset I_g, such that the local mapping $\mathfrak{D}/I(M, g) \times \mathfrak{S}_g \to \mathscr{M}_1$, $(\bar{\varphi}, \tilde{g}) \mapsto \chi(\bar{\varphi})^*\tilde{g}$ is a diffeomorphism onto a neighborhood of g in $\mathscr{M}_1$.

In particular, the induced map $I(M,g)\backslash\mathfrak{S}_g \to \mathfrak{D}\backslash\mathcal{M}_1$ is a homeomorphism onto a neighborhood of the Riemannian structure defined by g.

12.23 Definition. Let g be an Einstein metric on M. The subset of Einstein metrics in the slice $\mathfrak{S}_g$ at g is called the *premoduli space of Einstein structures around* g.

From now on, we shall work on this premoduli space. To recover a true neighborhood of the Einstein structure defined by g, one must still divide by the action of the isometry group of g, a compact Lie group.

12.24 Remark. When the scalar curvature s of g is negative or zero, the identity component $(I_g)^0$ of the isometry group of g acts trivially on the premoduli space of Einstein structures around g.

Therefore, the Moduli Space is obtained from the premoduli space by dividing by the action of the *finite* group $I(M,g)/(I_g)^0$.

Indeed, according to S. Bochner (cf. 1.84), if $s < 0$, the isometry group I_g is finite. If $s = 0$, we shall see in 12.52 that neighboring Einstein metrics $\tilde{g}$ also are Ricci flat. For such a $\tilde{g}$, Killing fields and harmonic 1-forms are all parallel, thus $\dim(I(M,\tilde{g})) = b_1(M) = \dim(I(M,g))$. If $\tilde{g}$ is in the slice $\mathfrak{S}_g$, 12.22 implies that $I(M,\tilde{g}) \subset I(M,g)$, and so $(I_g)^0 = (I_{\tilde{g}})^0$. □

12.25. As Remark 12.24 indicates, we shall often encounter spaces which are *locally* obtained by dividing a *local* manifold by the action of a finite group. Lacking of a widely accepted terminology, we shall call such an object an *orbifold*.

12.26. There are several ways to express the Einstein condition. Since Einstein metrics are critical points of the total scalar curvature functional

$$\mathbf{S}(g) = \int_M s_g \mu_g,$$

for our purposes, the most convenient expression seems to be the following:

$$\text{“}g \text{ is Einstein if and only if } r_g = \frac{\mathbf{S}(g)}{n\,\mathrm{vol}(g)} g\text{”}.$$

Definition. We call the differential operator $E\colon \mathcal{M}_1 \to \mathcal{S}^2 M$, $E(g) = r_g - \frac{1}{n}\mathbf{S}(g)g$, the *Einstein operator*.

D. Infinitesimal Einstein Deformations

12.27. Let g be an Einstein metric on M. If there exists a smooth curve, $g(t)$, in the premoduli space around g, with $g(0) = g$, its first jet $h = \left.\frac{dg(t)}{dt}\right|_{t=0}$ will satisfy the *linearized Einstein equation*

$$E'_g(h) = \frac{d}{dt} E(g(t))\Big|_{t=0} = 0,$$

where $E'_g(h)$ is computed with the help of Formula (1.182). Since g is an Einstein metric, $S'_g(h) = 0$, so that $E'_g(h) = r'_g(h) - \frac{s}{n}h$, that is,

(12.28) $$2E'_g(h) = D^*Dh - 2\delta^*\delta h - Dd(\operatorname{tr} h) - 2\mathring{R}h.$$

Or, in terms of the Lichnerowicz Laplacian Δ_L defined in 1.143:

(12.28′) $$2E'_g(h) = \Delta_L h - 2\delta^*\delta h - Dd(\operatorname{tr} h) - 2\frac{s}{n}h.$$

Therefore, a first step when investigating the Moduli Space is to study the linear differential operator E'_g on the tangent space to the slice.

12.29 Definition. An *infinitesimal Einstein deformation* of an Einstein metric g is a symmetric 2-tensor field h such that

$$E'_g(h) = 0, \quad \delta_g h = 0, \quad \text{and} \quad \int_M \operatorname{tr}_g h\mu_g = 0.$$

Their space is *denoted by* $\varepsilon(g)$.

12.30 Theorem (M. Berger-D.G. Ebin, [Be-Eb]). *A symmetric 2-tensor field h is an infinitesimal Einstein deformation of g if and only if it satisfies the following equations*:

$$(D^*_g D_g - 2\mathring{R}_g)h = 0, \qquad \delta_g h = 0, \quad \operatorname{tr}_g h = 0.$$

Proof. If h is an infinitesimal Einstein deformation of g, then $D^*Dh - Dd(\operatorname{tr} h) - 2\mathring{R}h = 0$. Taking the trace, one obtains $\Delta \operatorname{tr} h = \frac{s}{n}\operatorname{tr} h$. But $\frac{s}{n}$ cannot be a non zero eigenvalue of the Laplacian on functions, since, according to A. Lichnerowicz [Lic 1], the smallest non zero eigenvalue is not less than $\frac{s}{n-1}$. Therefore, either $\operatorname{tr}_g h = 0$, or $s = 0$ and $\operatorname{tr}_g h$ is constant. In either cases $\operatorname{tr}_g h = 0$. □

Remark. For a more geometric proof, see [Bou 5].

12.31 Corollary. *The space $\varepsilon(g)$ of infinitesimal Einstein deformations is finite dimensional.*

Clear from Appendix 31, since the operator $D^*D - 2\mathring{R}\colon \mathscr{S}^2M \to \mathscr{S}^2M$ is elliptic. □

E. Formal Integrability

12.32. Let g be an Einstein metric on M. Is the premoduli space around g a submanifold of the slice $\mathfrak{S}_g$, with tangent space $\varepsilon(g)$ at g?

12.33. In order to apply directly the implicit function theorem to the map E, we need its differential E'_g to be surjective. Unfortunately, the image of E'_g satisfies a differential equation, which comes from the differential Bianchi identity (cf. 1.14), namely $\delta r + \frac{1}{2}ds = 0$.

In fact, let us *denote by* $\beta_g = \delta_g + \frac{1}{2}d\,\mathrm{tr}_g$ the *Bianchi operator*, then 1.94 (see also 4.19) states that, for any metric g,

$$\beta_g(E(g)) = 0.$$

Differentiating with respect to the metric g, we obtain, for any h,

$$\beta'_g(h)(E(g)) + \beta_g(E'_g(h)) = 0.$$

Thus, when g is Einstein,

$$\beta_g \circ E'_g = 0. \tag{12.33}$$

12.34. Since E'_g is never surjective, we shall look for weaker conditions under which any infinitesimal Einstein deformation h integrates into a curve of Einstein metrics, at least at the formal level.

12.35. The following discussion is valid for any smooth map E between Banach spaces. It is classical in the theory of deformations of complex structures (cf [Gri]).

If $g(t)$ is a smooth curve, starting at g, with jet

$$h_k = \frac{d^k}{dt^k} g(t)\bigg|_{t=0} \quad \text{for } k = 1, 2, \ldots$$

we *set*

$$E_g^k(h_1, \ldots, h_k) = \frac{d^k}{dt^k} E(g(t))\bigg|_{t=0}.$$

The map E_g^k is polynomial in $h_1, \ldots, h_k$, of degree one in h_k. It allows us to define the action of E on formal power series.

12.36 Definition. For a formal series $g(t) = g + \sum_{k=1}^{+\infty} \frac{t^k}{k!} h_k$, we *define*

$$E(g(t)) = E(g) + \sum_{k=1}^{+\infty} \frac{t^k}{k!} E_g^k(h_1, \ldots, h_k).$$

A tangent vector $h_1 \in \operatorname{Ker} E_g^1$ is *formally integrable* if there exists a formal series $g(t) = g + th_1 + \sum_{k=2}^{+\infty} \frac{t^k}{k!} h_k$ such that $E(g(t)) = 0$.

12.37. Clearly,

$$E_g^{k+1}(h_1,\ldots,h_{k+1}) = \frac{d}{dt} E_{g(t)}^k\left(\frac{dg(t)}{dt},\ldots,\frac{d^k}{dt^k}g(t)\right)\Bigg|_{t=0}$$

$$= \sum_{j=0}^{k-1} \frac{\partial}{\partial hj} E_g^k(h_1,\ldots,h_k)\cdot h_{j+1} + E_g^k(h_1,\ldots,h_{k-1},h_{k+1}).$$

By induction, we obtain

$$E_g^k(h_1,\ldots,h_k) = P_g^k(h_1,\ldots,h_{k-1}) + E_g^1(h_k), \tag{12.37}$$

where P_g^k is a polynomial in $h_1,\ldots,h_{k-1}$ only. In the above formula, it is clear that, if E_g^1 is surjective, so is the map $h_k \mapsto E_g^k(h_1,\ldots,h_k)$ for any $h_1,\ldots,h_{k-1}$. Thus, given any h_1 in $\operatorname{Ker} E_g^1$, a formal solution of E may be constructed step by step.

12.38. If E_g^1 is not surjective, there exists a linear map B_g into some Banach space such that $\operatorname{Im} E_g^1 \subset \operatorname{Ker} B_g$. (For example, let B_g be the projection onto $(\operatorname{Im} E_g^1)^\perp$).

Assume that such a $B_{\tilde g}$ is defined for $\tilde g$ close to g, and depends smoothly on $\tilde g$. Then, for every formal series $g(t)$, the formal identity $B_{g(t)}\left(\frac{d}{dt}E(g(t))\right) = 0$ holds.

Let us construct inductively the formal solution $g(t)$ of E. Assume we have already found $h_1,\ldots,h_k$ such that, for $0 \leqslant j \leqslant k$,

$$E_g^j(h_1,\ldots,h_j) = 0.$$

For $g(t) = g + \sum_{j=0}^{k} \frac{t^j}{j!} h_j$, we have

$$0 = \frac{d^k}{dt^k} B_{g(t)}\left(\frac{d}{dt}E(g(t))\right) = \sum_{j=0}^{k} \binom{k}{j}\left(\frac{d^{k-j}}{dt^{k-j}} B_{g(t)}\right)\left(\frac{d^{j+1}}{dt^{j+1}} E(g(t))\right).$$

At $t = 0$,

$$0 = \sum_{j=0}^{k-1} \binom{k}{j}\left(\frac{d^{k-j}}{dt^{k-j}} B_{g(t)}\bigg|_{t=0}\right)(E_g^{j+1}(h_1,\ldots,h_{j+1})) + B_g(E_g^{k+1}(h_1,\ldots,h_k,0)),$$

that is,

$$B_g(P_g^{k+1}(h_1,\ldots,h_k)) = 0.$$

In view of (12.37), our next step h_{k+1} must satisfy

$$E_g^1(h_{k+1}) = -P_g^{k+1}(h_1,\ldots,h_k).$$

If $\operatorname{Im} E_g^1 = \operatorname{Ker} B_g$, this equation always has a solution, and any h_1 in $\operatorname{Ker} E_g^1$ is formally integrable. This is why the quotient space $\operatorname{Ker} B_g/\operatorname{Im} E_g^1$ is usually called the *obstruction space* of the equation E, subject to the *integrability condition* B.

12.39. Notice that the same discussion would apply to an integrability condition of the form $B_g(E(g)) = 0$.

12.40 Example. *Deformations of complex structures.* We refer to Chapter 2 for definitions and notation. Fix an integrable almost complex structure J on M. The first jets $I = \frac{d}{dt} J(t)\Big|_{t=0}$ of variations $J(t)$ of J satisfy the linearized equations

$$IJ + JI = 0, \qquad N'_J(I) = 0. \tag{12.40}$$

The first equation means that I may be considered as a section of $\bigwedge^{0,1} M \otimes T^{1,0}M$. One easily checks that

$$N'_J(I) = \tfrac{1}{2} J \circ d'' I \in \textstyle\bigwedge^{0,2} M \otimes T^{1,0} M$$

12.41. Obvious solutions of these equations (12.40) are the first jets of trivial deformations, these are of the form $L_X J$, where X is a real vector field. Write $X = Z + \bar{Z}$, where Z is a section of $T^{1,0}M$. Then

$$L_X J = d'' Z \in \textstyle\bigwedge^{0,1} M \otimes T^{1,0} M.$$

12.42. Both d'' operators appearing above are part of the *Dolbeault complex* of the holomorphic vector bundle $T^{1,0}M$: this complex of sheaves

$$0 \to \Theta \to \underline{T}^{1,0} \to \underline{\textstyle\bigwedge}^{0,1} \otimes \underline{T}^{1,0} \to \underline{\textstyle\bigwedge}^{0,2} \otimes \underline{T}^{1,0} \to \underline{\textstyle\bigwedge}^{0,3} \otimes \underline{T}^{1,0} \to \cdots$$

is a resolution of the sheaf Θ of germs of holomorphic vector fields of (M, J) (see [Ko-Mo]). The Dolbeault Theorem provides us with an isomorphism between our deformation space $\mathrm{Ker}(d'')/d''(C^\infty T^{1,0}M)$ and $H^1(M, \Theta)$.

12.43. The relation $d'' \circ d'' = 0$ expresses an integrability condition for the differential $d'' \colon \bigwedge^{0,1} M \otimes T^{1,0}M \to \bigwedge^{0,2} M \otimes T^{1,0}M$. The corresponding obstruction space is $\mathrm{Ker}(d'')/d''(C^\infty(\bigwedge^{0,1} M \otimes T^{1,0}M)) \simeq H^2(M, \Theta)$.

12.44 Example. *Deformations of Einstein structures.* The most natural integrability condition is the Bianchi operator β_g. A description of the obstruction space is provided by the following theorem, which we will not prove.

12.45 Theorem (N. Koiso, [Koi 4]). *Let g be an Einstein metric on M, of volume* 1. *The subspace* $\mathrm{Ker}\,\beta_g$ *of* $T_g\mathcal{M}_1$ *splits into an orthogonal sum* $\mathrm{Ker}\,\beta_g = \mathrm{Im}\,E'_g \oplus \varepsilon(g)$.

12.46. In other words, the obstruction space for the Einstein equation coincides with the deformation space itself. Consequently, the integrability criterion assuming that this space vanishes is meaningless. One has to investigate more closely the polynomials

$$\bar{P}^k \colon \varepsilon(g)^k \xrightarrow{P^k} \mathrm{Ker}\,\beta_g \to \mathrm{Ker}\,\beta_g/\mathrm{Im}\,E'_g \simeq \varepsilon(g).$$

N. Koiso worked this out for compact symmetric spaces. In particular, he discovered that, for the symmetric metric g on $\mathbb{C}P^{2k} \times \mathbb{C}P^1$, $\bar{P}^2$ does not vanish. This means that *g has infinitesimal Einstein deformations, none of which are formally integrable.* This example is of great importance, since it shows the complexity of the discussion of Einstein deformations. The difficulty is not of an analytic nature—see the next paragraphs. It stems from the algebraic structure of the system of equations E. We shall come back to this example in 12.51 and 12.77.

F. Structure of the Premoduli Spaces

12.47. We have just exhibited sufficient conditions under which the Einstein operator E has formal power series solutions. Here, we are concerned with the convergence of these series. The main analytic property used here is the fact that E'_g has a closed image. Of course, this follows from Theorem 12.45, but we will prove this corollary completely.

12.48 Lemma. *Let g be an Einstein metric on M. The linear operator E'_g: $T_g\mathfrak{S}_g \to \mathscr{S}^2M$ has a closed image.*

Proof. We replace the non elliptic operator E'_g by an elliptic one. We set

$$F: \mathscr{S}^2M \to \mathscr{S}^2M, \qquad F = 2E'_g + 2\delta^*\delta = D^*D - 2\mathring{R} - Dd\,\mathrm{tr}.$$

Then $\mathrm{Im}(F)$ is closed, and $F(T_g\mathfrak{S}_g) = E'_g(T_g\mathfrak{S}_g)$. Indeed, the relation

$$\mathrm{tr}\, F = 2\left(\Delta - \frac{s}{n}\right)\mathrm{tr}$$

shows that $F(T_g\mathscr{M}_1)$ is also closed. Formula 12.33 yields

$$\beta_g \circ F = 2\beta_g\delta^*\delta = G\delta_g, \tag{12.48}$$

where $G = 2\delta\delta^* - d\delta$ is elliptic.

Clearly, $F(T_g\mathfrak{S}_g) \subset \mathrm{Ker}\,\beta_g$. Conversely, let $h = Fk \in \mathrm{Ker}\,\beta_g$, where $k \in T_g\mathscr{M}_1$. We have $G\delta_g k = \beta_g F k = 0$, so $\delta_g k \in \mathrm{Ker}\, G$, a finite dimensional vector space. This proves the inclusions

$$F(T_g\mathfrak{S}_g) \subset F(T_g\mathscr{M}_1) \cap \mathrm{Ker}\,\beta_g \subset F(T_g\mathscr{M}_1 \cap \delta_g^{-1}\,\mathrm{Ker}\, G).$$

Since $T_g\mathfrak{S}_g$ is closed and has finite codimension in $T_g\mathscr{M}_1 \cap \delta_g^{-1}\,\mathrm{Ker}\, G$, the image $F(T_g\mathfrak{S}_g)$ is closed in $F(T_g\mathscr{M}_1 \cap \delta_g^{-1}\,\mathrm{Ker}\, G)$ (this is a standard argument in topological vector spaces, see [Pal 1]). We conclude that $F(T_g\mathfrak{S}_g)$ is closed in $F(T_g\mathscr{M}_1) \cap \mathrm{Ker}\,\beta_g$, which is closed in $\mathscr{S}^2M$. □

13.49 Theorem (N. Koiso, [Koi 4]). *Let g be an Einstein metric on M. In the slice $\mathfrak{S}_g$, there exists a finite dimensional real analytic submanifold Z such that*:

(i) *the tangent space of Z at g is $\varepsilon(g)$, i.e., consists of all infinitesimal Einstein deformations of g*;

(ii) *Z contains the premoduli space around g as a real analytic subset.*

Proof. Denote by p the orthogonal projection of $\mathscr{S}^2M$ onto $E'_g(T_g\mathfrak{S}_g)$. The real analytic map $p \circ E$: $\mathfrak{S}_g \to E'_g(T_g\mathfrak{S}_g)$ is a submersion at g. Thus $Z = (p \circ E)^{-1}(0)$ is a real analytic submanifold near g, with tangent space $\operatorname{Ker} E'_g \cap T_g\mathfrak{S}_g = \varepsilon(g)$ at g. On this real analytic manifold, the map E is real analytic; therefore, the premoduli space $E^{-1}(0)$ is a real analytic subset. □

12.50 Corollary. *An infinitesimal Einstein deformation integrates into a curve of Einstein metrics if and only if it is formally integrable.*

Proof. If $g(t)$ is an honest smooth curve in the premoduli space around g, with $g(0) = g$, $g'(0) = h$, its Taylor series is a formal solution to the equation $E = 0$, with h as first jet. Conversely, according to M. Artin ([Art]), any formal curve in a real analytic set can be approximated (in the sense of formal series) by convergent curves in this set. □

12.51 Remark. Corollary 12.50 shows that the integrability problem for infinitesimal Einstein deformations is not an analytic one.

Let us come back to N. Koiso's example 12.46. The symmetric structure $[g_0]$ on $\mathbb{C}P^1 \times \mathbb{C}P^{2k}$ is isolated in the Moduli Space. Indeed, if a real analytic set has an accumulation point, then it contains a curve. In this case, the dimension of the premoduli space is strictly smaller than $\dim \varepsilon(g_0) = 4(4k^2 - 1)$. The premoduli space around g_0 still is a manifold. Nevertheless, it seems that one should not hope that premoduli spaces always be manifolds. However, an example where a premoduli space is singular is still lacking.

G. The Set of Einstein Constants

Theorem 12.49 contains information about the topology of the Moduli Space.

12.52 Corollary. *The Moduli Space of Einstein structures on M is locally (real analytically) arcwise connected. The total scalar curvature* **S** *is a locally constant function, and takes (at most) countably many values on the Moduli Space.*

Proof. The local connectedness holds for the premoduli space around each point, hence for the Moduli Space. Since Einstein metrics are critical points of the functional **S**, it is constant on each premoduli space. Finally, the topology of the Moduli Space, as that of $\mathscr{M}_1$, has a countable basis. Therefore, this space has at most countably many connected components. □

12.53. This implies that certain Moduli Spaces are disconnected. In fact, the method of dilating the fibers in a Riemannian submersion with totally geodesic fibres 9.67, originally due to G. Jensen, yields two Einstein structures on spheres S^{4n+3}, with different total scalar curvatures. On S^{15}, two of the possible submersions

$$S^7 \to S^{15} \to \mathbb{C}aP^2 = S^8, \qquad S^3 \to S^{15} \to \mathbb{H}P^3,$$

yield three different total scalar curvatures. Hence, on a sphere S^{4n+3}, the Moduli

Space of Einstein structures has at least two connected components. On S^{15}, at least three.

12.54. Conversely, examples are known where the Moduli Space is connected. This is the case for surfaces. Consider 3-manifolds of constant sectional curvature. The Einstein structure is unique when the curvature is positive (J.A. Wolf [Wol 4]) or negative (G.D. Mostow [Mos]). In section B, we have seen that the Moduli Space of the flat manifolds of dimension 3 and 4 is connected. This is again the case for the underlying manifold of K3 surfaces, as we shall see in Section K.

12.55. For the sake of completeness, we mention that, since it is contained in the space of all Riemannian structures, the Moduli Space of Einstein structures is Hausdorff. This follows from the compactness properties of sets of isometries and quasi-isometries. On the other hand, there exist non-Hausdorff moduli spaces of complex structures (See 12.113).

12.56. According to Corollary 12.52, the set of Einstein constants of all Einstein metrics of volume one on a given manifold M is countable. For certain examples, it has only one element, for others, it can be infinite, see Add.A.

12.57. Let us turn to another question. Look at all Einstein manifolds of a given dimension n, of volume one. What is the set of Einstein constants? It is more convenient to normalize metrics in order that the Einstein constant λ (in 1.95) be equal to $\pm(n-1)$ (zero is ruled out)—as is usually done in the theory of spaces of constant sectional curvature. Then what can be said on the set of volumes of Einstein n-manifolds with Einstein constant $(n-1)$ (resp., $-(n-1)$)? Let us look at known examples.

12.58. If (M, g) is a surface of Euler characteristic $\chi(M)$, and $r = \pm g$, then

$$\text{volume}(g) = 2\pi|\chi(M)|.$$

More generally, if (M, J, g) is a $2m$-dimensional Kähler-Einstein manifold, $r = \pm(2m-1)g$ implies (notation of Chapter 2) $\rho = \pm(2m-1)\omega$, and

$$\text{vol}(g) = \left(\frac{+1}{2m-1}\right)^m \int_M |\rho^m| = \left(\frac{2\pi}{2m-1} c_1(J)\right)^m.$$

When M is biholomorphic to a hypersurface of degree $d > m + 2$ in $\mathbb{C}P^{m+1}$, Formula (11.10) yields $\text{vol}(g) = d\left(2\dfrac{d-m-2}{2m-1}\right)^m \text{vol}(\mathbb{C}P^m)$.

We observe that, thanks to the Aubin-Yau theorem, these hypersurfaces carry negative Einstein metrics, with arbitrarily large volume.

A manifold M may admit several complex structures, and thus, several Kähler-Einstein structures, but, in known examples, they have the same first Chern number, and so the same volume.

12.59. Since Einstein 3-manifolds have constant sectional curvature, we have fairly good information on the possible volumes. In case of curvature +1, the volumes are explicit. They form a decreasing sequence, starting at $2\pi^2$ (the standard sphere) and converging to zero (lens spaces, i.e., cyclic quotients of the standard sphere).

No exact value of the volume of a compact manifold with constant curvature -1 is known. However, approximate values may be numerically computed (see [Thu 2], such as $\simeq 0.98$, which is conjectured to be the smallest possible. In [Thu 1], summed up in [Gro 5], it is shown that the volumes of compact 3-manifolds of constant curvature -1 form a well-ordered subset of $\mathbb{R}_+$. It is *not discrete*, i.e., it contains a converging sequence. Whether the limit is the volume of a compact manifold of constant curvature is unknown (it is however the volume of a *complete* manifold of constant curvature).

12.60. We add two remarks about spaces of constant curvature.

In even dimensions, the generalized Gauss-Bonnet formula shows that volumes are proportional to Euler characteristics, and thus form a closed discrete set.

In odd dimensions, there are lens spaces, so arbitrarily small volumes in constant curvature +1. This does not happen in negative curvature: it is a theorem of D.A. Kazhdan and G.A. Margulis [Ka-Ma]. In fact, the set of volumes of compact n-dimensional manifolds of constant curvature -1 is discrete and closed, when $n \geqslant 4$, [Wng].

12.61. In higher dimensions, we have only one result. For positive Einstein manifolds, an upper bound on volume follows from R. Bishop's inequality, see 6.61, 0.65 or [Bis]. Indeed,

$$r = (n-1)g \quad \text{implies} \quad \mathrm{vol}(g) \leqslant \mathrm{vol}(S^n, \mathrm{can}).$$

Since we have so few examples, it seems very hard to make any conjecture. However, we single out two questions.

12.62 Question. On a given manifold M, does the volume of Einstein metrics take only finitely many values?

In fact, for negative Einstein, we have no example of more than one volume on a fixed manifold.

12.63 Question. For a given dimension n, is the set of volumes of negative (resp. positive) Einstein manifolds of dimension n discrete and closed? (resp. is zero the only accumulation point?)

As a kind of evidence, we suggest the following results. There exists a positive constant ε_n such that, if (M, g) satisfies $r = -(n-1)g$ and one of the following conditions, (+), (++), (+++); then $\mathrm{vol}(g) \geqslant \varepsilon_n$.

(+) (M, g) is locally symmetric, of non-compact type (D.A. Kazhdan, G.A. Margulis [Ka-Ma]).

(++) M has non zero simplicial volume. (M. Gromov, Isolation Theorem, [Gro 2] p. 14 and so). The simplicial volume is a topological invariant of manifolds, see 6.43. It does not vanish for manifolds which carry a metric of negative sectional curvature, and also for their products and connected sums, but it vanishes for all simply connected manifolds.

$(+++)$ M satisfies the homotopical condition introduced by M. Gromov in [Gro 3], Proposition 6.6.D p. 70. For example, M admits an irreducible metric of non positive sectional curvature.

H. Rigidity of Einstein Structures

12.64 Definition. We say that an Einstein structure is *rigid* if it is an isolated point of the Moduli Space.

12.65 Examples. We shall prove that metrics with constant, non-zero, sectional curvature, are rigid (as Einstein structures). On the contrary a flat metric on a torus is not rigid. A Kähler-Einstein K3 surface is not rigid.

Theorem 12.49 yields the following sufficient condition for rigidity.

12.66 Corollary. *If an Einstein structure has no infinitesimal Einstein deformations, it is rigid.*

In fact, the premoduli space, contained in Z, reduces to a point. □

This condition is realized under certain curvature conditions.

12.67 Theorem (N. Koiso [Koi 1]). *Let g be an Einstein metric on M. Denote by a_0 the largest eigenvalue of the zero order operator $\mathring{R}_g$ on trace-free symmetric 2-tensor fields, i.e.,*

$$a_0 = \sup\{\langle \mathring{R}h, h\rangle / \|h\|_2^2 \text{ for } h \in C^\infty(S_0^2M)\}.$$

If $a_0 < \max\left\{-\dfrac{s}{n}, \dfrac{s}{2n}\right\}$, then g has no infinitesimal Einstein deformation (and thus is rigid).

12.68 Remark. In fact, what is shown is that an Einstein metric with $a_0 < \max\left\{-\dfrac{s}{n}, \dfrac{s}{2n}\right\}$ is *stable* in the sense of 4.63.

12.69 *Proof* of Theorem 12.67. This follows from two Weitzenböck formulas (compare 1.151 and [Bes 2], p. 331).

First, let us consider a symmetric 2-tensor field h as a 1-form with values in the cotangent bundle T^*M. This bundle comes equipped with the Levi-Civita covariant derivative D, thus there is an induced exterior differential d^D on T^*M-valued differential forms. For a 0-form α and a 1-form h with values in T^*M, one has

$$d^D\alpha(x) = D_x\alpha, \qquad d^Dh(x \wedge y) = D_xh(y, .) - D_yh(x, .).$$

Let δ^D denote the adjoint operator: $C^\infty(\overset{2}{\Lambda}{}^1 T^*M \otimes T^*M) \to C^\infty(\overset{1}{\Lambda}{}^0 T^*M \otimes T^*M)$. An easy computation yields the Weitzenböck formula

$$(\delta^D d^D + d^D\delta^D)h = D^*Dh - \mathring{R}h + h \circ r.$$

Second, let S denote the symmetrized covariant derivative on symmetric p-tensor fields. If $h \in \mathscr{S}^p M$,

$$Sh(x_1, \dots, x_{p+1}) = D_{x1} h(x_2, \dots, x_{p+1}) + D_{x2} h(x_3, \dots, x_{p+1}, x_1) + \cdots$$
$$+ D_{x_{p+1}} h(x_1, \dots, x_p).$$

Let S^* denote the adjoint: $\mathscr{S}^{p+1}M \to \mathscr{S}^p M$. When $p = 1$, $S^* = \delta$ and $S = \delta^*$. An easy computation yields another Weitzenböck formula

$$(S^*S - \delta^*\delta)h = D^*Dh + 2\mathring{R}h - 2h \circ r.$$

12.70. If h is an infinitesimal Einstein deformation of g, then

$$0 \leqslant \|\delta^D h\|^2 + \|d^D h\|^2 = \left\langle \left(D^*D - \mathring{R} + \frac{s}{n} \right) h, h \right\rangle = \left\langle \left(\mathring{R} + \frac{s}{n} \right) h, h \right\rangle$$
$$\leqslant \left(a_0 + \frac{s}{n} \right) \|h\|^2,$$

$$0 \leqslant \|Sh\|^2 = \left\langle \left(D^*D + 2\mathring{R} - 2\frac{s}{n} \right) h, h \right\rangle = \left\langle \left(4\mathring{R} - 2\frac{s}{n} \right) h, h \right\rangle$$
$$\leqslant 4\left(a_0 - \frac{s}{2n} \right) \|h\|^2.$$

If $h \neq 0$, necessarily $a_0 \geqslant \max\left\{ -\frac{s}{n}, \frac{s}{2n} \right\}$. □

The largest eigenvalue a_0 may be estimated from the sectional curvature. First estimate a_0 by

$$\sup_{p \in M} \sup \{ (\mathring{R}_p \eta, \eta)/|\eta|^2 \quad \text{for } \eta \in S_0^2 T_p M \}.$$

Since the metric g is Einstein, $\operatorname{tr} \mathring{R}h = \frac{s}{n} \operatorname{tr} h$, and so $\mathring{R}(S_0^2) \subset S_0^2$.

12.71 Algebraic Lemma (T. Fujitani [Fuj]). *Let R be an Einstein curvature tensor on an n-dimensional euclidean space. If we denote by $K_{\min}$ and $K_{\max}$ the minimum and maximum of its sectional curvature, then*

$$a = \sup\{ (\mathring{R}\eta, \eta)/|\eta|^2 \quad \text{for } \eta \in S_0^2 \} \leqslant \min\left\{ (n-2)K_{\max} - \frac{s}{n}, \frac{s}{n} - nK_{\min} \right\}.$$

Proof. Choose η such that $\mathring{R}\eta = a\eta$. Choose a basis in which η is diagonal, with eigenvalues $\lambda_1, \dots, \lambda_n$ such that $\lambda_1 = \sup|\lambda_i|$ and $\sum_{i=1}^n \lambda_i = 0$. Then

$$a\lambda_1 = a\eta_{11} = (\mathring{R}\eta)_{11} = \sum_{i,k} R_{i1k1} \eta_{ik} = \sum_{i \neq 1} R_{i1i1} \lambda_i.$$

Thus

$$\begin{aligned} a\lambda_1 &= \sum_{i\neq 1} K_{\max}\lambda_i - \sum_{i\neq 1} (K_{\max} - R_{i1i1})\lambda_i \\ &\leqslant -\lambda_1 K_{\max} + \lambda_1 \sum_{i\neq 1} (K_{\max} - R_{i1i1}) \\ &= \left((n-2)K_{\max} - \frac{s}{n}\right)\lambda_1. \end{aligned}$$

In the other direction,

$$\begin{aligned} a\lambda_1 &= \sum_{i\neq 1} K_{\min}\lambda_i + \sum_{i\neq 1} (R_{i1i1} - K_{\min})\lambda_i \\ &\leqslant -\lambda_1 K_{\min} + \lambda_1 \sum_{i\neq 1} (R_{i1i1} - K_{\min}) \\ &= \left(-nK_{\min} + \frac{s}{n}\right)\lambda_1. \end{aligned}$$ □

12.72 Corollary (J.P. Bourguignon, unpublished). *An Einstein structure with δ-pinched sectional curvature,* $\frac{n-2}{3n} < \delta$, *is rigid.*

Indeed, the assumption means that $2nK_{\min} > \frac{2}{3}(n-2)K_{\max}$. Therefore, either $\frac{s}{n} > \frac{2}{3}(n-2)K_{\max}$, or $\frac{s}{n} < 2nK_{\min}$. In either case, $a_0 < \frac{s}{2n}$. □

12.73 Corollary. *An Einstein structure with negative sectional curvature is rigid, provided* $n \geqslant 3$.

In fact, $K_{\max} < 0$ implies $a_0 < -\frac{s}{n}$. □

Earlier results may be found in [Ber] [Fuj] [Gal 1].

In 1962, A. Weil [Wei] proved that locally symmetric spaces of non-compact type are rigid as locally symmetric spaces. In fact, a refined version of Lemma 12.71 shows that these spaces have no Einstein deformations either.

12.74 Proposition (N. Koiso, [Koi 1]). *A locally symmetric Einstein space of non compact type is rigid, provided it has no local 2-dimensional factor.*

To deal with locally symmetric spaces of compact type, Theorem 12.67 is a poor tool. However, in this case, infinitesimal Einstein deformations may be computed directly, thanks to representation theory.

12.75 Proposition (N. Koiso, [Koi 3]). *Irreducible symmetric spaces of compact type are rigid, except possibly spaces from the following list:*

$SU(p+q)/S(U(p) \times U(q))$, $p \geqslant q \geqslant 2$; $SU(m)/SO(m)$; $SU(2m)/Sp(m)$; $SU(m)$ ($m \geqslant 3$); E_6/F_4.

12.76 Remark. The spaces listed above do have infinitesimal Einstein deformations, but we do not know whether they are rigid or not.

Looking at known examples leads to the following question: Is any Einstein manifold of positive scalar curvature rigid? As symmetric spaces show, it might not be sufficient to look at infinitesimal Einstein deformations. One has to investigate "second order deformations". In this direction, N. Koiso has been successful, as we already saw in 12.46.

12.77. The symmetric metric on $\mathbb{C}P^1 \times \mathbb{C}P^{2k}$ is rigid, although it has infinitesimal Einstein deformations.

I. Dimension of the Moduli Space

12.78. Let g be an Einstein metric on M. Let us *denote by* V its volume and by $K_{\min}$ the minimal value of its sectional curvature. As s (resp $K_{\min}$) is not invariant under homothetic changes of the metric, its exact value gives no more information than its sign. So we introduce $A(g)$ and $\kappa(g)$ which are the values of s and $K_{\min}$ normalized by the diameter, i.e. $A(g) = \dfrac{s}{n(n-1)} d^2$ and $\kappa(g) = K_{\min} \cdot d^2$.

Now set

$$\alpha(n) = n^{(n-2)/2n} \left(\frac{\operatorname{Vol} S^{n-1}}{\operatorname{Vol} S^n}\right)^{1/n} \left(\frac{2}{(n-2)^{1/2}}\right) + 2^{1-1/n}$$

$$\xi(x) = \prod_{i=0}^{\infty} \left[1 + \frac{\alpha(n)\beta^i}{(2\beta^i - 1)^{1/2}} \sqrt{x}\right]^{2/\beta^i} \quad \text{for } x \in [0, +\infty[,$$

where $$\beta = \frac{n}{n-2} \quad \text{if } n \geqslant 3,\ \beta = 100 \text{ if } n = 2.$$

12.79 Theorem ([Gal 4] et [Gal 2]). *For every n-dimensional Einstein manifold* (M, g), *the dimension of* $\varepsilon(g)$ *is smaller than*

$$N \cdot \xi\left[\frac{2(n-1)A - 2n\kappa}{C^2 \cdot d^2 \cdot V^{-2/n}}\right], \quad \text{where } N = \frac{n(n+1)}{2} - 1$$

and C is the isoperimetric constant $\operatorname{Inf}_{\Omega}\left(\dfrac{\operatorname{Vol} \partial\Omega}{(\operatorname{Vol} \Omega)^{1-1/n}}\right)$ *for all domains* Ω *such that* $\operatorname{Vol} \Omega \leqslant V/2$.

Before proving 12.79, we shall state more readable corollaries.

12.80 Corollary ([Gal 4]). *There exists an explicit continuous function* η *from* $\mathbb{N} \times \mathbb{R}$ *to* $\mathbb{R}^+$ *such that, for every n-dimensional Einstein manifold* (M, g) *satisfying* $\kappa(g) \geqslant k$, *the dimension of the space of infinitesimal Einstein deformations of g is smaller than* $\eta(n, k)$.

Proof. The main ingredient is the isoperimetric estimate of [Gal. 2] and [Gal. 3] which gives $C \cdot V^{-1/n} d \geqslant \Gamma(A(g))$, where

$$\Gamma(\alpha) = 2^{-1/n} \cdot H(\alpha) \quad \text{if } \alpha \geqslant 0,$$

$$\Gamma(\alpha) = |\alpha|^{1/2n} \left[\int^{|\alpha|^{1/2}} \left(\frac{cht}{H(\alpha)} + \frac{|\alpha|^{-1/2}}{n} sht \right)^{n-1} dt \right]^{-1/n} \quad \text{if } \alpha < 0.$$

with

$$H(\alpha) = |\alpha|^{1/2} \left(\int_0^{(1/2)|\alpha|^{1/2}} \cos t^{n-1}\, dt \right)^{-1} \quad \text{if } \alpha > 0,$$

$$H(\alpha) = 2 \quad \text{if } \alpha = 0,$$

$$H(\alpha) = |\alpha|^{1/2} \left(\int_0^{(1/2)|\alpha|^{1/2}} cht^{n-1}\, dt \right)^{-1} \quad \text{if } \alpha < 0.$$

As $k \leqslant \kappa \leqslant A \leqslant \pi^2$ by Myers' Theorem 6.51, 12.79 implies 12.80.

(12.81)
$$\eta(n, k) = N \cdot \xi \left[\frac{2(n-1)\pi^2 - 2nk}{\Gamma(k)^2} \right]. \qquad \square$$

12.82 Remark. It is impossible to bound the dimension of $\varepsilon(g)$ in terms of the dimension only. In fact, this dimension is arbitrarily large already for surfaces. Examples in higher dimensions are obtained by taking products.

Note that the dimension of $\varepsilon(g)$ for a flat torus is precisely equal to N (see 12.17). As $\xi(x)$ goes to 1 when x goes to zero, this gives by 12.80 and 12.81 the

12.83 Corollary ([Gal 2 et 4]). *For each dimension n, there exists some positive constant $\tilde{\alpha}(n)$ such that, for all Einstein manifolds (M, g) of dimension n satisfying*

$$(n-1)A(g) - n\kappa(g) \leqslant \tilde{\alpha}(n),$$

the dimension of the space $\varepsilon(g)$ of infinitesimal Einstein deformations is maximum for flat tori.

Remark. On K3 surfaces, one has $A(g) = 0$ and $N = 9$. As the dimension of the Moduli Space is there equal to 57 (see 12.106), it implies that $\kappa(g)$ must be far from zero.

Proof of 12.83. By 12.30, every element of $\varepsilon(g)$ is a section h of the fiber bundle of trace-free symmetric 2-tensors (whose fiber is N-dimensional) satisfying (in particular) $(D^*D + 2\mathring{R})(h) = 0$. As $(\mathring{R}h, h) \geqslant \left(nK_{\min} - \frac{s}{n} \right) |h|^2$, see 12.71, this gives

$$(D^*Dh, h) \leqslant 2((n-1)A(g) - n\kappa(g)) \cdot d^{-2} |h|^2. \qquad \square$$

The following theorem ends the proof of 12.79.

12.84 Theorem ([Gal 2, theorem 3.3], compare with [Li]). *Let E be a fibre bundle on M with some metric $(.,.)$ and a Riemannian connection D. Let l be the dimension* *of the fiber E and λ be any real number. Let $\mathcal{T}$ be a subspace of the space of C^2 sections of E. If the inequality $((D^*D - \lambda^2)(T), T) \leqslant 0$ is satisfied at each point of M and for every T in $\mathcal{T}$, then* $\dim \mathcal{T} \leqslant l \cdot \xi\left(\frac{\lambda^2}{C^2 \cdot V^{-2/n}}\right)$.

Idea of the proof: The Sobolev inequality related to the inclusion $H_1^2 \subset L^{2n/n-2}$ (see Appendix 8) admits an explicit expression.

From [Gal 3] we get, for every f in H_1^2 satisfying $\int_M f \cdot \mu_g = 0$,

$$\|f\|_{L^{2\beta}} \leqslant \frac{\alpha(n)}{C} \|df\|_{L^2} \tag{8}$$

where β and $\alpha(n)$ are given in 12.78. The hypothesis implies

$$|T| \cdot \Delta(|T|) \leqslant (D^*DT, T) \leqslant \lambda^2 |T|^2.$$

By integration, this gives

$$\|d(|T|^k)\|_{L^2}^2 \leqslant \frac{k^2}{2k-1} \cdot \lambda^2 \cdot \|T\|_{L^{2k}}^{2k}.$$

Using the same method as in [Li] and applying the Sobolev inequality to the function $|T|^{\beta^i}$, we get

$$\|T\|_{L^{2\beta^{i+1}}}^{\beta^i} \leqslant \left(1 + \alpha(n) \frac{\beta^i}{(2\beta^i - 1)^{1/2}} \frac{\lambda}{C \cdot V^{-1/n}}\right) V^{-1/n} \cdot \|T\|_{L^{2\beta^i}}^{\beta^i}.$$

As $\|T\|_{L^{2\beta^i}}$ goes to $\|T\|_{L^\infty}$ when i goes to $+\infty$, we get by iterating.

$$V \cdot \|T\|_{L^\infty}^2 \leqslant \xi\left(\frac{\lambda^2}{C^2 \cdot V^{-2/n}}\right) \cdot \|T\|_{L^2}^2.$$

As D^*D is elliptic, $\mathcal{T}$ is finite-dimensional (see Appendix 31). To each point m of M corresponds an orthonormal basis $(T_i)_i$ which is also orthogonal for the scalar product in the fibre E_m. There exists some point m such that $\dim \mathcal{T} \leqslant V \cdot \sum_i |T_i(m)|^2$. It follows

$$\begin{aligned} \dim \mathcal{T} &\leqslant V \cdot N \cdot \sup \|T_i\|_{L^\infty}^2 \\ &\leqslant N \cdot \xi\left(\frac{\lambda^2}{C^2 V^{-2/n}}\right). \end{aligned}$$ □

12.85 Some Open Problems:

—When the sectional curvature is nonnegative, the dimension of the Moduli Space is bounded by some constant $C(n)$ by 12.80. Can we improve $C(n)$ such that is would be the number of moduli of the flat torus?

—For given a, b and n, let $\mathcal{M}_{a,b}$ be the set of all n-dimensional manifolds satisfying $A(g) \geqslant a$ and $d \leqslant b$. M. Gromov (see [Gro 1]) proved that $\mathcal{M}_{a,b}$ is precompact for the Hausdorff metric (cf. 0.67). So it seems natural to ask whether it is possible to replace $\kappa(g)$ by $A(g)$ in 12.80. Note that, if it is true, the upper bound would not have the sharp property 12.83 if one only assumes that $A(g) > -\alpha(n)$.

For example, K3 surfaces satisfy $A(g) = 0$, but the dimension of their Moduli Space is greater than the dimension of the Moduli Space of a flat torus T^4.

—In the normalization, is it possible to replace the diameter by the volume, i.e., is it possible to bound the dimension of the Moduli Space of Einstein structures in terms of $-s \cdot V^{2/n}$ or at least $-K_{\min} \cdot V^{2/n}$?

Such a normalization should be more convenient because the functionals on the space of metrics on M are usually normalized by the volume (see Chapter 4).

J. Deformations of Kähler-Einstein Metrics

12.86. Chapter 11 already has emphasized the link between complex structures and Einstein metrics.

When the first Chern class is negative, the Calabi-Aubin theorem asserts that, to each complex structure J, there corresponds a unique Einstein metric g, such that the pair (g, J) is Kähler. Here, we shall use an easy local version of this theorem, to show that a deformation of the complex structure J produces a deformation of the Einstein metric.

Similarly, when a fixed complex structure has vanishing first Chern class, one may deform the prescribed Kähler class, this again will produce a deformation of the Einstein metric.

12.87. Recall the following facts about complex deformations. Let Θ *denote* the sheaf of germs of holomorphic vector fields of the complex manifold (M, J). An *infinitesimal complex deformation* of J is a vector of $H^1(M, \Theta)$. We call it *integrable* if it comes from a d''-closed $T^{1,0}M$-valued (0, 1)-form, which is formally integrable in the sense of 12.36. According to K. Kodaira, L. Nirenberg and D. Spencer [Ko-Ni-Sp], such an infinitesimal deformation is the first jet of a curve of complex structures $J(t)$. In particular, if all infinitesimal complex deformations of J are integrable, the moduli space of complex structures on M is an orbifold (see 12.25) near J, with $H^1(M, \Theta)$ as tangent space.

The main result is the following theorem (proved in 12.99 and 12.100).

12.88 Theorem (N. Koiso [Koi 4]). *Let (g, J) be a Kähler-Einstein structure on M. Assume that*:

(i) *the first Chern class of J is negative or zero*;

(ii) *all infinitesimal complex deformations of J are integrable.*

Then the premoduli space of Einstein metrics around g is a manifold with $\varepsilon(g)$ as tangent space at g. Moreover, any metric $\tilde{g}$ in it is Kähler with respect to some complex structure $\tilde{J}$ close to J.

In the Moduli Space of Einstein Structures, these Kähler-Einstein structures form an open subset, which is an orbifold. (see 12.25).

12.89 Example. Hypersurfaces of $\mathbb{C}P^m$, $m \geqslant 3$. As already seen in 11.10, the first Chern class of a hypersurface of degree $d \geqslant m + 2$ is negative. According to K.

Kodaira and D. Spencer [Ko-Sp], assumption (ii) is also satisfied. Moreover, all complex deformations of a given hypersurface are hypersurfaces of the same projective space. This means that the moduli space of complex structures of a manifold biholomorphically equivalent to a hypersurface is, locally, an orbifold of complex dimension $\binom{m+d}{d} - (m+2)^2$. Theorem 12.88 states that an open subset of the Moduli Space of Einstein metrics on the underlying manifold is a manifold of dimension $2(\binom{m+d}{d} - (m+2)^2)$.

12.90. Before we prove Theorem 12.88, let us show how the fact that an Einstein metric be Kähler with respect to some complex structure on M enlightens the discussion of infinitesimal Einstein deformations.

Contravariant 2-tensors split into hermitian

$$h(Jx, Jy) = h(x, y),$$

and skew-hermitian ones

$$h(Jx, Jy) = -h(x, y).$$

In this context, each species has a new interpretation. For a 2-tensor h and an endomorphism I, we shall *denote by* $h \circ I$ the tensor

$$h \circ I(x, y) = h(x, Iy).$$

Then setting

$$\psi = h \circ J$$

turns a symmetric hermitian 2-form h into a real differential 2-form φ of type (1, 1). Setting

$$g \circ I = h \circ J$$

turns a symmetric skew-hermitian 2-form h into a real, symmetric endomorphism I, which anticommutes with J, i.e., $IJ + JI = 0$. As already seen in 12.40, one may consider I as a $T^{1,0}M$-valued 1-form of type (0, 1).

12.91. Under these identifications, the system of equations for infinitesimal Einstein deformations involving $D^*D - 2\mathring{R}$, δ, tr, is related to more familiar operators.

If h is a symmetric hermitian 2-form on M, then $\operatorname{tr} h$ is the scalar product of ψ with the Kähler form ω,

$$\delta h = -\delta\psi \circ J, \tag{12.92}$$

and the usual (real) Hodge-De Rham Laplacian Δ satisfies the following Weitzenböck formula.

$$\Delta\psi = (D^*D - 2\mathring{R})h \circ J + 2\frac{s}{n}\psi. \tag{12.92'}$$

If h is a skew-hermitian symmetric 2-form on M, then

$$\operatorname{tr} h = 0, \tag{12.93}$$

$$\delta h = -J \circ (\delta'' I),$$

and the complex Laplacian $\Delta'' = d''\delta'' + \delta''d''$ satisfies a Weitzenböck formula

(12.93') $$g \circ (\Delta'' I) = (D^*D - 2\mathring{R})h \circ J.$$

12.94 Lemma. *Let g be a Kähler-Einstein metric on M. If h is an infinitesimal Einstein deformation of g, so are its hermitian and skew-hermitian parts h_H and h_A.*

Proof. The Kähler property $D_X JY = D_Y JX$ and its consequence $R_{JX,JY} = R_{X,Y}$ imply that the operator $D^*D - 2\mathring{R}$ preserves hermitian and skew-hermitian forms. Thus $(D^*D - 2\mathring{R})h_H = (D^*D - 2\mathring{R})h_A = 0$. Formula (12.93') now implies that $\Delta'' I = 0$, and $\delta h_A = -J_0(\delta'' I) = 0$. Since $\operatorname{tr} h_A = 0$, then h_A (and necessarily $h_H = h - h_A$) is an infinitesimal Einstein deformation. □

12.95. Formula (12.92') immediately yields the following interpretation of hermitian infinitesimal Einstein deformations. If the scalar curvature (or, equivalently, the first Chern class) is negative, there are none. If the first Chern class is zero, *Hermitian infinitesimal Einstein deformations are in one to one correspondance with the real* (1, 1)-*differential forms which are harmonic and orthogonal to the Kähler form. Thus they form a space of dimension* $\dim H^{1,1}_{\mathbb{R}}(M, J) - 1$.

In other words, to (the first jet of) a deformation of the cohomology class of the Kähler form, we can associate an infinitesimal Einstein deformation.

12.96. On the other hand, *a skew-hermitian infinitesimal Einstein deformation corresponds* to a symmetric harmonic I, i.e., *to a symmetric infinitesimal complex deformation.* This correspondance is one to one, since, by Hodge theory, $H^1(M, \Theta) \simeq \operatorname{Ker}(d'')/d''(C^\infty(\wedge^{0,1}M \otimes T^{1,0}M)) \simeq \operatorname{Ker}(\Delta'') \cap C^\infty(\wedge^{0,1} \otimes T^{1,0})$.

What about skew-symmetric infinitesimal complex deformations? A section I of $\wedge^{0,1}M \otimes T^{1,0}M$ is skew-symmetric if and only if $g \circ I$ is a differential 2-form of type (2, 0), and I is harmonic if and only if $g \circ I$ is harmonic, and thus holomorphic. Therefore, *skew-symmetric infinitesimal complex deformations are in one to one correspondance with holomorphic 2-forms.* The dimension of this space is equal to $2 \dim_{\mathbb{C}} H^{2,0}(M, J)$.

12.97 Remark. A classical vanishing theorem of S. Bochner (see 11.24) asserts that, if M admits a non-zero holomorphic 2-form ψ, then the Ricci curvature is zero, and ψ is parallel. Assume M has irreducible holonomy representation (see 10.41). Then ψ has to be non degenerate. In general, existence of ψ implies that one of the irreducible factors will have holonomy in Sp(right dimension) that is, is hyperkählerian. This is a rather strong condition, though not fully understood (see Chapter 14).

To sum up, we have shown

12.98 Proposition (N. Koiso [Koi 4], J.P. Bourguignon in case of a K3 surface). *Let (g, J) be a Kähler-Einstein structure on M.*

If $c_1(J) < 0$, $\dim \varepsilon(g) = 2 \dim_{\mathbb{C}} H^1(M, \Theta)$.

If $c_1(J) = 0$, $\dim \varepsilon(g) = \dim H^{1,1}_{\mathbb{R}}(M, J) - 1 + 2 \dim_{\mathbb{C}} H^1(M, \Theta) - 2 \dim_{\mathbb{C}} H^{0,2}(M, J)$.

If $c_1(J) > 0$, $\dim \varepsilon(g) \geqslant 2 \dim_{\mathbb{C}} H^1(M, \Theta)$.

In the third case, the inequality may be strict. On $\mathbb{C}P^1 \times \mathbb{C}P^{2k}$, the canonical complex structure is rigid (R. Bott [Bot]), whereas the Kähler symmetric metric has infinitesimal Einstein deformations (12.46). Of course, all of them are hermitian.

12.99 *Proof of Theorem* 12.88, *case* $c_1(J) < 0$. Let (g, J) be a Kähler-Einstein metric on M, which satisfies the assumptions of Theorem 12.88. The assumption on J ensures that there exists a manifold $\mathscr{C}$ of complex structures $\tilde{J}$ around J, with $H^1(M, \Theta_J)$ as tangent space.

According to K. Kodaira and D. Spencer [Ko-Sp] there exist Kähler metrics $\omega(\tilde{J})$ depending smoothly on $\tilde{J}$, with $\omega(J) = g \circ J$. As in 11.40, let us seek a Kähler-Einstein metric in the class of $\omega(\tilde{J})$, i.e., of the form

$$\omega(\tilde{J}) - \tfrac{1}{2} dd^c_{\tilde{J}} \varphi,$$

for a function $\varphi \in \mathscr{C}M$. The equation

$$\mathrm{Cal}^-_{\tilde{J}, \omega(\tilde{J})}(\varphi) = f_{\omega(\tilde{J})}$$

to be solved is written explicitely in 11.41. There, it is checked that the derivative of Cal^- with respect to φ is an isomorphism. Thus, there exists a smooth solution $\varphi(\tilde{J}, \omega(\tilde{J}))$, that is, a metric $g(\tilde{J})$. We may assume that $g(\tilde{J})$ is in the slice $\mathfrak{S}_g$, in particular, in the manifold Z of Theorem 12.49, whose tangent space is $\varepsilon(g)$.

It remains to check that the mapping $\tilde{J} \mapsto g(\tilde{J})$ is a submersion. Given a curve $\tilde{J}(t)$ with first jet $I = I^s + I^a$, let $g(\tilde{J}(t))$ have first jet $h = h_A + h_H$. Since $g(\tilde{J}(t))$ is in the slice, h_A and h_H are infinitesimal Einstein deformations. Differentiating the Kähler form $g(t) \circ \tilde{J}(t)$, we find that $h \circ J + g \circ I = h_A \circ J + g \circ I^s + h_H \circ J + g \circ I^a$ is a closed differential form. Clearly, the symmetric path $h_A \circ J + g \circ I^s$ has to be zero (which implies, by (12.93′), that I^s is Δ''-harmonic). Thus the derivative of the mapping $\tilde{J} \mapsto g(\tilde{J})$ is just the isomorphism pointed out in 12.96 (remember that there are no h_H's nor I^a's when $c_1 < 0$). □

12.100 *Proof of Theorem* 12.88, *case* $c_1(J) = 0$. Again, let $\mathscr{C}$ be a manifold of complex structures $\tilde{J}$ around J of maximal dimension, and $\omega(\tilde{J})$ be Kähler metrics depending smoothly on $\tilde{J}$. Let $\mathscr{H}^{1,1}$ denote the smooth vector bundle over $\mathscr{C}$ whose fiber at $\tilde{J}$ is the space of real harmonic (1, 1)-forms relative to the metric $\omega(\tilde{J})$.

Given a $(\tilde{J}, \kappa)$ in $\mathscr{H}^{1,1}$, that is, a complex structure $\tilde{J}$ and a harmonic (1,1)-form κ, let us seek a Kähler-Einstein metric in the cohomology class of $\omega(\tilde{J}) + \kappa$,

$$\omega(\tilde{J}) + \kappa - \tfrac{1}{2} dd^c_{\tilde{J}} \varphi.$$

As checked in 11.38, the implicit function theorem applies, and there exists a smooth solution $\varphi(\tilde{J}, \kappa)$, that is, a metric $g(\tilde{J}, \kappa)$. Again, we assume $g(\tilde{J}, \kappa)$ is in the slice $\mathfrak{S}_g$.

In order to check, that $(\tilde{J}, \kappa) \mapsto g(\tilde{J}, \kappa)$ is a submersion, let $(\tilde{J}(t), \kappa(t))$ be a curve in $\mathscr{H}^{1,1}$; the first jet consists in a base component $I^s + I^a$ and a fiber component $\psi \in \mathscr{H}^{1,1}_{\mathbb{R}}(\tilde{J})$. Let $h_A + h_H$ be the first jet of the corresponding curve of metrics $g(t)$. Differentiating the Kähler form $g(t) \circ \tilde{J}(t)$, we find that $h_A \circ J = -g \circ I^s$, and $h_H \circ J + g \circ I^a$ is a closed differential form in the cohomology class of ψ. Since h_H is an infinitesimal Einstein deformation, $h_H \circ J$ is harmonic (by 12.92′). Type considerations then show that $h_H \circ J = \psi$. The derivative $(I, \kappa) \mapsto h$ is surjective, and the mapping $(\tilde{J}, \kappa) \mapsto g(\tilde{J}, \kappa)$ is a submersion. □

12.101 Remarks. (i) As already mentionned in 11.47, the proof in 12.99 also works when $c_1(J) > 0$, with the additional assumption that (M, J) has no holomorphic vector fields. The conclusion is that the Kähler-Einstein metrics form a submanifold in the premoduli space around g, with tangent space consisting in all skew-hermitian infinitesimal Einstein deformations. However, we are unable to illustrate this statement, since all known examples of Kähler-Einstein manifolds with $c_1 > 0$ have holomorphic vector fields.

(ii) If $c_1 = 0$, § 12.100 yields more than expressed by Theorem 12.88.

12.102 Definition. The *space* of *Kähler-Einstein structures* on M is the quotient of the subset of Kähler-Einstein pairs (J, g) in $C^\infty(\text{End}\, TM) \times \mathscr{M}_1$ under the action of $\mathfrak{D}$. We *denote* it *by* $\mathscr{K}(M)$.

12.103 Theorem. *Let* $[J, g]$ *be a Kähler-Einstein structure on M such that* $c_1(J) = 0$, *and all infinitesimal complex deformations of J are integrable. Then the space* $\mathscr{K}$ *is, up to a finite action, a manifold around* $[J, g]$. *The map to the complex moduli space is a submersion. The map to the Einstein Moduli Space is a locally trivial fibration, with compact fibers.* (again, up to a finite action).

The example of K3 surfaces shows that the map $\mathscr{K} \to \mathscr{C}$ may not be a fibration (see § 12.113).

K. The Moduli Space of the Underlying Manifold of K3 Surfaces

12.104 Definition. A K3 *surface* is a complex surface with vanishing first Chern class and no global holomorphic 1-forms.

12.105. These surfaces appear as one special category in Kodaira's classification of complex surfaces, see [Kod 3]. A lot is known about them, in particular, that they all are diffeomorphic to a manifold M of real dimension 4, which justifies the title.

Moreover, according to Todorov-Siu [Siu 2], all of them admit Kähler metrics, and so Kähler-Einstein metrics, by the Calabi-Yau Theorem. These surfaces will provide a pretty good illustration of our results on the moduli space of Kähler-Einstein structures.

12.106 Local Deformations. All necessary numerical invariants may be computed as follows. Since $c_1 = 0$, the Riemann-Roch formula yields $\chi(M) = c_2 = 24$. Since the Betti numbers $b_3(M) = b_1(M) = 2h^{0,1} = 0$, one gets $b_2 = 22$. One shows that the canonical line bundle K is trivial, that is, there is a nowhere vanishing holomorphic 2-form λ. Clearly, any holomorphic 2-form has to be proportional to λ; this implies that $h^{2,0} = 1$, and thus $h^{1,1} = 20$.

The holomorphic 2-form λ, a "complex symplectic structure", induces an isomorphism of sheaves $\Theta \simeq \Omega^1$, thus

$$\dim H^2(M, \Theta) = \dim H^2(M, \Omega^1) = h^{2,1} = h^{0,1} = 0,$$
$$\dim H^1(M, \Theta) = \dim H^1(M, \Omega^1) = h^{1,1} = 0.$$

Consequently, the moduli spaces $\mathscr{C}$ of complex structures, $\mathscr{K}$ of Kähler-Einstein structures, $\mathscr{E}$ of Einstein structures on M are, up to discrete actions, manifolds of real dimensions 40, 59 and 57.

A description of infinitesimal Einstein deformations in terms of the Hodge $*$ on harmonic 2-forms is given in [Bes 2] p. 174. Since our aim is rather to give a global description of the space $\mathscr{E}(M)$, we will not develop this point of view. We leave it to the reader to find his way through the jungle of isomorphisms $\Theta \simeq \Omega^1$ and in 12.95, 12.96, and [Bes 2], Chapter XVI, in order to recognize hermitian and skew-hermitian infinitesimal Einstein deformations, in Besse's setting.

12.107 Global Description of $\mathscr{E}(M)$. Recall that $\mathscr{E}(M)$ is the quotient of Einstein metrics of volume one modulo all diffeomorphisms. Let us *denote by* $\mathfrak{D}'$ the group of diffeomorphisms of M which induce the identity on the cohomology group $H^2(M, \mathbb{Z})$. We *define* the *Teichmüller space* $\tilde{\mathscr{E}}(M)$ as the quotient of Einstein metrics of volume one modulo $\mathfrak{D}'$. In the following paragraph, we shall construct an homeomorphism of $\tilde{\mathscr{E}}(M)$ with an open subset T of the symmetric space

$$SO(3,19)/SO(3) \times SO(19),$$

More precisely, T is the complement of a countable union of submanifolds of codimension 3 (compare [Tod 2]).

As a consequence, the Moduli Space $\mathscr{E}(M)$ is homeomorphic to a quotient of T modulo a discrete subgroup of $SO(3,19)$.

12.108 The "Period Mapping" for Einstein Structures on M. First observe that all K3 complex structures on M induce the same orientation. Indeed, they have the same Hodge numbers $h^{p,q}$, and thus, by the Hodge Index Formula, they have the same signature

$$\tau = \sum_{p,q=1}^{2} (-1)^q h^{p,q} = -16.$$

For this natural orientation, the *intersection form*—denoted by a dot—on $H^2(M, \mathbb{R})$ has 3 plus and 19 minus signs. It means that, for any Riemannian metric g on M, the subspace $\mathscr{H}^+(g) \subset H^2(M, \mathbb{R})$ of self-dual—i.e., $*\alpha = \alpha$—harmonic forms has dimension 3, and the intersection form is positive definite on it.

Moreover, if g is Einstein, the 3-space $\mathscr{H}^+(g)$ carries a natural orientation. It is a consequence of Hitchin's Theorem 6.40. Self-dual harmonic—in fact, parallel—forms α of unit norm are in 1-1 correspondance with complex structures I, compatible with the metric and the orientation. Now, given any $I \neq J$, the basis (I, J, IJ) is a direct basis for $\mathscr{H}^+(g)$.

Therefore, the *period mapping* p_e

$$g \bmod \mathfrak{D}' \mapsto p_e[g] = \mathscr{H}^+(g) \in U$$

is well defined on $\tilde{\mathscr{E}}(M)$, where U is the space of oriented positive definite 3-planes in $H^2(M, \mathbb{R})$.

12.109. Neither the injectivity of this mapping nor to determine its image are obvious matter. We shall use the deep results on the period mapping for Kähler K3 surfaces.

Let $\mathscr{K}$ denote the space of pairs (J, κ)—where J is a K3 almost complex structure on M and $\kappa \in H^2(M, \mathbb{R})$ is a cohomology class containing a Kähler form (relative to J) and such that $\kappa \cdot \kappa = 1$ – modulo $\mathfrak{D}'$. The Calabi-Yau Theorem 11.15 yields a map $\mathscr{K}(M) \to \mathscr{E}(M)$. Hitchin's Theorem 6.37 implies that this map is surjective.

12.110. Let J be a K3 almost complex structure on M. The subspace $H^{2,0}(J) \subset H^2(M, \mathbb{C})$ is a line $\mathbb{C}\lambda$, such that $\lambda \cdot \lambda = 0$ and $\lambda \cdot \bar{\lambda} > 0$. The *period* $p_c(J)$ is the positive definite 2-plane π in $H^2(M, \mathbb{R})$ with oriented basis $\left(\lambda + \bar{\lambda}, \frac{\lambda - \bar{\lambda}}{i}\right)$. Clearly p_c is well-defined on the space $\mathscr{C}$ of complex structures $[J]$ modulo $\mathfrak{D}'$.

We define the *period mapping* p_k on the space $\mathscr{K}$ by

$$(J, \kappa) \bmod \mathfrak{D}' \mapsto p_k([J], \kappa) = (\pi, \kappa).$$

Let us try to guess what its image should be. For type reasons—$H^{1,1}(J)$ is orthogonal to $H^{2,0}(J)$ in $H^2(M, \mathbb{C})$—the Kähler class κ is in $\pi^\perp$. Furthermore, if γ is the first Chern class of a holomorphic curve—or, more generally, of an effective divisor C—the intersection $\kappa \cdot \gamma = \text{area}(C) > 0$ for any metric in the class κ. According to the Lefschetz Theorem on (1, 1)-classes, any integral class γ in $H^{1,1}(J) = \pi^\perp$ is the first Chern class of some divisor C. In general, one cannot tell whether C is effective or not. At least, the Riemann-Roch formula ([Hir], p. 149), implies that, if $\gamma \cdot \gamma = -2$, either C or $-C$ is effective. Such a class is called a *root*, and we can claim that $\gamma \cdot \kappa \neq 0$ for any Kähler class κ. Thus it appears that the image of the period mapping is contained in the space

$$\begin{aligned} K\Pi = \{(\pi, \kappa)/\pi &\text{ is an oriented, positive definite 2-plane in } H^2(M, \mathbb{R}), \kappa \in \pi^\perp, \\ &\kappa \cdot \kappa = 1 \text{ and } \kappa \cdot \gamma \neq 0 \text{ for any } \gamma \in \Gamma\}, \end{aligned}$$

where $\Gamma = \{\gamma \in H^2(M, \mathbb{Z})/\gamma \cdot \gamma = -2\}$ denotes the set of roots.

12.111. We claim that p_k is a homeomorphism of $\mathscr{K}(M)$ onto one of the two connected components of $K\Pi$. This is a restatement of D. Burns and M. Rapoport's Torelli Theorem [Bu-Ra] and A. Todorov's Surjectivity Theorem [Tod 1] (see also [Siu 1]), slightly improved by E. Looijenga in [Loo]. Indeed, our space $\mathscr{K}(M)$ coïncides with E. Looijenga's Moduli Space of Marked Kähler K3 Surfaces KM: $\mathscr{K}(M)$ is clearly homeomorphic to an open and closed subset of KM; since KM is connected, $\mathscr{K}(M) = KM$. This proves incidentally that $\mathfrak{D}^+/\mathfrak{D}'$ has index two in $\text{Aut}(H^2(M, \mathbb{Z}))$, where $\mathfrak{D}^+$ is the group of orientation preserving diffeomorphisms.

12.112. We finally have a commuting diagram

$$\begin{array}{ccc} \mathscr{K}(M) & \to & \mathscr{E}(M) \\ \downarrow p_k & & \downarrow p_e \\ K\Pi & \to & T \end{array}$$

where T is the image of $K\Pi$ in U, that is,

$$T = \{\text{oriented positive definite 3-planes } \tau \text{ in } H^2(M, \mathbb{R}) \text{ such that } \tau^\perp \cap \Gamma = \varnothing\}.$$

By construction, p_e is surjective onto T. If g, g' are Einstein metrics on M such that $p_e[g] = \tau = p_e[g']$, let π be an oriented plane in τ, let κ denote its unit oriented normal vector in τ. Then the g- (resp g'-) harmonic form in κ corresponds to a complex structure J (resp. J') such that (g, J) is Kähler with Kähler class κ (resp. (g', J') is Kähler with Kähler class κ). Since $p_k[J, \kappa] = p_k[J', \kappa] = (\pi, \kappa)$, we have $[J, \kappa] = [J', \kappa]$ in $\tilde{\mathscr{K}}(M)$ and $[g] = [g']$ in $\tilde{\mathscr{E}}(M)$, so p_e is injective.

Since an Einstein metric on M has a canonical orientation, isometric Einstein metrics differ by a diffeomorphism in $\mathfrak{D}^+$. Thus $\mathscr{E}(M)$ equals $\tilde{\mathscr{E}}(M)$ modulo $\mathfrak{D}^+/\mathfrak{D}'$, and is homeomorphic to T modulo the discrete subgroup $\mathrm{Aut}(H^2(M, \mathbb{Z})) \cap SO(3, 19)$, of index two in $\mathrm{Aut}(H^2(M, \mathbb{Z}))$.

12.113 Remarks. (i) Let $\mathscr{C}(M)$ be the quotient of K3 complex structures on M modulo $\mathfrak{D}'$. The Burns-Rapoport theorem [Bu-Ra] indicates that the relevant period mapping is p_c: $J \bmod \mathfrak{D}' \mapsto (\pi, K_J)$ where K_J is the connected component of the open set $C_\pi = \{\kappa \in \pi^\perp / \kappa \cdot \kappa > 0, \kappa \cdot \gamma \neq 0 \text{ for all } \gamma \in \Gamma\}$ containing all Kähler classes of J. Indeed, p_e is a homeomorphism of $\mathscr{C}(M)$ onto one of the two connected components of the space

$\Pi = \{(\pi, K)/\pi$ is an oriented positive definite 2-plane in $H^2(M, \mathbb{R})$, K is a connected component of $C_\pi\}$.

Clearly, the projection $K\Pi \to \Pi$ is a submersion but not a fibration, since the number of components in C_π (generically 2, always finite) varies. In fact, the space Π is not Hausdorff.

(ii) A lot is known about the lattice $H^2(M, \mathbb{Z})$. It is even (see [Bes 2], p. 161) which suffices to determine its structure. It breaks down into 5 irreducible even factors: two negative rank 8 E_8 factors, and 3 indefinite rank 2 factors. Its automorphism group is rather "big", see [Wll], where generators are given.

(iii) The uniqueness statement in the Burns-Rapoport theorem [Bu-Ra] implies that the automorphism group of a K3 complex structure J on M, and, a fortiori, the isometry group of an Einstein metric g, is imbedded in $\mathrm{Aut}(H^2(M, \mathbb{Z})) \cap SO(3, 19)$ as the isotropy subgroup of its period $p_c[J] \in \Pi$ (resp. $p_e[g] \in T$). The possible isometry groups—i.e., the finite subgroups of $\mathrm{Aut}(H^2(M, \mathbb{Z})) \cap SO(3, 19)$—are listed in [Nik] and [Muk]. A lot of information on the automorphism groups (finiteness, for example) is to be found in [Nik].

(iv) The Moduli Space of Einstein Structures on the K3 manifold M appears as a good (symmetric) space with holes in it. One may converge to a hole, and be Kähler with respect to a fixed complex structure. Then, it is the Kähler class κ which leaves from the Kähler chamber K_J. It means that some effective divisor has area converging to zero. Can one tell more about this collapsing? In other words, can points in the holes be interpreted as geometric objects?

Chapter 13. Self-Duality

A. Introduction

13.1. Oriented Riemannian geometry in four dimensions has some special features most of which may be derived from the action of the Hodge $*$-operator. Using it one may split 2-forms into *self-dual* and *anti-self-dual* forms. This can be applied in particular to the middle cohomology of a compact four-manifold or to the curvature form of any bundle with connection over an oriented four-manifold.

13.2. Connections with self-dual curvature on an oriented four-manifold turn out to be automatically absolute minima of the Yang-Mills functional on the space of connections. Moreover, the Einstein condition can be interpreted in this setting since it is equivalent to the self-duality of the induced Levi-Civita connection on the bundle of self-dual 2-forms.

13.3. On oriented four-dimensional manifolds the Weyl conformal curvature tensor W decomposes into two irreducible components, W^+ and W^-, under the action of the special orthogonal group. Manifolds for which W^+ or W^- vanishes are called *half-conformally flat*. Typical examples include the standard sphere, the complex projective plane or K3 surfaces with their Ricci-flat metrics. The first two examples are in fact the only half conformally flat Einstein spaces with positive scalar curvature. A new proof of this theorem using index theorems for twisted Dirac operators on appropriate bundles is given.

13.4. Half conformally flat manifolds are the spaces for which the twistor construction due to R. Penrose works. Over an oriented Riemannian manifold one considers the space of compatible complex structures, the so called twistor space. This space has a natural almost complex structure which is integrable if and only if W^+ vanishes. The complex structure of the twistor space encodes the conformal structure of the base space. Anti-self-dual connections on vector bundles are pulled back to $(1, 0)$ connections on holomorphic vector bundles on the twistor space. One can also express the Einstein condition in holomorphic terms on the twistor space. So far, this construction has been especially successful for non-compact manifolds, or in providing Yang-Mills fields on the standard sphere. (In this case the twistor space is identified with $\mathbb{C}P^3$). It is worth pointing out that the Taub-NUT metrics are Lorentzian counterparts of half conformally flat Einstein metrics on $\mathbb{R}^4$ constructed here by twistor methods (see Chapter 3).

B. Self-Duality

13.5. Recall that the Hodge map $*$ is defined as follows. If α, β are exterior p-forms and ω_g is the volume form of an oriented Riemannian manifold (M, g) (see 1.50), then

$$\alpha \wedge *\beta = (\alpha, \beta)\omega_g.$$

If $\dim M = 4$ and $p = 2$, then the operator defines an endomorphism

$$*: \textstyle\bigwedge^2 T^*M \to \bigwedge^2 T^*M \text{ such that } *^2 = Id.$$

The two eigenspaces corresponding to the eigenvalues $+1$ and -1 of the $*$ operator are denoted respectively by $\bigwedge^+ T^*M$ and $\bigwedge^- T^*M$. They define 3-dimensional vector bundles of the so-called *self-dual* and *anti-self-dual* 2-forms on M.

Clearly we have

$$\textstyle\bigwedge^2 = \bigwedge^+ \oplus \bigwedge^-.$$

13.6. The curvature operator $R: \bigwedge^2 \to \bigwedge^2$ may be written in block diagonal form relative to the direct sum decomposition:

$$R = \begin{pmatrix} A & B \\ C & D \end{pmatrix}$$

By the Bianchi identity R is self-adjoint hence $A = A^*$, $C = B^*$, $D = D^*$, and under the decomposition of R into its irreducible components (1.126)

$$\operatorname{Tr} A = \operatorname{Tr} D = s/4, \quad B = r - \tfrac{1}{4}sg \quad \text{and}$$

$$\begin{pmatrix} A & 0 \\ 0 & D \end{pmatrix} - s/12 = W, \text{ the Weyl tensor.}$$

The two components of the Weyl tensor $W^+ = A - s/12$, $W^- = D - s/12$ are called the *self-dual* and *anti-self-dual* parts respectively.

13.7. The Hodge $*$ operator commutes with the Laplacian on forms. Hence, corresponding to the decomposition $\bigwedge^2 = \bigwedge^+ \oplus \bigwedge^-$, there is a decomposition of harmonic forms. If M is compact then by Hodge theory there is a decomposition of the 2-dimensional cohomology group

$$H^2(M, \mathbb{R}) = \mathscr{H}^+ \oplus \mathscr{H}^-.$$

If α is a self-dual 2-form, then $\alpha \wedge \alpha$ is a positive multiple of the volume form (see 1.50) ω_g by (13.5) and if α is anti-self-dual $\alpha \wedge \alpha$ is negative. Thus the signature τ of M is given by

$$\tau = b_+ - b_-,$$

where $b_\pm = \dim \mathscr{H}^\pm$.

13.8. By the Hirzebruch signature Theorem [Hir],

$$\tau(M) = \tfrac{1}{3}p_1(M),$$

where p_1 is the first Pontryagin class, and by the Chern-Weil theorem [Ko-No]:

$$p_1(M) = \frac{-1}{8\pi^2}\int_M \operatorname{Tr}(R \wedge R)$$

where R is considered as a matrix of 2-forms. Since B and B^* are acting on orthogonal spaces,

$$\begin{aligned}\operatorname{Tr}(R \wedge R) &= \operatorname{Tr}(A \wedge A) + \operatorname{Tr}(D \wedge D)\\ &= -2(|W^+|^2 - |W^-|^2)\omega_g,\end{aligned}$$

because $\alpha \wedge \alpha = |\alpha|^2\omega_g$ if α is self-dual. Hence the signature theorem gives

(13.9)
$$\tau = b_+ - b_- = \frac{1}{12\pi^2}\int_M (|W^+|^2 - |W^-|^2)\omega_g,$$

which we used already in 6.34.

The Gauss-Bonnet Theorem represents the Euler characteristic as the integral

$$\begin{aligned}\chi(M) &= \frac{1}{8\pi^2}\int_M \operatorname{Tr}({*R})^2\mu_g\\ &= \frac{1}{8\pi^2}\int_M \operatorname{Tr}(A^2 - 2BB^* + D^2)\mu_g.\end{aligned}$$

In particular, if M is an Einstein manifold $B = 0$, so

(13.10)
$$\begin{aligned}\chi(M) &= \frac{1}{8\pi^2}\int_M \operatorname{Tr}(A^2 + D^2)\mu_g\\ &= \frac{1}{8\pi^2}\int_M \left(|W^+|^2 + |W^-|^2 + \frac{1}{24}s^2\right)\mu_g.\end{aligned}$$

As in 6.35, we derive the inequality $\tau \leqslant \frac{2}{3}\chi$, with equality if and only if $W^- = 0$ and $s = 0$.

13.11. Part of the motivation for physical interest in Riemannian Einstein manifolds is derived from the study of Yang-Mills fields. In differential geometric terms, such a field is the curvature of a connection ∇ on a principal G-bundle P over a 4-manifold M, where G is a compact group.

13.12 Definition. A Yang-Mills "*instanton*" is a critical point of the functional

$$\mathbf{YM}(\nabla) = \int_M |F|^2\mu_g$$

where $F \in C^\infty(\mathfrak{g} \otimes \bigwedge^2)$ is the curvature of the connection ($\mathfrak{g}$ is the Lie algebra of G). At such a critical point the curvature F is harmonic:

$$d^\nabla {*F} = 0, \qquad d^\nabla F = 0$$

where d^∇ is an exterior differential as in 1.12.

By the Chern-Weil theorem, the second Chern class $c_2(V)$ of some vector bundle associated with P via a representation of G is defined by

$$c_2(V) = \frac{1}{8\pi^2}\int_M \mathrm{Tr}(F \wedge F)$$

$$= \frac{-c}{8\pi^2}\int_M (|F^+|^2 - |F^-|^2)\omega_g$$

for some universal constant c where $F^\pm$ are the $\mathfrak{g} \otimes \bigwedge^\pm$ components of F. Thus

$$\mathbf{YM}(\nabla) \geqslant \frac{8\pi^2}{c}|c_2(V)|$$

and the minimum is attained if and only if $F^+ = 0$ or $F^- = 0$.

13.13 Definition. A *self-dual connection* is a connection whose curvature satisfies $F^- = 0$
It is in particular a critical point of the Yang-Mills functional.

There is a close analogy between the Einstein equations and the Yang-Mills equations. Indeed, on an Einstein manifold, the curvature of the Levi-Civita connections is harmonic, and therefore is a critical point of the Yang-Mills functional.
There is a further relation:

13.14 Theorem (Singer, [At-Hi-Si]). *A 4-manifold M^4 is Einstein if and only if the Levi-Civita connection on $\bigwedge^+$ is self-dual.*

Proof. The curvature of $\bigwedge^+$ is given by the first row of the block decomposition (13.6) of the curvature tensor. Thus the anti-self-dual part is B^* and this vanishes if and only if M is Einstein. □

However, the analogy between the Einstein equations and the Yang-Mills equations fails at the level of functionals, see 4.82.

C. Half-Conformally Flat Manifolds

13.15. In this section we examine within a differential geometric framework those Einstein manifolds which we call half conformally flat. They bear the same relationship to a general Einstein manifold that self-dual solutions bear to a general solution of the Yang-Mills equations.

13.16 Definition. The oriented 4-manifold (M, g) is *half-conformally flat* if $W^- = 0$.
The condition is clearly conformally invariant.

13.17 Examples. (1) Every conformally flat 4-manifold satisfies this condition, hence

$$M = \mathbb{R}^4, S^4, T^4, S^1 \times S^3$$

and connected sums of these all have half conformally flat structures [Kul 1].

(2) The projective plane $\mathbb{C}P^2$ is half conformally flat with respect to the orientation defined by the complex structure, and the Fubini-Study metric. This is clear because the curvature operator is invariant under the action of the isotropy group $U(2)$. Since $U(2) \subseteq SO(4)$ acts irreducibly on $\wedge^-$ it follows from Schur's Lemma that the component D in 13.6 must be a scalar.

(3) A K3-surface with the opposite orientation and the Calabi-Yau metric is half conformally flat. This follows in particular from $\tau \leqslant \frac{2}{3}\chi$ since with the canonical orientation $\tau = -16$ and $\chi = 24$.

13.18. Before we proceed, it will be useful to extend our techniques in analysing the special properties of 4-dimensional Riemannian geometry. The principal bundle of oriented orthonormal frames P has structure group $SO(4)$. The 3-dimensional bundle $\wedge^\pm$ has an induced metric and is associated to P by a representation

$$\lambda^+ : SO(4) \to SO(3),$$

where kernel is $SU(2)$. Similarly there is a representation $\lambda^- : SO(4) \to SO(3)$ giving $\wedge^-$. The two kernels are commuting subgroups of $SO(4)$ isomorphic to $SU(2)$ which intersect in -1, and there is an isomorphism

$$SO(4) \cong SU(2) \times SU(2)/\mathbb{Z}_2,$$

where $\mathbb{Z}_2$ is generated by the diagonal element $(-1, -1)$. The non-trivial double covering Spin(4) is thus isomorphic to $SU(2) \times SU(2)$.

13.19. Let us denote by Σ^+, Σ^- the two complex vector bundles on M associated to the defining 2-dimensional representations of the two factors. These will only exist globally if M is a spin manifold, i.e. the 2nd Stiefel-Whitney class $w_2(M) = 0$. Denote by $S^m\Sigma^\pm$ the mth symmetric power bundle of $\Sigma^\pm$. This is a complex bundle of dimension $(m+1)$. Then it follows from the theory of group representations that *every* bundle associated to an irreducible representation of Spin(4) is of the form

$$S^m\Sigma^+ \otimes S^n\Sigma^- \qquad \text{[Var]},$$

and is a representation of $SO(4)$ if $(-1, -1)$ acts trivially i.e. if $m + n$ is even.

13.20. The basic representation Σ^+ has a symplectic structure (because $SU(2) \leqslant SL(2, \mathbb{C})$) and a unitary structure (because $SU(2) \leqslant U(2)$). These define respectively isomorphisms

$$\Sigma^+ \cong (\Sigma^+)^*$$
$$\Sigma^+ \cong (\bar{\Sigma}^+)^*,$$

and the composite is a quaternionic structure

$$j \colon \Sigma^+ \to \Sigma^+,$$

which is antilinear ($j(\lambda v) = \bar{\lambda} j(v)$) and satisfies $j^2 = -1$.

These two structures induce symplectic forms and quaternionic structures on $S^m\Sigma^+$ for m odd and metrics and real structures on $S^m\Sigma^+$ if m is even.

13.21. The 3-dimensional representations $\lambda^\pm$ give rise to a natural isomorphism

$$\Lambda^\pm \cong S^2\Sigma^\pm$$

furthermore the unique 4-dimensional real irreducible representation of $SO(4)$ gives the isomorphism

$$\Lambda^1 \cong \Sigma^+ \otimes \Sigma^-.$$

The final ingredient to this formalism is the formula for decomposing a tensor product:

$$S^m\Sigma^+ \otimes S^n\Sigma^+ \cong \bigoplus_{k=0}^{\min(m,n)} S^{m+n-2k}\Sigma^+ \qquad \text{[Var]}. \tag{13.22}$$

This arises from the repeated contraction with respect to the symplectic form. These properties of irreducible representations make it possible to minimize complicated tensor calculations.

13.23. If we consider the components in the block decomposition of the curvature from this point of view we have

$$B \in C^\infty(\Lambda^+ \otimes \Lambda^-) \cong C^\infty(S^2\Sigma^+ \otimes S^2\Sigma^-),$$

which is an irreducible representation of $SO(4)$ and

$$A \in C^\infty(\Lambda^+ \otimes \Lambda^+) \cong C^\infty(S^2\Sigma^+ \otimes S^2\Sigma^+).$$

From (13.22), $S^2\Sigma^+ \otimes S^2\Sigma^+ \cong S^4\Sigma^+ \oplus S^2\Sigma^+ \oplus S^0\Sigma^+$ and the skew symmetric part is $S^2\Sigma^+$. Thus since A is self-adjoint, $A \in C^\infty(S^0\Sigma^+ \oplus S^4\Sigma^+)$. The bundle $S^0\Sigma^+$ is the trivial bundle consisting of multiples of the identity, so the trace-free part, in which W^+ lies, is $S^4\Sigma^+$, which is irreducible. Similarly $W^- \in C^\infty(S^4\Sigma^-)$.

13.24. Consider now the bundle $S^m\Sigma^-$. The covariant derivative D of the Levi-Civita connection defines a differential operator

$$D\colon C^\infty(S^m\Sigma^-) \to C^\infty(S^m\Sigma^- \otimes \Lambda^1).$$

Now $S^m\Sigma^- \otimes \Lambda^1 \cong S^m\Sigma^- \otimes \Sigma^+ \otimes \Sigma^- \cong (S^{m+1}\Sigma^- \otimes \Sigma^+) \oplus (S^{m-1}\Sigma^- \otimes \Sigma^+)$ and these are the unique irreducible components. By projecting onto them we obtain two first order differential operators

$$\begin{aligned} D_m\colon C^\infty(S^m\Sigma^-) &\to C^\infty(S^{m-1}\Sigma^- \otimes \Sigma^+),\\ \bar{D}_m\colon C^\infty(S^m\Sigma^-) &\to C^\infty(S^{m+1}\Sigma^- \otimes \Sigma^+). \end{aligned} \tag{13.25}$$

We can also form their formal adjoints

$$\begin{aligned} D_m^*\colon C^\infty(S^{m-1}\Sigma^- \otimes \Sigma^+) &\to C^\infty(S^m\Sigma^-),\\ \bar{D}_m^*\colon C^\infty(S^{m+1}\Sigma^- \otimes \Sigma^+) &\to C^\infty(S^m\Sigma^-). \end{aligned} \tag{13.26}$$

13.27 Proposition. *If M is a half conformally flat 4-manifold then the sequence of differential operators*

$$C^\infty(S^m\Sigma^-)\xrightarrow{\bar{D}_m} C^\infty(S^{m+1}\Sigma^-\otimes\Sigma^+)\xrightarrow{D^*_{m+2}} C^\infty(S^{m+2}\Sigma^-)$$

forms an elliptic complex.

Proof. First we must show that this is a complex—that $D^*_{m+2}\bar{D}_m = 0$.

Now each operator factors through the covariant derivative D, hence

$$D^*_{m+2}\bar{D}_m = \sigma D^2$$

for some linear map $\sigma \in C^\infty(\mathrm{Hom}(S^m\Sigma^-\otimes\Lambda^1\otimes\Lambda^1, S^{m+2}\Sigma^-))$. But σ is defined by $SO(4)$-invariant projections so σ itself must be invariant. If we restrict σ to symmetric forms, then it lies in $\mathrm{Hom}(S^m\Sigma^-\otimes(S^2\Sigma^+\otimes S^2\Sigma^- + 1), S^{m+2}\Sigma^-)$. By expanding the tensor product it is clear that there are no invariants here. Thus $D^*_{m+2}\bar{D}_m$ is a linear invariant function of the skew part of D^2 – the curvature. But there is only one such function, the symmetrization map $S^m\Sigma^-\otimes\Lambda^- \cong S^m\Sigma^-\otimes S^2\Sigma^- \to S^{m+2}\Sigma^-$, and this depends on the component $W^-\in S^4\Sigma^-$ of the curvature. If M is half conformally flat this is zero, so $D^*_{m+2}\bar{D}_m = 0$.

To prove that the complex is elliptic is equivalent to showing that

$$C^\infty(S^{m+1}\Sigma^-\otimes\Sigma^+)\xrightarrow{\bar{D}^*_m\oplus D^*_{m+2}} C^\infty(S^m\Sigma^-\oplus S^{m+2}\Sigma^-)$$

is an elliptic operator. But

$$S^m\Sigma^-\oplus S^{m+2}\Sigma^- \cong S^{m+1}\Sigma^-\otimes\Sigma^-,$$

and we already have a natural elliptic operator—the *Dirac operator* $\mathscr{D}^D$ with coefficient bundle $S^{m+1}\Sigma^-$.

$$\mathscr{D}^D\colon C^\infty(S^{m+1}\Sigma^-\otimes\Sigma^+)\to C^\infty(S^{m+1}\Sigma^-\otimes\Sigma^-).$$

Since this again factors through the covariant derivative with an invariant projection we see that $\mathscr{D}^D = C_1\bar{D}^*_m\oplus C_2D^*_{m+2}$. Since $\mathscr{D}^D$ is elliptic, C_1 and C_2 must be non-zero and it follows that the original complex is elliptic.

For each of these complexes there is a dual complex

$$C^\infty(S^{m+2}\Sigma^-)\xrightarrow{D_{m+2}} C^\infty(S^{m+1}\Sigma^-\otimes\Sigma^+)\xrightarrow{\bar{D}^*_m} C^\infty(S^m\Sigma^-). \tag{13.28}$$

As an example, if $m = 0$ there is the complex

$$C^\infty \xrightarrow{d} C^\infty(\Lambda^1)\xrightarrow{d^-} C^\infty(\Lambda^-),$$

where

$$d^- = \tfrac{1}{2}(d - *d).$$

The dual complex is

$$C^\infty(\Lambda^-)\xrightarrow{d} C^\infty(\Lambda^3)\xrightarrow{d} C^\infty(\Lambda^4),$$

which is half the de Rham complex.

If m is odd, then of course we need M to be a spin manifold to obtain a globally defined complex in (13.28).

13.29. The preceding discussion can be applied to produce some powerful restrictions on half conformally flat manifolds. In particular, we shall apply them to obtain the following:

13.30 Theorem (Hitchin). *Let M be a compact half conformally flat Einstein manifold. Then*

(1) *If $s > 0$, M is isometric to S^4 or CP^2 with their canonical metrics.*

(2) *If $s = 0$, M is either flat or its universal covering is a K3 surface with the Calabi-Yau metric.*

Proof. (1) Suppose $s > 0$, and consider the complex

$$C^\infty(S^2\Sigma^-) \overset{\bar{D}_2}{\to} C^\infty(S^3\Sigma^- \otimes \Sigma^+) \overset{D_4^*}{\to} C^\infty(S^4\Sigma^-).$$

By Hodge theory the index of the complex is given by

$$\text{index} = \dim\ker \bar{D}_2 + \dim\ker D_4 - \dim\ker D_4^* \oplus \bar{D}_2^*,$$

and this, as we have seen, is the index of the Dirac operator

$$\mathscr{D}^D\colon C^\infty(\Sigma^- \otimes S^3\Sigma^-) \to C^\infty(\Sigma^+ \otimes S^3\Sigma^-).$$

By the Atiyah-Singer index theorem ([At-Si], [Sha]) this is given by

$$\begin{aligned}\text{index}\,\mathscr{D}^D &= -ch(S^3)\hat{A}(M)\\ &= (10c_2(S_-) - 4)(1 - p_1/24)[M]\\ &= 5\chi - 7\tau\end{aligned}$$

using the signature theorem.

13.31. We now use the Weitzenböck formula (see 1.151) for the Dirac operator with coefficient bundle $S^3\Sigma^-$. This gives

$$(\mathscr{D}^D)^2 = D^*D - 4(\sigma^2 \otimes 1 + \sigma \otimes \rho)(R), \tag{13.32}$$

where σ is the representation Σ and ρ the representation $S^3\Sigma^-$.

Since the metric is half conformally flat and Einstein, then the curvature of the Levi-Civita connection on $S^3\Sigma^-$ involves only the scalar curvature s (13.6).

We already know (1.152) that $\sigma^2(R) = -\frac{1}{16}s$ so the curvature term in the Weitzenböck decomposition may be evaluated by considering a curvature tensor of the form $R = cg \otimes g$. Now if $\{X_i\}$ is an orthonormal basis for $\mathfrak{o}(n)$ with respect to the Killing form the *Casimir operator* (see [Var]) of a representation $\rho\colon \mathfrak{o}(n) \to \operatorname{End} V$ is

$$C(V) = \rho^2\left(\sum X_i \otimes X_i\right) = \sum_i \rho(X_i)^2 \in \operatorname{End} V.$$

On an irreducible representation it is a scalar by Schur's lemma and for $SO(4)$

$$C(S^m\Sigma^+ \otimes S^n\Sigma^-) = -\tfrac{1}{8}(m(m+2) + n(n+2)) \qquad \text{[Var]}.$$

We have

$$C(\Sigma \otimes S^3\Sigma^-) = K(\sigma^2 \otimes 1 + 2\sigma \otimes \rho + 1 \otimes \rho^2)(g \otimes g)$$

for some positive constant K and so

$$(\sigma^2 \otimes 1 + \sigma \otimes \rho)(R) = \frac{Kc}{2}(C(\Sigma \otimes S^3\Sigma^-) + C(\Sigma) - C(S^3\Sigma^-))$$

$$= -\frac{3Kc}{8} \quad \text{on } \Sigma^+ \otimes S^3\Sigma^- \subset \Sigma \otimes S^3\Sigma^-$$

and $$= -\frac{3Kc}{4} \quad \text{on } S^4\Sigma^- \subset \Sigma \otimes S^3\Sigma^-$$

thus by the vanishing argument (1.137), if $c > 0$,

$$\ker D_4 = 0,$$

$$\ker D_4^* \oplus \bar{D}_2^* = 0.$$

Hence from (13.31),

$$\dim \ker \bar{D}_2 = \operatorname{ind} \mathscr{D}^D = 5\chi - 7\tau, \tag{13.33}$$

If we apply the analogous vanishing theorem to the complex for $m = 0$, i.e.

$$C^\infty \xrightarrow{d} C^\infty(\Lambda^1) \xrightarrow{d^-} C^\infty(\Lambda^-)$$

we deduce that $b_1 = 0$ and $b_- = 0$. Hence

$$\chi = 2 + b_+,$$

$$\tau = b_+,$$

and so

$$\dim \ker \bar{D}_2 = 10 - 2b_+. \tag{13.34}$$

Now if we finally use the integral inequality

$$\tau \leqslant \tfrac{2}{3}\chi$$

we see that $b_+ \leqslant 4$, and since $s > 0$ equality does not occur, so $b_+ \leqslant 3$. Hence we obtain the inequality

$$\dim \ker \bar{D}_2 \geqslant 4. \tag{13.35}$$

We now need the following:

13.36 Lemma. *Let M be an Einstein 4-manifold and suppose $\alpha \in C^\infty(\Lambda^-)$ satisfies $\bar{D}_2\alpha = 0$. Let X be the vector field such that $d\alpha = i_X\omega_g$. Then X is a Killing vector field.*

Proof. Let $\pi_1 \colon \Lambda^- \otimes \Lambda^1 \otimes \Lambda^1 \to \Lambda^- \otimes \Lambda^+$ denote the invariant projection and π_2: $\Lambda^1 \to \Lambda^- \otimes \Lambda^1$ the inclusion

$$\Sigma^+ \otimes \Sigma^- \to (\Sigma^+ \otimes \Sigma^-) \oplus (\Sigma^+ \otimes S^3\Sigma^-).$$

Then the composition

$$\pi_1(\pi_2 \otimes id) \colon \Lambda^1 \otimes \Lambda^1 \to \Lambda^- \otimes \Lambda^+$$

is either zero or a multiple of the unique invariant map. In fact by evaluating on a decomposable element it is easily seen to be non-trivial.

Now consider $D^2\alpha \in C^\infty(\Lambda^- \otimes \Lambda^1 \otimes \Lambda^1)$. The projection $\pi_1 D^2\alpha$ is skew in $\Lambda^1 \otimes \Lambda^1$ and hence is $\pi_1(R\alpha)$. But since M is Einstein R has no $\Lambda^- \otimes \Lambda^+$ component, so $\pi_1(R\alpha) = 0$. However, since $\bar{D}_2\alpha = 0$, the projection of $D\alpha$ onto the $\Sigma^+ \otimes S^3\Sigma^-$ component of $S^2\Sigma^- \otimes \Lambda^1$ is zero, hence $D\alpha \in C^\infty(\Sigma^+ \otimes \Sigma^-)$. Under the isomorphism $\Sigma^+ \otimes \Sigma^- \cong \Lambda^3$ this element is $d\alpha$, or using the metric to define a vector field, X.

Thus $\pi_1 D^2\alpha = \pi_1(\pi_2 \otimes id)DX$, and since $\pi_1 D^2\alpha = 0$, X satisfies the unique invariant equation $\pi_1(\pi_2 \otimes id)DX = 0$ i.e. the trace-free symmetric part of DX vanishes. Since $d^2\alpha = 0$, $\operatorname{div} X = 0$ hence the whole symmetric part of DX is zero, so X is a Killing vector field. □

13.37. Now if $\bar{D}_2\alpha = 0$ and $d\alpha = 0$ then α is covariant constant, and so harmonic, but since $b_1 = 0$ this is impossible hence if G denotes the compact Lie group of isometries of M, from (13.35)

(13.38) $$\dim G \geqslant \dim \ker \bar{D}_2 \geqslant 4.$$

13.39. Now since $0 \leqslant b_+ \leqslant 3$, there are just four cases to consider. Suppose first $b_+ \geqslant 2$, then there are two linearly independent harmonic self-dual 2-forms α_1, α_2. Linear independence here is of course with respect to real constant coefficients. Suppose in some open set $\alpha_2 = f\alpha_1$ for some C^∞ function f. Then

$$0 = d\alpha_2 = df \wedge \alpha_1.$$

Hence $\alpha_1 = \theta \wedge df$ whenever $df \neq 0$, and so $\alpha_1 \wedge \alpha_1 = 0$. But α_1 is self-dual so $\alpha_1 \wedge \alpha_1 = |\alpha_1|^2\omega_g$, thus f must be constant. But since α_1 and α_2 are harmonic, if $\alpha_2 - f\alpha_1$ vanishes on an open set it vanishes everywhere [Aro], so we deduce that if α_1 and α_2 are linearly independent over $\mathbb{R}$, there is some point $x \in M$ at which the forms are linearly independent.

Every element in the identity component G_0 of G leaves fixed each harmonic form. Let $H \leqslant G_0$ be the stabilizer of x, then $H \leqslant SO(4)$ and H leaves fixed α_1 and α_2. Since they are linearly independent at x this implies that

$$H \leqslant SU(2) = \text{kernel}\,\lambda^+.$$

If $\dim H = 0$, then since $\dim G \geqslant 4$ and G is compact, then $M \cong G/H$, a finite quotient of G. Since $\chi(M) \neq 0$ this is impossible so $\dim H = 1$ or 3.

Let T_1 be the maximal torus of H and T_2 a circle group in G which commutes with T_1. Then if $T_2 \neq T_1$, the tangent direction of the orbit of T_2 at x is non-zero and preserved by T_1. But the only elements in $SU(2)$ which have real eigenvectors are ± 1, so $T_2 = T_1$. This means that rank $G = 1$, but this contradicts $\dim G \geqslant 4$. Hence we must have $b_+ = 0$ or 1.

13.40. If $b_+ = 0$, $\dim G \geqslant 10$ from (13.38). But for an oriented 4-manifold this implies that M is isometric to S^4. [Kob 4].

13.41. If $b_+ = 1$, $\dim G \geqslant 8$. If H is the stabilizer of a general point, then $H \leqslant U(2)$ since it leaves fixed a non-zero self-dual 2-form. Thus $\dim H = 4$ and $\dim G = 8$.

An argument with maximal tori exactly analogous to the above shows that rank $G = 2$. This means that G_0 is a quotient of $SU(3)$, so M is a quotient of $\mathbb{C}P^2$. Since $\tau = 1$, there are no such quotients so M is isometric to $\mathbb{C}P^2$ with a homogeneous metric. There is only one such Einstein metric, the Fubini-Study metric, (see Chapter 8) so part (1) of the theorem is proved.

13.42. Suppose $s = 0$. Then from [Hit 2] $\chi(M) \geqslant 0$ and vanishes iff M is flat. If $b_1 \neq 0$, then by Bochner's Theorem 1.155 there must exist a parallel and hence non-vanishing 1-form and so $\chi = 0$. Thus if M is not flat, $b_1 = 0$ and likewise for every finite covering. It follows from the splitting Theorem 6.65 for manifolds with $r \geqslant 0$, that $\pi_1(M)$ is finite.

Thus the universal cover $\tilde{M}$ is compact and simply connected. Now if $r = 0$ and $W^- = 0$ the curvature of Λ^- is zero (13.6), so since $\tilde{M}$ is simply connected we have a reduction of the holonomy group from $SO(4)$ to $\ker \lambda^- = SU(2)$. Thus (see 10.29) $\tilde{M}$ is a Kähler manifold with zero Ricci tensor. In particular $b_1 = 0$ and $c_1 = 0$ so $\tilde{M}$ is a K3 surface (cf. 12.104). □

D. The Penrose Construction

13.43. Perhaps the most remarkable aspect of self-duality is its link with holomorphic geometry. The methods of algebraic geometry may be used to prove results in 4-dimensional Riemannian geometry. More specifically, one can construct self-dual Einstein metrics from the geometry of a family of holomorphic curves in a complex manifold. Moreover, the basic geometry is defined not by a tensorial object like a metric but by the infinitesimal intersection properties of those curves. The originator of this point of view is Penrose [Pen 2] specifically in order to solve problems in mathematical physics.

13.44. The starting point for this construction is to consider a certain fibre bundle $\pi\colon P \to M$. We take $P = S(\Lambda^- M)$, the unit sphere bundle of the 3-dimensional real vector bundle Λ^-. This is a 6-manifold whose fibres are 2-spheres. Using the metric to identify 2-forms and skew-adjoint endomorphisms of TM, a unit anti-self-dual 2-form at $x \in M$ becomes an endomorphism J such that

$$J^* = -J,$$
$$J^2 = -1,$$

which thus defines a complex structure on the tangent space $T_x M$. The canonical orientation with respect to J is the opposite one to that of M.

Using the Levi-Civita connection we may split the tangent bundle of P:

$$TP \cong TF \oplus \pi^* TM,$$

where TF is the tangent bundle along the fibre.

Then at a point $z = J_x$ of P we define a complex structure on $T_z P$ by taking J_x

on the horizontal space $\pi^* T_x M$ and the standard one (the Riemann sphere) in the fibre.

13.45 Definition. The *twistor space* of M is the manifold $P = S(\bigwedge^- M)$ with the almost complex structure defined above.

13.46 Theorem (R. Penrose [Pen 2], M.F. Atiyah, N.J. Hitchin and I.M. Singer [At-Hi-Si]). *Let (M, g) be an oriented Riemannian 4-manifold and $P = S(\bigwedge^- M)$ be its twistor space. Then the almost complex is integrable iff (M, g) is half conformally flat.*

13.47 *Proof.* It is convenient to adopt a slightly different description of the almost complex structure on P, and to introduce complex numbers at a more fundamental level. In fact

$$P \cong \mathbb{P}(\Sigma^-) \qquad \text{(see [At-Hi-Si])}.$$

This is the projective bundle of Σ^-. We shall work locally on the 8-manifold $\Sigma^- \backslash 0$ (the total space of Σ^- without the zero-section) whose quotient by complex scalar multiplication is $\mathbb{P}(\Sigma^-)$.

On the bundle $\pi\colon \Sigma^- \backslash 0 \to M$ there is a tautological non-vanishing section Φ of $\pi^* \Sigma^-$. We define the almost complex structure on $\Sigma^- \backslash 0$ by splitting with the connection and then at each point defining the $(1, 0)$ forms in the horizontal space $\pi^* T_x M$ to be of the form $\psi \otimes \Phi$ where ψ runs through $(\Sigma^+)_x$. Here we have used the isomorphism $\bigwedge^1 \cong \Sigma^+ \otimes \Sigma^-$. In the vertical direction we take the standard complex structure on the fibre, which is a punctured complex vector space.

Now choose a local orthonormal frame $\{e_i\}$ for TM and corresponding frames $\{\psi_\alpha\}$, $\{\varphi_\alpha\}$ for Σ^+ and Σ^-. We parametrize $\Sigma^- \backslash 0$ locally by

$$(x, \lambda) \to \sum_{\alpha=1}^{2} \lambda_\alpha \varphi_\alpha(x),$$

and the tautological section Φ is given by

$$\Phi = \sum_{\alpha=1}^{2} \lambda_\alpha \varphi_\alpha.$$

If $\omega_{\alpha\beta}$ is the connection form of Σ^- with respect to this frame, then the space of $(1, 0)$ forms on $\Sigma^- \backslash 0$ is spanned by

$$\textbf{(13.48)} \qquad \begin{cases} \theta_\alpha = d\lambda_\alpha - \sum_\gamma \omega_{\alpha\gamma} \lambda_\gamma & \alpha = 1, 2, \\ \sigma_\beta = \psi_\beta \otimes \Phi & \beta = 1, 2. \end{cases}$$

The connection defines a projection onto the tangent space of the fibres of $\Sigma^- \backslash 0$. In local coordinates this is

$$\pi_1 = \sum_\alpha \theta_\alpha \frac{\partial}{\partial \lambda_\alpha}.$$

Now if $\pi_2\colon \Sigma^- \backslash 0 \to P$ denotes the canonical projection

$$\pi_2\pi_1 = (\lambda_2\theta_1 - \lambda_1\theta_2)/\lambda_2^2\frac{\partial}{\partial\mu} = \theta/\lambda_2^2\frac{\partial}{\partial\mu},$$

where $\mu = \lambda_1/\lambda_2$ is an affine coordinate for $\lambda_2 \neq 0$ on the projective bundle $\mathbb{P}(\Sigma^-)$.

The space of (1, 0) forms on P is then spanned by

(13.49)
$$(\theta/\lambda_2^2, \sigma_1/\lambda_2, \sigma_2/\lambda_2) = (\theta', \sigma_1', \sigma_2'),$$

which are all homogeneous of degree 0 and hence functions of $\mu = \lambda_1/\lambda_2$.

13.50. By the Newlander-Nirenberg theorem [Ne-Ni] (see also 2.12) the integrability condition is that the ideal $\mathscr{I}$ generated by $(\theta', \sigma_1', \sigma_2')$ be closed under the exterior derivative d. To simplify calculations let us suppose that the frame is chosen to be geodesic at a fixed point $x \in M$. Then $\omega_{\alpha\beta}$ vanishes at x. From this we see that $\theta' = \theta/\lambda_2^2 = d\mu$ at x.

Taking exterior derivatives at x,

(13.51)
$$\begin{cases} d(\sigma_\beta') = d\mu \wedge (\psi_\beta \otimes \varphi_1) \\ d(\theta') = \omega(\Phi, R^-(\Phi))/\lambda_2^2 \end{cases},$$

where $R^- \in C^\infty(\operatorname{End}\Sigma^- \otimes \bigwedge^2)$ is the curvature tensor of Σ^- and ω is the symplectic form on Σ^-.

Hence $d(\sigma_\beta')$ is in $\mathscr{I}$ and since $d\theta'$ contains no $d\mu$ term, it will be in $\mathscr{I}$ iff

(13.52)
$$d\theta' \wedge \sigma_1' \wedge \sigma_2' = \frac{1}{\lambda_2^4}\omega(\Phi, R^-(\Phi)) \wedge (\Phi \otimes \Phi) = 0.$$

Here $\sigma_1 \wedge \sigma_2 = \Phi \otimes \Phi$ under the isomorphism $\bigwedge^- \cong S^2\Sigma^-$.

13.53. Now since $\alpha \wedge \beta = 0$ if $\alpha \in \bigwedge^+$ and $\beta \in \bigwedge^-$, clearly the $\bigwedge^+$ part of the curvature (B in the notation of 13.6) has no effect on this condition. Neither does the scalar curvature for that contributes

$$C_1 s(\Phi \otimes \Phi) \wedge (\Phi \otimes \Phi) = 2C_1 s\omega(\Phi, \Phi) = 0.$$

There remains $W^- \in C^\infty(S^4\Sigma^-)$ and if we consider W^- as a homogeneous symmetric form of degree 4, the integrability condition is

$$W^-(\Phi, \Phi, \Phi, \Phi) = 0.$$

As (λ_1, λ_2) vary this vanishes iff $W^- = 0$. Thus the almost complex structure on P is integrable iff $W^- = 0$, and the theorem is proved. □

13.54. Let us consider now the bundle $\Sigma^-\backslash 0$ over P. We shall show that this is not an extra piece of data, but is canonically associated to P.

If $W^- = 0$, then it is easy to see that $\{\theta_\alpha, \sigma_\beta\}$ span the (1, 0) space of an integrable complex structure on $\Sigma^-\backslash 0$, such that $\pi_2\colon \Sigma^-\backslash 0 \to P$ is holomorphic. Consider the globally defined section

(13.55)
$$\alpha = \theta \wedge \sigma_1 \wedge \sigma_2 \in C^\infty(\Sigma^-\backslash 0, \pi_2^* K),$$

where $K \cong \bigwedge^{3,0}P$ is the canonical line bundle over P.

Locally, we may write

$$\alpha = \lambda_2^4 \theta' \wedge \sigma'_1 \wedge \sigma'_2.$$

Now from (13.52)

$$d\theta' \wedge \sigma'_1 \wedge \sigma'_2 = 0,$$

and at x, in a geodesic frame $\theta' = d\mu$, so

$$\theta' \wedge d\sigma'_\beta = 0.$$

Moreover, again at x, $d\lambda_2 = \theta_2 \in \bigwedge_x^{1,0} P$, so

$$d''\lambda_2 = 0.$$

Hence

(13.56) $$d''\alpha = 0.$$

Thus α defines a holomorphic trivialization of $\pi_2^* K$ on $\Sigma^- \backslash 0$. Furthermore, if $\lambda \in \mathbb{C}^*$ then

$$\alpha(\lambda z) = \lambda^4 \alpha(z),$$

since α is homogeneous of degree 4 in (λ_1, λ_2).

Thus $\alpha\colon \Sigma^- \backslash 0/\mathbb{Z}_4 \to K\backslash 0$ is a biholomorphic map, commuting with the $\mathbb{C}^*$ action of scalar multiplication. In other words $\Sigma^- \backslash 0$ with its $\mathbb{C}^*$ action is the principal bundle of a holomorphic bundle $K^{1/4}$ such that

$$\bigotimes^4 K^{1/4} \cong K.$$

This line bundle exists globally on P iff M is a spin manifold. In all cases, there is a distinguished holomorphic line bundle $K^{1/2} = \Sigma^- \backslash 0 \times_{\mathbb{Z}_2} \mathbb{C}$ such that

$$K^{1/2} \otimes K^{1/2} \cong K.$$

Note for future reference that $\theta = \lambda_2 \theta_1 - \lambda_1 \theta_2$ with this interpretation is an element of $C^\infty(K^{-1/2} \otimes \bigwedge^{1,0})$.

13.57. The integrability condition in Theorem 13.46 is conformally invariant, so it is not surprising to find that the almost complex structure on P is independent of the metric within the conformal equivalence class. Suppose, then, that the metric g is conformally related to g by $g = e^{2f} g$. Let $\{e_i\}$ denote an oriented orthonormal basis for the tangent bundle of M. Then [Hit 4] the connection matrices relative to this basis of the two Levi-Civita connections are related by

(13.58) $$\tilde{\omega}_{ij} = \omega_{ij} + e_j(df, e_i) - e_i(df, e_j)$$

identifying tangent vectors and 1-forms with the metric g. On the spinor bundle, we obtain

(13.59) $$\tilde{\omega}_{\alpha\beta} = \omega_{\alpha\beta} - \frac{1}{4} \sum_{\gamma=1}^{2} \langle df \cdot \varphi_\alpha, \psi_\gamma \rangle \sigma_\gamma \otimes \varphi_\beta,$$

and so from (13.48)

(13.60)
$$\tilde{\theta}_\alpha = \theta_\alpha - \frac{1}{4}\sum_{\gamma=1}^{2} \langle df \cdot \varphi_\alpha, \psi_\gamma \rangle \sigma_\gamma,$$

and moreover

(13.61)
$$\tilde{\sigma}_\alpha = e^f \sigma_\alpha.$$

Thus the space of (1, 0) forms is unchanged since $\tilde{\theta}'$, $\tilde{\sigma}'_1$, $\tilde{\sigma}'_2$ (see 13.49) forms another basis.

We have chosen here not to change the symplectic form on the spin bundles when the metric changes conformally. This, although being sufficient to prove the conformal invariance of the almost complex structure, has the disadvantage that the holomorphic form $\alpha = \theta \wedge \sigma_1 \wedge \sigma_2$ is not conformally invariant. Indeed, from (13.60) and (13.61)

$$\tilde{\theta} \wedge \tilde{\sigma}_1 \wedge \tilde{\sigma}_2 = e^{2f}\theta \wedge \sigma_1 \wedge \sigma_2.$$

If the symplectic form on spinors changes by

$$\langle , \rangle \to e^f \langle , \rangle,$$

when the metric changes by

$$g \to e^{2f} g$$

then α is conformally invariant and (13.60) and (13.61) give

(13.62)
$$\tilde{\theta}_\alpha = e^{-1/2f}(\theta_\alpha + \sum a_{\alpha\beta}\sigma_\beta),$$
$$\tilde{\sigma}_\alpha = e^{1/2f}\sigma_\alpha.$$

This is the conformal weighting which makes the operator $\overline{D_1}$ conformally invariant and is the basis for the more invariant treatment of the subject in [At-Hi-Si].

13.63 Proposition. (1) *The fibres of $\pi: P \to M$ are holomorphic curves whose normal bundle is $H \oplus H$ where H is the holomorphic line bundle over $\mathbb{C}P^1$ with $c_1(H) = 1$.*

(2) *P possesses a free antiholomorphic involution $\tau: P \to P$ which transforms each fibre to itself.*

Proof. (1) We shall work in $\Sigma^-\backslash 0$. The forms σ_1, σ_2 pulled back to a fibre vanish and in a geodesic frame at x, $\theta_\alpha = d\lambda_\alpha$, which is a (1, 0) form in the standard complex structure on the fibre $(\Sigma^-)_x$. Thus each fibre is holomorphic, and hence each fibre of $\mathbb{P}(\Sigma^-) = P$ is likewise a holomorphic curve.

The forms σ_1 and σ_2 span the conormal bundle of the fibre $(\Sigma^-)_x$ and are holomorphic there, since $d\sigma_\alpha$ contains no $d\bar{\lambda}_\beta$ terms. Hence they trivialize the conormal bundle in $\Sigma^-\backslash 0$. However, they are *linear* in λ and so in the quotient space they trivialize $H \otimes N^*$ where H is the (positive) Hopf bundle and N^* the conormal bundle, thus

$$N \cong H \oplus H.$$

More invariantly,

$$N \cong H \otimes (\Sigma^+)_x,$$

since the indices of σ_α refer to a basis of Σ^+.

(2) The quaternionic structure $j\colon \Sigma^- \to \Sigma^-$ defines a free involution τ on P (note that $j^2 = -1$ which is a scalar so $\tau^2 = id$). Since the Levi-Civita connection on Σ^- has holonomy group $SU(2)$, it preserves the structure and hence $j^*\bar{\theta}, j^*\bar{\sigma}$ are again $(1,0)$ forms. Thus τ is antiholomorphic, and clearly preserves the fibres. □

13.64 Examples. Take $M = \mathbb{R}^4$, with the flat Euclidean metric. Then $P \cong S^2 \times \mathbb{R}^4$ and in the notation of (13.46) the complex structure is defined by the 1-forms

$$\theta' = d\mu,$$
$$\sigma'_1 = \psi_1 \otimes (\mu\varphi_1 + \varphi_2),$$
$$\sigma'_2 = \psi_2 \otimes (\mu\varphi_1 + \varphi_2).$$

Now $\{\varphi_\alpha, \psi_\beta\}$ are $SU(2)$ bases, so using the quaternionic structure j, we have $j\psi_1 = \psi_2, j\varphi_1 = \varphi_2$. If we fix φ_1 and consider the complex structure on $\mathbb{R}^4$ defined by the map $\psi \to \mathrm{Re}(\psi \otimes \varphi_1)$, $\psi \in \Sigma^+ \cong \mathbb{C}^2$ then with respect to linear coordinates (z_1, z_2) on $\mathbb{C}^2$, the forms above become

$$\theta' = d\mu,$$
$$\sigma'_1 = \mu dz_1 - d\bar{z}_2,$$
$$\sigma'_2 = \mu dz_2 + d\bar{z}_1.$$

Clearly, this is not the product structure.

Hence
$$w_1 = \mu z_1 - \bar{z}_2,$$
$$w_2 = \mu z_2 + \bar{z}_1$$

are local holomorphic functions on P, defined where $\lambda_2 \neq 0$. Similarly

$$\tilde{w}_1 = z_1 - \frac{1}{\mu}\bar{z}_2,$$

$$\tilde{w}_2 = z_2 + \frac{1}{\mu}\bar{z}_1$$

are well defined on $\lambda_1 \neq 0$. Since $\mu\tilde{w}_\alpha = w_\alpha$ it follows that $P(\mathbb{R}^4)$ has the holomorphic structure of the complex vector bundle $H \oplus H$ over $\mathbb{C}P^1$.

The real structure τ is defined by the quaternionic structure on the fibre of P, hence

$$\tau(\mu, z_1, z_2) = \left(-\frac{1}{\bar{\mu}}, z_1, z_2\right),$$

and
$$\tau(\mu, w_1, w_2) = \left(-\frac{1}{\bar{\mu}}, -\frac{\bar{w}_2}{\bar{\mu}}, \frac{\bar{w}_1}{\bar{\mu}}\right)$$

is the antiholomorphic involution.

The fibres of the projection are the curves $z_1 = c_1, z_2 = c_2$ i.e.

$$w_1 = \mu c_1 - \bar{c}_2,$$
$$w_2 = \mu c_2 + \bar{c}_1,$$

and these are precisely the holomorphic sections of $H \oplus H$ left fixed by τ.

13.65. Take $M = S^4$. We may consider the constant curvature metric on S^4 as the metric of the symmetric space $\mathbb{H}P^1 = Sp(2)/Sp(1) \times Sp(1)$, the quaternion projective line. The bundle Σ^- is then a homogeneous vector bundle obtained from the defining representation of one of the $Sp(1)$ factors—it is the quaternionic Hopf bundle.

If we consider a 2-dimensional quaternionic vector space $\mathbb{H}^2$ as a complex vector space $\mathbb{C}^4$ with a j-operator, then a point in the Hopf bundle consists of an arbitrary non-zero vector. Hence the projective bundle P is $\mathbb{C}P^3$.

The fibres of P are the projective lines in $\mathbb{C}P^3$ which are invariant under the quaternionic structure j, corresponding to the fact that a point of $\mathbb{H}P^1$ represents a 2-dimensional subspace of $\mathbb{C}^4$ invariant under j.

Since the complement of a point in S^4 is conformally equivalent to $\mathbb{R}^4$, then the complement of a line in $\mathbb{C}P^3$ should, by Example 13.64, be holomorphically equivalent to $H \oplus H$. In fact projection onto a line disjoint from the one removed defines such an equivalence.

13.66. Take $M = T^2$, a flat torus. Then T^4 is isometric to the quotient of $\mathbb{R}^4$, by the action of a lattice. Hence $P(T^4)$ is the quotient of $H \oplus H$ by a corresponding lattice action. It is a compact complex manifold, holomorphically fibred over $\mathbb{C}P^1$ by complex tori, and has been considered by Blanchard [Bla.]

13.67. Take $M = S^3 \times S^1$. This is the quotient of $\mathbb{R}^4 \backslash 0$ by the conformal action of $\mathbb{Z}$ generated by $x \to \lambda x$ $(\lambda \in \mathbb{R}\backslash\{0, \pm 1\})$. The complex manifold $P(S^3 \times S^1)$ is a compact complex manifold diffeomorphic to $S^1 \times S^2 \times S^3$. It is fibred holomorphically over $\mathbb{C}P^1$ by Hopf surfaces. Neither of the last two examples is Kählerian, and in fact it may be shown that only S^4 and $\mathbb{C}P^2$ have compact Kählerian twistor spaces [Hit 4]. This provides an alternative proof of Theorem (13.30) (see [Fr-Ku]).

E. The Reverse Penrose Construction

13.68. The power of the Penrose approach is that it gives a method of constructing half flat structures, and subsequently half conformally flat Einstein manifolds. This construction is provided by the following (converse to Theorem 13.46).

13.69 Theorem (R. Penrose [Pen]). *Let Q be a complex 3-manifold such that*

(1) Q is fibred by projective lines whose normal bundles are isomorphic to $H \oplus H$.

(2) Q possesses a free antiholomorphic involution τ which transforms each fibre to itself. Then Q is holomorphically equivalent to $P(M)$ for some self-dual manifold M.

Proof. Condition (1) defines the manifold M—the base space of the fibration. Since we know that the complex structure on $P(M)$ is defined by the conformal structure on M and not a particular choice of metric, the theorem will yield naturally a conformal structure.

13.70. Since we begin with holomorphic data we must use holomorphic methods. We shall use the extension theory of Griffiths [Gri] to produce a conformal structure on M, defined as a second-order G-structure (see [Kob 4]).

First let $Y \subset Q$ be one of the projective lines in the theorem, a fibre of $p: Q \to M$, and let $Z = P(\mathbb{R}^4) = H \oplus H$ be the twistor space of $\mathbb{R}^4$. Choose an isomorphism onto the zero section of Z

$$f: Y \to Z.$$

The obstruction to extending this to the first order is an element of the sheaf cohomology group $H^1(Y, f^*\underline{TZ} \otimes \underline{N}^*)$. However, on the zero section $\mathbb{C}P^1 \subset Z$

$$TZ \cong H^2 \oplus H \oplus H, \tag{13.71}$$

and by hypothesis

$$N \cong H \oplus H. \tag{13.72}$$

Since $H^1(\mathbb{C}P^1, \underline{H}^k) = 0$ for $k \geqslant -1$ [Wel], it follows that $H^1(Y, f^*\underline{TZ} \otimes N^*) = 0$, and so f may be extended to a map $f^{(1)}$ of the first order neighbourhood of $Y \subset Q$.

The extension $f^{(1)}$ has a derivative

$$f' \in H^0(Y, \mathrm{Hom}(\underline{TQ}, f^*\underline{TZ})). \tag{13.73}$$

Let $\pi: H \oplus H \to \mathbb{C}P^1$ denote the projection. Then $\pi f: Y \to \mathbb{C}P^1$ is an isomorphism so $\mathrm{Ker}(\pi f)' = E \subset TQ|_y$ is a rank 2 holomorphic subbundle, transverse to TY and hence isomorphic to the normal bundle N.

Since the difference of any two extensions lies in $H^0(Y, f^*\underline{TZ} \otimes \underline{N}^*)$ it is clear from (13.71) and (13.72) that there are enough extensions to make f' an isomorphism. We choose such an extension.

13.74. The obstruction to extending $f^{(1)}$ to the second order is an element of $H^1(Y, f^*\underline{TZ} \otimes S^2 N^*)$ (see [Gri]). This again vanishes so we have an extension $f^{(2)}$. Now since $(\pi f)'$ vanishes in the directions $E \subset TQ$, there is a well-defined second derivative map

$$(\pi f)'' \in H^0(Y, \mathrm{Hom}(S^2\underline{E}, (\pi f)^*\underline{T}\mathbb{C}P^1)).$$

The difference of two extensions lies in

$$H^0(Y, f^*\underline{TZ} \otimes S^2\underline{N}^*) \quad \text{but now}$$

$$\pi: H^0(Y, f^*\underline{TZ} \otimes S^2N^*) \to H^0(Y, (\pi f)^*\underline{T}\mathbb{C}P^1 \otimes S^2\underline{N}^*)$$

is an isomorphism since $H^0(\mathbb{C}P^1, \underline{H}^{-1}) = 0$. We may therefore choose the second extension so that $(\pi f)''$ vanishes.

13.75. The real structure τ on Q and the equivalent one on $P(\mathbb{R}^4)$ (see 13.64) make all the cohomology groups considered real vector spaces if the initial isomorphism f was compatible with τ. We may therefore choose a real extension to the second order.

13.76. By the conformal invariance of the twistor space construction (13.57) the group of orientation preserving conformal transformations of S^4, namely $SO(5,1)$, acts as holomorphic transformations of $\mathbb{C}P^3$ preserving the real structure τ. But the 15-dimensional group $SO(5,1)$ is a real form of $SL(4,\mathbb{C})$, hence any biholomorphic transformation of $\mathbb{C}P^3$ compatible with τ is obtained this way.

13.77. Consider now the ambiguity in the choice of the extension $f^{(1)}$. We have

$$\dim H^0(Y, f^*\underline{T}Z \otimes \underline{N}^*) = 8$$

from (13.71) and (13.72).

However any two extensions which are isomorphisms are related by an isomorphism of the first order neighbourhood of the zero section $f(Y) \subset Z$ to itself, inducing the identity on $f(Y)$. Also, the subgroup of $SL(4,\mathbb{C})$ leaving fixed every point on a line in $\mathbb{C}P^3$ is a connected 8-dimensional group. Since Z is an open set in $\mathbb{C}P^3$ with the zero-section going to a line, we deduce from 13.76 that any two real extensions of f to the first order are equivalent by an element of the conformal group acting on $P(\mathbb{R}^4)$. This element leaves fixed every point on the line. Different choices of initial real isomorphism f differ by an element of $SU(2)$ acting on the line so finally we obtain a holomorphic isomorphism to the second order along Y:

$$f^{(2)}: Q \to P(\mathbb{R}^4)$$

well-defined modulo the action of

$$\mathscr{L}_0 \subseteq SO(5,1) \quad \text{on } P(\mathbb{R}^4) \text{ where } \mathscr{L}_0$$

is the subgroup leaving fixed a point on S^4 (or equivalently leaving stable a line in $\mathbb{C}P^3$).

13.78. Finally consider the map

$$\mathbb{R}^4 \xrightarrow{(id,\, pt)} \mathbb{R}^4 \times S^2 \cong P(\mathbb{R}^4) \xrightarrow{(f^{(2)})^{-1}} Q \xrightarrow{p} M.$$

This is a 2-frame at $p(Y) = x \in M$, well-defined modulo $\mathscr{L}_0$. As x varies it defines a conformal structure [Kob 4] and by construction we have an isomorphism $Q \cong P(M)$.

The distinguished 2-frames in the G-structure approach to conformal structures are the 2-frames of geodesic coordinates for all the metrics in the conformal equivalence class. The complex frame θ', σ_1', σ_2' on $P(M)$ is equivalent in geodesic coordinates to the flat frame on $P(\mathbb{R}^4)$. Hence

$$\theta'_M = f^{(2)*}\theta'_{\mathbb{R}^4},$$

$$\sigma'_{\beta,M} = f^{(2)*}\sigma'_{\beta,\mathbb{R}^4},$$

but these are of type $(1,0)$ with respect to the given complex structure on Q, which is by hypothesis integrable, so from Theorem 13.46 Q is the twistor space of a half-conformally flat structure on M.

13.79 Remark. There is a more concrete method of obtaining the conformal structure in practice. The above theorem guarantees that it is half flat and inverts the construction of 13.46.

We begin with a theorem of Kodaira [Kod 2]. If X is a complex manifold and $Y \subset X$ a compact complex submanifold such that $H^1(Y, \underline{N}) = 0$, then Y belongs to a locally complete family parametrized by a complex manifold Z. Furthermore there is a natural isomorphism $T_z \cong H^0(Y_z, \underline{N})$, where N denotes the normal bundle of the manifold Y_z of the family.

Take $Y = \pi^{-1}(x)$ and $X = Q$ then since $N \cong H \oplus H$ and $H^1(\mathbb{C}P^1, \underline{H} \oplus \underline{H}) = 0$, Kodaira's theorem holds. Moreover $\dim H^0(\mathbb{C}P^1, \underline{H} \oplus \underline{H}) = 4$, so Z is a complex 4-manifold.

Since M is connected, we have an inclusion $M \subset Z$ and we have constructed a complexification of M. In fact if Y_z is a curve of the family sufficiently close to Y_x, $(x \in M)$, then by local completeness $\tau(Y_z)$ also belongs to the family. Thus a neighbourhood of M admits an anti-holomorphic involution of which M is the fixed point set, since by hypothesis τ takes Y_x to itself. Thus M is a real form of the complex manifold of all curves of the family generated by the fibres of π.

Now consider the isomorphism

$$T_z \cong H^0(Y_z, \underline{N}),$$

and define

$$C = \{e \in H^0(Y_z, \underline{N}) | e(y) = 0 \text{ for some } y \in Y_z\}.$$

Since $N \cong H \oplus H$, a holomorphic section e is given by a pair of linear forms $(a\mu + b, c\mu + d)$ and this vanishes somewhere iff $ad - bc = 0$. Thus C defines a quadratic form on T_z, up to a scalar multiple—a complex conformal structure. Over $x \in M$ we obtain a real conformal structure. To see that this is the same structure as is produced by the theorem, it is enough to check on $\mathbb{R}^4$ (use normal coordinates).

On $\mathbb{R}^4$, from (13.64) a curve invariant by τ is given by

$$w_1 = \mu c_1 - \bar{c}_2,$$

$$w_2 = \mu c_2 + \bar{c}_1.$$

A tangent vector is then given by the section of $H \oplus H$, $e = (\mu c_1' - \bar{c}_2', \mu c_2' + \bar{c}_1')$. So the conformal structure is defined by the quadratic form $c_1'\bar{c}_1' + c_2'\bar{c}_2'$ which is the Euclidean metric.

13.80 Example. Let Q be the flag manifold $F_3 = SU(3)/T$. This is the space consisting of pairs (m, l) where $m \in \mathbb{C}P^2$, l is a line on $\mathbb{C}P^2$ and $m \in l$. Equivalently it is the projective holomorphic tangent bundle of $\mathbb{C}P^2$.

Define a map $\pi: F_3 \to \mathbb{C}P^2$ by

$$\pi(m, l) = l \cap m^{\perp}.$$

Here we use a hermitian inner product on $\mathbb{C}^3$ to define the 2-dimensional orthogonal complement of the 1-dimensional subspace defined by m. The corresponding projective line is denoted by $m^\perp$.

Define $\tau: F_3 \to F_3$ by

$$\tau(m, l) = (l^\perp, m^\perp).$$

Then τ is antiholomorphic, $\tau^2 = id$ and τ has no fixed points since $m \notin m^\perp$. Furthermore π clearly commutes with τ.

Although π is not holomorphic the fibre $\pi^{-1}(x)$ is a holomorphic curve. Indeed

$$\pi^{-1}(x) = \{(m, l) | x \in l \text{ and } m \in x^\perp\}.$$

That is we consider all lines through x (a copy of $\mathbb{C}P^1$) and a distinguished point on each given by its intersection with $x^\perp$. This defines a holomorphic projective line in F_3, invariant under τ.

To calculate the conformal structure on $\mathbb{C}P^2$ defined by this complex manifold, we first consider the full holomorphic family of curves Z. In this case we take

$$Z = \{(z, w) \in \mathbb{C}P^2 \times \mathbb{C}P^{2^*} | z \notin w\},$$

where $\mathbb{C}P^{2^*}$ is the dual projective space of lines. The real structure induced on Z is

$$\tau(z, w) = (w^\perp, z^\perp).$$

In homogeneous coordinates, a point of Z is a pair $(z, w) \in \mathbb{C}^3 \times \mathbb{C}^3$ and $\tau(z, w) = (\bar{w}, \bar{z})$.

Now if a section of the normal bundle of a curve Y_z vanishes at some point, the corresponding tangent vector in T_z is tangent to a one parameter family of curves passing through this point. The condition for the curves defined by (z, w) and $(\tilde{z}, \tilde{w}) \in Z$ to intersect is that the intersection of the lines w and $\tilde{w}$ should lie on the line joining z and $\tilde{z}$ i.e.

$$\det \begin{pmatrix} z_0 & z_1 & z_2 \\ \tilde{z}_0 & \tilde{z}_1 & \tilde{z}_2 \\ w_1\tilde{w}_2 - w_2\tilde{w}_1 & w_2\tilde{w}_0 - w_0\tilde{w}_2 & w_0\tilde{w}_1 - w_1\tilde{w}_0 \end{pmatrix} = 0,$$

hence the infinitesimal condition, in affine coordinates is

$$\det \begin{pmatrix} 1 & z_1 & z_2 \\ 0 & dz_1 & dz_2 \\ w_1 dw_2 - w_2 dw_1 & -dw_2 & dw_1 \end{pmatrix} = 0,$$

i.e.

$$dz_1 dw_1 + dz_2 dw_2 + (z_1 dz_2 - z_2 dz_1)(w_1 dw_2 - w_2 dw_1) = 0.$$

On the real part $\mathbb{C}P^2$ of Z, $w = \bar{z}$, so the conformal structure is defined by the metric

$$g = dz_1 d\bar{z}_1 + dz_2 d\bar{z}_2 + (z_1 dz_2 - z_2 dz_1)(\bar{z}_1 d\bar{z}_2 - \bar{z}_2 d\bar{z}_1),$$

and this is conformally equivalent to the Fubini-Study metric.

F. Application to the Construction of Half-Conformally Flat Einstein Manifolds

Let us turn now to the problem of finding an Einstein metric within a given half flat conformal equivalence class. Recall that we have a well defined element

$$\theta = \lambda_2\theta_1 - \lambda_1\theta_2 \in C^\infty(K^{-1/2} \otimes \Lambda^{1,0})$$

on the twistor space P, depending on the metric.

13.81 Theorem. *Let (M, g) be a Riemannian 4-manifold whose conformal structure is half flat. Then*

(1) *g is an Einstein metric iff $\theta \in C^\infty(P(M), K^{-1/2} \otimes \Lambda^{1,0})$ is holomorphic.*

(2) *If θ is holomorphic then the scalar curvature s of the metric is defined by*

$$s = \theta \wedge d\theta \in H^0(P, K^{-1} \otimes K) = H^0(P, \mathbb{C}) = \mathbb{C}.$$

Proof. A geodesic frame at $x \in M$ defines a trivialization of $K^{-1/2}$ for $\lambda_2 \neq 0$, which by the construction (13.46) is holomorphic along the fibre P_x. Relative to this trivialization we can represent θ by the (1, 0) form θ' of (13.44). Now

$$d\theta' = \omega(\Phi, R^-(\Phi))/\lambda_2^2$$

so $d''\theta' = 0$ if $\omega(\Phi, R^-(\Phi))^{(1,1)} = 0$.

The scalar part of R^- contributes $Cs\Phi \otimes \Phi$ to $d\theta'$ (see (13.51)) but this is a (2, 0) form and hence contributes nothing to $d''\theta'$. Since $W^- = 0$, the only contribution is the Λ^+ part of R^-, i.e. the component B. Now every such form is of type (1, 1) (J acts trivially on it) and hence as (λ_1, λ_2) varies, $d''\theta' = 0$ iff $B = 0$. Thus $d''\theta = d''(\lambda_2^4\theta') = 0$ iff g is Einstein.

If g is Einstein, then

$$\begin{aligned} d\theta' &= Cs\Phi \otimes \Phi/\lambda_2^2 \quad \text{at } x,\\ &= Cs\sigma_1' \wedge \sigma_2'. \end{aligned}$$

Hence

$$\begin{aligned} \theta \wedge d\theta &= \lambda_2^8\theta' \wedge d\theta',\\ &= Cs\lambda_2^8\theta' \wedge \sigma_1' \wedge \sigma_2',\\ &= Cs\theta \wedge \sigma_1 \wedge \sigma_2,\\ &= Cs, \end{aligned}$$

where we have used the isomorphism between $\Sigma^-\backslash 0$ and $K^{1/4}\backslash 0$ defined by $\theta \wedge \sigma_1 \wedge \sigma_2$ (13.54). The constant C can be evaluated by considering the 4-sphere. □

There is a converse to the theorem which enables us to construct Einstein metrics:

13.82 Theorem. *Let Q be a complex manifold as in Theorem 13.69 and let φ be a holomorphic section of $K^{-1/2} \otimes \Lambda^{1,0}$. Suppose*

(1) *on each fibre of* Q, φ *restricts to a non-zero form.*

(2) $\tau^*\varphi = \varphi$. *Then, using the isomorphism* $Q \cong P(M)$ *of Theorem* 13.69, *there is a unique Einstein metric on* M *within the conformal equivalence class, such that* $\theta = \varphi$.

Proof. First note that on each fibre P_x, $N \cong H \oplus H$, and $TP_x \cong H^2$, so that $K_Q \cong H^{-4}$. Hence $K_Q^{-1/2} \otimes \Lambda^{1,0}P_x$ is a trivial holomorphic line bundle, so from condition (1) if φ is non-zero at some point of the fibre, it is non-zero everywhere. If we choose any metric within the conformal equivalence class, then the form θ defines for each fibre P_x a trivialization of $K^{-1/2} \otimes \Lambda^{1,0}P_x$ holomorphic in the direction of P_x, but not necessarily in other directions. Thus

$$\varphi = e^f\theta + b_1\sigma_1 + b_2\sigma_2$$

for some function $f \in C^\infty(M)$.
From 13.57 there exists a conformally equivalent metric such that

$$\varphi = \theta + a_1\sigma_1 + a_2\sigma_2,$$

where a_1 and a_2 are local C^∞ sections of $K^{-1/4}$.

Now since by hypothesis $d''\varphi = 0$, we have

$$0 = \omega(\varphi, R^-(\varphi))^{(1,1)} + d''(\textstyle\sum a_\alpha\sigma_\alpha).$$

Restricting the forms to a fibre P_x, the curvature term vanishes since it is horizontal and $d''\sigma_\alpha = 0$ from (13.52). Thus $d''a_\alpha = 0$ and so a_α defines a holomorphic section of $K^{-1/4} \cong H$ on $P_x \cong \mathbb{C}P^1$. Hence each a_α is linear in (λ_1, λ_2) and we may write

$$\varphi = \theta + \textstyle\sum a_{\alpha\beta}\lambda_\alpha\sigma_\beta,$$

where $a_{\alpha\beta} \in C^\infty(M, \Sigma^+ \otimes \Sigma^-)$ is a 1-form. Now $d''\theta$ and $d''a_{\alpha\beta}$ are horizontal forms. Moreover in a geodesic frame $d''\lambda_\alpha = 0$ at x. However from (13.51), in affine coordinates $d''\sigma_\beta'$ has a $d\mu$ factor. Thus from the equation $d''\varphi = 0$ we obtain

$$\sum_\beta (a_{1\beta}\mu + a_{2\beta})(\psi_\beta \otimes \varphi_1)^{0,1} = 0.$$

However, from the definition of the almost complex structure (13.49), $(\psi \otimes \varphi_1)^{0,1} = 0$ only if $\mu = \infty$ or $\psi = 0$. Thus

$$\textstyle\sum (a_{1\beta}\mu + a_{2\beta})\psi_\beta = 0 \quad \forall \mu,$$

and hence $a_{\alpha\beta} = 0$.

Consequently $\varphi = \theta$ and the theorem is proved. For an alternative treatment of this result see [Leb]. □

Examples. 13.83. Let $Q = H \oplus H$, the twistor space of $\mathbb{R}^4$. Then $\theta = \lambda_2 d\lambda_1 - \lambda_1 d\lambda_2$ defines the flat Euclidean metric. Since $K_Q \cong p^*(H^{-4})$, $\theta \in H^0(Q, \underline{K}^{-1/2} \otimes \underline{\Lambda}^{1,0})$ is the pull back of the canonical trivialization of

$$H^2 \otimes K \text{ on } \mathbb{C}P^1.$$

13.84. Let $Q = \mathbb{C}P^3$ with homogeneous coordinates (z_0, z_1, z_2, z_3) and involution $(z_0, z_1, z_2, z_3) = (\bar{z}_1, -\bar{z}_0, \bar{z}_3, -\bar{z}_2)$.

Take $\theta = z_0 dz_1 - z_1 dz_0 + z_2 dz_3 - z_3 dz_2$. Since $K \cong H^{-4}$, H being the positive Hopf bundle, θ is a well-defined section of $K^{-1/2} \otimes \Lambda^{1,0}$.

A quaternionic subspace of $\mathbb{C}^4$ is defined by

$$\begin{pmatrix} z_2 \\ z_3 \end{pmatrix} = \begin{pmatrix} a & b \\ -\bar{b} & \bar{a} \end{pmatrix} \begin{pmatrix} z_0 \\ z_1 \end{pmatrix},$$

and on this subspace

$$\theta = (1 + |a|^2 + |b|^2)(z_0 dz_1 - z_1 dz_0)$$

which is non-vanishing.

This describes the metric of constant positive curvature on S^4.

If we take

$$\theta = z_0 dz_1 - z_1 dz_0 - z_2 dz_3 + z_3 dz_2,$$
$$\theta = (1 - |a|^2 - |b|^2)(z_0 dz_1 - z_1 dz_0).$$

Thus θ defines a metric on the ball $|a|^2 + |b|^2 < 1$. This is the hyperbolic metric.

13.85. Let $Q = F_3$, the twistor space of $\mathbb{C}P^2$. Then $F_3 \cong \mathbb{P}(T^*\mathbb{C}P^2)$ and on any projective cotangent bundle there is a canonical contact form. This defines the Fubini-Study metric on $\mathbb{C}P^2$.

13.86. From Theorem 13.81, $\theta \wedge d\theta = 0$ iff the Ricci tensor vanishes. But this is the Frobenius integrability condition for the distribution of holomorphic 2-planes defined by θ. This corresponds to the fact that $W^- = 0$ and $r = 0$ implies that Σ^- is flat, and the integral submanifolds of the connection form are the complex surfaces defined by θ.

If M is simply connected then the connection trivializes the bundle $\mathbb{P}(\Sigma^-)$ and so differentiably $P \cong M \times S^2$. The holomorphic structure of P is such that the projection $\pi: P \to S^2$ is holomorphic and the fibres (the leaves of the distribution defined by θ) are complex surfaces diffeomorphic to M. In this case, the form θ defines a non-vanishing holomorphic section of $K_Q^{-1/2} \otimes \pi^* K$ or equivalently an isomorphism $K_Q \cong \pi^*(H^{-4})$.

As we have seen the conditions $W^- = 0$, $r = 0$ reduce the holonomy group to $SU(2)$ and we have a Kähler manifold with zero Ricci tensor. The theorems above show that any such metric may be obtained from a complex 3-manifold fibring over $\mathbb{C}P^1$, with a distinguished family of holomorphic sections, each with normal bundle $H \oplus H$, an isomorphism $K = \pi^*(H^{-4})$ and an antiholomorphic involution τ.

The only non-trivial compact such manifold is covered by a K3 surface. Thus, the existence of the Calabi/Yau metric implies the existence of a compact complex 3-manifold P with the properties outlined above. It has yet to be described from any other viewpoint, but it must have the following properties:

(i) P is diffeomorphic to a product of a K3 surface and S^2.

(ii) The Chern numbers are given by

$$c_3 = 48, \qquad c_1 c_2 = 96, \qquad c_1^3 = 0.$$

(iii) P has Kodaira dimension $-\infty$.
(iv) P has algebraic dimension 1.

13.87. As a final non-trivial example of this construction, we shall produce a family of complete Kähler metrics with zero Ricci tensor on $\mathbb{C}^2$, by deforming the flat metric. Since we are aiming to produce half conformally flat metrics with zero scalar curvature we must seek a complex 3-manifold fibring over $\mathbb{C}P^1$, whose fibres are $\mathbb{C}^2$. For flat space this is $P = H \oplus H$. Another description of this twistor space is

$$\tilde{P} = \{(x, y, z) \in H \oplus H \oplus H^2 | xy = z\}.$$

The map $x = z_1$, $y = z_2$, $z = z_1 z_2$ defines the isomorphism $\tilde{P} \cong P$.
Take the standard open sets

$$U_0 = \{(z_0, z_1) \in \mathbb{C}P_1 | z_0 \neq 0\},$$
$$U_1 = \{(z_0, z_1) \in \mathbb{C}P_1 | z_1 \neq 0\},$$

and let $u = z_1/z_0$ be an affine coordinate on U_0.

The total space of the bundle H^2 is covered by two corresponding coordinate neighbourhoods. Indeed, since $H^2 \cong T\mathbb{C}P^1$, the map

$$(z, u) \to z\frac{d}{du}$$

gives a holomorphic coordinate system on the set $\tilde{U}_0$ covering U_0.

Consider the line bundle L over H^2 defined by the transition function $\varphi = e^{cz/u}$ on $\tilde{U}_0 \cap \tilde{U}_1$, and define

$$Q = \{(x, y) \in (L \oplus L^*) \otimes \pi^* H | xy = z\}.$$

Here z is the tautological section of $\pi^* H^2$ over H^2 and $\pi\colon H^2 \to \mathbb{C}P^1$ the projection. Then Q is a complex 3-manifold fibring over $\mathbb{C}P^1$ with each fibre holomorphically equivalent to $\mathbb{C}^2$. It is in fact a locally trivial $\mathbb{C}^2$ bundle over $\mathbb{C}P^1$, with a distinguished section $x = y = z = 0$ but which is *not* a vector bundle. Furthermore the form $dx \wedge dy$ defines an isomorphism $\bigwedge^2 T^*F \cong \pi^* H^{-2}$ or equivalently $K \cong \pi^* H^{-4}$.

Now H^2 has an antiholomorphic involution

$$\tau(z, u) = (\bar{z}/\bar{u}^2, -1/\bar{u}),$$

and if c is real then τ induces an antiholomorphic map

$$\tau\colon L \to L^*.$$

The bundle H has an antiholomorphic quaternionic map $\sigma\colon H \to H (\sigma^2 = -1)$ and we may assemble this data into an involution

$$\tau(x, y) = (-\tau\sigma y, \tau\sigma x)$$

with no fixed points.

To find the sections, first take $z(u) = au^2 + 2bu - \bar{a}$ where $b = \bar{b}$. This is a section of H^2 invariant by τ. Let α, β be the roots of $z(u) = 0$ and set

$$x(u) = Ae^{c(au+b)}(u-\alpha),$$
$$y(u) = Be^{-c(au+b)}(u-\beta),$$

relative to the local trivialization of L on $\tilde{U}_0$ for which $e^{cz/u}$ is the transition function. Then over $\tilde{U}_1$,

$$x(u) = A\exp\left(c(au+b) - c\left(\frac{au^2+2bu-\bar{a}}{u}\right)\right)(u-\alpha)\frac{1}{u},$$
$$= A\exp\left(-c\left(b-\frac{\bar{a}}{u}\right)\right)\left(1-\frac{\alpha}{u}\right),$$

which is regular at $u = \infty$.

Since $\alpha, \beta = \dfrac{-b \pm \sqrt{b^2+|a|^2}}{a}$ we can unambiguously set $\alpha = \dfrac{-b+\sqrt{b^2+|a|^2}}{a}$ and $\beta = \dfrac{-b-\sqrt{b^2+|a|^2}}{a}$.

Since $AB = a$ if $xy = z$, the invariance under τ of the curve defined by x and y implies that

$$A\bar{A} = b + \sqrt{b^2+|a|^2}.$$

Thus the curves we have found are parametrized by a point $\underline{r} = (b, \Im\mathrm{m}\, a, \Re\mathrm{e}\, a) \in \mathbb{R}^3$ and an angular variable $\eta = \arg A$. Since $H^1(\mathbb{C}P^1, \underline{\mathrm{End}\, N}) = 0$. It actually follows from Kodaira's deformation theorem on vector bundles that these new curves for small enough c being deformations of the sections of $H \oplus H$ have the same normal bundle. We may now compute the metric.

The tangent (A', a', b') to a deformation defines a section of the normal bundle of a curve by means of the equations.

$$z'(u) = a'u^2 + 2b'u - \bar{a}',$$
$$\frac{x'(u)}{x(u)} = \frac{A'}{A} + c(a'u+b') - \frac{\alpha'}{u-\alpha},$$
$$= \frac{A'}{A} + c(a'u+b') - \frac{(\alpha a'u + \bar{a}')}{2ur},$$

where $r = \sqrt{b^2+|a|^2}$

The conformal structure is defined by the condition that $z'(u) = 0$ and $x'(u) = 0$ should have a common root. This becomes

$$\left(cr - \frac{1}{2}\right)^2 (a'\bar{a}' + b'^2) = \left(\frac{\alpha a'}{2} + \frac{b'}{2} - \frac{A'r}{A}\right)^2.$$

However, since $A\bar{A} = b + r = -\alpha a$, this last term is the square of an imaginary number, and the positive definite conformal structure is defined by

$$\left(cr - \frac{1}{2}\right)^2 (a'\bar{a}' + b'^2) + \left(\mathrm{Im}\left(\frac{A'r}{A} - \frac{\alpha a'}{2}\right)\right)^2. \tag{13.88}$$

At $u = 0$, the local coordinates y, z on the fibre are given by

$$z = z(0) = -\bar{a}$$

$$y = y(0) = Be^{-cb}\left(\frac{b+r}{a}\right) = \bar{A}e^{-cb}.$$

Hence

$$\frac{A'r}{A} - \frac{\alpha a'}{2} = \frac{\bar{y}'r}{\bar{y}} + cb'r + \frac{\alpha \bar{z}'}{2},$$

and we know that

$$Re\left(\frac{A'r}{A} - \frac{\alpha a'}{2}\right) = \frac{b'}{2}.$$

Thus

$$\left\| \frac{\bar{y}'r}{\bar{y}} + \frac{\alpha \bar{z}'}{2} \right\|^2 = \left(cr - \frac{1}{2}\right)^2 b'^2 + \left(\mathrm{Im}\left(\frac{A'r}{A} - \frac{\alpha a'}{2}\right)\right)^2.$$

To fix the metric within the conformal class the length of the holomorphic 2-form on the fibre, $dx \wedge dy = dz \wedge dy/y$ must be 1. If we take $c = -1/m$, then the metric is given in the form (13.88) by

$$\left(1 + \frac{2m}{r}\right) d\underline{r} \cdot d\underline{r} + \left(1 + \frac{2m}{r}\right)^{-1} \left(d\tau + \frac{x_2\, dx_3 - x_3\, dx_2}{r(r - x_1)}\right)^2,$$

which is the Taub-NUT metric [Eg-Gi-Ha] (compare 3.38 and Add.12).

In Kählerian form with respect to local coordinates on $\mathbb{C}^2$ given by $z = z_1 z_2$, $y = z_1$ it is

$$g = \gamma dz d\bar{z} + \gamma^{-1}\left(\frac{2dy}{y} + \frac{\bar{\delta} dz}{z}\right)\left(\frac{2d\bar{y}}{\bar{y}} + \frac{\delta d\bar{z}}{\bar{z}}\right),$$

where

$$\gamma = 1 + 2m/\sqrt{b^2 + |z|^2},$$

$$\delta = 1 - b/\sqrt{b^2 + |z|^2},$$

and b is determined implicitly by the equation

$$|y|^2 = (b + \sqrt{b^2 + |z|^2})e^{2b/m}.$$

(see [Hit 4] for a related construction).

Chapter 14. Quaternion-Kähler Manifolds

A. Introduction

14.1. In this chapter, we study $4n$-dimensional Riemannian manifolds with holonomy group contained in $Sp(n)$ or $Sp(n)Sp(1)$ (see Chapter 10). These two cases are in fact quite different, more different for example than $SU(n)$ from $U(n)$. More precisely, $Sp(n)$ is included in $SU(2n)$, so Riemannian manifolds with holonomy contained in $Sp(n)$ are particular cases of Kähler manifolds with zero Ricci curvature.

On the other hand, $Sp(n)Sp(1)$ is not a subgroup of $U(2n)$. In fact, this follows from E.B. Dynkin's classification of subgroups of semi-simple Lie groups [Dyn 2]

14.2 Proposition. *The subgroup $Sp(n)Sp(1)$ is a maximal subgroup of $SO(4n)$.*

(See for example A. Gray [Gra 7] for a direct proof.)

We study separately the two cases $Sp(n)$ and $Sp(n)Sp(1)$, even if they have some common features. To begin with, we give them different names.

14.3 Definition. A $4n$-dimensional Riemannian manifold is called

a) *hyperkählerian* if its holonomy group is contained in $Sp(n)$,

b) *locally hyperkählerian* if its restricted holonomy group is contained in $Sp(n)$.

The denomination "hyperkählerian" is due to E. Calabi, who gave the first complete non trivial examples in dimensions $4n \geqslant 8$ ([Cal 5]).

We study hyperkählerian manifolds in § B, as an application of the theory of Kähler manifolds with vanishing Ricci curvature. Notice that non trivial compact examples (for $n \geqslant 2$) appeared only recently (A. Fujiki [Fjk], S. Mukai [Muk], A. Beauville [Bea 2]). We give some indications on the construction of those compact examples in § C, and we postpone the complete non compact Calabi examples until Chapter 15.

14.4 Definition. A $4n$-dimensional Riemannian manifold is called

a) *quaternion-Kähler* if its holonomy group is contained in $Sp(n)Sp(1)$,

b) *locally quaternion-Kähler* if its restricted holonomy group is contained in $Sp(n)Sp(1)$.

The main result, which justifies our interest for these manifolds in this book, is the following

14.5 Theorem (M. Berger [Ber 7], see 14.39). *If $n \geqslant 2$, a quaternion-Kähler manifold is Einstein.*

This will be proved in § D, where we show also that the theory of quaternion-Kähler manifold splits naturally in two different parts, because of the

14.6 Theorem (M. Berger [Ber 1], see 14.45). a) *A quaternion-Kähler manifold with vanishing Ricci tensor is locally hyperkählerian.*

b) *A quaternion-Kähler manifold with non-vanishing Ricci tensor is (even locally) de Rham irreducible (if $n \geqslant 2$).*

Up to now, the theory of non-hyperkählerian quaternion-Kähler manifolds seems far from complete. In particular, there are very few *complete* examples, and all the known examples are homogeneous. Among those, there are some nonsymmetric ones, due to D.V. Alekseevskii [Ale 3, 4]; we indicate their construction in § I. In the *compact* case, the only known examples are symmetric; they were classified by J.A. Wolf [Wol 2], see § E.

14.7. In § F, we introduce the definition of a *quaternionic manifold*; this kind of structure is a quaternionic equivalent for the notion of a complex manifold, which was intensely used in the study of manifolds with holonomy contained in $U(n)$. But the quaternionic theory is quite different from the complex one; in particular, a quaternion-Kähler manifold is not necessarily integrable as a quaternionic manifold.

14.8. We describe in § G a new tool for the study of these quaternionic manifolds, which was introduced recently independently by S. Salamon ([Sal 1, 2, 3, 4] and L. Bérard Bergery ([BéBer 7] and [BB-Oc]). This is a generalization of the celebrated "twistor space theory" of R. Penrose and it converts the study of quaternionic manifolds into problems on complex manifolds (of real dimension $4n + 2$).

In § H, we give some applications of this twistor space theory to the study of quaternion-Kähler manifolds. For example, S. Salamon gave a very nice characterization of the quaternionic projective space (see 14.89). A key result is

14.9 Theorem (S. Salamon [Sal 2], L. Bérard Bergery [BéBer 7], see 14.80). *Given a $4n$-dimensional ($n \geqslant 2$) compact quaternion-Kähler manifold (M, g) with positive scalar curvature, there exists on its twistor space a Kähler-Einstein metric such that the projection on M is a Riemannian submersion with totally geodesic fibres.*

Unfortunately, the theorem is not true with negative scalar curvature. And on the other hand, Kähler-Einstein manifolds with positive scalar curvature are not yet classified; this is the difficult case of the generalized Calabi conjectures, see 11.D. So only partial results may be deduced from 14.9 for quaternion-Kähler manifolds.

B. Hyperkählerian Manifolds

From the theory of holonomy groups, we deduce easily the following characterization of hyperkählerian manifolds in terms of invariants of the Riemannian metric.

14.10 Proposition. *A Riemannian manifold (M, g) is hyperkählerian if and only if there exist on M two complex structures I and J such that*

a) *I and J are* parallel (*i.e. g is a Kähler metric for each*)

b) $IJ = -JI$.

14.11. Notice that $K = IJ$ is still a parallel complex structure on M and more generally, given (x, y, z) in $\mathbb{R}^3$ with $x^2 + y^2 + z^2 = 1$, then the complex structure $xI + yJ + zK$ on M is still parallel. So there is a whole manifold (isomorphic to S^2) of parallel complex structures on M.

14.12. In the following, we *choose* once and for all one of these complex structures (say I) and we consider M as a complex manifold for I. Notice that there is no "natural" choice in general, but we have to make one if we want to apply complex geometry. (Another way to make this choice is to pick up some particular subgroup $SU(2n)$ of $SO(4n)$ containing $Sp(n)$). We recall that (M, g, I) is a Kähler manifold with holonomy group contained in particular in $SU(2n)$, so we deduce from 10.67

14.13 Theorem (M. Berger [Ber 1]). *A hyperkählerian manifold is Ricci-flat.*

As a corollary, if M is compact, the first (real) Chern class $c_1(M)$ vanishes (this is of course true also for $J, K, \ldots$).

In the following, we want to characterize hyperkählerian manifolds among Kähler manifolds. We introduce for that an extra structure.

14.14 Definition. A (*complex-*)*symplectic structure* on a complex manifold (M, I) is a *closed holomorphic* 2-*form* on M, which is *non degenerate* at each point of M.

An obvious example of a complex-symplectic structure is the constant 2-form $\sum_{j=1}^{n} dz^j \wedge dz^{j+n}$ on $\mathbb{C}^{2n}$. We will see other examples below (some compact ones in 14.25 and 14.28, and some non-compact ones in 14.33). Notice that, if M has a complex-symplectic structure ω, then M is $2n$-dimensional as a complex manifold (and $4n$-dimensional as a real manifold). Moreover, $\bigwedge^n \omega$ is a complex *volume form* on M (i.e., in local coordinates $(z^1, \ldots, z^{2n})$ on M, then $\bigwedge^n \omega = f(z)\, dz^1 \wedge \cdots \wedge dz^{2n}$, with an everywhere non-zero function f).

The following is an easy consequence of Proposition 14.10

14.15 Proposition. *Let (M, g) be a hyperkählerian manifold and I, J two anti-commuting parallel complex structures as in* 14.10. *Then the complex 2-form*

$$\omega(X, Y) = g(JX, Y) + ig(IJX, Y) \qquad \text{(14.15)}$$

is parallel and I-holomorphic, so ω is in particular a complex-symplectic structure on (M, I).

So any hyperkählerian manifold is a complex-symplectic manifold in a very precise way (once we have choosen some I). Now the converse is true, at least in the *compact* case.

14.16 Theorem (A. Beauville [Bea 2]). *Let* (M, I) *be a compact complex manifold of Kähler type admitting a complex-symplectic structure* ω.

Then, for any Kähler class α *in* $H^2(M, \mathbb{R})$, *there exists on M a unique Riemannian metric g such that*

a) *the metric g is hyperkählerian,*

b) *the complex structure I of M is g-parallel,*

c) *the Kähler class of* (g, I) *is precisely* α.

Proof. Since $\bigwedge^n \omega$ is a holomorphic form of type $(2n, 0)$ which never vanishes, the canonical line bundle of M is trivial, and its first real Chern class is zero. So we may apply Calabi-Yau Theorem 11.15; for any Kähler class α on M, there exists a unique Riemannian metric g which is Kähler for I, has vanishing Ricci tensor and Kähler class α. Then a) follows from the

14.17 Lemma (S. Bochner [Bo-Ya]). *Let* (M, g, I) *be a Kähler manifold with zero Ricci tensor. Then any holomorphic tensor field on M is g-parallel.*

Proof. For any holomorphic tensor field A on M, we have

$$\Delta(\|A\|^2) = \|DA\|^2 \qquad \text{(see 1.156)}$$

so by integrating on M, we get $DA = 0$. □

(For some modern references and generalizations, see [Mic] or [Kob 5].)

From the lemma, it follows that any complex-symplectic form on M is parallel. This implies in particular that the holonomy group is contained in $Sp(n)$. □

14.18. Notice that the Riemannian product of two hyperkählerian manifolds is obviously hyperkählerian. Conversely, we look at the de Rham decomposition of hyperkählerian manifolds. We recall that any Ricci-flat homogeneous Riemannian manifold is flat (see 7.61), so the symmetric factors in the decomposition are flat ones. A glance at the other possible groups in Berger's list gives the

14.19 Theorem. a) *Let* (M, g) *be a* complete *hyperkählerian manifold. Then its universal covering* $(\tilde{M}, \tilde{g})$ *is the Riemannian product of* $\mathbb{H}^k$ $(=\mathbb{R}^{4k})$ *with its flat metric and a finite number of complete simply-connected* $4r_i$*-dimensional hyperkählerian manifolds* M_i *with holonomy group* $Sp(r_i)$.

b) *If M is furthermore* compact, *then the factors* M_i *above are also compact; and there exists a finite covering* $(\hat{M}, \hat{g})$ *of* (M, g) *such that* $(\hat{M}, \hat{g})$ *is the Riemannian product of a flat hyperkählerian torus and the* M_i*'s as above.*

Notice that b) follows from the Cheeger-Gromoll Theorem 6.65.

It is possible also to characterize (de Rham-)irreducible compact hyperkählerian manifolds in terms of complex-symplectic structures.

14.20 Theorem (A. Beauville [Bea 2]). *Let (M, I) be a 4n-dimensional complex compact manifold of Kähler type. Then M admits a Kähler metric g whose holonomy group is exactly $Sp(n)$ if and only if M is simply connected and admits a unique complex-symplectic structure (up to a scaling factor).*

Proof. Since all the complex-symplectic structures are g-parallel in the proof of Theorem 14.16, the holonomy theorem shows that there is only a one-dimensional (complex) vector space of them if M is holonomy-irreducible. The only difficult point is to show that M is necessarily simply-connected. From 14.19b), we know that $\pi_1(M)$ is finite. Then the result follows by considering the universal (hyperkählerian) covering together with the following

14.21 Lemma. *Let (M, I) be a 4n-dimensional compact hyperkählerian manifold with holonomy $Sp(n)$. Then the spaces of (global) holomorphic forms on M are*

$$H^0(M, \Omega^p) = \begin{cases} 0 & \text{if } p \text{ is odd} \\ \mathbb{C}\omega^q & \text{if } p = 2q \text{ with } 0 \leqslant q \leqslant n, \end{cases}$$

where ω is the "unique" complex-symplectic structure.

In particular the Euler characteristic of the canonical sheaf $\chi(\mathcal{O}_M) = n + 1$ depends only on the dimension. It follows that (M, I) is simply connected.

Proof. From Lemma 14.17, all holomorphic forms are parallel. Since the holonomy is precisely $Sp(n)$, they may only be (up to a complex factor) multiples of the complex symplectic structure. Now

$$\chi(\mathcal{O}_M) = \sum_i (-1)^i \dim H^0(M, \Omega^i) = n + 1.$$

And the last result follows from the fact that the universal covering $\tilde{M}$ of M is still compact and hyperkählerian. But $\chi(\mathcal{O}_{\tilde{M}}) = n + 1 = \chi(\mathcal{O}_M)$ so the covering must be one-sheeted and $\tilde{M} = M$. □

C. Examples of Hyperkählerian Manifolds

14.22. In the particular case $n = 1$, then $Sp(1) = SU(2)$ in $SO(4)$, so a 4-dimensional Riemannian manifold is hyperkählerian exactly when it is Kähler and Ricci flat. In particular, this shows that any compact complex surface M of Kähler type with vanishing first Chern class is either a torus or simply connected and admits a unique complex-symplectic structure, i.e. is a so-called "K3-*surface*".

The so-called *Enriques surfaces* are $\mathbb{Z}_2$-quotients of some K3-surfaces. They are examples of locally hyperkählerian Kähler manifolds which are not hyperkählerian. Their restricted holonomy group is $Sp(1)$, but their holonomy group is $Sp(1) \rtimes_{\mathbb{Z}_2} \mathbb{Z}_4 (\subset U(2))$. (This follows for example from [Bou 2]).

14.23. If $n \geqslant 2$, then $Sp(n)$ is strictly smaller than $SU(2n)$. Now, in any dimension $4n$, there are examples of compact hyperkählerian manifolds. The first examples were given by A. Fujiki in dimension 8 ($n = 2$) [Fjk] and examples in all dimensions $4n$ (generalizing Fujiki's example) were given recently by S. Mukai [Muk] and A. Beauville [Bea 2]. We describe below Beauville's examples, and we refer to [Bea 2] for the corresponding proofs, that we will *not* give here.

14.24. Let S be a compact complex surface. We denote $S^{(r)}$ the r-th symmetric product of S (i.e. the quotient of S^r by the symmetric group $\mathfrak{S}_r$). The complex space $S^{(r)}$ is not a smooth manifold but, *because S is 2-dimensional*, its Douady space $S^{[r]}$ (or its Hilbert scheme in the algebraic case) is a (*smooth*) complex manifold, of complex dimension $2r$. In technical terms. "$S^{[r]}$ parametrizes the finite analytic subspaces Z of S with $lg(\mathbb{O}_Z) = r$ (here we consider $S^{(r)}$ as the variety of effective zero-cycles of degree r)". Then Beauville proves the following

14.25 Theorem ([Bea 2]). *If the canonical line bundle of S is trivial, then $S^{[r]}$ has a complex-symplectic structure.*

If S is a K3-surface, Beauville shows also that $S^{[r]}$ is simply connected, with $b_2 = 23$. Moreover, he states that $S^{[r]}$ is always Kählerian (i.e. admits a Kähler metric), according to recent results of J. Varouchas.

14.26 Theorem ([Bea 2]). *Let S be a* K3-*surface. Then the manifold $S^{[r]}$ is simply connected and admits hyperkählerian metrics with holonomy group exactly $Sp(r)$.*

14.27. Now let T be a complex 2-dimensional torus, and $r \geqslant 2$. Then if T is Kähler, the manifold $T^{[r+1]}$ admits hyperkählerian metrics, but it is not simply connected. We consider the map $u\colon T^{(r+1)} \to T$ defined by $u([p_0] + \cdots + [p_1]) = [\sum_{i=0}^{r} p_i]$. Composing with the projection $T^{[r+1]} \to T^{(r+1)}$, we get a map $v\colon T^{[r+1]} \to T$ which is furthermore T-equivariant for the action of T on $T^{[r+1]}$ by translation and on T by $(t, x) \mapsto x + (r+1)t$. So v is a complex fibre bundle. We *denote* K_r the fibre of v. Then Beauville proves

14.28 Proposition ([Bea 2]). *The symplectic structure of $T^{[r+1]}$ induces a symplectic structure on K_r.*

Furthermore, Beauville shows that K_r is simply-connected with $b_2 = 7$ (if $r \geqslant 2$), and that K_r and $S^{[r]}$ are not even bimeromorphically equivalent. Notice that K_1 is the so-called Kummer surface of T, and in particular is a K3-surface. Finally

14.29 Theorem ([Bea 2]). *Let T be a Kählerian complex 2-dimensional torus. Then the manifold K_r is simply connected and admits hyperkählerian metrics with holonomy group exactly $Sp(r)$.*

14.30. In the paper [Bea 2], Beauville studies also the moduli space of these manifolds. There are many deformations of those $S^{[r]}$ and K_r into other complex manifolds of Kähler type with a complex-symplectic structure. Of course they are still diffeomorphic to $S^{[r]}$ or K_r, but different as complex manifolds. In particular, most of them may not be obtained by the constructions which gives $S^{[r]}$ or K_r. A key ingredient here is the following

14.31 Theorem (V.A. Bogomolov [Bog]). *Let M be a compact complex manifold of Kähler type admitting a "unique" complex-symplectic structure. Then the Kuranishi space of moduli is smooth at the base point.*

14.32 Remarks. a) Be careful that Theorem 2 of [Bog] is false; there is a mistake in n° 8. But its Theorem 1 is true.

b) The complete classification of complex-symplectic manifolds is an open problem, even in the compact case.

c) Recently, A.N. Todorov has announced that every compact complex-symplectic manifold admits a Kähler metric [Tod 3].

14.33. In the non-compact case, many examples of complex-symplectic manifolds are known. The simplest ones are the following. Let N be any complex manifold (compact or not). We denote by $M = \bigwedge^{(1,0)}N$ its complex cotangent bundle, i.e., the bundle of 1-forms of complex type $(1,0)$. Then M is naturally a complex (even-dimensional) manifold in the following way: let $(z^1, \ldots, z^n)$ be local complex coordinates on some open subset U in N; then $(dz^1, \ldots, dz^n)$ form a local basis of $(1,0)$-forms. For any $(1,0)$-form ξ on U, we denote by $\xi_1, \ldots, \xi_n$ its coordinates with respect to that basis, and we take $(\xi_1, \ldots, \xi_n, z^1, \ldots, z^n)$ as local complex coordinates on the open subset $\bigwedge^{(1,0)}U$ of $\bigwedge^{(1,0)}N = M$.

Moreover, the 1-form $\sum_{j=1}^n \xi_j \, dz^j$ is the restriction of a well-defined 1-form θ over M. And $d\theta = \omega$ is a well-defined complex-symplectic structure on M; indeed, in the above local coordinates, $\omega = \sum_{j=1}^n d\xi_j \wedge dz^j$.

It is not known under which conditions such an M admits a hyperkählerian metric and in particular a complete one. But at least one example is known in any (even) dimension. These examples, due to E. Calabi, were indeed the first known examples of complete manifolds with holonomy precisely $Sp(n)$.

14.34 Theorem (E. Calabi [Cal 5]). *The complex cotangent bundle of the canonical complex projective space (of any dimension) admits a complete hyperkählerian metric.*

See 15.24 for some more details on Calabi's construction and Add.E for more examples.

D. Quaternion-Kähler Manifolds

14.35. The 4-dimensional Riemannian manifolds with holonomy $Sp(1)Sp(1)$ are simply the oriented Riemannian manifolds, since $Sp(1)Sp(1) = SO(4)$. We will not consider this case and in the following we will only consider

$$\boxed{n \geqslant 2}\,.$$

(A few results are still valid if $n = 1$, but as we will see, the precise analog of quaternion-Kähler manifolds in dimension 4 are Einstein self dual manifolds, which were studied already in Chapter 13).

Now, using as before the holonomy theorem and the invariants of $Sp(n)Sp(1)$, we may give another characterization of quaternion-Kähler manifolds.

14.36 Proposition. *A Riemannian manifold (M, g) is quaternion-Kähler if and only if there exists a covering of M be open sets U_i and, for each i, two almost complex structures I and J on U_i such that*

a) *g is Hermitian for I and J on U_i,*

b) *$IJ = -JI$,*

c) *the Levi-Civita derivatives of I and J are linear combinations of I, J and $K = IJ$,*

d) *for any x in $U_i \cap U_j$, the vector space of endomorphisms of T_xM generated by I, J and K is the same for i and j.*

Proof. This follows from the fact that the subgroup $Sp(n)Sp(1)$ of $SO(4n)$ may be characterized by the fact that it preserves the 3-dimensional vector space of endomorphisms of $\mathbb{R}^{4n} = \mathbb{H}^n$ generated by the product with i, j, k on the right on each component (here $Sp(n)$ acts on the left by (n, n)-quaternion matrices and $Sp(1)$ acts on the right by the product with a quaternion with norm 1). □

14.37. Notice that in general, it is not possible to define I, J, K everywhere on M. For example, the canonical quaternionic projective space $\mathbb{H}P^n$ is quaternion-Kähler since its holonomy group coïncides with its isotropy subgroup which is exactly $Sp(n)Sp(1)$; but $\mathbb{H}P^n$ admits no almost complex structure (for topological reasons).

On the other hand, the vector space generated by I, J and K is well defined at each point of M, (condition d) of 14.36) so we get a 3-dimensional subbundle E of $\mathrm{End}(TM)$ and the condition c) means that the subbundle E is in fact "globally parallel" (but not parallelized) under the Levi-Civita connection D of g (more precisely, the orthogonal projection from $\mathrm{End}(TM)$ to E is a parallel tensor).

14.38. Now the main result is the following theorem, of which we will give two proofs, the first one "elementary" (i.e. using only 14.36 and the symmetry properties of R), and the other one (due to D.V. Alekseevskii [Ale 1]), more sophisticated, using representation theory.

14.39 Theorem (M. Berger [Ber 7]). *A quaternion-Kähler manifold (M, g) is Einstein (in dimension $4n \geqslant 8$).*

First proof. (after S. Ishihara [Ish 2]). Since the Einstein condition is local, we may choose an open set U with two almost complex structures I and J as in Proposition 14.36. We differentiate I, J and K twice by the Levi-Civita connection and use the Ricci formula; this gives, for any vector fields X, Y on U,

$$[R(X, Y), I] = \gamma(X, Y)J - \beta(X, Y)K, \tag{14.39a}$$

$$[R(X, Y), J] = -\gamma(X, Y)I + \alpha(X, Y)K, \tag{14.39b}$$

$$[R(X, Y), K] = \beta(X, Y)I - \alpha(X, Y)J, \tag{14.39c}$$

where α, β, γ are 2-forms on U.

Then these forms are related to the Ricci curvature by the following

14.40 Lemma (If $n \geqslant 2$):

(14.40a) $$\alpha(X,Y) = \frac{2}{n+2} r(IX,Y),$$

(14.40b) $$\beta(X,Y) = \frac{2}{n+2} r(JX,Y),$$

(14.40c) $$\gamma(X,Y) = \frac{2}{n+2} r(KX,Y).$$

Proof (*of the Lemma*). We apply (14.39c) to a vector field Z and evaluate the result with JZ by g. We get

(14.40d) $$\alpha(X,Y)|Z|^2 = g(R(X,Y)Z,IZ) + g(R(X,Y)JZ,KZ).$$

Now we choose a local orthonormal basis (X_i) such that for any X_i, then (up to a $\pm$sign) IX_i, JX_i, KX_i are also in the basis. In particular the set of pairs (X_i, IX_i) and (JX_i, KX_i) are the same. By summing on such a basis we get

(14.40e) $$4n\alpha(X,Y) = 2\sum_{i=1}^{4n} g(R(X,Y)X_i, IX_i).$$

We apply Bianchi's first identity on the last three terms, which give

$$2n\alpha(X,Y) = \sum_{i=1}^{4n} [g(R(X,X_i)Y,IX_i) - g(R(X,IX_i)Y,X_i)]$$

Here also, the sets of pairs (X_i, IX_i) and $(IX_i, -X_i)$ give the same result so

$$n\alpha(X,Y) = \sum_{i=1}^{4n} g(R(X,X_i)Y,IX_i) = -\sum_{i=1}^{4n} g(IR(X,X_i)Y,X_i).$$

We apply (14.39a) and we get

$$n\alpha(X,Y) = -\sum_{i=1}^{4n} [(g(R(X,X_i)IY,X_i) - \gamma(X,X_i)g(JY,X_i) + \beta(X,X_i)g(KY,X_i)]$$

so $n\alpha(X,Y) = -r(X,IY) + \gamma(X,JY) - \beta(X,KY)$.

We apply that formula to IY instead of Y. This gives

$$n\alpha(X,IY) + \beta(X,JY) + \gamma(X,KY) = r(X,Y).$$

The same computation for β or γ gives

$$\alpha(X,IY) = \beta(X,JY) = \gamma(X,KY) = \frac{2}{n+2} r(X,Y).$$

Since r is symmetric and α, β, γ alternating, we get the result. □

End of the First Proof. By using (14.40d) four times, we get easily

$$g(R(X,IX)Z,IZ) + g(R(X,IX)JZ,KZ) + g(R(JX,KX)Z,IZ) + g(R(JX,KX)JZ,KZ) = \frac{4}{n+2}r(X,X)\|Z\|^2 = \frac{4}{n+2}r(Z,Z)\|X\|^2$$

for any X and Z, so

$$r(X,X) = \lambda\|X\|^2 \quad \text{and } M \text{ is Einstein.} \qquad \square$$

14.41 *Second Proof* (after S. Salamon [Sal 2]). We shall use representation theory to describe the decomposition of the curvature tensor of a quaternion-Kähler manifold into irreducible components under the action of $Sp(n)Sp(1)$, and compare it with the decomposition under $O(4n)$ of 1.101. This idea is due to D.V. Alekseevskii who gave this decomposition (without proof) for all possible holonomy groups in [Ale 1]. For a complete and detailed proof in the quaternion-Kähler case, we refer to [Sal 1] or [Sal 2]. Here we will only sketch the method of [Sal 2].

We know from 10.53 that, in the quaternion-Kähler case, the (4,0)-curvature tensor R must belong to the subspace $S^2(\mathfrak{a}) \subset S^2(\wedge^2 T_xM)$ where $\mathfrak{a}$ is the Lie algebra of $Sp(n)Sp(1)$, viewed as a subspace of skew-symmetric endomorphisms of T_xM, which we identified with $\wedge^2 T_xM$. Since $Sp(n)Sp(1)$ is not simple, $\mathfrak{a}$ and $S^2(\mathfrak{a})$ are not irreducible under the action of $Sp(n)Sp(1)$.

We denote by λ the canonical representation of $Sp(n)$ acting on $\mathbb{H}^n$ on the left and by ρ the canonical representation of $Sp(1)$ acting on $\mathbb{H}$ on the right. They are *complex* representations (of dimension $2n$ and 2 respectively) with a quaternionic structure, hence the tensor product representation $\lambda \otimes_{\mathbb{C}} \rho$ is a complex representation (of dimension $4n$) with a real structure, i.e., it is the complexification of a real representation, which is precisely the canonical representation of $Sp(n)Sp(1)$ into $\mathbb{H}^n$. In the following, we will argue on complex representations instead of real ones. In that setting, any irreducible representation of $Sp(n) \times Sp(1)$ is a subspace of some tensor product $(\bigotimes^p \lambda) \otimes (\bigotimes^q \rho)$ and it factors to a representation of $Sp(n)Sp(1)$ if and only if $p+q$ is even. Now the adjoint representation of $Sp(n)Sp(1)$ into $\mathfrak{a}$ decomposes into the direct sum $(S^2\lambda) \oplus (S^2\rho)$. We deduce that

$$S^2(\mathfrak{a}) = S^2(S^2\lambda) \oplus (S^2\lambda \circ S^2\rho) \oplus S^2(S^2\rho)$$

(where $\circ$ is the symmetrized product), but only $S^2\lambda \circ S^2\rho$ is indeed an irreducible factor. We see easily for example that $S^2(S^2\rho) = \mathbb{R}B_1 \oplus S^2\rho$, where $B_1 \in S^2(S^2\rho)$ is the Killing form of the Lie algebra $S^2\rho$ of $Sp(1)$. But the decomposition of $S^2(S^2\lambda)$ into irreducible factors is more complicated and we will not give it here.

Now we have to take care of the first Bianchi identity, and we will use the projector b of 1.107. We get the following result:

(14.41) $S^2(\mathfrak{a}) \cap \mathrm{Ker}(b) = \mathbb{R}R_0 \oplus S^4\lambda$,

where R_0 is the curvature tensor of the canonical quaternionic projective space $\mathbb{H}P^n$.

The simplest way to prove that, is to exhibit "generic" terms in $S^2\lambda \circ S^2\rho$ and $S^2\rho$ which are not annihilated by b (for example $(S^2x)\circ(S^2y)$ or (S^2y) for any x in $\mathrm{Im}(\lambda)$ and y in $\mathrm{Im}(\rho)$). Now the holonomy group of $\mathbb{H}P^n$ is exactly $Sp(n)Sp(1)$, hence R_0

is not included into $S^2(S^2\lambda)$ and it has a non-zero component on $\mathbb{R}B_1$. It follows that we may write, for any curvature tensor R of a quaternion-Kähler manifold, $R = \alpha R_0 + R_1$, where α is some function on M and R_1 belongs to $S^2(S^2\lambda)$. Now R_1 is at each point an algebraic curvature tensor whose corresponding holonomy is in $Sp(n) \subset SU(2n)$, hence satisfies $c(R_1) = 0$, where c is the Ricci contraction of 1.109. Since $c(R) = \alpha c(R_0)$ is then proportional to the Riemannian metric g, we see that M is Einstein. One should take care that R_0 is *not* in the irreducible component $\mathscr{S}$ of 1.114 (since $\mathbb{H}P^n$ has non-constant curvature), but it is in $\mathscr{S} \oplus \mathscr{W}$ (since $\mathbb{H}P^n$ is Einstein). In fact, we may show with 7.72 that $R_0 = -(B_1 + B_2)$, where B_2 is the Killing form of the Lie algebra $S^2\lambda$ of $Sp(n)$.

Finally, one needs more computation to show that $S^2(S^2\lambda) \cap \operatorname{Ker}(b) = S^4\lambda$ and we will not do it here since we do not need it (see [Sal 1]). □

14.42 Remark. There are some useful relations between the components of the curvature tensor. For example, for any X,

$$g(R(X,IX)X,IX) + g(R(X,JX)X,JX) + g(R(X,KX)X,KX) = \frac{3\lambda}{n+2}\|X\|^4. \tag{14.42a}$$

Also if

$$g(X,Y) = g(X,IY) = g(X,JY) = g(X,KY) = 0,$$

$$g(R(X,Y)X,Y) + g(R(X,IY)X,IY) + g(R(X,JY)X,JY) + g(R(X,KY)X,KY) = \frac{\lambda}{n+2}\|X\|^2\|Y\|^2. \tag{14.42b}$$

We mention among others one application (without proof).

14.43 Theorem (M. Berger [Ber 8]). *A compact orientable locally quaternion-Kähler manifold with positive sectional curvature is isometric to the canonical quaternion projective space.*

14.44 Exercise. The curvature tensor R_0 of the canonical quaternion projective space satisfies

$$\begin{aligned} 4R_0(X,Y)Z = {} & g(X,Z)Y - g(Y,Z)X + 2g(IX,Y)IZ + g(IX,Z)IY \\ & - g(IY,Z)IX + 2g(JX,Y)JZ + g(JX,Z)JY \\ & - g(JY,Z)JX + 2g(KX,Y)KZ + g(KX,Z)KY \\ & - g(KY,Z)KX. \end{aligned} \tag{14.44}$$

As a by-product of the preceding study, we get the following result, which shows that the theory of quaternion-Kähler manifold splits naturally into two completely different cases, the hyperkählerian one and the other.

14.45 Theorem (M. Berger [Ber 7]). *Let (M, g) be a quaternion-Kähler manifold*

$(n \geqslant 2)$. Then

a) (M,g) is Ricci-flat if and only if it is locally hyperkählerian.

b) If (M,g) is not Ricci-flat, it is (even locally) de Rham-irreducible.

Proof. a) With the notation of 14.40, if $\lambda = 0$ then $\alpha = \beta = \gamma = 0$ and it is possible to choose three independent linear combinations of I, J, K which are parallel. And this is true everywhere, so at least the restricted holonomy group is contained in $Sp(n)$. For the proof of b), we begin with a

14.46 Lemma. *If (M,g) is quaternion-Kähler and not Ricci-flat (with $n \geqslant 2$), then the holonomy group of (M,g) contains the $Sp(1)$ factor of $Sp(n)Sp(1)$.*

Proof (of the Lemma). We use (14.40e) and (14.40a). This gives

$$\sum_{i=1}^{n} g(R(X_i, IX_i)X, Y) = 2n\alpha(X, Y) = \frac{4n}{n+2} r(IX, Y)$$

$$= \frac{4n\lambda}{n+2} g(IX, Y)$$

so $I = \frac{n+2}{4n\lambda} \sum_{i=1}^{n} R(X_i, IX_i)$ since $\lambda \neq 0$.

But we know that all $R(X, Y)$ belong to the Lie algebra of the holonomy group, so I also and similarly J and K and all the $\mathfrak{sp}(1)$-factor of the Lie algebra $\mathfrak{sp}(n) \oplus \mathfrak{sp}(1)$ of $Sp(n)Sp(1)$. Finally $Sp(1)$ is the connected subgroup of $SO(4n)$ whose Lie algebra is precisely that $\mathfrak{sp}(1)$. □

Proof of 14.45b). Let us suppose that locally $TM = E_1 \oplus E_2$ with two orthogonal distributions E_1 and E_2 invariant by the holonomy group, and let X and Y be non-zero vectors of E_1 and E_2 respectively. Since I, J, K are in the holonomy group, IY, JY and KY are also in E_2. In particular $g(R(X, IX)Y, IY) = 0$ and $g(R(X, IX)JY, KY) = 0$. But their sum is $\alpha(X, IX)|Y|^2 = \frac{2\lambda}{n+2}|X|^2|Y|^2 \neq 0$, a contradiction.

So (M,g) is irreducible. □

14.47 Corollary (M. Berger [Ber 7]). *If the restricted holonomy group H of a $4n$-dimensional Riemannian manifold $(n \geqslant 2)$ is contained into $Sp(n)U(1)$, then it is contained into $Sp(n)$.*

Proof. Since $Sp(n)U(1) \subset U(2n)$, we see easily that there exist locally I, J and K as in Proposition 14.36, where furthermore I is *parallel*. Then, with the notations of Theorem 14.40, $\beta = \gamma = 0$, so $r = 0$ and we apply 14.45a). □

14.48 Remark. In the other case $n = 1$, then $Sp(1)U(1) = U(2) \subset SO(4)$, so non-Ricci-flat Kähler (complex-) surfaces are examples of 4-manifolds with holonomy $Sp(1)U(1)$. Here I is parallel (since a complex structure is always parallel for the Levi-Civita connection of a Kähler metric), hence $\beta = \gamma = 0$ (with the notations of

Theorem 14.39), but α and r may be non zero. The proof of Lemma 14.40 gives $r(x, Y) = \alpha(X, IY)$ and nothing more in this case if $n = 1$. Of course, the case $\alpha = r = 0$ corresponds to the case where the restricted holonomy group is indeed $Sp(1) = SU(2) \subset SO(4)$.

Notice that considering the $Sp(1)$ part of $Sp(1)U(1)$ gives another quaternionic structure on the manifold M^4. The corresponding I_1, J_1, K_1 all commute with the preceding I, J, K. But this does not seem to throw any new light on these manifolds.

E. Symmetric Quaternion-Kähler Manifolds

14.49. As first examples, we describe below those symmetric spaces which are quaternion-Kähler. We recall that homogeneous Riemannian manifolds with zero Ricci curvature are flat, so the homogeneous hyperkählerian Riemannian manifolds are the $4n$-dimensional products of flat Euclidean spaces and tori.

Now let (M, g) be a symmetric Riemannian manifold which is quaternion-Kähler and not Ricci flat; we denote G the connected component of the identity in the isometry group and K the isotropy subgroup. From Theorem 14.45, we deduce that M is an irreducible symmetric space. Moreover, if we *denote* by $M^* = G^*/K$ the dual symmetric space of M, it is obvious that M^* is quaternion-Kähler also. So we may consider only the case where M is compact (so G is compact).

14.50. We recall that the holonomy subgroup of M coïncides with the isotropy subgroup K; so we deduce from Lemma 14.46 that K contains a simple normal subgroup isomorphic to $Sp(1)$, whose isotropy representation is isomorphic to the representation of the normal subgroup $Sp(1)$ of $Sp(n)Sp(1)$ in $\mathbb{H}^n = \mathbb{R}^{4n}$.

Conversely, if the isotropy subgroup K of a symmetric space contains a normal subgroup $Sp(1)$ with such a representation, then we may write $K = K_1 \cdot Sp(1)$, with a normal subgroup K_1 such that $K_1 \cap Sp(1)$ is contained in the center $\{\pm 1\}$ of $Sp(1)$. Furthermore, the isotropy representation of K_1 commutes with that of $Sp(1)$, so K_1 must be contained into the commutator $Sp(n)$ of $Sp(1)$ and M is quaternion-Kähler.

Now it is not difficult to detect such K's in E. Cartan's list of irreducible symmetric spaces (Table 7.102 of Chapter 7). We deduce immediately

14.51 Theorem (J.A. Wolf [Wol 2]). *Let (M, g) be a quaternion-Kähler symmetric Riemannian manifold with non-zero Ricci curvature. Then M is irreducible, simply connected and*

a) *either the Ricci curvature is positive and M is compact and one of the G/K of the following Table* 14.52.

b) *or the Ricci curvature is negative and M is non-compact and one of the G^*/K of the following table* 14.52.

14.52 Table. Quaternion-Kähler symmetric Riemannian manifolds with non-zero Ricci curvature.

G	G^*	K	$\dim M$	Notes
$Sp(n+1)$	$Sp(n,1)$	$Sp(n)Sp(1)$	$4n\ (n \geqslant 1)$	$M = \mathbb{H}P^n$
$SU(n+2)$	$SU(n,2)$	$S(U(n)U(2))$	$4n\ (n \geqslant 1)$	$Sp(1) = SU(2)$ and see 14.53b) below
$SO(n+4)$	$SO(n,4)$	$SO(n)SO(4)$	$4n\ (n \geqslant 3)$	Here
G_2	G_2^2	$SO(4)$	8	$SO(4) = Sp(1)Sp(1)$
F_4	F_4^{-20}	$Sp(3)Sp(1)$	28	
E_6	E_6^2	$SU(6)Sp(1)$	40	
E_7	E_7^{-5}	$\mathrm{Spin}(12)Sp(1)$	64	
E_8	E_8^{-24}	$E_7Sp(1)$	112	

14.53 Remarks. a) We recall that by convention, we allow G to be only nearly effective. For example, the isometry group of $\mathbb{H}P^n$ is $Sp(n+1)/\{\pm 1\}$.

b) The symmetric space $SU(n+2)/S(U(n)U(2))$ admits also a complex structure J_0 such that (M, g, J_0) is Kähler, and more precisely is a Hermitian symmetric space. But that J_0 is different from all the local complex structures $xI + yJ + zK$ (with $x^2 + y^2 + z^2 = 1$) which define the quaternionic structure. The Lie algebra $\mathfrak{su}(n) \oplus \mathfrak{su}(2) \oplus \mathbb{R}$ of K, viewed as the holonomy algebra at eK, contains J_0 as a generator of $\mathbb{R}$ and (I, J, K) as generators of $\mathfrak{su}(2) = \mathfrak{sp}(1)$.

c) Be careful that the Paragraph 3 of Wolf's paper [Wol 2] is false (in particular Theorems 3.5 and 3.7); the mistake is in Lemma 3.4 (about quaternionic structure with "complex scalar part" in Wolf's terminology). But the rest of the paper is correct.

14.54. It appears on Table 14.52 that there is exactly one compact quaternion-Kähler symmetric space for each type of compact simple Lie group (except $SU(2)$). In his paper [Wol 2], J.A. Wolf gives an a priori proof of that fact, by studying quaternion-Kähler symmetric spaces from the point of view of roots.

Let $\mathfrak{g}$ be a compact simple Lie algebra, $\mathfrak{t}$ a Cartan subalgebra, and α a root *with maximal length.* There is a corresponding 3-dimensional Lie subalgebra $\{H_\alpha, X_\alpha, Y_\alpha\} = \mathfrak{sp}(1)$. We consider the centralizer $\mathfrak{k}_1$ of $\mathfrak{sp}(1)$ and the subalgebra $\mathfrak{k} = \mathfrak{k}_1 \oplus \mathfrak{sp}(1)$, which contains $\mathfrak{t}$. If G is the simply-connected compact simple Lie group with Lie algebra $\mathfrak{g}$, and K the compact subgroup of G generated by $\mathfrak{k}$, then G/K is a compact quaternion-Kähler symmetric space.

14.55. We recall once again that, up to now, these spaces are the only known examples of compact quaternion-Kähler manifolds. In the non-compact case, D.V. Alekseevskii proved in [Ale 3, 4] that there exist non-symmetric homogeneous (hence complete) quaternion-Kähler manifolds (see § I below). But he proved also

14.56 Theorem (Alekseevskii [Ale 2, 3]). *Let M be a quaternion-Kähler Riemannian manifold which is G-homogeneous under a* reductive *G. Then M is symmetric.*

In particular, any compact quaternion-Kähler homogeneous Riemannian manifold is symmetric.

We refer to [Ale 2] for the proof.

F. Quaternionic Manifolds

14.57. As the name tries to suggest, there are some analogies between the theory of Kähler manifolds and that of quaternion-Kähler manifolds. We recall that a Kähler manifold (M, g, J) may be viewed as a Riemannian manifold (M, g) together with an almost complex structure J such that (g, J) is almost-Hermitian and J is parallel for the Levi-Civita connection D of g. It turns out (since D is torsion-free) that $DJ = 0$ implies that J is in fact a complex structure. We want to consider quaternion-Kähler manifolds in a similar way.

To begin with, we need quaternionic analogs of the notions of almost-complex structure and complex structure, but there is more than one possible choice here. Notice first that there is no good notion of "holomorphic functions" in the quaternionic case, because of the

14.58 Theorem (C. Ehresmann [Ehr 2]). *Let $f: U \to \mathbb{R}^{4n}$ be a differentiable map on some connected open set U of $\mathbb{R}^{4n} = \mathbb{H}^n$. If the differential $f'(x)$ of f belongs to $Gl(n, \mathbb{H})$ for any x in U, then f is an affine map.*

There exists nevertheless a "quaternionic holomorphic calculus", developed by R. Fueter ([Fue]), see also the recent survey by A. Sudbery [Sud], but it does not fit our purposes.

14.59. We may look at these structures from the point of view of G-structures. Then an equivalent formulation of Theorem 14.58 is the proposition that the subgroup $Gl(n, \mathbb{H})$ of $Gl(4n, \mathbb{R})$ has "order 1", i.e. its first prolongation (as defined by E. Cartan) is zero; this implies that any integrable $Gl(n, \mathbb{H})$-structure is in fact an affine structure. In any case, $Gl(n, \mathbb{H})$ is not the "good" group to consider, since even the quaternionic projective space $\mathbb{H}P^n$ has no $Gl(n, \mathbb{H})$-structure (even non-integrable) for topological reasons. For the study of $Gl(n, \mathbb{H})$-structures, see A. Sommese [Som] or E. Bonan [Bon 2].

14.60. So we consider the subgroup $Gl(n, \mathbb{H})Sp(1)$ of $Gl(4n, \mathbb{R})$, which we define in the same way as $Sp(n)Sp(1)$. We identify $\mathbb{R}^{4n}$ with the (right) vector space $\mathbb{H}^n$ over $\mathbb{H}$; then $Gl(n, \mathbb{H})$ is the linear group (acting on the left) of the vector space $\mathbb{H}^n$ (over $\mathbb{H}$), and $Sp(1)$ is as before the group of unit quaternions acting on $\mathbb{H}^n$ by scalar multiplications *on the right*. Of course, $Gl(n, \mathbb{H})$ and $Sp(1)$ commute and their intersection is the center $\{\pm 1\}$ of $Sp(1)$ (notice that the center of $Gl(n, \mathbb{H})$ is precisely the subgroup $\mathbb{R}^*$ of homotheties in $Gl(4n, \mathbb{R})$). We see easily that $Sp(n)Sp(1)$ is the maximal compact subgroup of $Gl(n, \mathbb{H})Sp(1)$.

14.61 Definitions (E. Bonan [Bon 2]). a) An *almost-quaternionic* manifold is a $4n$-dimensional manifold with a $Gl(n, \mathbb{H})Sp(1)$-structure (i.e. a reduction of its tangent principal bundle to the subgroup $Gl(n, \mathbb{H})Sp(1)$ of $Gl(4n, \mathbb{R})$).

b) A Riemannian metric on an almost-quaternionic manifold is called *quaternion-Hermitian* if there exists a common reduction of the $Gl(n, \mathbb{H})Sp(1)$-bundle and the $SO(4n)$-bundle to $Sp(n)Sp(1)$.

Equivalently, an almost-quaternionic structure may be defined by a covering (U_i) of M with two almost-complex structures I, J on each U_i such that $IJ = -JI$ and the 3-dimensional vector spaces of endomorphisms generated by I, J and IJ is the same on all of M.

Then a Riemannian metric g is quaternion-Hermitian if moreover g is Hermitian for each of the I and J above. See [Bon 2] for various results on these structures.

In analogy with the complex case, we set

14.62 Definition (S. Salamon [Sal 1, 3]). *A quaternionic manifold* is an almost-quaternionic manifold admitting a torsion-free $Gl(n, \mathbb{H})Sp(1)$-connection.

Notice that such a connection is never unique. Since the Levi-Civita connection is torsion-free, we deduce immediately

14.63 Proposition. *A quaternion-Kähler manifold is quaternion-Hermitian and quaternionic.*

We recall that an almost-complex structure with a complex *torsion-free* connection is integrable, and hence a complex structure (Newlander-Nirenberg theorem, see 2.12). *This is not true for quaternionic manifolds* and our analogy breaks down here definitely.

By definition, a G-structure is integrable if there exist local coordinates (U_i, φ_i) on M such that the coordinate changes $\psi_{ij} = \varphi_j \circ \varphi_i^{-1}$ on $\varphi_i(U_i \cap U_j)$ have a differential in G ($\subset Gl(4n, \mathbb{R})$) at each point. Now $Gl(n, \mathbb{H})Sp(1)$ has order 2 in E. Cartan's sense (as subgroup of $Gl(4n, \mathbb{R})$), and this imposes some rigidity. We deduce from the general theory of integrable G-structures of order 2

14.64 Theorem (R. Kulkarni [Kul 3], see also [Sal 1, 3]). *Let M be an integrable quaternionic manifold. Then there exists an etale mapping f from the universal cover $\tilde{M}$ of M to $\mathbb{H}P^n$, which is compatible with the quaternionic structures.*

In particular, if M is furthermore compact and simply connected it is diffeomorphic to $\mathbb{H}P^n$.

In this theorem, the canonical quaternionic projective space $\mathbb{H}P^n$ is endowed with its canonical quaternionic structure, which may also be described in the following way. We consider $\mathbb{H}P^n$ as the homogeneous space $Gl(n+1, \mathbb{H})/Gl(1, n; \mathbb{H})$, where $Gl(1, n; \mathbb{H})$ is the closed subgroup of $Gl(n+1, \mathbb{H})$ which leaves globally invariant $\mathbb{H} \subset \mathbb{H} \times \mathbb{H}^n = \mathbb{H}^{n+1}$; i.e., $Gl(1, n; \mathbb{H})$ is the subgroup of matrices of $Gl(n+1, \mathbb{H})$ of the following form: $\begin{pmatrix} a & {}^tX \\ 0 & A \end{pmatrix}$, where $a \in \mathbb{H}_*$, $X \in \mathbb{H}^n$ and A is an invertible (n, n)-matrix. Notice that $Gl(n+1, \mathbb{H})$ does not act effectively on $\mathbb{H}P^n$ since its center $Z = \{a\mathrm{Id}, a \in \mathbb{R}_*\}$ is contained in $Gl(1, n; \mathbb{H})$. (If we want effectivity, we need to consider $PGl(n+1, \mathbb{H}) = Gl(n+1, \mathbb{H})/Z$, but we will not do it here; one may also consider other subgroups of $Gl(n+1, \mathbb{H})$ which act transitively and nearly effectively on $\mathbb{H}P^n$, such as $Sl(n+1, \mathbb{H})$ or $Sp(n+1)$).

Now the isotropy representation ρ of $Gl(1, n; \mathbb{H})$ is not faithful, it has a kernel N which is the *normal* subgroup of matrices of the following form: $\begin{pmatrix} a & {}^tX \\ 0 & aI \end{pmatrix}$ where $a \in \mathbb{R}_*$, $X \in \mathbb{H}^n$ and I is the (n, n)-identity matrix. Then the image of ρ is precisely

$Gl(n, \mathbb{H})Sp(1)$ and the fibration $P = Gl(n+1, \mathbb{H})/N \to \mathbb{H}P^n$ is a *principal* $Gl(n, \mathbb{H})Sp(1)$-*bundle*; moreover the vector bundle associated with the representation ρ is the tangent bundle, hence P is a reduction of the principal tangent bundle and a quaternionic structure. Now it is easy to see, in the canonical local coordinates for the projective space $\mathbb{H}P^n$, that this quaternionic structure is integrable.

Finally, we want to point out that $\mathbb{H}P^n$ has *no finite quotient* (in fact $\mathbb{H}P^n$ cannot be the total space of any non-trivial fibration of manifolds: this is due to J.C. Becker, D.H. Gottlieb and R. Schultz, see [Be-Go]). As a corollary, any compact integrable quaternionic manifold with finite fundamental group is isomorphic to $\mathbb{H}P^n$.

14.65. Now the compact quaternion-Kähler symmetric spaces different from $\mathbb{H}P^n$ are compact simply connected quaternionic manifolds and it is not difficult to see from their topology that they are not diffeomorphic to $\mathbb{H}P^n$. In particular, they are examples of *non-integrable* quaternionic manifolds.

In some sense, quaternionic manifolds are examples of "semi-integrable" G-structures. In the following chapter, we describe a new tool for the study of these manifolds, which was introduced recently by S. Salamon and L. Bérard Bergery independently.

G. The Twistor Space of a Quaternionic Manifold

14.66. Let M be a quaternionic manifold (as defined in 14.61 and 14.62). We *denote* by E the 3-dimensional vector subbundle of $\mathrm{End}(TM)$ generated by I, J and $K = IJ$ on each local chart U_i. We insist that, by definition, E is a well-defined vector bundle over all of M (but it may be non-trivial; in particular, there may be no global almost-complex structure on M, as for example on $\mathbb{H}P^n$). We may also view E as the vector bundle associated with the given $Gl(n, \mathbb{H})Sp(1)$-bundle for the 3-dimensional representation of $Gl(n, \mathbb{H})Sp(1)$ on $\mathbb{R}^3$ such that $Gl(n, \mathbb{H})$ acts trivially and $Sp(1)$ acts through the adjoint representation on its Lie algebra $\mathfrak{sp}(1)$ $(= \mathbb{R}^3)$.

In particular, the vector bundle E carries a natural Euclidean structure, such that (I, J, K) forms an orthonormal basis of E over each U_i. Then the unit-sphere subbundle of E has as fibre at m in U_i precisely the set of those linear combinations of I, J, K which are complex structures on $T_m M$.

14.67 Definition (S.Salamon [Sal 1, 2, 3, 4], L. Bérard Bergery [BéBer 7]). We call this unit-sphere subbundle of E, the *bundle of twistors* Z or *twistor space* of the quaternionic manifold M.

The name is chosen in order to emphasize that this theory is a direct generalization of Penrose's theory of twistor space for self-dual manifolds (for other generalizations, see [BB-Oc]).

14.68 Theorem (S. Salamon [Sal 1, 2]). *Let M be a quaternionic manifold (with dimension $4n \geqslant 8$). Then its bundle of twistors Z admits a natural complex struc-*

ture such that the fibres of the projection $\pi\colon Z \to M$ *are compact complex curves of genus* 0.

14.69 Remark. This is not true in dimension 4. In fact $Gl(1, \mathbb{H})Sp(1)$ is isomorphic to the oriented conformal group $\mathbb{R}_+^* SO(4)$, so a 4-dimensional quaternionic manifold is nothing but an oriented manifold with a conformal structure. In this case, there exists also a natural *almost*-complex structure on Z, but it is integrable only if the manifold is self-dual (as we saw in Chapter 13).

14.70 *Proof.* We will not give Salamon's proof which is beautiful and quite conceptual, but involves quite a lot of representation theory. We only sketch a more elementary proof (due to L. Bérard Bergery).

We choose first a torsion-free $Gl(n, \mathbb{H})Sp(1)$-connection θ on the given $Gl(n, \mathbb{H})Sp(1)$-bundle Q, which induces a torsion-free linear connection D on M. Since E and Z are associated bundles, D induces a *metric* connection on E and also a splitting of the tangent-bundle $TZ = \mathscr{V} \oplus \mathscr{H}$, where $\mathscr{V}$ is the vertical distribution (tangent to the fibres of π) and $\mathscr{H}$ a supplementary horizontal distribution (defined by θ). Notice that, by definition, the horizontal transport associated to $\mathscr{H}$ as in 9.40 preserves the canonical metric of the fibres S^2 of π and also their orientation; as a corollary, it preserves also the canonical complex structure of S^2, which is uniquely determined on S^2 in terms of the canonical metric and orientation. Since the vertical distribution $\mathscr{V}$ is tangent to the fibres, this complex structure induces an endomorphism $\hat{J}$ with $\hat{J}^2 = -Id$ on $\mathscr{V}_z$, for each z in Z.

14.71. On the other hand, each point z in Z is by definition a complex structure on the vector space $T_{\pi(z)}M$. Since the tangent map π_* identifies $\mathscr{H}_z$ with $T_{\pi(z)}M$, we may lift up this complex structure z into an endomorphism $\bar{J}$ on $\mathscr{H}_z$ with $\bar{J}^2 = -Id$.

Finally, we *define* the natural almost complex structure J on Z as follows (see Figure (14.71)):

$$\text{we take}\quad J(\mathscr{V}) = \mathscr{V} \quad\text{and}\quad J\restriction \mathscr{V} = \hat{J},$$
$$J(\mathscr{H}) = \mathscr{H} \quad\text{and}\quad J\restriction \mathscr{H} = \bar{J}.$$

It is not difficult to see that in fact J does not depend on the particular torsion-free $Gl(n, \mathbb{H})Sp(1)$-connection that we chose.

14.72. In order to prove the integrability of J, we consider its Nijenhuis tensor NJ (see 2.11), and the various components of it in the splitting $TZ = \mathscr{V} \oplus \mathscr{H}$.

Obviously, if U, V are vertical vector fields, then so are JU, JV and all the brackets of them; since $\hat{J}$ is a complex structure on each fibre, we get $NJ(U, V) = N\hat{J}(U, V) = 0$.

Now let X, Y be basic vector fields (see 9.22). Since the horizontal transport of $\mathscr{H}$ respects $\hat{J}$, we have $[X, JU] = J[X, U]$ (since $[X, U]$ is vertical if U is). Notice that JX is *not* a basic vector field, but we have still $\mathscr{V}[JX, JU] = J\mathscr{V}[JX, U]$ since this is obviously tensorial for a horizontal X. To study the remaining (horizontal) term of $NJ(X, U)$, we project it onto the base. We see easily that

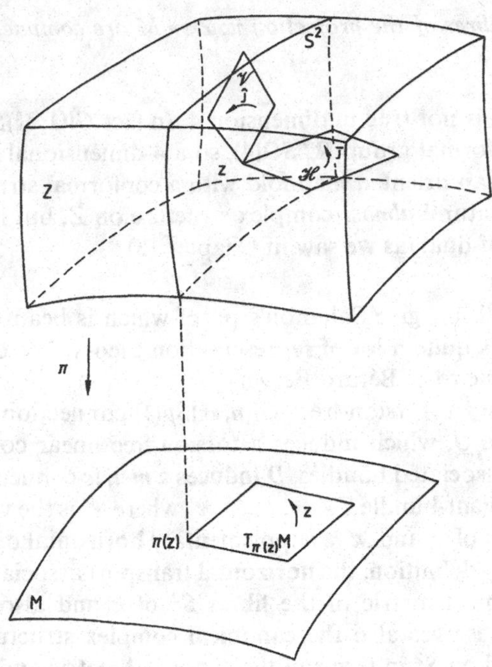

Fig. 14.71

$$\pi_*(\mathscr{H}[JX, U]) = -\varphi_*(U)\bar{X}, \tag{14.72}$$

where φ is the map from the fibre Z_m (of Z at m in M) into $\operatorname{End}(T_mM)$ which associates to z the corresponding complex structure on $T_{\pi(z)}M$.

Then $NJ(X, U) = 0$ follows from the following result (which is easy for example in local coordinates U_i).

14.73 Lemma. *$\varphi_*(\hat{J}U) = \varphi(z) \circ \varphi_*(U)$ for any vertical U at z in Z.*

For any z in Z, we consider a local section s of Z on U around $m = \pi(z)$, such that $Ds = 0$ at m (where s is viewed in E). This gives a local almost complex structure S on U and the projection of $\mathscr{H}NJ(\mathscr{H}, \mathscr{H})$ (restricted to $s(U)$) is the Nijenhuis tensor NS of S. Since D is torsion-free and $DS = 0$ at m, then $NS = 0$ at m and so $\mathscr{H}NJ(\mathscr{H}, \mathscr{H}) = 0$.

The most delicate problem is the vanishing of $\mathscr{V}NJ(\mathscr{H}, \mathscr{H})$. As shown in 9.53, $\mathscr{V}[\mathscr{H}, \mathscr{H}]$ is a tensor, related to the curvature R of the connection D. Through that identification the vanishing of $\mathscr{V}NJ(\mathscr{H}, \mathscr{H})$ is equivalent to the following

14.74 Lemma. *If I, J, K are as in 14.61, then the curvature R of D satisfies*

$$[R(SX, SY), S] - S[R(SX, Y), S] - S[R(X, SY), S] - [R(X, Y), S] = 0 \tag{14.74a}$$

for any $S = xI + yJ + zK$, where $x^2 + y^2 + z^2 = 1$.

Proof. Since D is a $Gl(n, \mathbb{H})Sp(1)$-connection, the covariant derivatives of I, J, K are linear combinations of I, J, K as in 14.36c) and, differentiating twice, we get the same formulas than in 14.39.

We denote $I(R)(X, Y, Z)$ the first member of (14.74a), (for $S = I$), applied to a vector Z. A direct computation using Bianchi identity gives immediately

(14.74b) $$I(R)(X, Y, Z) + I(R)(Y, Z, X) + I(R)(Z, X, Y) = 0.$$

Now a glance at Formulas 14.39 shows that $I(R)(X, Y, Z)$ is a linear combination of JZ and KZ only. So we deduce from (14.74b) that if Y is not a linear combination of X and IX, then $I(R)(X, Y, Z) = 0$. Now a direct computation shows that if $Y = aX + bIX$, then $I(R)(X, Y, Z) = 0$. □

14.75 Examples. a) The quaternionic projective space $\mathbb{H}P^n$ is the quotient of $\mathbb{H}^{n+1} - \{0\}$ by $\mathbb{H}^*$ acting on the right. Also $\mathbb{H}P^n$ is the homogeneous space $Gl(n+1, \mathbb{H})/Gl(1, n; \mathbb{H})$, and the fibration $P \to \mathbb{H}P^n$ of 14.64 is the principal $Gl(n, \mathbb{H})Sp(1)$-bundle which defines the quaternionic structure. Now the fibration $\mathbb{H}^{n+1} - \{0\} \to \mathbb{H}P^n$ is the associated fibration for the quotient group $Gl(n, \mathbb{H})Sp(1)/Sl(n, \mathbb{H}) = \mathbb{H}^*$.

Since $Gl(n, \mathbb{H})U(1)/Sl(n, \mathbb{H}) = \mathbb{C}^*$ (if we identify $\mathbb{H}^{n+1} = \mathbb{C}^{2n+2}$), then the twistor space Z of $\mathbb{H}P^n$ is the complex projective space $\mathbb{C}P^{2n+1} = (\mathbb{C}^{2n+2} - \{0\})/\mathbb{C}^*$ and the S^2-fibration $\mathbb{C}P^{2n+1} = Z \to \mathbb{H}P^n$ is the Hopf fibration of 9.83.

b) Similarly, let $M = G/K$ (where $K = HSp(1)$) be one of the symmetric quaternion-Kähler manifolds of 14.52. Then the twistor space of M is the complex homogeneous manifold $Z = G/HU(1)$ (corresponding to the inclusion $U(1) \subset Sp(1)$).

14.76 *$Gl(n, \mathbb{H})$-manifolds.* Let M be a $Gl(n, \mathbb{H})$-manifold with a torsion-free $Gl(n, \mathbb{H})$-connection. Then the bundle E is trivial and furthermore we may choose I, J, K to be parallel. Obviously, the twistor space Z of M is the product $M \times S^2$ as a real differentiable manifold, but Z is *not* a product as a complex manifold, even if $M = \mathbb{H}^n$. The trivialization by I, J, K gives a holomorphic fibration $q\colon Z \to S^2$ whose fibre at $S = xI + yJ + zK$ in S^2 is the complex manifold (M, S), so q is in general non-trivial.

For example, the twistor space $Z(\mathbb{H}^n)$ of $\mathbb{H}^n$ is the total space of the complex vector bundle $L \otimes_{\mathbb{C}} \mathbb{C}^{2n}$ over $S^2 = \mathbb{C}P^1$, where L is the complex line bundle induced by the Hopf fibration (see [Som]); in particular $Z(\mathbb{H}^n)$ is not isomorphic to $\mathbb{C}^{2n} \times \mathbb{C}P^1$ (as a complex manifold).

In the case where M is a complex torus $T^{4n} = \mathbb{H}^n/\Gamma$ for some lattice Γ, then its twistor space is a compact complex manifold which admits no compatible Kähler metric (this example is due to Blanchard [Bla]).

H. Applications of the Twistor Space Theory

14.77. The theory of twistor space gives a way to translate any problem on quaternionic manifolds into a problem on complex manifolds. For example, since any complex manifold is analytic, we deduce that any quaternionic manifold is analytic.

Unfortunately, the twistor space is often a complicated complex manifold, as we saw for $\mathbb{H}^n$. We consider now the case of nonhyperkählerian quaternion-Kähler manifolds. The only general result is the following

14.78 Theorem (S. Salamon [Sal 1]). *Let (M,g) be a quaternion-Kähler manifold with non-zero Ricci curvature. Then the twistor space Z admits a* complex contact structure [i.e. *a family of holomorphic* 1*-forms ω_i on the open sets U_i of a covering of Z such that*

(a) $\omega_i \wedge (d\omega_i)^{2n} \neq 0$ *and*

(b) *on $U_i \cap U_j$ (if non void), there exists a holomorphic function f_{ij} such that* $\omega_i = f_{ij}\omega_j$].

Proof. The Levi-Civita connection D of M is a torsion-free $Sp(n)Sp(1)$-connection, so it may be used as in 14.70 to give a splitting $TZ = \mathscr{V} \oplus \mathscr{H}$ compatible with the complex structure. Now $\mathscr{V}$ is a complex line bundle on Z, so the projection $\mathscr{H}\colon TZ \to \mathscr{V}$ extends to a holomorphic 1-form ω on Z with values in $\mathscr{V}$. Also $\mathscr{V}$ has a natural connection $\hat{D}$ given by the canonical connection of S^2. We consider the differential $d^{\hat{D}}\omega$. Then the restriction of $d^{\hat{D}}\omega$ to $\mathscr{H} \times \mathscr{H}$ is related to the obstruction to integrability A of $\mathscr{H}$ (see 9.24) by

$$d^{\hat{D}}\omega(X,Y) = -2\omega(A_X Y)$$

for any X, Y in $\mathscr{H}$. And now A may be expressed in terms of the curvature R of g by Formula (9.53b), using the map φ of 14.72. If I, J, K are a local orthonormal basis of E, then at the point $S = xI + yJ + zK$ (with $x^2 + y^2 + z^2 = 1$) of Z, we get

$$\varphi(A_X Y) = -\tfrac{1}{2}[R(\pi_* X, \pi_* Y), S]. \tag{14.78}$$

Now Formulas 14.39 and 14.40 give easily the result. □

14.79 Remark. If M is furthermore *compact*, a theorem by S. Kobayashi [Kob 1] gives the following relation on first Chern classes (since Z is compact)

$$c_1(Z) = (n+1)c_1(\mathscr{V}). \tag{14.79}$$

We have seen in 14.76 that the twistor space of a compact quaternionic manifold may be non-kählerian. So the following result is of particular importance.

14.80 Theorem (S. Salamon [Sal 1, 2], L. Bérard Bergery [BéBer 7]). *Let (M,g) be a quaternion-Kähler manifold with* positive *Ricci curvature. Then the twistor space Z of M admits a Kähler-Einstein metric with positive Ricci curvature such that the projection $\pi\colon Z \to M$ is a Riemannian submersion with totally geodesic fibres.*

Proof. We recall that (M,g) is Einstein so, up to a homothety, we may choose $r = (n+2)g$ (as for the canonical $\mathbb{H}P^n$). We choose on Z the unique metric $\tilde{g}$ such that π becomes a Riemannian submersion with totally geodesic fibres isometric to the canonical S^2 (with constant curvature *one*) and horizontal distribution $\mathscr{H}$ induced by the Levi-Civita connection D as in 14.78. Obviously, $\tilde{g}$ is Hermitian for J. We claim that $\tilde{g}$ is Kähler and Einstein. This follows from the following properties of the corresponding O'Neill tensor A.

14.81 Lemma. *Under the hypothesis above, if U is a vertical and X, Y are basic vector fields, then we have*

$$\text{(14.81)}\qquad \text{a) } D_U(JX) = JA_XU \qquad \text{b) } JA_XU = A_X(JU)$$
$$\text{c) } A_X(JY) = J(A_XY) \qquad \text{d) } \|A_X\|^2 = \tfrac{1}{2}\|X\|^2$$
$$\text{e) } \|AU\|^2 = n\|U\|^2.$$

Proof of the Lemma. Formulas 14.81b) to e) follow easily from Formulas 14.39 and 14.40 under the identification (14.78). The crucial point is (14.81a)).

With the same local basis than in 14.78, if we consider at I in Z a vertical vector U such that $\varphi(U) = uJ + vK$, then $\pi_*(D_U(JX)) = uJ\bar{X} + vK\bar{X}$ (where $\bar{X} = \pi_*(X)$).

On the other hand $\pi_*(J(A_XU)) = I\pi_*(A_XU) = I(-uK\bar{X} + vJ\bar{X}) = uJ\bar{X} + vK\bar{X}$, because $\tilde{g}(A_XU, Y) = -\tilde{g}(U, A_XY) = -ug(K\bar{X}, \bar{Y}) + vg(J\bar{X}, \bar{Y})$. □

14.82 *End of the Proof of the Theorem* 14.80. If U, V are vertical, then $(\tilde{D}_UJ)V = 0$ since this is true on each fibre. If X is a basic vector field $\mathscr{V}D_UX = 0$ and $\mathscr{H}D_UX = A_XU$, so $(D_UJ)X = D_U(JX) - J(D_UX) = 0$. Also $[X, U]$ is vertical and $\mathscr{V}(D_XU) = [X, U]$ so $\mathscr{V}(D_XJ)U = 0$ follows from 14.72. And $\mathscr{H}(D_XJ)U = 0$ follows from (14.81b). Similarly (14.81c) gives $\mathscr{V}(D_XJ)Y = 0$. Finally, $\mathscr{H}(D_XJ)Y = 0$ follows from the same argument as in 14.74.

We compute the Ricci curvature $\tilde{r}$ by Formulas (9.61). Then Formulas (14.79d and e) give $\tilde{r} = (n+1)\tilde{g}$, so $(Z, J, \tilde{g})$ is Kähler-Einstein. □

14.83 Corollary 1 (S. Salamon [Sal 1, 2]). *If (M, g) is a compact quaternion-Kähler manifold with* positive *Ricci curvature, then M is simply connected and its odd-dimensional Betti numbers vanish.*

Proof. By Kobayashi's Theorem 11.26, we have $\pi_1(Z) = 0$, so $\pi_1(M) = 0$. The second assertion follows from the fact that Z has only (p, p)-cohomology (see [Sal 2] for details). □

14.84 Corollary 2 (L. Bérard Bergery, unpublished). *If (M, g) is a compact quaternion-Kähler manifold with* positive *Ricci curvature, then the twistor space Z admits a non-Kähler Einstein Hermitian metric such that π is a Riemannian submersion with totally geodesic fibres.*

Proof. This follows immediately from Theorem 9.73. □

In the same direction, we may also consider the principal $SO(3)$-bundle $\pi'\colon Y \to M$, associated to Q for the quotient $Sp(n)Sp(1)/Sp(n) = SO(3)$. One may show easily that there is a principal S^1-fibration $\pi''\colon Y \to Z$ and that Y is a real contact manifold. Then computations similar to those in 14.80 give

14.85 Proposition. *Let (M, g) be a quaternion-Kähler manifold with* positive *Ricci curvature. Then the bundle Y above admits two different Einstein metrics such that π' is a Riemannian submersion with totally geodesic fibres, one of which being such that π'' is a Riemannian submersion with totally geodesic fibres over $(Z, \tilde{g})$.*

14.86 Remarks. a) Unfortunately, since the only known compact examples of quaternion-Kähler manifolds with positive Ricci curvature are the symmetric spaces $G/HSp(1)$ of 14.52, we obtain (up to now) only homogeneous metrics over $Z = G/HU(1)$ and $Y = G/H$.

b) If (M, g) is a quaternion-Kähler manifold with *negative* Ricci curvature, the construction in 14.80 gives a Riemannian metric $\tilde{g}$ on Z which is neither Kähler nor Einstein (but still Hermitian). But, if we consider instead of $\tilde{g}$ the pseudo-Riemannian metric $\tilde{h}$ with signature $(4n, 2)$ such that $\tilde{h}(\mathscr{V}, \mathscr{H}) = 0$, $\tilde{h} \restriction \mathscr{V} = -\tilde{g} \restriction \mathscr{V}$, $\tilde{h} \restriction \mathscr{H} = \tilde{g} \restriction \mathscr{H}$ [and if we normalize g such that $r = -(n+2)g$], then $\tilde{h}$ is pseudo-Kähler and Einstein, and there is also another pseudo-Riemannian metric with signature $(4n, 2)$ which is Einstein and non-Kähler. Similarly, Y possesses two pseudo-Riemannian Einstein metrics with signature $(4n, 3)$.

14.87. We consider now some characteristic classes and characteristic forms of a *compact* quaternion Kähler manifold M. Since the bundle E is a $SO(3)$-bundle, it has essentially 2 interesting characteristic class, namely its second Stiefel-Whitney class $w_2(E)$ in $H^2(M, \mathbb{Z}_2)$ (since $w_1(E) = 0$) and its first Pontryagin class $p_1(E)$ in $H^4(M, \mathbb{Z})$.

14.88 Definition (S. Marchiafava and G. Romani [Ma-Ro]). We call $\xi = w_2(E)$ the *fundamental class* of M.

The fundamental result on ξ is the following

14.89 Theorem (S. Salamon [Sal 1, 2]). *If (M, g) is a compact quaternion-Kähler manifold with* positive *Ricci curvature and $\xi(M) = 0$, then (M, g) is isometric to the canonical $\mathbb{H}P^n$ (up to a constant factor).*

Proof. Obviously $\pi^*(E) = \mathscr{V} \oplus \mathbb{R}$, so $w_2(\mathscr{V}) = 0$, which means that there exists a complex line bundle L over Z such that $\mathscr{V} = L \otimes L$ (over $\mathbb{C}$).

Together with (14.79), it follows that

$$c_1(Z) = (2n+2)c_1(L).$$

Now a theorem by S. Kobayashi and T. Ochiai ([Ko-Oc]) insures that Z is isomorphic to $\mathbb{C}P^{2n+1}$.

From Matsushima's Theorems 11.52, 8.2, 8.95 and Theorem 14.80, it follows that Z is even isometric to the canonical $\mathbb{C}P^{2n+1}$. Finally, the O'Neill Formulas 9.29 relating the curvature of Z and M insure that M has positive sectional curvature so by Berger's Theorem 14.43, M is isometric to the canonical $\mathbb{H}P^n$. □

14.90 Remark. The original proof by Salamon was slightly different at the end: he gets the dimension of the isometry group of M using various complexes on Z; this has applications even if $\xi \neq 0$ (see [Sal 1, 2]).

14.91. Since D induces a connection on E, the first Pontryagin class $p_1(E)$ may be realized by some 4-form Ω on M. In fact, this form Ω is a quaternionic analog of the Kähler form in complex geometry. It is not difficult to see that Ω may be computed as follows

14.92 Proposition (V.Y. Kraines [Kra]). *Let (M, g) be a quaternion-Kähler manifold with local I, J, K. We consider the local 2-forms $\alpha(X, Y) = g(IX, Y), \beta(X, Y) = g(JX, Y), \gamma(X, Y) = g(KX, Y)$. Then the 4-form*

$$\Omega = \alpha \wedge \alpha + \beta \wedge \beta + \gamma \wedge \gamma \tag{14.92}$$

is well-defined over all of M, parallel, non degenerate (i.e. $\Omega^n \neq 0$) *and*

$$[\Omega] = 8\pi^2 p_1(E) \text{ in } H^4(M, \mathbb{R}).$$

14.93 Definition (V.Y. Kraines [Kra]). We call Ω the *fundamental* 4-*form* of M.

The form Ω may be used to get some informations on the cohomology of M in the same way as the Kähler class in Kähler geometry. We obtain for example

14.94 Theorem (V.Y. Kraines [Kra], E. Bonan [Bon 2, 3]). *Let (M, g) be a compact quaternion-Kähler manifold with non-zero Ricci curvature. Then the Betti numbers b_i of M satisfy*

$$b_{2p} \leqslant b_{2p+4} \quad \text{and} \quad b_{2p-1} \leqslant b_{2p+1} \quad \text{for any } p < n = \frac{\dim(M)}{4}.$$

(The reference [Bon 3] refines and extends earlier results of [Kra] and [Bon 2]). Compare with 14.83 when M has positive scalar curvature.

Finally, the Leray-Hirsch theorem relates the real cohomology of Z and M. We get

14.95 Theorem (S. Salamon [Sal 1]). *The real cohomology of the twistor space Z of a compact quaternion-Kähler manifold M with non-zero Ricci curvature is the $H^*(M, \mathbb{R})$-module generated by one 2-form l such that $l^2 = [\Omega]$.*

More precisely, l is the real class corresponding to $\frac{1}{2}c_1(\mathscr{V})$. (If $\xi = 0$, l is indeed an integral class).

I. Examples of Non-Symmetric Quaternion-Kähler Manifolds

14.96. Up to now, we have given only one class of examples of quaternion-Kähler manifolds with non-zero Ricci curvature, namely the symmetric spaces G/K or G^*/K of Table 14.52. We insist that the (compact) G/K of Table 14.52 are the *only known examples* of quaternion-Kähler manifolds with positive Ricci curvature (notice that they do not have any quotient by 14.83). In the case of negative Ricci curvature, the non-compact G^*/K have many quotients, some of them being compact (indeed, a theorem by A. Borel [Bor 5] asserts that any symmetric space has compact quotients). All those quotients are of course complete quaternion-Kähler manifolds.

14.97. But there do exist some examples of *non locally-symmetric* complete quaternion-Kähler manifolds (with negative Ricci curvature). They were discovered by D.V. Alekseevskii when he classified in [Ale 3, 4] the quaternion-Kähler manifolds with a transitive solvable group of isometries. He obtained the following result, where $N(p)$ denotes the well-known Radon-Hurwitz number, which satisfies $N(8p+q) = 8^p N(q)$ and $N(1) = 1$, $N(2) = 2$, $N(3) = N(4) = 4$, $N(5) = N(6) = N(7) = N(8) = 8$.

14.98 Theorem (D.V. Alekseevskii [Ale 3, 4]). (a) *There exist two distinct families of simply-connected solvable groups, denoted by*

$W(p,q)$ for all $0 \leqslant p \leqslant q$, of dimension $4(4+p+q)$ and

$V(p,q)$ for all $1 \leqslant p$ and $0 \leqslant q$, of dimension $4(4+q+2pN(q))$

admitting a quaternion-Kähler left-invariant Riemannian metric (unique up to homotheties).

These metrics are non locally-symmetric, *except for the following cases, where we get the symmetric spaces given in the table below*:

$W(0,q)$	$SO(q+4,4)/SO(q+4)SO(4)$
$V(p,0)$	$Sp(p+4,1)/Sp(p+4)Sp(1)$
$V(1,1)$	$F_4^4/Sp(3)Sp(1)$
$V(1,2)$	$E_6^2/SU(6)Sp(1)$
$V(1,4)$	$E_7^{-5}/\mathrm{Spin}(12)Sp(1)$
$V(1,8)$	$E_8^{-24}/E_7Sp(1)$

b) *Conversely, let (M,g) be a quaternion-Kähler Riemannian manifold admitting a transitive solvable group of isometries.*

Then (M,g) is either symmetric or isometric to one of the $W(p,q)$ or $V(p,q)$ with the metric above.

14.99. We will not give here the proof of Theorem 14.98, since it is quite difficult, long and technical. We will not even give any precise definition of the groups $W(p,q)$ and $V(p,q)$, since this involves the delicate techniques of the classification of homogeneous non-compact Kähler manifolds. We want only to mention why the number $N(p)$ appears; the crucial point is that the family $V(p,q)$ is defined via the theory of $\mathbb{Z}_2$-graded Clifford modules of Clifford algebras. We recall that the Clifford algebra C_p admits a unique irreducible $\mathbb{Z}_2$-graded module, with dimension precisely $N(p)$; now $V(p,q)$ is associated with the action of C_p on q copies of its irreducible module. For more details, see [Ale 4].

14.100 Remarks. (a) We recall that if G^*/K is a non-compact irreducible symmetric space, there exists a solvable subgroup R of G^* which acts simply-transitively on G^*/K (this follows directly from the classical Iwasawa decomposition of G^*).

(b) In [Ale 3], D.V. Alekseevskii states that, for the metric of Theorem 14.98, the $W(p,q)$ and the $V(p,q)$ (these ones only if $q \neq 0$) admit a unique invariant complex structure (which is not Kähler) and also an almost-Kähler invariant almost-complex structure (which is not integrable). Here "invariant" means R-invariant and not G^*-invariant in the symmetric cases.

(c) In [Ale 4], D.V. Alekseevskii works only in the case of "completely solvable" groups and asserts that the calculations are the same (but still more complicated) in the general case.

(d) The author does not know if any of the non-symmetric $W(p,q)$ or $V(p,q)$ admits a quotient, and in particular a compact quotient (as a Riemannian manifold).

Chapter 15. A Report on the Non-Compact Case

A. Introduction

15.1. Since the main emphasis of the book is on compact spaces, this chapter on non-compact examples is only meant as a report.

In some sense there are many more non-compact Einstein spaces than compact ones. Notice though that it follows from Myers' theorem (cf. Theorem 6.52) that complete non compact Einstein metrics have a non-positive Einstein constant. Except for this numerical constraint, there is no known obstruction to the existence of a complete Einstein metric on a non-compact manifold even in dimension 4.

Aside from the local examples due to J. Gasqui [Gas] described in Chapter 5 which are germs of metrics, there are some more global examples arising most of the time from bundle constructions. The most interesting ones are of course complete.

15.2. The case of homogeneous Einstein metrics on non-compact spaces has already been treated in Chapter 7. For those spaces the most striking fact is the theorem of Alekseievski and Kim'elfeld (Theorem 7.61) which reduces the Ricci-flat locally homogeneous spaces to the flat ones. This is due to the fact that Ricci-flat spaces lie on the borderline between positive and negative Ricci curvature. In this case the main argument comes from the Cheeger-Gromoll theorem (cf. Theorem 6.65) interacting with the isometry group. For the Kähler case, see 15.47.

15.3. In the first section B, we mention an old construction of nonhomogeneous Einstein metrics due to E. Calabi although it is by now covered by more recent results. It provided the first nonhomogeneous examples of complete Kähler-Einstein metrics. It shows a nice interaction between real and complex techniques and allows us to introduce some notions we need later. The metrics that E. Calabi actually constructs are in fact cohomogeneity one metrics (cf. Chapter 9).

15.4. In Section C, we review some bundle constructions providing other examples of complete Einstein metrics. One family due to L. Bérard Bergery has already been explained in Chapter 9. We only quote results from it. The other family that we present is due to E. Calabi. It is the Kähler analogue of the preceding one. It is quite interesting, providing examples of geometric interest as (non-compact) spaces with holonomy group $Sp(m)$. It contains some examples due to Eguchi-Hanson and N. Hitchin which can be obtained by the Penrose construction (see Chapter 13).

15.5. A special section (Section D) is devoted to the deep result characterizing bounded domains admitting a complete Kähler-Einstein metric as the bounded domains of holomorphy. This was obtained by N. Mok and S.T. Yau after some earlier work by S.Y. Cheng and S.T. Yau. The techniques involved are heavily analytical and have many points in common with the solution of the Calabi conjecture (see Chapter 11).

B. A Construction of Nonhomogeneous Einstein Metrics

15.6. Before we describe the examples due to E. Calabi we must introduce a few notions of complex geometry that we will need in this section and in Section D.

A domain $\mathscr{D}$ in $\mathbb{C}^m$ (i.e., a connected open set) is called *a domain of holomorphy* if there is a function holomorphic in $\mathscr{D}$ which cannot be extended at any point p of the boundary $\partial\mathscr{D}$ into a holomorphic function defined in a neighbourhood of p.

As is well known, any domain in $\mathbb{C}$ is a domain of holomorphy. In higher dimensions the notion becomes rather restrictive. Domains of holomorphy include for example the polycylinders.

15.7. In this section we will consider complex tubular domains $M = D \times i\mathbb{R}^m$ where D is a domain in $\mathbb{R}^m$. It is well known (cf. [Bo-Ma] pp. 90–92) that M is a domain of holomorphy if and only if D is convex. This will become clear later in this paragraph.

In [Cal 4], E. Calabi equips M with Kähler metrics which are invariant under the group of purely imaginary translations. For some examples, the metrics will even be invariant under the full Euclidean group $E_m = SO(m) \times \mathbb{R}^m$. These Kähler metrics are derived from special Kähler potentials Φ (cf. 2.11) adapted to the domain D. For z in D, one sets

$$\Phi(z) = f(\mathfrak{Re}(z)),$$

where f is a real-valued function defined on D.

In order that $\omega = dd^c\Phi$ be a Kähler metric, one needs the matrix $\left(\dfrac{\partial^2\Phi}{\partial z^\alpha \partial \bar{z}^\beta}\right)$ to be positive definite (one then says that Φ is *strictly plurisubharmonic*). Because of the relation between f and Φ, this is so if and only if f is strictly convex (indeed $\left(\dfrac{\partial^2\Phi}{\partial z^\alpha \partial \bar{z}^\beta}\right) = \left(\dfrac{\partial^2 f}{\partial x^\alpha \partial x^\beta}\right)$ if $z^\alpha = x^\alpha + iy^\alpha$ are the complex coordinates on $D \times i\mathbb{R}^m$).

15.8. It follows then from 2.100 that the volume element ω^m can be expressed as

$$\omega^m = i^m \det\left(\frac{\partial^2 f}{\partial x^\alpha \partial x^\beta}\right) dz^1 \wedge d\bar{z}^1 \wedge \cdots \wedge dz^m \wedge d\bar{z}^m.$$

In particular, ω^m coincides with the volume element of the flat metric on $M \subset \mathbb{C}^m$ as soon as $\det\left(\dfrac{\partial^2 f}{\partial x^\alpha \partial x^\beta}\right) = 1$.

One easily sees that this scalar equation has many local solutions. On the other hand, as was explained in 2.101, if f satisfies this equation the Kähler metric ω is Ricci-flat. By checking directly the value of the full curvature tensor of ω, one sees that ω is flat if and only if f is quadratic.

The bad point about these examples is that, except for the trivial solution, this metric cannot be complete, nor defined on the whole space (i.e., when $D = \mathbb{R}^m$) as was found along the study of improper affine hyperspheres (see [Cal 3] and [Pog 3]).

15.9. If f is solution of the generalized real Monge-Ampère equation,

$$e^{\lambda f} \det\left(\frac{\partial^2 f}{\partial x^i \partial x^j}\right) = 1, \tag{15.9}$$

then the metric g is Kähler-Einstein with Einstein constant λ. To get a complete metric, λ must be negative.

Examples arising in this way are products of irreducible homogeneous Bergman manifolds after adjusting properly the Einstein constants of the factors.

Other examples can be constructed by taking D to be the open ball of radius r and the function f to depend only on the distance to the center. We set $f(x) = \varphi(r)$. In this way we come back to a cohomogeneity one construction. Notice that then the full Euclidean group of the imaginary factor acts as a group of isometries.

If φ is a solution of

$$\left(\frac{\varphi'}{r}\right)^{m-1} \varphi'' = e^{\varphi}, \tag{15.10}$$

then f satisfies (15.9).

To ensure smoothness at the origin of the metric, one must assume

$$\varphi'(0) = 0, \qquad m\varphi''(0) = e^{\varphi(0)}. \tag{15.10$'$}$$

To prove that f is indeed a smooth convex solution of (15.9) which diverges at the boundary and that the metric is complete requires the use of barrier functions which are sub- and supersolutions of (15.10) and (15.10′). This type of tool will also be used in Section D.

In [Wol 6] J.A. Wolf gives a nice proof of the fact that this metric is not locally homogeneous. It is based on group theoretic arguments using the E_m-invariance of the metric constructed.

C. Bundle Constructions

15.11. The main idea behind bundle constructions of Einstein metrics is to break the Einstein equation into pieces using the geometric structure. On some of the pieces (e.g., the base and the fibres), the equation is known to be fulfilled because the building blocks satisfy some geometric properties (like being already an Einstein metric of lower dimension or a slight modification of it). In some sense, the problem is then reduced to controlling how the base and fibres fit together.

15.12. If the base is one-dimensional, this means solving an ordinary differential system. If the total space is compact, the base space must be either a circle or an interval. The second case is more interesting but leads to exceptional fibres over which one gets two boundary conditions for the differential system. The system becomes overdetermined. It is therefore no surprise that on non compact spaces it is easier to get examples. Over a half-line the system has only boundary conditions at one end and the whole line has no boundary.

15.13. The first result illustrating this fact is probably classical. It can be found in [Koi 5]. It relies on the use of the Einstein equation as an evolution equation, a Cauchy datum on a hypersurface being given.

Theorem [Koi 5, Theorem A]. *Let (M, g) be a real analytic Riemannian manifold with constant scalar curvature. Then, there exists an Einstein manifold $(\overline{M}, \overline{g})$ such that (M, g) is isometrically embedded into $(\overline{M}, \overline{g})$ as a totally geodesic hypersurface.*

15.14. Of course, if one looks for complete Einstein metrics one must work harder. L. Bérard Bergery proved for example (cf [BéBer 3] Proposition 5.8) that $G/K \times \mathbb{R}$ admits no G-invariant complete metrics as soon as all G-invariant metrics on G/K are proportional.

15.15. The first easy example recalled in [BéBer 3], Proposition 9.3, is given by $M \times \mathbb{R}$ with metric $\dfrac{3k}{n-2}(\cosh t)^2 g + dt^2$ so long as g is an Einstein metric on the $(n-1)$-dimensional manifold M with negative constant λ.

15.16. For a 2-dimensional non-compact base space, aside from trivial product situations or constant curvature spaces, one has the following examples

Proposition [BéBer 3, Proposition 9.5]. *If (M, g) is a complete Einstein manifold with constant λ, then*

i) *if $\lambda > 0$, $M \times \mathbb{R}^2$ admits a complete Ricci-flat metric;*

ii) *if $\lambda < 0$ (resp. $\lambda = 0$), $M \times S^1 \times \mathbb{R}$ admits a complete Einstein metric with negative constant (resp. with vanishing constant);*

iii) *for all values of λ, $M \times \mathbb{R}^2$ admits a one-parameter family of complete Einstein metrics with negative constant.*

15.17. It is interesting to notice that this proposition carries over to Lorentzian metrics (as many of the bundle constructions). As an example, the Schwarzschild (cf. Chapter 3) can be found in case *i*) by taking $(M, g) = (S^2, \text{can})$ and a Lorentzian metric on $\mathbb{R}^2$ thought as a base space.

15.18. We now come to the bundle construction of Kähler-Einstein complete metrics on non-compact spaces. We first follow E. Calabi [Cal 5]. The starting point is to notice that the construction of Riemannian submersions with totally geodesic fibres (cf. Chapter 9) although not adapted to holomorphic bundles can be adapted to the Kähler situation. To require that the differential of the projection map be an isometry on each horizontal space prevents the metric to be Kählerian except in trivial product situations. If $\pi\colon L \to (M, \omega_M)$ is a holomorphic vector bundle over a Kähler manifold (we use the letter L to denote the total space in that case) equipped with a hermitian fibre metric t and its canonical Hermitian connection ∇, one then requires the metric condition to hold only along the zero section. Then, there is at most one Kähler metric ω_L on L which induces the given Hermitian fibre metric t and for which horizontal vectors for the connection ∇ are orthogonal to vertical vectors. It is sometimes convenient to think of t as a quadratic function on each fibre. If one does so, one easily sees for example that $t + \Phi \circ \pi$ is a candidate Kähler potential for ω_L for any Kähler potential Φ for the base metric ω_M. By computing in a well chosen frame (differentials of holomorphic coordinates on the base, covariant differentials of holomorphic coordinates on the bundle), one can see that, at the point ζ of the bundle,

(15.19) $$\omega_L = \omega_M + \pi^* R^\nabla(\zeta, \bar{\zeta}) + \tilde{t}$$

where the curvature form R^∇ is considered as a 2-form on the base with values in the skew-hermitian forms on the fibre using the vertical Hermitian metric and $\tilde{t}$ is the extension of t to tangent vectors of the total space by using the orthogonality between horizontal and vertical vectors. It follows from Formula (15.19) that ω_L is positive in a neighborhood of the zero section. Moreover, if the Hermitian bundle (L, t) is positive, ω_L is also globally positive.

15.20. This suggests to generalize the preceding considerations to an open (hence non-compact) sub-bundle E of L. We will endow the fibres with a not necessarily flat Hermitian metric which we suppose to be invariant under the structure group of the Hermitian bundle. It is clear that the Kähler potential of such a metric can be of the type $u \circ t$ where $u\colon [0, +\infty[\to \mathbb{R}$. In that case we have

$$\omega_E = \pi^* \omega_M + dd^c(u \circ t),$$

which combined with (15.19) gives, at the point ζ of E,

(15.21) $$\omega_E = \pi^* \omega_M + u' R^\nabla(\zeta, \bar{\zeta}) + u' t(\nabla\zeta, \nabla\bar{\zeta}) + u'' t \otimes t(\zeta, \overline{\nabla\zeta}, \nabla\zeta, \bar{\zeta}).$$

In vertical directions, ω_E is positive if and only if, for $x \geqslant 0$,

$$u'(x) \geqslant 0, \qquad u'(x) + x u''(x) \geqslant 0.$$

15.22. We want ω_E to be an Einstein metric. For that purpose, we suppose first that ω_M is a Kähler-Einstein metric. Computations are easier to conduct when the fibres are one-dimensional. In that case the curvature form R^∇ of the bundle can be considered as an ordinary 2-form on the base. We *suppose* moreover that R^∇ is proportional to the Kähler form, i.e., $R^\nabla = l\omega_M$. This condition can be fulfilled only if $[l\omega_M]$ is an integral form. Since we already assumed ω_M to be Kähler-Einstein with constant λ_M, we know that $[\lambda_M \omega_M]$ is an integral class.

If $\lambda_M \neq 0$, l must belong to a discrete subgroup of $\mathbb{R}$ if $[\omega_M] \neq 0$. If $[\omega_M] = 0$ (hence M is non-compact), l can be any real number.

If $\lambda_M = 0$, one can conduct the same discussion, but ω_M must be a Hodge metric up to a constant. In the sequel, we will suppose that this condition is fulfilled.

Then the metric ω_E can be written in the form

$$\omega_E = (1 + l|\zeta|^2 u')\pi^*\omega_M + (u' + u''|\zeta|^2)t(\nabla\zeta, \nabla\bar{\zeta}) \tag{15.23}$$

15.24 Theorem [Cal 5, Proposition 4.1]. *Let $\pi: L \to M$ be a holomorphic Hermitian line bundle with curvature l over an $(m - 1)$-dimensional Kähler manifold. Let u be a C^∞ real-valued function defined on $[0, +\infty[$. On the subdomain $E = t^{-1}(I)$ of L where I is the interval of $\mathbb{R}$ on which*

$$1 + lxu'(x) > 0 \quad \text{and} \quad u'(x) + xu''(x) > 0, \tag{15.25}$$

ω_E is a Kähler-Einstein metric with constant λ_E if and only if u satisfies the differential equation

$$(1 + lxu'(x))^{m-1}(xu''(x) + u'(x))\exp(\lambda_E u(x)) = cx^{(l+\lambda_E-\lambda_M)/l} \tag{15.26}$$

for some positive constant c.

If $l = 0$, we must have $\lambda_M = \lambda_E = \lambda$. If $\lambda \neq 0$,

$$\lambda u(x) = 2\log(1 + \alpha\lambda x^a) + \beta\log x + \gamma$$

with $\alpha, \beta, \gamma \in \mathbb{R}$, $a \neq 0$, and if $\lambda = 0$,

$$u(x) = \alpha^2 x^a + \beta\log x + \gamma$$

are solutions.

15.27. The main point about solving Equation (15.26) is the existence of the integrating factor $\lambda_M - \lambda_E - l\lambda_E u'(x)$. A geometric interpretation of this integrating factor is yet to be found.

A Kähler-Einstein metric can be extended to the zero section for $l \neq 0$ if one has $\lambda_E = \lambda_M - l$ and $u'(0) > 0$ and for $l = 0$ if there is no logarithmic term. The equation is then reduced to the following first order nonlinear differential equation

$$(m + 1)\lambda_M[(1 + lxu'(x))^m - 1] - m\lambda_E[(1 + lxu'(x))^{m+1} - 1] = \alpha x^l e^{-ku(x)} + \beta. \tag{15.28}$$

What are even more important are criteria to ensure that the Kähler metric ω_E is complete.

15.29 Theorem [Cal 5, Theorem 4.3]. *The Kähler metric ω_E is complete on its maximal domain of definition if and only if*

i) *ω_M is complete;*
ii) *$\lambda_E = \lambda_M - l$;*
iii) *$l \geqslant 0$ and $\lambda_E < 0$.*

15.30. E. Calabi notices that a solution of (15.26) can be given in closed form for the case $\lambda_E = 0$ and $l > 0$. The solution is then defined on the total space of the linear bundle L we started from. This construction applies in particular to the total space of the canonical bundle of a Kähler-Einstein manifold with positive constant which as a consequence admits a complete Ricci-flat Kähler metric. This remark applies in particular to $\mathbb{C}P^1$. The Kähler-Einstein metric obtained on $T^*\mathbb{C}P^1$ coincides with the metric given by the Penrose construction by T. Eguchi-A.J. Hanson and N. Hitchin (cf [Eg-Ha], [Hit 4]). This other point of view is treated in the Chapter 13 where a translation into twistor terms of the Einstein condition is given. Notice that this other construction has so far only provided non-compact examples (compare 13.87).

15.31. Using similar constructions on holomorphic cotangent bundles, E. Calabi obtains a complete Kähler-Einstein metric on the total space of the holomorphic cotangent bundle of $\mathbb{C}P^m$. This metric on $T^*\mathbb{C}P^m$ has holonomy group $Sp(m)$ (cf. Chapter 14 and Add.E).

15.32. In Chapter 9, one finds also Kähler-Einstein metrics on the total spaces of some (almost) bundles as cohomogeneity one metrics. See 9.129 for details.

D. Bounded Domains of Holomorphy

15.33. In this section we describe results on non-compact complex manifolds which tie together holomorphic properties with differential geometric ones.

When M is a compact complex manifold, the first Chern class $c_1(M)$ in $H^2(M, \mathbb{Z})$ is a cohomological invariant attached to the complex structure of M. One way of introducing differential geometric assumptions on this cohomology class is to suppose that the image of $c_1(M)$ in $H^2(M, \mathbb{R})$ admits a nowhere degenerate closed differential 2-form, hence a Kähler form or the opposite of a Kähler form, as representative.

When $c_1(M)$ is negative, it is known that there exists a unique Kähler-Einstein metric up to a scale (see 11.17).

15.34. When M is non-compact, the assumption on the first Chern class can be replaced by the following: "on M, there exists a complete Hermitian metric with Ricci form bounded from above by a negative multiple of the metric". The assumption is obviously satisfied if the manifold has a complete Kähler-Einstein metric.

The question is then to relate complex properties of M with the existence of a complete Kähler-Einstein metric.

15.35. We begin by reviewing some basic notions of complex geometry that we need later. A domain in $\mathbb{C}^m$ (i.e., a connected open set) with C^2 boundary is said to be *pseudo-convex* (resp. *strictly pseudo-convex*) if one can find a nowhere vanishing C^2 function φ vanishing on the boundary which is *plurisubharmonic* (resp. *strictly plurisubharmonic*), that is such that its Levi form $dd^c\varphi$ is non-negative (resp. positive).

15.36. Another notion of great importance is that of a domain of holomorphy. One says that D is *a domain of holomorphy* if there is a function holomorphic in D which cannot be extended holomorphically at any point of the boundary by a function defined in a neighbourhood of this point. Any domain in $\mathbb{C}$ verifies this property. In higher dimensions the notion is much more restrictive. On the other hand, it includes for example the polycylinders.

In fact Oka's theorem (cf [Oka 1] and [Oka 2]) states that D is a domain of holomorphy if and only if $-\log \rho_{\partial D}$ (where $\rho_{\partial D}$ denotes the Euclidean distance to the boundary of D) is plurisubharmonic. If one extends the notion of pseudoconvexity appropriately to domains with non-C^2 boundaries, this says that a domain D in $\mathbb{C}^m$ is a domain of holomorphy if and only if D is pseudoconvex.

15.37. Moreover, to get a satisfactory theory domains in $\mathbb{C}^m$ have to be generalized to coverings of domains in $\mathbb{C}^m$, the so-called *Riemann domains over* $\mathbb{C}^m$. This generalizes what one does in one complex variable to study the Riemann surface of $z \mapsto z^{1/k}$ $(k \in \mathbb{N}^*)$. Thanks to the covering map, the notions of pseudo-convexity and of domains of holomorphy extend to Riemann domains over $\mathbb{C}^m$ in a straightforward manner (cf [Fr-Gr] p. 66 for example).

15.38. The main theorem is now the following

Theorem [Mo-Ya, Main theorem]. *The following statements are equivalent*:

i) *D is a bounded Riemann domain of holomorphy over $\mathbb{C}^m$*;

ii) *the bounded Riemann domain D over $\mathbb{C}^m$ admits a complete Kähler-Einstein metric*;

iii) *the bounded Riemann domain D over $\mathbb{C}^m$ admits a complete Hermitian metric with bounded non-positive Ricci curvature.*

15.39. This theorem elaborates on previous work of S.Y. Cheng-S.T. Yau (cf. [Ch-Ya 2]) who proved the existence of complete Kähler-Einstein metrics on a large class of pseudoconvex domains containing those with C^2 boundaries.

We only give a sketch of the proof. To show that i) implies iii) relies on solving an appropriate Monge-Ampère equation. The background Kähler metric ω is taken to be $id'd''(-\log \rho_{\partial D})$ if D is a strictly peusdoconvex domain in $\mathbb{C}^m$ with smooth boundary. Further complications arise either if the boundary is not smooth or not strictly pseudoconvex. We refer the reader to [Ch-Ya 2] for details.

15.40. The nice feature of this metric is that it is almost Einstein near the boundary. The general form of the Monge-Ampère equation that one obtains is as in the compact case (cf. 11.41)

$$\log((\omega - \tfrac{1}{2}dd^c\varphi)^m/\omega^m) - \lambda\varphi = F.$$

In our situation $\lambda < 0$. The crucial property here is the boundedness of F on $\bar{D}$, the closure of D.

To get existence by the continuity method, one needs to find the right function space to show openness of the interval on which the family of equations obtained by taking tF instead of F is solvable. To prove closedness one must get estimates up to the third derivatives to be able to apply Schauder estimates on functions having $C^{2,\alpha}$-Hölder regularity. To get the estimates one works as in the compact case, the generalized maximum principle being the only new ingredient. This is carried out in [Ch-Ya]. The survey article [Be-Fe-Gr] gives also a sketch of the proof. Notice that boundedness of the geometry has to be assumed or derived here, unlike the compact case.

Earlier work by H. Grauert [Gra], C. Fefferman [Fef] and K. Diederich and P. Pflug [Di-Pf] should be quoted.

15.41. There is a unique solution $\tilde{\omega}$ which is metrically equivalent to the background metric, i.e., which satisfies $C_1\omega < \tilde{\omega} < C_2\omega$ for some positive real constants C_1 and C_2. A nice consequence of this fact is the following theorem tying up differential geometric and complex properties. In some sense it means that the complete Kähler-Einstein metric is really the "best" metric one can find on manifolds which admit one.

Theorem [Ch-Ya, Proposition 5.5]. *Any biholomorphic map between complex manifolds is an isometry of complete negative Kähler-Einstein metrics.*

15.42. To show that the metric obtained is complete is a delicate point. This was the missing part of [Ch-Ya 2] where only almost completeness was proved. The proof is given in [Mo-Ya]. The main estimate is a lower bound of the distance in the new metric between two points z_0 and z_1 of D in terms of $\log(-\log \rho_{\partial D}(z_0))$ where $\rho_{\partial D}$ denotes the distance to the boundary in the Euclidean metric. Such an estimate follows from application of the following generalization of the Schwarz lemma which is an extension of [Yau 2].

15.43 Theorem [Mo-Ya, § 1]. *Let M_1 and M_2 be m-dimensional complex manifolds. If M_1 has a Kähler metric with scalar curvature bounded from below by $-k_1$ and M_2 a volume form v_2 such that its Ricci form is negative definite with $(dd^c \log v_2)^m \geqslant k_2 v_2$, for any non everywhere singular map f one has*

$$\sup \frac{f^* v_2}{v_1} \leqslant \frac{k_1^m}{m^m k_2}.$$

15.44. The proof relies on a maximum principle for complete Riemannian manifolds which is of general interest and that we state below

Maximum Principle for Complete Riemannian Manifolds [Yau 1]. *On a complete Riemannian manifold with Ricci curvature bounded from below, let f be a C^2 function which is bounded from below on M. Then, for any $\varepsilon > 0$, there exists a point p in M such that*

$$|df(p)| < \varepsilon, \qquad \Delta f(p) < \varepsilon, \qquad f(p) < \inf_M f + \varepsilon.$$

15.45. At this point we want to mention that the Bergman metric is not necessarily complete on a bounded domain of holomorphy (compare [Kob 3]). This remark makes the Kähler-Einstein metric that we now know exists on a bounded domain of holomorphy more interesting than the Bergman metric.

In fact the two metrics have some relations. In [Ch-Ya 2] § 6 it is shown that the boundary behaviour of the Kähler-Einstein metric is connected with the regularity of the Bergman kernel at the boundary. In particular S.Y. Cheng and S.T. Yau obtain almost optimal boundary regularity for a solution of a Dirichlet problem considered by C. Fefferman in his study of the Bergman kernel. (One may also consult [Be-Fe-Gr].)

15.46. In [Mo-Ya] one also finds an extension of Theorem 15.38 to certain Stein manifolds (and some unbounded domains of holomorphy). We recall that a *Stein manifold* is an m-dimensional complex manifold which admits a finite holomorphic mapping to $\mathbb{C}^m$ and in which the holomorphic convex hull of a compact set is compact.

15.47. *Non-compact homogeneous Kähler-Einstein manifolds.* On a homogeneous bounded domain, there is one and only one invariant Kähler-Einstein metric, the Bergmann metric. It has negative Ricci curvature. This has been first observed by E. Cartan [Car 13].

The converse is true: all non-compact homogeneous Kähler manifolds with negative Ricci curvature (this includes negative Einstein manifolds) are bounded domains in their Bergmann metric. This appears in unpublished work of J.L. Koszul [Koz 2], see also [Nak], theorem 13.

Chapter 16. Generalizations of the Einstein Condition

A. Introduction

In this chapter we discuss some *generalizations of Einstein metrics*, that is, a few classes of Riemannian manifolds characterized by tensorial conditions, which are consequences of the Einstein metric equation. Among such generalizations, we restrict our consideration to those which have been studied in the differential geometric literature and can be illustrated by interesting examples.

Since the Einstein condition is an algebraic linear equation in r, it is to some extent natural that we first consider linear differential equations generalizing it. The simplest one is $Dr = 0$, providing a local characterization of products of Einstein manifolds. Its most immediate consequences consist in turn in imposing on Dr the *natural linear conditions*, which correspond to vanishing of certain irreducible components of Dr under the action of the orthogonal group. The bundle where Dr takes its values splits into 3 irreducible invariant subbundles, giving rise, besides $Dr = 0$, to 6 conditions of this type, presented (in dimensions $n \geqslant 3$) by the following table:

Condition	Known examples of compact manifolds of this type, other than Einstein or locally product ones
$dr = 0$, i.e., r is a Codazzi tensor. Dr takes its values in the invariant subbundle denoted by S. Equivalent conditions: 1) $\delta R = 0$ (harmonic curvature) 2) If $n \geqslant 4$: $\delta W = 0$ and constant scalar curvature	1) Compact conformally flat manifolds with constant scalar curvature 2) Compact quotients of $(\mathbb{R} \times \overline{M}, dt^2 + f^{4/n}(t) \cdot \overline{g})$, where $(\overline{M}, \overline{g})$ is Einstein with scalar curvature $\overline{s} > 0$ and f is a positive solution of $d^2f/dt^2 - \frac{1}{4}n(n-1)^{-1} \cdot \overline{s} f^{1-4/n} = cf$ with a constant $c < 0$ 3) Compact quotients of $(M_1 \times M_2, f^{q-2}(g_1 \times g_2))$, where (M_1, g_1) has constant curvature $K_1 < 0$, (M_2, g_2) is Einstein with scalar curvature $s_2 = -n_2(n_2-1)K_1$, $n_i = \dim M_i \geqslant 2$, $q = 2(n_1+n_2)(n_1+n_2-2)^{-1}$, $n_2 > n_1 - K_1^{-1}\lambda_1$, λ_1 being the lowest positive eigenvalue of the Laplacian Δ of (M_1, g_1), and $f\colon M_1 \to (0, \infty)$ is a C^∞ solution to $\Delta f + \frac{1}{4}(n_1 - n_2)(n_1 + n_2 - 2)K_1 f = cf^{q-1}$ with a constant $c > 0$ 4) For further examples, see 16.40.

Condition	Known examples of compact manifolds of this type, other than Einstein or locally product ones
$2(n-1)(n+2)D[r-(2n-2)^{-1}sg] = (n-2)\,ds\circ g$. *Dr* is a section of the invariant subbundle denoted by *Q*. (This implies $\delta W = 0$)	Compact quotients of $(\mathbb{R}\times\overline{M}, dt^2 + f^{-2}(t)\cdot\overline{g})$, where $(\overline{M},\overline{g})$ is Einstein with scalar curvature $\overline{s}<0$ and f is a positive solution of $$d^2f/dt^2 - 2(n-1)^{-1}(n-2)^{-1}\cdot\overline{s}f^3 = cf$$ with a constant $c>0$
$(D_Xr)(X,X) = 0$ for all vectors X. *Dr* is a section of the invariant subbundle denoted by *A*. (This implies constant scalar curvature)	1) Compact quotients of naturally reductive homogeneous Riemannian manifolds 2) Nilmanifolds covered by the generalized Heisenberg groups of A. Kaplan
$d[r-(2n-2)^{-1}sg] = 0$, i.e., $r-(2n-2)^{-1}sg$ is a Codazzi tensor. $Dr\in C^\infty(Q\oplus S)$. Equivalent condition (if $n\geqslant 4$): $\delta W = 0$ (harmonic Weyl tensor)	1) Compact conformally flat manifolds 2) Compact manifolds locally isometric to $(M_1\times M_2, f\cdot(g_1\oplus g_2))$, where $\dim M_i = n_i\geqslant 1$, (M_i,g_i) has scalar curvature s_i and either a) (M_1,g_1) of constant curvature, (M_2,g_2) Einstein with $n_1(n_1-1)s_2 + n_2(n_2-1)s_1 = 0$ (e.g., $n_1 = 1$), and f is an *arbitrary* positive function on M_1, or b) $n_1 = 2$, (M_2,g_2) two-dimensional or of constant curvature, $f = \lvert 2s_2 + n_2(n_2-1)s_1\rvert^{2/(3-n)} > 0$
$\delta r = 0$ (constant scalar curvature) $Dr\in C^\infty(S\oplus A)$.	Every compact manifold admits such a metric. For details, see Chapter 4.
$(D_X[r-2(n+2)^{-1}sg])(X,X) = 0$ for each vector X. $Dr\in C^\infty(Q\oplus A)$	See the examples for $Dr\in C^\infty(Q)$ and $Dr\in C^\infty(A)$.

In Section D through G we are concerned with four of these conditions, except for $Dr\in C^\infty(S\oplus A)$, dealt with in Chapter 4, and $Dr\in C^\infty(Q\oplus A)$, since very little is known about it.

Section C is devoted to Codazzi tensors, discussed separately for reasons explained in 16.6. Finally, in Section H we study oriented Riemannian four-manifolds satisfying $\delta W^+ = 0$, which is a natural linear condition on *Dr* relative to the *special* orthogonal group.

B. Natural Linear Conditions on *Dr*

16.1. In order to discuss the irreducible components of *Dr* under the orthogonal group action, let us first recall that, for any Riemannian manifold (M,g), *Dr* is a 3-tensor field having two additional algebraic properties, the one coming from the symmetry of r and the latter from the Bianchi identity $\delta r = -\frac{1}{2}ds$ (see 1.94). Thus, Dr is a section of the vector bundle $H = H(M,g) \subset T^*M \otimes S^2M \subset \bigotimes^3 T^*M$ the fibre of which, at any point $x \in M$, consists of all 3-linear maps ξ of T_xM into $\mathbb{R}$ such that $\xi(X,Y,Z) = \xi(X,Z,Y)$ and $\sum_{i=1}^n [\xi(X_i,X_i,X) - \frac{1}{2}\xi(X,X_i,X_i)] = 0$ for any X, Y, $Z \in T_xM$ and any orthonormal basis $(X_1,\ldots,X_n)$ of T_xM, $n = \dim M$.

A discussion of the irreducible components of Dr can also be found in A. Gray's article [Gra 3].

16.2. Given a Riemannian manifold (M,g), $\dim M = n \geqslant 3$, one has the following natural vector bundle homomorphisms associated with $\bigotimes^3 T^*M$: the *contraction* $\gamma\colon \bigotimes^3 T^*M \to T^*M$, the *partial alternation* $\alpha\colon \bigotimes^3 T^*M \to \Lambda^2 M \otimes T^*M$, the *partial symmetrization* $\sigma\colon \bigotimes^3 T^*M \to \bigotimes^3 T^*M$ and the mapping $\varphi\colon T^*M \to H(M,g)$, given by

$$(\gamma(\xi))X = \sum_{i=1}^n \xi(X_i,X_i,X),$$

$$(\alpha(\xi))(X,Y,Z) = \tfrac{1}{2}[\xi(X,Y,Z) - \xi(Y,X,Z)],$$

$$(\sigma(\xi))(X,Y,Z) = \tfrac{1}{3}[\xi(X,Y,Z) + \xi(Y,Z,X) + \xi(Z,X,Y)].$$

$$(\varphi(\zeta))(X,Y,Z) = (X,Y)\cdot\zeta(Z) + (X,Z)\cdot\zeta(Y) + 2n(n-2)^{-1}(Y,Z)\cdot\zeta(X),$$

for $\xi \in \bigotimes^3 T_x^*M$, $X,Y,Z \in T_xM$, $\zeta \in T_x^*M$ and any orthonormal basis $(X_1,\ldots,X_n)$ of T_xM, $x \in M$. Since $(n-2)\gamma\circ\varphi = (n-1)(n+2)\cdot Id_{T^*M}$ and $(n-2)(\varphi(\zeta),\xi) = (7n-6)(\gamma(\xi),\zeta)$ for any $\zeta \in T^*M$ and $\xi \in H(M,g)$, it follows that the n-dimensional invariant subbundle $Q = Q(M,g) = \operatorname{Im}\varphi$ of $H = H(M,g)$ coincides with the orthogonal complement of $H \cap \operatorname{Ker}\gamma$ in H. The subbundles $S = S(M,g) = H \cap \operatorname{Ker}\alpha \subset \operatorname{Ker}\gamma$ and $A = A(M,g) = H \cap \operatorname{Ker}\sigma \subset \operatorname{Ker}\gamma$ of H are mutually orthogonal. It is now easy to verify that

$$H = Q \oplus S \oplus A$$

is an orthogonal decomposition of H into a direct sum of invariant (i.e., naturally defined) subbundles (explicitly, and $\xi \in H$ has the components $\xi_Q = (n-2)(n-1)^{-1}(n+2)^{-1}\varphi(\gamma(\xi))$, $\xi_S = \sigma(\xi - \xi_Q)$). Using standard arguments of invariant theory one can prove (cf. [Gra 5], compare with 1.114) that this is the *unique irreducible orthogonal decomposition of H*. Moreover, the pairwise direct sums of the subbundles Q, S, A are easily seen to admit the following characterizations: for any $\xi \in H$,

i) $\xi \in S \oplus A$ if and only if $\gamma(\xi) = 0$.

ii) $\xi \in Q \oplus S$ if and only if $\alpha[\xi - (n-1)^{-1}\gamma(\xi)\otimes g] = 0$.

iii) $\xi \in Q \oplus A$ if and only if $\sigma[\xi - 4(n+2)^{-1}\gamma(\xi)\otimes g] = 0$.

16.3. Let (M,g) be an n-dimensional Riemannian manifold, $n \geqslant 3$. For a tensor field $T \in \mathscr{T}T^{k+1}$, we have defined in 1.58 its *divergence* $\delta T \in \mathscr{T}^k T^*M$ by

$$(\delta T)(X_1,\ldots,X_k) = -tr_g[(Y,Z) \to (D_Y T)(Z,X_1,\ldots,X_k)].$$

Thus, the divergences δR and δW of R and W are sections of $T^*M \otimes \Lambda^2 M$, and, using an obvious switch of the arguments, we may view them as sections of $\wedge^2 M \otimes T^*M$. Under this identification, the differential Bianchi identity gives

$$\delta R = -dr, \qquad \delta W = -\frac{n-3}{n-2} d[r - (2n-2)^{-1} sg],$$

where, for any symmetric 2-tensor field b, db denotes the *exterior derivative* of b (viewed as a T^*M-valued 1-form cf. 1.12), i.e., $db = 2\alpha(Db)$. We shall say that (M,g) has *harmonic curvature* (resp., *harmonic Weyl tensor*) if $\delta R = 0$ (resp., if $\delta W = 0$). A symmetric 2-tensor field b on (M,g) will be called a *Codazzi tensor* if $db = 0$, i.e., if b satisfies the *Codazzi equation* $(D_X b)(Y,Z) = (D_Y b)(X,Z)$ for arbitrary tangent vectors X, Y, Z.

16.4. For a Riemannian manifold (M,g), $\dim M = n \geqslant 3$, the natural linear conditions that can be imposed on Dr can be characterized, in view of 16.2 and 16.3, as follows:

(i) $Dr \in C^\infty(Q)$ if and only if
$D[r - (2n-2)^{-1} sg] = \frac{1}{2}(n-2)(n+2)^{-1}(n-1)^{-1} ds \odot g$.

(ii) $Dr \in C^\infty(S)$ is equivalent to each of the following conditions:
- a) $\alpha(Dr) = 0$;
- b) $dr = 0$, i.e., r is a Codazzi tensor;
- c) $\delta R = 0$, i.e., (M,g) has harmonic curvature;
- d) $n \geqslant 4$: (M,g) has harmonic Weyl tensor and constant scalar curvature;
- e) $n = 3$: (M,g) is conformally flat and has constant scalar curvature (cf. 1.164).

(iii) $Dr \in C^\infty(A)$ if and only if $\sigma(Dr) = 0$, i.e., $(D_X r)(X,X) = 0$ for any tangent vector X.

(iv) $Dr \in C^\infty(S \oplus A)$ if and only if s is constant.

(v) $Dr \in C^\infty(Q \oplus S)$ is equivalent to any of the following conditions:
- a) $\alpha(D[r - (2n-2)^{-1} sg]) = 0$;
- b) $d[r - (2n-2)^{-1} sg] = 0$, i.e., $r - (2n-2)^{-1} sg$ is a Codazzi tensor;
- c) $n \geqslant 4$: $\delta W = 0$, i.e., (M,g) has harmonic Weyl tensor;
- d) $n = 3$: (M,g) is conformally flat (cf. 1.170).

(vi) $Dr \in C^\infty(Q \oplus A)$ if and only if $\sigma(D[r - 2(n+2)^{-1} sg]) = 0$, i.e., if $(D_X[r - 2(n+2)^{-1} sg])(X,X) = 0$ for any vector X.

16.5. For an *oriented four-dimensional* Riemannian manifold (M,g), the decomposition $\wedge^2 M = \wedge^+ M \oplus \wedge^- M$ (see 1.122) gives rise to additional subbundles of $H = H(M,g)$, invariant under the action of the *special* orthogonal group (i.e., naturally determined by the metric and the orientation). Namely, $A = A^+ \oplus A^-$, where $A^\mp = A \cap \operatorname{Ker}(\pi^\pm \circ \alpha)$, $\pi^\pm : \wedge^2 M \otimes T^*M \to \wedge^\pm M \otimes T^*M$ being the projection. For instance, in terms of the decomposition $W = W^+ + W^-$, it is easy to see that (M,g) satisfies the natural linear condition $Dr \in C^\infty(Q \oplus S \oplus A^-)$ (i.e., $d(r - sg/6) \in C^\infty(\wedge^- M \otimes T^*M)$) if and only if $\delta W^+ = 0$. Note that $\delta W^\pm$ are the $(T^*M \otimes \wedge^\pm M)$-components of $\delta W = -\frac{1}{2} d(r - sg/6)$ (cf. 16.3), which follows immediately from the fact that $\wedge^\pm M$ are invariant under parallel transport.

C. Codazzi Tensors

16.6. Since Codazzi tensors occur in a natural way in the study of Riemannian manifolds with harmonic curvature or with harmonic Weyl tensor (cf. 16.4ii)b), 4v)b)), and a considerable part of known results on such manifolds are easy consequences of theorems on Codazzi tensors, we discuss these tensors separately in this section. After presenting various examples, we state some theorems relating Codazzi tensors to the structure of the curvature operator (see 16.9, 16.14, 16.21). Throughout this section, for a given Codazzi 2-tensor b on (M, g), we shall *denote* by B the *Codazzi tensor of type* $(1, 1)$, corresponding to b via g.

16.7. The simplest examples of Codazzi tensors (for further examples, see 16.12 and 16.48, 16.49).

(i) Parallel tensors.

(ii) Second fundamental forms of hypersurfaces in spaces of constant curvature, see 1.72.

(iii) For any function f on a space (M, g) of constant curvature K, the Ricci identity 1.21 implies that $b = Ddf + Kfg$ is a Codazzi tensor (as shown by D. Ferus [Frs], these are, locally, the only Codazzi tensors in such spaces). Similarly, for any surface (M, g) with Gaussian curvature K, $b = 2DdK + K^2 g$ is a Codazzi tensor.

(iv) By 16.4v), a Riemannian manifold (M, g) has harmonic Weyl tensor if and only if $b = r - (2n-2)^{-1} sg$ is a Codazzi tensor ($n = \dim M \geqslant 4$; see examples given in Section D). Hence (cf. (iii)), every conformally flat manifold admits a Codazzi tensor which is not a constant multiple of the metric.

16.8 Proposition (Y. Matsushima [Mat 3], S. Tanno [Tan 1], cf. [Gra 5]. *In a Kähler manifold (M, J, g), any Codazzi tensor B which is Hermitian (i.e., commutes with J), must be parallel.*

In fact, the composite $\zeta = J \circ B$ is then skew-adjoint and hence the expression $((D_X \zeta) Y, Z) = (J((D_X B) Y), Z)$ is symmetric in X, Y and skew-symmetric in Y, Z, so that it is zero.

16.9 Theorem (M. Berger, cf. [Be-Eb], [Rya], [Sio 2], [Weg], [Bou 8], [Gra 5]). *Every Codazzi tensor b with constant trace on a compact Riemannian manifold (M, g) with non-negative sectional curvature K is parallel. If, moreover, $K > 0$ at some point, then b is a constant multiple of g.*

Proof. For any Codazzi tensor b, the Weitzenböck formula 1.136 can be rewritten as

$$\delta Db + Dd(tr_g b) = \mathring{R}(b) - b \circ r. \tag{16.9}$$

For any $x \in M$ and some orthonormal basis $(X_1, \ldots, X_n)$ of $T_x M$ ($n = \dim M$), we have $b_x(X_i, X_j) = \lambda_i \delta_{ij}$ and, at x, $(b, \mathring{R}(b) - b \circ r) = -\sum_{i<j} R(X_i, X_j, X_i, X_j)(\lambda_i - \lambda_j)^2 \leqslant 0$. On the other hand, $\int_M (b, \delta Db) v_g = \int_M |Db|^2 v_g \geqslant 0$, so that our assertion follows from Formula (16.9). □

16.10. Let b be any symmetric 2-tensor field on a Riemannian manifold (M, g). Given

$x \in M$ and an eigenvalue λ of b_x, we shall *denote* by $V_\lambda(x) \subset T_xM$ the corresponding eigenspace. In every connected component of the open dense subset M_b of M, consisting of points at which the number of distinct eigenvalues of b is locally constant, the eigenvalues of b form mutually distinct smooth *eigenvalue functions* and, for such a function λ, the assignment $x \mapsto V_{\lambda(x)}(x)$ defines a smooth *eigenspace distribution* V_λ of b. If λ and μ are such eigenvalue functions, then, for any vector fields X, Y, Z with $X \in C^\infty(V_\lambda)$, $Y \in C^\infty(V_\mu)$, the Leibniz rule yields

(16.10) $$(D_Zb)(X,Y) = (X,Y)\cdot Z\lambda + (\lambda - \mu)(D_ZX, Y).$$

16.11 Proposition (cf. [Der 2], [Hi-Re]). *Given a Codazzi tensor b on a Riemannian manifold (M,g) and an eigenvalue function λ of b, defined in a component of M_b, we have*

(i) *the eigenspace distribution V_λ is integrable;*

(ii) *each integral manifold N of V_λ is umbilical in (M,g). More precisely, for any eigenvalue function $\mu \neq \lambda$ and any sections X, Z of V_λ, Y of V_μ, the Y-component h^Y of the second fundamental form h of N is given by* $h^Y(Z,X) = -(D_ZX, Y) = (\mu - \lambda)^{-1}(X,Z)\cdot Y\lambda$;

(iii) *if* $\dim V_\lambda > 1$, *then λ is constant along V_λ.*

Proof. For μ, X, Y, Z as in (ii), $(X,Y) = (Z,Y) = 0$ and hence (16.10) implies $(\lambda - \mu)([Z,X],Y) = (D_Zb)(X,Y) - (D_Xb)(Z,Y)$ and $(\lambda - \mu)(D_ZX,Y) - (X,Z)\cdot Y\lambda = (D_Zb)(X,Y) - (D_Yb)(X,Z)$, so that (i) and (ii) follow from the Codazzi equation. If $\dim V_\lambda > 1$ and $X \in C^\infty(V_\lambda)$, we can find, locally, a non-zero $Y \in C^\infty(V_\lambda)$ with $(X,Y) = 0$. Applying (16.10) with $\lambda = \mu$, we obtain $|Y|^2 \cdot X\lambda = (D_Xb)(Y,Y) = (D_Yb)(X,Y) = 0$, which completes the proof. □

16.12. As an application of 16.11, we shall now derive a *local classification of Codazzi tensors b having exactly two distinct eigenvalue functions* λ, μ (with $\dim V_\lambda \leqslant \dim V_\mu$). For simplicity, we assume in addition that $\dim V_\lambda > 1$ or $\operatorname{tr}_g b$ is constant (cf. [Der 2] for the latter case); a similar argument works without this hypothesis. Let $\dim M = n \geqslant 3$.

By 16.11, V_λ (resp., V_μ) is integrable and has umbilical leaves with mean curvature vector $H_\lambda = (\mu - \lambda)^{-1}(D\lambda)_\mu$ (resp., $H_\mu = (\lambda - \mu)^{-1}(D\mu)_\lambda$), the subscript convention being that $X = X_\lambda + X_\mu \in V_\lambda \oplus V_\mu = TM$.

(i) If $\dim V_\lambda > 1$ (more generally, if λ is constant along V_λ, cf. 16.11iii)), we have $(D\lambda)_\lambda = (D\mu)_\mu = 0$ and so $H_\lambda = -(D\log|\lambda - \mu|)_\mu$, $H_\mu = -(D\log|\lambda - \mu|)_\lambda$. For the conformally related metric $\bar{g} = (\lambda - \mu)^2 g$, it is now easy to verify that V_λ and V_μ are totally geodesic in $(M,\bar{g})$, i.e., they are $\bar{g}$-parallel along each other and along themselves, and so the splitting $TM = V_\lambda \oplus V_\mu$ comes from a local Riemannian product decomposition of $(M,\bar{g})$. Therefore, we have locally

$$M = M_1 \times M_2, g = (\lambda - \mu)^{-2}(\bar{g}_1 \times \bar{g}_2), b = (\lambda - \mu)^{-2}(\lambda\bar{g}_1 + \mu\bar{g}_2),$$

where $\lambda\colon M_2 \to \mathbb{R}$, $\mu\colon M_1 \to \mathbb{R}$ have disjoint ranges. Conversely, for Riemannian manifolds $(M_i, \bar{g}_i)$, $i = 1, 2$, and functions λ, μ with these properties, the above formula defines a Riemannian manifold (M,g) with a Codazzi tensor b satisfying our conditions.

(ii) Let b have constant trace and assume that b is not parallel. By (i), $\dim V_\lambda = 1$

(since, in (i), λ and μ depend on separate variables). In view of 16.11iii), ii), μ and λ are constant along V_μ and the integral curves of V_λ are geodesics, i.e., for a fixed local unit section X of V_λ, $D_X X = 0$. Each leaf of V_μ has mean curvature $\eta = (X, H_\mu) = (\lambda - \mu)^{-1} X\mu$, which is constant along the leaf. In fact, for any $Y \in C^\infty(V_\mu)$, $Y\eta = (\lambda - \mu)^{-1} YX\mu$ and $YX\mu = [Y, X]\mu$, while $([Y, X], X) = -(D_X Y, X) = (Y, D_X X) = 0$, so that $[Y, X] \in C^\infty(V_\mu)$ and $YX\mu = 0 = Y\eta$. For $x \in M$ we can find (cf. 16.11i)) local coordinates $(t, y_1, \ldots, y_{n-1})$ at x with $\partial_t = \partial/\partial t \in V_\lambda$, $\partial_i = \partial/\partial y_i \in V_\mu$, $1 \leqslant i < n$. Since $[\partial_i, \partial_t] = 0$, 16.11ii) yields $\partial_i \cdot |\partial_t|^2 = 2(\mu - \lambda)^{-1} |\partial_t|^2 \partial_i \lambda = 0$, i.e., making a substitution in t, we may assume that $\partial_t = X$. Similarly, $\partial_t(\partial_i, \partial_j) = -2(D_{\partial_i}\partial_j, \partial_t) = 2\eta \cdot (\partial_i, \partial_j)$. Since $\partial_i \eta = 0$, we obtain $(\partial_i, \partial_j) = e^{2\Psi} \cdot \bar{g}_{ij}$ with $\partial_i \Psi = 0$ and $d\Psi/dt = \eta = (\lambda - \mu)^{-1} \partial_t \mu$, and $\partial_t \bar{g}_{ij} = 0$. This, together with the fact that $\operatorname{tr}_g b = C_0$ is constant, gives, locally,

$$\text{(16.13)} \qquad M = I \times \overline{M}, \quad g = dt^2 + e^{2\Psi(t)} \cdot \bar{g}, \quad b = \lambda dt^2 + \mu e^{2\Psi(t)} \cdot \bar{g},$$
$$\lambda = C_0/n + (1 - n)Ce^{-n\Psi(t)}, \quad \mu = C_0/n + Ce^{-n\Psi(t)},$$

where I is an interval, $(\overline{M}, \bar{g})$ an $(n-1)$-dimensional Riemannian manifold, and C a real constant. Conversely, for any such data, and for an arbitrary function Ψ on I, (16.13) defines a Riemannian manifold (M, g) with a Codazzi tensor b of the type discussed above.

16.14 Theorem (A. Derdziński and C.-L. Shen, [De-Sh]). *Let B be a Codazzi tensor of type $(1,1)$ on a Riemannian manifold (M, g), x a point of M, λ and μ eigenvalues of B_x. Then the subspace $V_\lambda(x) \wedge V_\mu(x) \subset \bigwedge^2 T_x M$, spanned by all exterior products of elements of $V_\lambda(x)$ and $V_\mu(x)$, is invariant under the curvature operator $R_x \in \operatorname{End} \bigwedge^2 T_x M$.*

16.15 *Proof*. Adding a constant multiple of Id to B, we may assume that B is nondegenerate in a neighborhood M' of x. The automorphism B of TM' transforms g and D into the metric $G = B^* g$ and the connection $\bar{\nabla} = B^* D$ on M' (so that $G(X, Y) = g(BX, BY)$, $B(\bar{\nabla}_X Y) = D_X(BY)$). Clearly, $\bar{\nabla} G = 0$ and the curvature tensor $\bar{R}$ of $\bar{\nabla}$ satisfies $\bar{R} = B^* R$, i.e., $G(\bar{R}(X, Y)Z, U) = g(R(X, Y)BZ, BU)$. As observed by N. Hicks [Hic], the Codazzi equation for B means that $\bar{\nabla}$ is torsion-free. Thus, the Riemannian connection D_G of G and its curvature 4-tensor R^G are given by $D_G = \bar{\nabla}$ and

$$\text{(16.16)} \qquad R^G(X, Y, Z, U) = R(X, Y, BZ, BU).$$

Let $X \in V_\lambda(x)$, $Y \in V_\mu(x)$, $Z \in V_\nu(x)$, $U \in V_\xi(x)$. Using abbreviated notations like $R_{XYZU} = R(X, Y, Z, U)$, we have, by (16.16), $\nu\xi R_{XYZU} = R^G_{XYZU} = R^G_{ZUXY} = \lambda\mu R_{XYZU}$ and, similarly, $(\mu\xi - \lambda\nu) R_{XZUY} = (\mu\nu - \lambda\xi) R_{XUYZ} = 0$, while the Bianchi identity for R^G yields $0 = \nu\xi R_{XYZU} + \mu\xi R_{XZUY} + \mu\nu R_{XUYZ}$. Combining these equalities, we obtain the matrix equation

$$\begin{bmatrix} \lambda & \xi & \nu \\ \xi & \lambda & \mu \\ \nu & \mu & \lambda \\ 1 & 1 & 1 \end{bmatrix} \cdot \begin{bmatrix} R(X, Y, Z, U) \\ R(X, Z, U, Y) \\ R(X, U, Y, Z) \end{bmatrix} = 0.$$

If $R_{XYZU} \neq 0$, the coefficient matrix satisfies the cofactor relations $(\lambda - \xi)(\lambda + \xi - \mu - \nu) = (\nu - \lambda)(\nu + \lambda - \mu - \xi) = (\lambda - \mu)(\lambda + \mu - \nu - \xi) = 0$, which easily imply that λ is equal to one of μ, ν, ξ. Therefore, evaluating R_x on four eigenvectors of B_x yields zero if more than two eigenspaces are involved. Hence $R_{XY}Z = 0$ if λ, μ, ν are mutually distinct. On the other hand, if $\lambda = \mu \neq \nu$, (16.16) gives $0 = B(R^G_{XY}Z + R^G_{YZ}X + R^G_{ZX}Y) = R_{XY}(BZ) + R_{YZ}(BX) + R_{ZX}(BY) = (\nu - \lambda)R_{XY}Z$, which completes the proof. □

16.17 Corollary (J.P. Bourguignon, [Bou 9]). *Let b be a Codazzi tensor on a Riemannian manifold (M, g). Then*

(i) *b commutes with r, $\mathring{R}(b)$ and with $Dd(\mathrm{tr}_g b) + \delta Db$.*

(ii) *The endomorphisms R, $g \owedge r$ and W of $\wedge^2 TM$ commute with $g \owedge b$.*

Proof. By 16.14, $r(X, Y) = \mathring{R}(b)(X, Y) = 0$ for eigenvectors X, Y of b corresponding to distinct eigenvalues, so that r and $\mathring{R}(b)$ commute with b. Hence (i) follows from Formula (16.9) (which also directly implies that r and b commute). Since $\wedge^2 T_xM$ is spanned by the subspaces $V_\lambda(x) \wedge V_\mu(x)$ and $(g \owedge b)_x$ restricted to such a subspace is $\lambda + \mu$ times the identity, (ii) is immediate from 16.14. □

16.18. An orthonormal basis $(X_1, \ldots, X_n)$ of a Euclidean space E is said to *diagonalize* an algebraic curvature tensor $R \in \mathscr{C}E$ if all exterior products $X_i \wedge X_j$, $i < j$, are eigenvectors of R (viewed as an endomorphism of $\wedge^2 E$), i.e., if $R(X_i, X_j)X_k = 0$ whenever i, j, k are mutually distinct. Following H. Maillot ([Mai 1], [Mai 2]) we shall call $R \in \mathscr{C}E$ *pure* if it is diagonalized by some orthonormal basis of E. A Riemannian manifold (M, g) will be said to have *pure curvature operator* (resp., *pure Weyl tensor*) if for any $x \in M$, R_x (resp., W_x) is pure. If $n = \dim E = 2$, every orthonormal basis of E diagonalizes any $R \in \mathscr{C}E$, while, for $n \geqslant 3$, such a basis diagonalizes R if and only if it diagonalizes $W(R)$ and $c(R)$. Thus, each of the following conditions implies that the Riemannian manifold (M, g) has pure curvature operator:

(i) $\dim M = 2$, or $W = 0$ (e.g., $\dim M = 3$, cf. 1.119);

(ii) (M, g) is a hypersurface in a space of constant curvature (cf. the Gauss equation (1.72c));

(iii) (M, g) is a Riemannian product of manifolds with pure curvature operators.

Moreover, it is obvious that

(iv) the property of having pure Weyl tensor is conformally invariant.

16.19. If $R \in \mathscr{C}E$ has pure Weyl component $W = W(R)$ (e.g., if R is pure), then all the Pontryagin forms $p_i(R) \in \wedge^{4i}E^*$, $i \geqslant 1$, are zero (see [Mai 1]). In fact, it is easy to see (cf. [Gre]) that $p_i(R) = p_i(W)$ and the subalgebra $p(R) = p(W)$ of $\wedge E^*$ generated by the $p_i(W)$ has another system of generators $\Omega_i \in \wedge^{4i}E^*$, Ω_i being obtained by alternating the map $(Y_1, \ldots, Y_{4i}) \to \mathrm{tr}[W(Y_1 \wedge Y_2) \circ \cdots \circ W(Y_{4i-1} \wedge Y_{4i})]$. However, if an orthonormal basis $(X_1, \ldots, X_n)$ diagonalizes W, then $W(X_i \wedge X_j) \circ W(X_k \wedge X_l) = 0$ for mutually distinct i, j, k, l, which implies $p(W) = \wedge^0 E^* = \mathbb{R}$.

16.20 Lemma. *Let E be an oriented 4-dimensional Euclidean space and let $W \in \mathscr{W}(E)$*

be pure. Then W^+ and W^-, viewed as endormorphisms of $\bigwedge^+ E$ and $\bigwedge^- E$, respectively, have equal spectra, i.e., $|W^+| = |W^-|$ and det $W^+ = \det W^-$.

Proof. Since W is pure, we have $W(X_i \wedge X_j) = \lambda_{ij} X_i \wedge X_j$ for some orthonormal basis $X_1, \ldots, X_4$ of E and real numbers λ_{ij}. If $\{i,j,k,l\} = \{1,2,3,4\}$, then $*(X_i \wedge X_j) = \pm X_k \wedge X_l$, and so $\lambda_{ij} = \lambda_{kl}$, since W and $*$ commute by 1.128. Elements of the form $X_i \wedge X_j \pm X_k \wedge X_l$ now give rise to bases of $\bigwedge^+ E$ and $\bigwedge^- E$, realizing equal spectra for W^+ and W^-. □

16.21 Theorem (cf. [De-Sh], [Bou 9]). *Let b be a Codazzi tensor on a Riemannian manifold (M, g), $\dim M = n$.*

(i) *If b has n distinct eigenvalues at all points of a dense subset of M, then, for any $x \in M$, R_x is diagonalized by some orthonormal basis of $T_x M$, diagonalizing b_x. The Pontryagin forms of (M, g) and the real Pontryagin classes of M are all zero.*

(ii) *If $n = 4$ and $b - (\mathrm{tr}_g b)g/4 \neq 0$ in a dense subset of M, then, at each $x \in M$, W_x is diagonalized by some orthonormal basis of $T_x M$, diagonalizing b_x. The Pontryagin form p_1 of (M, g) vanishes identically and $p_1(M, \mathbb{R}) = 0$. At each $x \in M$, W_x^+ and W_x^- have equal spectra.*

16.22 *Proof.* (i) is obvious from 16.14 and 16.19 together with a continuity argument. By 16.19 and 16.20, (ii) will follow if we show that W_x is diagonalized by an orthonormal eigenbasis of b_x, for any $x \in M_b$. In a neighbourhood M' of x, b has m distinct eigenvalue functions $\lambda(1), \ldots, \lambda(m)$, $2 \leqslant m \leqslant 4$, their multiplicities being $k_1, \ldots, k_m$ with $k_1 \leqslant \cdots \leqslant k_m$. Four cases are possible: I. $m = 2$, $k_1 = k_2 = 2$; II. $m = 2$, $k_1 = 1$, $k_2 = 3$; III. $m = 3$, $k_1 = k_2 = 1$, $k_3 = 2$; IV. $m = 4$. In case I, 16.12i) implies that near x, g is conformal to a product of surface metrics, compatibly with the b-eigenspace decomposition of TM', and our assertion on W_x follows from 16.18i), iii), iv). Assume now case II (resp., case III). Since $V_{\lambda(1)} \wedge V_{\lambda(2)}$ (resp., $V_{\lambda(1)} \wedge V_{\lambda(2)}$ and $V_{\lambda(1)} \wedge V_{\lambda(3)}$) are invariant, by 16.14 and 16.17i), under the self-adjoint endomorphism W of $\bigwedge^2 TM'$, we can choose an orthonormal b_x-eigenvector basis $(X_1, \ldots, X_4)$ of $T_x M$ with $W(X_1 \wedge X_i) = \mu_i X_1 \wedge X_i$ for some μ_i, $2 \leqslant i \leqslant 4$. The fact that W commutes with $*$ for any orientation (cf. 1.128) implies that all $X_i \wedge X_j$ are eigenvectors of W_x, as required. Finally, in case IV our assertion is immediate from (i) and 16.18, which completes the proof. □

D. The Case $Dr \in C^\infty(Q \oplus S)$: Riemannian Manifolds with Harmonic Weyl Tensor

16.23. The n-dimensional Riemannian manifolds (M, g) for which Dr is a section of $Q \oplus S$, i.e., $r - (2n-2)^{-1} sg$ is a Codazzi tensor, can also be characterized by the condition $\delta W = 0$ (when $n \geqslant 4$), or by conformal flatness (when $n = 3$; see 16.4.v)). In this section, we discuss these manifolds, always assuming that $n \geqslant 4$. They are then said to have harmonic Weyl tensor. This terminology is justified by the fact that $\delta W = 0$ implies the "Bianchi identity" $dW = 0$, so that W, viewed as a $\bigwedge^2 M$-valued 2-form, is both closed and co-closed (see 16.41 for details).

After describing various examples of manifolds with harmonic Weyl tensor, we give some general theorems on the structure of their curvature operator (16.28, 16.31) and a local classification result (16.32).

16.24. The simplest examples of manifolds with harmonic Weyl tensor:

(i) Manifolds with $Dr = 0$, locally isometric to products of Einstein manifolds.

(ii) Conformally flat manifolds ($W = 0$). This class contains many *compact* examples, apart from spaces of constant curvature: the conformal inversion map $X \mapsto |X|^{-2}X$ in the model space $\mathbb{R}^n$ immediately gives rise to a connected sum operation for such manifolds ([Kul 1]). Note that a Riemannian product is conformally flat if and only if both factors have constant sectional curvatures and either one of them is 1-dimensional, or the sum of their curvatures is zero, see 1.167.

(iii) The condition $DW = 0$ gives no new examples of manifolds with $\delta W = 0$: by a result of W. Roter (cf. [De-Ro]), it implies that $DR = 0$ or $W = 0$ (see also [Miy] and, for $n = 4$, 16.75iii)). Moreover, the class of manifolds with $\delta W = 0$ is not closed under taking Riemannian products, unless both factors have constant scalar curvatures.

16.25. Other examples of manifolds with $\delta W = 0$ can be obtained by conformal deformations. Under a conformal change $g' = e^{2f}g$ of metric in dimension n, δW transforms like

$$\delta_{g'} W_{g'} = \delta W - (n-3)\, W(Df, ., ., .) \tag{16.25}$$

(cf. 1.159). Thus, we can proceed by taking a Riemannian manifold with $Dr = 0$ and finding on it a function f with $W(Df, ., ., .) = 0$. An easy computation gives the following

Lemma. *Let (M, g) be a Riemannian product of two Einstein manifolds (M_i, g_i). For a non-zero vector X tangent to M we have $W(X, ., ., .) = 0$ if and only if $n_2(n_2-1)s_1 + n_1(n_1-1)s_2 = 0$ and $W_i(X_i, ., ., .) = 0, i = 1, 2$, where $n_i = \dim M_i \geqslant 1$, s_i is the scalar curvature of g_i, X_i denotes the M_i-component of X and W_i is the Weyl tensor of g_i (defined to be zero if $n_i \leqslant 3$).*

16.26. The following constructions of examples are immediate from 16.25:

(i) For an Einstein manifold $(\overline{M}, \overline{g})$, a 1-dimensional manifold (I, dt^2) and a positive function F on I, the metric $g = dt^2 + F^2(t)\cdot\overline{g} = e^{2\log F(t)}[F^{-2}(t)\,dt^2 \oplus \overline{g}]$ on the product manifold $M = I \times \overline{M}$ satisfies $\delta W = 0$. Suppose now that $\overline{M}$ is compact, $I = \mathbb{R}$, F is periodic, with period T, and let Φ be an isometry of $(\overline{M}, \overline{g})$. Clearly, the mapping $(t, y) \mapsto (t + T, \Phi(y))$ of $\mathbb{R} \times \overline{M}$ is an isometry of (M, g), generating a properly discontinuous $\mathbb{Z}$-action on M. The quotient manifold $(M, g)/\mathbb{Z}$ (cf. [Ko-No 1], p. 44 and 9.11), with the "*twisted*" *warped product* metric determined by g, is then an example of a *compact Riemannian manifold with harmonic Weyl tensor*, diffeomorphic to a bundle with fibre $\overline{M}$ over the circle. In general, it is neither conformally flat (unless $\overline{g}$ is of constant curvature, cf. 16.24ii)), nor does it have parallel Ricci tensor (non-constant functions F for which this happens cannot be periodic and positive everywhere on $\mathbb{R}$).

(ii) Let (M_1, g_1) be a space of constant sectional curvature K_1, (M_2, g_2) an Einstein manifold with scalar curvature $s_2 = -n_2(n_2 - 1)K_1$, where $n_2 = \dim M_2$. For an *arbitrary* positive function F on M_1, *the warped product metric* $g = F^2 g_1 + F^2 g_2 = e^{2 \log F}(g_1 \oplus g_2)$ *on* $M_1 \times M_2$ *has harmonic Weyl tensor*. As in (i), g is, in general, not of type 16.24i) or ii). Choosing (M_1, g_1) to be a sphere and (M_2, g_2) a simply connected compact complex manifold having $c_1 < 0$, endowed with a Kähler-Einstein metric (cf. 11.17), we obtain here *examples of compact simply connected manifolds satisfying* $\delta W = 0$ *and neither* $W = 0$, *nor* $Dr = 0$.

16.27. Further examples obtained by conformal deformations. Let a Riemannian manifold (M, g) have *recurrent conformal curvature* ([Ad-Mi]) in the sense that $2|W|^2 \cdot DW = d(|W|^2) \otimes W$. By 16.25, the metric $g' = |W|^{2/(3-n)} \cdot g$, defined wherever $W \neq 0$ ($n = \dim M \geqslant 4$), satisfies $\delta_{g'} W_{g'} = 0$. As shown by W. Roter ([Rot]), locally, in dimensions $n \geqslant 5$, the only analytic manifolds with recurrent conformal curvature are those with $W = 0$, or $DR = 0$, or products of surfaces with spaces of constant curvature. In dimension 4 there are more such examples, e.g., all Riemannian products of surfaces. Thus, *for* $(M, g) = (M_1, g_1) \times (M_2, g_2)$, *where* $\dim M_1 = 2$ *and* g_2 *is of constant curvature or* $\dim M_2 = 2$, *the metric* g' *defined above has harmonic Weyl tensor*. Since, for such g, $|W|$ is proportional to $|K_1 + K_2|$, where K_i is the Gaussian (resp., constant sectional) curvature of g_i, this construction gives *many examples of compact manifolds with harmonic Weyl tensor, including simply connected ones*. In fact, if M_1 and M_2 are compact and $K_2 \neq 0$ everywhere, we may rescale g_1 to obtain $|W_{g_1 \oplus g_2}| > 0$, so that g' is defined everywhere on M.

16.28. As an immediate consequence of 16.4v), 16.21 and 16.18, we obtain the following

Theorem. *Let* (M, g) *be a Riemannian manifold with harmonic Weyl tensor*, $\dim M = n \geqslant 4$.

(i) *If* r *has* n *distinct eigenvalues at all points of a dense subset of* M, *then* (M, g) *has pure curvature operator and* $p_i(M, \mathbb{R}) = 0$ *for* $i \geqslant 1$.

(ii) *If* $n = 4$ *and* $r - sg/4 \neq 0$ *in a dense subset of* M, *then* (M, g) *has pure curvature operator*, $p_1(M, \mathbb{R}) = 0$ *and, at each point* $x \in M$, W_x^+ *and* W_x^- *have equal spectra.*

16.29 Corollary (cf. [De-Sh], [Der 3], [Bou 9]). *Let an oriented Riemannian four-manifold* (M, g) *satisfy* $W^- = 0$ *and have harmonic Weyl tensor. Then* $W \otimes (r - sg/4) = 0$ *everywhere in* M, *so that* (M, g) *is conformally flat or Einstein.*

In fact, by 16.28ii), $|W^+| = |W^-| = 0$ wherever $r - sg/4 \neq 0$. For the last statement, cf. proof of 16.31. □

16.30 Proposition (cf. [Mat 3], [Tan 1], [Gra 5]). *Let a Kähler manifold* (M, J, g) *of real dimension* $n \geqslant 4$ *have harmonic Weyl tensor. Then its Ricci tensor is parallel.*

In fact, since r commutes with J (cf. 2.42), our assertion is immediate from 16.4v) and 16.8. Note that in this case (M, J, g) is, locally, a product of Kähler-Einstein manifolds. Using 16.24ii), one can now easily conclude that *a conformally flat Kähler manifold which is not flat must be 4-dimensional and locally isometric to a product of surfaces with mutually opposite constant curvatures* (cf. [Ya-Mo] and 1.167).

16.31 Theorem (D. DeTurck and H. Goldschmidt, [DT-Go]). *Let (M, g) be a Riemannian manifold with harmonic Weyl tensor,* $\dim M = n \geqslant 4$. *Suppose that, at some point $x \in M$, $W_x = 0$ and r_x has n distinct eigenvalues. Then (M, g) is conformally flat.*

Proof. In a neighborhood of x, we can find an orthonormal frame field $(X_1, \ldots, X_n)$ diagonalizing r. By 16.4v), 16.14 and 16.17i), it also diagonalizes W, so that the essential components of W are $w_{ij} = w_{ji} = W(X_i, X_j, X_i, X_j)$, $i \neq j$. Condition $\delta W = 0$ and its consequence $dW = 0$ (cf. 16.23, 16.42) now easily give

$$D_k w_{ik} = \sum_{j \neq i, k} (w_{ik} - w_{ij}) \Gamma_{jj}^k, \qquad i \neq k,$$

$$D_k w_{ij} = (w_{ij} - w_{kj}) \Gamma_{ii}^k + (w_{ij} - w_{ki}) \Gamma_{jj}^k, \qquad i \neq j \neq k \neq i,$$

where D_i is the directional derivative along X_i and $\Gamma_{ij}^k = (D_{X_i} X_j, X_k)$. Thus, the w_{ij} satisfy a first order linear system of equations solved for the derivatives. As $w_{ij}(x) = 0$ by hypothesis, we have $W = 0$ near x, and therefore everywhere, since W satisfies the elliptic system $(\delta d + d\delta) W = 0$ [Aro]. □

16.32. In dimension 4, we can state the following local classification result. For the proof, see 16.70.

Proposition. *Let (M, g) be an oriented Riemannian four-manifold with harmonic Weyl tensor. Suppose that W^+, viewed as an endomorphism of $\bigwedge^+ M$, has less than 3 distinct eigenvalues at any point. In a neighborhood of each point $x \in M$ at which $W \neq 0$ and $r \neq sg/4$, g is obtained by a conformal deformation of a product of surface metrics as in* 16.27. *In terms of g, that product metric equals $|W|^{2/3} \cdot g$.*

E. Condition $Dr \in C^\infty(S)$: Riemannian Manifolds with Harmonic Curvature

16.33. The study of Riemannian manifolds for which $Dr \in C^\infty(S)$ (i.e., r is a Codazzi tensor, which is also equivalent to $\delta R = 0$, cf. 16.4ii)) is additionally motivated by their relationship with the Yang-Mills connections (see 16.34). In view of the differential Bianchi identity $dR = 0$ (Formula 1.14), it is natural to call them manifolds with harmonic curvature.

By a recent result of D. DeTurck and H. Goldschmidt [DT-Go], *all manifolds with $\delta R = 0$ are analytic in suitable local coordinates.* On the other hand, in dimensions $n \geqslant 4$, these manifolds are characterized by having harmonic Weyl tensor and constant scalar curvature. Thus, *all arguments of Section D remain valid for manifolds with harmonic curvature.* Therefore we have, for such manifolds, Bourguignon's theorems 16.28ii) and 16.29 ([Bou 9]), the Matsushima-Tanno Theorem 16.30, Theorem 16.31 due to DeTurck and Goldschmidt, as well as 16.28i) and 16.32, with the obvious simplifications: in 16.28 we may replace the "dense subset" by "a point".

In this section we describe the known examples of compact Riemannian manifolds with harmonic curvature and give two further results (16.36, 16.38) on such manifolds, including a classification theorem. Finally, we discuss the harmonicity of arbitrary algebraic curvature tensor fields.

16.34. Condition $\delta R = 0$ also appears in the following context. Given a vector bundle E over a compact Riemannian manifold (M, g) and a fibre metric g_E in E, one assigns to any connection ∇ in E, compatible with g_E, the *Yang-Mills integral* $\mathbf{YM}(\nabla) = \int_M |R^\nabla|^2 \cdot v_g$, R^∇ being the curvature of ∇. A connection ∇ is a critical point of this *Yang-Mills functional* if and only if $\delta^\nabla R^\nabla = 0$ (cf. [Par], 4.82 and 13.11). For $E = TM$ and $g_E = g$, this is just condition $\delta R = 0$.

16.35 Examples. According to 16.33, many (usually non-compact) manifolds with harmonic curvature can be constructed by imposing on the examples described in 16.24, 16.26 and 16.27 the additional condition that their scalar curvature be constant. This gives the following *compact manifolds with harmonic curvature*:

(i) *Compact manifolds with $Dr = 0$.*

(ii) *Compact conformally flat manifolds with constant scalar curvature.*

Since the Yamabe conjecture is also true in the conformally flat case (R. Schoen, cf. [Sce]), this yields a metric with $\delta R = 0$ on every compact conformally flat manifold, e.g. on connected sums of tori, cf. 16.24ii) and 6.77.

(iii) *Compact manifolds locally isometric to products of manifolds with $\delta R = 0$.*

(iv) Given a compact Einstein manifold $(\overline{M}, \overline{g})$ with scalar curvature $\overline{s} > 0$, $\dim \overline{M} = n - 1 \geqslant 2$, a non-constant positive periodic function F on $\mathbb{R}$ and an isometry Φ of $(\overline{M}, \overline{g})$, it is easy to verify that the manifold $(M, g)/\mathbb{Z}$, defined as in 16.26i), has constant scalar curvature if and only if we have, for $f = F^{n/2}$ and a constant C_0,

$$\textbf{(16.35iv)} \qquad d^2f/dt^2 - \frac{1}{4}n(n-1)^{-1} \cdot \overline{s} f^{1-4/n} = \frac{n}{4} C_0 f.$$

Choosing f to be a non-constant positive periodic solution of this equation (for its existence, see 16.37), we obtain from our construction examples of *compact Riemannian manifolds with harmonic curvature*, namely, *bundles with fibre $\overline{M}$ over S^1 endowed with twisted warped product metrics* (cf. [Der 1]). These manifolds never satisfy $Dr = 0$ and are not conformally flat unless $\overline{g}$ has constant sectional curvature (cf. 16.26i)).

(v) Let (M_1, g_1) be a compact manifold of constant sectional curvature $K_1 < 0$, (M_2, g_2) a compact Einstein manifold with scalar curvature $s_2 = -n_2(n_2 - 1)K_1$, where $n_i = \dim M_i \geqslant 2, i = 1, 2$, and $n_2 > n_1 - K_1^{-1}\lambda_1$, λ_1 being the lowest positive eigenvalue of the Laplace operator Δ of (M_1, g_1) acting on functions. Set $q = 2(n_1 + n_2)(n_1 + n_2 - 2)^{-1}$ and $a = \frac{1}{4}(n_1 - n_2)(n_1 + n_2 - 2)K_1$, so that $q > 2$, $a > \lambda_1(q-2)^{-1}$ and $q < 2n_1(n_1 - 2)^{-1}$ (in the case where $n_1 > 2$). For any positive C^∞ function f on M_1, the warped product metric $g = f^{q-2}(g_1 \oplus g_2)$ on $M = M_1 \times M_2$ satisfies $\delta R = 0$ if and only if its scalar curvature s is constant (cf. 16.26ii)), which (see 1.161) is in turn equivalent to $\Delta f + af = cf^{q-1}$ with a *constant* c (in fact, $c = \frac{1}{4}s(n_1 + n_2 - 2)(n_1 + n_2 - 1)^{-1}$). A *non-constant* positive solution f to this equation (with any $c > 0$) must exist on (M_1, g_1) (see 16.37), which, consequently, gives rise to examples of *metrics g with harmonic curvature on the compact manifolds $M = M_1 \times M_2$ with (M_i, g_i) as above* ([Der 4]); note that $\dim M \geqslant 5$ and $\pi_1 M$ is infinite. If g_2 is not of constant curvature, the manifolds (M, g) constructed in this way are not locally isometric to any of the manifolds described in (i)–(iv) above (see [Der 4] for details). This construction can be slightly generalized if one takes, instead of $M_1 \times M_2$, certain bundles M with fibre M_2 over M_1, obtained by using an arbitrary homomorphism from $\pi_1(M_1)$ into the isometry group of (M_2, g_2); such M then carry *twisted warped product metrics with harmonic curvature.*

16.36 Theorem (M. Berger, cf. [Be-Eb], [Sio 2], [Bou 8], [Gra 5]). *Every compact Riemannian manifold with harmonic curvature and non-negative sectional curvature K satisfies $Dr = 0$. If, moreover, $K > 0$ at some point, then the manifold is Einstein.*

In fact, this is immediate from 16.4ii) and 16.9.

16.37 Lemma. (i) *Given a C^2 function v on a closed interval I and $f_1, f_2 \in I$, the existence of a non-constant periodic C^2 function f on $\mathbb{R}$ with range $[f_1, f_2]$, satisfying the equation $d^2f/dt^2 = \frac{1}{2}v'(f)$, is equivalent to the conditions $f_1 < f_2$, $v(f_1) = v(f_2)$, $v'(f_1) \cdot v'(f_2) \neq 0$ and $v(f_0) > v(f_1)$ for all $f_0 \in (f_1, f_2)$.*

(ii) (H. Yamabe [Yam 2], cf. [Aub 6], p. 115–119). *For any compact Riemannian manifold (M, g) and real numbers a, q, c with $q > 2$, $c > 0$, $a > \lambda_1(q-2)^{-1}$ (where λ_1 is the lowest positive eigenvalue of the Laplace operator) and $q < 2n(n-2)^{-1}$ (if $n = \dim M > 2$), or $q < \infty$ (if $n = 2$), the equation*

(16.37ii) $$\Delta f + af = cf^{q-1}$$

admits a non-constant positive C^∞ solution $f: M \to \mathbb{R}$.

Proof. (i): an easy exercise.

(ii): Using Yamabe's method ([Aub 6], p. 115–119), one finds a positive C^∞ solution f to (16.37ii), with some $c > 0$, by minimizing the functional

$$I_q(f) = \left(\int_M |df|^2 v_g + a\int_M f^2 v_g\right) \cdot \left(\int_M f^q v_g\right)^{-2/q}$$

in the class of all functions f in the first Sobolev space L_1^2 with $f \geqslant 0$ and $\int_M f v_g > 0$; by rescaling f, any $c > 0$ can be attained. If this minimizing f were constant, the second variation of I_q at f would give $\int_M |d\psi|^2 v_g \geqslant a(q-2)\int_M \psi^2 v_g$ for all C^∞ functions ψ with $\int_M \psi v_g = 0$ and hence ([Be-Ga-Ma] p. 186) $a(q-2) \leqslant \lambda_1$, contrary to our hypothesis. □

16.38 Theorem (A. Derdziński, [Der 1]). *Let (M, g) be a compact Riemannian manifold with harmonic curvature,* $\dim M = n \geqslant 3$. *If its Ricci tensor is not parallel and has, at each point, less than three distinct eigenvalues, then (M, g) is isometrically covered by one of the compact manifolds constructed in* 16.35iv). *Conversely, each of those manifolds has the stated properties.*

16.39 *Proof.* Fix $x \in M$ with $r_x \neq s(x)g_x/n$ and $(Dr)_x \neq 0$. By 16.4ii) and 16.12ii), near x, g is given by (16.13) with some $(n-1)$-dimensional Riemannian manifold $(\overline{M}, \overline{g})$, a function Ψ on an interval I and with $b = r$. Computing, in (16.13), r from g and comparing it with b, we see that $(\overline{M}, \overline{g})$ is Einstein and $f = e^{n\Psi/2}$ satisfies equation (16.35iv), where $\overline{s}$ is the scalar curvature of $\overline{g}$. The elementary symmetric functions of the eigenvalues of r are analytic on M (cf. 16.33) and, by (16.13), one of them is non-constant, since $(Dr)_x \neq 0$. A suitable regular level of such a function gives an extension of $(\overline{M}, \overline{g})$ to a *compact* Einstein manifold, whose universal covering space we denote by $(\hat{M}, \hat{g})$.

In terms of (16.13), one easily verifes that the curves

$$I \ni t \mapsto c(t) = (t, y) \in I \times \overline{M} \subset M$$

are geodesics and satisfy

$$d^2 f^{2/n}/dt^2 = (1-n)^{-1} r(\dot{c}, \dot{c}) f^{2/n}. \tag{16.39}$$

Since (M, g) is complete and analytic (cf. 16.33), this *linear* differential equation implies that $f^{2/n}$ has an analytic extension to $\mathbb{R}$. This extension is non-zero everywhere. In fact, if it vanished at $t_0 \in \mathbb{R}$, equation (16.35iv), rewritten for $f^{2/n}$, would imply that $\overline{s} > 0$ (since $df^{2/n}/dt \neq 0$ at $t = t_0$ by the uniqueness of solutions of (16.39)) and would determine, up to a sign, the power series expansion of $f^{2/n}$ at t_0. Hence $f^{2/n}$ would be one of the "obvious" solutions (linear, trigonometric or hyperbolic) of the rewritten equation, which would give $Dr = 0$ near x, contradicting our hypothesis. Therefore f can be extended to a positive analytic function on $\mathbb{R}$, again denoted by f.

The warped product manifold $(M', g') = (\mathbb{R} \times \hat{M}, dt^2 + f^{4/n}(t) \cdot \hat{g})$ is analytic, complete (see [Bi-ON], p. 23) and has an open subset isometric to a subset of (M, g). The universal covering space of (M, g) is therefore isometric to (M', g') (cf. [Ko-No 1], p. 252) and so $(M, g) = (M', g')/\Gamma$, Γ being a discrete group of isometries. Clearly, Γ preserves the product foliations of $M' = \mathbb{R} \times \hat{M}$ (tangent to the eigenspace distributions of r) and, passing to a finite covering space, we may assume that Γ preserves the orientation in the $\mathbb{R}$-direction. Using equation 16.35iv) with $(Dr)_x \neq 0$, one easily concludes that all elements of Γ operate on $\mathbb{R} \times \hat{M}$ as product maps of a translation of $\mathbb{R}$, keeping f invariant, with an isometry of $(\hat{M}, \hat{g})$. For some element $T \times \Phi$ of Γ, the translation $T \in \mathbb{R}$ must be non-zero, for otherwise the projection $M' \to \mathbb{R}$ would define an unbounded function on M. Thus, f is periodic, and, by 16.37, $\overline{s} > 0$ and $C_0 < 0$. The $\mathbb{Z}$-action on M' generated by $T \times \Phi$ defines a finite covering space $(M', g')/\mathbb{Z}$ of (M, g), which is of the type described in 16.35iv). This completes the proof. □

16.40. *Further examples.* $S^2 \times S^2$ and other products of compact surfaces admit non-standard metrics with harmonic curvature (of type 16.27), which can be obtained by a bifurcation argument, see [Der 6].

16.41. Let $\bar{R}$ be an algebraic curvature tensor field on a Riemannian manifold (M, g), $\dim M = n \geqslant 4$. We say that $\bar{R}$ is *closed* (as a $\bigwedge^2 M$-valued 2-form) if it satisfies the "differential Bianchi identity" $dR = 0$, where $(dR)(X, Y, Z, U, V) = (D_X R)(Y, Z, U, V) + (D_Y R)(Z, X, U, V) + (D_Z R)(X, Y, U, V)$, and that $\bar{R}$ is *co-closed* if $\delta \bar{R} = 0$. Let $\bar{R} = \bar{U} + \bar{Z} + \bar{W}$ be the decomposition of $\bar{R}$ in the sense of 1.114, and set $\bar{r} = c(\bar{R})$ and $\bar{s} = \mathrm{tr}_g \bar{r}$.

Proposition (K. Nomizu [Nom 1], J.P. Bourguignon [Bou 4]). *In the above notations,*

(i) *Conditions $d\bar{U} = 0$, $\delta \bar{U} = 0$ and $d\bar{s} = 0$ are mutually equivalent.*

(ii) *Conditions $d\bar{Z} = 0$ and $\delta \bar{Z} = 0$ are mutually equivalent. They are satisfied if and only if $d\bar{r} = 0$, i.e., if $\bar{r}$ is a Codazzi tensor.*

(iii) *If, moreover, $d\bar{R} = 0$, then $d\bar{W} = 0$ if and only if $\delta \bar{W} = 0$, which is in turn equivalent to $d[\bar{r} - (2n-2)^{-1} \bar{s} g] = 0$.*

(iv) *If $d\bar{R} = 0$, then $\bar{R}$ is co-closed if and only if so are $\bar{U}, \bar{Z}, \bar{W}$.*

Proof. (i) and (ii) are obvious from (1.116) together with the easily verified fact that, for any symmetric 2-tensor field a, conditions $\delta(a \owedge g) = 0$ and $da = 0$ are equivalent provided $\mathrm{tr}_g a$ is constant, and $d(a \owedge g) = 0$ if and only if $da = 0$. The latter equivalence, together with $\bar{R} = (n-2)^{-1} g \owedge [\bar{r} - (2n-2)^{-1} \bar{s} g] + \bar{W}$ (cf. (1.116)) and the formula $\delta \bar{W} = -(n-3)(n-2)^{-1} d[\bar{r} - (2n-2)^{-1} \bar{s} g]$ (obtained from $d\bar{R} = 0$ by contraction, cf. 16.3), yields (iii). Since $\delta \bar{R} = -dr$ and $2\delta \bar{r} = -d\bar{s}$ whenever $d\bar{R} = 0$ (cf. 16.3 and 1.94), condition $d\bar{R} = \delta \bar{R} = 0$ implies that $\bar{u}$ is constant. Hence (iv) is immediate from (i) and (ii), which completes the proof. □

F. The Case $Dr \in C^\infty(Q)$

16.42. The class of Riemannian manifolds with $Dr \in C^\infty(Q)$ (see 16.4i) for an equivalent condition) has already been discussed in the literature ([Gra 5], [Sin], [Siv]). In dimensions $n \geqslant 4$, all these manifolds have harmonic Weyl tensor (cf. 16.4) and so *all results of Section* D *remain valid for them.* In this section we construct examples of compact manifolds with $Dr \in C^\infty(Q)$ and $Dr \neq 0$, and prove a pinching theorem. We also discuss some questions related to the local classification problem for such manifolds.

16.43 Examples. By 16.42, an obvious construction procedure is to impose condition $Dr \in C^\infty(Q)$ on the manifolds described in Section D, which immediately gives rise to the following *examples of compact manifolds with* $Dr \in C^\infty(Q)$:

(i) *compact manifolds with* $Dr = 0$;

(ii) *bundles with fibre* $\overline{M}$ over S^1 *with twisted warped product metrics constructed as in* 16.26i), *where the compact Einstein manifold* $(\overline{M}, \overline{g})$ *has scalar curvature* $\overline{s} < 0$ *and* $f = F^{-1}$ *is a non-constant positive periodic solution of* $d^2f/dt^2 = 2(n-1)^{-1}(n-2)^{-1} \cdot \overline{s}f^3 + Cf$ *with a constant* $C > 0$ (such an f exists by 16.37). These examples never satisfy $Dr = 0$. Some of them (those for which $\overline{g}$ is of constant curvature) are conformally flat.

16.44 Proposition. *Let* (M, g) *be a compact Riemannian manifold with* $Dr \in C^\infty(Q)$ *and*

$$(16.44) \qquad (n-1)(n+6)r \leqslant 4s, \qquad n = \dim M \geqslant 3$$

(the latter condition holds, e.g., if r is sufficiently C^0 *close to* φg *for a function* $\varphi < 0$). *Then* $Dr = 0$.

16.45 Corollary. *Let* g_0 *be an Einstein metric with negative scalar curvature on a compact manifold M. If a metric g on M satisfies* $Dr \in C^\infty(Q)$ *and is sufficiently* C^2 *close to* g_0, *then g has parallel Ricci tensor.*

Proof of the Proposition. From 16.4i) and 16.3, we obtain $2(n-1)(\delta R)(X, Y, Z) = (X, Y)\, ds(Z) - (X, Z)\, ds(Y)$. The Weitzenböck formula (16.9), applied to 16.4i) implies, after taking divergence and using the formula $r(Ds) = d\Delta s - \delta D ds$ (cf. 1.155) that $(n-2)\delta D ds = (n+6)r(Ds) - 4(n-1)^{-1} s\, ds$, which easily yields (cf. [Sin])

$$(n-2)\Delta |Ds|^2 = 2(n+6)r(Ds, Ds) - 8(n-1)^{-1}s|Ds|^2 - 2(n-2)|Dds|^2.$$

Integrating this and using the inequality (16.44), we conclude that $Dds = 0$, so that s is constant and, by 16.4i), $Dr = 0$, as required. □

16.46. By 16.4i), condition $Dr \in C^\infty(Q)$ can be rewritten as

$$(16.47) \qquad Db = \zeta \circ g, \qquad \zeta = (n+2)^{-1} d(\mathrm{tr}_g b), \qquad n = \dim M,$$

where $b = r - (2n-2)^{-1}sg$. To determine the local structure of all solutions (M, g, b) of (16.47), let us first note that for any such solution, any eigenvalue functions λ, ν of b with $\lambda \neq \nu$ and any $X \in C^\infty(V_\lambda)$, $Z \in C^\infty(V_\nu)$, we have

(i) $X \cdot \lambda = 3\zeta(X)$, $Z \cdot \lambda = \zeta(Z)$;

(ii) if λ is constant along V_λ (e.g., if $\dim V_\lambda > 1$, cf. 16.11iii)), then all eigenvalues of b are constant along V_λ, ζ annihilates V_λ and $\zeta = d\lambda$.

In fact, setting in formula (16.10) first $\lambda = \mu = \nu$ and $Z = Y = X$ and then $\lambda = \mu \neq \nu$ and $Y = X$, we obtain (i) from (16.47). If λ is constant along V_λ, (i) gives $\zeta(X) = 0$, $X \cdot \nu = \zeta(X) = 0$ and $Z \cdot \lambda = \zeta(Z)$, which proves (ii). □

16.48. *Local classification of the solutions* (M, g, b) *of* (16.47) *which are of generic type* in the sense that, for each $x \in M$, b_x has n distinct eigenvalues and the restriction of ζ_x to each eigenspace is non-zero: Set $x_i = (n+2)\lambda_i - \mathrm{tr}_g b$, where λ_i are the eigenvalue functions of b, $1 \leqslant i \leqslant n$. By 16.46i), $X \cdot x_i = 2X \cdot (\mathrm{tr}_g b) = 2(n+2)\zeta(X) \neq 0$ and $Y \cdot x_i = 0$ whenever $X \in V_i, X \neq 0$ and $Y \in (V_i)^\perp$, V_i being the eigenspace distribution of λ_i. Hence the x_i form, locally, a coordinate system and

$$g = \sum_j \psi_i \, dx_i^2, \qquad b = \sum_i \lambda_i \psi_i \, dx_i^2, \qquad \lambda_i = (2n+4)^{-1}(2x_i + \sum_j x_j)$$

with $\psi_i = |\partial_i|^2$, $\partial_i = \partial/\partial x_i$. Since $[\partial_i, \partial_j] = 0$, 16.11ii) yields $\partial_j \psi_i = -2(D_{\partial_i}\partial_i, \partial_j) = 2\psi_i(\lambda_j - \lambda_i)^{-1}\partial_j \lambda_i$ for $j \neq i$, which implies

$$\psi_i = \exp(q_i(x_i)) \cdot \prod_{j \neq i} |x_j - x_i|,$$

where q_i are functions of one variable. Conversely, one easily verifies that in any domain M of $\mathbb{R}^n$ where the Cartesian coordinates x_i are mutually distinct, the above formulae, with arbitrary functions $q_i(x_i)$, define a generic type solution (M, g, b) of (16.47).

16.49. There exist solutions (M, g, b) of (16.47) which are *not of generic type*, other than those with $Db = 0$:

Example. Let (M_1, g_1, b_1) satisfy (16.47) with the 1-form ζ_1, $\dim M_1 = n_1 \geqslant 1$, and denote by D_1, etc., the Riemannian connection, etc., of g_1. Assume that M_1 is simply connected and that, for some $\mu_0 \in \mathbb{R}$, $T = (b_1 - \mu g_1)^{-1}$ exists, where $\mu = (n_1 + 2)^{-1} \mathrm{tr}_{g_1} b_1 + \mu_0$ and b_1 is viewed as a (1, 1) tensor; locally, these hypotheses are always satisfied. For any Riemannian manifold (M_2, g_2), we shall construct a solution (M, g, b) of (16.47) which is an *extension of* (M_1, g_1, b_1) *by* (M_2, g_2), that is

(16.49) $$M = M_1 \times M_2, g = g_1 + e^\varphi g_2, b = b_1 + \mu e^\varphi g_2$$

for some function φ on M_1. Condition (16.47) for this (M, g, b) is equivalent to $2T(d\mu) = d\varphi$. To prove that such a φ exists, i.e., that $d(T(d\mu)) = 0$, note that $d\mu = \zeta_1$ by (16.47) and so $((D_1)_X T)(Y, Z) = -T(d\mu, Y)T(X, Z) - T(d\mu, Z)T(X, Y)$, while (16.47) gives $D_1 d(\mathrm{tr}_{g_1} b_1) + \delta_1 D_1 b_1 = n_1 D_1 d\mu + \Delta_1 \mu \cdot g_1$. Thus, by 16.17i), $D_1 d\mu$ commutes with b_1 and hence also with T, which implies $d(T(d\mu)) = 0$, as required.

16.50. *Local classification of arbitrary solutions* (M, g, b) *of* (16.47) (at points in general position). Assume that $M = M_b$ and ζ annihilates a fixed eigenspace distribution $\mathbf{V}_\mu$ (cf. 16.46ii) and 16.47). We claim that, *if b is not parallel, (M, g, b) is, locally, obtained from a generic type solution by a finite number of extension procedures as described in* 16.49. In fact, for $X \in C^\infty(V_\lambda)$, $Y \in C^\infty(V_\mu)$, $Z \in C^\infty(V_\nu)$ with $\lambda \neq \mu \neq \nu$, (16.10) and (16.47) yield $(\lambda - \mu)(D_Z X, Y) = (X, Z) \cdot \zeta(Y) = 0$, i.e., $D_Z X \in C^\infty((\mathbf{V}_\mu)^\perp)$. Thus, by 16.11, $(\mathbf{V}_\mu)^\perp$ and V_μ are integrable and the leaves of $(\mathbf{V}_\mu)^\perp$ are totally geodesic, while those of V_μ are umbilical. Hence we have, locally, formula (16.49), (M_i, g_i) being Riemannian manifolds with $b_1 \in S^2 M_1$ (b_1 is "constant along M_2" in view of (16.47)), and φ is a function on $M_1 \times M_2$, while (M_1, g_1, b_1) satisfies (16.47) with $\zeta_1 = \zeta$, and $b_1(d_1\varphi) - \mu d_1 \varphi = 2d\mu$, d_i being the differential along M_i. Since, by 16.46i), $d_2 \mu = 0$, this gives $(b_1 - \mu g_1)(d_2 d_1 \varphi) = 0$, and hence (general position!) $\varphi = \varphi_1 + \varphi_2$ for some functions φ_i on M_i. Replacing g_2 by $\exp(\varphi_2) g_2$, we obtain relation (16.49) with $\varphi\colon M_1 \to \mathbb{R}$ and so (M, g, b) is an extension as in 16.49. Repeating this procedure, we can successively split off all eigenspace distributions of b annihilated by ζ, which eventually leads to a generic type situation.

16.51. There is the obvious problem of finding a local classification (at points in general position) of all Riemannian manifolds with $Dr \in C^\infty(Q)$. An obvious idea would be to take our classification 16.48, 16.50 of the solutions (M, g, b) of (16.47) and impose on them the condition $b = r - (2n-2)^{-1} sg$. This direct approach leads, however, to hopeless computations.

G. Condition $Dr \in C^\infty(A)$: Riemannian Manifolds such that $(D_X r)(X, X) = 0$ for all Tangent Vectors X

16.52. A Riemannian manifold (M, g) satisfies the condition $Dr \in C^\infty(A)$ if and only if $\sigma(Dr) = 0$ (see 16.4iii)), which is clearly equivalent to $(D_X r)(X, X) = 0$ for all $X \in TM$. This condition occurs as a consequence of some geometrical hypotheses. Namely, it holds (see [DA-Ni 1] and 16.55i)) for all *D'Atri spaces* [Va-Wi], i.e., Riemannian manifolds (M, g) such that, for each $x \in M$, the *local geodesic symmetry* at x (assigning $\exp_x(-X)$ to $\exp_x X$, for $X \in T_x M$ close to 0) preserves, up to a sign, the volume element v_g. Moreover, all Riemannian manifolds (M, g) for which the operator $f \mapsto (r, Ddf)$, acting on functions, commutes with Δ, satisfy $\sigma(Dr) = 0$ (see [Sum]).

In the present section we describe various examples of manifolds with $\sigma(Dr) = 0$ and state a theorem on them.

16.53 *Examples of Riemannian manifolds with $\sigma(Dr) = 0$, i.e., with $(D_X r)(X, X) = 0$ for all vectors X.*

(i) *Manifolds with $Dr = 0$.*

(ii) All D'Atri spaces (cf. 16.55i)). This class contains (see [DA-Ni 2]) *all homogeneous Riemannian manifolds which are naturally reductive* (for some G *and* $\mathfrak{p}$, cf. 7.84), in particular, *the normal homogeneous Riemannian manifolds.* Thus, all such manifolds satisfy $\sigma(Dr) = 0$, which gives rise to a large variety of examples of *compact locally homogeneous manifolds with $(D_X r)(X, X) = 0$ for all vectors X*, including all metric spheres in rank one symmetric spaces (as shown directly by Chen and Vanhecke [Ch-Va], cf. also [Zil 1]), all nearly Kählerian 3-symmetric spaces (by Gray's result [Gra 5], cf. [Gra 2], [Tr-Va 2]) and many compact Lie groups with suitable left-invariant metrics constructed by D'Atri and Ziller [DA-Zi]. Most of these examples have $Dr \neq 0$. Moreover, there exist homogeneous D'Atri spaces which are not naturally reductive (for any G and $\mathfrak{p}$). Namely, A. Kaplan constructed ([Kap 1], [Kap 2], [Kap 3], cf. [Tr-Va 2]) a class of left-invariant metrics on certain 2-step nilpotent Lie groups, all of which have the D'Atri space property without being, in general, naturally reductive. Many of Kaplan's examples admit compact quotient manifolds ([Kap 2]).

(iii) *Riemannian products of manifolds with $\sigma(Dr) = 0$.*

16.54 Theorem (U. Simon [Sio 1], cf. [Gra 5]). *Let (M,g) be a compact Riemannian manifold with $(D_X r)(X,X) = 0$ for all $X \in TM$ and with non-positive sectional curvature K. Then $Dr = 0$. If, moreover, $K < 0$ at some point, then (M,g) is Einstein.*

Proof. For any symmetric 2-tensor field b on (M,g) such that $(D_X b)(X,X) = 0$ for all vectors X, the Weitzenböck Formula 1.143 can easily be rewritten as

$$\delta Db + Dd(\mathrm{tr}_g b) = b \circ r + r \circ b - 2\mathring{R}(b).$$

If $b = r$, then $\mathrm{tr}_g b$ is constant (cf. 16.4) and our assertion can be obtained by integration, exactly as in 16.9. □

16.55 Theorem (J.E. D'Atri and H.K. Nickerson, [DA-Ni 1], [DA-Ni 2], [DAt]).

(i) *Every D'Atri space (M,g) satisfies the condition $(D_X r)(X,X) = 0$ for all vectors X.*

(ii) *Every naturally reductive Riemannian homogeneous manifold is a D'Atri space.*

Proof. (i) For any Riemannian manifold (M,g), $x \in M$ and $X \in T_x M$,

$$(D_X r)(X,X) = \frac{d^3}{dt^3} \log \theta(\exp_x tX)\bigg|_{t=0},$$

$\theta = \theta_x$ being the *normal volume function* centered at x (see 1.48). This follows from a direct computation in normal coordinates at x, using the easy equality $\sum_{i,j} g^{ij}\partial_k g_{ij} = \partial_k \log\det g_{ij}$ and relations obtained by differentiating (1.92). In a D'Atri space, the function $t \mapsto \theta_x(\exp_x tX)$ is even, which completes the proof.

(ii) See [DA-Ni 2] or [DAt]. □

16.56 Some *open problems.* (i) Find examples of Riemannian manifolds with $(D_X r)(X,X) = 0$ for all vectors X, which are neither locally homogeneous, nor locally isometric to Riemannian products and have non-parallel Ricci tensor.

(ii) Do there exist D'Atri spaces which are not locally homogeneous? (cf. [Tr-Va 1]).

H. Oriented Riemannian 4-Manifolds with $\delta W^+ = 0$

16.57. For oriented Riemannian four-manifolds, relation $\delta W^+ = 0$ is equivalent to a linear condition on Dr, namely, to $d(r - sg/6) \in C^\infty(\bigwedge^- M \otimes T^*M)$ (cf. 16.5). In the present section we discuss the manifolds satisfying this condition and show that some of them are in a natural conformal relationship with Kähler manifolds, which allows us to construct many examples of compact manifolds of this type. We also give some results concerning the behavior of the function $\det W^+$ on such manifolds.

16.58. In our discussion of oriented Riemannian four-manifolds (M,g), it is convenient to use local trivializations of $\bigwedge^+ M$ by eigenvectors of W^+. In the open dense subset M_{W^+} of M, consisting of points at which the number of distinct eigenvalues of $W^+ \in C^\infty(\text{End}\,\Lambda^+ M)$ is locally constant, we can, locally, choose C^∞ functions μ_i and mutually orthogonal C^∞ sections ω_i of Λ^+ with $|\omega_i|^2 = 2$ and $W^+(\omega_i) = \mu_i\omega_i$, $1 \leqslant i \leqslant 3$. Note that, for $x \in M$, every oriented orthonormal basis $X_1, \ldots, X_4$ of T_xM gives rise to orthonormal bases $2^{-1/2}(\mp X_1 \wedge X_2 - X_3 \wedge X_4)$, $2^{-1/2}(\mp X_1 \wedge X_3 - X_4 \wedge X_2)$, $2^{-1/2}(\mp X_1 \wedge X_4 - X_2 \wedge X_3)$ of $\Lambda_x^\pm M$. This construction can easily be verified to be equivariant relative to the two-fold covering homomorphism $SO(4) \to SO(3) \times SO(3)$, so that each pair of suitably oriented orthonormal bases of $\Lambda_x^+ M$ and $\Lambda_x^- M$ is obtained in this way from an oriented orthonormal basis of T_xM. Hence, by changing the signs of some of our ω_i, we get (cf. 1.128)

(16.59)
$$\omega_i\omega_j = \omega_k = -\omega_j\omega_i \quad \text{if } \varepsilon_{ijk} = 1, \qquad \omega_i^2 = -Id,$$
$$W^+ = \frac{1}{2}\sum_i \mu_i\omega_i \otimes \omega_i, \qquad \sum_i \mu_i = 0,$$

where ε_{ijk} is the *Ricci symbol* (skew-symmetric in i, j, k, with $\varepsilon_{123} = 1$). The invariance of $\Lambda^+ M$ under parallel displacements implies

(16.60)
$$D\omega_i = \zeta_k \otimes \omega_j - \zeta_j \otimes \omega_k, \quad \text{if } \varepsilon_{ijk} = 1$$

for certain C^∞ 1-forms ζ_i defined (locally) in M_{W^+}. Calculating the curvature of the Riemannian connection in $\Lambda^+ M$ in terms of the connection forms ζ_i, i.e., applying to (16.60) the Ricci identity 1.21, we obtain

(16.61)
$$d\zeta_i + \zeta_j \wedge \zeta_k = (\mu_i - s/6)\cdot\omega_i + \tfrac{1}{2}(r\omega_i + \omega_i r) \quad \text{if } \varepsilon_{ijk} = 1.$$

16.62 Proposition (cf. [Gra 3], [Pol 2], [Der 3]). *Let (M, J, g) be a Kähler manifold of real dimension four, oriented in the natural way (so that the Kähler form ω is a section of $\Lambda^+ M$). Then $W^+(\omega) = s\cdot\omega/6$ and $W^+(\eta) = -s\cdot\eta/12$ for any 2-form $\eta \in \Lambda^+ M$ orthogonal to ω.*

Proof. The Weitzenböck formula for 2-forms η (cf. 1.154),

$$2W(\eta) = D^*D\eta - \Delta\eta + 2(n-1)^{-1}(n-2)^{-1}s\cdot\eta + (n-4)(n-2)^{-1}(r\eta + \eta r)$$

implies, in dimension $n = 4$ and for $\eta = \omega$ with $D\omega = 0$, that $W(\omega) = s\cdot\omega/6$. In M_{W^+} (notation of 16.58), set $\omega_i = \omega$ for a fixed i, so that $\mu_i = s/6$. From (16.60) and (16.61), $\zeta_l = 0$ and $r\omega_l + \omega_l r = (s/3 - 2\mu_l)\omega_l$, if $l \neq i$. Since $r\omega_i = \omega_i r$ (cf. 2.42), this gives, by (16.59), $(s/3 - 2\mu_j)\omega_k = \omega_i[(s/3 - 2\mu_j)\omega_j] = \omega_i(r\omega_j + \omega_j r) = r\omega_k + \omega_k r = (s/3 - 2\mu_k)\omega_k$ for j, k such that $\varepsilon_{ijk} = 1$. Hence (cf. (16.59)) $\mu_j = \mu_k = -s/12$, as required. □

16.63 Remark. By 16.62, every Kähler manifold (M, J, g) of real dimension four satisfies the relation

(16.64)
$$\#\,\text{Spec}_{(\Lambda^+)}\, W^+ \leqslant 2$$

i.e., the endomorphism W^+ of $\Lambda^+ M$ has, at each point, less than three distinct eigenvalues. By the conformal invariance of W, (16.64) will also hold for any metric conformal to our Kähler metric g. On the other hand, 16.62 implies for (M, J, g) the equality $W^+ = s \cdot T$ for some tensor field T, which is *parallel*, since it is naturally determined by ω and $\bigwedge^+ M$. Therefore, $s \cdot \delta W^+ + W^+(Ds, ., ., .) = 0$. Consequently (cf. (16.25) and 16.5), for any Kähler four-manifold (M, J, g), the metric $g' = g/s^2 = g/(24|W^+|^2)$, defined wherever $W^+ \neq 0$, satisfies the conditions $\delta_{g'} W_{g'}^+ = 0$ and $\sharp\,\mathrm{Spec}_{(\Lambda^+)} W_{g'}^+ \leqslant 2$; clearly, $g = (24 g'(W_{g'}^+, W_{g'}^+))^{1/3} \cdot g'$. For a converse statement, see 16.67.

16.65 *Examples of compact oriented Riemannian four-manifolds with* $\delta W^+ = 0$.

(i) *Compact oriented four-manifolds with* $Dr = 0$ (cf. 16.57).

(ii) *Compact manifolds satisfying the conformally invariant condition* $W^+ = 0$. The only known examples of this type are *compact conformally flat 4-manifolds*, and (suitably) oriented *manifolds conformal to a quotient of a* K3 *surface with a Ricci-flat Kähler metric, or to the standard* $\mathbb{C}P^2$, *or to compact quotients of its dual* ($\mathbb{C}P^2$) (cf. [At-Hi-Si] and 13.3).

(iii) For any compact Kähler manifold (M, J, g) with $\dim_{\mathbb{R}} M = 4$ and $s \neq 0$ everywhere, the conformally related *compact manifold* $(M, g') = (M, s^{-2} g)$ *satisfies* $\delta_{g'} W_{g'} = 0$ (for the natural orientation; cf. 16.63). Using small Kähler deformations of g (see 2.113), we obtain a large variety of such examples (e.g., with the underlying manifold $\mathbb{C}P^2$).

16.66. In the notations of 16.58, Formulae (16.59) and (16.60) easily imply that *condition* $\delta W^+ = 0$ *is equivalent to*

$$d\mu_i = (\mu_i - \mu_j)\omega_k(\zeta_k) + (\mu_i - \mu_k)\omega_j(\zeta_j) \quad \text{if } \varepsilon_{ijk} = 1, \tag{16.66}$$

where we set $\omega(\zeta) = -\iota_\zeta \omega = \omega(., \zeta)$ for any 2-form ω and any 1-form ζ.

16.67. We have the following converse of the statement given in 16.63:

Theorem (A. Derdziński, [Der 3]). *Let an oriented Riemannian four-manifold* (M, g) *satisfy the conditions* (16.64) *and* $\delta W^+ = 0$. *Then the metric* $g' = (24|W^+|^2)^{1/3} \cdot g$, *defined wherever* $W^+ \neq 0$, *is (locally) Kähler with respect to some complex structure, defined explicitly (up to a sign) at points where* $W^+ \neq 0$ *and compatible with the original orientation. Moreover,* $s_{g'} = \varphi^{1/3}$, *where* $\varphi/6$ *is the simple eigenvalue of* W^+, *and* $g = s_{g'}^{-2} g'$.

Proof. In the notations of 16.58, let $6\mu_i = -12\mu_j = -12\mu_k = \varphi$ for some fixed i, j, k with $\varepsilon_{ijk} = 1$, the function φ being smooth wherever $W^+ \neq 0$. Clearly, $\varphi^2 = 24|W^+|^2$, $g' = \varphi^{2/3} g$ and, by (16.66),

$$d\varphi = 3\varphi\omega_k(\zeta_k) + 3\varphi\omega_j(\zeta_j). \tag{16.68}$$

We claim that

$$3\varphi \cdot D_X \omega_i + d\varphi \wedge \omega_i(X) + \omega_i(d\varphi) \wedge X = 0 \tag{16.69}$$

for any tangent vector X. In fact, let η be the 2-form constituting the left-hand side of (16.69). Any $\xi \in \bigwedge^- M$ is orthogonal to $D_X \omega_i$ and commutes with ω_i (cf. 16.58, 1.125), so that $(\eta, \xi) = 0$. On the other hand, (16.59), (16.60) and (16.68) yield $(\omega_j, \eta) =$

$(X, 6\varphi \cdot \zeta_k + 2\omega_k(d\varphi)) = 0$ and, similarly, $(\omega_k, \eta) = (\omega_i, \eta) = 0$, which proves (16.69). It is now easy to verify (cf. 1.159a)) that (16.69) means nothing but $D_{g'}(\varphi^{2/3}\omega_i) = 0$, so that g' is a Kähler metric with Kähler form determined by g up to a sign (since ω_i is a simple eigenvector of W^+). Our assertion is now immediate from the obvious relations between the spectra of W_g^+ and $W_{g'}^+$, acting on $\wedge^+ M$, together with 16.62. □

16.70 *Proof of Proposition* 16.32. By 16.28ii), relations $|W^+| = |W^-| > 0$ and $\# \mathrm{Spec}_{(\Lambda^-)} W^- \leqslant 2$ hold near x. Thus, 16.67 implies that the metric $g' = |W|^{2/3} g$ is Kählerian for *two complex structures* J^+, J^-, corresponding to different orientations in a neighborhood of x. The corresponding Kähler forms $\omega^\pm$ are sections of $\Lambda^\pm$, and therefore J^+ and J^- must commute (since $\omega^\pm = \mp X_1 \wedge X_2 - X_3 \wedge X_4$ for some local g'-orthonormal frame $X_1, \ldots, X_4$, cf. 16.58). Hence $a = J^+ J^-$ is a g'-parallel self-adjoint (1, 1) tensor field with $a^2 = Id$, $\det a = 1$ and $a \neq \pm Id$ (note that $(J^+)^{-1} = -J^+$). Consequently, the (± 1)-eigenspaces of a form g'-parallel plane fields near x. Together with the conformal transformation rule for $|W|$, this completes the proof. □

16.71. Let (M, g) be an oriented Riemannian four-manifold with $\delta W^+ = 0$. In the notations of 16.58, Formulae (16.59) and (16.61) yield $(\omega_i, d\zeta_i) = -\omega_i(\zeta_j, \zeta_k) + 2\mu_i + s/6$, $(\omega_j, d\zeta_i) = -\omega_j(\zeta_j, \zeta_k)$, $(\omega_k, d\zeta_i) = -\omega_k(\zeta_j, \zeta_k)$ and hence (cf. (16.60)), $\delta(\omega_i(\zeta_i)) = \omega_j(\zeta_i, \zeta_k) - \omega_k(\zeta_i, \zeta_j) + \omega_i(\zeta_j, \zeta_k) - 2\mu_i - s/6$, whenever $\varepsilon_{ijk} = 1$. Using (16.66), we obtain the equality (cf. [Der 3])

$$\textbf{(16.71)} \qquad \Delta\mu_i = 2\mu_i^2 + 4\mu_j\mu_k - s \cdot \mu_i/2 + 2(\mu_j - \mu_i)|\zeta_k|^2 + 2(\mu_k - \mu_i)|\zeta_j|^2,$$

whenever $\varepsilon_{ijk} = 1$.

16.72 Proposition. *Let an oriented Riemannian 4-manifold (M, g) satisfy the conditions $\delta W^+ = 0$ and $\det_{(\Lambda^+)} W^+ = 0$. Then $W^+ = 0$.*

Proof. Locally in M_{W^+}, we have $\mu_i = 0$ for a fixed i (notations of 16.58). Suppose that $W^+ \neq 0$ near $x \in M_{W^+}$. By (16.59), $\mu_j = -\mu_k \neq 0$, j, k being such that $\varepsilon_{ijk} = 1$, and so, by (16.66), $\omega_j(\zeta_j) = \omega_k(\zeta_k)$, which gives $|\zeta_j| = |\zeta_k|$. Hence (16.71) implies $0 = \Delta\mu_i = 4\mu_j\mu_k$. This contradiction completes the proof. □

16.73 Proposition. *Every oriented Riemannian 4-manifold with $\delta W^+ = 0$ satisfies the relation*

$$\textbf{(16.73)} \qquad \Delta|W^+|^2 = -s \cdot |W^+|^2 + 36 \det{}_{(\Lambda^+)} W^+ - 2|DW^+|^2.$$

Proof. In the notations of 16.58, set $Y_i = (\mu_j - \mu_k)\omega_i(\zeta_i)$, if $\varepsilon_{ijk} = 1$. By (16.59) and (16.60), $|DW^+|^2 = \sum_i (|d\mu_i|^2 + 2|Y_i|^2)$. Computing $\Delta|W^+|^2$ from (16.71), we now easily obtain (16.73). □

16.74 Theorem. *Let (M, g) be a compact oriented Riemannian 4-manifold with $\delta W^+ = 0$ and $s \geqslant 0$. Viewing W^+ as an endomorphism of $\wedge^+ M$, we have*

(16.74) $$\int_M \det W^+ \cdot v_g \geqslant 0,$$

the inequality being strict unless $W^+ = 0$ *identically.*

Proof. The weak inequality follows from (16.73) by integration. Equality occurs there if and only if $DW^+ = 0$ and $s \cdot |W^+| = 0$, which implies that $|W^+|$ is constant. Thus, by (16.73), equality in (16.74) yields $\det_{(\Lambda^+)} W^+ = 0$ and hence $W^+ = 0$ in virtue of 16.72. □

16.75 Remarks. (i) It follows immediately from (16.59) that every oriented Riemannian 4-manifold (M, g) satisfies the conditions $\check{W} = |W|^2 \cdot g$ and $\check{W}^\pm = |W^\pm|^2 \cdot g$, where, for an algebraic curvature tensor T, $\check{T}(X, Y) = (T(X, ., ., .), T(Y, ., ., .))$. cf. 1.131a), (4.72) and [Bac]. Consequently, the vector bundle homomorphism which maps X in TM to $W^+(X, ., ., .) \in T^*M \otimes \Lambda^+ M$ is injective wherever $W^+ \neq 0$. The transformation rule $\delta_{g'} W_{g'}^+ = \delta W^+ - W^+(Df, ., ., .)$ of δW^+ for conformally related metrics $g' = e^{2f} g$ on an oriented four-dimensional manifold (see (16.25) and 16.5) implies that in the open subset of such a manifold defined by $W^+ \neq 0$, the given conformal class cannot contain two essentially distinct (i.e., not proportional with a constant factor) metrics with $\delta W^+ = 0$.

(ii) An oriented Riemannian four-manifold (M, g) satisfies the condition $DW^+ = 0$ if and only if either $W^+ = 0$, or g is locally Kähler (Kähler with respect to a complex structure defined on a two-fold isometric covering) in a way compatible with the orientation and has non-zero constant scalar curvature. In fact, if $W^+ \in C^\infty(S_0^2 \Lambda^+ M)$ (cf. 1.128) is parallel and non-zero, it must have a simple eigenvalue and a local section ω of the corresponding line subbundle of $\Lambda^+ M$, normed by $|\omega|^2 = 2$, is parallel, so that our assertion follows from 16.63.

(iii) Roter's theorem saying that Riemannian manifolds (M, g) with $DW = 0$ must have $W = 0$ or $DR = 0$ (cf. 16.24iii)) can be proved in dimension four as follows. If $W \neq 0$, g is locally Kählerian by (ii), so that 16.30 implies $Dr = 0$, which, together with $DW = 0$, gives $DR = 0$ (cf. 1.1).

Appendix. Sobolev Spaces and Elliptic Operators

To discuss the Laplacian and related elliptic differential operators, one must introduce certain function spaces. It turns out that the spaces one thinks of first, namely C^0, C^1, C^2, etc... are, for better or worse, not appropriate; one is forced to use more complicated spaces. For example, if $\Delta u = f \in C^k$, one would like to have $u \in C^{k+2}$, a fact which is *false* for these C^k spaces [Mor, p.54], but which is true for the spaces to be introduced now.

For simplicity M will always denote a C^∞ compact Riemannian manifold without boundary, $n = \dim M$, and E and F are smooth vector bundles (with inner products) over M. Of course, there are related assertions if M has a boundary or if M is not C^∞.

A. Hölder Spaces [Gi-Tr]

Let $A \subset \mathbb{R}^n$ and $0 < \alpha < 1$. Then $f: A \to \mathbb{R}$ is *Hölder continuous with exponent* α if the following expression is finite

$$[f]_{\alpha,A} = \sup_{\substack{x,y\in A\\ x\neq y}} \frac{|f(x)-f(y)|}{|x-y|^\alpha}. \tag{1}$$

The simplest example of such a function is $f(x) = |x|^\alpha$ for real x (we could allow $\alpha = 1$ in (1), but must exclude it in Sections C and H through K below). Let $\Omega \subset \mathbb{R}^n$ be an open set. The *Hölder space* $C^{k+\alpha}(\Omega)$ is the Banach space of real valued functions, f defined on Ω all of whose k^{th} order partial derivatives are Hölder continuous with exponent α. The norm is

$$\|f\|_{k+\alpha} = \|f\|_{C^k(\bar{\Omega})} + \max_{|j|=k} [\partial^j f]_{\alpha,\bar{\Omega}}, \tag{2}$$

where $\| \ \|_{C^k(\bar{\Omega})}$ is the usual C^k norm, $j = (j_1, \dots, j_n)$, and $\partial^j = (\partial/\partial x_1)^{j_1} \cdots (\partial/\partial x_n)^{j_n}$. On a manifold, M, one obtains the space $C^{k+\alpha}(M)$ by using a partition of unity. Note that if $0 < \alpha < \beta < 1$, then $C^{k+\beta}(M) \hookrightarrow C^{k+\alpha}(M)$ and by the Arzela-Ascoli theorem, this embedding is compact (recall that for Banach spaces A, B, a continuous map $T: A \to B$ is *compact* if for any bounded set $Q \subset A$, the closure of its image $T(Q)$ is compact).

B. Sobolev Spaces

For $f \in C^\infty(M)$, $1 \leqslant p < \infty$, and an integer $k \geqslant 0$ define the norm

(3) $$\|f\|_{p,k} = \left[\int_M \sum_{0 \leqslant j \leqslant k} |D^j f|^p \mu_g\right]^{1/p},$$

where $|D^j f|$ is the pointwise norm of the jth covariant derivative and μ_g is the Riemannian element of volume. The *Sobolev space* $L_k^p(M)$ is the completion of $C^\infty(M)$ in this norm; equivalently, by using local coordinates and a partition of unity, one can describe $L_k^p(M)$ as equivalence classes of measurable functions all of whose partial derivatives up to order k are in $L^p(M)$. The space $L_k^p(M)$ is a Banach space. If $p = 2$ these spaces are Hilbert spaces with the obvious inner product. This simplest case, $p = 2$, is generally adequate for linear problems (such as Hodge theory); nonlinear problems make frequent use of arbitrary values of p. The notation $H_{p,k}$ and W_k^p are often used instead of L_k^p. For a vector bundle E with an inner product one defines $L_k^p(E)$ similarly.

Note that if one changes the metric on a compact Riemannian manifold (M, g), although the norms (and inner products) on the spaces $C^{k+\alpha}(M)$ and $L_k^p(M)$ do change, the new norms are equivalent to the old ones so the topologies do not change.

C. Embedding Theorems [Fri, § 8–11]

It is clearly useful to investigate relationships between these spaces $C^{k+\alpha}$ and L_k^p. For example, as we shall see shortly, there is the psychologically reassuring fact that if $f \in L_k^p$ for all k, then $f \in C^\infty$. One easy but useful observation is that if $f \in C^0(M)$ and if we write

$$\|f\|_\infty = \max_{x \in M} |f(x)|$$

then

(4) $$\lim_{p \to \infty} \|f\|_{L^p(M)} = \|f\|_\infty$$

[Proof: given $\varepsilon > 0$, let $M_\varepsilon = \{x \in M : |f(x)| \geqslant \|f\|_\infty - \varepsilon\}$. Then

$$(\|f\|_\infty - \varepsilon)\operatorname{Vol}(M_\varepsilon)^{1/p} \leqslant \|f\|_{L^p(M)} \leqslant \|f\|_\infty \operatorname{Vol}(M)^{1/p}].\ \square$$

Another elementary inequality—an immediate consequence of Hölder's inequality—states that if $1 \leqslant q \leqslant p$, then

(5) $$\|f\|_{L^q(M)} \leqslant \operatorname{Vol}(M)^{(p-q)/pq} \|f\|_{L^p(M)}.$$

This shows that if $l \leqslant k$ and $q \leqslant p$, then there is a continuous injection $L_k^p(M) \hookrightarrow$

$L^q_l(M)$. The Sobolev embedding theorems give many other such relationships. Recall that $n = \dim M$, and let $\delta(p, k) = k - \dfrac{n}{p}$.

6 Theorem (Sobolev inequality and embedding theorem). *Let $0 \leqslant l \leqslant k$ be integers and assume $f \in L^p_k(M)$.*

(a) *If $\delta(p, k) < l$, that is, $k - l < n/p$, and if q satisfies*

$$\delta(q, l) \leqslant \delta(p, k), \text{ that is } \frac{1}{p} - \frac{k-l}{n} \leqslant \frac{1}{q}, \tag{7}$$

then there is a constant $c > 0$ independent of f such that

$$\| f \|_{q,l} \leqslant c \| f \|_{p,k}. \tag{8}$$

Thus, there is a continuous inclusion $L^p_k \hookrightarrow L^q_l(M)$. Moreover, if $l < k$ and strict inequality holds in (7), then this inclusion is a compact operator.

(b) *If $l < \delta(p, k) < l + 1$, that is, $(k - l) - 1 < n/p < k - l$, let $\alpha = \delta(p, k) - l$ so $0 < \alpha < 1$. Then there is a constant c independent of f such that*

$$\| f \|_{\delta(p,k)} = \| f \|_{l+\alpha} \leqslant c \| f \|_{p,k}. \tag{9}$$

Thus, there is a continuous inclusion $L^p_k(M) \hookrightarrow C^{\delta(p,k)}(M)$ and a compact inclusion $L^p_k(M) \hookrightarrow C^\gamma(M)$ for $0 < \gamma < \delta(p, k)$.

Of course, in the inclusion $L^p_k \hookrightarrow C^{l+\alpha}$ we identify functions that differ only on sets of measure zero. The compactness assertions of part (a) in this theorem were proved by Rellich ($p = 2$) and generalized by Kondrashov.

Note that all of the above results are proved first for a smoothly bounded open set in $\mathbb{R}^n$ and then extended to vector bundles over compact manifolds using a partition of unity.

Some useful special cases of the theorem are:

i) if $f \in L^p_k(M)$ and $p > n$, then $f \in C^{k-1}(M)$,

ii) if $f \in L^p_k(M)$ and $pk > n$, then $f \in C^0(M)$,

iii) if $f \in L^2_1(M)$, then $f \in L^{2n/(n-2)}(M)$ (here $n \geqslant 3$), and there are constants A, $B > 0$ such that

$$\| f \|_{L^{2n/(n-2)}} \leqslant A \| Df \|_{L^2} + B \| f \|_{L^2}. \tag{10}$$

The value $q = 2n/(n - 2)$ in (10) is the largest number for which (8) holds in the case $k = 1$, $p = 2$. It is a "limiting case" of the Sobolev inequality. The smallest value of A for which there is some constant B such that (10) holds is known (see [Gi-Tr, p.151] and also [Aub 3]). This smallest constant is independent of the manifold M. On the other hand, for fixed $B > 0$ the smallest permissible value of A does depend on the geometry of M and is related to the isoperimetric inequality (see [Gal 4]). Related inequalities for limiting cases have been found ([Tru 2], [Br-Wa], [Aub 3]) and play an important role in several recent geometric problems.

Since the condition (7) and the related condition in part (b) may seem mysterious, it may be useful to point out that they are both optimal and easy to discover. To see this, for example with $l = 0$, let $\varphi \in C_0^\infty(|x| < 1)$, $\varphi \not\equiv 0$, and let $f_\lambda(x) = \varphi(\lambda x) \in C_0^\infty(|x| < 1)$ for $\lambda \geqslant 1$. Applying (8) to the f_λ and doing a brief computation, one obtains

$$\|\varphi\|_{L^q} \leqslant c\lambda^{k+n(1/q-1/p)}\|\varphi\|_{p,k}.$$

Letting $\lambda \to \infty$ there is a contradiction unless (7) holds. This example uses the conformal map $x \to \lambda x$; it leads one to suspect that conformal defomations of metrics lead one to the limiting case of the Sobolev inequality. This suspicion is verified in Chapter 4.

D. Differential Operators

Here and below we will use the notation $F(x, \partial^k u)$, $a(x, \partial^k u)$, etc. to represent any (possibly nonlinear) differential operator of order k. We use local coordinates on $\mathbb{R}^n$ and consider a differential operator on vector valued functions u:

$$F(x, \partial^k u) = \sum_{|\alpha|=k} a_\alpha(x, \partial^l u)\partial^\alpha u + f(x, \partial^l u),$$

where α is a multi-index and $l \leqslant k$. The equation $F(x, \partial^k u) = 0$ is *linear* if the coefficients a_α—which may be complex matrices—and f do not depend on u or its derivatives, while it is *quasilinear* if a_α and f depend only on the derivatives $\partial^\beta u$ of at most order $k-1$, so $l \leqslant k-1$. In this case, F is linear in its highest order derivatives. The usual formulas for the various curvatures of a Riemannian manifold are examples of second order quasilinear differential operators in the metric g. Other examples are the minimal surface equation—as well as all Euler-Lagrange equations for variational problems. On the other hand, the Gaussian curvature, K, of a surface $u\colon \mathbb{R}^2 \to \mathbb{R}$ in $\mathbb{R}^3$ (in the notation $u_x = \partial u/\partial x$ etc) satisfies

(12) $$u_{xx}u_{yy} - u_{xy}^2 - K(x, y)(1 + u_x^2 + u_y^2)^2 = 0,$$

which is not quasilinear, it is fully nonlinear. Another fully nonlinear equation is the Monge-Ampere equation for Kähler-Einstein metrics [it is customary to refer to nonlinear equations involving the determinant of the hessian as "Monge-Ampère equations"].

One often meets second order quasilinear equations in *divergence form*

$$\operatorname{div} A(x, u, Du) = f(x, u, Du),$$

where A is a vector field. For example, on $\mathbb{R}^n$ this equation is

$$-\sum_i \frac{\partial}{\partial x^i} A^i(x, u, Du) = f(x, u, Du).$$

Clearly, if $u \in C^2$ is a solution of this in an open set $\Omega \subset \mathbb{R}^n$, then for any $\varphi \in L_1^2$ with support in Ω we have

$$\int \sum_i A^i(x, u, Du) \frac{\partial \varphi}{\partial x^i} \mu_g = \int f(x, u, Du) \varphi \mu_g. \tag{13}$$

In many circumstances, this last formula makes sense even for $u \in L_1^2(\Omega)$, rather than $C^2(\Omega)$ (the model is where $A(x, u, Du) = \operatorname{grad} u$). Then we say that $u \in L_1^2(\Omega)$ is a *weak solution* of the original equation if (13) holds for all $\varphi \in L_1^2$ with support in Ω. Caution: there are several other useful, but not equivalent, definitions of "weak solution".

The main application of these notions is to calculus of variations problems, where one seeks a critical point of

$$J(u) = \int F(x, u, Du) \mu_g.$$

Then the Euler-Lagrange equation is automatically in divergence form, so one may think of (13) as a weak form of an Euler-Lagrange equation.

The virtue of enlarging the class of admissible solutions to allow weak solutions is that this may make it much easier to prove the existence of a solution of the equation. On the other hand, one is then faced with the often difficult regularity problem of determining to what extent this weak solution is actually smooth.

E. Adjoint

If E and F are smooth hermitian vector bundles over M and if $P: C^\infty(E) \to C^\infty(F)$ is a *linear* differential operator, then one can use the L^2 inner product to define the *formal adjoint*, P^*, by the usual rule

$$\langle Pu, v \rangle_F = \langle u, P^* v \rangle_E$$

for all smooth sections $u \in C^\infty(E)$, $v \in C^\infty(F)$. Since the supports of u and v can be assumed to be in a coordinate patch, one can compute P^* locally using integration by parts. For example if $u: \mathbb{R} \to \mathbb{R}^N$ and if $Pu = Du$ (where here $D = d/dx$), then $P^* v = -Dv$, while if

$$Pu = A(x) D^2 u, \quad \text{then} \quad P^* v = D^2 (A^*(x) v),$$

where A^* is the adjoint of the matrix A. We use the term "formal adjoint" to avoid confusion with the strict Hilbert space adjoint of an unbounded operator.

F. Principal Symbol

To a linear differential operator $P: C^\infty(E) \to C^\infty(F)$ of order k, at every point $x \in M$ and for every $\xi \in T_x^* M$ one can associate an algebraic object, the *principal symbol* $\sigma_\xi(P; x)$, often written simply $\sigma_\xi(P)$. If, in local coordinates,

(14) $$Pu = \sum_{|\alpha| \leqslant k} a_\alpha(x)\partial^\alpha u,$$

where the a_α are $\dim F \times \dim E$ matrices, then $\sigma_\xi(P; x)$ is the matrix

(15) $$\sigma_\xi(P; x) = i^k \sum_{|\alpha|=k} a_\alpha(x)\xi^\alpha \qquad (i = \sqrt{-1}).$$

Here we use the notation $\xi^\alpha = \prod_j \xi^{\alpha_j}$.
(It is often convenient to delete the factor i^k here and in (16). While this slightly complicates the property (iii) below, it eliminates using awkward factors of i in Example 17, in which M could be a real manifold so complex numbers would be out of place.) To define this invariantly, let E_x and F_x be the fibres of E and F at $x \in M$, let $u \in C^\infty(E)$ with $u(x) = z$, and let $\varphi \in C^\infty(M)$ have $\varphi(x) = 0$, $d\varphi(x) = \xi$, then $\sigma_\xi(P; x)\colon E_x \to F_x$ is the following endomorphism

(16) $$\sigma_\xi(P; x)z = \frac{i^k}{k!} P(\varphi^k u)|_x.$$

The point is that many of the properties of P depend only on the highest order derivatives appearing in P; the principal symbol is a simple invariant way to refer to this highest order part of P (although it is occasionally useful to use the whole symbol involving the lower order derivatives in P, not just its principal part). For example, by verifying an algebraic property of $\sigma_\xi(P)$ we will define an elliptic differential operator (see below), the *algebraic* property then implies *analytic* conclusions.

The symbol has several obvious, but useful, algebraic properties

(i) $\sigma_\xi(P + Q) = \sigma_\xi(P) + \sigma_\xi(Q)$,

(ii) $\sigma_\xi(PQ) = \sigma_\xi(P)\sigma_\xi(Q)$,

(iii) $\sigma_\xi(P^*) = \sigma_\xi(P)^*$ (hermitian adjoint of $\sigma_\xi(P)$).

Note that without the factor i^k in (15)–(16), the property (iii) would need an extra factor $(-1)^k$ since $(\partial/\partial x)^* = -\partial/\partial x$.

17 Example. The exterior derivative $d\colon \Omega^p(M) \to \Omega^{p+1}(M)$ has the property $d(\varphi\alpha) = d\varphi \wedge \alpha + \varphi d\alpha$ for any $\varphi \in C^\infty(M)$, $\alpha \in \Omega^p(M)$. Thus

(18) $$\sigma_\xi(d)\alpha = i\xi \wedge \alpha.$$

Similarly, for any vector bundle E over M, the covariant derivative $D\colon \Omega^0(E) \to \Omega^1(E)$ satisfies $(D(\varphi v) = d\varphi \otimes v + \varphi Dv$ for any $\varphi \in C^\infty(M)$, $v \in \Omega^0(E)$. Consequently

$$\sigma_\xi(D)v = i\xi \otimes v.$$

G. Elliptic Operators

A linear differential operator $P\colon C^\infty(E) \to C^\infty(F)$ is *elliptic at a point* $x \in M$ if the symbol $\sigma_\xi(P; x)$ is an isomorphism for every real $\xi \in T_x^*M - \{0\}$. Of course a necessary condition for this is that $\dim E_x = \dim F_x$. Note that there is a more general definition of ellipticity for systems that allows different orders in the various dependent variables (see [Do-Ni] and [Ag-Do-Ni 2]).

19 Example. Consider the second order scalar equation

$$Pu = \sum_{i,j} a_{ij}(x) \frac{\partial^2 u}{\partial x_i \partial x_j} + \sum_j b_j(x) \frac{\partial u}{\partial x_j} + c(x)u, \tag{20}$$

where u and the coefficients are real-valued functions. Then for each x and ξ

$$\sigma_\xi(P; x) = -\sum_{i,j} a_{ij}(x)\xi_i\xi_j.$$

Hence P is elliptic at x if and only if the matrix $(a_{ij}(x))$ is positive (or negative) definite.

21 Example. The Cauchy-Riemann equation in $\mathbb{R}^2 = \mathbb{C}^1$ is $\bar{\partial}u = \frac{1}{2}(\partial/\partial x + i\partial/\partial y)u = 0$. It is ellipitic.

22 Example. Let $C^\infty(E) \xrightarrow{P} C^\infty(F) \xrightarrow{Q} C^\infty(G)$, where P and Q are first order linear differential operators and E, F, G are hermitian vector bundles over M. The second order operator

$$L = PP^* + Q^*Q\colon C^\infty(F) \to C^\infty(F) \tag{23}$$

is elliptic at x if the following symbol sequence is exact at F_x for every $\xi \in T_x^*M - \{0\}$

$$E_x \xrightarrow{\sigma_\xi(P;x)} F_x \xrightarrow{\sigma_\xi(Q;x)} G_x$$

(*Proof*: just use (30) below). □

The rule (23) is a common construction of an elliptic operator. If $E = \bigwedge^{p-1}$, $F = \bigwedge^p$, and $G = \bigwedge^{p+1}$ and P and Q are exterior differentiation, then the Hodge-de Rham Laplacian on p-forms is of this form. Another special case is if $P = 0$; in this situation we see that if $\sigma_\xi(Q)$ is injective then Q^*Q is elliptic.

The operator P is *underdetermined elliptic at* x if $\sigma_\xi(P; x)$ is surjective for every $\xi \in T_x^*M - \{0\}$ (the simplest example is the divergence of a vector field on $\mathbb{R}^n$). In this case PP^* is elliptic at x. Similarly, P is *overdetermined elliptic at* x *if* $\sigma_\xi(P; x)$ is injective for every $\xi \in T_x^*M - \{0\}$ (the simplest example is the gradient of a real-valued function; another example is the Cauchy-Riemann equations for an analytic function of several complex variables). In this case P^*P is elliptic at x.

For a nonlinear differential operator $F(x, \partial^k u)$, its *linearization at* u is the linear operator

$$Pv = \frac{d}{dt} F(x, \partial^k(u + tv))\bigg|_{t=0}. \tag{24}$$

Thus, the quasilinear operator (11) (recall $l \leqslant k - 1$) has as its linearization at u the kth order operator

$$Pv = \sum_{|\alpha|=k} a_\alpha(x, \partial^l u)\partial^\alpha v + \text{lower order terms}, \tag{25}$$

while the linearization of (12) at u is

$$P v=u_{yy} v_{xx}-2 u_{xy} v_{xy}+u_{xx} v_{yy}+\text{lower order terms.} \tag{26}$$

The *nonlinear equation* $F(x, \partial^k u)=0$ is *elliptic at* (x, u) (that is, it is elliptic at x for the function u), if its linearization at u is elliptic at the point x, which a similar definition for *underdetermined* and *overdetermined elliptic*. From (26) it is clear that the equation (12) is elliptic precisely at those points where $K>0$. The equation $u u_{xx}+u_{yy}=0$ is evidently elliptic if $u>0$, but not if $u \leqslant 0$; thus $u(x) \equiv+1$ is an elliptic solution while $u(x) \equiv-1$ is a non-elliptic solution.

H. Schauder and L^p Estimates for Linear Elliptic Operators

After roughly one hundred years of hard work, we now know that the main technical step in the theory of linear elliptic operators is establishing the following inequality. Recall that E and F are vector bundles over M and that M is compact *without* boundary.[1])

27 Theorem. *Let $P\colon C^{\infty}(E) \rightarrow C^{\infty}(F)$ be a linear elliptic differential operator of order k. Given a constant $0<\alpha<1$ and an integer $l \geqslant 0$ there are constants $c_1, c_2, \ldots, c_6$ such that*

(a) *(Schauder estimates) for every $u \in C^{k+l+\alpha}(E)$,*

$$\|u\|_{k+l+\alpha} \leqslant c_1\|P u\|_{l+\alpha}+c_2\|u\|_{C(E)} \leqslant c_3\|u\|_{k+l+\alpha}, \tag{28}$$

(b) *(L^p estimates) for every $u \in L_{k+l}^p(E)$, $1<p<\infty$,*

$$\|u\|_{p, k+l} \leqslant c_4\|P u\|_{p, l}+c_5\|u\|_{L^1} \leqslant c_6\|u\|_{p, k+l}, \tag{29}$$

Moreover, if one restricts u so that it is orthogonal (in $L^2(E)$) to $\ker P$, *then we can let $c_2=c_5=0$ (with new constants c_1 and c_4).*

Conceivable $\ker P \mid L_{k+l}^p(E)$ could get smaller as l increases. In fact, the regularity part of Theorem 31 states that $\ker P \subset C^{\infty}(E)$ so there is no ambiguity. Note that the right-hand sides in (28)–(29) are obvious. The moral of (28)–(29) is that in these Hölder and Sobolev spaces, $\|P u\|$ defines a norm equivalent to the standard norm, except that one must add an extra term if $\ker P \neq 0$, since then $\|P u\|$ is only a seminorm. This theorem is proved—in greater generality—in [Do-Ni], [Ag-Do-Ni 1 and 2], and [Mor, Theorem 6.4.8]; since M has no boundary, all one really needs are the simpler "interior estimates" from these references coupled with a partition of unity argument. (In particular, one does not need the assumption of "proper ellipticity" here, or anywhere else in this appendix).

[1]) In the following we consider a linear differential operator. $P\colon C^{\infty}(E) \rightarrow C^{\infty}(F)$ of order k; clearly this operator can be extended uniquely to act on $C^{k+l+\alpha}(E)$ and $L_{k+l}^p(E)$. We presume this extension has been done wherever needed, in particular, in Theorems 27 and 31, and Corollary 32.

I. Existence for Linear Elliptic Equations

For a linear map $L: \mathbb{R}^n \to \mathbb{R}^k$ one knows that one can solve the equation $Lx = y$ if and only if y is orthogonal to $\ker L^*$. [Proof: we show that $(\operatorname{Im} L)^\perp = \ker L^*$. Now $z \perp \operatorname{Im} L \Leftrightarrow \langle Lx, z\rangle = 0$ for all $x \Leftrightarrow \langle x, L^*z\rangle = 0$ for all $x \Leftrightarrow L^*z = 0$]. This assertion can be summarized by saying that

$$\mathbb{R}^k = L(\mathbb{R}^n) \oplus \ker L^*, \tag{30}$$

and can also be formulated as an alternative: "either one can always solve $Lx = y$, or else $\ker L^* \neq 0$, in which case a solution exists is and only if y is orthogonal to $\ker L^*$."

If L is a continuous map between Hilbert spaces, the above reasoning is still valid and shows that $\operatorname{Im}(L)^\perp = \ker L^*$. Hence $\overline{\operatorname{Im}(L)} = (\ker(L^*))^\perp$. However, in order to pass to the analogue of (30) one needs that the image of L be a *closed* subspace.

The estimates of Theorem 27 allow one to prove that the image of P is closed. As a consequence, the existence theory for a linear elliptic operator on a compact manifold can be stated exactly as in the finite dimensional case. It is often called the *Fredholm alternative.*

31 Theorem (Existence). *Let $P: C^\infty(E) \to C^\infty(F)$ be a linear elliptic differential operator of order k. Then both $\ker P$ and $\ker P^* \subset C^\infty$ and they are also finite dimensional. If $f \in L^2_l(F)$, then there is a solution $u \in L^2_{k+l}(E)$ of $Pu = f$ if and only if f is orthogonal (in $L^2(F)$) to $\ker P^*$; this solution u is unique if one requires that u be orthogonal (in $L^2(E)$) to $\ker P$. If $E = F$, the eigenspace $= \ker(P - \lambda I)$ is therefore finite dimensional.*

Moreover, for $1 < p < \infty$, if $f \in L^p_l$, $C^{l+\alpha}$, or C^∞, then a solution u is in L^p_{k+l}, $C^{k+l+\alpha}$, or C^∞, respectively.

The proof for $f \in L^2_l$ or C^∞ can be found, for example in [War, Chapter 6], while the last sentence is a consequence of the regularity Theorem 40 below.

In the next corollary we extend part of Theorem 31 to underdetermined and overdetermined systems. The usefulness of part (b) of the corollary to geometric problems was first pointed out in [Be-Eb].

32 Corollary. *Let $P: C^\infty(E) \to C^\infty(F)$ be a linear differential operator of order k.*

(a) *If P is either elliptic or underdetermined elliptic, then $\ker P^* \subset C^\infty$ is finite dimensional and*

$$L^p_l(F) = P(L^p_{k+l}(E)) \oplus \ker P^* \qquad (1 < p < \infty), \tag{33i}$$

$$C^{l+\alpha}(F) = P(C^{k+l+\alpha}(E)) \oplus \ker P^*, \tag{33ii}$$

$$C^\infty(F) = P(C^\infty(E)) \oplus \ker P^*. \tag{33iii}$$

(b) *If P is overdetermined elliptic, then* (33) *are valid if one replaces* $\ker P^*$ *by the intersection of* $\ker P^*$ *with $L^p_l(F)$, $C^{l+\alpha}(F)$, and $C^\infty(F)$, respectively (if $l < k$, then* $\ker P^* \cap L^p_l(F)$ *are distributions).*

34 *Proof.* (a) If P is elliptic, this is immediate. If P is underdetermined elliptic, apply the theorem to $Q = PP^*$. Note that $\ker Q = \ker P^*$ since $\langle Qv, v\rangle = \langle PP^*v, v\rangle = \|P^*v\|^2$ in L^2.

(b) First we prove the analogue of (33i). Since P^*P is elliptic and—in the $L^2(F)$ inner product—$\operatorname{Im}(P^*)$ is orthogonal to $\ker P$ ($=\ker P^*P$), by (33i), for any $f \in L_l^p(F)$ there is a solution $u \in L_{l+k}^p$ of $P^*Pu = P^*f$. Thus $Pu - f \in \ker P^* \cap L_l^p(F)$. But Pu is orthogonal to $\ker P^* \cap L_l^p(F)$, since if $\Psi \in (\ker P^*) \cap L_l^p(F)$ then in L^2, $\langle Pu, \Psi\rangle = \langle u, P^*\Psi\rangle = 0$. Hence $Pu - f = 0$ if and only if f is orthogonal to $\ker P^* \cap L_l^p(F)$. The proof of the case $f \in C^{k+\alpha}$ is similar. □

One obtains the classical Hodge decomposition as an immediate consequence of (33) applied to the elliptic operator $L = d\delta + \delta d$ (see (23)) acting on differential forms (recall that in our notation $\delta = d^*$).

Here is an example, due to R.T. Seeley [See], of an elliptic operator P having *every* complex number λ as an eigenvalue. Let M be any compact Riemannian manifold with Laplacian Δ_M and let $0 \leqslant \theta < 2\pi$ be a coordinate on S^1. Then the operator

(35)
$$Pu = -\left(e^{-i\theta}\frac{\partial}{\partial\theta}\right)^2 u + e^{-2i\theta}\Delta_M u$$

(or take the real and imaginary parts if one prefers a pair of real equations) is elliptic on $S^1 \times M$. Making the change of variable $t = \sqrt{\lambda}e^{i\theta}$, we see that for any complex $\lambda \neq 0$, $u = \exp(\pm i\sqrt{\lambda}\exp i\theta)$ is an eigenfunction with eigenvalue λ, while for $\lambda = 0$, $u \equiv 1$ is an eigenfunction. This awkward situation does not occur for "strongly elliptic operators". To define them, let $P\colon C^\infty(E) \to C^\infty(E)$ and regard the symbol in local coordinates as a square matrix $[\sigma_\xi(P; x)]_{ij}$. *Strong ellipticity at* x means that for some (complex) number γ and some $c > 0$ (γ and c may depend on x, but not ξ on η) such that the following quadratic form is definite:

(36)
$$Q(\eta) = Re\left\{\gamma\sum_{i,j}[\sigma_\xi(P;x)]_{ij}\eta_i\bar{\eta}_j\right\} \geqslant c|\eta|^2$$

for all complex vectors η and all real vectors $\xi \in T_x^*M$ with $|\xi| = 1$. Replacing ξ by $-\xi$ reveals that P must have even order. The Laplacian on forms or tensors is strongly elliptic, for example, while the Cauchy-Riemann equations on $\mathbb{R}^2$, equation (35), and the second order operator $(\partial/\partial x + i\partial/\partial y)^2$ on $\mathbb{R}^2$ are not strongly elliptic.

37 Theorem [Mor, 6.5.4]. *If $P\colon C^\infty(E) \to C^\infty(E)$ is strongly elliptic, then it is elliptic* (clearly) *and its eigenvalues are discrete, having a limit point only at infinity.*

As a brief illustration of the use of both the L^p and Schauder theory for nonlinear equations, say $f(x, s)$ is a bounded C^∞ function on $M \times \mathbb{R}$ and say $u \in L_2^2(M)$ is a solution of

$$\Delta u = f(x, u).$$

We claim that, in fact, $u \in C^\infty(M)$. Now $f(x, u)$ is bounded, and hence in L^p for all

$p < \infty$. Thus, by Theorem 31, $u \in L_2^p$ for all $p < \infty$. Choosing some $p > n = \dim M$ the Sobolev embedding theorem then implies that $u \in C^{1+\alpha}$ for some $0 < \alpha < 1$, and therefore so is $\Delta u = f(x, u)$. By Theorem 31, again $u \in C^{3+\alpha}$. Thus $\Delta u = f(x, u) \in C^{3+\alpha}$, so $u \in C^{5+\alpha}$ etc. This reasoning is often called a "bootstrap argument," since the regularity of u is "raised by its own bootstraps."

In order to apply the existence portion of these results and solve $Pu = f$ on M, one needs to know that $\ker P^* = 0$. As an example, consider the scalar equation

$$Pu = \Delta u + c(x)u, \tag{38}$$

where $c(x) > 0$ (recall the sign convention $\Delta u = -u''$ on $\mathbb{R}$). We claim that $\ker P = 0$ and present two proofs. The first uses the obvious *maximum principle* that if $Pu \leqslant 0$ then $u \leqslant 0$, that is, u can not have a positive local maximum—since at such a point $\Delta u \geqslant 0$ and $cu > 0$ so $Pu > 0$ there. If $u \in \ker P$, then it can not have a positive maximum or negative minimum. Hence $u = 0$. (There is a stronger version of the maximum principle that holds if $c \geqslant 0$: "If $Pu \leqslant 0$, then at a local maximum $u < 0$ unless $u \equiv$ constant," see [Pr-We]).[1]) This basic idea is equally applicable to real second order nonlinear scalar equations.

For the second proof, multiply the equation $Pu = 0$ by u and integrate over M, integrating by parts (the divergence theorem) to obtain

$$0 = \int_M u(\Delta u + cu)\mu_g = \int_M (|Du|^2 + cu^2)\mu_g.$$

Since $c > 0$, then clearly $u = 0$.

This proof is still applicable for vector valued functions u with c a positive definite matrix—and similar equations on vector bundles. Bochner and others have used it effectively to prove "vanishing theorems" in geometry.

In this example, $P = P^*$. Thus $\ker P^* = 0$; we conclude that for any $f \in C^\infty(M)$ there is a unique solution of $\Delta u + cu = f$. Moreover, $u \in C^\infty$.

Both of these proofs show that on scalar functions $\ker \Delta$ is the constants. Consequently we can solve $\Delta u = f$ if and only if $\langle f, 1\rangle = 0$, that is, $\int_M f\mu_g = 0$. □

J. Regularity of Solutions for Elliptic Equations

In brief, solutions of elliptic equations are as smooth as the coefficients and data permit them to be. The results are, of course, local. First we consider the linear case of a system

$$Pu \equiv \sum_{|\alpha| \leqslant k} a_\alpha(x)\partial^\alpha u = f(x). \tag{39}$$

[1]) Warning: Because many mathematicians use the opposite sign convention on the Laplacian, it is very important to check the sign convention before using inequalities like $\Delta u \geqslant f$.

40 Theorem (Regularity). *Assume P is elliptic in an open set $\Omega \subset \mathbb{R}^n$, that in Ω for some integer $l \geqslant 0$ and some $0 < \sigma < 1$*

$$a_\alpha(x) \text{ is in } C^l, \text{ or } C^{l+\sigma}, \text{ or } C^\infty, \text{ or } C^\omega,$$

while

$$f(x) \text{ is in } L_l^p,\ 1 < p < \infty, \text{ or } C^{l+\sigma}, \text{ or } C^\infty, \text{ or } C^\omega,$$

respectively. If $u \in L_k^p(\Omega)$ satisfies $Pu = f$ (almost everywhere) in Ω, then in Ω

$$u \text{ is in } L_{k+l}^p, \text{ or } C^{k+l+\sigma}, \text{ or } C^\infty, \text{ or } C^\omega, \text{ respectively.}$$

This theorem is an amalgamation of [Ag-Do-Ni 2 Th.10.7], [Do-Ni, Th 4], [Mor Theorems 6.2.5 and 6.6.1] where a slightly more general result is proved.

Upon first seeing such results, one may wonder if there is any practical situation where the coefficients are not in C^∞. To answer this, one looks at nonlinear equations, where one often needs results for linear equations with minimal smoothness assumptions (indeed, one even needs results with bounded measurable coefficients).

As a specific example, let $u \in C^2$ be a solution of the quasilinear elliptic equation

$$\sum_{|\alpha| \leqslant 2} a_\alpha(x, \partial^l u)\partial^\alpha u = f(x, \partial^l u),$$

where $l \leqslant 1$ and the coefficients a_α and f are C^∞ functions of their variables. We claim that, in fact, $u \in C^\infty$. To prove this we use another bootstrap argument. Since $u \in C^2$ then the functions $a_\alpha(x, \partial^l u(x))$ and $f(x, \partial^l u(x))$ are in C^1 as functions of x, and hence in C^σ for all $0 < \sigma < 1$. By Theorem 40 then $u \in C^{2+\sigma}$ for all $0 < \sigma < 1$. Thus a_α and f are in $C^{1+\sigma}$ so $u \in C^{3+\sigma}$ etc. The same proof works if we had assumed only that $u \in L_2^p$ for some $p > n$.

For a nonlinear system $F(x, \partial^k u) = 0$ the local results concerning regularity are similar.

41 Theorem. *Assume that $F(x, s)$ is a C^1 (or $C^{l+\sigma}$ for some integer $l \geqslant 1$, $0 < \sigma < 1$, or C^∞, or C^ω) function of all its variables for x in an open set $\Omega \subset \mathbb{R}^n$ and all s, and that $u \in C^k(\Omega)$ is an elliptic solution of the k-th order equation $F(x, \partial^k u) = 0$. Then $u \in C^{k+\lambda}(\Omega)$ for any $0 < \lambda < 1$ (or in $C^{k+l+\sigma}(\Omega)$, or $C^\infty(\Omega)$, or $C^\omega(\Omega)$ respectively).*

See [Ag-Do-Ni 2, Theorem 12.1] and [Mor Theorems 6.7.6 and 6.8.1] for a proof. The key ingredient is Theorem 40 above. Regularity for overdetermined elliptic systems near the point x_0 is obtained by considering the elliptic system $P_0^* F(x, \partial^l u) = 0$, where P_0 is the linearization of F at $(x_0, j_k u)$ with $j_k u = k$-jet of u at x_0.

K. Existence for Nonlinear Elliptic Equations

At the present time there is no adequate general existence theory for nonlinear equations. The subject essentially consists of some significant examples and several techniques that have been useful. We will limit ourselves to stating two results, both of which are consequences of the implicit function theorem, and then giving a list of some methods.

The first result (from [Ag-Do-Ni 2, § 12]) considers the question of solving a nonlinear equation (or system) of order k

$$F(x, \partial^k u; t) = 0, \tag{42}$$

where t is a parameter. Say one has a solution at $t = t_0$. Can one always find a solution for t near t_0? It is reasonable that the implicit function theorem (in Banach spaces) is the key to this. Recall that we are on a compact manifold M without boundary so the question of boundary conditions does not enter.

43 Theorem (Perturbation). *In (42), let F be a C^∞ function of all of its arguments. Assume that*

(i) *$u_0 \in C^k$ is a solution of (42) for $t = t_0$,*
(ii) *the linearization, P, at u_0 is elliptic, and*
(iii) *the linearized equation $Pv = f$ has a unique solution for any $f \in C^\sigma$ (for some $0 < \sigma < 1$).*

Then there is a solution $u \in C^{k+\alpha}$ of (42) if $|t - t_0|$ is small enough.

44 *Proof.* By Theorem 41, we know that $u_0 \in C^{k+\alpha}(M)$, indeed $u_0 \in C^\infty(M)$. Let $T(u, t) = F(x, \partial^k u; t)$, so $T\colon C^{k+\alpha}(M) \times \mathbb{R} \to C^\alpha(M)$ is a smooth map. Note that $T(u_0, t_0) = 0$, while $T_u(u_0, t_0) = P\colon C^{k+\alpha}(M) \to C^\alpha(M)$ is bijective. The desired conclusion now follows from the implicit function theorem [Nir 2]. Because we assumed $F \in C^\infty$, then we also know that $u \in C^\infty$, but it is obvious that this proof requires only very mild smoothness of F to obtain a solution $u \in C^{k+\alpha}$. □

This theorem is often used in the "continuity method" (see below and Chapter 11) as well as in a variety of perturbation situations. If the linearization, P, is elliptic but *not* invertible, one can investigate the higher order terms in the Taylor series of $T(u, t)$ near (u_0, t_0) and (attempt to) determine when the nonlinear equation $T(u, t) = 0$ is solvable. This is called "bifurcation theory" (see [Nir 2]).

Several straightforward modifications of this perturbation Theorem 43 are often required in practice.

1. In the frequently occuring case when (42) is quasilinear, one can use Sobolev spaces instead of Hölder spaces. One uses $P\colon L^p_k \to L^p$ where $p > \dim M$, since then by the Sobolev embedding theorem the coefficients in (11) with $|\beta| \leqslant k - 1$ are continuous. An example is [Ka-Wa 2, 3] and Chapter 4 above.

2. Underdetermined elliptic systems can often be treated by combining the perturbation theorem with the device in the proof of Corollary 32(a).

3. There is a simple situation where the perturbation theorem can be used with P not invertible. This is when P is invertible on a subspace and this subspace is "invariant" under F. For example, if $A = \{f \in C^0(M)\colon \int_M f = 0\}$, and if $F\colon C^{k+\alpha} \cap A \to C^\alpha \cap A$ with $P\colon C^{k+\alpha} \cap A \to C^\alpha \cap A$ invertible (which is the case if $P = \Delta$), then the Perturbation Theorem applies to yield a solution $u \in C^{k+\alpha} \cap A$ (presuming $u_0 \in A$ too). See [Aub 5], [Yau 3] and Chapter 11 where this occurs.

Next we discuss the *local solvability* of a nonlinear equation (or system)

$F(x, \partial^k u) = 0$ of order k. The question is quite modest, since we seek a solution only in some neighborhood of a point x_0, not on a compact manifold and do not impose any boundary conditions. The well-known Lewy example (see [Nir 4]) shows that this may be impossible, even for a linear equation with smooth coefficients. However if one assumes ellipticity there is no difficulty as long as the equation is solvable at one point 'to avoid obviously unsolvable examples as finding a real solution of $(\Delta u)^2 = -1$).

45 Theorem (Local solvability). *Assume $F(x, \partial^k u)$ is a C^∞ function of all of its arguments and that $u_0(x) \in C^k$ is an elliptic solution of $F(x, \partial^k u_0) = 0$ at $x = x_0$. Then in some neighborhood of x_0 there is a solution $u \in C^k$ (and hence C^∞) of $F(x, \partial^k u) = 0$. Moreover, one can also specify that $\partial^\alpha u = \partial^\alpha u_0$ for $|\alpha| \leqslant k - 1$ at x_0.*

See [Mal 2 Section 9], [Nir 4], and [DeT] for a proof. One can call u_0 an "infinitesimal solution" in which case the theorem states that if $F(x, \partial^k u)$ is elliptic at u_0, then infinitesimal solvability implies local solvability. Again, using the device of the proof of Corollary 32 (a) one can prove the local solvability of underdetermined elliptic equations; one application is in 1.169, another is Malgrange's proof of the Newlander-Nirenberg theorem (see [Nir 4] or [Mal 2]). A different extension can be made if the linearization of F at (x_0, u_0) is strongly elliptic; then one can solve the Dirichlet problem in a small disc, instead of asking that $\partial^\alpha u = \partial^\alpha u_0, |\alpha| \leqslant k - 1$, at the origin.

Finally, we list various methods which have been used to prove existence for nonlinear elliptic problems, and give some references (see also the survey article [Nir 1]).

1. *Calculus of Variations.* One proves that an equation has a solution by showing that it is the Euler-Lagrange equation of a variational problem (and hence quasi-linear), and then proving that this variational problem has a critical point. See [Cou], [Ka-Wa 1], [Lem], [Sa-Uh], and Chapter 4 above.

2. *Continuity Method.* To solve the k-th order equation $F(x, \partial^k u) = 0$, one considers a family of problems

$$F(x, \partial^k u; t) = 0, \qquad 0 \leqslant t \leqslant 1,$$

where $F(x, \partial^k u; 1) = F(x, \partial^k u)$ is the problem we wish to solve, while at $t = 0$, $F(x, \partial^k u; 0) = 0$ is a simpler problem that we know how to solve. Let A be the set of $t \in [0, 1]$ such that the problem is solvable. We know that $t = 0 \in A$. One shows that A is open, usually by the Perturbation Theorem 43. The final step is to show A is closed, since then $A = [0, 1]$. Say $A \ni t_j \to t$. Then there are u_j satisfying $F(x, \partial^k u_j; t_j) = 0$. To prove that $t \in A$ (that is, A is closed) one needs to find a subsequence of the (u_j) that converges in C^k. The standard approach is to show that the sequence (u_j) is in a bounded set in $C^{k+\sigma}$ for some $0 < \sigma < 1$, and then apply the Arzela-Ascoli theorem. Thus, one has the (usually difficult) task of finding an "a priori estimate": if u is a solution $F(x, \partial^k u; t) = 0$ for some $0 \leqslant t \leqslant 1$, then $\|u\|_{k+\alpha} \leqslant$ constant, where the constant is independent of t. Two applications of this method are in [Nir 3], and [Aub 5], [Yau 3] (see also [Ast] and Chapter 11).

3. *Schauder Fixed-Point Theorem.* One proves that $F(x, \partial^k u) = 0$ has a solution by showing that a related equation involving a compact operator has a fixed point (this is an extension of the Brouwer Fixed Point Theorem). For example, given a function v, let $u = T(v)$ be the solution of the linear problem (so we are assuming *linear* solvability) for an elliptic equation

$$\sum_{|\alpha|=k} a_\alpha(x, \partial^l v)\partial^\alpha u = f(x, \partial^l v).$$

A fixed point $u = T(u)$ is then a solution of the corresponding quasi-linear equation. Note that $T: C^{k-1+\sigma} \to C^{k+\sigma} \hookrightarrow C^{k-1+\sigma}$ is a compact operator since the inclusion $C^{k+\sigma} \hookrightarrow C^{k-1+\sigma}$ is compact. To apply the method, one approach is to find a ball $B = \{u: \|u\| \leqslant c\}$ in $C^{k-1+\sigma}$ and show that $T: B \to B$; thus, one wants to show that if $\|v\| \leqslant c$ then $\|u\| = \|T(v)\| \leqslant c$, i.e., find an a priori bound on solutions of $T(v) = u$. For examples and some modified versions, see [Co-Hi], [Gi-Tr] Chapter 10], and [Nir 2].

4. *Leray-Schauder Degree.* This is similar to (but more complicated than) the fixed point approach. It is an extension of the Brouwer degree to Banach spaces. See [Nir 2].

5. *Monotonicity.* This method applies to some quasi-linear equations that do not quite fit in the calculus of variations approach. See [Mor § 5.12].

6. *Sub and Super Solutions.* While only applying to second order scalar equations, this method is often quite simple—when it works. For example u_- is called a *subsolution*[1]) of $\Delta u = f(x, u)$ if $\Delta u \leqslant f(x, u_-)$, with the inequality reversed for a *supersolution* u_+. If there are $u_\pm$ with $u_- \leqslant u_+$, then there is a solution u with $u_- \leqslant u \leqslant u_+$. As a simple illustration, one can use constants for $u_\pm$ in the equation $\Delta u = f(x) - g(x)e^u$, where f, $g > 0$, thus proving the existence of a solution (which is unique, since if $w = u - v$, where u and v are solutions, then $\Delta w = -c(x)w$ for some function $c > 0$; hence $w = 0$ by our discussion of (38)). See [Ka-Wa 1].

7. *Alexandrov's Method.* This applies only to second order scalar equations $F(x, \partial^2 u) = 0$, where u is a convex function. One uses the convexity to obtain approximate polyhedral solutions, and then pass to the limit. See [Pog 1 and 4].

8. *Heat Equation.* One solves the "heat equation" $\partial u/\partial t = F(x, \partial^2 u)$ and shows that as $t \to \infty$ the solution approaches "equilibrium", that is, $\partial u/\partial t \to 0$, to obtain a solution of $F(x, \partial^2 u) = 0$. A notable application is [Ee-Sa] (see also [Ee-Le]). Another application is recent result [Ham 2], which is also discussed in Chapter 5.

9. *Steepest Descent.* This is an alternate to the calculus of variations. To minimize a functional (and hence obtain a solution of the corresponding calculus of variations problem), one follows the gradient lines. There is a close resemblance to the heat equation procedure, see [Ino] for an example.

---

[1]) Recall our sign convention $\Delta u = -u''$ on $\mathbb{R}^1$.

Addendum

In this Addendum, we collect a choice of recent results (1985) which seem relevant to the contents of the book.

A. Infinitely many Einstein Constants on $S^2 \times S^{2m+1}$

Let (M_i, g_i), $i = 1, \ldots, m$, be Kähler-Einstein metrics with positive scalar curvature, and let $c_1(M_i) = q_i\alpha_i$ where $q_i \in \mathbb{Z}$ and α_i is an indivisible integer class. Set $M = M_1 \times \cdots \times M_m$ and denote by π_i the projection of M onto M_i. Let

$$\pi\colon P_{k_1,\ldots,k_m} \to M$$

be the principal circle bundle over M with Euler class $\sum_i k_i \pi_i^*(\alpha_i)$.

Add.1 **Theorem** (M. Wang, W. Ziller [Wa-Zi 4]). *For every choice of integers $k_1, \ldots, k_m$ which do not vanish simultaneously, the circle bundle $P_{k_1,\ldots,k_m}$ admits an Einstein metric with positive scalar curvature.*

If $m = 1$, this is Kobayashi's theorem (9.76). But in this case, the manifolds P that one obtains for different values of k are only coverings of each other, if $m > 1$, P is simply connected if and only if $k_1, \ldots, k_m$ are relatively prime. The metrics are obtained in a similar fashion as in Kobayashi's theorem, i.e., π is made into a Riemannian submersion with totally geodesic fibers such that the principal connection has the harmonic representative of the Euler form as curvature and such that the metric on the base is $\sum_i x_i g_i$ with appropriately chosen positive real numbers x_i.

If all M_i are homogeneous, so is P. But by choosing for M_i the inhomogeneous examples of N. Koiso and Y. Sakane [Ko-Sa], one can show

Add.2 **Corollary.** *For each positive integer l, there exists a compact Einstein manifold with positive scalar curvature of cohomogeneity l.*

By using $M_1 = \mathbb{C}P^1$ and $M_2 = \mathbb{C}P^m$, one can show

Add.3 **Corollary.** *There exist infinitely many non-isometric Einstein metrics of* *positive scalar curvature on* $S^2 \times S^{2m+1}$, $m \geqslant 1$, *on the unique non-trivial* S^{4k+1}*-bundle over* S^2, $k \geqslant 1$, *and on the unique non-trivial* $\mathbb{R}P^{2m+1}$*-bundle over* S^2, $m \geqslant 1$.

These metrics are all homogeneous, but with respect to different transitive actions. E.g., $S^2 \times S^3$ is diffeomorphic to $S^3 \times S^3/S^1$ for every embedding of S^1, and for each embedding, $S^3 \times S^3/S^1$ carries a unique $S^3 \times S^3$-invariant Einstein metric. But in general, the manifolds $P_{k_1,\ldots,k_m}$ will have different homotopy types for different choices of the integers $k_1, \ldots, k_m$.

If, for the infinitely many Einstein metrics in Add.3, we normalize the metric so that the volume is equal to 1, then the Einstein constant tends to zero. In particular, the Moduli Space of Einstein Structures on $S^2 \times S^3$ has infinitely many components.

The examples in Add.3 also show how the Palais-Smale condition fails for the total scalar curvature functional on the set of total volume 1 metrics, i.e., an infinite sequence of critical points does not have to have a convergent subsequence. But, if one uses convergence in Gromov's sense of Hausdorff convergence ([Gro 1], Chapter 3), then the sequence of Einstein metrics always has a subsequence that converges to a product Riemannian metric on $M_1 \times \cdots \times M_m$ which in general is not Einstein.

The examples one gets from Theorem Add.1 in dimension 7, i.e., circle bundles over $\mathbb{C}P^2 \times \mathbb{C}P^1$ and over $\mathbb{C}P^1 \times \mathbb{C}P^1 \times \mathbb{C}P^1$, were independantly discovered by some physicists [Ca-OA-Fr], [OA-Fr-VN], [Pa-Po 1], who constructed homogeneous Einstein metrics on these spaces. They were used in the theory of Kaluza-Klein supergravity, which is a unified field theory modeled on $M^4 \times M^7$ where M^4 is Minkowski space and M^7 is an Einstein manifold with positive scalar curvature. As was observed in [Ca-Ro-Wa], this shows that every compact simply connected homogeneous space in dimension 7 (or in fact in dimension $\leqslant 7$) carries a homogeneous Einstein metric with positive scalar curvature. Besides the above circle bundles over $\mathbb{C}P^2 \times \mathbb{C}P^1$ and over $\mathbb{C}P^1 \times \mathbb{C}P^1 \times \mathbb{C}P^1$, one has the examples of M. Wang on $SU(3)/S^1$ (9.102) (see also [Ca-Ro] and [Pa-Po 2]), the Jensen metric on S^7 (9.82), the Kobayashi metric on the Stiefel manifold $T_1 S^4$ (9.77b), the isotropy irreducible spaces $Sp(2)/Sp(1)$ and S^7 (7.107); and the obvious symmetric space products $S^2 \times S^2 \times S^3$, $S^4 \times S^3$, $S^5 \times S^2$, $\mathbb{C}P^2 \times S^3$ and $S^2 \times SU(3)/SO(3)$.

B. Explicit Metrics with Holonomy G_2 and Spin(7)

Add.4 **Theorem** (R. Bryant [Bry 2]). *There are explicit cohomogeneity* 1 *metrics with holonomy* G_2 *and* Spin(7).

The two examples are "cones" on homogeneous spaces. By a cone, we mean $M = \mathbb{R}_+ \times N$ endowed with a metric $g = dt^2 \oplus t^2 g_N$. Note that such a metric is Ricci-flat if and only if g_N is Einstein with Einstein constant $\dim N - 1$.

In the example with holonomy G_2, one takes the normal $SU(3)$-invariant metric g_{N^6} on $N^6 = SU(3)/T^2$ where T^2 is a maximal torus in $SU(3)$. In the example with holonomy Spin(7), one takes the $Sp(2)$-invariant metric g_{N^7} on $N^7 = Sp(2)/K$ where $K \subset Sp(2)$ is isomorphic to $SU(2)$ but is *not* the standard embedding of $SU(2)$ in

$Sp(2)$ (in fact, up to conjugacy, K is characterized by the fact that $K = SU(2)$ and $Sp(2)/K$ is isotropy irreducible).

Recall that the previously known examples of metrics with exceptional holonomy (see 10.96 (VII) and (VIII)) had been obtained using an existence theorem in partial differential equations. The present examples are really explicit, as we now explain.

Denote by Φ the canonical G_2-invariant 3-form on the imaginary Cayley numbers $\operatorname{Im}\mathbb{C}a$. Since the stabilizer of Φ in $Gl(7,\mathbb{R})$ is G_2, which is irreducible, the form Φ determines the Euclidean structure and orientation on $\operatorname{Im}\mathbb{C}a$. Recall that, in order to construct a metric with holonomy contained in G_2, it is sufficient to produce a differential 3-form ψ in dimension 7 such that

(i) at each point, φ is linearly equivalent to Φ;

then φ determines a Riemannian metric g_φ and a Hodge operator $*_\varphi$

(ii) $d\varphi = 0$, $d*_\varphi \varphi = 0$

Similarly, there is a 4-form Ψ on the Cayley numbers $\mathbb{C}a$ whose stabilizer is Spin(7). In order to construct a metric with holonomy contained in Spin(7), it is sufficient to produce a closed differential 4-form ψ, linearly equivalent to Ψ at each point (note that $*_\psi\psi = \psi$ automatically).

The 7-dimensional Example

Let κ be the canonical $\mathfrak{su}(3)$-valued left invariant one form on $SU(3)$. We may write it in the form

$$\kappa = \begin{pmatrix} -i\theta_3 & -\bar{\omega}^1 & i\omega^2 \\ \omega^1 & -i\theta_2 & -\bar{\omega}^3 \\ i\bar{\omega}^2 & \omega^3 & -i\theta_1 \end{pmatrix}.$$

The real and imaginary parts of the ω^i's constitute a trivialisation $F\colon TN^6 \to \mathbb{R}^6$, $N^6 = SU(3)/T^2$. Let

$$G = dt \oplus tF\colon TM^7 \to \mathbb{R}^7$$

where $M^7 = \mathbb{R}_+ \times N$. This is a trivialization of TM^7. We decide that $\mathbb{R}^7 = \operatorname{Im}\mathbb{C}a$. Then the differential form $\psi = G^*\Phi$ satisfies (i). One checks that ψ and $*_\psi\psi$ are closed. Indeed,

$$\psi = d\left(\frac{t^3}{3}\Omega\right),$$

$$*_\psi\psi = d\left(-\frac{t^4}{4}\Xi\right)$$

where

$$\Omega = \frac{i}{2}(\omega^1 \wedge \bar{\omega}^1 + \omega^2 \wedge \bar{\omega}^2 + \omega^3 \wedge \bar{\omega}^3)$$

$$I = \operatorname{Im}(\omega^1 \wedge \omega^2 \wedge \omega^3).$$

Since ψ is homogeneous of degree 3 under the homotheties in the $\mathbb{R}_+$ factor, the metric g_ψ is homogeneous of degree 2, so it is a cone metric. There remains to check that the holonomy is not a proper subgroup of G_2, see [Bry 2]. □

The 8-dimensional Example

The irreducible complex 4-dimensional representation of $SU(2)$ is unitary and fixes a symplectic form. It thus gives an embedding

$$SU(2) \to K \subset Sp(2).$$

The corresponding representation ρ of K on $\mathfrak{sp}(2)/\mathfrak{K}$ is 7-dimensional, irreducible, and $\rho(K) \subset G_2$, thus one can identify $\mathfrak{sp}(2)/\mathfrak{k}$ with $\mathrm{Im}\,\mathbb{C}\mathrm{a}$. Since the canonical 3-form Φ is K-invariant, it gives rise to a differential 3-form φ on $N^7 = Sp(2)/K$. One checks that

$$d\varphi = 4 *_\varphi \varphi.$$

If $F\colon TN^7 \to \mathrm{Im}\,\mathbb{C}\mathrm{a}$ is a local trivialization such that $F^*\Phi = \varphi$, one uses

$$G = dt \oplus tF\colon TM^8 \to \mathbb{R} \oplus \mathrm{Im}\,\mathbb{C}\mathrm{a} = \mathbb{C}\mathrm{a}$$

($M^8 = \mathbb{R}_+ \times N^7$) to construct a differential 4-form $\psi = G^*\Psi$ which satisfies (i). Then $*_\psi \psi = \psi$ and

$$\psi = d\left(\frac{t^4}{4}\varphi\right),$$

so g_ψ is a cone metric with holonomy contained in Spin(7). As before, one checks that the holonomy is not a proper subgroup of Spin(7), see [Bry 2]. □

C. Inhomogeneous Kähler-Einstein Metrics with Positive Scalar Curvature

Let N be a compact Kähler-Einstein manifold with positive first Chern class. Write $c_1(N) = q\alpha$ where α is an indivisible integer class. Let $L \to N$ be the holomorphic line bundle over N such that $c_1(L) = k\alpha$, and let $L \times L \to N \times N$ be the $\mathbb{C} \oplus \mathbb{C}$-bundle. *Define* $N_k = \mathbb{P}(L \times L)$ which is a $\mathbb{C}P^1$-bundle over $N \times N$. Then $c_1(N_k) > 0$ if and only if $k < q$.

Extending previous results by Y. Sakane [Sae 2], N. Koiso and Y. Sakane have shown

Add.5 **Theorem** ([Ko-Sa]). *For $k < q$, N_k carries a Kähler-Einstein metric.*

The construction is a variant of [Béber 3]. Let $P \to M$ be a principal circle bundle over a complex manifold M, endowed with a principal connection θ. Let S^1 act isometricly on $\mathbb{C}P^1$ in the canonical metric. Denote by t the distance to one of the fixed points. Let

$$\overline{M} = P \times_{S^1} \mathbb{C}P^1.$$

Then $\pi: \overline{M} \to M$ is a holomorphic $\mathbb{C}P^1$-bundle over M equipped with a horizontal distribution and a function t. One searches a Kähler-Einstein metric in the form

$$\bar{\omega} = \pi^* \omega_t \oplus u(t) \omega_{\mathbb{C}^*}$$

where ω_t is a 1-parameter family of Kähler metrics on M, and $u(t)$ is some positive function of t.

Then $\bar{\omega}$ is Kähler if and only if

$$\omega_t = \omega_0 - U(t)\rho$$

where ρ is the curvature of the connection θ, and the function U is determined by u.

Now we are left with 3 parameters in the construction: a Kähler metric ω_0 on M, a connection θ on P, a real function U of one real variable. We assume

(i) the metric ω_0 is Kähler-Einstein, i.e., its Ricci form satisfies $\rho_0 = \omega_0$;

(ii) the form ρ has constant eigenvalues with respect to ω_0. In the examples, this will be achieved by a choice of ρ (and thus of θ) in its cohomology class.

Then Einstein's equations for $\bar{\omega}$ boil down to a second order ODE for the function U, which always has solutions in $]0, \frac{\pi}{2}[$. One of these solutions extends smoothly to a Kähler-Einstein metric on $\overline{M}$ if and only if a certain integral involving the function U vanishes. This integral is nothing but *Futaki's functional* $F(X)$ computed on the generator X of the $\mathbb{C}^*$ action on the fibers (complexification of the S^1 action).

When (M, ω_0) is homogeneous, one chooses $\rho = k\rho_0$. The metric $\bar{\omega}$ has cohomogeneity one. The corresponding manifolds $\overline{M}$ can be characterized by their automorphism groups [Hu-Sn]. This yields

Add.6 Theorem ([Ko-Sa]). *Consider the compact complex manifolds M whose automorphism group has an open orbit which separates M. Among these, vanishing of Futaki's functional is a necessary and sufficient condition for the existence of a Kähler-Einstein metric.*

Still, it is hard to find examples of manifolds where Futaki's functional vanishes. Theorem Add.5 is obtained by taking $M = N \times N$ and the bundle P whose associated line bundle is $L \otimes L^*$. For symmetry reasons, Futaki's functional vanishes automatically.

For various approaches to Futaki's functional, see [Fut 2], [Fu-Mo], [Bou 11] and Add.9.

D. Uniqueness of Kähler-Einstein Metrics with Positive Scalar Curvature

Add.7 Theorem (S. Bando and T. Mabuchi [Ba-Ma]). *Let M be a complex manifold with positive first Chern class. If M admits two Kähler-Einstein metrics $\omega_0, \omega_1 \in 2\pi c_1(M)$, then there exists an automorphism $a \in \mathfrak{A}^0(M)$ such that*

$$\omega_1 = a^* \omega_0.$$

One starts with an attempt by T. Aubin ([Aub 7]) to prove the existence of Kähler-Einstein metrics via the continuity method (compare App.45). Let us *denote by* $\mathfrak{K}$ the space of Kähler metrics in the cohomology class $2\pi c_1(M)$. Given $\omega \in \mathfrak{K}$, let us write its Ricci form ρ in the form

$$\rho = \omega - \tfrac{1}{2}dd^c f,$$

(compare 2.75, 2.110). The following 1-parameter family of equations

$$\text{(Add.8)} \qquad \log((\omega - \tfrac{1}{2}dd^c\psi_t)^m/\omega^m) = t\psi_t + f$$

interpolates between the Calabi problem of finding a metric $\tilde{\omega}$ whose Ricci form is ω at $t = 0$, compare 2.101 and 11.34, and Einstein' equations at $t = 1$, compare 11.41. For any initial data ω, there exists a solution ψ_0 at $t = 0$ [Yau 3], and thus, for t in an interval $[0, \tau[$. The solution is *unique*. However, one has not been able to prove yet that ψ_t exists for all $t \in [0, 1]$. In fact, the functionals (see below) which could be used to obtain a priori estimates tend to increase with t.

Conversely, *assuming* that there exists an Einstein metric θ in $\mathfrak{K}$, S. Bando and T. Mabuchi show that, for a generic $\omega \in K$, there exists a solution

$$\omega_t = \omega - \tfrac{1}{2}dd^c\psi_t$$

of (Add.8) such that $\omega_0 = \omega$ and ω_1 is in the $\mathfrak{A}^0(M)$-orbit of θ.

Let us identify the tangent space of $\mathfrak{K}$ at a point ω with $\mathscr{C}M$. Then the equation

$$\alpha_\omega(\varphi) = \int_M \varphi\omega^m$$

defines a differential 1-form α on $\mathfrak{K}$. It is closed, thus exact, and we denote by L a primitive of α. If

$$\omega' = \omega - \tfrac{1}{2}dd^c\varphi,$$

let

$$\mathbf{N}(\omega, \omega') = L(\omega') - L(\omega) - \alpha_\omega(\varphi).$$

The function $\mathbf{N}$ on $\mathfrak{K} \times \mathfrak{K}$ behaves somewhat like the square of a distance. For fixed ω, $\mathbf{N}(\omega, .)$ attains a unique minimum on the orbit of θ under $\mathfrak{A}^0(M)$. If ω is generic, its Hessian is positive definite at this minimum ω_1, which allows to construct a solution ω_t of equation (Add.8) for $t \in]\varepsilon, 1]$. Now $\mathbf{N}$ itself provides an a priori bound on ω_t, so that ω_t exists for all $t \in]0, 1]$. Unfortunately, this bound blows up when t tends to 0, so that extra estimates, relying on [Gal 5] and [Yau 3], have to be found to prove existence up to $t = 0$.

If there were several distinct orbits of Einstein metrics in $\mathfrak{K}$, then, for a suitable choice of ω, there would be one curve of solutions of equation (Add.8) joining ω to each orbit. Since equation (Add.8) has only one solution for t small, one concludes that there is at most one orbit of Einstein metrics under $\mathfrak{A}^0(M)$. □

Add.9 In [Mab 1] and [Mab 2], T. Mabuchi gives a beautiful picture of the space $\mathfrak{K}$. He observes (see also [Bou 11]) that the Futaki functional 2.153 can be interpreted as a differential 1-form γ on $\mathfrak{K}$, i.e., for a (real) holomorphic vector field

X, and any $\omega \in \mathfrak{R}$,

$$F(X) = \gamma_\omega\left(\frac{d}{dt}(\exp(tX)^*\omega)\right).$$

Since γ is closed (this is the content of the Calabi-Futaki Theorem 2.160), it admits a primitive μ, called a "K-energy map". Then the critical points of μ are exactly the Einstein metrics.

The natural Riemannian metric on $\mathfrak{R}$ has non-positive sectional curvature, and μ is a convex function. Furthermore, if μ is constant on a geodesic, it has to be entirely contained in a single orbit of $\mathfrak{A}^0(M)$. In particular, if a geodesic joins two Einstein metrics, the convex function μ has to be constant on it, so its endpoints are in the same $\mathfrak{A}^0(M)$-orbit. However, these properties are far from proving Theorem Add.7. Indeed, there is no reason why $\mathfrak{R}$ be geodesically complete. This fails even for finite dimensional moduli spaces, see [Wop].

E. Hyperkählerian Quotients

A method for producing non-compact but complete hyperkählerian manifolds by a quotient construction has been given by N. Hitchin, A. Karlhede, U. Lindström and M. Rocek [H-K-L-R]. This is based on the symplectic quotient construction of J. Marsden and A. Weinstein, see [Gu-St].

Add.10 If (M, ω) is a symplectic manifold and a Lie group G acts by symplectic transformations then each element $A \in \mathfrak{g}$ defines a vector field X_A such that $L_{X_A}\omega = 0$. Since

$$0 = L_{X_A}\omega = d(i(X_A)\omega) + i(X_A)\,d\omega$$

and ω is closed, then if $H^1(M, \mathbb{R}) = 0$ there is a function f_A such that

$$df_A = i(X_A)\omega$$

and this function is well defined up to the addition of a constant. Considering all elements $A \in \mathfrak{g}$ one defines a moment map

$$\mu\colon M \to \mathfrak{g}^*$$

by

$$\langle \mu(m), A \rangle = f_A(m).$$

If G is compact or semi-simple, the ambiguities in the choices of f_A may be adjusted to make μ equivariant. The remaining ambiguity in μ is simply the addition of an abelian character in $\mathfrak{g}^*$.

If G acts freely on $\mu^{-1}(0)$ with Hausdorff quotient manifold $\mu^{-1}(0)/G$, then this quotient inherits a natural symplectic structure. This is the Marsden-Weinstein quotient.

Suppose (M, g) is a hyperkählerian manifold with Kähler forms ω_1, ω_2, ω_3

corresponding to the complex structures I, J, K, and G is a Lie group of isometries preserving the symplectic forms $\omega_1, \omega_2, \omega_3$. One obtains three moment maps as above μ_1, μ_2, μ_3 or alternatively a single map

$$\mu: M \to \mathfrak{g}^* \otimes \mathbb{R}^3.$$

The quotient construction consists of the

Add.11 Theorem [H-K-L-R]. *Let M be a hyperkählerian manifold acted on by a group G of isometries preserving the hyperkählerian structure and let $\mu: M \to \mathfrak{g}^* \otimes \mathbb{R}^3$ be the corresponding moment map. Then, if G acts freely on $\mu^{-1}(0)$ with Hausdorff quotient, the quotient manifold $\mu^{-1}(0)/G$ with its induced metric is hyperkählerian.*

Note that if M is complete and G compact, then $\mu^{-1}(G)$ is complete.

Add.12 Many examples may be constructed by taking M to be a quaternionic vector space with an inner product and the group G to be a subgroup of the group of affine linear transformations. In particular, any subgroup $G \subset Sp(m)$ can be used, although it is necessary to have a non-trivial centre to construct a moment map for which $\mu^{-1}(0) \neq \emptyset$: a necessary condition for the free action of G.

Examples

1. $M = \mathbb{H}^2 = \mathbb{C}^2 \times \mathbb{C}^2$ and $G = \mathbb{R}$ acting by

$$t(z_1, z_2, w_1, w_2) = (e^{it} z_1, e^{-it} z_2, w_1 + t, w_2).$$

This gives the Taub-NUT metric, compare 3.38, 13.87.

2. $M = \mathbb{H}^{m+1} = \mathbb{C}^{m+1} \times \mathbb{C}^{m+1}$ and $G = S^1$ acting by

$$e^{it}(z, w) = (e^{it} z, e^{-it} w).$$

This yields Calabi's metric [Cal 5] on $T^*\mathbb{C}P^m$.

3. $M = \mathbb{H}^m$ and $G = T$, the maximal torus of $SU(m)$ acting on $\mathbb{C}^m \times (\mathbb{C}^m)^*$. The resulting quotient is the gravitational multi-instanton of Gibbons and Hawking [Hit 4].

4. A generalization of the previous example has been given by P. Kronheimer, producing a hyperkählerian metric on the resolution of a rational double point [Kro]. In this case, if $\Gamma \subset SU(2)$ is a finite subgroup with regular representation V then taking $M = (\mathrm{End}(V) \otimes \mathbb{C}^2)^\Gamma$, the Γ-invariant part, and $G = SU(V)^\Gamma$ gives the required metric.

5. Infinite dimensional generalizations of the construction where M is a space of connections and G is the group of gauge transformations yield hyperkählerian metrics on finite dimensional moduli spaces, [At-Hi], [Hit 6].

Bibliography

[Ab-Ma] R. Abraham and J. Marsden, Fondations of Mechanics, Benjamin, New York (1967).

[Ada] F. Adams, Lectures on Lie groups, W.A. Benjamin, Inc. New York (1969).

[Ad-Mi] T. Adati and T. Miyazawa, On a Riemannian space with recurrent conformal curvature. Tensor, N. S. **18**, 348–354 (1967).

[Ag-Do-Ni 1] S. Agmon, A. Douglis, L. Nirenberg, Estimates near the boundary for the solutions of elliptic partial differential equations satisfying general boundary conditions, I, Comm. Pure. Appl. Math **12**, 623–727 (1959).

[Ag-Do-Ni 2] S. Agmon, A. Douglis, and L. Nirenberg, Estimates near the boundary for the solutions of elliptic partial differential equations satisfying general boundary conditions, II. Comm. Pure Appl. Math. **17**, 35–92 (1964).

[Ahl] L.V. Ahlfors, Some remarks on Teichmüller's space of Riemann surfaces, Annals of Math. **74**, 171–191 (1961).

[Ale 1] D.V. Alekseevskii, Riemannian manifolds with exceptional holonomy groups, Funktional Anal. i Priložen... **2**(2), 1–10 (1968), (English translation: Functional Anal. Appl. **2**, 97–105 (1968)).

[Ale 2] D.V. Alekseevskii, Compact quaternion spaces, Funktional Anal. i Priložen... **2**(2), 11–20 (1968), (English translation: Functional Anal. Appl. **2**, 106–114 (1968)).

[Ale 3] D.V. Alekseevskii, Quaternion Riemannian spaces with transitive reductive or solvable group of motions, Funktional Anal. i Priložen... **4**(4), 68–69 (1970), (English translation: Functional Anal. Appl. **4**, 321–322 (1970)).

[Ale 4] D.V. Alekseevskii, Classification of quaternionic spaces with a transitive solvable group of motions, Izv. Akad. Nauk SSSR Ser. Mat. **9**(2), 315–362 (1975), (English translation: Math. USSR-Izv. **9**, 297–339 (1975)).

[Al-Ki 1] D.V. Alekseevskii and B.N. Kimel'fel'd, Structure of homogeneous Riemannian spaces with zero Ricci curvature, Funktional Anal. i Priložen... **9**(2), 5–11 (1975), (English translation: Functional Anal. Appl. **9**, 97–102 (1975)).

[Al-Ki 2] D.V. Alekseevskii and B.N. Kimel'fel'd, Classification of homogeneous conformally flat Riemannian manifolds (in Russian). Mat. Zametki **24**, 103–110 (1978), (English translation: Math. Notes **24**, 559–562 (1978).)

[Am-Si] W. Ambrose and I.M. Singer, A theorem on holonomy, Trans. Amer. Math. Soc. **79**, 428–443 (1953).

[An-Bu-Ka] P.L. Antonelli, D. Burghelea and P.J. Kahn, The non-finite homotopy type of some diffeomorphism groups, Topology **11**, 1–49 (1972)

[Apt] M. Apte, Sur certaines classes caractéristiques des variétés kählériennes compactes, C.R. Acad. Sci. Paris **240**, 149–151 (1955).

[Arn] V.I. Arnold, Méthodes Mathématiques de la mécanique classique., Mir, Moscow (1974).

[Aro] N. Aronszjan, A unique continuation theorem for solutions of elliptic partial differential equations or inequalities of second order. J. Math. Pures Appl. **35**, 235–249 (1957).

[Art] M. Artin, On the solutions of analytic equations, Inventiones Math. **5**, 277–291 (1968).

[Ast] Première classe de Chern et courbure de Ricci: preuve de la conjecture de Calabi. Séminaire Palaiseau 1978, Astérisque 58, Société Math. de France (1978).

[At-Bo-Sh] M.F. Atiyah, R. Bott and A. Shapiro, Clifford Modules, Topology **3** (Supp. 1), 3–38 (1964).

[At-Hi] M.F. Atiyah, N. Hitchin, Low-energy scattering of non-abelian magnetic monopoles, Phil. Trans. R. Soc. Lond. A **315**, 459–469 (1985).

[At-Hi-Si] M.F. Atiyah, N.J. Hitchin and I.M. Singer, Self-duality in four-dimensional Riemannian geometry, Proc. Roy. Soc. London A **362**, 425–461 (1978).

[Ati 1] M.F. Atiyah, Complex fibre bundles and ruled surfaces, Proc. London Math. Soc. **5** 407–434 (1955).

[Ati 2] M.F. Atiyah, Geometry of Yang-Mills fields, Fermi Lectures Notes, Scuola Normale Sup. di Pisa, Pisa (1979).

[At-Si 1] M.F. Atiyah and I.M. Singer, The index of elliptic operators III, Ann. of Math. **87**, 546–604 (1968).

[At-Si 2] M.F. Atiyah, I.M. Singer, The index of elliptic operators IV, Ann. of Math. **92**, 119–138 (1970).

[At-Si 3] M.F. Atiyah and I.M. Singer, The index of elliptic operators V, Ann. of Math. **93**, 139–149 (1971).

[Aub 1] T. Aubin, Métriques riemanniennes et courbure, J. Diff. Geom. **4**, 383–424 (1970).

[Aub 2] T. Aubin, Equations différentielles non linéaires et problème de Yamabe concernant la courbure scalaire, J. Math. Pures Appli. **55**, 269–296 (1976).

[Aub 3] T. Aubin, Meilleurs constantes dans le théorème d'inclusion de Sobolev et un théorème de Fredholm non linéaire pour la transformation conforme de la courbure scalaire, J. Funct. Anal. **32**, 148–174 (1979).

[Aub 4] T. Aubin, Problèmes isopérimétriques et espaces de Sobolev. J. Diff. Geom. **11**, 573–598 (1976).

[Aub 5] T. Aubin, Equations du type de Monge-Ampère sur les variétés kähleriennes compactes, C.R. Acad. Sci. Paris **283**, 119–121 (1976).

[Aub 6] T. Aubin, Non-linear analysis on Manifolds, Monge-Ampère Equations. Springer Verlag, New-York (1982).

[Aub 7] T. Aubin, Réduction du cas positif de l'équation de Monge-Ampère sur les variétés kähleriennes compactes à la démonstration d'une inégalité, J. Funct. Anal. **57**, 143–153 (1984).

[Avz 1] A. Avez, Valeur moyenne du scalaire de courbure sur une variété compacte. Applications relativistes, C.R. Acad. Sci, Paris **256**, 5271–5273 (1963).

[Avz 2] A. Avez, Remarques sur les variétés de dimension 4, C. R. Acad. Sci. Paris A **264**, 738–740 (1967).

[Ba-Ma] S. Bando, T. Mabuchi, Uniqueness of Einstein Kähler metrics modulo connected group actions, Preprint Tôhoku University, Sendai (1985).

[Bac] R. Bach, Zur Weylschen Relativitätstheorie und der Weylschen Erweiterung des Krümmungstensorbegriffs. Math. Z. **9**, 110–135 (1921).

[BB-Bo-La] L. Bérard Bergery, J.P. Bourguignon and J. Lafontaine, Déformations localement triviales des métriques riemanniennes, Proc. Symp. Amer. Math. Soc. **27**, 3–32 (1973).

[BB-Oc] L. Bérard-Bergery and T. Ochiai, On some generalizations of the construction of twistor spaces, in "Global Riemannian Geometry", ed. T.J. Willmore and N. Hitchin, page 52–58, Ellis Horwood, Chichester (1984).

[Bea 1] A. Beauville, Surfaces Algébriques complexes, Asterisque **54**, Soc. Math. France, Paris (1978).

[Bea 2] A. Beauville, Variétés Kähleriennes dont la lère classe de Chern est nulle, J. Diff. Geom. **18**, 755–782 (1983).

[BéBer 1] L. Bérard Bergery, Sur la courbure des métriques riemannienness invariantes des groupes de Lie et des espaces homogènes, Annales Sc. Ec. Norm. Sup. **11**, 543–576 (1978).

[BéBer 2] L. Bérard Bergery, La courbure scalaire des variétés riemanniennes. Séminaire Bourbaki, 1979–80, exposé n° 556. Lecture Notes 842, Springer, Berlin (1981).

[BéBer 3] L. Bérard Bergery, Sur de nouvelles variétés riemanniennes d'Einstein. Publications de l'Institut E. Cartan n° 4 (Nancy), 1–60 (1982).

[BéBer 4] L. Bérard Bergery, Scalar curvature and isometry groups., in "Spectra of Riemannian manifolds", ed. M. Berger, S. Murakami, T. Ochiai, Kagai, Tokyo, pp. 9–28 (1983).

[BéBer 5] L. Bérard Bergery, Sur certaines fibrations d'espaces homogene riemanniens, Compositio

Math. **30**, 43–61 (1975).

[BéBer 6] L. Bérard Bergery, Certains fibrés à courbure de Ricci positive, C.R. Acad. Sci. Paris, A **286**, 929–931 (1978).

[BéBer 7] L. Bérard Bergery, Variétés quaternionniennes, Notes d'une conférence à la table ronde "Variétés d'Einstein", Espalion (1979). (Unpublished).

[Be-Eb] M. Berger and D. Ebin, Some decompositions of the space of symmetric tensors on a Riemannian manifold, J. Diff. Geom. **3**, 379–392 (1969).

[Be-Eh] J. Beem, P. Ehrlich, Global Lorentzian Geometry, Dekker, New York (1981).

[Be-Fe-Gr] M. Beals, C. Fefferman, R. Grossman, Strictly pseudoconvex domains in $\mathbb{C}^n$, Bull. Amer. Math. Soc. **8**, 125–322 (1983).

[Be-Ga-Ma] M. Berger, P. Gauduchon, E. Mazet, Le spectre d'une variété riemannienne, Lecture Notes in Math. n° 194 Springer, Berlin (1971).

[Be-Go] J.C. Becker and D.H. Gottlieb, Applications of the evaluation map and transfer map theorems, Math. Ann. **211**, 277–288 (1974).

[Be-La] M. Berger and A. Lascoux, Variétés kähleriennes compactes, Lecture Notes n° 154, Springer-Verlag, Berlin (1971).

[Ber 1] M. Berger, Sur les groupes d'holonomie des variétés à connexion affine et des variétés riemanniennes, Bull. Soc. Math. France **83**, 279–330 (1955).

[Ber 2] M. Berger, Sur quelques variétés d'Einstein compactes, Annali di Math. Pura e Appl. **53**, 89–96 (1961).

[Ber 3] M. Berger, Sur les variétés d'Einstein compactes, C.R. de la III° réunion du groupement des mathématiciens d'expression latine, Namur (1965).

[Ber 4] M. Berger, Quelques formules de variation pour une structure riemannienne, Ann. Sc. Ec. Norm. Sup. **3**, 285–294 (1970).

[Ber 5] M. Berger, Sur la première valeur propre des variétés riemanniennes, Compositio Math. **26**, 29–149 (1973).

[Ber 6] M. Berger, Rapport sur les variétés d'Einstein, in Analyse sur les variétés (Metz 1979), Astérisque 80, Soc. Math. France, Paris (1980).

[Ber 7] M. Berger, Remarques sur le groupe d'holonomie des variétés Riemanniennes, C. R. Acad. Sci. Paris **262**, 1316–1318 (1966).

[Ber 8] M. Berger, Trois remarques sur les variétés riemannienes à courbure positive, C. R. Acad. Sci. Paris **263**, 76–78 (1966).

[Bes 1] A.L. Besse, Manifolds all of whose geodesics are closed, Ergebnisse der Math. n° 93, Springer (1978).

[Bes 2] A.L. Besse, Géométrie riemannienne en dimension 4, Cedic-Fernand Nathan, Paris (1981).

[Bie] L. Bieberbach, Uber die Bewegungsgruppen des Euklidischen Räume I, II, Math. Ann. **70**, 297–336 (1911), **72**, 400–412 (1912).

[Bi-On] R.L. Bishop and B. O'Neill, Manifolds of negative curvature, Trans. Amer. Math. Soc. **145**, 1–49 (1969).

[Bis] R.L. Bishop, A relation between volume, mean curvature and diameter, Notices Am. Math. Soc. **10**, 364 (1963).

[Bla] A. Blanchard, Sur les variétés analytiques complexes, Ann. Sc. Ecole Norm. Sup. **73**, 157–202 (1956).

[Ble] D.D. Bleecker, Critical Riemannian manifolds, J. Diff. Geom. **14**, 599–608 (1979).

[Bln] E. Blancheton, Equations aux variations de la Relativité Générale, C.R. Acad. Sci. Paris **245**, 284–287 (1957).

[Boc 1] S. Bochner, Vectors fields and Ricci curvature, Bull. Am. Math. Sc. **52**, 776–797 (1946)

[Boc 2] S. Bochner, Curvature in Hermitian metric, Bull. Amer. Math. Soc. **53**, 179–195 (1947).

[Bo-Eb-Ma] J.P. Bourguignon, D. Ebin, and J.E.D. Marsden, Sur le noyau des opérateurs différentiels à symbole injectif et non surjectif. C.R. Acad. Sc. Paris **282**, 867–870 (1976).

[Bog] V.A. Bogomolov, Hamiltonian Kähler manifolds, Dokl. Akad. Nauk SSSR (Mat.), **243**, 1101–1104 (1978). English translation: Soviet Math. Dokl. **19**, 1462–1465 (1978).

[Bo-Hi] A. Borel and F. Hirzebruch, Characteristic classes and homogeneous spaces, I, II, III, Amer. J. Math. **80**, 315–382 (1958), **81**, 315–382 (1959), **82**, 491–504 (1960).

[Bo-Ka] J.P. Bourguignon and H. Karcher, Curvature operators: pinching estimates and geometric examples. Ann. Sc. Ec. Norm. Sup. Paris **11**, 71–92 (1978).

[Bo-La] J.P. Bourguignon and H.B. Lawson, Yang Mills theory: its physical origins and differential geometric aspects, in Seminar on Diff. Geometry (ed. S.T. Yau), Princeton University Press, Princeton, 395–422 (1982).

[Bo-Li] A. Borel and A. Lichnerowicz, Groupes d'holonomie des variétés riemanniennes, C.R. Acad. Sci. Paris **234**, 1835–1837 (1952).

[Bo-Ma] S. Bochner and W.T. Martin, Several complex variables, Princeton Math. Ser. 10, Princeton Univ. Press, Princeton (1948).

[Bon 1] E. Bonan, Sur des variétés riemanniennes à groupe d'holonomie G_2 ou Spin (7), C.R. Acad. Sci. Paris **262**, 127–129 (1966).

[Bon 2] E. Bonan, Sur les G-structures de type quaternionien, Cahiers Top. et Geom. Diff. **9**, 389–461 (1967).

[Bon 3] E. Bonan, Sur l'algèbre extérieure d'und variété presque hermitienne quaternionique, C.R. Acad. Sci. Paris **295**, 115–118 (1982).

[Bo-Po] J.P. Bourguignon and A. Polombo, Integrands des nombres caractéristiques et courbure: rien ne va plus dès la dimension 6, J. of Diff. Geom. **16**, 537–550 (1981).

[Bor 1] A. Borel, Some remarks about Lie groups transitive on spheres and tori, Bull. A.M.S. 55, 580–587 (1949).

[Bor 2] A. Borel, Le plan projectif des octaves et les sphères comme espaces homogènes, C.R. Acad. Sc. Paris **230**, 1378–1380 (1950).

[Bor 3] A. Borel, Sur la cohomologie des espaces fibrés principaux et des espaces homogènes de groupes de Lie compacts, Ann. of Math. **57**, 115–207 (1953).

[Bor 4] A. Borel, Kählerian coset spaces of semi-simple Lie groups, Proc. Nat. Acad. Sci. US **40**, 1147–1151 (1954).

[Bor 5] A. Borel, Compact Clifford-Klein forms of symmetric spaces, Topology **2**, 111–122 (1963).

[Bo-Re] A. Borel, R. Remmert, Über Kompakte homogene Kählersche Mannigfaltigkeiten, Math. Annalen **145**, 429–439 (1962).

[Bot 1] R. Bott, Homogeneous vector bundles, Ann. of Math. **66**, 203–248 (1957).

[Bot 2] R. Bott, The geometry and representation theory of compact Lie groups, in: Representation theory of Lie groups, London Math. Society. Lecture Notes Series 34.

[Bou 1] J.P. Bourguignon, Une stratification de l'espace des structures riemanniennes, Compositio Math. **30**, 1–41 (1975).

[Bou 2] J.P. Bourguignon, Sur les géodésiques fermées des variétés quaternioniennes de dimension 4, Math. Ann. **221**, 153–165 (1976).

[Bou 3] J.P. Bourguignon, Premières formes de Chern des variétés Kähleriennes compactes, Séminaire Bourbaki, n° 507, vol. 1977–79, Lecture Notes 710, Springer, Berlin (1979)

[Bou 4] J.P. Bourguignon, Sur les formes harmoniques du type de la courbure, C.R. Acad. Sci. Paris A **286**, 337–339 (1978).

[Bou 5] J.P. Bourguignon, Déformations des variétés d'Einstein, in Analyse sur les variétés (Metz 1979), Astérisque 80, Soc. Math. France, Paris (1980).

[Bou 6] J.P. Bourguignon, Les surfaces K3, in Géométrie riemannienne en dimension 4, Cedic, Fernand Nathan, Paris (1981).

[Bou 7] J.P. Bourguignon, Formules de Weitzenböck en dimension 4, Séminaire A. Besse sur la géométrie riemannienne en dimension 4, Cedic, Fernand Nathan, Pairs (1981).

[Bou 8] J.P. Bourguignon, On harmonic forms of curvature type. (Unpublished).

[Bou 9] J.P. Bourguignon, Les variétés de dimension 4 à signature non nulle dont la courbure est harmonique sont d'Einstein. Invent. Math. **63**, 263–286 (1981).

[Bou 10] J.P. Bourguignon, A review of Einstein manifolds, in DD2 Proceedings (Shanghai-Hefei 1981), Science Press, Beijing (1984).

[Bou 11] J.P. Bourguignon, Invariants intégraux fonctionnels pour des équations aux dérivées partielles d'origine géométrique, to appear in "Proc. Conference on Differential Geometry and Global Analysis, Peniscola (1985)", Lecture Notes in Math., Springer, Berlin (1986).

[Bou 12] L'équation de la chaleur associée à la courbure de Ricci (d'après R.S. Hamilton), Sémi-

naire Bourbaki, exposé n° 653, 1985–1986.
[Bo-Ya] S. Bochner and K. Yano, Curvature and Betti numbers, Ann. Math. Studies n° 32 Princeton Univ. Press, Princeton (1953).
[Bre] G.E. Bredon, Introduction to compact transformation groups, Academic Press, New York, London (1972).
[Br-Gr] R. Brown and A. Gray, Riemannian manifolds with holonomy group Spin(9), in Diff. Geometry in honor of K. Yano, Kinokuniya, Tokyo, 41–59 (1972).
[Bri] H.W. Brinkmann, Einstein spaces which are mapped conformally on each other, Math. Ann. **94**, 119–145 (1925).
[Brs] L. Bers, Finite dimensional Teichmüller space and generalizations, Bull. Amer. Math. Soc. **5**, 131–172 (1981).
[Br-Wa] H. Brezis and S. Wainger, A note on limiting cases of Sobolev Imbedding and Convolution Inequalities, Comm. Part. Diff. Eq. **5**, 773–789 (1980).
[Bry 1] R. Bryant, Local existence of Riemannian metrics with holonomy G_2 or Spin(7), to appear.
[Bry 2] R. Bryant, Explicit metrics with holonomy G_2 and Spin(7), Preprint I.H.E.S., Bures-sur-Yvette (1985).
[Bse] J. Besse, Sur le domaine d'existence d'une fonction analytique, Comment. Math. Helv. **10**, 302–305 (1938).
[Bsr] P. Buser, Riemannsche Flächen and Längenspectrum vom Trigonometrischen Standpunkt aus, Habilitations-schript, Universität Bonn (1980).
[Bu-Ra] D. Burns and M. Rapoport, On the Torelli problem for Kählerian K3 surfaces, Ann. Sci. Ec. Norm. Sup., **8**, 235–274 (1975).
[Bus 1] H. Busemann, The Geometry of Geodesics, Academic Press, London (1955).
[Bus 2] H. Busemann, Recent Synthetic Differential Geometry, Ergebnisse bd 54, Springer, Berlin (1970).
[Ca-Ho-Ko] J. Carrell, A. Howard and C. Kosniowski, Holomorphic vector fields on complex surfaces, Math. Annalen **204**, 73–81 (1973).
[Cal 1] E. Calabi, The space of Kähler metrics, Proc. Internat. Congress. Math. Amsterdam **2**, 206–207 (1954).
[Cal 2] E. Calabi, On Kähler manifolds with vanishing canonical class, in Algebraic geometry and topology, a symposium in honor of S. Lefschetz, Princeton Univ. Press, 78–89 (1955).
[Cal 3] E. Calabi, Improper affine hyperspheres of convex type and a generalization of a theorem by K. Jörgens, Michigan Math. J. **5**, 105–126 (1958).
[Cal 4] E. Calabi, A construction of non homogeneous Einstein metrics. Differential Geometry, Proc. Symp. Pure Math. **27**, 17–24 (1975).
[Cal 5] E. Calabi, Métriques kähleriennes et fibrés holomorphes, Ann. Ecol. Norm. Sup. **12**, 269–294 (1979).
[Cal 6] E. Calabi, Extremal Kähler metrics, in Seminar on Differential Geometry, ed. S.T. Yau, Annals of Math. Studies 102, Princeton Univ. Press, Princeton, 259–290 (1982).
[Cal 7] E. Calabi, Extremal Kähler metrics II, in "Differential geometry and complex analysis", ed. I. Chavel, H.M. Farkas, Springer, Berlin, 95–114 (1985).
[Cal 8] E. Calabi, An extension of E. Hopf's maximum principle with an application to Riemannian geometry, Duke Math. J. **25**, 45–56 (1957).
[Car 1] E. Cartan, Les groupes projectifs qui ne laissent invariante aucune multiplicité plane, Bull. Soc. Math. Fr. **41**, 53–96, (1913) ou Oeuvres complétes t. I. Vol 1, p. 355–398.
[Car 2] E. Cartan, Les groupes projectifs continus réels qui ne laissent invariante aucune multiplicité plane, J. Math. pures et appliquées **10**, 149–186 (1914) ou Oeuvres complètes tome I, volume 1, 493–530.
[Car 3] E. Cartan, Les groupes réels simples finis et continus, Ann. Scient. Ecol. Norm. Sup. **31**, 263–355 (1914) ou Oeuvres complètes tome 1, Vol. 1. 493–530.
[Car 4] E. Cartan, La déformation des hypersurfaces dans l'espace conforme réel à $n \geqslant 5$ dimensions, Bull. Soc. Math. France **45**, 57–121 (1917) ou Oeuvres complètes t. III, Vol. 1, 221–286.
[Car 5] E. Cartan, Sur les variétés à connexion affine et la théorie de la relativité généralisée I et II,

Ann. Sci. Ecol. Norm. Sup. **40**, 325–412 (1923) et **41**, 1–25 (1924) ou Oeuvres complètes t. III, 659–746 et 799–824.

[Car 6] E. Cartan, Sur une classe remarquable d'espaces de Riemann, Bull. Soc. math. France **54**, 214–264 (1926), **55**, 114–134 (1927) ou Oeuvres complètes t. I, Vol. 2, 587–659.

[Car 7] E. Cartan, "Sur la possibilité de plonger un espace riemannien donné dans un espace euclidien", Ann. Soc. Pol. Math. **6**, 1–17 (1927) ou Oeuvres complètes t. III, Vol. 2, 1091–1098.

[Car 8] E. Cartan, Les groupes d'holonomie des espaces généralisés, Acta. Math. **48**, 1–42 (1926) ou Oeuvres complètes, tome III, volume 2, 997–1038.

[Car 9] E. Cartan, Sur certaines formes riemanniennes remarquables de géométries à groupe fondamental simple, Annales Sci, Ecol. Norm. Sup. **44**, 345–467 (1927) ou Oeuvres complètes t. I, Vol. I, 867–990.

[Car 10] E. Cartan. Les espaces de Finsler, Exposés de Géométrie, Hermann, Paris, 1934.

[Car 11] E. Cartan, La topologie des espaces représentatifs des groupes de Lie, Act. Sci. Ind. n° 358, Hermann, Paris (1936). ou Oeuvres complètes, tome I, vol. 2, 1307–1330.

[Car 12] E. Cartan, Sur la détermination d'un système orthogonal complet dans un espace de Riemann symétrique clos, Rendic. Circ. Mat. Palermo **53**, 217–252 (1929) ou Oeuvres complètes, tome I, vol 2, 1045–1080.

[Car 13] E. Cartan, Sur les domaines bornés homogènes de l'espace de *n* variables complexes, Abh. Math. Sem. Univ. Hamburg **11**, 110–162 (1936) ou Oeuvres complètes t. I, Vol. 1, 1259–1306.

[Ca-OA-Fr] L. Castellani, R.O'Auria, P. Fre, $SU(3) \otimes SU(2) \otimes U(1)$ from $D = 11$ supergravity, Nuclear Phys. **239B**, 610–652 (1984).

[Ca-Ro] L. Castellani, L.J. Romans, $N = 3$ and $N = 1$ supersymmetry in a new class of solutions for $D = 11$ supergravity, Nuclear Phys. **238B**, 683–701 (1984).

[Ca-Ro-Wa] L. Castellani, L.J. Romans, N.P. Warner, A classification of compactifying solutions for $D = 11$ supergravity, Nuclear Phys. **241B**, 429–462 (1984).

[Cas] L. Castellani, Einstein metrics on homogeneous spaces and $D = 11$ supergravity, Phys. Lett. **149B**, 103–106 (1984).

[Ca-Wa] M. Cahen and N. Wallach, Lorentzian symmetric spaces, Bull AMS **76**, 585–591 (1970).

[CB-DW-DB] Y. Choquet-Bruhat, C. De Witt-Morette, M. Dillard-Bleick, Analysis, Manifolds and Physics, North-Holland, Amsterdam (1977).

[CB-Le] Y. Choquet-Bruhat and J. Leray, Le problème de Dirichlet quasi linéaire d'ordre 2, C.R. Acad. Sc. Paris **214**, 81–85 (1972).

[Cha] I. Chavel, Riemannian symmetric spaces of rank one, Marcel Dekker, New York (1972).

[Che] J. Cheeger, Some examples of manifolds of non-negative curvature, J. Diff. Geom. **8**, 623–628 (1973).

[Ch-Eb] J. Cheeger and D. Ebin, Comparisons theorems in Riemannian Geometry, North-Holland, Amsterdam (1975).

[Ch-Gr 1] J. Cheeger, D. Gromoll, The splitting theorem for manifolds of non-negative Ricci curvature, J. of Diff. Geometry **6**, 119–128 (1971).

[Ch-Gr 2] Cheeger, D. Gromoll, On the structure of complete manifolds of nonnegative curvature, Annals of Math. **96**, 413–443 (1972).

[Chv] C. Chevalley, Theory of Lie groups, Princeton Math. Series, N° 8 (1946).

[Ch-Va] B.Y. Chen and L. Vanhecke, Differential geometry of geodesic spheres, J. reine angew. math. **325**, 28–67 (1981).

[Ch-Ya 1] S.Y. Cheng and S.T. Yau, On the regularity of the Monge-Ampère equation $\det(\partial^2 u/\partial x_i \partial y_j) = F(x, u)$, Comm. Pure and Appl. Math. **30**, 41–68 (1977).

[Ch-Ya 2] S.Y. Cheng, S.T. Yau, On the existence of a complete Kähler metric on non compact complex manifolds and the regularity of Fefferman's equation, Comm. Pure Applied Math. **33**, 507–544 (1980).

[Chr 1] S.S. Chern, An elementary proof of the existence of isothermal parameters on a surface. Proc. Amer. Math. Soc. **6**, 771–782 (1955).

[Chr 2] S.S. Chern, Complex manifolds without potential theory, Second Edition, Universitext,

Springer, Berlin (1979).

[Chr 3] S.C. Chern, Geometry of characteristic classes. Proc. 13th Biennal Seminar, Canadian Math. Congress, 1–40 (1972), now an appendix in [Chr 2].

[Coh] S. Cohn-Vossen, Totalkrümmung und geodätische Linien auf einfach Zusammenhängenden, offenen, vollständigen Flächenstücken, Mat. Sb. **43**, 139–163 (1936).

[Co-Hi] R. Courant and D. Hilbert, Methods of Mathematical Physics, II, Interscience-Wiley, New-York (1962).

[Cou] R. Courant, Dirichlet's principle, conformal mapping, and minimal surface, Interscience-Wiley, New York (1950).

[Crr] R. Carr, 4-manifolds of positive scalar curvature, PhD Thesis, Stony Brook (1984).

[DA-Ni 1] J.E. D'Atri and H.K. Nickerson, Divergence-preserving geodesic symmetries, J. Differential Geometry **3**, 467–476 (1969).

[DA-Ni 2] J.E. D'Atri and H.K. Nickerson, Geodesic symmetries in spaces with special curvature tensors, J. Differential Geometry **9**, 251–262 (1974).

[DAt] J.E. D'Atri, Geodesic spheres and symmetries in naturally reductive spaces, Michigan Math. J. **22**, 71–76 (1975).

[DA-Zi] J.E. D'Atri and W. Ziller, Naturally reductive metrics and Einstein metrics on compact Lie groups, Memoirs Amer. Math. Soc. **18**, No. 215 (1979).

[De-Ka] D. DeTurck and J. Kazdan, some regularity theorems in Riemannian geometry, Ann. Scient. Ec. Norm. Sup. 4° série, **14**, 249–260 (1981).

[De-Ko] D. DeTurck and N. Koiso, Uniqueness and non-existence of metrics with prescribed Ricci curvature, Ann. Sci. Ec. Norm. Sup. **14**, 249–260 (1984).

[Del] E.D. Deloff, Naturally reductive metrics and metrics with volume preserving geodesic symmetries on NC algebras, Thesis, Rutgers (1979).

[Der 1] A. Derdzinski, On compact Riemannian manifolds with harmonic curvature, Math. Ann. **259**, 145–152 (1982).

[Der 2] A. Derdzinski, Some remarks on the local structure of Codazzi tensors, in Global differential geometry and global analysis, Lecture Notes in Math, No. 838, Springer-Verlag, Berlin, p. 251–255 (1981).

[Der 3] A. Derdzinski, Self-dual Kähler manifolds and Einstein manifolds of dimension four, Comp. Math. **49**, 405–433 (1983).

[Der 4] A. Derdzinski, An easy construction of new compact Riemannian manifolds with harmonic curvature (Preliminary report). Preprint SFB 40–MPI n° 21, Bonn (1983).

[Der 5] A. Derdzinski, Hermitian Einstein metrics, in "Global Riemannian Geometry", ed. N. Hitchin and T.J. Willmore, Ellis Horwood, Chichester (1984).

[Der 6] A. Derdzinski, Riemannian metrics with harmonic curvature on 2-sphere bundles over compact surfaces, Preprint Math. Sci. Res. Inst., Berkeley (1985).

[De-Ri] A. Derdzinski and A. Rigas, Unflat connections in 3-sphere bundles over S^4, Trans. A.M.S. **265**, 485–493 (1981).

[De-Ro] A. Derdzinski and W. Roter, On conformally symmetric manifolds with metrics of indices 0 and 1, Tensor, N.S. **31**, 255–259 (1977).

[De-Sh] A. Derdzinski and C.L. Shen, Codazzi tensor fieds, curvature and Pontryagin forms, Proc. London Math. Soc. **47**, 15–26 (1983).

[DeT 1] D. DeTurck, Metrics with prescribed Ricci curvature, in Seminar on Differential Geometry, ed S.T. Yau, Annals of Math Studies 102, Princeton University Press, Princeton, 525–537 (1982).

[DeT 2] D. DeTurck, Existence of metrics with prescribed Ricci curvature: Local theory, Inventiones Math. **65**, 179–201 (1981).

[DeT 3] D. DeTurck, The Cauchy problem for Lorentz metrics with prescribed Ricci curvature, Compositio Math. **48**, 327–349 (1983).

[DeT 4] D. DeTurck, Deforming metrics in the direction of their Ricci tensors, J. Differential Geometry **18**, 157–162 (1983), (also see the corrected version, preprint, U. of Pennsylvania).

[DeT 5] D. DeTurck, Non linear Douglis-Nirenberg systems in geometry, lecture notes, Universita di Roma (1982).

[Die] J. Dieudonne, Fondements de l'Analyse Moderne, Cahiers Sci. **28**, Gauthier-Villars, Paris (1963).

[Di-Pf] K. Diedrich-P. Pflug, Über Gebiete mit vollständiger Kählermetrik, Math. Ann. **257**, 191–198 (1981).

[D-G-M-S] P. Deligne-Ph. Griffiths–J. Morgan and D. Sullivan, Real homotopy theory of Kähler manifolds, Inventiones math. **29**, 245–274 (1975).

[DoM] I. Doti-Miatello, Ricci curvature of left-invariant metrics on solvable unimodular Lie groups, Math. Zeit. **180**, 257–263 (1982).

[Don] S. Donaldson, An application of Gauge Theory to four-dimensional topology, J. Diff. Geom. **18**, 279–315 (1983).

[Do-Ni] A. Douglis and L. Nirenberg, Interior estimates for elliptic systems of partial differential equations, Comm. Pure and Appl. Math. **8**, 503–538 (1955).

[DT-Go] D. DeTurck and H. Goldschmidt, Preprint Univ. Pennsylvania, Philadelphia, PA (1984).

[Du-Ni-Po] M.J. Duff, B.E.W. Nilsson, C.N. Pope, Kaluza-Klein supergravity, Phys. Rep. **130**, 1–142 (1986).

[Dyn 1] E.B. Dynkin, Semi-simple subalgebras of semi-simple Lie algebras, Mat. Sb. **72**, 349–462 (1952). English translation: Amer. Math. Soc. Transl. **2**, 111–244 (1957).

[Dyn 2] E.B. Dynkin, Maximal subgroups of the classical groups, Trudy Moscow Mat. Obshch. **1**, 39–166 (1952). English translation: Amer. Math. Soc. Transl. **2**, 245–378 (1957).

[Ea-Ee] C.J. Earle and J. Eells, A fibre bundle description of Teichmüller theory, J. Differential Geom. **3**, 19–43 (1969).

[Ebi 1] D.G. Ebin, Espace des métriques riemanniennes et mouvement des fluides via les variétés d'applications, Ecole Polytechnique, Paris (1972).

[Ebi 2] D.G. Ebin, The space of Riemannian metrics, Proc. Symp. AMS n° 15, Global analysis, 11–40 (1968).

[Ee-Le] J. Eells and L. Lemaire, A report on harmonic maps, Bull. London Math. Soc. **10**, 1–68 (1978).

[Ee-Sa] J. Eells and J.M. Sampson, Harmonic mappings of Riemannian manifolds, Am. J. Math. **86**, 109–160 (1964).

[Eg-Gi-Ha] T. Eguchi, P. Gilkey, A.J. Hanson, Gravitation, Gauge Theories and Differential Geometry. Physical Reports **66**, 213–393 (1980).

[Eg-Ha] T. Eguchi and A.J. Hanson, Asymptotically flat self-dual solutions to Euclidean Gravity, Phys. Lett. **237**, 249–251 (1978).

[Ehr 1] C. Ehresman, les connexions infinitésimales dans un espace fibré différentiable, Colloque de Topologie, Bruxelles, 29–55 (1950).

[Ehr 2] C. Ehresmann, Sur la théorie des espaces fibrés, Colloque international du CNRS n° 2, Paris 1947, CNRS, Paris, 3–15 (1949).

[Ein] A. Einstein, Näherungsweise Integration der Feldgleichungen der Gravitation, S.B. Preuss. Akad. Wiss., 688–696 (1916).

[Eis] L.P. Eisenhart, Riemannian Geometry, Princeton University Press, Princeton (1966).

[Eli] H.T. Eliasson, On variations of metrics, Math. Scand. **29**, 317–327 (1971).

[Eps] D.B.A. Epstein, Natural tensors on Riemannian manifolds, J. Differential Geometry **10**, 631–645 (1975).

[Es-OS] J-H. Eschenburg and J. O'Sullivan, Jacobi tensors and Ricci curvature, Math. Ann. **252**, 1–26 (1980).

[Esc] J.H. Eschenburg, New examples of manifolds with strictly positive curvature, Invent. Math. **66**, 496–480 (1982).

[Es-He] J.H. Eschenburg and E. Heintze, An elementary proof of the Cheeger-Gromoll splitting theorem, Preprint, Math. Institut der Universität Münster.

[Eva] L.C. Evans, Classical solutions of fully non linear, convex 2nd order elliptic equations, Comm. Pure Appl. Math **35**, 333–363 (1982).

[Fa-La-Po] A. Fathi, F. Laudenbach and V. Poenaru, Travaux de Thurston sur les surfaces, Astérisque 66–67, Soc. Math. de France, Paris (1980).

[Fef] C. Fefferman, Monge-Ampère equations, the Bergman kernel, and geometry of pseudo-

convex domains, Ann. of Math. **103**, 395–416 (1976).
[Fer] J. Ferrand, les géodésiques des structures conformes, C.R. Acad. Sci. Paris **294**, 629–632 (1982).
[Fjk] A. Fujiki, On primitively symplectic compact Kähler V-manifolds of dimension four, in "Classification of algebraic and analytic manifolds", ed. Ueno, Progress in Math. n° 39, Birkhäuser (1983).
[Fra] T. Frankel, Gravitational curvature, Freeman, Boston (1979).
[Fre] H. Freudenthal, Bericht über die Theorie der Rosenfeldschen elliptischen Ebenen, in Proc. Coll. "Algebraic and Topological Foundations of Geometry", Utrecht 1959, Pergamon Press, Oxford, p. 35–37 (1962).
[Fr-Gr] K. Fritsche and H. Grauert, Several complex variables, Graduate texts in Math. 38, Springer Verlag, New York (1976).
[Fri] A. Friedman, Partial Differential equations, Holt, New York (1969).
[Fr-Ku] T. Friedrich and H. Kurke, Compact four-dimensional self dual Einstein manifolds with positive scalar curvature, Math. Nachr **106**, 271–299 (1982).
[Frs] D. Ferus, A remark on Codazzi tensors in constant curvature spaces, Global Differential Geometry and Global Analysis, Lectures Notes in Math. N° 838, Springer-Verlag, Berlin, p. 257 (1981).
[Fue] R. Fueter, Die Funktionentheorie des Differentialgleichungen $\Delta u = 0$ und $\Delta\Delta u = 0$ mit vier reellen Variabeln, Comment. Math. Helv. **7**, 307–330 (1935).
[Fuj] T. Fujitani, Compact suitably pinched Einstein manifolds, Bull. Faculty Liberal Arts, Nagasaki Univ. **19**, 1–5 (1979).
[Fu-Mo] A. Futaki, S. Morita, Invariant polynomials of the automorphism group of a compact complex manifold, J. Differential Geom. **21**, 135–142 (1985).
[Fut 1] A. Futaki, An obstruction to the existence of Einstein Kähler Metrics, Inven. Math. **73**, 437–443 (1983).
[Fut 2] A. Futaki, On compact Kähler manifolds of constant scalar curvature, Proc. Japan Acad. Sci. **59**, 401–402 (1983).
[Gal 1] S. Gallot, Variétés dont le spectre ressemble à celui de la sphère, Astérisque 80, 33–52, Soc. Math. de France, Paris (1980).
[Gal 2] S. Gallot, Estimées de Sobolev quantitatives sur les variétés riemanniennes et applications, Comptes rendus Acad. Sci. Paris, **292**, 375–377 (1981).
[Gal 3] S. Gallot, Minorations sur le λ_1 des variétés riemanniennes, Séminaire Bourbaki n° 569, Lecture Notes 901, Springer, Berlin (1981).
[Gal 4] S. Gallot, Inégalités isopérimétriques, courbure de Ricci et invariants géométriques I, Comptes rendus Acad. Sc. Paris, **296**, 365–368 (1983).
[Gal 5] S. Gallot, Inégalités isopérimétriques, courbure de Ricci et invariants géométriques II, Comptes rendus Acad. Sc. Paris, **296**, 365–368 (1983).
[Gao] L.Z. Gao, The construction of negative Ricci curved manifolds, Math. Ann. **271**, 185–208 (1985).
[Gas 1] J. Gasqui, Sur la courbure de Ricci d'une connexion linéaire, Comptes rendus Acad. Sc. Paris, Série A **281**, 283–288 (1975).
[Gas 2] J. Gasqui, Connexions à courbure de Ricci donnée, Math. Z. **168**, 167–179 (1979).
[Gas 3] J. Gasqui, Sur la résolubilité locale des équations d'Einstein, Compositio Math. **47**, 43–69 (1982).
[Gau] C.F. Gauss, Allgemeine Auflösung des Aufgabe die Theile einer gegebenen Fläche auf einer andern gegebenen Fläche so abzubilden, dass die Abbildung dem abgebildeten in den kleinsten Theilen ähnlich wird, Opera omnia t. IV, p. 193 (1825).
[Ga-Ya] L.Z. Gao, S.T. Yau, The existence of negatively Ricci curved metrics on three-manifolds, to appear in Inventiones Math.
[Ger] R. Geroch, Positive sectional curvature does not imply positive Gauss-Bonnet integrand, Proc. Am. Math. Soc **54**, 267–270 (1976).
[Gil] P. Gilkey, The index theorem and the heat equation, Math. Lect. Series, Publish or Perish Inc., Boston (1974).
[Gi-Tr] S. Gilbarg and N.S. Trudinger, Elliptic partial differential equations of second order,

second edition, Grundlehren series vol. 224, Springer-Verlag, Berlin (1983).

[Go-Ke] J.N. Goldberg and R.V. Kerr, Some applications of the infinitesimal holonomy group to the Petrov classification of Einstein spaces, J. of Math. Physics **2**, 327–332 (1961) and R.V. Kerr and J.N. Goldberg Einstein spaces with 4 parameter holonomy group, J. of Math. Physics **2**, 332–336 (1961).

[Gol] S.I. Goldberg. Integrability of almost Kähler manifolds, Proc. A.M.S. **21**, 96–100 (1969).

[Got] M. Goto, On algebraic homogeneous spaces, Amer. J. Math. **76**, 811–818 (1954).

[Gra 1] A. Gray, Weak holonomy groups, Math. Z. **123**, 290–300 (1971).

[Gra 2] A. Gray, Riemannian manifolds with geodesic symmetries of order 3, J. Differential Geometry **7**, 343–369 (1972).

[Gra 3] A. Gray, Invariants of curvature operators of four-dimensional Riemannian manifolds, Proc. 13th biennal Seminar Canadian Math. Congress, Vol. 2, 42–65 (1972).

[Gra 4] A. Gray, Compact Kähler manifolds with non-negative sectional curvature, Inventiones Math. **41**, 33–43 (1977).

[Gra 5] A. Gray, Einstein-like manifolds which are not Einstein, Geometriae Dedicata **7**, 259–280 (1978).

[Gra 6] A. Gray, Pseudo-Riemannian almost product manifolds and submersions, J. Math. Mech. **16**, 715–737 (1967).

[Gra 7] A. Gray, A note on manifolds whose holonomy group is a subgroup of Sp(n) Sp(1), Mich. Math. J. **16**, 125–128 (1965).

[Gre] W.H. Greub, Pontrjagin classes and Weyl tensors, Comptes rendus Math. Rep. Acad. Sci. Canada **3**, 177–182, (1981).

[Grf] F.P. Greenleaf, Invariant means on topological groups, Van Nostrand, New York (1969).

[Gr-Ha] PH. Griffiths and J. Harris, Principles of algebraic geometry, Wiley-Interscience, New York (1978).

[Gr-Ha-Va] W. Greub, S. Halperin and R. Vanstone, Connections, Curvature and cohomology (3 volumes), Academic Press, London (1972-73-76).

[Gr-He] A. Gray and L. Hervella, The sixteen classes of almost Hermitean manifolds, Ann. Mat. Pura e App. **123**, 35–58 (1980).

[Gri] P. Griffiths, The extension problem in complex analysis II, Amer. J. Math. **88**, 366–446 (1966).

[Gr-La 1] M. Gromov and H.B. Lawson, Spin and scalar curvature in the presence of a fundamental group, Ann. of Math. **111**, 209–230 (1980)

[Gr-La 2] M. Gromov and H.B. Lawson, The classification of simply connected manifolds of positive scalar curvature, Ann. of Math **111**, 423–434 (1980).

[Gr-La 3] M. Gromov and H.B. Lawson, Positive scalar curvature and the Dirac operator on complete Riemannian manifolds, Publ. Math. I.H.E.S. **58**, 83–196 (1983).

[Gr-Me] D. Gromoll and W. Meyer, An exotic sphere with nonnegative sectional curvature, Ann. of Math. **100**, 401–406 (1974).

[Gr-Kl-Me] D. Gromoll, W. Klingenberg and W. Meyer, Riemannsche Geometrie im Grossen, Lecture Notes n° 55, Springer Verlag, Berlin (1975).

[Gro 1] M. Gromov, Structures métriques pour les variétés riemanniennes Cédic Nathan, Paris (1981).

[Gro 2] M. Gromov, Volume and bounded cohomology, Publ. Math. I.H.E.S. **56**, 5–99 (1982).

[Gro 3] M. Gromov, Filling Riemannian manifolds, J. Diff. Geo. **18**, 1–147 (1983).

[Gro 4] M. Gromov, Oral communication (1983).

[Gro 5] M. Gromov, Hyperbolic manifolds according to Thurston and Jørgensen, Séminaire Bourbaki n° 546, Lecture Notes 842, Springer, Berlin (1981).

[Gro 6] M. Gromov, Hyperbolic manifolds, groups and actions, in Seminar on Diff. Geometry, ed. S.T. Yau, Ann. of Math. Studies 102, Princeton Univ. Press, Princeton (1982).

[Grr] J.Ch. Gerretsen, Tensor calculus and differential geometry. P. Noordhoff N.V., Groningen (1962).

[Grt] H. Grauert, Charakterisierung der Holomorphiegebiete durch die Vollständige Kählersche Metrik, Math. Am. **131**, 38–75 (1956).

[Gr-Va] A. Gray and L. Vanhecke, Riemannian geometry as determined by the volume of small

geodesics balls, Acta Math. **142**, 157–198 (1979).
[Gu-St] V. Guillemin, S. Sternberg, Symplectic techniques in physics, Cambridge Univ. Press, Cambridge (1984).
[Hae] A. Haefliger, Variétés feuilletées, Ann. Scuola Norm. Sup. Pisa **16**, 367–397 (1962).
[Ha-El] S. Hawking and G. Ellis, The large scale structure of space-time, Cambridge University Press, Cambridge (1973).
[Ham 1] R.S. Hamilton, The inverse function theorem of Nash and Moser, Bull. Amer. Math. Soc. (New series) **7**, 65–222 (1982).
[Ham 2] R.S. Hamilton, Three manifolds with positive Ricci curvature, J. Diff. Geom. **17**, 255–306 (1982).
[Ham 3] R.S. Hamilton, The Ricci curvature equation, Lecture notes, Math. Sci. Research Inst. Berkeley, (1983).
[Ham 4] R.S. Hamilton, Four-manifolds with positive curvature operator, Preprint U.C. San Diego (1985).
[Ha-Oz] J. Hano and H. Ozeki, On the holonomy group of linear connections, Nagoya Math. J. **10**, 97–100 (1956).
[Hei] E. Heintze, Riemannsche solvmanigfaltigkeiten, Geom. Dedicata **1**, 141–147 (1972).
[He-Ka] E. Heintze and H. Karcher, A general comparison theorem with applications to volume estimates for submanifolds, Ann. Scient. Ec. Norm. Sup., **11**, 451–470 (1978).
[Hel 1] S. Helgason, Differential geometry and symmetric spaces, Academic Press New-York and London, (1962), second edition (1978).
[Hel 2] S. Helgason, The Radon transform, Birkhaüser, Basel (1980).
[Her] R. Hermann, A sufficient condition that a mapping of Riemannian manifolds be a fibre bundle, Proc. A.M.S. **11**, 236–242 (1960).
[Hic] N. Hicks, Linear perturbations of connections, Michigan Math. J. **12**, 389–397 (1965).
[Hi-Re] S. Hiepko and H. Reckziegel, Uber sphärische Blätterungen und die Vollständigkeit ihrer Blätter, Manuscripta Math. **31**, 269–283 (1980).
[Hil] D. Hilbert, Die grundlagen der physik, Nach. Ges. Wiss., Göttingen, 461–472 (1915).
[Hir] F. Hirzebruch, New topological methods in algebraic topology, Springer-Verlag, Berlin Heidelberg, New York (1966).
[Hit 1] N. Hitchin, On compact four-dimensional Einstein manifolds, J. of Diff. Geometry **9**, 435–442 (1974).
[Hit 2] N. Hitchin, Harmonic spinors, Adv. in Math. **14**, 1–55 (1974).
[Hit 3] N. Hitchin, On the curvature of rational surfaces, in "Differential Geometry", Proc. of symp. Pure Math. **27**, 65–80 (1975).
[Hit 4] N. Hitchin, Polygons and Gravitons, Math. Proc. Camb. Phil. Soc. **85**, 465–476 (1979).
[Hit 5] N. Hitchin, Kählerian twistor spaces, Proc. London Math. Soc. **43**, 133–150 (1981).
[Hit 6] N. Hitchin, Metrics on moduli spaces, Proceedings of Lefschetz Centennial Conference, Mexico City (1984), to appear in Contemporary Mathematics.
[H-K-L-R] N.J. Hitchin, A. Karlhede, U. Lindström, M. Rocek, Hyperkähler metrics and supersymmetry, to appear.
[Hoc] G. Hochschild, The structure of Lie Groups, Holden-Day, San Francisco (1965).
[Hop] E. Hopf, Elementare Bemerkungen über die Lösungen partieller Differentialgleichungen zweiter Ordnung vom elliptischen Typ, Sitz. Ber. Preuss. Ak. d. Wiss. **19**, 147–152 (1927).
[Hör 1] L. Hörmander, The Frobenius-Nirenberg Theorem, Arkiv för Mathematik **5**, 425–432 (1964).
[Hör 2] L. Hörmander, Pseudo differential operators and non-elliptic boundary problems, Ann. of Math. **83**, 129–209 (1966).
[Hu-Sn] A.T. Huckleberry, D.M. Snow, Almost homogeneous manifolds with hypersurface orbit, Osaka Math. J. **19**, 763–786 (1982).
[Ino] A. Inoue, On Yamabe's problem by a modified direct method, Tôhoku Math. J. **34**, 499–507 (1982).
[Ish 1] S. Ishihara, Homogeneous Riemannian spaces of four dimensions, J. Math. Soc. Japan **7**, 345–370 (1955).
[Ish 2] S. Ishihara, Quaternion Kählerian manifolds, J. of Diff. Geom. **4**, 483–500 (1974).

[Iwa] H. Iwamoto, On the structure of Riemannian spaces whose holonomy fix a null system, Tôhoku Math. J. **1**, 109–135 (1950).

[Jac] R. Jackiw, Quantum meaning of classical field theory, Rev. Modern Phys. **49**, 681–706 (1977).

[Jan] M. Janet, Sur la possibilité de plonger un espace riemannien donné dans un espace euclidien, Ann. Soc. Pol. Math. **5**, 38–42 (1926).

[Jen 1] G.R. Jensen, Homogeneous Einstein spaces of dimension 4, J. of Diff. Geom. **3**, 309–349 (1969).

[Jen 2] G.R. Jensen, The scalar curvature of left invariant Riemannian metrics, Indiana Univ. Math. J. **20**, 1125–1143 (1971).

[Jen 3] G.R. Jensen, Einstein metrics on principal fibre bundles, J. Diff. Geom. **8**, 599–614 (1973).

[Jen 4] G.R. Jensen, Imbeddings of Stiefel manifolds into Grassmanians, Duke Math. J. **42**, 397–407 (1975).

[Jin] Z. Jin, Conformal deformation of Riemannian metrics on a class of non-compact complete Riemannian manifolds and the Yamabe problem, Preprint (1986).

[Joh] K. Johannson, Homotopy equivalence of 3-manifolds with boundaries, Lecture Notes 761, Springer, Berlin (1979).

[Jør] T. Jørgensen, Compact 3-manifolds of constant negative curvature fibring over the circle, Ann. of Math. **106**, 61–72 (1977).

[Ka-Kr] J.L. Kazdan and R. Kramer, Invariant criteria for existence of solutions to second order quasilinear elliptic equations, Comm. Pure Appl. Math. **31**, 619–645 (1978).

[Ka-Ma] D. Kazhdan and G.A. Margulis, A proof of Selberg's hypothesis, Math. Sb. (117) **75**, 163–168 (1968).

[Kap 1] A. Kaplan, Fundamental solutions for a class of hypoelliptic PDE generated by composition of quadratic forms Trans. Amer. Math. Soc. **258**, 147–153 (1980).

[Kap 2] A. Kaplan, Riemannian nilmanifolds attached to Clifford modules, Geom. Dedicata **11**, 127–136 (1981).

[Kap 3] A. Kaplan, On the geometry of groups of Heisenberg type, bull. London Math. Soc. **15**, 35–42 (1983).

[Kat] T. Kato, Perturbation theory for linear differential operators. Grundlehren n° 132, Springer, Berlin (1966).

[Ka-Wa 1] J.L. Kazdan and F.W. Warner, Curvature functions for compact 2-manifolds, Ann. of Math. **99**, 14–47 (1974).

[Ka-Wa 2] J.L. Kazdan and F.W. Warner, Existence and Conformals Deformation of metrics with Prescribed Gaussian and Scalar curvature, Ann. of Math. **101**, 317–331 (1979).

[Ka-Wa 3] J.L. Kazdan and F.W. Warner, A direct approach to the determination of Gaussian and scalar curvature functions. Inventiones Math. **28**, 227–230 (1975).

[Ka-Wa 4] J.L. Kazdan and F.W. Warner, Prescribing curvature, in "Differential Geometry", Proc. Amer. Math. Soc. Symp. Pure Math. **27**, 219–226 (1975).

[Ka-Wa 5] J.L. Kazdan, F.W. Warner, Curvature functions for open 2-manifolds, Ann. Math. **99**, 203–219 (1974).

[Kaz 1] J. Kazdan, A remark on the preceding paper of Yau, Comm. Pure and Appl. Math **31**, 413–414 (1978).

[Kaz 2] J.L. Kazdan, Another proof of Bianchi's identity in Riemannian Geometry, Proc. A.M.S. **81**, 341–342 (1981).

[Kaz 3] J.L. Kazdan, Deformation to positive scalar curvature, on complete manifolds, Math. Ann. **261**, 227–234 (1982).

[Kir] A Kirillov, Eléments de la théorie des représentations, Editions Mir, Moscou (1974).

[Kle] F. Klein, Vergleichende Berachtungen über neuere geometrische Forschungen, Math. Ann. 43, 83–100 (1893), or Klein's Gesammelte Mathematische Abhandlungen Springer Berlin, 1921–23, Vol 1, 460–497, or Bull. N.J. Math Soc. **2**, 215 (1892), or Annales ENS **8**, 87–102 et 173–199 (1891).

[Kob 1] S. Kobayashi, Remarks on complex contact manifolds, Proc. AMS **10**, 164–167 (1959).

[Kob 2] S. Kobayashi, On compact Kähler manifolds with positive definite Ricci tensor, Ann. of Math. **74**, 570–574 (1961).

[Kob 3] S. Kobayashi, Geometry of bounded domains, Trans. Amer. Math. Soc. **92**, 267–290 (1959).

[Kob 4] S. Kobayashi, "Transformations groups in differential geometry", Ergebnisse der Mathematik, und Ihre Grenzgebiete, 70, Springer-Verlag, Berlin (1972).

[Kob 5] S. Kobayashi, First Chern class and holomorphic tensor fields, Nagoya Math. J. **77**, 5–11 (1980).

[Kob 6] S. Kobayashi, Recent results in complex differential geometry, J. ber. d. Dt. Math-Verein **83**, 147–158 (1981).

[Kob 7] S. Kobayashi, Recent results in complex differential geometry, preprint, Berkeley (1982).

[Kob 8] S. Kobayashi, Principal fiber bundles with 1-dimensional toroidal group, Tôhoku Math. J. **8**, 29–45 (1956).

[Kob 9] S. Kobayashi, Topology of positively pinched Kähler manifolds, Tôhoku Math. J. **15**, 121–139 (1963).

[Kod 1] K. Kodaira, On Kähler varieties of restricted type (an intrinsic characterisation of algebraic varieties), Ann. of Math. **60**, 28–46 (1954).

[Kod 2] K. Kodaira, A theorem of completenes. of characteristic systems for analytic families of compact submanifolds of complex manifolds, Ann. Math. **84**, 146–162 (1962).

[Kod 3] K. Kodaira, On the structure of compact complex analytic surfaces I, Amer. J. Math. **86**, 751–798 (1964).

[Ko-Hi] K. Kodaira and F. Hirzebruch, On the complex projective spaces, J. Math. pures et App. **36**, 201–206 (1957).

[Koh] J.J. Kohn, Harmonic integrals on strongly pseudo-convex manifolds, I, Ann. of Math. **78**, 206–213 (1963).

[Koi 1] N. Koiso, A decomposition of the space of Riemannian metrics on a manifold, Osaka, J. of Math. **16**, 423–429 (1979).

[Koi 2] N. Koiso, Non deformability of Einstein metrics, Osaka, J. of Math. **15**, 419–433 (1978).

[Koi 3] N. Koiso, On the second derivative of the total scalar curvature, Osaka, J. of Math. **16**, 413–421 (1979).

[Koi 4] N. Koiso, Rigidity and stability of Einstein metrics. The case of compact symetric spaces, Osaka J. Math. **17**, 51–73 (1980).

[Koi 5] N. Koiso, Hypersurfaces of Einstein manifolds, Ann. Sci. de l'Ecole Norm. Sup., Paris, **14**, 433–443 (1981).

[Koi 6] N. Koiso, Rigidity and infinitesimal deformability of Einstein metrics, Osaka J. Math. **19**, 643–668 (1982).

[Koi 7] N. Koiso, Einstein metrics and complex structures, Inventiones Math. **73**, 71–106 (1983).

[Ko-Mo] K. Kodaira and J. Morrow, Complex manifolds, Holt, Rinehart and Winston Inc., New York (1971).

[Ko-Ni-Sp] K. Kodaira, L. Nirenberg, D.C. Spencer, On the existence of deformations of complex analytic structures, Ann. of Math. **68**, 450–459 (1958).

[Ko-No 1] S. Kobayashi, K. Nomizu, Foundations of differential geometry, Vol. I, Interscience, Wiley, New York (1963).

[Ko-No 2] S. Kobayashi and K. Nomizu, Foundations of differentiable geometry, Vol. II, Interscience, Wiley, New York (1969).

[Koo] O. Kobayashi, A differential equation arising from scalar curvature, J. Math. Soc. of Japan **34**, 665–675 (1982).

[Ko-Oc] S. Kobayashi, T. Ochiai, Characterizations of complex projective spaces and hyperquadrics, J. Math. Kyoto Univ. **13**, 31–47 (1973).

[Kor] A. Korn, Zwei Anwendungen der Methode der suksessiven Annäherungen, Scwarz Festschrift, Berlin, 215–229 (1919).

[Kos] B. Kostant, On differential geometry and homogeneous spaces II, Proc. N.A.S. U.S.A. **42**, 354–357 (1956).

[Ko-Sa] N. Koiso, Y. Sakane, Non-homogeneous Kähler-Einstein metrics on compact complex manifolds, Preprint Univ. Osaka (1985).

[Ko-Sp] K. Kodaira, D.C. Spencer, On deformations of complex analytic structures, I–II, Ann. of Math **67**, 328–466 (1958), III, Ann. of Math. **71**, 43–76 (1960).

[Koz 1] J.L. Koszul, Sur la forme hermitienne canonique des espaces homogènes complexes, Can. J. of Math. **7**, 562–576 (1955).

[Koz 2] J.L. Koszul, Sur les j-algèbres propres, Preprint (1966).

[Kra] V.Y. Kraines, Topology of quaternionic manifolds, Trans. AMS 122, 357–367 (1966).

[Kro] P. Kronheimer, Gravitational instantons and Kleinian singularities, to appear.

[Kui] N.H. Kuiper, On conformally flat spaces in the large, Ann. of Math. **50**, 916–924 (1949).

[Kul 1] R.S. Kulkarni, Conformally flat manifolds, Proc. Mat. Acad. Sci. USA **69**, 2675–2676 (1972).

[Kul 2] R.S. Kulkarni, On the Bianchi identities, Math. Ann. **199**, 175–204 (1972).

[Kul 3] R.S. Kulkarni, On the principle of uniformization, J. of Diff. Geom. **13**, 109–138 (1978).

[Laf 1] J. Lafontaine, Remarques sur les variétés conformément plates, math. Ann. **259**, 313–319 (1982).

[Laf 2] J. Lafontaine, Courbure de Ricci et fonctionnelles critiques, C.R. Acad. Sc. Paris. **295**, 687–690 (1982).

[Laf 3] J. Lafontaine, Sur la géométrie d'une généralisation de l'équation différentielle d'Obata, J. Math. Pures Appliquées **62**, 63–72 (1983).

[Lan 1] C. Lanczos, Ein vereinfachendes Koordinatensystem fur die Einsteinschen Gravitationsgleichungen, Phys. Z. **23**, 537–539 (1922).

[Lan 2] C. Lanczos, A remarkable property of the Riemann-Christoffel tensor in four dimensions, Ann. Math. **39**, 842–850 (1938).

[Lan 3] C. Lanczos, The splitting of the Riemann tensor, Rev. Modern Phys. **34**, 379–389 (1962).

[La-Sp-Ur] O.A. Ladyzhenskaja, V.A. Solonnikov, and N.N. Uraltseva, Linear and quasilinear equations of parabolic type, Nauka, Moscow (1967) Trans. of Math. Monographs **23**, Amer. Math. Soc., Providence R.I. (1968)

[Leb] C.R. Lebrun, H-spaces with a cosmological constant, Proc. Roy. Soc. Lond. **A 380**, 171–185 (1982).

[Le-DM] M.L. Leite and I. Doti Miatello, Metrics of negative Ricci curvature on SL(n, $\mathbb{R}$), $n \geqslant 3$, J. of Diff. Geom. **17**, 635–641 (1982).

[Lem] L. Lemaire, Applications harmoniques de surfaces riemanniennes, J. Diff. Geom. **13**, 51–78 (1978).

[Lev] M. Levine, A remark on extremal Kähler metrics, J. Differential Geom. **21**, 73–77 (1985).

[Li] P. Li, On the Sobolev Constant and the p-spectrum of a compact Riemannian manifold, Ann. Sc. E.N.S. Paris **13**, 451–469 (1980).

[Lic 1] A. Lichnerowicz, Géométrie des groupes de transformation, Dunod (1958).

[Lic 2] A. Lichnerowicz, Propagateurs et commutateurs en relativité générale, Inst. Hautes Etudes Sci. Publ. Mat. **10**, 293–344 (1961).

[Lic 3] A. Lichnerowicz, Théorie globale des connexions et des groupes d'holonomie, Dunod, Paris (1955).

[Lic 4] A. Lichnerowicz, Spineurs harmoniques, C.R. Acad. Sci. Paris **257**, 7–9 (1963).

[Lic 5] A. Lichnerowicz, Espaces homogènes kähleriens, in "Coll. Intern. de géom. différentielle, Strasbourg (1953)", CNRS, Paris, 171–184 (1953).

[Lic 6] A. Lichnerowicz, Analytic transformations of Kähler manifolds, Mimeo. notes Univ. Illinois (1960).

[Lie] Séminaire S. Lie (lère année 1954/55) Ecole Normale Supérieure, Paris.

[Lit] L. Lichtenstein, Neuere Entwicklung des Theorie partieller Differenzialgleichungen zweiter Ordnung vom elliptischen Typus, in Ens. d. Math. Wiss. 2.3.2, 1277–1334 (1924) Teubner, Leipzig (1927).

[Loo] E. Looijenga, A Torelli theorem for Kähler-Einstein K3 surfaces, in "Geometry Symposium, Utrecht (1980)" Lecture Notes n° 894, Springer-Verlag, New York, 107–112 (1982).

[Los] O. Loos, Symmetric spaces, Benjamin, New-York-Amsterdam (1969).

[Mab 1] T. Mabuchi, A functional integrating Futaki's invariant, Proc. Japan Acad. Sci. **61**, 119–120 (1985).

[Mab 2] T. Mabuchi, Some symplectic geometry on compact Kähler manifolds, I, Preprint Max-Planck-Institut, Bonn (1986).

[Mag] Ch. Margerin, Pointwise pinched manifolds are space-forms, in "Geometric Measure Theory and Calculus of Variations, Arcata 1983", Proc. Amer. Math. Soc. Symp. in Pure Math. **44**, 307–328 (1986).

[Mai 1] H. Maillot, Sur les variétés riemanniennes à opérateur de courbure pur. C.R. Acad. Sci. Paris A **278**, 1127–1130 (1974).

[Mai 2] H. Maillot, Sur l'opérateur de courbure d'une variété riemannienne, Thèse (3° cycle), Université Claude-Bernard, Lyon (1974).

[Mal 1] B. Malgrange, Pseudo-groupes de Lie elliptiques, Séminaire Leray, Collège de France (1969–70).

[Mal 2] B. Malgrange, Equations de Lie, I, II, J. Differential geometry **6**, 503–522, **7**, 117–141 (1972).

[Ma-Mo] R. Mandelbaum and B. Moishezon, The topological type of algebraic surfaces: Hypersurfaces in $P_3(C)$, Proc. Symp. Pure Math. **30**, Part 1, 277–283 (1977).

[Man 1] O.V. Manturov, Homogeneous asymmetric Riemannian spaces with an irreducible group of rotations, Dokl. Akad. Nauk. SSSR **141**, 792–795 (1961).

[Man 2] O.V. Manturov, Riemannian spaces with orthogonal and symplectic groups of motions and an irreducible group of rotations, Dokl. Akad. Nauk. SSSR **141**, 1034–1037 (1961).

[Man 3] O.V. Manturov, Homogeneous Riemannian manifolds with irreducible isotropy group, Trudy Sem. Vector and Tensor Analysis **13**, 68–145 (1966).

[Ma-Ro] S. Marchiafava and G. Romani, Sui fibrati con struttura quaternionale generalizzata, Ann. Mat. Pura Appl. **107**, 131–157 (1976).

[Mat] H. Matsumoto, Quelques remarques sur les espaces riemanniens isotropes, C.R. Acad. Sc. Paris **272**, 316–319 (1971).

[Mar] S. Marchiafava, Alcune osservazioni riguardanti i gruppi di Lie G_2 e Spin (7), candidati a gruppi di olonomia, Annali di Mat. pura ed appl. **129**, 247–264 (1981).

[Mat 1] Y. Matsushima, Sur les espaces homogènes Kählériens d'un groupe réductif, Nagoya Math. J. **11**, 53–60 (1957).

[Mat 2] Y. Matsushima, Sur la structure du groupe d'homéomorphismes analytiques d'une certaine variété kählérienne, Nagoya Math. J. **11**, 145–150 (1957).

[Mat 3] Y. Matsushima, Remarks on Kähler-Einstein manifolds, Nagoya Math. J. **46**, 161–173 (1972).

[Mic] M.L. Michelsohn, Clifford and spinor cohomology of Kähler manifolds, Amer. J. Math. **102**, 1084–1146 (1980).

[Mil 1] J. Milnor, A note on curvature and fundamental group, J. of Diff. Geom. **2**, 1–7 (1968).

[Mil 2] J. Milnor, Remarks concerning spin manifolds, Differential and combinatorial Topology, a symposium in Honor of Marston Morse, Princeton Univ. Press, Princeton, 55–62 (1965).

[Mi-Th-Wh] C. Misner, K. Thorne, J. Wheeler, Gravitation, Freeman, San Francisco (1973).

[Miy] T. Miyazawa, Some theorems on conformally symmetric spaces, Tensor, N.S. **32**, 24–26 (1978).

[Moi] S. Mori, Projective manifolds with ample tangent bundles, Ann. of Math. **110**, 593–606 (1973).

[Mon] D. Montgomery, Simply connected homogeneous spaces, Proc. Amer. Math. Soc. **1**, 457–469 (1950).

[Mo-Sa 1] D. Montgomery and H. Samelson, Transformation groups of spheres, Annals of Math. **44**, 454–470 (1943).

[Mo-Sa 2] D. Montgomery and H. Samelson, Groups transitive on the n-dimensional torus, Bull A.M.S. **49**, 455–456 (1943).

[Mor] C.B. Morrey, Multiple Integrals in the Calculus of Variations, Grundlehren Series Vol 130, Springer-Verlag, Berlin (1966).

[Mos] G.D. Mostow, Quasiconformal mappings in *n*-space and the strong rigidity of hyperbolic space forms, Publ. Math. de l'I.H.E.S. **34**, 53–104 (1968).

[Mo-Si] D. Mostow and T. Siu, A compact Kähler surface of negative curvature not covered by the ball, Ann. of Math. **112**, 321–360 (1980).

[Mo-Ya] N. Mok and S.T. Yau, Completeteness of the Kähler-Einstein metric on bounded Riemann domains and the characterization of domains of holomorphy by curvature conditions, in The mathematical heritage of Henri Poincaré, Proc. Symp. Pure Math. **39**, Part 1, 41–60 (1983).

[Msr] J. Moser, On the volume element on a manifold, Trans. AMS **120**, 286–294 (1965).

[Muk] S. Mukai, Symplectic structure of the moduli space of sheaves on an abelian or K3 surface, Preprint

[Mum 1] D. Mumford, Algebraic geometry I, complex projective varieties, Springer-Verlag, Grundlehren 221, Springer, Berlin (1976).

[Mut 1] Y. Muto, On Einstein metrics, J. Diff. Geom. **9**, 521–530 (1974).

[Mut 2] Y. Muto, Riemannian submersions and critical Riemannian metrics, J. Math. Soc. Japan **29**, 493–511 (1977).

[Mye 1] S.B. Myers, Riemannian manifold with positive mean curvature, Duke Math. J. **8**, 401–404 (1941).

[Mye 2] S.B. Myers, Riemannian manifolds in the large, Duke Math. J. **1**, 39–49 (1935).

[My-St] S.B. Myers and N. Steenrod, The group of isometries of a Riemannian manifold, Annals of Math. **40**, 400–416 (1939).

[Nab] P. Nabonnand, Sur les variétés riemanniennes complètes à courbure de Ricci positive, C.R.A.S. **291**, 591–593 (1980).

[Nag] T. Nagano, On fibred Riemannian manifolds, Sc. Papers of the college of General Education, Univ. of Tokyo **10**, 17–27 (1960).

[Nai] M.A. Naimark, Theory of group representations (translated from Russian), Grundlehren n° 246, Springer, Berlin (1982).

[Nak] K. Nakajima, On *j*-algebras and homogeneous Kähler manifolds, Hokkaido Math. J. **15**, 1–20 (1986).

[Nas] J.C. Nash, Positive Ricci curvature on fiber bundles, J. of Diff. Geom. **14**, 241–254 (1979).

[Ne-Ni] A. Newlander and L. Nirenberg, Complex analytic coordinates in almost complex manifolds, Ann. of Math. **65**, 391–404 (1957).

[Nij 1] A. Nijenhuis, On the holonomy group of linear connections, Indag. Math. **15**, 233–249, (1953) **16**, 17–25 (1954).

[Nij 2] A. Nijenhuis, A note on infinitesimal holonomy groups, Nagoya Math. J. **12**, 145–147 (1957).

[Nik] V. Nikulin, Finite groups of automorphisms of Kählerian surfaces of type K3, Proc. Moscow Math. Soc. **38**, 75–137 (1979).

[Nir 1] L. Nirenberg, Variational and topological methods in non linear problems, Bull. Am. Math. Soc. (New series) **4**, 267–302 (1981).

[Nir 2] L. Nirenberg, Topics in non linear functionnal analysis, New York University Lecture Notes (1974).

[Nir 3] L. Nirenberg. The Weyl and Minskowski problems in differential geometry in the large, Comm. Pure Appli. Math. **6**, 337–394 (1953).

[Nir 4] L. Nirenberg, Lectures on linear partial differential equations, Amer. Math. Soc. 1973.

[Nom 1] K. Nomizu, On the decomposition of generalized curvature tensor fields, Differential geometry, conference in honor of K. Yano, Kinokuniya, Tokyo, 335–345 (1972).

[Nom 2] K. Nomizu, Un théorème sur les groupes d'holonomie, Nagoya Math. J. **10**, 101–103 (1956).

[Nom 3] K. Nomizu, Lie Groups and Differential Geometry, Publ. Math. Soc. Japan n° 2 (1956).

[Oba] M. Obata, Certain conditions for a Riemannian manifold to be isometric with a sphere, J. Math. Soc., Japan **14**, 333–340 (1962).

[Oka 1] K. Oka, Domaines finis sans point critique intérieur, Japanese J. Math. **23**, 97–115 (1953).

[Oka 2] K. Oka, Domaines pseudoconvexes, Tôhoka Math. J. **49**, 15–52 (1942).

[Oa-Fr-Vn] R. O'Auria, P. Fre, P. Van Nieuwenhuizen, $N = 2$ matter coupled supergravity from compactification on a coset G/H possessing an additional Killing vector, Phys. Lett. **136B**, 347–353 (1984).

[Omo] H. Omori, On the group of diffeomorphisms of a complete manifold, Global Analysis, Proc. Sympos. Pure Math. **15**, 167–183 (1968).

[One 1] B. O'Neill, The fundamental equations of a submersion, Mich Math. J. **13**, 459–469 (1966).

[One 2] B. O'Neill, Submersions and geodesics, Duke Math. J. **34**, 363–373 (1967).

[One 3] B. O'Neill, Semi-riemannian geometry, Academic Press, New York (1983).

[Oni] A. L. Oniscik, Transitive compact transformation groups, Math. Sb. **60**, (102), 447–485 (1963), (English translation: Amer. Math. Soc. Trans. **55**(2), 153–194 (1966)).

[Pag] D. Page, A compact rotating gravitational instanton, Phys. Lett. **79** B, 235–238 (1979).

[Pal 1] R.S. Palais, Foundations of global non-linear analysis. New-York, Benjamin (1968).

[Pal 2] R.S. Palais, The principle of symmetric criticality, Comm. Math. Phys. **69**, 19–30 (1979).

[Pal 3] R.S. Palais, Critical point theory and the minimax principle, in global Analysis (Proc. Symp. Pure Math.), Amer. Math. Soc., Providence, Rhode Island. XX, 185–212 (1970).

[Pal 4] R.S. Palais, Applications of the symmetric criticality principle to mathematical physics and differential geometry, in DD2 Proceedings (Shanghai-Hefei 1981), Science Press, Beijing (1984).

[Pal 5] R.S. Palais, The classification of G-spaces, Memoirs of the A.M.S. n° 36 (1960).

[Pa-Po 1] D.N. Page, C.N. Pope, Which compactifications of $D = 11$ supergravity are stable? Phys. Lett. **144B**, 346–350 (1984).

[Pa-Po 2] D.N. Page, C.N. Pope, New squashed solutions of $D = 11$ supergravity, Phys. Lett. **147B**, 55–60 (1984).

[Par] T.H. Parker, Gauge theories on four dimensional Riemannian manifolds, Comm. Math. Phys. **85**, 563–602 (1982).

[Pat] E.H. Patterson, A class of critical riemannian metrics, J. London Math. Soc. **12**, 349–358 (1981).

[Pen 1] R. Penrose, Nonlinear gravitons and curved twistor theory, Gen. Relativ. Grav. **I**, 31–52 (1976).

[Pen 2] R. Penrose, Techniques of differential topology in relativity, SIAM (1972).

[Per] M.J. Perry, Gravitational instantons, in Seminar on Diff. Geometry, ed. S.T. Yau, Ann. Math. Studies n° 102, Princeton Univ. Press, Princeton, 603–630 (1982).

[Pet] A.Z. Petrov, Einstein spaces, Pergamon, Oxford (1969).

[Pir] F. Pirani, Survey of gravitational radiation theory, in Recent Developments in General Relativity, Pergamon, Oxford (1962).

[PiS] I.I. Piatetskii-Shapiro, Geometry of classical domains and the theory of automorphic functions, Moscow (1961), English translation New-York (1969).

[Pog 1] A.V. Pogorelov, The multidimensional Minkowski problem, English translation, Winston and sons, Washington (1978).

[Pog 2] A.V. Pogorelov, Monge-Ampère equations of elliptic type, Noordhoff, Groningen (1964).

[Pog 3] A.V. Pogorelov, On the improper convex affine hyperspheres, Geometriae Dedicata **1**, 33–46 (1972).

[Pog 4] A. V. Pogorelov, Extrinsic geometry of convex surfaces, Nauka, Moscow, English: translation of Math. Monographs, Vol 35, Amer. Math. Soc., Providence (1973).

[Pol 1] A. Polombo, Nombres caractéristiques d'une variété riemannienne de dimension 4, J. of Diff. Geom. **13**, 145–162 (1978).

[Pol 2] A. Polombo, Nombres caractéristiques d'une surface Kälherienne compacte, C.R. Acad. Sci. Paris A **283**, 1025–1028 (1976).

[Poo 1] W.A. Poor, Differential Geometric structures, Mc Graw Hill, New York (1981).

[Poo 2] W.A. Poor, Some exotic spheres with positive Ricci curvature, Math. Annalen **216**, 245–252 (1975).

[Pr-We] M.H. Protter and H.F. Weinberger, Maximum principles in Differential Equations, Prentice Hall, Englewood Cliffs N.J. (1967).

[Rau] H.E. Rauch, A transcendental view of the space of algebraic Riemann surfaces, Bull. Amer. Math. Soc. **71**, 1–39 (1965).

[Rbg 1] J. Rosenberg, C*-algebras, positive scalar curvature, and the Novikov Conjecture, Publ. I.H.E.S. **58**, 197–212 (1983).

[Rbg 2] J. Rosenberg, C*-algebras, positive scalar curvature, and the Novikov Conjecture II, Preprint University of Maryland (1984), (to appear in Proc. U.S.-Japan Seminar on "Geometric Methods in Operator Algebras" Kyoto, Japan, 1983, E. Effros and H. Araki, eds).

[Rha] G. De Rham, Sur la réductibilité d'un espace de Riemann, Comm. Math. Helv. **26**, 328–344 (1952).

[Rig] A. Rigas, Some bundles of nonnegative curvature, Math. Annalen **232**, 187–193 (1978).

[Rol] A. Rolfsen, Knots and links, Math. Lect. Ser 7, Publish or Perish, Boston (1976).

[Ros] B.A. Rosenfeld, Einfache Lie-Gruppen und Nichteuklidische Geometrien, in Proc. Coll. "Algebraic and Topological Foundations of Geometry, Utrecht 1959, Pergamon Press, Oxford, p. 135–155 (1962).

[Rot] W. Roter, Some existence theorems on conformally recurrent manifolds, in preparation.

[Run] H. Rund, The differential geometry of Finsler spaces, Springer (1959).

[Rya] P. Ryan, A note on conformally flat spaces with constant scalar curvature, Proc. 13th biennal Seminar Canadian Math. Congress, Vol 2, 115–124 (1972).

[Sae 1] Y. Sakane, On compact Einstein-Kähler manifolds with abundant holomorphic transformations, in "Manifolds and Lie groups" (Notre-Dame, Ind. 1980), pp. 337–35!, Progr. in Math. 14, Birkhaüser Boston (1981).

[Sae 2] Y. Sakane, Examples of compact Einstein Kähler manifolds with positive Ricci tensor, Preprint Univ. Osaka (1985).

[Sag] A.A. Sagle, Some homogeneous Einstein manifolds, Nagoya Math. J. **39**, 81–106 (1970).

[Sak] K. Sakamoto, On the topology of quaternion Kähler manifolds, Tôhoku Math. J. **26**, 389–405 (1974).

[Sal 1] S.M. Salamon, Quaternionic manifolds, D. Phil. Thesis, Oxford (1980).

[Sal 2] S.M. Salamon, Quaternionic Kähler manifolds, Invent. Math. **67**, 143–171 (1982).

[Sal 3] S.M. Salamon, Quaternionic manifolds, Symposia Mathematica, **26**, 139–151 (1982).

[Sal 4] S.M. Salamon, Quaternionic structures and twistor space, in "Global Riemannian Geometry", ed. T.J. Willmore and N. Hitchin, Ellis Horwood, Chichester, p. 65–74 (1984).

[Sa-Sh] J. Sabitov and S. Shefel', The connections between the order of smoothness of a surface and its metric, Sibirsk. Mat. Zh. **17**, 916–925 (1976). English translation: Siberian Math. J. **17**, 687–694 (1976).

[Sa-Uh] J. Sacks and K. Uhlenbeck, The existence of minimal immersions of 2-spheres, Ann. of Math. **113**, 1–24 (1981).

[Sa-Wu] R. Sachs and H. Wu, General relativity for Mathematicians, Springer (1977).

[Sch] J.F. Schell, Classification of 4 dimensional Riemannian spaces, J. of Math. Physics **2**, 202–206 (1961).

[Sce] R. Schoen, Conformal deformation of a Riemannian metric to constant scalar curvature, J. Differential Geom. **20**, 479–495 (1985).

[Sco] P. Scott, The geometries of 3-manifolds, Bull. London Math. Soc. **15**, 401–487 (1983).

[Sc-Ya 1] R. Schoen and S.T. Yau, Existence of incompressible minimal surfaces and the topology of three dimensional manifolds with non negative scalar curvature, Ann. Math. **110**, 127–142 (1979).

[Sc-Ya 2] R. Schoen and S.T. Yau, Incompressible minimal surfaces, three dimensional manifolds with non-negative scalar curvature and the positive mass conjecture in general relativity, Proc. N.A.S. U.S.A. **75**, 2567 (1978).

[Sc-Ya 3] R. Schoen and S.T. Yau, On the structure of manifolds with positive scalar curvature, Manuscripta Math. **28**, 159–183 (1979).

[Sc-Ya 4] R. Schoen and S.T. Yau, Complete three dimensional manifolds with positive Ricci curvature and scalar curvature, in "Seminar on Differential Geometry", ed. S.T. Yau, p. 209–227, Princeton Univ. Press, Princeton (1982).

[See] R.T. Seeley, in Notices of Amer. Math. Soc. **11**, 570 (1964).

[Ser 1] J.P. Serre, Représentations linéaires et espaces homogènes kähleriens des groupes de Lie compacts, (d'après résultats non publiés de A. Borel et A. Weil) Séminaire Bourbaki, n° 100, 1953/54, Benjamin, New York (1966).

[Ser 2] J.P. Serre, Algèbres de Lie semi-simples complexes, W.A. Benjamin, Inc. (1966).

[Sev] F. Severi, Some remarks on the topological caracterization of algebraic surfaces, In studies presented to R. Von Mises, Academic Press, New-York 54–61 (1954).

[Sha] P. Shahanan, The Atiyah-Singer Index theorem: An introduction, Lecture Notes in Math. 638, Springer Verlag, Berlin (1976).

[Sim] J. Simons, On transitivity on holonomy systems, Ann. Math. **76**, 213–234 (1962).

[Sin] E.N. Sinyukova, Geodesic mappings of the spaces L_n (in Russian), Izvestiya Vyssh. Ucheb. Zaved, (Mat.), n° **3**, 57–61 (1982).

[Sio 1] U. Simon, On differential operators of second order on Riemannian manifolds with nonpositive curvature, Colloq. Math. **31**, 223–229 (1974).

[Sio 2] U. Simon, Compact conformally symmetric Riemannian spaces, Math. Z. **132**, 173–177 (1973).

[Si-Th] I.M. Singer and J.A. Thorpe, The curvature of 4-dimensional Einstein spaces, in "Global Analysis", Papers in Honour of K. Kodaira, Princeton University Press, Princeton, 355–365, (1969).

[Siu 1] Y.T. Siu, A simple proof of the surjectivity of the period map of K3 surfaces, Manuscripta Math. 35, 225–255 (1981).

[Siu 2] Y.T. Siu, Every K3 surface is Kähler, Inventiones Math. 73, 139–150 (1983).

[Siu 3] Y.T. Siu, Pseudocenvexity and the problem of Levi, Bull. Amer. Math. Soc. **84**, 481–512 (1978).

[Siv] N.S. Sinyukov, Geodesic mappings of Riemannian manifolds (in Russian), Nauka, Moscow [1979).

[Si-Ya] Y.T. Siu and S.T. Yau, Compact Kähler manifolds of positive bisectional curvature, Inv. Math. **59**, 189–204 (1980).

[Si-Yg] Y.T. Siu and P. Yang, Compact Kähler-Einstein surfaces of non-positive bisectional curvature., Invent. Math. **64**, 471–487 (1981).

[Smy] B. Smyth, Differential geometry of Hypersurfaces, Annals of Math. **85**, 246–266 (1967).

[Som] A. Sommese, Quaternionic manifolds, Math. Ann. **212**, 191–214 (1975).

[Spi] M. Spivak, A comprehensive introduction to differential geometry, Publish or Perish (1975).

[Stp] H. Stephani, Allgemeine Relativitätstheorie, VEB Deutschen Verlag der Wissenschaften, Berlin (1980).

[Str] P. Stredder, Natural differential operators on Riemannian manifolds and representations of the orthogonal and special orthogonal groups, J. Differential Geometry **10**, 647–660 (1975).

[Sud] A. Sudbery, Quaternionic analysis, Math. Proc. Camb. Phil. Soc. **85**, 199–225 (1979).

[Sul] D. Sullivan, Travaux de Thurston sur les groupes quasi-Fuchsiens et les variétés hyperboliques de dimension 3 fibrées sur S^1, Séminaire Bourbaki n° 554, Lecture Notes 842, Springer, Berlin (1981).

[Sum] T. Sumitomo, On the commutator of differential operators, Hokkaido Math. J. 1, 30–42 (1972).

[Tan 1] S. Tanno, Curvature tensors and covariant derivatives, Ann. Mat. Pura Appl. **96**, 233–241 (1973).

[Tan 2] S. Tanno, The automorphism groups of almost hermitian manifolds, Trans. Amer. Math. Soc. **137**, 269–275 (1969).

[Ter] C.L. Terng, Natural vector bundles and natural differential operators, Amer. J. Math. **100**, 775–828 (1978).

[Thu 1] W. Thurston, Geometry and Topology of 3-manifolds, preprint Inst. for Adv. Study, Princeton (1980).

[Thu 2] W. Thurston, Three dimensional manifolds, Kleinian groups and hyperbolic geometry, Bull. Amer. Math. Soc. **6**, 357–381 (1982).

[Thu 3] W. Thurston, Hyperbolic structures on 3-manifolds I, deformations of acylindrical manifolds, preprint Inst. Adv. Study, Princeton.

[Tho 1] J. Thorpe, Curvature and the Petrov canonical forms, Journal of Mathematical Physics **10**, 1 (1969).

[Tho 2] J. Thorpe, Curvature invariants and space-time singularities, Journal of Mathematical Physics **18**, 960 (1977).

[Tho 3] J.A. Thorpe, Some remarks on the Gauss-Bonnet integral, J. Math. Mec. **18**, 779–786 (1969).

[Tod 1] A. Todorov, Applications of the Kähler-Einstein Calabi-Yau metrics to moduli of K3 surfaces, Inventiones Math. **61**, 251–265 (1980).

[Tod 2] A. Todorov, How many Kähler metrics has a K3 surface in "Arithmetic and Geometry", papers dedicated to I.R. Shafarevitch on the occasion of his sixtieth birthday, volume 2, page 451–464, Birkhäuser, Boston-Basel-Stuttgart (1983).

[Tod 3] A.N. Todorov, Every holomorphic symplectic manifold admits a Kähler metric, Preprint Univ. Sofia (1986).

[Top] V.A. Toponogov, Riemannian spaces which contain straight lines, Amer. Math. Soc. Translations (2) **37**, 287–290 (1964).

[Tr-Va 1] F. Tricerri and L. Vanhecke, Naturally reductive homogeneous spaces and generalized Heisenberg groups, to appear

[Tr-Va 2] F. Tricerri and L. Vanhecke, Homogeneous Structures on Riemannian Manifolds, London Math. Soc. Lecture Notes Series, N° 83, Cambridge Univ. Press, Cambridge (1983).

[Tru 1] N.S. Trudinger, Remarks concerning the conformal deformation of Riemannian structures on compact manifolds, Ann. Scuola Norm. Sup. Pisa **22**, 265–274 (1968).

[Tru 2] N.S. Trudinger, On imbeddings into Orlicz spaces and some applications, J. Math. Mech. **17**, 473–483 (1967).

[Va-Wi] L. Vanhecke and T.J. Willmore, Interaction of tubes and spheres, Math. Ann. **263**, 31–42 (1983).

[Var] V.S. Varadarajan, Lie groups, Lie algebras & their representations, Prentice Hall, Englewood Cliffs, NJ (1965).

[Vie] M. Ville, Sur le volume des variétés riemanniennes pincées, to appear in Bull. Soc. Math. France (1985).

[Vil] J. Vilms, Totally geodesic maps, J. of Diff. Geom. **4**, 73–79 (1970).

[Wak] H. Wakakuwa, Holonomy groups, Public. Study Group of Geometry, 6, Okayama Univ., Okayama (1971).

[Wal] N.R. Wallach, Compact homogeneous Riemannian manifolds with strictly positive curvature, Ann. of Math. **96**, 277–295 (1972).

[Wan] M. Wang, Some examples of homogeneous Einstein manifolds in dimension seven, Duke Math. J. **49**, 23–28 (1982).

[War] F. Warner, Foundations of differential manifolds and Lie groups, Scott, Foresman, Chicago (1971).

[Wa-Zi 1] M. Wang and W. Ziller, On the isotropy representation of a symmetric space, in Proc. Conf. "Differential geometry on homogeneous spaces, Torino (1983)", Rend. Sem. Mat. Univ. Politec. Torino, 253–261 (1983).

[Wa-Zi 2] M. Wang and W. Ziller, On normal homogeneous Einstein manifolds, Ann. Sci. Ec. Norm. Sup. **18**, 563–633 (1985).

[Wa-Zi 3] M. Wang and W. Ziller, Existence and Non-existence of Homogeneous Einstein Metrics, Inventiones Math. **84**, 177–194 (1986).

[Wa-Zi 4] M. Wang, W. Ziller, Einstein metrics with positive scalar curvature, preprint I.H.E.S., Bures-sur-Yvette (1985).

[Wa-Zi 5] M. Wang, W. Ziller, New examples of Einstein metrics on principal bundles, preprint (1986).

[Wa-Zi 6] M. Wang, W. Ziller, Isotropy irreducible spaces, symmetric spaces, and maximal subgroups of classical Lie groups, in preparation.

[Web] S. Weinberg, Gravitation and Cosmology, Wiley, New York (1972).

[Weg] B. Wegner, Codazzi-Tensoren und Kennzeichnungen sphärischer Immersionen, J. Differential Geometry **9**, 61–70 (1974).

[Wei 1] A. Weil, Introduction à l'étude des variétés kählériennes, Hermann, Paris (1958).

[Wei 2] A. Weil, Sur les modules des surfaces de Riemann, Sém. Bourbaki, Benjamin (1958).

[Wei 3] A. Weil, On discrete subgroups of Lie groups II, Annals of Math. **75**, 578–602 (1962).

[Wel] R.O. Wells, Differential analysis on complex manifolds, Graduate texts in Mathematics, Springer Verlag, Berlin (1979).

[Wey] H. Weyl, Classical groups, their invariants and representations, Princeton University Press, Princeton (1946).

[Whi] White, Hail to Chern, Choral communication, Oberwolfach (1981).

[Wl 1] C.T.C. Wall, On the orthogonal groups of unimodular quadratic forms, Math. Ann. **147**, 328–338 (1962).

[Wol 1] J.A. Wolf, Differentiable fibre spaces and mappings compatible with Riemannian metrics, Mich. Math. J. **11**, 65–70 (1964).

[Wol 2] J.A. Wolf, Complex homogeneous contact manifolds and quaternionic symmetric spaces, J. of Math. and Mech. **14**, 1033–1047 (1965).

[Wol 3] J. Wolf, The geometry and structure of isotropy irreductible homogeneous spaces, Acta Math. **120**, 59–148 (1968) and Erratum. Acta Math. **152**, 141–142 (1984).

[Wol 4] J. Wolf, Spaces of constant curvature, Publish or Perish, Boston, Mass. (1974).

[Wol 5] J.A. Wolf, Discrete groups, symmetric spaces, and global holonomy, Amer. J. Math. **84**, 527–542 (1962).

[Wol 6] J.A. Wolf, On Calabi's inhomogeneous Einstein-Kähler manifolds, Proc. Amer. Math Soc. **63**, 287–288 (1977).

[Wop] S. Wolpert, Non-completeness of Weil-Petersson's metric on Teichmüller space, Pacif. Math. J. **61**, 573–577 (1977).

[Wng] H.C. Wang, Topics in totally discontinuous groups, in Symmetric Spaces, ed. Boothby-Weiss, New York, 460–485 (1972).

[Wns 1] A. Weinstein, Fat bundles and symplectic manifolds, Adv. in Math **37**, 239–250 (1980).

[Wns 2] A. Weinstein, Distance spheres in complex projective spaces, Proc. A.M.S. **39**, 649–650 (1973).

[Wu 1] H. Wu, Decomposition of Riemannian manifolds, Bull. A.M.S., **70**, 610–617 (1964).

[Wu 2] H. Wu, On the de Rham decomposition theorem, Illinois J. Math. **8**, 291–311 (1964).

[Wu 3] H. Wu, Holonomy groups of indefinite metrics, Pacific J. of Math. **20**, 351–352 (1967).

[Wu 4] H. Wu, An elementary method in the study of nonnegative curvature, Acta Math. **142**, 57–78 (1979).

[Yam 1] H. Yamabe, On an arcwise connected subgroup of a Lie group, Osaka Math. J. **2**, 13–14 (1950).

[Yam 2] H. Yamabe, On a deformation of Riemannian structures on compact manifolds, Osaka Math. J. **12**, 21–37 (1960)

[Ya-Mo] K. Yano and I. Mogi, On real representations of Kählérian manifolds, Ann. Math. **61**, 170–189 (1955).

[Yan] C.N. Yang, Fibre bundles and the physics of the magnetic monopole, in The Chern Symposium 1979, ed. W.Y. Hsiang, Springer, Berlin (1980).

[Yau 1] S.T. Yau, Harmonic functions on complete Riemannian manifolds, Comm. Pure and Appl. Math. **28**, 201–228 (1975).

[Yau 2] S.T. Yau, A general Schwarz lemma for Kähler manifolds, Amer. J. Math. **100**(1), 197–203 (1978).

[Yau 3] S.T. Yau, On the Ricci curvature of a compact Kähler manifold and the complex Monge-Ampère equation I, Com. Pure and Appl. Math **31**, 339–411 (1978).

[Yau 4] S.T. Yau, On the curvature of compact Hermitian manifolds, Inventiones Math. **25**, 213–239 (1974).

[Yau 5] S.T. Yau, On Calabi's conjecture and some new results in algebraic geometry, Proc. Nat. Acad. Sci. USA **74**, 1798–1799 (1977).

[Yau 6] S.T. Yau, Remarks on the group of isometries of a Riemannian manifold, Topology **16**, 239–247 (1977).

[Yau 7] S.T. Yau, Problem section, in Seminar on Differential Geometry, ed. S.T. Yau, Ann. of Math. Study n° 102, Princeton University Press, Princeton, 669–706 (1982).

[Zil 1] W. Ziller, The Jacobi equation on naturally reductive compact Riemannian homogeneous spaces, Comment. Math. Helvetici **52**, 573–590 (1977).

[Zil 2] W. Ziller, Homogeneous Einstein metrics on spheres and projective spaces, Math. Ann. **259**, 351–358 (1982).

[Zil 3] W. Ziller, Homogeneous Einstein metrics, in "Global Riemannian Geometry", ed. N. Hitchin and T. Willmore, Ellis Horwood, Chichester (1984).

Notation Index

A	239	submersion: O'Neill tensor A
$A(g)$	358	normalized scalar curvature
$\mathfrak{A}(M)$	86	automorphism group of complex M
$\mathfrak{a}(M)$	86	Lie algebra of holomorphic vector fields
$b_i(M)$		Betti numbers of M
B	135	Bach tensor
B	185	Killing-Cartan form of a Lie algebra
$c_i(M)$	79	Chern classes of M
$C^{k+\alpha}(\Omega)$	456	Hölder space on domain Ω
$C_{\rho,q}$	197	Casimir operator of representation ρ
$\mathfrak{C}$	116	space of metrics with constant scalar curvature
C	359	isoperimetric constant
$\mathscr{C}$	365	space of complex structures
Cal	326	operator solving Einstein's equations
$\mathscr{C}E$	46	space of algebraic curvature tensors
$\mathscr{C}M$	21	space of smooth functions on M
$\mathbb{C}P^m$	35	complex projective m-space
d	24	exterior differential
d', d''	68	complex exterior differentials
d^c	68	complex exterior differential made real
d^∇	24	exterior differential of ∇
$d^{\nabla'}, d^{\nabla''}$	69	complex exterior differentials of ∇
D^m	320	unit ball in $\mathbb{C}^m$
D	30	Levi-Civita connection
$\mathscr{D}$	55	Dirac operator
$\mathscr{D}^\nabla$	55	Dirac operator on twisted bundle
Ddf	34	Hessian of function f
Df	34	gradient of function f
D^+, D^-	74	complex splitting of D on 1-forms
$\mathfrak{D}$	117	diffeomorphism group
$\mathfrak{D}_0$	344	identity component in Diff group

$E_{\mathbb{C}}$		complexification of vector space E
$\mathscr{E}(M)$	340	moduli space of Einstein metrics on M
$\tilde{\mathscr{E}}(M)$	340	Teichmüller space of M
End E		endomorphisms of vector space E
Exp		exponential map of a Lie group
$\exp_p$	28	exponential map at point p
F	90	Ricci potential of a Kähler metric
$\mathscr{F}$	90	Futaki functional
$\hat{g}_b$	236	submersion: fiber metric
$\check{g}$	236	submersion: base metric
grad	119	gradient of a Riemannian functional
$h^{p,q}$	84	Hodge numbers
H	38	mean curvature vector
$H^+(g)$	366	space of positive harmonic 2-forms
H^n	29	hyperbolic n-space
$H_p(M, .)$		cohomology groups of M
H	75	holomorphic sectional curvature
$\mathscr{H}$		submersion: horizontal projector
$\mathscr{H}_x$		submersion: horizontal space
$\mathfrak{hol}$	291	Lie algebra of holonomy group
Hol(p)	280	holonomy group at point p
$\mathrm{Hol}_0(p)$	280	restricted holonomy at p
$I(M, g)$	41	isometry group of (M, g)
ILH	127	inverse limit of Hilbert
II	38	second fundamental form
ker		kernel of operator
K	42	sectional curvature
K	82	canonical line bundle
K^*	82	anti-canonical line bundle
$\mathscr{K}$	365	space of Kähler-Einstein structures
$K_{\min}$, $K_{\max}$	356	range of sectional curvature
$L_k^p(M)$	457	Sobolev space on M
L_X	35	Lie derivative of vector field X
$\mathscr{M}_1^G$	121	Riemannian metrics (G-invariant)
$\mathscr{M}_1^K$	92	Kahler metrics of unit volume
$\mathscr{M}_1$	117	space of metrics of volume 1
$\mathscr{M}_{-1}$	344	space of surfaces of constant curvature -1
$\tilde{M}$		universal covering space of M
$\mathscr{M}$	117	space of Riemannian metrics on M

$\hat{M}$ 321 blowing up of complex manifold M
$M \# N$ 157 connected sum of manifolds M and N

N 68 complex torsion (or Nijenhuis) tensor
$\mathcal{N}_\mu$ 118 space of metrics given volume element μ

P 379 twistor space of 4-manifold

r 42 Ricci curvature
R 23 Riemann curvature tensor
$\check{R}$ 51 curvature acting on exterior 2-forms
$\mathring{R}$ 51 curvature acting on symmetric 2-tensors
$\mathbb{R}P^n$ real projective n-space
R'_g, r'_g, s'_g 62 derivatives of various quantities
Ric 137 Ricci as an operator

s 43 scalar curvature
S 119 total scalar curvature
S^n 30 sphere of dimension n
$\mathcal{S}^2 M$ space of symmetric 2-forms on M
$S^k E$ 45 symmetric powers of vector space E
$S_0^2 E$ 45 tracefree symmetric 2-tensors on E
$\mathfrak{S}_k$ 46 symmetric group on k letters
$Sp(A)$ spectrum of operator A
Ss, **S**r, **S**R, **S**W, **S**Z 127 quadratic functionals

T 26 torsion of connection D
T 96 stress/energy tensor
T 238 submersion: O'Neill tensor T
T^∇ 26 torsion of connection ∇
T^n n-torus
$T^{0,1}, T^{1,0}$ 66 tangent bundle split into types
$T^{(r,s)}M$ 21 bundle of (r, s)-tensors on M
$T_p M$ tangent space to M at point p
TM tangent bundle of M
$\mathrm{tr}_{\mathbb{C}}$ 72 complex trace of a 2-form

U, Z, W 48 irreducible components of curvature
$\mathcal{U}E$; $\mathcal{Z}E$; $\mathcal{W}E$ 47 irreducible subspaces of curvature tensors
UM unit tangent bundle of M
V_λ 437 eigenvalue distribution of Codazzi tensor
$\mathcal{V}_x$ 236 submersion: vertical space
$\mathcal{V}$ 236 submersion: vertical projector

W 48 Weyl curvature tensor
W^+, W^- 50 Weyl curvature (positive, negative part)

X_U	209	fundamental vectorfield attached to U
$\mathbf{YM}(\nabla)$	371	Yang-Mills functional
Z	412	twistor space of a quaternionic manifold
Z		trace free Ricci curvature
β_g	348	Bianchi operator
δ	34	codifferential (on forms)
δ	35	divergence (of symmetric tensors)
$\check{\delta}$	243	submersion: horizontal divergence
$\tilde{\delta}$	243	submersion: mixed divergence
$\hat{\delta}$	243	submersion: vertical divergence
δ^*	35	adjoint of divergence
$\delta^{\nabla'}$, $\delta^{\nabla''}$	70	complex codifferentials of ∇
δ_i^j	31	Kronecker symbol
Δ	34	Laplacian
Δ', Δ''	83	complex Laplacians
Δ_L	54	Lichnerowicz Laplacian
$\varepsilon(g)$	347	space of infinitesimal Einstein deformations
ε_{ijk}	452	Ricci symbol
κ	49	Gauss curvature
$\kappa(g)$	356	normalized minimum of sectional curvature
λ	44	Einstein constant
Λ	71	contraction by Kähler form
$\mu(g)$	122	Yamabe invariant
μ_g	32	volume element
$\pi_1(M)$		fundamental group of M
ρ	73	Ricci form
ρ_0	77	primitive part of Ricci form
σ_A	460	symbol of operator A
Σ_r	321	Hirzebruch surfaces
Σ^+M, Σ^-M	54	bundles of spinors over M
τ_γ	245	parallel translation along γ
τ	81	signature of a manifold
θ	32	volume element in polar coordinates
θ	382	twistor space (a canonical form on)
Θ	362	sheaf of holomorphic vectorfields
χ	178	Isotropy representation of homogeneous space
χ		Euler characteristic
ω_g	33	volume form of oriented (M, g)
ω	70	Kähler form
$\Omega^p M$	21	space of differential p-forms on M
$\bigwedge^+ E$	49	space of self-dual 2-forms
$\bigwedge_0^{1,1} M$	77	bundle of trace-free (1, 1)-forms
$\bigwedge^p M$	21	bundle of exterior p-forms on M

Symbol	Page	Meaning
$\bigwedge^{p,q}$	67	differential forms split into types
$[f]_{\alpha,A}$	456	Hölder norm of function f on domain A
$\|f\|_{k+\alpha}$	456	$C^{k+\alpha}$ norm of f
$\|f\|_{L^p}$	457	L^p-norm of f
$\|f\|_{p,k}$	457	Sobolev norm of f
$\|M\|$	161	simplicial volume of manifold M
$\otimes$		tensor product
$h \circ k$	45	symmetric product of symmetric tensors h and k
$\owedge$	47	Kulkarni-Nomizu product of symmetric tensors
$[\alpha]$		cohomology class of a closed form α
$[X, Y]$		bracket of vector fields X and Y
$\cup$	80	cup product of cohomology classes
$\flat$ $\sharp$	30	musical isomorphisms

Generic Notations

Symbol	Page	Meaning
α		differential form
α_0	72	primitive part of form α
$\dot{c}$		derivative of a curve c
A^*	34	adjoint of operator A
$\mathbf{F}$	118	Riemannian functional
g		Riemannian metric
$[g]$	344	orbit of metric g under Diff group
$\mathfrak{g}, \mathfrak{h}, \mathfrak{k}$		Lie algebra
γ		genus of a surface
J		almost complex structure
m	66	complex dimension of complex manifold
n		dimension of manifold
p		point on manifold
V, W, X		vectorfield
v, w, x		vector
∇	22	connection

Subject Index

3-momentum 95
4-velocity 95
$\hat{A}$-genus 170
absolutely simple 195, 202
adjoint 460
– representation *182, 188, 194, 199, 208, 297*
affine connection 278
Albanese torus 320
algebraic curvature tensor 46
almost complex structure 66, 284
– Kähler structure 255
– quaternionic manifold 410
Ambrose-Singer theorem 291
amenable 158
ample 322
anti-canonical line bundle 82, 320
anti-self-dual connection 369, 372
Apte formula 80
associated 2-form 70
– connection 248
Aubin-Calabi-Yau metric 323
– theorem 323
automorphism group 86, 320, 322, 329, 331, 368

Bach tensor 135
basic vector field 240
bending of light 109
Berger-Simons theorem 300
Bergmann manifold 424
– metric 77, 320
Betti number 161, 166, 167, 302, 306, 323
Bian(g, r) 138
Bianchi identity 27, 41, 120, 138, 286, 293, 294, 300, 348, 434
– map 46
Bishop's theorem 16
black hole 112
blowing up 321, 331, 334, 338
Bochner method 53, 166, 324, 399, 466
– theorem 233, 57, 166, 378
Borel-Lichnerowicz theorem 290
Busemann function 172

Calabi conjecture 83, 322
Calabi-Eckmann manifold 84
Calabi-Yau metric 323
canonical line bundle 82, 284, 320, 365, 401
– sheaf 400
– variation (of a Riemannian submersion) 252
capture of light 109, 110
Cartan subalgebra 298, 409
Cartan-Janet Theorem 145
Cartan-Kähler theorem 302
Casimir operator 197, 376
causal geodesic 108
Cayley numbers 298
Cheeger-Gromoll theorem 171, 308, 323
Chern class 78, 302, 320, 326, 371
– connection 82
– number 321, 392
Chern-Weil homomorphism 78
Christoffel symbol 23, 30, 144
Clifford algebra 169
closed trapped surface 115
Codazzi equation 434
– tensor 434
Codazzi-Mainardi equation 38
codifferential 34
cohomogeneity 275, 471, 472, 475
complementary roots 218
complete 36
– (Ehresmann-) 244
– intersection 320
– orbit 209
complex connection 25
– dimension 66
– exterior derivative 68
– – – of vector-valued form 69
– Laplacian 54
– manifold 68
– projective space 76
– symplectic structure 365, 398, 402
– torus 76
– volume form 215, 284
conformal 121, 122, 337

– change of metric 58
conformally flat 60, 122, 136
– – (half) 135, 136, 372, 380
conjugate point 28
connected sum 157, 161, 163, 164, 171
connection 22
– (anti self-dual) 369, 372
– (associated) 248
– (Ehresmann-) 244
– (*G*-) 247
– (*Gl*(*F*)-) 247
– (Levi-Civita) 30
– (principal) 248
contraction (on 3-tensors) 434
– by Kähler form 71
cosmological constant 96
cotangent bundle 21
covariant derivative 22, 25, 282
critical 116, 119, 121, 132, 134, 135
curvature 24
– acting on exterior 2-forms 51
– – – symmetric 2-tensors 51
– endomorphism 153, 290, 291, 296, 303
– tensor 30, 41
– – (algebraic) 46

D'Atri spaces 450, 451
dd^c-Lemma 326
de Rham's holonomy theorem 240, 288
decomposition of Kähler curvature 77
Dehn surgery 159
Del Pezzo surfaces 321
diameter 36, 165, 359
diffeomorphism group 116, 117, 340
differential form 21
Dirac Laplacian 54
– operator 55, 369, 375
direct sum of connections 24
distance 36
– spheres 257, 258
divergence 125, 35
– form (in) 459
divisor 367
Dolbeault complex 350
– theorem 84, 350
domain of holomorphy 423, 429

effective divisor 367
Ehresmann connection 244
– theorem 244
eigenspace distribution 436
eigenvalue function 436
Einstein equation 96, 104, 113
– – (linearized) 346
– manifold 3
– operator 346
– structure 340
Elie Cartan theorem 295
elliptic complex 375, 462
– operator 123, 327, 462
– (overdetermined) 462, 463
– (strongly) 465
– (underdetermined) 462, 463
Enriques surface 162, 400
Euclidean connection 25
Euler characteristic 161, 325, 371, 400
– number (of complex manifold) 86
Euler-Lagrange equation 122
exceptional divisor 321
exponential map 28
exterior derivative (complex) 68
– differential 21
– – of a vector valued form 24, 434
extremal metric 333

faithful 178
fat bundle 242
Finsler metric 1
first Chern class 320
– variation 62
flag manifold 211
flat 30
– manifold 345
formal series 349, 352
formally integrable 349, 352, 361
Fredholm alternative 464
Fröhlicher spectral sequence 84
Fubini-Study metric 256
functional (Riemannian) 118, 119
fundamental class in homology 164
– group 157, 164, 167, 171, 177, 189
– principle of holonomy 282
– vector field 209
Futaki functional 92, 332, 375, 475

G-connection 247
Gauss curvature 49
– equation 38, 102
– Lemma 32
Gauss-Bonnet 161, 371
geodesic 27, 36
– normal coordinates 32, 143, 144, 145
– symmetry 191, 295, 296
geodesics (first integrals) 108
– (Lorentz incomplete) 112, 113
Gl(*F*)-connection 247
gradient 34, 119
gravitational potential 96
growth function 167

Haken 159
half conformally flat 135, 136, 372

hamiltonian vector field 330, 477
harmonic coordinates 143
– curvature 434
– map 243
– Weyl tensor 434
Hermann theorem 245, 249
hermitian (infinitesimal Einstein deformation) 362, 363
– connection 25
– metric 69
– symmetric spaces (compact) 224
Hessian 34
Hirzebruch surfaces 321
Hodge $*$-operator 33, 49, 99, 369, 341
– index formula 366
Hölder spaces 143, 327, 429, 456
holomorphic line bundle 81
– sectional curvature 75
– vector field 87, 329, 330, 332, 334
holomorphy potential 88
holonomy group 280
– representation 280
homogeneous Kähler manifold (compact) 208, 224
– – – (non-compact) 431
– Lorentz manifold 205
homotopically atoroidal 157
Hopf bundle 332, 335, 383
– fibration 257, 340, 353
– – (generalized) 258
horizontal 236
– lift 244
hyperbolic space 29, 155, 158, 180
hyperkählerian 163, 284, 285, 294, 302, 324, 379, 396, 398
hypersurface 320, 362, 423

immersion 29, 37
incompressible 159
Index Theorem 170, 369, 376
infinitesimal automorphism 86
– complex deformation 361
– – – (symmetric) 363
– Einstein deformation 347, 352, 363
– – – (hermitian) 362, 363
– observer 95
instanton 371
integrability condition 349, 350
integrable complex structure 68
intersection form 161, 367, 368
involution 295, 296, 299
involutive distribution 282
irreducible 357, 363
– decomposition of curvature tensor 45
– isotropy 119
– symmetric space 119, 297
isometry 39
– group 39, 41, 178, 180, 181, 192
isotropy irreducible 119
– – space 7
– – spaces (tables of compact non symmetric strongly) 203
– – – (tables of non-compact non symmetric strongly) 204
– representation 178, 182, 187, 197, 204
– subgroup 178, 191, 195, 296, 298
Iwamoto-Lichnerowicz theorem 284
Iwasawa decomposition 180

Jacobi equations 107
Jacobi field 28, 102, 107, 296

K3 surface 162, 342, 365, 369, 373, 376, 400
Kähler class 85
– curvature 73
– – operator 73
– form 70
– metric 70
– potential 85
Kähler-Einstein metric 319
– structures (space of) 365, 367
kählerian 322
Kaluza-Klein 250, 472
Killing vector field 35, 40, 183, 334, 346, 377
Killing-Cartan form 184, 185, 198, 298
Klein bottle 283
Kruskal model 111
Kulkarni-Nomizu product 47
Kummer surface 401

Lagrangian 116
– (natural) 118
Laplacian 34
– (complex) 54
– (Dirac) 54
– (Hodge-de Rham) 35
– (Lichnerowicz) 54
– (rough) 52
– (twisted Dirac) 56
lattice 368
left-invariant Riemannian metric 182, 190
length 36
Levi form 85
Levi-Civita connection 30, 182
Lichnerowicz Laplacian 54, 132, 133, 347
Lie derivative 21
light 109, 110
line 37, 168, 172
linear connection 22, 25
linearization 62, 346, 462
linearized Einstein equation 346
link 159

local solvability 469
locally isometricly embeddable 145
– symmetric Lorentz manifold 206
– – space 193, 198, 297, 300, 302, 357
loop 280
Lorentz Einstein metric 256
– geometry 96
– manifold 309, 310, 425
– metric 29
– symmetric space 206
L^p estimate 463

mass/energy 96
maximum principle 173, 327, 328, 430, 466
mean curvature vector 38, 243
minimal submanifold 38
Minkowski space 205
Moduli Space of Einstein Structures 9, 340, 352, 361, 365, 471, 475
moment map 380, 477
Monge-Ampère equation 327, 424, 429
monodromy 288
motion 178
musical isomorphism 30
Myers theorem 146, 149, 165

Nash-Moser implicit function theorem 147
natural Lagrangian 118
naturally reductive 193, 196, 450
nearly effectively 179, 180
negative curvature (compact space of) 354
– – (constant) 155, 158
Newlander-Nirenberg theorem 68, 381, 414
Nijenhuis tensor 67
noncharacteristic directions 142
normal coordinates 32, 143, 144, 145
– homogeneous metric 197, 210, 221, 374
null curve 95

O'Neill tensors 238
obstruction space 349, 351
one parameter subgroup 181, 193
orbifold 346, 361
orbit 208
oriented cobordism ring 171
orthogonal group 180
overdetermined 462, 463

Page metric 273, 337, 338
pair of pants 156
Palais-Smale condition 151, 373
parallel 25
– translation 280
partial alternation (on 3-tensors) 434
– symmetrization (on 3-tensors) 434
past (see your own) 107
path 244
Penrose construction 379
perihelion procession 107
period mapping on complex structures 367
– – – Einstein structures 366
– – – Kähler-Einstein structures 367
photon sphere 106
pinched sectional curvature 357
planetary orbits 105, 107
plurisubharmonic (strictly) 423
Poincaré's conjecture 147
Poisson bracket 330
Pontryagin class 370
– number 6
positive (negative) spin bundle 372, 380
– 2-form 70
– cohomology class 319
– spin bundle 169, 346, 352
premoduli space of Einstein Structures 361
prime 157
primitive form 71
– part of a form 71, 334
– roots 225
principal connection 248
– orbit 237
– pressure 96
– symbol 169
product 29, 44
projectable (vector field) 210, 239
projective algebraic manifold 322
pseudo-Anosov 158
pseudo-Riemannian immersion 37
pseudo-Riemannian metric 29
pseudoconvex (strictly) 428
pseudosphere 145
pure curvature operator 439
– Weyl tensor 439

quadratic functional 133, 333, 337
quasilinear 459
quaternion-Hermitian manifold 410
quaternion-Kähler manifold 285, 307, 396, 403
quaternion-Kähler symmetric spaces (table of) 409
quaternionic manifold 411

rank 298, 299
ray 37, 172
reducible Kähler manifold 288
– Riemannian manifold 285
reductive 331
regular 211
restricted holonomy group 280, 287, 294
reverse Penrose construction 385

Ricci contraction 46
– curvature 2, 15, 42
– equation 38
– flat 124, 128, 135, 191, 209
– form 73, 216, 218, 284, 319, 428
– formula 26
– potential 90
– symbol 452
Riemann domain 429
Riemann-Roch-Hirzebruch theorem 86, 365, 367
Riemannian metric 1, 29
– product 285
– structure 1
– structures (spaces of) 341
– submersion 236
rigid 355
roots (complementary) 218
– (positive) 212
– (primitive) 225
rough Laplacian 52

saturated holonomy system 304
scalar curvature 15, 43, 185, 319
Schauder estimate 463
Schwarzschild metric 106, 113, 425
– model 106, 113, 425
second fundamental form 38, 60
sectional curvature 42
self-dual connection (anti) 369, 372
– form 370
– part of Weyl tensor 50
semi-simple 195, 198, 199, 331
Serre duality 84
sharp 30
signature 140, 142, 145, 161, 256, 325, 366, 371
– theorem (Hirzebruch) 161, 370
simple 195, 199, 201, 202
– components 209
simplicial volume 164
simply transitively 178, 180, 190
slice 116, 341, 345
smooth function 21
Sobolev class 457
– inequality 120, 146, 151, 458
space of Kähler-Einstein structures 365, 368
– – Riemannian structures 341
space-like curve 95
spacetime singularities 115
spectrum of the Laplacian 128, 130, 347
speed of light 94
sphere 29, 128, 130, 132, 354, 355
spheres (groups acting transitively on) 179, 300
spin 169
– bundle (positive) 169, 373, 380
– cobordism ring 170
standard homogeneous metric 198
Stein manifold 431
stress tensor 96
stress/energy tensor 96
strongly elliptic 465
– isotropy irreducible space 187
structural group (of a fibre bundle) 246
subharmonic 174
support function 172
symbol 460
symmetric infinitesimal complex deformation 363
– Lorentz space 206
– space 186, 294
– spaces (compact hermitian) 224
– – (tables of compact irreducible) 201, 312, 313, 314, 317
– – (tables of non-compact irreducible) 202, 315, 316, 317
– – (tables of quaternion-Kähler) 409
symplectic structure 220, 383, 477
– – (complex) 398

tangent bundle 21
Taub-NUT metric 104, 369, 395, 480
tautological section 211
Teichmüller space of Einstein structures 340, 343, 365, 367
test particle 95
tidal stress 97
timelike curve 95
torsion 26, 279
torus 180, 191, 226, 281, 286, 287
– (complex) 76
total scalar curvature 4, 116, 119, 128, 346, 352
totally geodesic 38, 285, 286, 288
– – fibers (submersion with) 235, 240, 249
trace 71, 72
transgression 222, 223
translations (group of) 178
transitive action 178
travel time of light 109, 110
trivial connection 23
twisted Dirac Laplacian 56
– – operator 56
twistor space 380
– – (quaternionic style) 397, 412
two-point homogeneous spaces 186
type of a form 67
– – a vector 67
types (of irreducible symmetric spaces) 195

umbilic point 38, 437
underdetermined 462, 463

unimodular 184, 190
universal covering 165, 191, 193, 205, 323

vacuum space time 96
vector field 21
– – (basic) 240
– – (fundamental) 209
– – (hamiltonian) 330
– – (holomorphic) 87, 329, 330, 332, 334
– – (Killing) 334, 346
vertical 236
very ample 323
Vilms theorem 249
volume 319, 353
– element 32
– form 33
– – (complex) 215, 284

Wang theorem 265
warped product 237
weak solution 460
weight 212, 217
Weingarten equation 38
Weitzenbock formula 53, 74, 166, 169, 178, 341, 362, 376, 399, 436, 448
Weyl chamber 227
– curvature tensor 337
– group 227
– tensor 48, 60, 78, 370
– – (harmonic) 434, 443
– – (self-dual part of) 50, 337, 370, 451
Weyl-Schouten tensor 62

Yamabe invariant 122
Yang-Mills 136, 243, 371, 444
Young symmetrizer 46

Errata

Page 21, line 9 *Replace the second paragraph of 1.4 by the following*
We denote by $\Lambda^s M$ the vector bundle of exterior s-forms on M. Hence $\Lambda^s_x M = \Lambda^s(T^*_x M)$, this vector space being as usual a quotient space of $\bigotimes^s(T^*_x M)$ (sometimes we will consider forms as elements of $\bigotimes^s(T^*_x M)$ through the $Gl(T^*_x M)$-invariant splitting of the canonical projection). More generally, $\wedge$ will denote any alternating procedure and similarly S or $\odot$ will denote symmetrisation.

Page 21, line -5 *Replace the first line of Formula (1.5a) by*

$$\alpha(X_0, X_1, \ldots, X_p) = \sum_{i=0}^{p} (-1)^i X_i \cdot \alpha\left(X_0, \ldots, \hat{X}_i, \ldots, X_p\right)$$

Page 24, line -7 *Replace the second line of the first formula of 1.12 by*

$$+\sum_{i<j} (-1)^{i+j} \alpha\left(\left[\tilde{X}_i, \tilde{X}_j\right], X_0, \ldots, \hat{X}_i, \ldots, \hat{X}_j, \ldots, X_p\right).$$

Page 24, line -6 *Replace line -6 ("Then") by the following*
Notice that, if $p = 0$ and s is a section of E, then $d^\nabla s$ is nothing but what we called ∇s in 1.8 Remark a). And then the curvature R of the connection ∇ is given through the formula ...

Page 25, line 10 *Replace the proof of Theorem 1.14 by the following*
Sketch of proof : This follows from Formula (1.12) and the fact that

$$(d^\nabla \circ d^\nabla) \circ d^\nabla = d^\nabla \circ (d^\nabla \circ d^\nabla),$$

provided we put on the vector bundle $End(E)$ the "natural" (see 1.15) connection, still denoted by ∇, induced by the connection ∇ on E, and consider the corresponding d^∇.

Page 25, line 12 *Replace paragraph 1.15 by the following*
1.15. Some more definitions. (a) Let E be a vector bundle with a linear connection ∇. Then, on any vector bundle "naturally associated" with E, like E^* or $End(E)$, there exists a "natural" connection associated with ∇, which

will be also denoted by ∇. We will not give the general definition here and refer to [Ko-No 1] for details. We give only one example here. If s is a section of E and α a section of E^*, then, for any tangent vector X, we have

$$(\nabla_X \alpha)(s) = X(\alpha(s)) - \alpha(\nabla_X s).$$

(b) A section s of a vector bundle E, equipped with a connection ∇ is called parallel if $\nabla s = 0$. Note that, given a point x in M, there always exists a neighborhood U and a finite family $(s_i)_{i \in I}$ of sections of E over U such that $(s_i(y))_{i \in I}$ is a basis of the fiber E_y for each y in U and $\nabla s_i(x) = 0$. (Consider a local trivialization of E centered at x. Then to find such s_i near x amounts to finding functions with appropriate first jet at x).
(c) If the vector bundle E is equipped with some additional structure (such as a Euclidean fiber metric h, a complex structure J, a symplectic form ω or a Hermitian triple (g, J, ω), then a linear connection ∇ is called respectively Euclidean (or metric), complex, symplectic or Hermitian, if respectively h, J, ω or (g, J, ω) are parallel (when the vector bundle in which they live respectively is endowed with the "natural" connection induced by ∇). In such a case, the curvature R^∇ of ∇ satisfies additional properties, namely, for each X, Y in T_xM, the linear map $R^\nabla_{X,Y}$ on E_x is respectively skew-symmetric, complex linear, skew-symplectic or skew-Hermitian.
In the particular case where E is the tangent bundle of the base manifold M, further considerations can be developed.

Page 25, line -9 *Replace "(see Remark(1.8.a))" by*
(see Remark (1.8.a) and 1.12)

Page 27, line 10 *At the end of 1.23, add the following*
Notation : We denote by $S_{X,Y,Z}$ the "cyclic sum" with respect to X, Y, Z, i.e., if $A(X, Y, Z)$ is any expression involving X, Y and Z, we write

$$S_{X,Y,Z} A(X,Y,Z) = A(X,Y,Z) + A(Y,Z,X) + A(Z,X,Y)$$

Page 30, line 5 *Replace paragraph 1.38 by the following*
1.38. Let (M, g) be a pseudo-Riemannian manifold. Then at each point x of M, the nondegenerate quadratic form g_x induces a canonical isomorphism $T_xM \to T_x^*M$. This isomorphism is often denoted by $\flat$ ("flat") and its inverse by $\sharp$ ("sharp") since in classical tensor notation, they correspond to lowering (resp. raising) indices, see below 1.42. More generally, g_x induces various isomorphisms between any $T_x^{(p,q)}M$ and $T_x^{(r,s)}M$, as soon as $p+q = r+s$, these isomorphisms depending of which indices are lowered or raised.
If we consider the unique isomorphism $T_x^{(p,q)}M \to T_x^{(q,p)}M$ which lowers the p "covariant" indices and raises the q "contravariant" indices, and compose it with the "evaluation map" (pairing any vector space with its dual, respecting

the order of the indices), we get b a nondegenerate quadratic form, still denoted by g_x on any $T_x^{(p,q)}M$. Note that if g_x is positive definite, then it is also positive definite on any $T_x^{(p,q)}M$. Similarly, g_x induces an isomorphism between $\Lambda_x^s M = \Lambda_x^s (T_x^* M)$ and $\Lambda_x^s (T_x M)$, or $S^s (T_x^* M)$ and $S^s (T_x M)$, and (through pairing) g_x induces a nondegenerate quadratique form on $\Lambda_x^s M$ and $S^s (T_x M)$, also denoted by g_x. Notice that if (e^i) is an orthonormal basis for $T_x M$, the "dual basis" (e_i^*) for T_x^M, which is defined by $e_i^* (e^j) = \delta_i^j$ (Kronecker symbol, see 1.42), is also an orthonormal basis, and furthermore $e_{i_1}^* \otimes \cdots \otimes e_{i_p}^* \otimes e^{j_1} \otimes \cdots \otimes e^{j_q}$ is an orthonormal basis for $T_x^{(p,q)}M, e_{i_1}^* \wedge \cdots \wedge e^* i_s$ an orthonormal basis for $\Lambda_x^s M$ $(i_1 < \cdots < i_s)$, and $e_{i_1}^* \odot \cdots \odot e_{i_s}^*, i_1 < \cdots < i_s$, an orthonormal basis for $S^s (T_x^* M)$.

Page 32, line 14 *Replace the beginning of the first line of 1.45 by*
1.45. Theorem (see [Ru-Wa-Wi] or [Eps]). Local...

Page 32, line 18 *Replace by*

$$\sum_{j=1}^{n} g_{ij} \left(x^1, \ldots, x^n\right) x^j = \sum_{j=1}^{n} g_{ij} (0)\, x^j .$$

Page 34, line 5 *Replace lines 5 and 6 by*

a) the *gradient* of f is the vector field $grad\, f = \sharp df$ (or $df^\sharp$), i.e., $grad\, f$ satisfies $g\,(grad\ \mathrm{f,X}) = X\,(f) = df\,(X)$ for any X in TM ; we will sometimes denote $grad\ f$ by Df to keep the notations shorter, despite the possible confusion with df.

Page 34, line -8 *Replace the formula by*

$$\delta = \begin{cases} - *_g \circ f \circ *_g, & \text{if } n \text{ is even ;} \\ (-1)^p *_g \circ d \circ *_g & \text{if } n \text{ is odd.} \end{cases}$$

Page 34, line -7 *Replace b) by*

b) δ may be viewed as the restriction of D^* for a suitable embedding of $\Lambda^{p+1} M$ into $\Lambda^1 M \otimes \Lambda^p M$

Page 35, line 4 *Replace paragraph 1.59 by*

1.59. Instead of forms, we may also consider symmetric tensors. Il we consider the covariant derivative

$$D = \mathscr{S}^p M \to \Omega^1 M \otimes \mathscr{S}^p M = \mathscr{S}^1 M \otimes \mathscr{S}^p M,$$

and compose with the "symmetrization" sym

$$sym : \mathscr{S}^1 M \otimes \mathscr{S}^p M \to \mathscr{S}^{p+1} M$$

given by

$$sym\,(\alpha, \omega)\,(X_0, \ldots, X_p) = \frac{1}{p+1} \sum_{i=0}^{p} \alpha\,(X_i)\,\omega\left(X_0, \ldots, \hat{X}_i, \ldots, X_p\right),$$

we obtain a differential operator, *denoted* by δ^*,

$$\delta^* = \mathscr{S}^p M \to \mathscr{S}^{p+1} M,$$

whose formal adjoint is called the divergence, and denoted by δ,

$$\delta : \mathscr{S}^{p+1} M \to \mathscr{S}^p M.$$

Notice that δ may be viewed as the restriction of D^* for a suitable embedding of $\mathscr{S}^{p+1} M$ into $T^{(p+1,0)} M$.

Page 35, line 16 *Change a sign in the formula. The correct sign is*

$$\delta^* \alpha = \frac{1}{2} L_{\alpha^\sharp} g,$$

Page 35, line 23 *Change a sign in the formula. The correct sign is*

$$= \frac{1}{2}\,(L_{\alpha^\sharp} g)\,(X, Y)$$

Page 43, line 16 *Replace formula (1.92) by*

$$(n-1)\,g\,(\dot{c}, \dot{c})\,\ddot{\phi} + r\,(\dot{c}, \dot{c})\,\phi \leq 0$$

Page 52, line 11 *In the first line of (1.133), the equation is*

$$\check{R} = \frac{1}{4} q\,(R, R)\,q$$

Page 53, lines -4 and -1, page 54, lines 3 and 4 *Replace E by E_p*

Page 54, (1.144) *The correct formula is*

$$\Gamma T = -2c_\rho^2(R) T$$

Page 55, line 11 *Replace line 11 by*

Thus, since σ is parallel,

Page 55, after (1.149) *Add*
Although σ is not a representation, c_σ can be given the same meaning as c_ρ in 1.139.

Page 65, line 7 *Change a sign in the formula. The correct sign is*

$$\mu'_g h = \frac{1}{2}(\mathrm{tr}_g h)\,\mu_g.$$

Page 65, line 10 *Change a sign in the formula. The correct sign is*

$$Vol\,(M)'_g\,h = \int_M \mu'_g h = \frac{1}{2}\int_M (\mathrm{tr}_g h)\mu_g$$

Page 74, (2.46) *The correct formula is*

$$\omega = \sum_{\alpha=1}^{n} e_\alpha^\flat \wedge J e_\alpha^\flat$$

Page 77, (2.64) *The correct formula is*

$$B = \frac{\mathrm{tr}B}{(m^2-1)} Id_{|\Lambda_0^{1,1}M} + B_0$$

i.e. $\Lambda_0^{1,1}$ *instead of* $\Lambda^{1,1}$

Page 92, (2.151) *The correct formula is*

$$\frac{1}{2}\Delta f + (dd^c f, \rho) = 0.$$

Page 166, 6.56 *Replace the first three lines of the statement by*
Let (M, g) be a compact connected Riemannian manifold. If the Ricci curvature of M is non-negative then

$$\dim H^1(M, R) = b_1 \leq \dim M$$

Moreover, the universal cover of M is a Riemannian product of the form $R^{b_1} \times \overline{M}$, where $\overline{M}$ is a manifold of dimension $\dim M - b_1$.
Page 310, line 2 *Replace by*
for $A^2 : f(z)\,dx^2 + 2dydz + g(y)\,dz^2$, for nonconstant positive functions f and g.

Page 310, line 15 *Replace the expressions for* B_θ^3 and B^4 by

$$B^3_\theta = \begin{pmatrix} e^{t\theta} & b \\ 0 & e^{-t\theta} \end{pmatrix} \; t \in \mathbb{R}, b \in \mathbb{C} \text{ where } \theta \in \mathbb{C} \text{ has norm } 1,$$

$$B^4 = \begin{pmatrix} a & b \\ 0 & 1/a \end{pmatrix}, \; a \in \mathbb{C}^*, b \in \mathbb{C}.$$

Page 310, line -7 *Replace the formula by*

$$dx^2 + dy^2 + 2dudv + 2\rho dxdu + \left(\omega - \frac{\partial \rho}{\partial x}\right) du^2.$$

Page 379, line 6 The reference should be [Hit 1]

Page 380, line 7 the almost complex *structure* is integrable...

Page 382, line -6 The reference should be [Hit 2]

Page 385, line -9 The reference should be [Hit 5]

Page 459, line 7 conformal *deformations*...

Page 478, line 10 *Replace* $\mu^{-1}(G)$ by $\mu^{-1}(0)$

Page 489, line [Hit 6] *the complete reference is*

[Hit 6] N. Hitchin, Metrics on moduli spaces, Proceedings of Lefschetz Centennial Conference, Mexico City (1984), Contemporary mathematics **58,** Part I, 157-178, Amer. Math. Soc., Providence (1986).

Page 489, line [H-K-L-R] *the complete reference is*

[H-K-L-R] N.J. Hitchin, A.Karlhede, U. Lindström, M. Rocek, Hyperkähler metrics and supersymmetry, Commun. Math. Phys. **108,** 535-549 (1986).

Page 492, line [Kro] *the complete reference is*

[Kro] P. Kronheimer, Instantons gravitationnels et singularités de Klein, C. R. Acad. Sci. Paris **303,** 53-55 (1986).

Page 496, line [Run] *Between references [Run] and [Rya], insert*

[Ru-Wa-Wi] H.S. Ruse, A.G. Walker and T.J. Willmore, Harmonic spaces, Consiglio Nacionale delle Ricerce, Monografie Matematiche 8, Edizioni Cremonese, Roma (1961).

Page 499, line [Wu3] The last page of that reference is 392 instead of 352.

Page 506 (index), line Futaki No reference to page 375.

CIM

M. Aigner Combinatorial Theory ISBN 978-3-540-61787-7
A. L. Besse Einstein Manifolds ISBN 978-3-540-74120-6
N. P. Bhatia, G. P. Szegő Stability Theory of Dynamical Systems ISBN 978-3-540-42748-3
J. W. S. Cassels An Introduction to the Geometry of Numbers ISBN 978-3-540-61788-4
R. Courant, F. John Introduction to Calculus and Analysis I ISBN 978-3-540-65058-4
R. Courant, F. John Introduction to Calculus and Analysis II/1 ISBN 978-3-540-66569-4
R. Courant, F. John Introduction to Calculus and Analysis II/2 ISBN 978-3-540-66570-0
P. Dembowski Finite Geometries ISBN 978-3-540-61786-0
A. Dold Lectures on Algebraic Topology ISBN 978-3-540-58660-9
J. L. Doob Classical Potential Theory and Its Probabilistic Counterpart ISBN 978-3-540-41206-9
R. S. Ellis Entropy, Large Deviations, and Statistical Mechanics ISBN 978-3-540-29059-9
H. Federer Geometric Measure Theory ISBN 978-3-540-60656-7
S. Flügge Practical Quantum Mechanics ISBN 978-3-540-65035-5
L. D. Faddeev, L. A. Takhtajan Hamiltonian Methods in the Theory of Solitons ISBN 978-3-540-69843-2
I. I. Gikhman, A. V. Skorokhod The Theory of Stochastic Processes I ISBN 978-3-540-20284-4
I. I. Gikhman, A. V. Skorokhod The Theory of Stochastic Processes II ISBN 978-3-540-20285-1
I. I. Gikhman, A. V. Skorokhod The Theory of Stochastic Processes III ISBN 978-3-540-49940-4
D. Gilbarg, N. S. Trudinger Elliptic Partial Differential Equations of Second Order ISBN 978-3-540-41160-4
H. Grauert, R. Remmert Theory of Stein Spaces ISBN 978-3-540-00373-1
H. Hasse Number Theory ISBN 978-3-540-42749-0
F. Hirzebruch Topological Methods in Algebraic Geometry ISBN 978-3-540-58663-0
L. Hörmander The Analysis of Linear Partial Differential Operators I – Distribution Theory and Fourier Analysis ISBN 978-3-540-00662-6
L. Hörmander The Analysis of Linear Partial Differential Operators II – Differential Operators with Constant Coefficients ISBN 978-3-540-22516-4
L. Hörmander The Analysis of Linear Partial Differential Operators III – Pseudo-Differential Operators ISBN 978-3-540-49937-4
L. Hörmander The Analysis of Linear Partial Differential Operators IV – Fourier Integral Operators ISBN 978-3-642-00117-8
K. Itô, H. P. McKean, Jr. Diffusion Processes and Their Sample Paths ISBN 978-3-540-60629-1
T. Kato Perturbation Theory for Linear Operators ISBN 978-3-540-58661-6
S. Kobayashi Transformation Groups in Differential Geometry ISBN 978-3-540-58659-3
K. Kodaira Complex Manifolds and Deformation of Complex Structures ISBN 978-3-540-22614-7
Th. M. Liggett Interacting Particle Systems ISBN 978-3-540-22617-8
J. Lindenstrauss, L. Tzafriri Classical Banach Spaces I and II ISBN 978-3-540-60628-4
R. C. Lyndon, P. E Schupp Combinatorial Group Theory ISBN 978-3-540-41158-1
S. Mac Lane Homology ISBN 978-3-540-58662-3
C. B. Morrey Jr. Multiple Integrals in the Calculus of Variations ISBN 978-3-540-69915-6
D. Mumford Algebraic Geometry I – Complex Projective Varieties ISBN 978-3-540-58657-9
O. T. O'Meara Introduction to Quadratic Forms ISBN 978-3-540-66564-9
G. Pólya, G. Szegő Problems and Theorems in Analysis I – Series. Integral Calculus. Theory of Functions ISBN 978-3-540-63640-3
G. Pólya, G. Szegő Problems and Theorems in Analysis II – Theory of Functions. Zeros. Polynomials. Determinants. Number Theory. Geometry ISBN 978-3-540-63686-1
W. Rudin Function Theory in the Unit Ball of $\mathbb{C}^n$ ISBN 978-3-540-68272-1
S. Sakai C*-Algebras and W*-Algebras ISBN 978-3-540-63633-5
C. L. Siegel, J. K. Moser Lectures on Celestial Mechanics ISBN 978-3-540-58656-2
T. A. Springer Jordan Algebras and Algebraic Groups ISBN 978-3-540-63632-8
D. W. Stroock, S. R. S. Varadhan Multidimensional Diffusion Processes ISBN 978-3-540-28998-2
R. R. Switzer Algebraic Topology: Homology and Homotopy ISBN 978-3-540-42750-6
A. Weil Basic Number Theory ISBN 978-3-540-58655-5
A. Weil Elliptic Functions According to Eisenstein and Kronecker ISBN 978-3-540-65036-2
K. Yosida Functional Analysis ISBN 978-3-540-58654-8
O. Zariski Algebraic Surfaces ISBN 978-3-540-58658-6